Civil Engineering
Reference Guide

The McGraw-Hill
Engineering Reference Guide Series

This series makes available to professionals and students a wide variety of engineering information and data available in McGraw-Hill's library of highly acclaimed books and publications. The books in the Series are drawn directly from this vast resource of titles. Each one is either a condensation of a single title or a collection of sections culled from several titles. The Project Editors responsible for the books in the Series are highly respected professionals in the engineering areas covered. Each Editor selected only the most relevant and current information available in the McGraw-Hill library, adding further details and commentary where necessary.

Hicks • PLUMBING DESIGN AND INSTALLATION REFERENCE GUIDE

Covers the fundamentals of plumbing design and installation for a wide variety of industrial buildings and structures. Culled by Tyler G. Hicks from several McGraw-Hill books.

Hicks • POWER PLANT EVALUATION AND DESIGN REFERENCE GUIDE

Provides concise evaluation and design information for power plants serving many different needs—utility, industrial, and commercial. Culled by Tyler G. Hicks from several McGraw-Hill books and magazine articles.

Johnson & Jasik • ANTENNA APPLICATIONS REFERENCE GUIDE

Includes practical information and guidelines to antenna applications in all areas of communication. Comprised of one full section of Johnson and Jasik's Antenna Engineering Handbook, *Second Edition. Prepared by Richard C. Johnson.*

Markus and Weston • CLASSIC CIRCUITS REFERENCE GUIDE

Collects in one source hundreds of electronic circuits immediately useful in a wide variety of applications. Culled by Charles D. Weston from Markus's Sourcebook of Electronic Circuits, Electronics Circuits Manual, *and* Guidebook of Electronics Circuits.

Merritt • CIVIL ENGINEERING REFERENCE GUIDE

Offers quick reference to major civil engineering fields: structural design, surveying, geotechnical, environmental, and water engineering. A condensation by Max Kurtz of Merritt's Standard Handbook for Civil Engineers, *Third Edition.*

Woodson • HUMAN FACTORS REFERENCE GUIDE FOR ELECTRONICS AND
         COMPUTER PROFESSIONALS

Presents all essential data on human factors (ergonomics) relevant to the electronics and computer fields. Compiled by Wesley E. Woodson from his Human Factors Design Handbook.

Woodson • HUMAN FACTORS REFERENCE GUIDE FOR PROCESS PLANTS

Makes available to engineers and specialists all essential data on human factors (ergonomics) relevant to the process industries. Compiled by Nicholas P. Chopey from Woodson's Human Factors Design Handbook.

Civil Engineering
Reference Guide

Frederick S. Merritt *Editor*

Consulting Engineer, West Palm Beach, FL

Max Kurtz, P.E. *Project Editor*

Consulting Engineer, New York, NY

McGraw-Hill Book Company

New York St. Louis San Francisco Auckland
Bogotá Hamburg Johannesburg London Madrid
Mexico Montreal New Delhi Panama Paris
São Paulo Singapore Sydney Tokyo Toronto

Library of Congress Cataloging-in-Publication Data

Merritt, Frederick S.
    Civil engineering reference guide.

    (The McGraw-Hill engineering reference guide series)
    Includes index.
    1. Civil engineering—Handbooks, manuals, etc.
I. Kurtz, Max, 1920-   . II. Title.   III. Series.
TA151.M44   1986   624   86-7291
ISBN 0-07-041522-6

1234567890   DOC/DOC   8932109876

ISBN 0-07-041522-6

This book is a condensation of *Standard Handbook for Civil Engineers,* 3d ed.,
Frederick S. Merritt (ed.), copyright © 1983 by McGraw-Hill, Inc., all rights reserved.
Printed and bound by R. R. Donnelley & Sons, Inc.

Contents

Section 1

STRUCTURAL THEORY *by Frederick S. Merritt* **1-1**

EQUILIBRIUM

1-1. Types of Load;   1-2. Static Equilibrium

STRESS AND STRAIN

1-3. Unit Stress;   1-4. Unit Strain;   1-5. Stress-Strain Relations;   1-6. Constant Unit Stress;   1-7. Poisson's Ratio;   1-8. Thermal Stresses;   1-9. Axial Stresses in Composite Members;   1-10. Stresses in Pipes and Pressure Vessels;   1-11. Strain Energy

STRESSES AT A POINT

1-12. Stress Notation;   1-13. Stress Components;   1-14. Two-Dimensional Stress;   1-15. Principal Stresses;   1-16. Maximum Shearing Stress at a Point;   1-17. Mohr's Circle

STRAIGHT BEAMS

1-18. Beam-and-Girder Framing;   1-19. Types of Beams;   1-20. Reactions;   1-21. Internal Forces;   1-22. Shear Diagrams;   1-23. Bending-Moment Diagrams;   1-24. Shear-Moment Relationship;   1-25. Moving Loads and Influence Lines;   1-26. Maximum Bending Moment;   1-27. Bending Stresses in a Beam;   1-28. Moment of Inertia;   1-29. Section Modulus;   1-30. Shearing Stresses in a Beam;   1-31. Combined Shear and Bending Stress;   1-32. Beam Stresses in the Plastic Range;   1-33. Beam Deflections;   1-34. Combined Axial and Bending Loads;   1-35. Eccentric Loading;   1-36. Unsymmetrical Bending;   1-37. Beams with Unsymmetrical Sections

GRAPHIC-STATICS FUNDAMENTALS

1-38. Representation of a Force;   1-39. Parallelogram of Forces;   1-40. Resolution of Forces;   1-41. Force Polygons;   1-42. Equilibrium Polygon;   1-43. Beam and Truss Reactions by Graphics;   1-44. Shear and Moment Diagrams by Graphics

STEEL BRIDGES

CONCRETE BRIDGES

Section 6

SOIL AND ROCK CLASSIFICATION

SOIL PROPERTIES AND PARAMETERS

SITE INVESTIGATIONS

SHALLOW FOUNDATIONS

HAZARDOUS SITE AND FOUNDATION CONDITIONS

EARTH AND WATER PRESSURE ON FOUNDATIONS

SOIL IMPROVEMENT

Preface

This book is a condensation of the highly popular and widely acclaimed *Standard Handbook for Civil Engineers* (Frederick S. Merritt, editor; 1983). The criterion applied in selecting material from the source book was a simple one: to satisfy the needs of the largest possible number of practicing civil engineers.

The material in this book was prepared by individuals who are eminent authorities in their respective fields and who have a keen understanding of the needs of practitioners in those fields. The book presents a comprehensive review of civil engineering principles, it demonstrates how those principles are applied in solving practical problems, and it stresses the use of simplified procedures. Thus, the book combines theory and practice.

The following features of this book are particularly noteworthy:

1. The book contains a host of time-saving aids in the form of tables and charts. For example, Table 6-7 presents a direct solution of the rather intricate Eq. (6-44), using the parameters that exist most frequently in practice, and Table 8-4 presents a direct solution of the Manning formula as applied to fluid flow in pipes.

2. The book has an extensive system of cross-references to enhance its utility. As a result, the reader who enters a section at some intermediate point and then encounters an unfamiliar expression or principle can turn to the point in the book at which that expression or principle is introduced. This system of cross-references also links articles which are located in different parts of the book but are logically related, so the reader can obtain a broader perspective on each subject.

3. Illustrations are used extensively to convey information clearly and effectively. Thus, Fig. 7-46 exhibits all possible profiles of fluid flow in an open channel, and Figs. 4-9 and 4-10 show how a timber beam is connected to each possible type of support. Many illustrations are accompanied by highly descriptive legends that explain in detail the significance of the illustrations.

4. Section 7 gives step-by-step solutions to problems in fluid mechanics that arise frequently in practice. As an illustration, Example 7-8 analyzes the water hammer in a pipe caused by the gradual closing of a valve. The solution is performed very systematically with the use of Table 7-10.

5. References to other books appear at strategic locations throughout this book. These references enable the reader to delve more deeply into a particular subject if he or she wishes to do so.

Structural-design codes are subject to periodic revision, and therefore it is imperative that the designer have the current codes available. However, these code revisions are often relatively minor, and they can readily be assimilated by the engineer with a thorough grasp of design principles.

The index is comprehensive, and it was compiled assiduously to enable the reader to locate a required subject as quickly as possible.

Max Kurtz
Project editor

Contributors

Russell C. Brinker, Visiting Professor of Civil Engineering, New Mexico State University, University Park, NM *(Surveying)*

Roger L. Brockenbrough, Research Consultant, Research Laboratory, United States Steel Corporation, Monroeville, PA *(Structural-Steel Design)*

William S. Gardner, Consulting Engineer, Executive Vice President, Woodward-Clyde Consultants, Plymouth Meeting, PA *(Geotechnical Engineering)*

William T. Ingram, Consulting Engineer, Whitestone, NY *(Environmental Engineering)*

John J. Kozak, Former Chief, Division of Structures, California Department of Transportation, Sacramento, CA *(Bridge Engineering)*

I. Paul Lew, Consulting Engineer, Vice President, Lev Zetlin Associates, Inc., New York, NY *(Concrete Design and Construction)*

Frederick S. Merritt, Consulting Engineer, West Palm Beach, FL *(Structural Theory, Geotechnical Engineering)*

Samuel B. Nelson, Former Director of Public Works, State of California; Retired General Manager and Chief Engineer, Department of Water and Power, City of Los Angeles; Former General Manager, Southern California Rapid Transit District; Vice President, Daniel, Mann, Johnson, and Mendenhall; Vice Chairman, Board of Directors, Metropolitan Water District of Southern California; Member, California Water Commission *(Water Engineering)*

Maurice J. Rhude, President, Sentinel Structures, Inc., Peshtigo, WI *(Wood Design and Construction)*

James E. Roberts, Project Director, Sacramento Transit Development Agency, Sacramento, CA *(Bridge Engineering)*

Charles H. Thornton, Consulting Engineer; President, Lev Zetlin Associates, Inc., New York, NY *(Concrete Design and Construction)*

Section 9

SURVEYING *by Russell C. Brinker* 9-1

Section 7

WATER ENGINEERING *by Samuel B. Nelson* **7-1**

Section 8

ENVIRONMENTAL ENGINEERING *by William T. Ingram* **8-1**

Structural theory describes the behavior of structures under various types of loads and predicts the strength and deformations of structures. Design formulas and methods based on structural theory, when verified by laboratory and field tests and observations of structures under service conditions, insure that a structure subjected to specified loads will not suffer structural damage. Such damage exists when any portion of a structure is unable to function satisfactorily and may be indicated by excessive elastic deformation, inelastic deformation or yielding, fracture, or collapse.

To serve design and analysis needs, structural theory relates properties and arrangements of materials and the behavior of structures made of such materials. But if structural theory were to take into account every variable involved, it would become too complicated for practical use in most cases, so general practice is to make simplifying assumptions that yield consistent and sufficiently accurate results. Experience, experiments, and basic understanding often are required to determine whether a given theory or method is applicable to a particular structure.

Section 1

Structural Theory

Frederick S. Merritt

Consulting Engineer
West Palm Beach, FL

EQUILIBRIUM

1-1. Types of Load

Loads are the external forces acting on a structure. Stresses are the internal forces that resist the loads.

Tensile forces tend to stretch a component, **compressive forces** tend to shorten it, and **shearing forces** tend to slide parts of it past each other.

Loads also may be classified as static or dynamic. **Static loads** are forces that are applied slowly and then remain nearly constant, such as the weight, or dead load, of a floor system. **Dynamic loads** vary with time. They include repeated loads, such as alternating forces from oscillating machinery; moving loads, such as trucks or trains on bridges; impact loads, such as that from a falling weight striking a floor or the shock wave from an explosion impinging on a wall; and seismic loads or other forces created in a structure by rapid movements of supports.

Loads may be considered distributed or concentrated. **Uniformly distributed loads** are forces that are, or for practical purposes may be considered, constant over a surface of the supporting member; dead weight of a rolled-steel beam is a good example. **Concentrated loads** are forces that have such a small contact area as to be negligible compared with the entire surface area of the supporting member. For example, a beam supported on a girder, may, for all practical purposes, be considered a concentrated load on the girder.

In addition, loads may be axial, eccentric, or torsional. An **axial load** is a force whose resultant passes through the centroid of a section under consideration and is perpendicular to the plane of the section. An **eccentric load** is a force perpendicular to the plane of the section under consideration but not passing through the centroid of the section, thus bending the supporting member. **Torsional loads** are forces that are offset from the shear center of the section under consideration and are inclined to or in the plane of the section, thus twisting the supporting member.

Also, loads are classified according to the nature of the source. For example: **Dead loads** include materials, equipment, constructions, or other elements of weight supported in, on, or by a structural element, including its own weight, that are intended to remain permanently in place. **Live loads** include all occupants, materials, equipment, constructions, or other elements of weight supported in, on, or by a structural element that will or are likely to be moved or relocated during the expected life of the structure. **Impact loads** are a fraction of the live loads used to account for additional stresses and deflections resulting from movement of the live loads. **Wind loads** are maximum forces that may be applied to a structural element by wind in a mean recurrence interval, or a set of forces that will produce equivalent stresses. Mean recurrence intervals generally used are 25 years for structures with no occupants or offering negligible risk to life, 50 years for ordinary permanent structures, and 100 years for permanent structures with a high degree of sensitivity to wind and an unusually high degree of hazard to life and property in case of failure. **Snow loads** are maximum forces that may be applied by snow accumulation in a mean recurrence interval. **Seismic loads** are forces that produce maximum stresses or deformations in a structural element during an earthquake, or equivalent forces.

Probably maximum loads should be used in design. For buildings, minimum design load should be that specified for expected conditions in the local building code or, in the absence of an applicable local code, in the American National Standard "Building Code Requirements for Minimum Design Loads in Buildings and Other Structures," A58.1, American National Standards Institute, New York. For highways and highway bridges, minimum design loads should be those given in "Standard Specifications for Highway Bridges," American Association of State Highway and Transportation Officials, Washington, D.C. For railways and railroad bridges, minimum design loads should be those given in "Manual for Railway Engineering," American Railway Engineering Association, Chicago.

1-2. Static Equilibrium

If a structure and its components are so supported that after a small deformation occurs no further motion is possible, they are said to be in equilibrium. Under such circumstances, external forces are in balance and internal forces, or stresses, exactly counteract the loads.

Since there is no translatory motion, the vector sum of the external forces must be zero. Since there is no rotation, the sum of the moments of the external forces about any point must be zero. For the same reason, if we consider any portion of the structure and the loads on it, the sum of the external and internal forces on the boundaries of that section must be zero. Also, the sum of the moments of these forces must be zero.

In Fig. 1-1, for example, the sum of the forces R_L and R_R needed to support the truss is equal to the 20-kip load on the truss (1 **kip** = 1 kilopound = 1000 lb = 0.5 ton). Also, the sum of the moments of the external forces is zero about any point; about the right end, for instance, it is $40 \times 15 - 30 \times 20 = 600 - 600$.

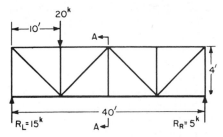

Fig. 1-1. Truss in equilibrium under load. Upward-acting forces, or reactions, R_L and R_R, equal the 20-kip downward-acting force.

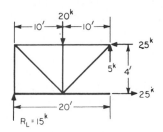

Fig. 1-2. Section of the truss shown in Fig. 1-1 is kept in equilibrium by stresses in the components.

Figure 1-2 shows the portion of the truss to the left of section AA. The internal forces at the cut members balance the external load and hold this piece of the truss in equilibrium.

When the forces act in several directions, it generally is convenient to resolve them into components parallel to a set of perpendicular axes that will simplify computations. For example, for forces in a single plane, the most useful technique is to resolve them into horizontal and vertical components. Then, for a structure in equilibrium, if H represents the horizontal components, V the vertical components, and M the moments of the components about any point in the plane,

$$\Sigma H = 0 \qquad \Sigma V = 0 \qquad \text{and} \qquad \Sigma M = 0 \qquad (1\text{-}1)$$

These three equations may be used to determine three unknowns in any nonconcurrent coplanar force system, such as the truss in Figs. 1-1 and 1-2. They may determine the magnitude of three forces for which the direction and point of application already are known, or the magnitude, direction, and point of application of a single force. Suppose, for the truss in Fig. 1-1, the reactions at the supports are to be computed. Take the sum of the moments about the right support and equate them to zero to determine the left reaction: $40R_L - 30 \times 20 = 0$, from which $R_L = 600/40 = 15$ kips. To find the right reaction, take moments about the left support and equate the sum to zero: $10 \times 20 - 40R_R = 0$, from which $R_R = 5$ kips. As an alternative, equate the sum of the vertical forces to zero to obtain R_R after finding R_L: $20 - 15 - R_R = 0$, from which $R_R = 5$ kips.

STRESS AND STRAIN

1-3. Unit Stress

It is customary to give the strength of a material in terms of unit stress, or internal force per unit of area. Also, the point at which yielding starts generally is expressed as a *unit stress*. Then, in some design methods, a safety factor is applied to either of these stresses to determine a unit stress that should not be exceeded when the member carries design loads. That unit stress is known as the *allowable stress,* or *working stress.*

In working-stress design, to determine whether a structural member has adequate load-carrying capacity, the designer generally has to compute the maximum unit stress produced by design loads in the member for each type of internal force—tensile, compressive, or shearing—and compare it with the corresponding allowable unit stress.

When the loading is such that the unit stress is constant over a section under consideration, the stress may be computed by dividing the force by the area of the section. But, generally, the unit stress varies from point to point. In those cases, the unit stress at any point in the section is the limiting value of the ratio of the internal force on any small area to that area, as the area is taken smaller and smaller.

1-4. Unit Strain

Sometimes in the design of a structure, the designer may be more concerned with limiting deformation or strain than with strength. Deformation in any direction is the total change in the dimension of a member in that direction. *Unit strain* in any direction is the deformation per unit of length in that direction.

When the loading is such that the unit strain is constant over the length of a member, it may be computed by dividing the deformation by the original length of the member. In general, however, unit strain varies from point to point in a member. Like a varying unit stress, it represents the limiting value of a ratio (Art. 1-3).

1-5. Stress-Strain Relations

When a material is subjected to external forces, it develops one or more of the following types of strain: linear elastic, nonlinear elastic, viscoelastic, plastic, and anelastic. Many structural materials exhibit linear elastic strains under design loads. For these materials, unit strain is proportional to unit stress until a certain stress, the proportional limit, is exceeded (point *A* in Fig. 1-3*a* to *c*). This relationship is known as **Hooke's law.**

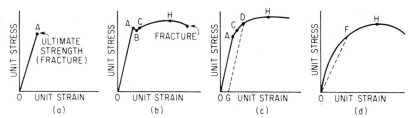

Fig. 1-3. Relationship of unit stress and unit strain for various materials. (*a*) Brittle. (*b*) Linear elastic with a distinct proportional limit. (*c*) Linear elastic with an indistinct proportional limit. (*d*) Nonlinear.

For axial tensile or compressive loading, this relationship may be written

$$f = E\epsilon \quad \text{or} \quad \epsilon = \frac{f}{E} \tag{1-2}$$

where f = unit stress
  ϵ = unit strain
  E = Young's modulus of elasticity

Within the elastic limit, there is no permanent residual deformation when the load is removed. Structural steels have this property.

In nonlinear elastic behavior, stress is not proportional to strain, but there is no permanent residual deformation when the load is removed. The relation between stress and strain may take the form

$$\epsilon = \left(\frac{f}{K}\right)^n \tag{1-3}$$

where K = pseudoelastic modulus determined by test
  n = constant determined by test

Viscoelastic behavior resembles linear elasticity. The major difference is that in linear elastic behavior, the strain stops increasing if the load does; but in viscoelastic behavior, the strain continues to increase although the load becomes constant and a residual strain remains when the load is removed. This is characteristic of many plastics.

Anelastic deformation is time-dependent and completely recoverable. Strain at any time is proportional to change in stress. Behavior at any given instant depends on all prior stress changes. The combined effect of several stress changes is the sum of the effects of the several stress changes taken individually.

Plastic strain is not proportional to stress, and a permanent deformation remains on removal of the load. In contrast with anelastic behavior, plastic deformation depends primarily on the stress and is largely independent of prior stress changes.

When materials are tested in axial tension and corresponding stresses and strains are plotted, stress-strain curves similar to those in Fig. 1-3 result. Figure 1-3a is typical of a brittle material, which deforms in accordance with Hooke's law up to fracture. The other curves in Fig. 1-3 are characteristic of ductile materials; because strains increase rapidly near fracture with little increase in stress, they warn of imminent failure, whereas brittle materials fail suddenly.

Figure 1-3b is typical of materials with a marked proportional limit A. When this is exceeded, there is a sudden drop in stress, then gradual stress increase with large increases in strain to a maximum before fracture. Figure 1-3c is characteristic of materials that are linearly elastic over a substantial range but have no definite proportional limit. And Fig. 1-3d is a representative curve for materials that do not behave linearly at all.

Modulus of Elasticity ▪ E is given by the slope of the straight-line portion of the curves in Fig. 1-3a to c. It is a measure of the inherent rigidity or stiffness of a material. For a given geometric configuration, a material with a larger E deforms less under the same stress.

At the termination of the linear portion of the stress-strain curve, some materials, such as low-carbon steel, develop an upper and lower **yield point** (A and B in Fig. 1-3b). These points mark a range in which there appears to be an increase in strain with no increase or a small decrease in stress. This behavior may be a consequence of inertia effects in the testing machine and the deformation characteristics of the test specimen. Because of the location of the yield points, the yield stress sometimes is used erroneously as a synonym for proportional limit and elastic limit.

The **proportional limit** is the maximum unit stress for which Hooke's law is valid. The **elastic limit** is the largest unit stress that can be developed without a permanent set remaining after removal of the load (C in Fig. 1-3). Since the elastic limit is always difficult to determine and many materials do not have a well-defined proportional limit, or even have one at all, the offset yield strength is used as a measure of the beginning of plastic deformation.

The **offset yield strength** is defined as the stress corresponding to a permanent deformation, usually 0.01% (0.0001 in/in) or 0.20% (0.002 in/in). In Fig. 1-3c the yield strength is the stress at D, the intersection of the stress-strain curve and a line GD parallel to the straight-line portion and starting at the given unit strain. This stress sometimes is called the **proof stress.**

For materials with a stress-strain curve similar to that in Fig. 1-3d, with no linear portion, a **secant modulus,** represented by the slope of a line, such as OF, from the origin to a specified point on the curve, may be used as a measure of stiffness. An alternative measure is the **tangent modulus,** the slope of the stress-strain curve at a specified point.

Ultimate tensile strength is the maximum axial load observed in a tension test divided by the original cross-sectional area. Characterized by the beginning of necking down, a decrease in cross-sectional area of the specimen, or local instability, this stress is indicated by H in Fig. 1-3.

Ductility is the ability of a material to undergo large deformations without fracture. It is measured by elongation and reduction of area in a tension test and expressed as a percentage. Ductility depends on temperature and internal stresses as well as the characteristics of the material; a material that may be ductile under one set of conditions may have a brittle failure at lower temperatures or under tensile stresses in two or three perpendicular directions.

Modulus of rigidity, or shearing modulus of elasticity, is defined by

$$G = \frac{v}{\gamma} \qquad (1\text{-}4)$$

where G = modulus of rigidity
v = unit shearing stress
γ = unit shearing strain

It is related to the modulus of elasticity in tension and compression E by the equation

$$G = \frac{E}{2(1 + \mu)} \qquad (1\text{-}5)$$

where μ is a constant known as Poisson's ratio (Art. 1-7).

Toughness is the ability of a material to absorb large amounts of energy. Related to the area under the stress-strain curve, it depends on both strength and ductility. Because of the difficulty of determining toughness analytically, often toughness is measured by the energy required to fracture a specimen, usually notched and sometimes at low temperatures, in impact tests. Charpy and Izod, both applying a dynamic load by pendulum, are the tests most commonly used.

Hardness is a measure of the resistance a material offers to scratching and indention. A relative numerical value usually is determined for this property in such tests as Brinell, Rockwell, and Vickers. The numbers depend on the size of an indentation made under a standard load. Scratch resistance is measured on the Mohs scale by comparison with the scratch resistance of 10 minerals arranged in order of increasing hardness from talc to diamond.

Creep is a gradual flow or change in dimension under sustained constant load. **Relaxation** is a decrease in load or stress under a sustained constant deformation.

If stresses and strains are plotted in an axial tension test as a specimen enters the inelastic range and then is unloaded, the curve during unloading, if the material was elastic, descends parallel to the straight portion of the curve (for example, DG in Fig. 1-3c). Completely unloaded, the specimen has a permanent set (OG). This also will occur in compression tests.

If the specimen now is reloaded, strains are proportional to stresses (the curve will practically follow DG) until the curve rejoins the original curve at D. Under increasing load, the reloading curve coincides with that for a single loading. Thus, loading the specimen into the inelastic range, but not to ultimate strength, increases the apparent elastic range. The phenomenon, called **strain hardening,** or work hardening, appears to increase the yield strength.

But if the reloading is in compression, the compressive yield strength is decreased, which is called the **Bauschinger effect.** This, however, is present only for relatively small strains. For large inelastic tensile strains initially, reloading in compression increases yield strength to some extent. But if this reloading is continued to a higher stress than that reached in the initial loading in tension, the yield strength will not show any increase in subsequent tension loading.

1-6. Constant Unit Stress

The simplest cases of stress and strain are those in which the unit stress and strain are constant. Stresses caused by an axial tension or compression load, a centrally applied shear, or a bearing load are examples. These conditions are illustrated in Figs. 1-4 to 1-7.

For constant unit stress, the equation of equilibrium may be written

$$P = Af \qquad (1\text{-}6)$$

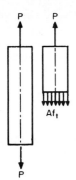

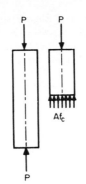

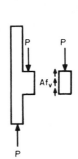

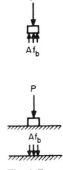

Fig. 1-4.
Tension member axially
loaded.

Fig. 1-5.
Compression member
axially loaded.

Fig. 1-6.
Bracket in shear.

Fig 1-7.
Bearing load.

where P = load, lb

A = cross-sectional area (normal to load) for tensile or compressive forces, or area on
which sliding may occur for shearing forces, or contact area for bearing loads, in^2

f = tensile, compressive, shearing, or bearing unit stress, psi

For torsional stresses, see Art. 1-76.

Unit strain for the axial tensile and compressive loads is given by

$$\epsilon = \frac{e}{L} \tag{1-7}$$

where ϵ = unit strain, in/in

e = total lengthening or shortening of member, in

L = original length of the member, in

Application of Hooke's law and Eq. (1-6) to Eq. (1-7) yields a convenient formula for the
deformation:

$$e = \frac{PL}{AE} \tag{1-8}$$

where P = load on member, lb

A = its cross-sectional area, in^2

E = modulus of elasticity of material, psi

[Since long compression members tend to buckle, Eq. (1-6) to (1-8) are applicable only to
short members. See Arts. 1-72 to 1-74.]

Although tension and compression strains represent a simple stretching or shortening of a
member, shearing strain is a distortion due to a small rotation. The load on the small rectangular portion of the member in Fig. 1-6 tends to distort it into a parallelogram. The unit shearing strain is the change in the right angle, measured in radians. (See also Art. 1-5.)

1-7. Poisson's Ratio

When a material is subjected to axial tensile or compressive loads, it deforms not only in the
direction of the loads but normal to them. Under tension, the cross section of a member
decreases, and under compression, it increases. The ratio of the unit lateral strain to the unit
longitudinal strain is called *Poisson's ratio*.

Within the elastic range, Poisson's ratio is a constant for a material. For materials such as concrete, glass, and ceramics, it may be taken as 0.25; for structural steel, 0.3. It gradually increases beyond the proportional limit and tends to approach a value of 0.5.

Assume, for example, that a steel hanger with an area of 2 in² carries a 40-kip (40,000-lb) load. The unit stress is 40,000/2 or 20,000 psi. The unit tensile strain, with modulus of elasticity of steel $E = 30,000,000$, is 20,000/30,000,000, or 0.00067 in/in. With Poisson's ratio as 0.3, the unit lateral strain is -0.3×0.00067, or a shortening of 0.00020 in/in.

1-8. Thermal Stresses

When the temperature of a body changes, its dimensions also change. Forces are required to prevent such dimensional changes, and stresses are set up in the body by these forces.

If α is the coefficient of expansion of the material and T the change in temperature, the unit strain in a bar restrained by external forces from expanding or contracting is

$$\epsilon = \alpha T \tag{1-9}$$

According to Hooke's law, the stress f in the bar is

$$f = E\alpha T \tag{1-10}$$

where E = modulus of elasticity.

When a circular ring, or hoop, is heated and then slipped over a cylinder of slightly larger diameter d than d_1, the original hoop diameter, the hoop will develop a tensile stress on cooling. If the diameter is very large compared with the hoop thickness, so that radial stresses can be neglected, the unit tensile stresses may be assumed constant. The unit strain will be

$$\epsilon = \frac{\pi d - \pi d_1}{\pi d_1} = \frac{d - d_1}{d_1}$$

and the hoop stress will be

$$f = \frac{(d - d_1)E}{d_1} \tag{1-11}$$

1-9. Axial Stresses in Composite Members

In a homogeneous material, the centroid of a cross section lies at the intersection of two perpendicular axes so located that the moments of the areas on opposite sides of an axis about that axis are zero. To find the centroid of a cross section containing two or more materials, the moments of the products of the area A of each material and its modulus of elasticity E should be used, in the elastic range.

Consider now a prism composed of two materials, with moduius of elasticity E_1 and E_2, extending the length of the prism. If the prism is subjected to a load acting along the centroidal axis, then the unit strain ϵ in each material will be the same. From the equation of equilibrium and Eq. (1-8), noting that the length L is the same for both materials,

$$\epsilon = \frac{P}{A_1 E_1 + A_2 E_2} = \frac{P}{\Sigma A E} \tag{1-12}$$

where A_1 and A_2 are the cross-sectional areas of each material and P the axial load. The unit stresses in each material are the products of the unit strain and its modulus of elasticity:

$$f_1 = \frac{PE_1}{\Sigma AE} \qquad f_2 = \frac{PE_2}{\Sigma AE} \tag{1-13}$$

1-10. Stresses in Pipes and Pressure Vessels

In a cylindrical pipe under internal radial pressure, the circumferential unit stresses may be assumed constant over the thickness of the pipe if the diameter is relatively large compared with the thickness (at least 15 times as large). Then, the circumferential unit stress, in pounds per square inch, is given by

$$f = \frac{pR}{t} \tag{1-14}$$

where p = internal pressure, psi
   R = average radius of pipe, in
   In a closed cylinder, the pressure against the ends will be resisted by longitudinal stresses in the cylinder. If the cylinder is thin, these stresses, psi, are given by

$$f_z = \frac{pR}{2t} \tag{1-15}$$

Equation (1-15) also holds for the stress in a thin spherical tank under internal pressure p with R the average radius.

In a thick-walled cylinder, the effect of radial stresses f_r becomes important. Both radial and circumferential stresses may be computed from Lamé's formulas:

$$f_r = p \frac{r_i^2}{r_o^2 - r_i^2} \left(1 - \frac{r_o^2}{r^2} \right) \tag{1-16}$$

$$f = p \frac{r_i^2}{r_o^2 - r_o^2} \left(1 + \frac{r_o^2}{r^2} \right) \tag{1-17}$$

where r_i = internal radius of cylinder, in
   r_o = outside radius of cylinder, in
   r = radius to point where stress is to be determined, in
The equations show that if the pressure p acts outward, the circumferential stress f will be tensile (positive) and the radial stress compressive (negative). The greatest stresses occur at the inner surface of the cylinder ($r = r_i$):

$$\text{Max } f_r = -p \tag{1-18}$$

$$\text{Max } f = \frac{k^2 + 1}{k^2 - 1} p \tag{1-19}$$

where $k = r_o/r_i$. Maximum shear stress is given by

$$\text{Max } f_v = \frac{k^2}{k^2 - 1} p \tag{1-20}$$

For a closed cylinder with thick walls, the longitudinal stress is approximately

$$f_z = \frac{p}{r_i(k^2 - 1)} \tag{1-21}$$

But because of end restraints, this stress will not be correct near the ends.

   (S. Timoshenko and J. N. Goodier, "Theory of Elasticity," McGraw-Hill Book Company, New York.)

1-11. Strain Energy

Stressing a bar stores energy in it. For an axial load P and a deformation e, the energy stored is

$$U = \frac{1}{2} Pe \qquad (1\text{-}22\,a)$$

assuming the load is applied gradually and the bar is not stressed beyond the proportional limit. The equation represents the area under the load-deformation curve up to the load P. Application of Eqs. (1-2) and (1-6) to Eq. (1-22a) yields another useful equation for energy, in inch-pounds:

$$U = \frac{f^2}{2E} AL \qquad (1\text{-}22\,b)$$

where f = unit stress, psi
    E = modulus of elasticity of material, psi
    A = cross-sectional area in^2
    L = length of bar, in
    Since AL is the volume of the bar, the term $f^2/2E$ gives the energy stored per unit of volume. It represents the area under the stress-strain curve up to the stress f.

Modulus of resilience is the energy stored per unit of volume in a bar stressed by a gradually applied axial load up to the proportional limit. This modulus is a measure of the capacity of the material to absorb energy without danger of being permanently deformed. It is important in designing members to resist energy loads.

Equation (1-22a) is a general equation that holds true when the **principle of superposition** applies (the total deformation produced at a point by a system of forces is equal to the sum of the deformations produced by each force). In the general sense, P in Eq. (1-22a) represents any group of statically interdependent forces that can be completely defined by one symbol, and e is the corresponding deformation.

The strain-energy equation can be written as a function of either the load or the deformation. For axial tension or compression, strain energy, in inch-pounds, is given by

$$U = \frac{P^2 L}{2AE} \qquad U = \frac{AEe^2}{2L} \qquad (1\text{-}23\,a)$$

where P = axial load, lb
    e = total elongation or shortening, in
    L = length of member, in
    A = cross-sectional area, in^2
    E = modulus of elasticity, psi
    For pure shear:

$$U = \frac{V^2 L}{2AG} \qquad U = \frac{AGe^2}{2L} \qquad (1\text{-}23\,b)$$

where V = shearing load, lb
    e = shearing deformation, in
    L = length over which deformation takes place, in
    A = shearing area, in^2
    G = shearing modulus, psi
    For torsion:

$$U = \frac{T^2 L}{2JG} \qquad U = \frac{JG\phi^2}{2L} \qquad (1\text{-}23\,c)$$

where T = torque, in-lb
ϕ = angle of twist, rad
L = length of shaft, in
J = polar moment of inertia of cross section, in^4
G = shearing modulus, psi
For pure bending (constant moment):

$$U = \frac{M^2 L}{2EI} \qquad U = \frac{EI\theta^2}{2L}$$ (1-23 d)

where M = bending moment, in-lb
θ = angle of rotation of one end of beam with respect to other, rad
L = length of beam, in
I = moment of inertia of cross section, in^4
E = modulus of elasticity, psi

For beams carrying transverse loads, the total strain energy is the sum of the energy for bending and that for shear. (See also Art. 1-55.)

STRESSES AT A POINT

Tensile and compressive stresses sometimes are referred to as *normal stresses* because they act normal to the cross section. Under this concept, tensile stresses are considered positive normal stresses and compressive stresses negative.

1-12. Stress Notation

Consider a small cube extracted from a stressed member and placed with three edges along a set of x, y, z coordinate axes. The notations used for the components of stress acting on the sides of this element and the direction assumed as positive are shown in Fig. 1-8.

For example, for the sides of the element perpendicular to the z axis, the normal component of stress is denoted by f_z. The shearing stress v is resolved into two components and requires two subscript letters for a complete description. The first letter indicates the direction of the normal to the plane under consideration; the second letter gives the direction of the component of stress. Thus, for the sides perpendicular to the z axis, the shear component in the x direction is labeled v_{zx} and that in the y direction v_{zy}.

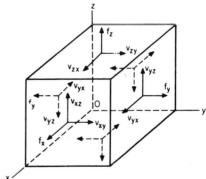

Fig. 1-8.   Stresses at a point in a rectangular coordinate system.

1-13. Stress Components

If, for the small cube in Fig. 1-8, moments of the forces acting on it are taken about the x axis, and assuming the lengths of the edges as dx, dy, and dz, the equation of equilibrium requires that

$$(v_{zy} \, dx \, dy) \, dz = (v_{yz} \, dx \, dz) \, dy$$

(Forces are taken equal to the product of the area of the face and the stress at the center.) Two similar equations can be written for moments taken about the y and z axes. These equations show that

$$v_{xy} = v_{yx} \qquad v_{zx} = v_{xz} \qquad v_{zy} = v_{yz} \tag{1-24}$$

Thus, components of shearing stress on two perpendicular planes and acting normal to the intersection of the planes are equal. Consequently, to describe the stresses acting on the coordinate planes through a point, only six quantities need be known: the three normal stresses f_x, f_y, f_z and three shearing components $v_{xy} = v_{yx}, v_{zx} = v_{xz}$, and $v_{zy} = v_{yz}$.

If only the normal stresses are acting, the unit strains in the x, y, and z directions are

$$\epsilon_x = \frac{1}{E} [f_x - \mu(f_y + f_z)]$$

$$\epsilon_y = \frac{1}{E} [f_y - \mu(f_x + f_z)] \tag{1-25}$$

$$\epsilon_z = \frac{1}{E} [f_z - \mu(f_x + f_y)]$$

where μ = Poisson's ratio. If only shearing stresses are acting, the distortion of the angle between edges parallel to any two coordinate axes depends only on shearing-stress components parallel to those axes. Thus, the unit shearing strains are (see Art. 1-5)

$$\gamma_{xy} = \frac{1}{G} v_{xy} \qquad \gamma_{yz} = \frac{1}{G} v_{yz} \qquad \gamma_{zx} = \frac{1}{G} v_{zx} \tag{1-26}$$

1-14. Two-Dimensional Stress

When the six components of stress necessary to describe the stresses at a point are known (Art. 1-13), the stresses on any inclined plane through the same point can be determined. For two-dimensional stress, only three stress components need be known.

Assume, for example, that at a point O in a stressed plate, the components f_x, f_y, and v_{xy} are known (Fig. 1-9). To find the stresses on any other plane through the z axis, take a plane parallel to it close to O, so that this plane and the coordinate planes form a tiny trianglular prism. Then, if α is the angle the normal to the plane makes with the x axis, the normal and shearing stresses on the inclined plane, to maintain equilibrium, are

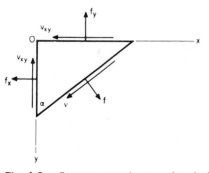

$$f = f_x \cos^2 \alpha + f_y \sin^2 \alpha + 2v_{xy} \sin \alpha \cos \alpha \tag{1-27}$$

$$v = v_{xy}(\cos^2 \alpha - \sin^2 \alpha) + (f_y - f_x) \sin \alpha \cos \alpha \tag{1-28}$$

(See also Art. 1-17.)

Fig. 1.9. Stresses at a point on a plane inclined to the axes.

1-15. Principal Stresses

If a plane at a point O in a stressed plate is rotated, it reaches a position for which the normal stress on it is a maximum or a minimum. The directions of maximum and minimum normal stress are perpendicular to each other, and on the planes in those directions, there are no shearing stresses.

The directions in which the normal stresses become maximum or minimum are called *principal directions,* and the corresponding normal stresses are called *principal stresses.* To find the principal directions, set the value of v given by Eq. (1-28) equal to zero. Then, the normals to the principal planes make an angle with the x axis given by

$$\tan 2\alpha = \frac{2v_{xy}}{f_x - f_y} \tag{1-29}$$

If the x and y axes are taken in the principal directions, $v_{xy} = 0$. In that case, Eqs. (1-27) and (1-28) simplify to

$$f = f_x \cos^2 \alpha + f_y \sin^2 \alpha \tag{1-30}$$

$$v = \frac{1}{2}(f_y - f_x) \sin 2\alpha \tag{1-31}$$

where f_x and f_y are the principal stresses at the point, and f and v are, respectively, the normal and shearing stress on a plane whose normal makes an angle α with the x axis.

If only shearing stresses act on any two perpendicular planes, the state of stress at the point is said to be one of pure shear or simple shear. Under such conditions, the principal directions bisect the angles between the planes on which these shearing stresses act. The principal stresses are equal to in magnitude to the pure shears.

1-16. Maximum Shearing Stress at a Point

The maximum unit shearing stress occurs on each of two planes that bisect the angles between the planes on which the principal stresses at a point act. The maximum shear equals half the algebraic difference of the principal stresses:

$$\text{Max } v = \frac{f_1 - f_2}{2} \tag{1-32}$$

where f_1 is the maximum principal stress and f_2 the minimum.

1-17. Mohr's Circle

As explained in Art. 1-14, if the stresses on any plane through a point in a stressed plate are known, the stresses on any other plane through the point can be computed. This relationship between the stresses may be represented conveniently on Mohr's circle (Fig. 1-10). In this diagram, normal stress f and shear stress v are taken as rectangular coordinates. Then, for each plane through the point there will correspond a point on the circle, the coordinates of which are the values of f and v for the plane.

Given the principal stresses f_1 and f_2 (Art. 1-15), to find the stresses on a plane making an angle α with the plane on which f_1 acts: Mark off the principal stresses on the f axis (points A and B in Fig. 1-10). Measure tensile stresses to the right of the v axis and compressive stresses to the left. Construct a circle passing through A and B and having its center on the f axis. This is the Mohr's circle for the given stresses at the point under consideration. Draw a radius making an angle 2α with the f axis, as indicated in Fig. 1-10. The coordinates of the intersection with the circle represent the normal and shearing stresses acting on the plane, f and v.

Given the stresses on any two perpendicular planes f_x, f_y, and v_{xy}, but not the principal stresses f_1 and f_2, to draw the Mohr's circle: Plot the two points representing the known stresses with respect to the f and v axes (points C and D in Fig. 1-11). The line joining these points is

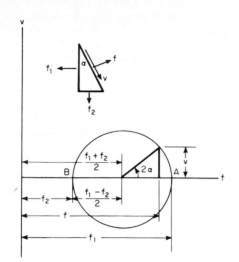

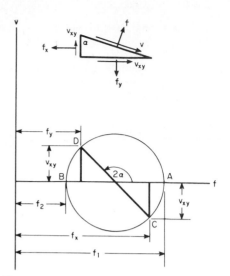

Fig. 1-10. Mohr's circle for stresses at a point—constructed from known principal stresses f_1 and f_2 in a plane.

Fig. 1-11. Stress circle constructed from two known normal positive stresses f_x and f_y and a known shear v_{xy}.

a diameter of the circle, so bisect CD to find the center of the circle and draw the circle. Its intersections with the f axis determine f_1 and f_2.

(S. Timoshenko and J. N. Goodier, "Theory of Elasticity," McGraw-Hill Book Company, New York.)

STRAIGHT BEAMS

1-18. Beam-and-Girder Framing

Bridge decks and building floors and roofs frequently are supported on a rectangular grid of flexural members. Different names often are given to the components of the grid, depending on the type of structure and the part of the structure supported on the grid. In general, though, the members spanning between main supports are called **girders** and those they support are called **beams** (Fig. 1-12). Hence, this type of framing is known as beam-and-girder framing.

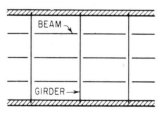

Fig. 1-12. Beam-and-girder framing.

In bridges, the smaller structural members parallel to the direction in which traffic moves may be called **stringers** and the transverse members **floor beams.** In building roofs, the grid components may be referred to as **purlins** and **rafters;** and in floors, they may be called **joists** and **girders.**

Beam-and-girder framing usually is used for relatively short spans and where shallow members are desired to provide ample headroom underneath.

1-19. Types of Beams

There are many ways in which beams may be supported. Some of the most common methods are shown Fig. 1-13 to 1-19. The beam in Fig 1-13 is called a simply supported beam, or

Fig. 1-13. Simple beam, both ends free to rotate.

Fig. 1-14. Cantilever beam.

Fig. 1-15. Beam with one end fixed.

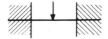

Fig. 1-16. Fixed-end beam.

Fig. 1-17. Beam with overhangs.

Fig. 1-18. Continuous beam.

Fig. 1-19. Hung-span (suspended-span) construction.

simple beam. It has supports near its ends that restrain it only against vertical movement. The ends of the beam are free to rotate. When the loads have a horizontal component, or when change in length of the beam due to temperature may be important, the supports may also have to prevent horizontal motion, in which case horizontal restraint at one support generally is sufficient. The distance between the supports is called the **span.** The load carried by each support is called a **reaction.**

The beam in Fig. 1-14 is a **cantilever.** It has a support only at one end. The support provides restraint against rotation and horizontal and vertical movement. Such support is called a **fixed end.** Placing a support under the free end of the cantilever produces the beam in Fig. 1-15. Fixing the free end yields a **fixed-end beam** (Fig. 1-16); no rotation or vertical movement can occur at either end. In actual practice, however, a fully fixed end can seldom be obtained. Most support conditions are intermediate between those for a simple beam and those for a fixed-end beam.

Figure 1-17 shows a beam that overhangs both its simple supports. The overhangs have a free end, like a cantilever, but the supports permit rotation.

Two types of beams that extend over several supports are illustrated in Figs. 1-18 and 1-19. Figure 1-18 shows a **continuous beam.** The one in Fig. 1-19 has one or two hinges in certain spans; it is called **hung-span,** or suspended-span, construction. In effect, it is a combination of simple beams and beams with overhangs.

Reactions for the beams in Figs. 1-13, 1-14, and 1-17 and the type of beam in Fig. 1-19 with hinges suitably located may be found from the equations of equilbrium, which is why they are classified as **statically determinate beams.**

The equations of equilibrium, however, are not sufficient to determine the reactions of the beams in Figs. 1-15, 1-16, and 1-18. For those beams, there are more unknowns than equations. Additional equations must be obtained based on a knowledge of the deformations, for example, that a fixed end permits no rotation. Such beams are classified as **statically indeterminate.** Methods for finding the stresses in that type of beam are given in Arts. 1-52 to 1-70.

1-20. Reactions

As pointed out in Art. 1-19, the loads imposed by a simple beam on its supports can be found by application of the equations of equilibrium [Eq. (1-1)]. Consider, for example, the 60-ft-long beam with overhangs in Fig. 1-20. This beam carries a uniform load of 200 lb/lin ft over its entire length and several concentrated loads. The span is 36 ft.

To find reaction R_1, take moments about R_2 and equate the sum of the moments to zero (assume clockwise rotation to be positive, counterclockwise, negative):

$$-2000 \times 48 + 36R_1 - 4000 \times 30 - 6000 \times 18 + 3000 \times 12 - 200 \times 60 \times 18 = 0$$
$$R_1 = 14{,}000 \text{ lb}$$

In this calculation, the moment of the uniform load was found by taking the moment of its resultant, 200×60, which acts at the center of the beam.

To find R_2, proceed in a similar manner by taking moments about R_1 and equating the sum to zero, or equate the sum of the vertical forces to zero. Generally it is preferable to use the moment equation and apply the other equation as a check.

As an alternative procedure, find the reactions caused by uniform and concentrated loads separately and sum the results. Use the fact that the reactions due to symmetrical loading are equal, to simplify the calculation. To find R_2 by this procedure, take half the total uniform load

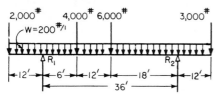

Fig. 1-20. Beam with overhangs loaded with both uniform and concentrated loads.

$$0.5 \times 200 \times 60 = 6000 \text{ lb}$$

and add it to the reaction caused by the concentrated loads, found by taking moments about R_1, dividing by the span, and summing:

$$-2000 \times \frac{12}{36} + 4000 \times \frac{6}{36} + 6000 \times \frac{18}{36} + 3000 \times \frac{48}{36} = 7000 \text{ lb}$$
$$R_2 = 6000 + 7000 = 13{,}000 \text{ lb}$$

Check to see that the sum of the reactions equals the total applied load:

$$14{,}000 + 13{,}000 = 2000 + 4000 + 6000 + 3000 + 200 \times 60$$
$$27{,}000 = 27{,}000$$

Reactions for simple beams with various loads are given in Figs. 1-33 to 1-38.

To find the reactions of a continuous beam, first determine the end moments and shears (Arts. 1-60 to 1-70); then if the continuous beam is considered as a series of simple beams with these applied as external loads, the beam will be statically determinate and the reactions can be determined from the equations of equilibrium. (For an alternative method, see Art. 1-59.)

1-21. Internal Forces

At every section of a beam in equilibrium, internal forces act to prevent motion. For example, assume the beam in Fig. 1-20 cut vertically just to the right of its center. Adding the external forces, including the reaction, to the left of this cut (see Fig. 1-21 a) yields an unbalanced down-

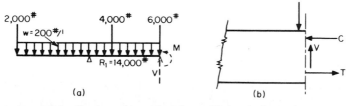

Fig. 1-21. Sections of beam kept in equilibrium by internal stresses.

ward load of 4000 lb. Evidently, at the cut section, an upward-acting internal force of 4000 lb must be present to maintain equilibrium. Also, taking moments of the external forces about the section yields an unbalanced moment of 54,000 ft-lb. To maintain equilibrium, there must be an internal moment of 54,000 ft-lb resisting it.

This internal, or resisting, moment is produced by a couple consisting of a force C acting on the top part of the beam and an equal but opposite force T acting on the bottom part (Fig. 1-21b). For this type of beam and loading, the top force is the resultant of compressive stresses acting over the upper portion of the beam, and the bottom force is the resultant of tensile stresses acting over the bottom part. The surface at which the stresses change from compression to tension—where the stress is zero—is called the **neutral surface.**

1-22. Shear Diagrams

As explained in Art. 1-21, at a vertical section through a beam in equilibrium, external forces on one side of the section are balanced by internal forces. The unbalanced external vertical force at the section is called the shear. It equals the algebraic sum of the forces that lie on either side of the section. For forces on the left of the section, those acting upward are considered positive and those acting downward negative. For forces on the right of the section, signs are reversed.

A shear diagram represents graphically the shear at every point along the length of a beam. The shear diagram for the beam in Fig. 1-20 is shown in Fig. 1-22b. The beam is drawn to scale and the loads and reactions are located at the points at which they act. Then, a convenient zero axis is drawn horizontally from which to plot the shears to scale. Start at the left end of the beam, and directly under the 2000-lb load there, scale off -2000 from the zero axis. Next, determine the shear just to the left of the next concentrated load, the left support: $-2000 - 200 \times 12 = -4400$ lb. Plot this downward under R_1. Note that in passing from just to the left of the support to just to the right, the shear changes by the magnitude of the reaction, from -4400 to $-4400 + 14,000$ or 9600 lb, so plot this value also under R_1. Under the 4000-lb load, plot the shear just to the left of it, 9600 $-$

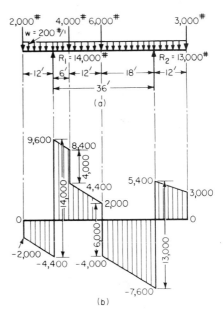

Fig. 1-22.
Shear diagram for beam in Fig. 1-20.

200×6, or 8400 lb, and the shear just to the right, $8400 - 4000$, or 4400 lb. Proceed in this manner to the right end, where the shear is 3000 lb, equal to the load on the free end.

To complete the diagram, the points must be connected. Straight lines can be used because shear varies uniformly for a uniform load (see Fig. 1-24b).

1-23. Bending-Moment Diagrams

About a vertical section through a beam in equilibrium, there is an unbalanced moment due to external forces, called *bending moment*. For forces on the left of the section, clockwise moments are considered positive and counterclockwise moments negative. For forces on the right of the section, the signs are reversed. Thus, when the bending moment is positive, the bottom of a simple beam is in tension and the top is in compression.

A bending-moment diagram represents graphically the bending moment of every point along the length of the beam. Figure 1-23c is the bending-moment diagram for the beam with concentrated loads in Fig. 1-23a. The beam is drawn to scale, and the loads and reactions are located at the points at which they act. Then, a horizontal line is drawn to represent the zero axis from which to plot the bending moments to scale. Note that the bending moment at both supports for this simple beam is zero. Between the supports and the first load, the bending moment is proportional to the distance from the support since the bending moment in that region equals the reaction times the distance from the support. Hence, the bending-moment diagram for this portion of the beam is a sloping straight line.

To find the bending moment under the 6000-lb load, consider only the forces to the left of it, in this case only the reaction R_1. Its moment about the 6000-lb load is 7000 × 10, or 70,000 ft-lb. The bending-moment diagram, then, between the left support and the first concentrated load is a straight line rising from zero at the left end of the beam to 70,000, plotted, to a convenient scale, under the 6000-lb load.

To find the bending moment under the 9000-lb load, add algebraically the moments of the forces to its left: 7000 × 20 − 6000 × 10 = 80,000 ft-lb. (This result could have been obtained more easily by considering only the portion of the beam on the right, where the only force present is R_2, and reversing the sign convention: 8000 × 10 = 80,000 ft-lb.) Since there are no other loads between the 6000- and 9000-lb loads, the bending-moment diagram between them is a straight line.

If the bending moment and shear are known at any section, the bending moment at any other section can be computed if there are no unknown forces between the sections. The rule is:

The bending moment at any section of a beam equals the bending moment at any section to the left, plus the shear at that section times the distance between sections, minus the moments of intervening loads. If the section with known moment and shear is on the right, the sign convention must be reversed.

For example, the bending moment under the 9000-lb load in Fig. 1-23a also could have been determined from the moment under the 6000-lb load and the shear just to the right of that load. As indicated in the shear diagram (Fig. 1-23b), that shear is 1000 lb. Thus, the moment is given by 70,000 + 1000 × 10 = 80,000 ft-lb.

Bending-moment diagrams for simple beams with various loadings are shown in Figs. 1-33 to 1-38. To obtain bending-moment diagrams for loading conditions that can be represented as a sum of the loadings shown, sum the bending moments at corresponding locations on the beam as given on the diagram for the component loads.

For a simple beam carrying a uniform load, the bending-moment diagram is a parabola (Fig. 1-24c). The maximum moment occurs at the center and equals $wL^2/8$ or $WL/8$, where w is the load per linear foot and $W = wL$ is the total load on the beam.

The bending moment at any section of a simply supported, uniformly loaded beam equals one-half the product of the load per linear foot and the distances to the section from both supports:

$$M = \frac{w}{2} x(L - x) \qquad (1\text{-}33)$$

1-24. Shear-Moment Relationship

The slope of the bending-moment curve at any point on a beam equals the shear at that point. If V is the shear, M the moment, and x the distance along the beam,

$$V = \frac{dM}{dx} \qquad (1\text{-}34)$$

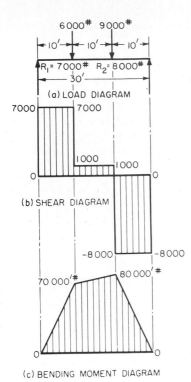

Fig. 1-23. Shear and moment diagrams for beam with concentrated loads.

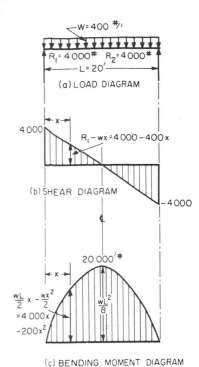

Fig. 1-24. Shear and moment diagrams for uniformly loaded beam.

Since maximum bending moment occurs when the slope changes sign, or passes through zero, maximum moment (positive or negative) occurs at the point of zero shear.

Integration of Eq. (1-34) yields

$$M_1 - M_2 = \int_{x2}^{x1} V\,dx \qquad (1\text{-}35)$$

Thus, the change in bending moment between any two sections of a beam equals the area of the shear diagram between ordinates at the two sections.

1-25. Moving Loads and Influence Lines

Influence lines are a useful device for solving problems involving moving loads. An influence line indicates the effect at a given section of a unit load placed at any point on the structure.

For example, to plot the influence line for bending moment at a point on a beam, compute the moments produced at that point as a unit load moves along the beam and plot these moments under the corresponding positions of the unit load. Actually, the unit load need not be placed at every point along the beam. The equation of the influence line can be determined in many cases by placing the load at an arbitrary point and computing the bending moment in general terms. (See also Art. 1-58.)

To draw the influence line for reaction at A for a simple beam AB (Fig. 1-25a), place a unit load at an arbitrary distance xL from B. The reaction at A due to this load is $1(xL/L) = x$. Then, $R_A = x$ is the equation of the influence line. It represents a straight line sloping down-

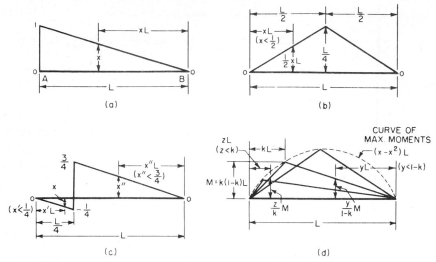

Fig. 1-25. Influence lines for (a) reaction at A, (b) midspan bending moment, (c) quarter-point shear, and (d) bending moments at several points in a beam.

ward from unity at A, when the unit load is at that end of the beam, to zero at B, when the load is at B (Fig. 1-25a).

Figure 1-25b shows the influence line for bending moment at the center of a beam. It resembles in appearance the bending-moment diagram for a load at the center of the beam, but its significance is entirely different. Each ordinate gives the moment at midspan for a load at the location of the ordinate. The diagram indicates that if a unit load is placed at a distance xL from one end, it produces a bending moment of xL/2 at the center of the span.

Figure 1-25c shows the influence line for shear at the quarter point of a beam. When the load is to the right of the quarter point, the shear is positive and equal to the left reaction. When the load is to the left, the shear is negative and equals the right reaction. Thus, to produce maximum shear at the quarter point, loads should be placed only to the right of the quarter point, with the largest load at the quarter point, if possible. For a uniform load, maximum shear results when the load extends from the right end of the beam to the quarter point.

Suppose, for example, that a 60-ft crane girder is to carry wheel loads of 20 and 10 kips, 5 ft apart. For maximum shear at the quarter point, place the 20-kip wheel there and the 10-kip wheel 5 ft to the right. The corresponding ordinates of the influence line (Fig. 1-25c) are ¾ and 40/45 × ¾. Hence, the maximum shear is 20 × ¾ + 10 × 40/45 × ¾ = 21.7 kips.

Figure 1-25d shows influence lines for bending moment at several points on a beam. The apexes of the triangular diagrams fall on a parabola, as indicated by the dashed line. From the diagram, it can be concluded that the maximum moment produced at any section by a single concentrated load moving along a beam occurs when the load is at that section. And the magnitude of the maximum moment increases when the section is moved toward midspan, in accordance with the equation for the parabola given in Fig. 1-25d.

1-26. Maximum Bending Moment

When a span is to carry several moving concentrated loads, an influence line is useful when determining the position of the loads for which bending moment is a maximum at a given section (see Art. 1-25). For a simple beam, maximum bending moment will occur at a section C as loads move across the beam when one of the loads is at C. The load to place at C is the one for which the expression $W_a/a - W_b/b$ (Fig. 1-26) changes sign as that load passes from one side of C to the other. (W_a is the sum of the loads on one side of C and W_b the sum of the loads on the other side of C.)

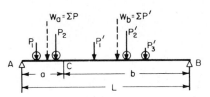

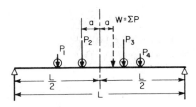

Fig. 1-26. Moving loads on simple beam *AB* placed for maximum moment at *C*.

Fig. 1-27. Moving loads placed for maximum moment in a simple beam.

When several concentrated loads move along a simple beam, the maximum moment they produce in the beam may be near but not necessarily at midspan. To find the maximum moment, first determine the position of the loads for maximum moment at midspan. Then, shift the loads until the load P_2 (Fig. 1-27) that was at the center of the beam is as far from midspan as the resultant of all the loads on the span is on the other side of midspan. Maximum moment will occur under P_2. When other loads move on or off the span during the shift of P_2 away from midspan, it may be necessary to investigate the moment under one of the other loads when it and the new resultant are equidistant from midspan.

1-27. Bending Stresses in a Beam

The commonly used flexure formula for computing bending stresses in a beam is based on the following assumptions:

1. The unit stress parallel to the bending axis at any point of a beam is proportional to the unit strain in the same direction at the point. Hence, the formula holds only within the proportional limit.

2. The modulus of elasticity in tension is the same as that in compression.

3. The total and unit axial strain at any point are both proportional to the distance of that point from the neutral surface. (Cross sections that are plane before bending remain plane after bending. This requires that all fibers have the same length before bending, thus that the beam be straight.)

4. The loads act in a plane containing the centroidal axis of the beam and are perpendicular to that axis. Furthermore, the neutral surface is perpendicular to the plane of the loads. Thus, the plane of the loads must contain an axis of symmetry of each cross section of the beam. (The flexure formula does not apply to a beam with cross sections loaded unsymmetrically.)

5. The beam is proportioned to preclude prior failure or serious deformation by torsion, local buckling, shear, or any cause other than bending.

Equating the bending moment to the resisting moment due to the internal stresses at any section of a beam yields the **flexure formula**:

$$M = \frac{fI}{c}$$

(1-36)

where M = bending moment at section, in-lb

f = normal unit stress at distance c, in, from the neutral axis (Fig. 1-28), psi

I = moment of inertia of cross section with respect to neutral axis, in^4

Generally, c is taken as the distance to the outermost fiber to determine maximum f.

1-28. Moment of Inertia

The neutral axis in a symmetrical beam coincides with the centroidal axis; that is, at any section the neutral axis is so located that

$$\int y \, dA = 0 \tag{1-37}$$

where dA is a differential area parallel to the axis (Fig. 1-28), y is its distance from the axis, and the summation is taken over the entire cross section.

Moment of inertia with respect to the neutral axis is given by

$$I = \int y^2 \, dA \tag{1-38}$$

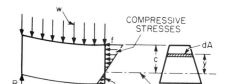

Fig. 1-28. Unit stresses on a beam section produced by bending.

Values for I for several common cross sections are in Fig. 1-29. Values for standard structural-steel sections are listed in the American Institute of Steel Construction "Steel Construction Manual." When the moments of inertia of other types of sections are needed, they can be computed directly by applying Eq. (1-38) or by breaking the section up into components for which the moment of inertia is known.

With the following formula, the moment of inertia of a section can be determined from that of its components:

$$I' = I + Ad^2 \tag{1-39}$$

where I = moment of inertia of component about its centroidal axis, in^4

I' = moment of inertia of component about parallel axis, in^4

A = cross-sectional area of component, in^2

d = distance between centroidal and parallel axes, in

The formula enables computation of the moment of inertia of a component about the centroidal axis of a section from the moment of inertia about the component's centroidal axis, usually obtainable from Fig. 1-29 or the AISC manual. By summing up the transferred moments of inertia for all the components, the moment of inertia of the section is obtained.

When the moments of inertia of an area with respect to any two perpendicular axes are known, the moment of inertia with respect to any other axis passing through the point of intersection of the two axes may be obtained by using Mohr's circle as for stresses (Fig. 1-11). In this analog, I_x corresponds with f_x, I_y with f_y, and the **product of inertia** I_{xy} with v_{xy} (Art. 1-17)

$$I_{xy} = \int xy \, dA \tag{1-40}$$

The two perpendicular axes through a point about which the moments of inertia are a maximum or a minimum are called the principal axes. The product of inertia is zero for the principal axes.

1-29. Section Modulus

The ratio $S = I/c$, relating bending moment and maximum bending stresses within the elastic range in a beam [Eq. (1-36)], is called the *section modulus*. I is the moment of inertia of the

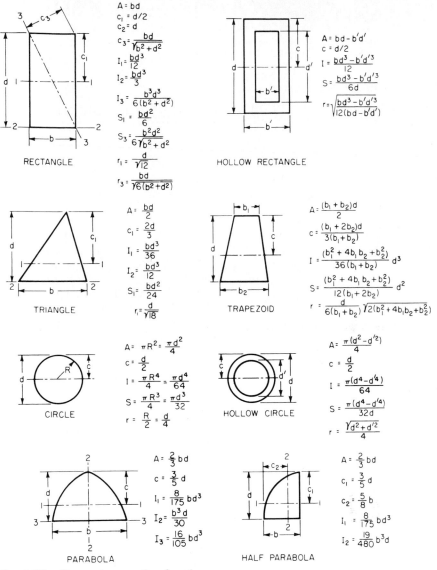

Fig. 1-29. Geometric properties of sections.

cross section about the neutral axis and c the distance from the neutral axis to the outermost fiber. Values of S for common types of sections are in Fig. 1-29. Values for standard structural-steel sections are listed in the American Institute of Steel Construction "Steel Construction Manual."

1-30. Shearing Stresses in a Beam

Vertical shear at any section in a beam is resisted by nonuniformly distributed, vertical unit stresses (Fig. 1-30). At every point in the section, there also is a horizontal unit stress, which is equal in magnitude to the vertical unit shearing stress there [see Eq. (1-24)].

At any distance y' from the neutral axis, both the horizontal and vertical shearing unit stresses are equal to

$$v = \frac{V}{It} A'\bar{y} \qquad (1\text{-}41)$$

where V = vertical shear at cross section, lb
$\quad t$ = thickness of beam at distance y' from neutral axis, in
$\quad I$ = moment of inertia of section about neutral axis, in⁴
$\quad A'$ = area between outermost surface and surface for which shearing stress is being computed, in²
$\quad \bar{y}$ = distance of center of gravity of this area from neutral axis, in

For a rectangular beam, with width $t = b$ and depth d, the maximum shearing stress occurs at middepth. Its magnitude is

$$v = \frac{V}{(bd^3/12)b} \frac{bd}{2} \frac{d}{4} = \frac{3}{2} \frac{V}{bd}$$

That is, the maximum shear stress is 50% greater than the average shear stress on the section. Similarly, for a circular beam, the maximum is one-third greater than the average. For an I or wide-flange beam, however, the maximum shear stress in the web is not appreciably greater than the average for the web section alone, assuming that the flanges take no shear.

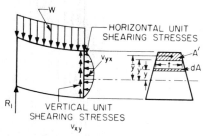

Fig. 1-30. Unit shearing stresses on a beam section.

1-31. Combined Shear and Bending Stress

For deep beams on short spans and beams with low tensile strength, it sometimes is necessary to determine the maximum normal stress f' due to a combination of shear stress v and bending stress f. This maximum or principal stress (Art. 1-15) occurs on a plane inclined to that of v and of f. From Mohr's circle (Fig. 1-11) with $f = f_x, f_y = 0$, and $v = v_{xy}$,

$$f' = \frac{f}{2} + \sqrt{v^2 + \left(\frac{f}{2}\right)^2} \qquad (1\text{-}42)$$

1-32. Beam Stresses in the Plastic Range

(Refer also to Art. 1-77.) When the bending stresses in a beam exceed the proportional limit and stress no longer is proportional to strain, the distribution of bending stresses over a cross section ceases to be linear. As the bending moment increases, the outer fibers deform with little change in stress, while the fibers not stressed beyond the proportional limit continue to take more stress.

If the actual stress distribution over the cross section is plotted then, for equilibrium, the area under the curve for the tensile stresses must equal the area under the curve for the compressive stresses. Also, the moments of the areas under these two curves about the neutral axis must equal the bending moment.

Modulus of rupture is the stress computed from the flexure formula [Eq. (1-36)] corresponding to the maximum bending moment a beam sustains at failure. Usually, it is considerably higher than the actual maximum unit stress in the beam, but it sometimes is used to compare the strength of beams with the same cross section and material.

1-33. Beam Deflections

The **elastic curve** is the position taken by the longitudinal centroidal axis of a beam when it deflects under load. The radius of curvature at any point of this curve is

$$R = \frac{EI}{M} \tag{1-43}$$

where M = bending moment at point
  E = modulus of elasticity
  I = moment of inertia of cross section about neutral axis
  Since the slope of the elastic curve is very small, $1/R$ is approximately d^2y/dx^2, where y is the deflection of the beam at a distance x from the origin of coordinates. Hence, Eq. (1-43) may be rewritten

$$M = EI\frac{d^2y}{dx^2} \tag{1-44}$$

To obtain the slope and deflection of a beam, this equation may be integrated, with M expressed as a function of x. Constants introduced during the integration must be evaluated in terms of known points and slopes of the elastic curve.
  After integration, Eq. (1-44) yields

$$\theta_B - \theta_A = \int_A^B \frac{M}{EI}\,dx \tag{1-45}$$

in which θ_A and θ_B are the slopes of the elastic curve at any two points A and B. If the slope is zero at one of the points, the integral in Eq. (1-45) gives the slope of the elastic curve at the other. The integral represents the area of the bending-moment diagram between A and B with each ordinate divided by EI.
  The **tangential deviation** t of a point on the elastic curve is the distance of this point, measured in a direction perpendicular to the original position of the beam, from a tangent drawn at some other point on the curve.

$$t_B - t_A = \int_A^B \frac{Mx}{EI}\,dx \tag{1-46}$$

  Equation (1-46) indicates that the tangential deviation of any point with respect to a second point on the elastic curve equals the moment about the first point of the area of the M/EI diagram between the two points. The moment-area method for determining beam deflections is a technique employing Eqs. (1-45) and (1-46).

  Moment-Area Method ▪ Suppose, for example, the deflection at midspan is to be computed for a beam of uniform cross section with a concentrated load at the center (Fig. 1-31). Since the deflection at midspan for this loading is the maximum for the span, the slope of the elastic curve at midspan is zero; that is, the tangent is parallel to the undeflected position of the beam. Hence, the deviation of either support from the midspan tangent equals the deflection at the center of the beam. Then, by the moment-area theorem [Eq. (1-46)], the deflection y_c is given by the moment about either support of the area of the M/EI diagram included between an ordinate at the center of the beam and that support

$$y_c = \left(\frac{1}{2}\frac{PL}{4EI}\frac{L}{2}\right)\frac{L}{3} = \frac{PL^3}{48EI}$$

Suppose now that the deflection y at any point D at a distance xL from the left support (Fig. 1-31) is to be determined. Note that from similar triangles, $xL/L = DE/t_{AB}$, where DE is the distance from the undeflected position of D to the tangent to the elastic curve at support A, and

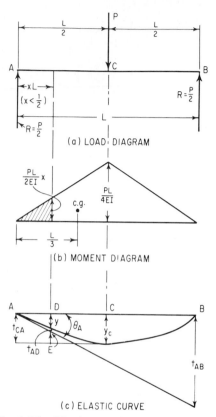

t_{AB} is the tangential deviation of B from that tangent. But DE also equals $y + t_{AD}$, where t_{AD} is the tangential deviation of D from the tangent at A. Hence,

$$y + t_{AD} = xt_{AB}$$

This equation is perfectly general for the deflection of any point of a simple beam, no matter how loaded. It may be rewritten to give the deflection directly:

$$y = xt_{AB} - t_{AB} \qquad (1\text{-}47)$$

But t_{AB} is the moment of the area of the M/EI diagram for the whole beam about support B, and t_{AD} is the moment about D of the area of the M/EI diagram included between ordinates at A and D. So at any point x of the beam in Fig. 1-31, the deflection is

$$y = x \left[\frac{1}{2} \frac{PL}{4EI} \frac{L}{2} \left(\frac{L}{3} + \frac{2L}{3} \right) \right]$$
$$- \frac{1}{2} \frac{PLx}{2EI} (xl) \frac{xL}{3} = \frac{PL^3}{48EI} x(3 - 4x^2)$$

It also is noteworthy that, since the tangential deviations are very small distances, the slope of the elastic curve A is given by

$$\theta_A = \frac{t_{AB}}{L} \qquad (1\text{-}48)$$

Fig. 1-31. Elastic curve for a simple beam and tangential deviations at ends.

This holds, in general, for all simple beams regardless of the type of loading.

Conjugate-Beam Method ▪ The procedure followed in applying Eq. (1-47) to the deflection of the loaded beam in Fig. 1-31 is equivalent to finding the bending moment at D with the M/EI diagram serving as the load diagram. The technique of applying the M/EI diagram as a load and determining the deflection as a bending moment is known as the conjugate-beam method.

The conjugate beam must have the same length as the given beam; it must be in equilibrium with the M/EI load and the reactions produced by the load; and the bending moment at any section must be equal to the deflection of the given beam at the corresponding section. The last requirement is equivalent to specifying that the shear at any section of the conjugate beam with the M/EI load be equal to the slope of the elastic curve at the corresponding section of the given beam. Figure 1-32 shows the conjugates for various types of beams.

Deflection Computations ▪ Deflections for several types of loading on simple beams are given in Figs. 1-33 and 1-35 to 1-38 and for cantilevers and beams with overhangs in Figs. 1-39 to 1-44.

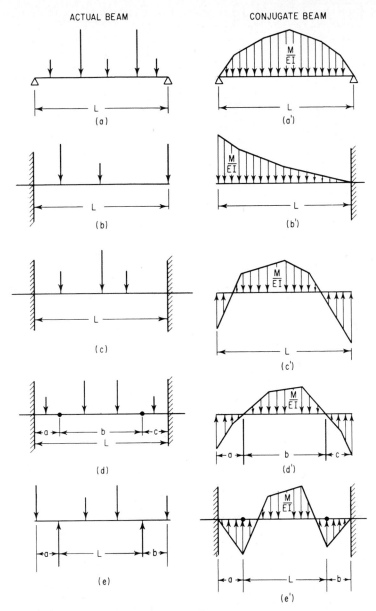

Fig. 1-32. Conjugate beams.

When a beam carries several different types of loading, the most convenient method of computing its deflection usually is to find the deflections separately for the uniform and concentrated loads and add them.

For several concentrated loads, the easiest method of obtaining the deflection at a point on a beam is to apply the reciprocal theorem (Art. 1-58). According to this theorem, if a concentrated load is applied to a beam at a point A, the deflection the load produces at point B equals

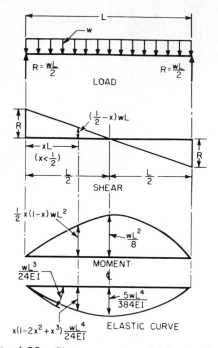

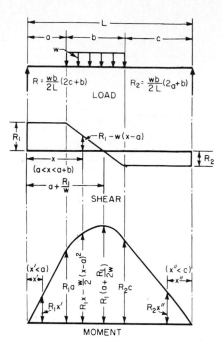

Fig. 1-33. Shears, moments, and deflections for full uniform load on a simply supported, prismatic beam.

Fig. 1-34. Shears and moments for a uniformly distributed load over part of a simply supported beam.

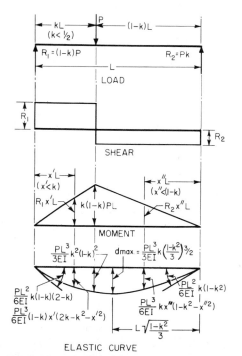

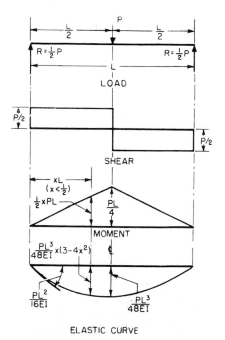

Fig. 1-35. Shears, moments, and deflections for a concentrated load at any point of a simply supported, prismatic beam.

Fig. 1-36. Shears, moments, and deflections for a concentrated load at midspan of a simply supported, prismatic beam.

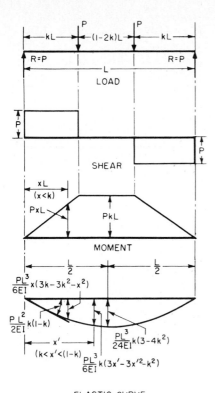

Fig. 1-37. Shears, moments, and deflections for two equal concentrated loads on a simply supported, prismatic beam.

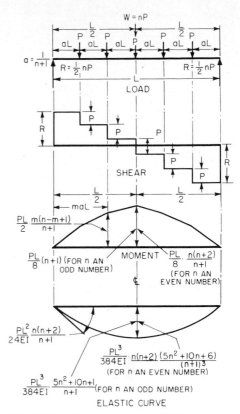

Fig. 1-38. Shears, moments, and deflections for several equal loads equally spaced on a simply supported, prismatic beam.

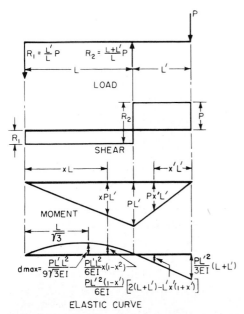

Fig. 1-39. Shears, moments, and deflections for a concentrated load on a beam overhang.

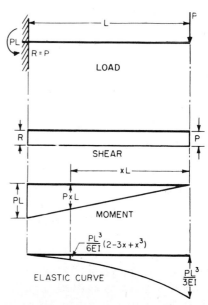

Fig. 1-40. Shears, moments, and deflections for a concentrated load on the end of a prismatic antilever.

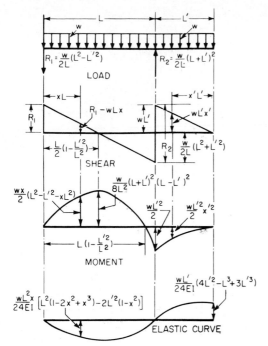

Fig. 1-41. Shears, moments, and deflections for a uniform load over the full length of a beam with overhang.

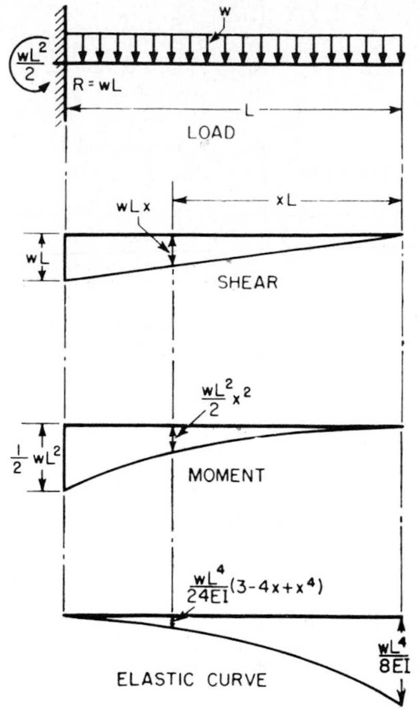

Fig. 1-42. Shears, moments, and deflections for a uniform load over the length of a cantilever.

the deflection at A for the same load applied at B ($d_{AB} = d_{BA}$). So place the loads one at a time at the point for which the deflection is to be found, and from the equation of the elastic curve determine the deflections at the actual location of the loads. Then, sum these deflections.

Suppose, for example, the midspan deflection is to be computed. Assume each load in turn applied at the center of the beam and compute the deflection at the point where it originally was applied from the equation of the elastic curve given in Fig. 1-36. The sum of these deflections is the total midspan deflection.

Another method for computing deflections is presented in Art. 1-55. This method also may be used to determine the deflection of a beam due to shear.

1-34. Combined Axial and Bending Loads

For short beams, subjected to both transverse and axial loads, the stresses are given by the principle of superposition if the deflection due to bending may be neglected without serious error. That is, the total stress is given with sufficient accuracy at any section by the sum of the axial stress and the bending stresses. The maximum stress, psi, equals

$$f = \frac{P}{A} + \frac{Mc}{I} \qquad (1\text{-}49)$$

where P = axial load, lb

A = cross-sectional area, in^2

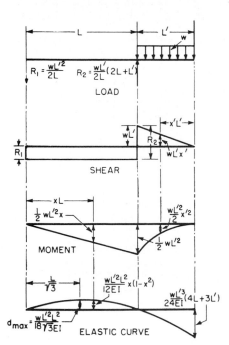

Fig. 1-43. Shears, moments, and deflections for uniform load on a beam overhang.

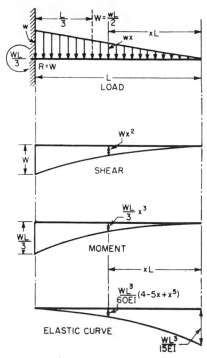

Fig. 1-44. Shears, moments, and deflections for triangular loading on a prismatic cantilever.

M = maximum bending moment, in-lb
c = distance from neutral axis to outermost fiber at section where maximum moment occurs, in
I = moment of inertia about neutral axis at that section, in^4

When the deflection due to bending is large and the axial load produces bending stresses that cannot be neglected, the maximum stress is given by

$$f = \frac{P}{A} + (M + Pd)\frac{c}{I} \qquad (1\text{-}50)$$

where d is the deflection of the beam. For axial compression, the moment Pd should be given the same sign as M, and for tension, the opposite sign, but the minimum value of $M + Pd$ is zero. The deflection d for axial compression and bending can be obtained by applying Eq. (1-44). (S. Timoshenko and J. M. Gere, "Theory of Elastic Stability," McGraw-Hill Book Company, New York; Friedrich Bleich, "Buckling Strength of Metal Structures," McGraw-Hill Book Company, New York.) But it may be closely approximated by

$$d = \frac{d_o}{1 - (P/P_c)} \qquad (1\text{-}51)$$

where d_o = deflection for transverse loading alone, in
P_c = critical buckling load, $\pi^2 EI/L^2$ (see Art. 1-72), lb

1-35. Eccentric Loading

If an eccentric longitudinal load is applied to a bar in the plane of symmetry, it produces a bending moment Pe, where e is the distance, in, of the load P from the centroidal axis. The total unit stress is the sum of the stress due to this moment and the stress due to P applied as an axial load:

$$f = \frac{P}{A} \pm \frac{Pec}{I} = \frac{P}{A}\left(1 \pm \frac{ec}{r^2}\right) \tag{1-52}$$

where A = cross-sectional area, in^2
c = distance from neutral axis to outermost fiber, in
I = moment of inertia of cross section about neutral axis, in^4
r = **radius of gyration** = $\sqrt{I/A}$, in
Figure 1-29 gives values of the radius of gyration for several cross sections.

If there is to be no tension on the cross section under a compressive load, e should not exceed r^2/c. For a rectangular section with width b and depth d, the eccentricity, therefore, should be less than $b/6$ and $d/6$; i.e., the load should not be applied outside the middle third. For a circular cross section with diameter D, the eccentricity should not exceed $D/8$.

When the eccentric longitudinal load produces a deflection too large to be neglected in computing the bending stress, account must be taken of the additional bending moment Pd, where d is the deflection, in. This deflection may be computed by using Eq. (1-44) or closely approximated by

$$d = \frac{4eP/P_c}{\pi(1 - P/P_c)} \tag{1-53}$$

P_c is the critical buckling load $\pi^2 EI/L^2$ (see Art. 1-72), lb.

If the load P does not lie in a plane containing an axis of symmetry, it produces bending about the two principal axes through the centroid of the section. The stresses, psi, are given by

$$f = \frac{P}{A} \pm \frac{Pe_x c_x}{I_y} \pm \frac{Pe_y c_y}{I_x} \tag{1-54}$$

where A = cross-sectional area in^2
e_x = eccentricity with respect to principal axis YY, in
e_y = eccentricity with respect to principal axis XX, in
c_x = distance from YY to outermost fiber, in
c_y = distance from XX to outermost fiber, in
I_x = moment of inertia about XX, in^4
I_y = moment of inertia about YY, in^4
The principal axes are the two perpendicular axes through the centroid for which the moments of inertia are a maximum or a minimum and for which the products of inertia are zero.

1-36. Unsymmetrical Bending

When a beam is subjected to loads that do not lie in a plane containing a principal axis of each cross section, unsymmetrical bending occurs. Assuming that the bending axis of the beam lies in the plane of the loads, to preclude torsion (see Art. 1-37), and that the loads are perpendicular to the bending axis, to preclude axial components, the stress, psi, at any point in a cross section is

$$f = \frac{M_x y}{I_x} \pm \frac{M_y x}{I_y} \tag{1-55}$$

where M_x = bending moment about principal axis XX, in-lb

$\quad M_y$ = bending moment about principal axis YY, in-lb

$\quad x$ = distance from point where stress is to be computed to YY axis, in

$\quad y$ = distance from point to XX, in

$\quad I_x$ = moment of inertia of cross section about XX, in^4

$\quad I_y$ = moment of inertia about YY, in^4

If the plane of the loads makes an angle θ with a principal plane, the neutral surface will form an angle α with the other principal plane such that

$$\tan \alpha = \frac{I_x}{I_y} \tan \theta$$

1-37. Beams with Unsymmetrical Sections

The derivation of the flexure formula $f = Mc/I$ (Art. 1-27) assumes that a beam bends, without twisting, in the plane of the loads and that the neutral surface is perpendicular to the plane of the loads. These assumptions are correct for beams with cross sections symmetrical about two axes when the plane of the loads contains one of these axes. They are not necessarily true for beams that are not doubly symmetrical because in beams that are doubly symmetrical, the bending axis coincides with the centroidal axis, whereas in unsymmetrical sections the two axes may be separate. In the latter case, if the plane of the loads contains the centroidal axis but not the bending axis, the beam will be subjected to both bending and torsion.

The **bending axis** is the longitudinal line in a beam through which transverse loads must pass to preclude the beam's twisting as it bends. The point in each section through which the bending axis passes is called the **shear center,** or center of twist. The shear center also is the center of rotation of the section in pure torsion (Art. 1-75). Its location depends on the dimensions of the section.

If a beam has an axis of symmetry, the shear center lies on it. In doubly symmetrical beams, the shear center lies at the intersection of two axes of symmetry and hence coincides with the centroid.

For any section composed of two narrow rectangles, such as a T beam or an angle, the shear center may be taken as the intersection of the longitudinal center lines of the rectangles.

For a channel section with one axis of symmetry, the shear center is outside the section at a distance from the centroid equal to $e(1 + h^2A/4I)$, where e is the distance from the centroid to the center of the web, h is the depth of the channel, A the cross-sectional area, and I the moment of inertia about the axis of symmetry. (The web lies between the centroid and the shear center.)

Locations of shear centers for several other sections are given in Freidrich Bleich, "Buckling Strength of Metal Structures," chap. 3, McGraw-Hill Book Company, New York, 1952.

When the cross section of a beam is unsymmetrical, stress in the elastic range is given by

$$f = \frac{P}{A} \pm \frac{M_y - M_x(I_{xy}/I_x)}{I_y - (I_{xy}/I_x)I_{xy}} x \pm \frac{M_x - M_y(I_{xy}/I_y)}{I_x - (I_{xy}/I_y)I_{xy}} y \qquad (1\text{-}56)$$

where A = cross-sectional area

$\quad M_x, M_y$ = bending moment about x-x and y-y axes

$\quad I_x, I_y$ = moment of inertia about x-x and y-y axes

$\quad x, y$ = distances of point under consideration from y-y axes

$\quad I_{xy}$ = product of inertia [Eq. (1-40)]

GRAPHIC-STATICS FUNDAMENTALS

1-38. Representation of a Force

Since a force is completely determined when it is known in magnitude, direction, and point of application, any force may be represented by the length, direction, and position of a straight line. The length of line to a given scale represents the magnitude of the force. The position of the line parallels the line of action of the force, and an arrowhead on the line indicates the direction in which the force acts.

Graphically represented, a force may be designated by a letter, sometimes followed by a subscript, such as P_1 and P_2 in Fig. 1-45. Or each extremity of the line may be indicated by a

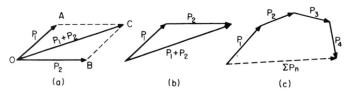

Fig. 1-45. Addition of forces by (*a*) parallelogram law, (*b*) triangle construction, and (*c*) polygon construction.

letter and the force referred to by means of these letters (Fig. 1-45*a*). The order of the letters indicates the direction of the force; in Fig. 1-45*a*, referring to P_1 as *OA* indicates it acts from *O* toward *A*.

Forces are concurrent when their lines of action meet. If they lie in the same plane, they are coplanar.

1-39. Parallelogram of Forces

The **resultant** of several forces is a single force that would produce the same effect on a rigid body. The resultant of two concurrent forces is determined by the **parallelogram law:**

If a parallelogram is constructed with two forces as sides, the diagonal represents the resultant of the forces (Fig. 1-45*a*).

The resultant is said to be equal to the sum of the forces, sum here meaning vectorial sum, or addition by the parallelogram law. Subtraction is carried out in the same manner as addition, but the direction of the force to be subtracted is reversed.

If the direction of the resultant is reversed, it becomes the **equilibrant,** a single force that will hold the two given forces in equilibrium.

1-40. Resolution of Forces

Any force may be resolved into two components acting in any given directions. To resolve a force into two components, draw a parallelogram with the force as a diagonal and sides parallel to the given directions. The sides then represent the components.

The procedure is: (1) Draw the given force. (2) From both ends of the force draw lines parallel to the directions in which the components act. (3) Draw the components along the parallels through the origin of the given force to the intersections with the parallels through the other end. Thus, in Fig. 1-45*a*, P_1 and P_2 are the components in directions *OA* and *OB* of the force represented by *OC*.

1-41. Force Polygons

Examination of Fig. 1-45a indicates that a step can be saved in adding forces P_1 and P_2. The same resultant could be obtained by drawing only the upper half of the parallelogram. Hence, to add two forces, draw the first force; then draw the second force at the end of the first one. The resultant is the force drawn from the origin of the first force to the end of the second force, as shown in Fig. 1-45b.

This diagram is called a force triangle. Again, the equilibrant is the resultant with direction reversed. If it is drawn instead of the resultant, the arrows representing the direction of the forces will all point in the same direction around the triangle. From the force triangle, an important conclusion can be drawn:

If three forces meeting at a point are in equilibrium, they form a closed force triangle.

To add several forces P_1, P_2, P_3, . . . , P_n, draw P_2 from the end of P_1, P_3 from the end of P_2, and so on. The force required to complete the force polygon is the resultant (Fig. 1-45c).

If a group of concurrent forces is in equilibrium, they form a closed force polygon.

1-42. Equilibrium Polygon

When forces are coplanar but not concurrent, the force polygon will yield the magnitude and direction of the resultant but not its point of application. To complete the solution, the easiest method generally is to employ an auxiliary force polygon, called an equilibrium, or funicular (string), polygon. Sides of this polygon represent the lines of action of certain components of the given forces; more specifically, they take the configuration of a weightless string holding the forces in equilibrium.

In Fig. 1-46a, the forces P_1, P_2, P_3, and P_4 acting on the given body are not in equilibrium.

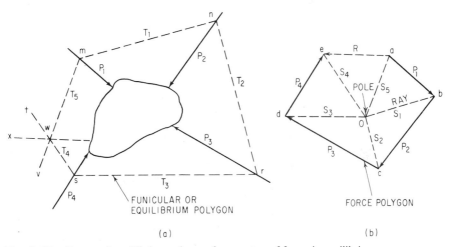

Fig. 1-46.   Force and equilibrium polygons for a system of forces in equilibrium.

The magnitude and direction of their resultant R are obtained from the force polygon *abcde*. The line of action of this resultant may be obtained as follows:

From any point O in the force polygon, draw a line to each vertex of the polygon. Since the lines Oa and Ob form a closed triangle with the force P_1, they represent two forces S_5 and S_1 that hold P_1 in equilibrium—two forces that may replace P_1 in a force diagram. So, as in Fig.

1-46*a*, at any point *m* on the line of action of P_1, draw lines *mn* and *mv* parallel to S_1 and S_5, respectively, to represent the lines of action of these forces. Similarly, S_1 and S_2 represent two forces that may replace P_2. The line of action of S_1 already is indicated by the line *mn*, and it intersects P_2 at *n*. So through *n* draw a line parallel to S_2, intersecting P_3 at *r*. Through *r*, draw *rs* parallel to S_3, and through *s*, draw *st* parallel to S_4. Lines *mv* and *st*, parallel to S_5 and S_4, respectively, represent the lines of action of S_5 and S_4. But these two forces form a closed force triangle with the resultant *ae* (Fig. 1-46*b*), and therefore the three forces must be concurrent. Hence, the line of action of the resultant must pass through the intersection *w* of the lines *mv* and *st*. The resultant of the four given forces is thus fully determined. A force of equal magnitude but acting in the opposite direction, from *e* to *a*, will hold P_1, P_2 P_3, and P_4 in equilibrium.

The polygon *mnrsw* is called an *equilibrium polygon*. Point *O* is called the *pole,* and $S_1 \ldots S_5$ are called the *rays of the force polygon.*

1-43. Beam and Truss Reactions by Graphics

Reactions of simple beams and trusses can be found with the aid of an equilibrium polygon (Art. 1-42). First, a force polygon is constructed to obtain the magnitude and direction to the sum of the reactions. Second, rays are drawn to the vertices of the polygon from a conveniently located pole. These rays are used to construct all but one side of an equilibrium polygon. The closing side is the common line of action of two equal but opposite forces that act with a pair of rays already drawn to hold the reactions in equilibrium. Therefore, draw a line through the pole parallel to the closing side. The intersection with the resultant in the force polygon separates it into two forces, which are equal to the reactions sought.

For example, suppose the reactions were to be obtained graphically for the beam (or truss) in Fig. 1-47*a*. As a first step, construct force polygon *ABCD* (Fig. 1-47*b*) with the loads P_1, P_2

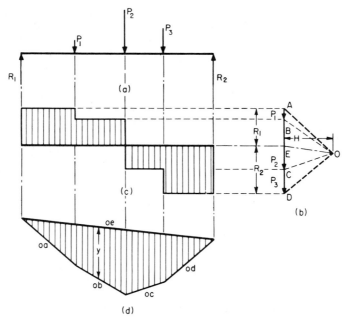

Fig. 1-47. Shear and moment diagrams obtained by graphic statics. (*a*) Loaded beam, (*b*) force polygon, (*c*) shear diagram, and (*d*) equilibrium polygon and bending-moment diagram.

and P_3. Since the loads are parallel, the force polygon is a straight line. Select a pole O, and draw rays from it to the extremities of the forces. Note that the sum of the reactions equals AD, the length of the line polygon.

Next, construct the equilbrium polygon as follows: Start with a convenient point on the line of action of R_1, the left reaction, and draw a line oa parallel to ray OA in Fig. 1-47b. Locate its intersection with the line of action of P_1 (Fig. 1-47d). Through this intersection, draw ob, a line parallel to OB. At the intersection with P_2, draw oc parallel to OC, and finally, thrcugh the intersection with P_3, a line od parallel to OD, which terminates on the line of action of R_2. Now, draw the closing line oe of the equilibrium polygon between the terminal and the starting point. The last step is to draw through the pole (Fig. 1-47b) OE a line parallel to oe, the closing line of the equilibrium polygon, cutting the force polygon resultant at E. Then, $DE = R_2$ and $EA = R_1$. (See Art. 1-20 for an analytical method of determining reactions.)

1-44. Shear and Moment Diagrams by Graphics

The shear at any section of a beam equals the algebraic sum of the loads and reactions on the left of the section, upward-acting forces being considered positive, downward forces negative. If the forces are arranged in the proper order, the shear diagram may be obtained directly from the force polygon after the reactions have been determined.

For example, the shear diagram for the beam in Fig. 1-47a can be easily obtained from the force polygon $ABCDE$ (in this case, a line, because the loads are parallel) in Fig. 1-47b. The zero axis is a line (Fig. 1-46c) parallel to the beam through E. As indicated in Fig. 1-47c, the ordinates of the shear are marked off, starting with R_1 along the line of action of the left reaction, by drawing lines parallel to the zero axis through the extremities of the forces in the force polygon. (See Art. 1-22 for an analytical method of determining shears.)

The moment of a force about a point can be obtained from the equilibrium and force polygons. In the equilibrium diagram, draw a line parallel to the force through the given point. Measure the intercept of this line between the two adjoining funicular-polygon sides (extended if necessary) that originate at the given force. The moment is the product of this intercept and the distance of the force-polygon pole from the force. The intercept should be measured to the same linear scale as the beam and load positions, and the pole distance to the same scale as the forces in the force polygon.

As a consequence of this relationship between the sides of the funicular polygon, each ordinate (parallel to the forces) multiplied by the pole distance equals the bending moment at the corresponding section of the beam or truss. In Fig. 1-47d, for example, the equilibrium polygon, to scale, is the bending-moment diagram for the beam in Fig. 1-47a. At any section, the bending moment equals the ordinate y multiplied by the pole distance H. (See Art. 1-23 for an analytical method.)

STRESSES IN TRUSSES

A truss is a coplanar system of structural members joined at their ends to form a stable framework. Usually, analysis of a truss is based on the assumption that the joints are hinged. Neglecting small changes in the lengths of the members due to loads, the relative positions of the joints cannot change. Stresses due to joint rigidity or deformations of the members are called **secondary stresses.**

Three bars pinned together to form a triangle represent the simplest type of truss. Some of the more common types of trusses are shown in Fig. 1-48.

The top members are called the **upper chord,** the bottom members the **lower chord,** and the verticals and diagonals **web members.**

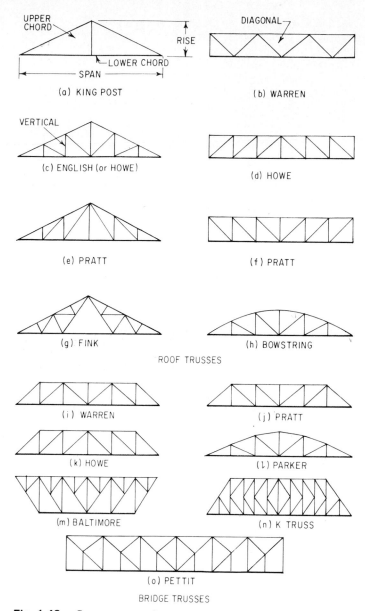

Fig. 1-48.   Common types of trusses.

Trusses act like long, deep girders with cutout webs. Roof trusses have to carry not only their own weight and the weight of roof framing but wind loads, snow loads, suspended ceilings and equipment, and a live load to take care of construction, maintenance, and repair loading. Bridge trusses have to support their own weight and that of deck framing and deck, live loads imposed by traffic (automobiles, trucks, railroad trains, pedestrians, and so on) and impact caused by live load, plus wind on structural members and vehicles. **Deck trusses** carry the live load on the upper chord and **through trusses** on the lower chord.

Loads generally are applied at the intersection of members, or panel points, so that the members will be subjected principally to direct stresses. To simplify stress analysis, the weight of the truss members is apportioned to upper- and lower-chord panel points. The members are assumed to be pinned at their ends, even though this may actually not be the case. However, if the joints are of such nature as to restrict relative rotation substantially, then the "secondary" stresses set up as a result should be computed and superimposed on the stresses obtained with the assumption of pin ends.

1-45. Bow's Notation

In analysis of trusses, especially in graphical analysis, Bow's notation is useful for identifying truss members, loads, and stresses. Capital letters are placed in the spaces between truss members and between forces; each member and load is then designated by the letters on opposite sides of it. For example, in Fig. 1-49a, the upper-chord members are AF, BH, CJ, and DL. The loads are AB, BC, and CD, and the reactions are EA and DE. Stresses in the members generally are designated by the same letters but in lowercase.

1-46. Graphical Analysis of Trusses

Stresses in trusses may be determined by the graphical methods explained in Arts. 1-38 to 1-43. The general method is based on the assumption that at every joint of a truss, the lines of action of loads and of stresses in members meet at a point. Then, since the loads and stresses are in equilibrium, the vectors, or arrows, representing them form a closed polygon. But in only a few cases are the stresses at a joint known initially; hence, the polygon for each joint usually is incomplete. The requirement that the forces form a polygon, however, permits determination of up to two unknowns. Thus, the procedure for analyzing trusses graphically is to start with those joints with only two unknown stresses (the lines of action are known because they lie along the longitudinal axes of the truss members). Solving those joints yields the magnitude of stresses that now can be used to solve other joints. By proceeding in this manner from joint to joint, all the stresses in a truss generally can be determined.

Figure 1-49b to e shows how the polygon of force may be applied at each joint of a truss to determine the two unknown stresses. The solution presumes that the reactions are known. They may be computed analytically or graphically, with load and funicular polygons, as explained in Arts. 1-20 and 1-43.

For convenience, use Bow's notation (Art. 1-45) to designate loads and truss members, as shown in Fig. 1-49a. Start with joint 1, where the reaction is known and there are only two unknown stresses, af in upper chord AF and fe in lower chord FE. Represent the reaction ea by an upward vertical arrow equal to 12 kips to a convenient scale. Through a draw a line parallel to AF and through e parallel to FE, to form the force triangle for joint 1 (Fig. 1-49b). The intersection is f, determining the stresses af and fe to the same scale as ea. Note that fe acts away from the joint and thus represents a tensile stress in FE; af acts toward the joint and represents a compressive stress in AF.

Move next to joint 2. There, the stress in the vertical FG is zero because there are no forces with a component in the direction of FG. Now, with the stresses af and fg known, solution of joint 3 by completing a force polygon becomes possible. There remain only two unknown stresses, bh in BH and hg in HG, the load $ab = 8$ kips being given. Figure 1-49c shows the solution for joint 3. Joints 4 and 5 are solved similarly in Fig. 1-49d and e.

Examination of these force polygons indicates that each stress occurs in two force polygons. Hence, the graphical solution can be shortened by combining the polygons. The combination of the various polygons for all the joints into one stress diagram is known as a Maxwell diagram.

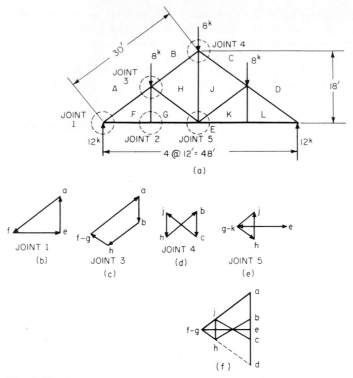

Fig. 1-49. Graphical determination of stresses at each joint of the truss in (*a*) may be expedited by constructing the single Maxwell diagram in (*f*).

Maxwell Diagram ▪ The procedure for a Maxwell diagram consists of first constructing the force polygon for the loads and reactions and then completing the force polygons for each joint on the same diagram. To make it easy to determine whether the stresses are compression or tension, plot loads and reactions in the force polygon in the order in which they are passed in going clockwise around the truss. Similarly, when drawing the force polygon for each joint, plot the forces in a clockwise direction around the joint. If these rules are followed, the order of the letters indicates the direction of the forces. Going around joint 1 clockwise, for example, we find, in Fig. 1-49*b*, the reaction *EA* as an upward-acting vertical force *ea* (*e* to *a*), the top-chord stress *af* acting toward the joint (*a* to *f*), and the bottom-chord stress *fe* acting away from the joint (*f* to *e*). Hence, *af* is compressive, *fe* tensile.

To construct a Maxwell diagram for the truss in Fig. 1-49*a*, lay off the loads and reactions clockwise (Fig. 1-49*f*). The force polygon *abcde* is a straight line because all the forces are vertical. To solve joint 1, draw a line through *a* in Fig. 1-49*f* parallel to *AF* and a second line through *e* parallel to *FE*. Their intersection is *f*. At joint 2, since the stress in the vertical *FG* is zero, *g* coincides with *f*.

To solve joint 3, start with the known stress *fa* and proceed clockwise around the joint. To complete the force polygon, draw a line through *b* parallel to *BH* and a line through *g* parallel to *HG*; the intersection of the lines locates *h*. Similarly, the force polygon for joint 4 is completed by drawing a line through *c* parallel to *CJ* and a line through *h* parallel to *JH*.

Wind loads on a roof truss with a sloping top chord are assumed to act normal to the roof, in which case the load polygon will be an inclined line or a true polygon. The reactions are computed generally on the assumption either that both are parallel to the resultant of the wind loads or that one end of the truss is free to move horizontally and therefore will not resist the horizontal components of the loads. The stress diagram is plotted in the same manner as for vertical loads after the reactions have been found.

Some trusses are complex and require special methods of analysis. For methods of solving these, see C. H. Norris and J. B. Wilbur, "Elementary Structural Analysis," McGraw-Hill Book Company, New York, which also presents a graphical method of obtaining the deflections of a truss. An analytical method is given in Art. 1-56.

1-47. Method of Sections for Truss Stresses

A convenient method of computing the stresses in truss members is to isolate a portion of the truss by a section so chosen as to cut only as many members with unknown stresses as can be evaluated by the laws of equilibrium applied to that portion of the truss. The stresses in the members cut by the section are treated as external forces and must hold the loads on that portion of the truss in equilibrium. Compressive forces act toward each joint or panel point, and tensile forces away from the joint.

Joint Isolation ▪ A choice of section that often is convenient is one that isolates a joint with only two unknown stresses. Since the stresses and load at the joint must be in equilibrium, the sum of the horizontal components of the forces must be zero, and so must be the sum of the vertical components. Since the lines of action of all the forces are known (the stresses act along the longitudinal axes of the truss members), we can therefore compute two unknown magnitudes of stresses at each joint by this method.

To apply it to joint 1 of the truss in Fig. 1-49a, first equate the sum of the vertical components to zero. This equation shows that the vertical component of the top chord must be equal and opposite to the reaction, 12 kips (see Fig. 1-49b). The stress in the top chord at this joint, then, must be a compression equal to $12 \times 30/18 = 20$ kips. Next, equate the sum of the horizontal components to zero. This equation indicates that the stress in the bottom chord at the joint must be equal and opposite to the horizontal component of the top chord. Hence, the stress in the bottom chord must be a tension equal to $20 \times 24/30 = 16$ kips.

Taking a section around joint 2 in Fig. 1-49a reveals that the stress in the vertical is zero since there are no loads at the joint and the bottom chord is perpendicular to the vertical. Also, the stress must be the same in both bottom-chord members at the joint since the sum of the horizontal components must be zero.

After joints 1 and 2 have been solved, a section around joint 3 cuts only two unknown stresses: S_{BH} in top chord BH (see Bow's notation, Art. 1-45) and S_{HG} in diagonal HG. Application of the laws of equilibrium to this joint yields the following two equations, one for the vertical components and the second for the horizontal components:

$$\Sigma V = 0.6 S_{FA} - 8 - 0.6 S_{BH} + 0.6 S_{HG} = 0$$
$$\Sigma H = 0.8 S_{FA} - 0.8 S_{BH} - 0.8 S_{HG} = 0$$

Both unknown stresses are assumed to be compressive, i.e., acting toward the joint. The stress in the vertical does not appear in these equations because it already was determined to be zero. The stress in FA, S_{FA}, was found from analysis of joint 1 to be 20 kips. Simultaneous solution of the two equations yields $S_{HG} = 6.7$ kips and $S_{BH} = 13.3$ kips. (If these stresses had come out with a negative sign, it would have indicated that the original assumption of their directions was incorrect; they would, in that case, be tensile forces instead of compressive forces.)

Parallel-Chord Trusses ▪ A convenient section for determining the stresses in diagonals of parallel-chord trusses is a vertical one, such as N-N in Fig. 1-50a. The sum of the forces acting

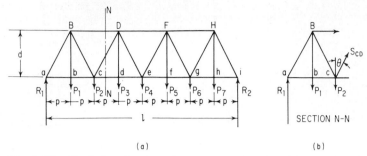

Fig. 1-50. Vertical section through the truss in (*a*) enables determination of stress in diagonal (*b*).

on that portion of the truss to the left of *N-N* equals the vertical component of the stress in diagonal *cD* (see Fig. 1-50*b*). Thus, if θ is the acute angle between *cD* and the vertical,

$$R_1 - P_1 - P_2 + S \cos \theta = 0$$

But $R_1 - P_1 - P_2$ is the algebraic sum of all the external vertical forces on the left of the section and is the vertical shear in the section. It may be designated as *V*. Therefore,

$$V + S \cos \theta = 0 \qquad \text{or} \qquad S = -V \sec \theta$$

From this it follows that for trusses with horizontal chords and single-web systems, the stress in any web member, other than the subverticals, equals the vertical shear in the member multiplied by the secant of the angle that the member makes with the vertical.

Nonparallel Chords ▪ A vertical section also can be used to determine the stress in diagonals when the chords are not parallel, but the previously described procedure must be modified.

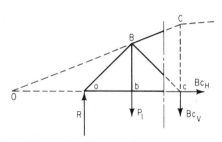

Fig. 1-51. Stress in truss diagonal determined by taking a vertical section and computing moments about the intersection of top and bottom chords.

Suppose, for example, that the stress in the diagonal *Bc* of the Parker truss in Fig. 1-51 is to be found. Take a vertical section to the left of joint *c*. This section cuts *BC*, the top chord, and *Bc*, both of which have vertical components, as well as the horizontal bottom chord *bc*. Now, extend *BC* and *bc* until they intersect, at *O*. If *O* is used as the center for taking moments of all the forces, the moments of the stresses in *BC* and *bc* will be zero since the lines of action pass through *O*. Since *Bc* remains the only stress with a moment about *O*, *Bc* can be computed from the fact that the sum of the moments about *O* must equal zero, for equilibrium.

Generally, the calculation can be simplified by determining first the vertical component of the diagonal and from it the stress. So resolve *Bc* into its horizontal and vertical components Bc_H and Bc_V, at *c*, so that the line of action of the horizontal component passes through *O*. Taking moments about *O* yields

$$(Bc_v \times Oc) - (R \times Oa) + (P_1 \times Ob) = 0$$

from which Bc_V may be determined. The actual stress in *Bc* is Bc_V multiplied by the secant of the angle that *Bc* makes with the vertical.

The stress in verticals, such as Cc, can be found in a similar manner. But take the section on a slope so as not to cut the diagonal but only the vertical and the chords. The moment equation about the intersection of the chords yields the stress in the vertical directly since it has no horizontal component.

Subdivided Panels ▪ In a truss with parallel chords and subdivided panels, such as the one in Fig. 1-52a, the subdiagonals may be either tension or compression. In Fig. 1-52a, the sub-

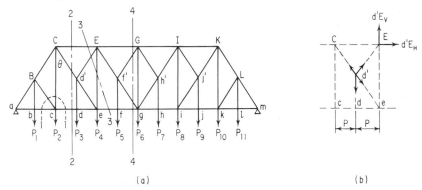

(a)   (b)

Fig. 1-52. Sections taken through truss with subdivided panels for finding stresses in web members.

diagonal Bc is in compression and $d'E$ is in tension. The vertical component of the stress in any subdiagonal equals half the stress in the vertical at the intersection of the subdiagonal and main diagonal.

This can be proved as follows: Take a circular section around d' in Fig. 1-52a. Both the upper and lower portions of the main diagonal Ce are in tension. The stress in the vertical $d'd$ also is tension and equal to the load applied at d. The direction of the stresses Cd', $d'e$, and $d'd$ with reference to joint d' is shown in Fig. 1-52b. Resolve the unknown stress $d'E$ into components at any point along its line of action, such as E. Then, if moments are taken about C, the moments of Cd', $d'e$, and $d'E_H$ are zero since the lines of action of these stresses pass through C. Equating the sum of the moments about C of the stresses d' to zero yields

$$(d'd \times p) - (d'E_V \times 2p) = 0 \qquad \text{or} \qquad d'E_V = \tfrac{1}{2}\,d'd$$

Generally, when determining the stresses in the long verticals and main diagonals, the stresses in the subdiagonals must be included. For example, take a section 1-1 about c (Fig. 1-52a) and equate to zero the sum of the vertical components of the forces. Since the stress in Bc is compressive and its vertical component equals $\tfrac{1}{2}P_1$, the equation for the stress in Cc is $Cc - \tfrac{1}{2}P_1 - P_2 = 0$. If you take a vertical section 2-2, the equation for the stress in Cd' is the same as for a diagonal of a truss with a single system of web members, that is, $V_2 \sec \theta$. If you take a sloping section 3-3, the equation for the stress in the vertical Ee is $Ee + \tfrac{1}{2}P_3 + V_3 = 0$. And if you take a vertical section 4-4, the equation for the stress in $f'g$ is $-f'g \cos \theta + \tfrac{1}{2}P_5 + V_4 = 0$.

For a truss with inclined chords and subdivided panels, the vertical component of the stress in a subdiagonal is not equal to half the stress in the vertical at the intersection with the main diagonal as for parallel-chord trusses. For example, the stress in $d'E$ for a truss with nonparallel chords is $d'd \times l/h$, where l is the length of $d'E$ and h is the length of Ee. The stress in $d'e$ can be found from its vertical component, which can be determined by taking a vertical section just to the left of E, resolving the stress in $d'E$ into horizontal and vertical components at E and the stress in $d'e$ into components at e, then equating to zero the moments about the intersection of the top and bottom chords.

1-48. Moving Loads on Trusses and Girders

To minimize bending stresses in truss members, framing is arranged to transmit loads to panel points. Usually, in bridges, loads are transmitted from a slab to stringers parallel to the trusses, and the stringers carry the load to transverse floor beams, which bring it to truss panel points. Similar framing generally is used for bridge girders.

In many respects, analysis of trusses and girders is similar to that for beams—determination of maximum end reaction for moving loads, for example, and use of influence lines. For girders, maximum bending moments and shears at various sections must be determined for moving loads, as for beams; and as indicated in Art. 1-47, stresses in truss members may be determined by taking moments about convenient points or from the shear in a panel. But girders and trusses differ from beams in that analysis must take into account the effect at critical sections of loads between panel points since such loads are distributed to the nearest panel points; hence, in some cases, influence lines differ from those for beams.

Stresses in Verticals ▪ The maximum total stress in a load-bearing stiffener of a girder or in a truss vertical, such as Bb in Fig. 1-53a, equals the maximum reaction of the floor beam at the panel point. The influence line for the reaction at b is shown in Fig. 1-53b and indicates that for maximum reaction, a uniform load of w lb/lin ft should extend a distance of $2p$, from a to c, where p is the length of a panel. In that case, the stress in Bb equals wp.

Maximum floor beam reaction for concentrated moving loads occurs when the total load between a and c, W_1 (Fig. 1-53c), equals twice the load between a and b. Then, the maximum live-load stress in Bb is

$$r_b = \frac{W_1 g - 2Pg'}{p} = \frac{W_1(g - g')}{p}$$

where g is the distance of W_1 from c, and g' is the distance of P from b.

Stresses in Diagonals ▪ For a truss with parallel chords and single-web system, stress in a diagonal, such as Bc in Fig. 1-53a, equals the shear in the panel multiplied by the secant of the angle θ the diagonal makes with the vertical. The influence diagram for stresses in Bc, then, is the shear influence diagram for the panel multiplied by sec θ, as indicated in Fig. 1-53d. For maximum tension in Bc, loads should be placed only in the portion of the span for which the influence diagram is positive (crosshatched in Fig. 1-54d). For maximum compression, the loads should be placed where the diagram is negative (minimum shear).

A uniform load, however, cannot be placed over the full positive or negative portions of the span to get a true maximum or minimum. Any load in the panel is transmitted to the panel points at both ends of the panel and decreases the shear. True maximum shear occurs for Bc when the uniform load extends into the panel a distance x from c equal to $(n - k)p/(n - 1)$, where n is the number of panels in the truss and k the number of panels from the left end of the truss to c.

For maximum stress in Bc caused by moving concentrated loads, the loads must be placed to produce maximum shear in the panel, and this may require several trials with different wheels placed at c (or, for minimum shear, at b). When the wheel producing maximum shear is at c, the loading will satisfy the following criterion: When the wheel is just to the right of c, W/n is greater than P_1, where W is the total load on the span and P_1 the load in the panel (Fig. 1-53a); when the wheel is just to the left of c, W/n is less than P_1.

For a truss with inclined chords and single-web system, stress in a diagonal, such as Bc in

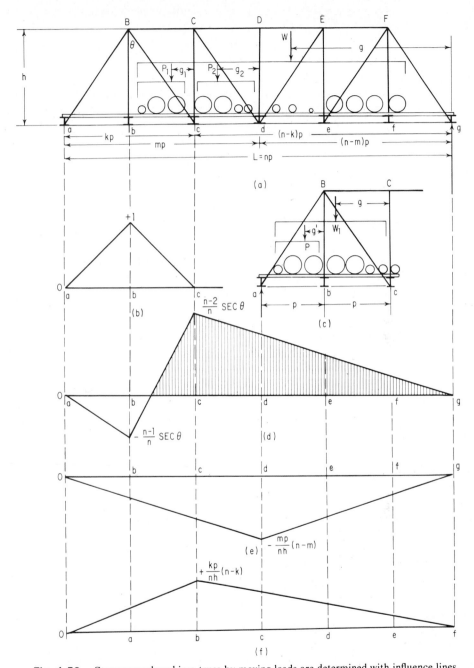

Fig. 1-53. Stresses produced in a truss by moving loads are determined with influence lines.

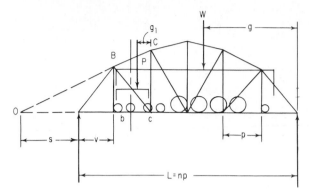

Fig. 1-54. Moving loads on a truss with inclined chords.

Fig. 1-54, is determined by taking moments about the intersection O of chord BC and the bottom chord. Maximum stress in Bc will occur with a heavy wheel at c, when the following criterion is satisfied: When the wheel is just to the right of c, W/n is greater than $P(1 + v/s)$, where W is the total load on the span, n the total number of panels in the truss, P the load in the panel through which the section is passed, v the distance from the left support to the left end of the panel through which the section cutting the web member is passed, and s the distance from O to the left support.

Stresses in Chords ▪ Stresses in truss chords, in general, can be determined from the bending moment at a panel point, so the influence diagram for chord stress has the same shape as that for bending moment at an appropriate panel point. For example, Fig. 1-53e shows the influence line for stress in upper chord CD (minus signifies compression). The ordinates are proportional to the bending moment at d since the stress in CD can be computed by considering the portion of the truss just to the left of d and taking moments about d. Figure 1-53f similarly shows the influence line for stress in bottom chord cd.

For maximum stress in a truss chord under uniform load, the load should extend the full length of the truss.

For maximum chord stress caused by moving concentrated loads, the loads must be placed to produce maximum bending moment at the appropriate panel point, and this may require several trials with different wheels placed at the panel point. Usually, maximum moment will be produced with the heaviest grouping of wheels about the panel point.

In all trusses with verticals, the loading producing maximum chord stress will satisfy the following criterion: When the critical wheel is just to the right of the panel point, Wm/n is greater than P, where mp is the distance of the panel point from the left end of the truss and P is the sum of the loads to the left of the panel point; when the wheel is just to the left of the panel point, Wm/n is less than P.

In a truss without verticals, the maximum stress in the loaded chord is determined by a different criterion. For example, the moment center for the lower chord bc (Fig. 1-55) is panel point C, at a distance c from b. When the critical load is at b or c, the following criterion will be satisfied: When the wheel is just to the right of b or c, Wk/L is greater than $P + Qc/p$; when the wheel is just to the left of b or c, Wk/L is less than $P + Qc/p$, where W is the total load on the span, Q the load in panel bc, P the load to the left of bc, and k the distance of the center of moments C from the left support. The moment at C is $Wgk/L - Pg_1 - Qcg_2/p$, where g is the distance of the center of gravity of the loads W from the right support, g_1 the distance of the center of gravity of the loads P from C, and g_2 the distance of the center of gravity of the loads Q from c, the right end of the panel.

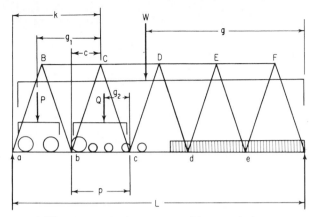

Fig. 1-55.   Moving loads on a truss without verticals.

(C. H. Norris and J. B. Wilbur, "Elementary Structural Analysis," McGraw-Hill Book Company, New York.)

1-49. Counters

For long-span bridges, it often is economical to design the diagonals of trusses for tension only. But in the panels near the center of a truss, maximum shear due to live loads plus impact may exceed and be opposite in sign to the dead-load shear, thus inducing compression in the diagonal. If the tension diagonal is flexible, it will buckle. Hence, it becomes necessary to place in such panels another diagonal crossing the main diagonal (Fig. 1-56). Such diagonals are called *counters.*

Designed only for tension, a counter is assumed to carry no stress under dead load because it would buckle slightly. It comes into action only when the main diagonal is subjected to compression. Hence, the two diagonals never act together.

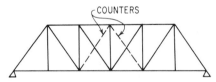

Fig. 1-56.   Truss with counters.

Although the maximum stresses in the main members of a truss are the same whether or not counters are used, the minimum stresses in the verticals are affected by the presence of counters. In most trusses, however, the minimum stresses in the verticals where counters are used are of the same sign as the maximum stresses and hence have no significance.

1-50. Stresses in Trusses Due to Lateral Forces

To resist lateral forces on bridge trusses, trussed systems are placed in the planes of the top and bottom chords, and the ends, or **portals,** also are braced as low down as possible without imping-ing on headroom needed for traffic (Fig. 1-57). When analyzing the lateral trusses, wind loads may be assumed as all applied on the windward chord or as applied equally on the two chords. In the former case, the stresses in the lateral struts are one-half panel load greater than if the latter assumption were made, but this is of no practical consequence.

Where the diagonals are considered as tension members only, counter stresses need not be computed since reversal of wind direction gives greater stresses in the members concerned than any partial loading from the opposite direction. When a rigid system of diagonals is used, the

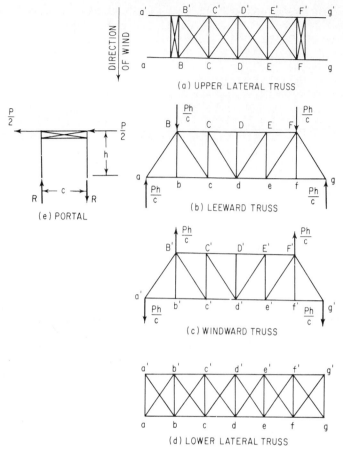

Fig. 1-57. Lateral trusses for bracing top and bottom chords of bridge trusses.

two diagonals of a panel may be assumed equally stressed. Stresses in the chords of the lateral truss should be combined with those in the chords of the main trusses due to dead and live loads.

When computing stresses in the lateral system for the loaded chords of the main trusses, the wind on the live load should be added to the wind on the trusses. Hence, the wind on the live load should be positioned for maximum stress on the lateral truss. Methods described in Art. 1-47 can be used to compute the stresses on the assumption that each diagonal takes half the shear in each panel.

When the main trusses have inclined chords, the lateral systems between the sloping chords lie in several planes, and the exact determination of all the wind stresses is rather difficult. The stresses in the lateral members, however, may be determined without significant error by considering the lateral truss flattened into one plane. Panel lengths will vary, but the panel loads will be equal and may be determined from the horizontal panel length.

Since some of the lateral forces are applied considerably above the horizontal plane of the end supports of the bridge, these forces tend to overturn the structure (Fig. 1-57e). The lateral forces of the upper lateral system (Fig. 1-57a) are carried to the portal struts, and the horizontal loads at these points produce an overturning moment about the horizontal plane of the supports. In Fig. 1-57e, P represents the horizontal load brought to each portal strut by the upper lateral bracing, h the depth of the truss, and c the distance between trusses. The overturning moment produced at each end of the structure is Ph, which is balanced by a reaction couple Rc. The value of the reaction R is then Ph/c. An equivalent effect is achieved on the main trusses if loads equal to Ph/c are applied at B and F and at B' and F', as shown in Fig. 1-57b and c. These loads produce stresses in the end posts and in the lower-chord members, but the web members are not stressed.

The lateral force on the live load also causes an overturning moment, which may be treated in a similar manner. But there is a difference as far as the web members of the main truss are concerned. Since the lateral force on the live load produces an effect corresponding to the position of the live load on the bridge, equivalent panel loads, rather than equivalent reactions, must be computed. If the distance from the resultant of the wind force to the plane of the loaded chord is h', the equivalent vertical panel load is Ph'/c, where P is the horizontal panel load due to the lateral force.

1-51. Complex Trusses

The method of sections may not provide a direct solution for some trusses with inclined chords and multiple-web systems. But if the truss is stable and statically determinate, a solution can be obtained by applying the equations of equilibrium to a section taken around each joint. The stresses in the truss members are obtained by solution of the simultaneous equations.

Since two equations of equilibrium can be written for the forces acting at a joint (Art. 1-47), the total number of equations available for a truss is $2n$, where n is the number of joints. If r is the number of horizontal and vertical components of the reactions, and s the number of stresses, $r + s$ is the number of unknowns.

If $r + s = 2n$, the unknowns can be obtained from solution of the simultaneous equations. If $r + s$ is less than $2n$, the structure is unstable (but the structure may be unstable even if $r + s$ exceeds $2n$). If $r + s$ is greater than $2n$, there are too many unknowns; the structure is statically indeterminate.

GENERAL TOOLS FOR STRUCTURAL ANALYSIS

For some types of structures, the equilibrium equations are not sufficient to determine the reactions or the internal stresses. These structures are called *statically indeterminate*.

For the analysis of such structures, additional equations must be written based on a knowledge of the elastic deformations. Hence, methods of analysis that enable deformations to be evaluated for unknown forces or stresses are important for the solution of problems involving statically indeterminate structures. Some of these methods, like the method of virtual work, also are useful in solving complicated problems involving statically determinate systems.

1-52. Virtual Work

A virtual displacement is an imaginary, small displacement of a particle consistent with the constraints upon it. Thus, at one support of a simply supported beam, the virtual displacement could be an infinitesimal rotation $d\theta$ of that end, but not a vertical movement. However, if the support is replaced by a force, then a vertical virtual displacement may be applied to the beam at that end.

Virtual work is the product of the distance a particle moves during a virtual displacement and the component in the direction of the displacement of a force acting on the particle. If the displacement and the force are in opposite directions, the virtual work is negative. When the displacement is normal to the force, no work is done.

Suppose a rigid body is acted on by a system of forces with a resultant R. Given a virtual displacement ds at an angle α with R, the body will have virtual work done on it equal to $R \cos \alpha \, ds$. (No work is done by internal forces. They act in pairs of equal magnitude but opposite direction, and the virtual work done by one force of a pair is equal and opposite in sign to the work done by the other force.) If the body is in equilibrium under the action of the forces, then $R = 0$, and the virtual work also is zero.

Thus, the principle of virtual work may be stated:

If a rigid body in equilibrium is given a virtual displacement, the sum of the virtual work of the forces acting on it must be zero.

As an example of how the principle may be used, let us apply it to the determination of the reaction R of the simple beam in Fig. 1-58a. First, replace the support by an unknown force R.

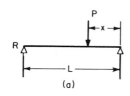

(a)

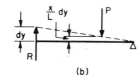

(b)

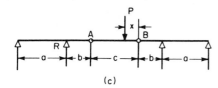

(c)

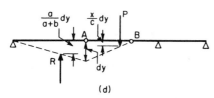

(d)

Fig. 1-58. Virtual work applied to determination of a simple-beam reaction (a) and (b) and the reaction of a beam with suspended span (c) and (d).

Next, move the end of the beam upward a small amount dy as in Fig. 1-58b. The displacement under the load P will be $x \, dy/L$, upward. Then, the virtual work is $R \, dy - Px \, dy/L = 0$, from which $R = Px/L$.

The principle also may be used to find the reaction R of the more complex beam in Fig. 1-58c. Again, the first step is to replace the support by an unknown force R. Next, apply a virtual downward displacement dy at hinge A (Fig. 1-58d). The displacement under the load P will be $x \, dy/c$ and at the reaction R will be $a \, dy/(a + b)$. According to the principle of virtual work, $-Ra \, dy/(a + b) + Px \, dy/c = 0$; thus, $R = Px(a + b)/ac$. In this type of problem, the method has the advantage that only one reaction need be considered at a time and internal forces are not involved.

Strain Energy ▪ When an elastic body is deformed, the virtual work done by the internal forces equals the corresponding increment of the strain energy dU, in accordance with the principle of virtual work.

Assume a constrained elastic body acted on by forces $P_1, P_2, \ldots$, for which the corresponding deformations are $e_1, e_2, \ldots$. Then, $\Sigma P_n \, de_n = dU$. The increment of the strain energy due to the increments of the deformations is given by

$$dU = \frac{\partial U}{\partial e_1} \, de_1 + \frac{\partial U}{\partial e_2} \, de_2 + \cdots$$

When solving a specific problem, a virtual displacement that is most convenient in simplifying the solution should be chosen. Suppose, for example, a virtual displacement is selected that affects only the deformation e_n corresponding to the load P_n, other deformations being unchanged. Then, the principle of virtual work requires that

$$P_n \, de_n = \frac{\partial U}{\partial e_n} \, de_n$$

This is equivalent to

$$\frac{\partial U}{\partial e_n} = P_n \qquad (1\text{-}57)$$

which states that the partial derivative of the strain energy with respect to a specific deformation gives the corresponding force.

Suppose, for example, the stress in the vertical bar in Fig. 1-59 is to be determined. All bars are made of the same material and have the same cross section A. If the vertical bar stretches an amount e under the load P, the inclined bars will each stretch an amount $e \cos \alpha$. The strain energy in the system is [From Eq. (1-23 a)]

$$U = \frac{AE}{2L}(e^2 + 2e^2 \cos^3 \alpha)$$

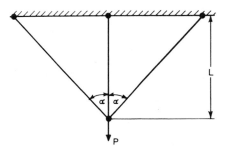

and the partial derivative of this with respect to e must be equal to P; that is,

$$P = \frac{AE}{2L}(2e + 4e \cos^3 \alpha) = \frac{AEe}{L}(1 + 2 \cos^3 \alpha)$$

Noting that the force in the vertical bar equals AEe/L, we find from the above equation that the required stress equals $P/(1 + 2 \cos^3 \alpha)$.

Fig. 1-59. Indeterminate truss.

1-53. Castigliano's Theorems

If strain energy is expressed as a function of statically independent forces, the partial derivative of the strain energy with respect to a force gives the deformation corresponding to that force:

$$\frac{\partial U}{\partial P_n} = e_n \qquad (1\text{-}58)$$

(See also Art. 1-52.)

This is known as Castigliano's first theorem. (His second theorem is the principle of least work.)

1-54. Method of Least Work

Castigliano's second theorem, also known as the principle of least work, states:

The strain energy in a statically indeterminate structure is the minimum consistent with equilibrium.

As an example of the use of the method of least work, an alternative solution will be given for the stress in the vertical bar in Fig. 1-59 (see Art. 1-52). Calling this stress X, we note that the stress in each of the inclined bars must be $(P - X)/2 \cos \alpha$. Using Eq. (1-23 a), we can express the strain energy in the system in terms of X:

$$U = \frac{X^2 L}{2AE} + \frac{(P - X)^2 L}{4AE \cos^3 \alpha}$$

Hence, the internal work in the system will be a minimum when

$$\frac{\partial U}{\partial X} = \frac{XL}{AE} - \frac{(P - X)L}{2AE \cos^3 \alpha} = 0$$

Solving for X gives the stress in the vertical bar as $P/(1 + 2 \cos^3 \alpha)$, as in Art. 1-52.

1-55. Dummy Unit-Load Method

The strain energy for pure bending is $U = M^2L/2EI$ [see Eq. (1-23d)]. To find the strain energy due to bending stress in a beam, we can apply this equation to a differential length dx of the beam and integrate over the entire span. Thus,

$$U = \int_0^L \frac{M^2\,dx}{2EI} \tag{1-59}$$

If we let M represent the bending moment due to a generalized force P, the partial derivative of the strain energy with respect to P is the deformation d corresponding to P. Differentiating Eq. (1-59) gives

$$d = \int_0^L \frac{M}{EI}\frac{\partial M}{\partial P}\,dx \tag{1-60}$$

The partial derivative in this equation is the rate of change of bending moment with the load P. It equals the bending moment m produced by a unit generalized load applied at the point where the deformation is to be measured and in the direction of the deformation. Hence, Eq. (1-60) can also be written as

$$d = \int_0^L \frac{Mm}{EI}\,dx \tag{1-61}$$

To find the vertical deflection of a beam, we apply a dummy unit load vertically at the point where the deflection is to be measured and substitute the bending moments due to this load and the actual loading in Eq. (1-61). Similarly, to compute a rotation, we apply a dummy unit moment.

As a simple example, let us apply the dummy unit-load method to the determination of the deflection at the center of a simply supported, uniformly loaded beam of constant moment of inertia (Fig. 1-60a). As indicated in Fig. 1-60b, the bending moment at a distance x from one end is $(wL/2)x - (w/2)x^2$. If we apply a dummy unit load vertically at the center of the beam (Fig. 1-60c), where the vertical deflection is to be determined, the moment at x is $x/2$, as indicated in Fig. 1-60d. Substituting in Eq. (1-61) gives

$$d = 2\int_0^{L/2}\left(\frac{wL}{2}x - \frac{w}{2}x^2\right)\frac{x}{2}\frac{dx}{EI} = \frac{5wL^4}{384EI}$$

As another example, let us apply the method to finding the end rotation at one end of a simply supported, prismatic beam produced by a moment applied at the other end. In other words, the problem is to find the end rotation at B, θ_B, in Fig. 1-61a, due to M_A. As indicated in Fig. 1-61b, the bending moment at a distance x from B due to M_A is M_Ax/L. If we apply a dummy unit moment at B (Fig. 1-61c), it will produce a moment at x of $(L - x)/L$ (Fig. 1-61d). Substituting in Eq. (1-61) gives

$$\theta_B = \int_0^L M_A\frac{x}{L}\frac{L - x}{L}\frac{dx}{EI} = \frac{M_AL}{6EI}$$

To determine the deflection of a beam due to shear, Castigliano's first theorem can be applied to the strain energy in shear:

$$U = \int\int\frac{v^2}{2G}\,dA\,dx$$

where v = shearing unit stress
G = modulus of rigidity
A = cross-sectional area

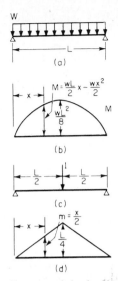

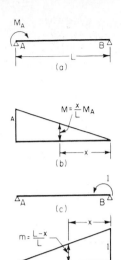

Fig. 1-60. Dummy unit-load method applied to a uniformly loaded beam (*a*) to find the midspan deflection; (*b*) moment diagram for the uniform load; (*c*) unit load at midspan; (*d*) moment diagram for unit load.

Fig. 1-61. End rotation at *B* in beam *AB* (*a*) caused by end moment at *A* is determined by dummy unit-load method; (*b*) moment diagram for end moment; (*c*) unit moment applied at beam end; (*d*) moment diagram for unit moment.

1-56. Truss Deflections by Dummy Unit-Load Method

Article 1-55 shows how the dummy unit-load method applies to determination of beam deflections. The method also may be adapted to computation of truss deformations.

The strain energy in a truss is given by

$$U = \sum \frac{S^2 L}{2AE} \tag{1-62}$$

which represents the sum of the strain energy for all the members of the truss. S is the stress in each member due to the loads, L the length of each, A the cross-sectional area, and E the modulus of elasticity. Application of Castigliano's first theorem (Art. 1-53) and differentiation inside the summation sign yield the deformation:

$$d = \sum \frac{SL}{AE} \frac{\partial S}{\partial P} \tag{1-63}$$

where, as in Art. 1-55, P represents a generalized load. The partial derivative in this equation is the rate of change of axial stress with P. It equals the axial stress u produced in each member of the truss by a unit load applied at the point where the deformation is to be measured and in the direction of the deformation. Consequently, Eq. (1-63) also can be written

$$d = \sum \frac{SuL}{AE} \tag{1-64}$$

To find the vertical deflection at any point of a truss, apply a dummy unit vertical load at the panel point where the deflection is to be measured. Substitute in Eq. (1-64) the stresses in each member of the truss due to this load and the actual loading. Similarly, to find the rotation of any joint, apply a dummy unit moment at the joint, compute the stresses in each member of the truss, and substitute in Eq. (1-64). When it is necessary to determine the relative movement of two panel points in the direction of a member connecting them, apply dummy unit loads in opposite directions at those points.

Note that members not stressed by the actual loads or the dummy loads do not enter into the calculation of a deformation.

As an example of the application of Eq. (1-64), let us compute the midspan deflection of the truss in Fig. 1-62a. The stresses in kips due to the 20-kip load at every lower-chord panel point

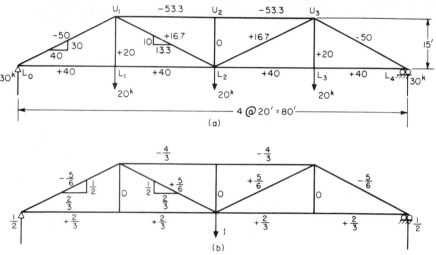

Fig. 1-62. Dummy unit-load method applied to a loaded truss (a) to find midspan deflection; (b) stresses produced by unit load applied at midspan.

are shown in Fig. 1-62a, and the ratios of length of members in inches to their cross-sectional areas in square inches are given in Table 1-1. We apply a dummy unit vertical load at L_2, where the deflection is required. Stresses u due to this load are shown in Fig. 1-62b and Table 1-1.

TABLE 1-1 Midspan Deflection of Truss of Fig. 1-62

Member	L/A	S	u	SuL/A
L_0L_2	160	$+40$	$+2/3$	4,267
L_0U_1	75	-50	$-5/6$	3,125
U_1U_2	60	-53.3	$-4/3$	4,267
U_1L_2	150	$+16.7$	$+5/6$	2,083
				13,742

$$d = \sum \frac{SuL}{AE} = \frac{2 \times 13,742,000}{30,000,000} = 0.916 \text{ in}$$

Table 1-1 also contains the computations for the deflection. Members not stressed by the 20-kip loads or the dummy unit loads are not included. Taking advantage of the symmetry of the truss, the values are tabulated for only half the truss and the sum is doubled. Also, to reduce the amount of calculation, the modulus of elasticity E, which is equal to 30,000,000, is not included until the very last step since it is the same for all members.

1-57.   Statically Indeterminate Trusses

A truss is statically indeterminate when the number of unknown quantities exceeds the number of independent equations of static equilibrium that may be written for the structure or portions of the structure. As noted in Art. 1-51, if n is the number of joints in a truss, r the number of horizontal and vertical components of the reactions, and s the number of stresses, the truss is statically indeterminate if $r + s$ is greater than $2n$.

The truss in Fig. 1-63a is statically indeterminate. It has 4 joints, 3 reaction components, and 6 members, so that $(r + s = 3 + 6) > (2n = 2 \times 4)$. To determine the stresses in this truss, it is necessary to add to the equations of equilibrium an equation based on a knowledge of the elastic deformations of the truss. For this purpose, remove member ac to make the truss statically determinate and instead apply unknown stresses S_1 at both a and c, as indicated in Fig. 1-63b. Then, determine S_1 by equating the sum of the horizontal deflection at c of the statically determinate truss and the elongation of ac to zero.

Let the stress in any member due to the load P and S_1 be $S' + uS_1$, where S' is the stress due to P, u the stress in any member due to unit force applied at a and c, and uS_1 the stress in any member due to S_1. Applying Eq. (1-64), we find that the horizontal displacement of c (assuming a fixed) due to both P and S_1 is

$$\delta = \sum_{2}^{n} \frac{(S' + uS_1)uL}{AE}$$

$$= \sum_{2}^{n} \frac{S'uL}{AE} + \sum_{2}^{n} \frac{S_1 u^2 L}{AE}$$

where L = length of member
A = cross-sectional area
E = modulus of elasticity

From Eq. (1-8), elongation of ac is $s_1 L_1/A_1 E = S_1 u^2 L_1/A_1 E$ since $u = 1$ in ac. Then, since the sum of δ and the elongation of ac is zero,

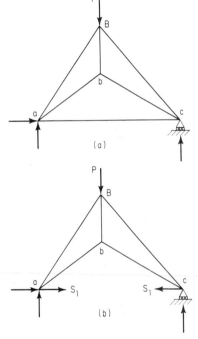

Fig. 1-63.   Statically indeterminate truss (a) is made determinate by cutting its tie (b).

$$\sum_{2}^{n} \frac{S'uL}{AE} + \sum_{2}^{n} \frac{S_1 u^2 L}{AE} + \frac{S_1 u^2 L_1}{A_1 E} = \sum_{2}^{n} \frac{S'uL}{AE} + S_1 \sum_{1}^{n} \frac{u^2 L}{AE} = 0$$

Solving for S_1 and assuming E constant, we obtain

$$S_1 = -\frac{\displaystyle\sum_{2}^{n} (S'uL/A)}{\displaystyle\sum_{1}^{n} (u^2 L/A)} \tag{1-65}$$

With S_1 known, the stress in each member of the truss can be determined by adding uS_1 for the member to S'.

As an example of the use of Eq. (1-65), let us determine the stresses in the truss of Fig. 1-64. The truss has 4 reaction components, 5 members (considering each of the main diagonals as one member), and 4 joints (not counting the intersection of the diagonals). Hence $(r + s = 4 + 5) > (2n = 2 \times 4)$. The truss is statically indeterminate, with one unknown more than equations for static equilibrium.

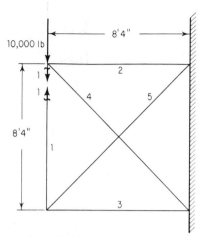

Assume member 1 as the redundant member and apply a pair of unit forces near the upper end of the member in such a direction as to cause tension in the member. Equation (1-65) is solved in Table 1-2 with E constant. From Table 1-2, $S_1 = -717,000/168.4 = -4300$ lb.

The true stress in each member of the complete frame is given in the last column of the table. (C. H. Norris and J. B. Wilbur, "Elementary Structural Analysis," McGraw-Hill Book Company, New York.)

Fig. 1-64. Truss with two diagonals is made determinate by cutting a member.

1-58. Reciprocal Theorem and Influence Lines

Consider a structure loaded by a group of independent forces A, and suppose that a second group of forces B is added. The work done by the forces A acting over the displacements due to B will be W_{AB}.

Now, suppose the forces B had been on the structure first and then load A had been applied. The work done by the forces B acting over the displacements due to A will be W_{BA}.

The reciprocal theorem states that $W_{AB} = W_{BA}$.

Some very useful conclusions can be drawn from this equation. For example, there is the reciprocal deflection relationship:

The deflection at a point A due to a load at B equals the deflection at B due to the same load applied at A. Also, the rotation at A due to load (or moment) at B equals the rotation at B due to the same load (or moment) applied to A.

Another consequence is that deflection curves also may be influence lines, to some scale, for reactions, shears, moments, or deflections (**Mueller-Breslau principle**). For example, suppose the influence line for a reaction is to be found; that is, we wish to plot the reaction R as a unit

TABLE 1-2 Computation of Stresses in Truss of Fig. 1-64

Member	A, in^2	L, in	S', lb	$S'L/A$	u, lb	$S'uL/A$	u^2L/A	S_1u	$S = S' + S_1u$, lb
1	4	100	0	....	+1.00	....	+25.0	−4,300	−4,300
2	4	100	+10,000	+250,000	+1.00	+250,000	+25.0	−4,300	+5,700
3	4	100	0	....	+1.00	....	+25.0	−4,300	−4,300
4	6	141	−14,100	−331,000	−1.41	+467,000	+46.7	+6,000	−8,100
5	6	141	0	....	−1.41	....	+46.7	+6,000	+6,000
				Σ		+717,000	+168.4		

load moves over the structure, which may be statically indeterminate. For loading condition A, we analyze the structure with a unit load on it at a distance x from some reference point. For loading condition B, we apply a dummy unit vertical load upward at the place where the reaction is to be determined, deflecting the structure off the support. At a distance x from the reference point, the displacement is d_{xR}, and over the support the displacement is d_{RR}. Hence, $W_{AB} = -1d_{xR} + Rd_{RR}$. On the other hand, W_{BA} is zero since loading condition A provides no displacement for the dummy unit load at the support in condition B. Consequently, from the reciprocal theorem, $W_{AB} = W_{BA} = 0$; hence

$$R = \frac{d_{xR}}{d_{RR}}$$

Since d_{RR}, the deflection at the support due to a unit load applied there, is a constant, R is proportional to d_{xR}. So the influence line for a reaction can be obtained from the deflection curve resulting from a displacement of the support (Fig. 1-65a). The magnitude of the reaction is obtained by dividing each ordinate of the deflection curve by d_{RR}.

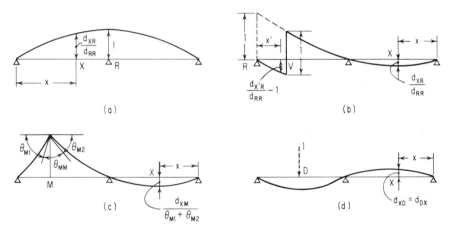

Fig. 1-65. Influence lines for a continuous beam are obtained from deflection curves. (a) Reaction at R; (b) shear at V; (c) bending moment at M; (d) deflection at D.

Similarly, the influence line for shear can be obtained from the deflection curve produced by cutting the structure and shifting the cut ends vertically at the point for which the influence line is desired (Fig. 1-65b).

The influence line for bending moment can be obtained from the deflection curve produced by cutting the structure and rotating the cut ends at the point for which the influence line is desired (Fig. 1-65c).

Finally, it may be noted that the deflection curve for a load of unity is also the influence line for deflection at that point (Fig. 1-65d).

CONTINUOUS BEAMS AND FRAMES

Continuous beams and frames are statically indeterminate. Bending moments in them are functions of the geometry, moments of inertia, and modulus of elasticity of individual members as well as of loads and spans. Although these moments can be determined by the methods described in Arts. 1-52 to 1-58, there are methods specially developed for beams and frames that often make analysis simpler. The following articles describe some of these methods.

1-59. General Method of Analysis

Continuous beams and frames consist of members that can be treated as simple beams, the ends of which are prevented by moments from rotating freely. Member LR in the continuous beam in Fig. 1-66a, for example, can be isolated, as shown in Fig. 1-66b, and the elastic restraints at

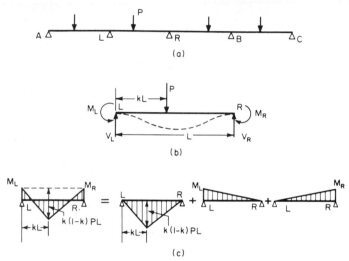

Fig. 1-66.   Any span of a continuous beam (a) can be treated as a simple beam, as shown in (b) and (c). In (c), the bending-moment diagram is decomposed into basic components.

the ends replaced by couples M_L and M_R. In this way, LR is converted into a simply supported beam acted upon by end moments and transverse loads.

The bending-moment diagram for LR is shown at the left in Fig. 1-66c. Treating LR as a simple beam, we can break down this diagram into three simple components, as shown at the right of the equals sign in Fig. 1-66c. Thus, the bending moment at any section equals the simple-beam moment due to the transverse loads plus the simple-beam moment due to the end moment at L plus the simple-beam moment due to the end moment at R.

Once M_L and M_R have been determined, the shears may be computed by taking moments about any section. Similarly, if the reactions or shears are known, the bending moments may be calculated.

A general method for determining the elastic forces and moments exerted by redundants, supports, and moments is as follows: Remove as many redundant supports or members as necessary to make the structure statically determinate. Compute, for the actual loads, the deflections or rotations of the statically determinate structure in the direction of the forces and couples exerted by the removed supports and members. Then, in terms of these forces and couples, compute the corresponding deflections or rotations the forces and couples produce in the statically determinate structure. Finally, for each redundant support or member, write equations that give the known rotations or deflections of the original structure in terms of the deformations of the statically determinate structure.

For example, one method of finding the reactions of the continuous beam in Fig. 1-67a is to remove the interior supports temporarily. The beam then will have deflections d_1, d_2, and d_3 at those supports (Fig. 1-67b). Next, place a unit load at the location of each support in succession (Fig. 1-67c, d, and e). Let y_{mn} denote the simple-beam deflection, where m represents the sup-

port where the deflection is measured and n the support at which the unit load is applied. Since the continuous beam has no deflections at its supports, we can now write one equation for each support equating the downward deflection of the loaded simple beam to its upward deflection due to the unknown reactions:

$$d_1 = y_{11}R_1 + y_{12}R_2 + y_{13}R_3$$
$$d_2 = y_{21}R_1 + y_{22}R_2 + y_{23}R_3 \quad (1\text{-}66)$$
$$d_3 = y_{31}R_1 + y_{32}R_2 + y_{33}R_3$$

Solution of these equations yields R_1, R_2, and R_3, and R_0 and R_4 can be obtained from these reactions by applying the equations of equilibrium.

For continuous beams and frames with a large number of redundants, this method becomes unwieldy because of the number of simultaneous equations. Special methods, like moment distribution, are preferable in such cases.

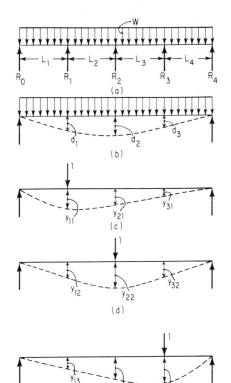

Fig. 1-67. Reactions of continuous beam (a) are found by making the beam statically determinate. (b) Deflections are computed with interior supports removed. (c), (d), and (e) Deflections are calculated for unit load over each removed support to obtain equations for each redundant.

1-60. Sign Convention

For moment distribution, the following sign convention is most convenient: A moment acting at an end of a member or at a joint is positive if it tends to rotate the end or joint, clockwise; negative if it tends to rotate the joint counterclockwise. Hence, in Fig. 1-66b, M_R is positive and M_L negative.

Similarly, the angular rotation at the end of a member is positive if in a clockwise direction, negative if counterclockwise. Thus, a positive end moment produces a positive end rotation in a simple beam. (See also Art. 1-59.)

For ease in visualizing the shape of the elastic curve under the action of loads and end moments, plot bending-moment diagrams on the tension side of each member. Hence, if an end moment is represented by a curved arrow, the arrow will point in the direction in which the moment is to be plotted.

1-61. Fixed-End Moments

A beam so restrained at its ends that no rotation is produced there by the loads is called a fixed-end beam, and the end moments are called fixed-end moments. Actually, it would be very difficult to construct a beam with ends that are truly fixed. The concept of fixed ends, however, is useful in determining the moments in continuous beams and frames.

Fixed-end moments may be expressed as the product of a coefficient and WL, where W is the total load on the span L. The coefficient is independent of the properties of other members of the structure. Thus, any member of a continuous beam or frame can be isolated from the rest of the structure and its fixed-end moments computed. Then, the actual moments in the beam can be found by applying a correction to each fixed-end moment.

Fixed-end moments may be determined conveniently by the moment-area method or the conjugate-beam method (Art. 1-33).

Fixed-end moments for several common types of loading on beams of constant moment of inertia (prismatic beams) are given in Fig. 1-68. Also, the curves in Fig. 1-70 enable fixed-end

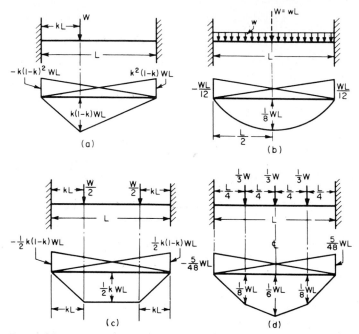

Fig. 1-68. Fixed-end moments for a prismatic beam: (*a*) for a concentrated load; (*b*) for a uniform load; (*c*) for two equal concentrated loads; (*d*) for three equal concentrated loads.

moments to be computed easily for any type of loading on a prismatic beam. Before the curves can be entered, however, certain characteristics of the loading must be calculated. These include $\bar{x}L$, the location of the center of gravity of the loading with respect to one of the loads; $G^2 = \Sigma b_n^2 P_n / W$, where $b_n L$ is the distance from each load P_n to the center of gravity of the loading (taken positive to the right); and $S^3 = \Sigma b_n^3 P_n / W$ (see case 8, Fig. 1-69). These values are given in Fig. 1-69 for some common types of loading.

The curves in Fig. 1-70 are entered at the bottom with the location a of the center of gravity of the loading with respect to the left end of the span. At the intersection with the proper G curve, proceed horizontally to the left to the intersection with the proper S line, then vertically to the horizontal scale indicating the coefficient m by which to multiply WL to obtain the fixed-end moment. The curves solve the equations:

$$m_L = \frac{M_L^F}{WL} = G^2[1 - 3(1 - a)] + a(1 - a)^2 + S^3 \qquad (1\text{-}67a)$$

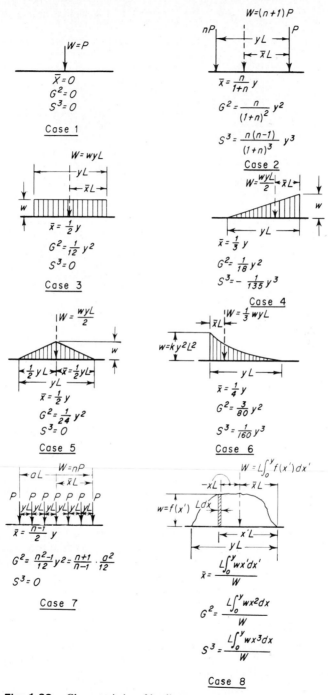

Fig. 1-69. Characteristics of loadings.

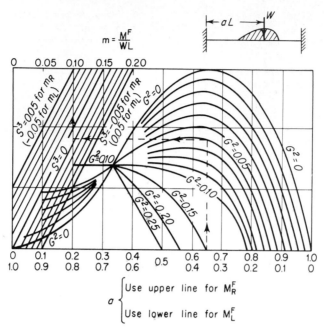

$$m = \frac{M^F}{WL}$$

Fig. 1-70. Chart for fixed-end moments caused by any type of loading.

$$m_R = \frac{M^F_R}{WL} = G^2(1 - 3a) + a^2(1 - a) - S^3 \qquad (1\text{-}67b)$$

where M^F_L is the fixed-end moment at the left support and M^F_R at the right support.

As an example of the use of the curves, find the fixed-end moments in a prismatic beam of 20-ft span carrying a triangular loading of 100 kips, similar to the loading shown in case 4, Fig. 1-69, distributed over the entire span, with the maximum intensity at the right support.

Case 4 gives the characteristics of the loading: $y = 1$; the center of gravity is $L/3$ from the right support; so $a = 0.67$, $G^2 = 1/18 = 0.056$, and $S^3 = -1/135 = -0.007$. To find M^F_R, we enter Fig. 1-70 at the bottom with $a = 0.67$ on the upper scale and proceed vertically to the estimated location of the intersection of the coordinate with the $G^2 = 0.06$ curve. Then we move horizontally to the intersection with the line for $S^3 = -0.007$, as indicated by the dashed line in Fig. 1-70. Referring to the scale at the top of the diagram, we find the coefficient m_R to be 0.10. Similarly, with $a = 0.67$ on the lower scale, we find the coefficient m_L to be 0.07. Hence, the fixed-end moment at the right support is $0.10 \times 100 \times 20 = 200$ ft-kips, and at the left support $-0.07 \times 100 \times 20 = -140$ ft·kips.

1-62. Fixed-End Stiffness

To correct a fixed-end moment to obtain the end moment for the actual conditions of end restraint in a continuous structure, the end of the member must be permitted to rotate. The amount it will rotate depends on its stiffness, or resistance to rotation.

The fixed-end stiffness of a beam is defined as the moment required to produce a rotation of unity at the end where it is applied, while the other end is fixed against rotation. It is represented by K^F_R in Fig. 1-71.

For prismatic beams, the fixed-end stiffnesses for both ends equal $4EI/L$, where E is the modulus of elasticity, I the moment of inertia of the cross section about the centroidal axis, and L the span (generally taken center to center of supports). When deformations need not be calculated, only the relative values of K^F for each member need be known; hence, only the ratio of I to L has to be computed. (For simply supported, prismatic beams, the actual stiffness is $3EI/L$, or three-fourths the fixed-end stiffness.)

For beams of variable moment of inertia, the fixed-end stiffness may be calculated by methods presented in Art. 1-69 or obtained from tables, such as those in the "Handbook of Frame Constants," EB034D, Portland Cement Association, Skokie, Ill. 60077; and J. M. Gere, "Moment Distribution Factors for Beams of Tapered *I*-Section," American Institute of Steel Construction, Chicago. Ill.

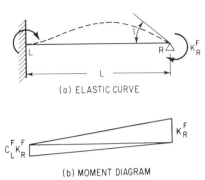

(a) ELASTIC CURVE

(b) MOMENT DIAGRAM

Fig. 1-71. Fixed-end stiffness.

1-63. Fixed-End Carry-Over Factor

When a moment is applied at one end of a continuous beam, a resisting moment is induced at the far end if that end is restrained by other beams or columns against rotation (Fig. 1-71). The ratio of the resisting moment at a fixed end to the applied moment is called the fixed-end carry-over factor C^F.

For prismatic beams, the fixed-end carry-over factor toward either end is 0.5. Note that the applied moment and the resisting moment have the same sign (Fig. 1-71*a*); that is, if the applied moment acts in a clockwise direction, the carry-over moment also acts clockwise.

For beams of variable moment of inertia, the fixed-end carry-over factor may be calculated by methods presented in Art. 1-69 or obtained from tables, such as those referred to in Art. 1-62.

1-64. Moment Distribution by Converging
Approximations

The frame in Fig. 1-72 consists of four prismatic members rigidly connected together at O and fixed at ends A, B, C, and D. If an external moment U is applied at O, the sum of the end moments in each member at O must be equal to U. Furthermore, all members must rotate at O through the same angle θ since they are assumed to be rigidly connected there. Hence, by the definition of fixed-end stiffness (Art. 1-62), the proportion of U induced in or "distributed" to the end of each member at O equals the ratio of the stiffness of that member to the sum of the stiffnesses of all the members at O. This ratio is called the distribution factor at O for the member.

Suppose a moment of 100 ft-kips is applied at O, as indicated in Fig. 1-72*b*. The relative stiffness (or I/L) is assumed as shown in the circle on each member. The distribution factors for the moment at O are computed from the stiffnesses and shown in the boxes. For example, the distribution factor for OA equals its stiffness divided by the sum of the stiffnesses of all the members at the joint: $3/(3 + 1 + 4 + 2) = 0.3$. Hence, the moment induced in OA at O is $0.3 \times 100 = 30$ ft-kips. Similarly, OB gets 10 ft-kips, OC 40 ft-kips, and OD 20 ft-kips.

Because the far ends of these members are fixed, one-half of these moments are carried over to them (Art. 1-63). Thus $M_{AO} = 0.5 \times 30 = 15$; $M_{BO} = 0.5 \times 10 = 5$; $M_{CO} = 0.5 \times 40 = 20$; and $M_{DO} = 0.5 \times 20 = 10$.

Most structures consist of frames similar to the one in Fig. 1-72, or even simpler, joined

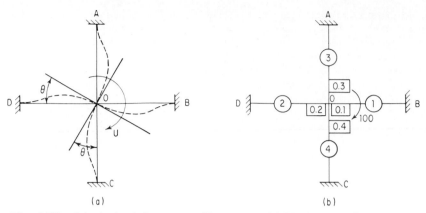

Fig. 1-72. Joint in simple frame rotated by moment. (*a*) Elastic curve; (*b*) stiffness and distribution factors.

together. Although the ends of the members may not be fixed, the technique employed for the frame in Fig. 1-72 can be applied to find end moments in such continuous structures.

Span with Simple Support ▪ Before the general method is presented, one shortcut is worth noting. Advantage can be taken when a member has a hinged end to reduce the work in distributing moments. This is done by using the true stiffness of the member instead of the fixed-end stiffness. (For a prismatic beam, the stiffness of a member with a hinged end is three-fourths the fixed-end stiffness; for a beam with variable moment of inertia, it is equal to the fixed-end stiffness times $1 - C_L^F C_R^F$, where C_L^F and C_R^F are the fixed-end carry-over factors to each end of the beam.) Naturally, the carry-over factor toward the hinge is zero.

Moment Release and Distribution ▪ When beam ends are neither fixed nor pinned but restrained by elastic members, moments can be distributed by a series of converging approximations. At first, all joints are locked against rotation. As a result, the loads will create fixed-end moments (Art. 1-61) at the ends of every loaded member. At each joint, the unbalanced moment, a moment equal to the algebraic sum of the fixed-end moments at the joint, is required to hold it fixed. But if the joint actually is not fixed, the unbalanced moment does not exist. It must be removed by applying an equal but opposite moment. One joint at a time is unlocked by applying a moment equal but opposite in sign to the unbalanced moment. The unlocking moment must be distributed to the members at the joint in proportion to their fixed-end stiffnesses. As a result, the far end of each member should receive a "carry-over" moment equal to the distributed moment times a carry-over factor (Art. 1-63).

After all joints have been released at least once, it generally will be necessary to repeat the process—sometimes several times—before the corrections to the fixed-end moments become negligible. To reduce the number of cycles, start the unlocking of joints with those having the greatest unbalanced moments. Also, include carry-over moments with fixed-end moments in computing unbalanced moments.

Example ▪ Suppose the end moments are to be found for the continuous beam *ABCD* in Fig. 1-73, given the fixed-end moments on the first line of the figure. The *I/L* values for all spans are given equal; therefore, the relative fixed-end stiffness for all members is unity. But since *A* is a hinged end, the computation can be shortened by using the actual relative stiffness,

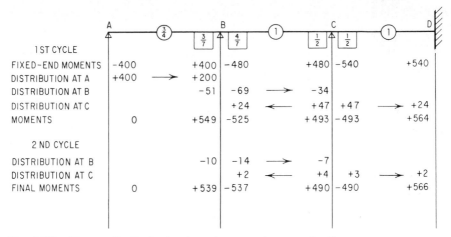

Fig. 1-73. Moment distribution in a beam by converging approximations.

which is ¾. Relative stiffnesses for all members are shown in the circle on each member. The distribution factors are shown in the boxes at each joint.

Begin the computation with removal of the unbalance in fixed-end moments (first line in Fig. 1-73). The greatest unbalanced moment, by inspection, occurs at hinged end A and is -400, so unlock this joint first. Since there are no other members at the joint, distribute the full unlocking moment of $+400$ to AB at A and carry over one-half to B. The unbalance at B now is $+400 - 480$ plus the carry-over of $+200$ from A, or a total of $+120$. Hence, a moment of -120 must be applied and distributed to the members at B by multiplying by the distribution factors in the corresponding boxes.

The net moment at B could be found now by adding the fixed-end and distributed moments at the joint. But it generally is more convenient to delay the summation until the last cycle of distribution has been completed.

After B is unlocked, the moment distributed to BA need not be carried over to A because the carry-over factor toward the hinged end is zero. But half the moment distributed to BC is carried over to C.

Similarly, unlock joint C and carry over half the distributed moments to B and D, respectively. Joint D should not be unlocked since it actually is a fixed end. Thus, the first cycle of moment distribution has been completed.

Carry out the second cycle in the same manner. Release joint B, and carry over to C half the distributed moment in BC. Finally, unlock C to complete the cycle. Add the fixed-end and distributed moments to obtain the final moments.

1-65. Continuous Frames

In continuous frames, maximum end moments and maximum interior moments are produced by different combinations of loadings. For maximum end moment in a beam, live load should be placed on that beam and on the beam adjoining the end for which the moment is to be computed. Spans adjoining these two should be assumed to be carrying only dead load.

For maximum midspan moments, the beam under consideration should be fully loaded, but adjoining spans should be assumed to be carrying only dead load.

The work involved in distributing moments due to dead and live loads in continuous frames in buildings can be greatly simplified by isolating each floor. The tops of the upper columns and the bottoms of the lower columns can be assumed fixed. Furthermore, the computations can be condensed considerably by following the procedure recommended in "Continuity in Concrete Building Frames," EB033D Portland Cement Association, Skokie, Ill. 60077, and illustrated in Fig. 1-74.

	A		B			C			D		E
1. STIFFNESS		1			1			1			1
2. DISTRIBUTION FACTOR	0.33		0.25	0.25		0.25	0.25		0.25	0.25		0.33
3. F.E.M. DEAD LOAD	—		+91	−37		+37	−70		+70	−59		—
4. F.E.M. TOTAL LOAD	−172	+99	+172	−78	+73	+78	−147	+85	+147	−126	+63	+126
5. CARRY OVER	−17	+11	+29	−1	+1	−2	−11	+7	+14	−21	+13	+7
6. ADDITION	−189	+18	+201	−79	−1	+76	−158	+9	+161	−147	+5	+133
7. DISTRIBUTION	+63		−30	−30		+21	+21		−4	−4		−44
8. MAX. MOMENTS	−126	+128	+171	−109	+73	+97	−137	+101	+157	−151	+81	+89

Fig. 1-74. Moment distribution in a continuous frame by converging approximations.

Figure 1-74 presents the complete calculation for maximum end and midspan moments in four floor beams *AB*, *BC*, *CD*, and *DE*. Columns are assumed to be fixed at the story above and below. None of the beam or column sections is known to begin with; so as a start, all members will be assumed to have a fixed-end stiffness of unity, as indicated on the first line of the calculation.

Column Moments ▪ The second line gives the distribution factors (Art. 1-64) for each end of the beams; column moments will not be computed until moment distribution to the beams has been completed. Then, the sum of the column moments at each joint may be easily computed since they are the moments needed to make the sum of the end moments at the joint equal to zero. The sum of the column moments at each joint can then be distributed to each column there in proportion to its stiffness. In this example, each column will get one-half the sum of the column moments.

Fixed-end moments at each beam end for dead load are shown on the third line, just above the horizontal line, and fixed-end moments for live plus dead loads on the fourth line. Corresponding midspan moments for the fixed-end condition also are shown on the fourth line, and like the end moments will be corrected to yield actual midspan moments.

Maximum End Moments ▪ For maximum end moment at *A*, beam *AB* must be fully loaded, but *BC* should carry dead load only. Holding *A* fixed, we first unlock joint *B*, which has a total-load fixed-end moment of +172 in *BA* and a dead-load fixed-end moment of −37 in *BC*. The releasing moment required, therefore, is −(172 − 37), or −135. When *B* is released, a moment of −135 × 0.25 is distributed to *BA*. One-half of this is carried over to *A*, or −135

$\times$ 0.25 $\times$ 0.5 = -17. This value is entered as the carry-over at A on the fifth line in Fig. 1-74. Joint B then is relocked.

At A, for which we are computing the maximum moment, we have a total-load fixed-end moment of -172 and a carry-over of -17, making the total -189, shown on the sixth line. To release A, a moment of $+189$ must be applied to the joint. Of this, 189 $\times$ 0.33, or 63, is distributed to AB, as indicated on the seventh line. Finally, the maximum moment at A is found by adding lines 6 and 7: $-189 + 63 = -126$.

For maximum moment at B, both AB and BC must be fully loaded, but CD should carry only dead load. We begin the determination of the maximum moment at B by first releasing joints A and C, for which the corresponding carry-over moments at BA and BC are $+29$ and $-(+78 - 70) \times 0.25 \times 0.5 = -1$, shown on the fifth line in Fig. 1-74. These bring the total fixed-end moments in BA and BC to $+201$ and -79, respectively. The releasing moment required is $- (201 - 79) = 122$. Multiplying this by the distribution factors for BA and BC when joint B is released, we find the distributed moments, -30, entered on line 7. The maximum end moments finally are obtained by adding lines 6 and 7: $+171$ at BA and -109 at BC. Maximum moments at C, D, and E are computed and entered in Fig. 1-74 in a similar manner. This procedure is equivalent to two cycles of moment distribution.

Maximum Midspan Moments ▪ The computation of maximum midspan moments in Fig. 1-74 is based on the assumption that in each beam the midspan moment is the sum of the simple-beam midspan moment and one-half the algebraic difference of the final end moments (the span carries full load but adjacent spans only dead load). Instead of starting with the simple-beam moment, however, we begin, for convenience, with the midspan moment for the fixed-end condition and then apply two corrections. In each span, these corrections equal the carry-over moments entered on line 5 for the two ends of the beam multiplied by a factor.

For beams with variable moment of inertia, the factor is $+ \frac{1}{2}(1/C^F + D - 1)$, where C^F is the fixed-end carry-over factor toward the end for which the correction factor is being computed and D the distribution factor for that end. The plus sign is used for correcting the carry-over at the right end of a beam and the minus sign for the carry-over at the left end. For prismatic beams, the correction factor becomes $\pm \frac{1}{2}(1 + D)$.

For example, to find the corrections to the midspan moment in AB, we first multiply the carry-over at A on line 5, $- 17$, by $- \frac{1}{2}(1 + 0.33)$. The correction, $+11$, also is entered on the fifth line. Then we multiply the carry-over at B, $+ 29$, by $+ \frac{1}{2}(1 + 0.25)$ and enter the correction, $+18$, on line 6. The final midspan moment is the sum of lines 4, 5, and 6: $+99 + 11 + 18 = +128$. Other midspan moments in Fig. 1-74 are obtained in a similar manner.

1-66. Moment-Influence Factors

For certain types of structures, particularly those for which different types of loading conditions must be investigated, it may be more convenient to find maximum end moments from a table of moment-influence factors. This table is made up by listing for the end of each member in a structure the moment induced in that end when a moment (for convenience, $+1000$) is applied to each joint successively. Once this table has been prepared, no additional moment distribution is necessary for computing the end moments due to any loading condition.

For a specific loading pattern, the moment at any beam end M_{AB} may be obtained from the moment-influence table by multiplying the entries under AB for the various joints by the actual unbalanced moments at those joints divided by 1000 and summing. (See also Art. 1-68 and Tables 1-3 and 1-4.)

1-67. Deflection of Supports

For some structures, it is convenient to know the effect of a movement of a support normal to the original position. But the moment-distribution method is based on the assumption that such movement of a support does not occur. The method, however, can be modified to evaluate end moments resulting from a support movement.

The procedure is to distribute moments as usual, assuming no deflection at the supports. This implies that additional external forces are exerted at the supports to prevent movement. These forces can be computed. Then, equal and opposite forces are applied to the structure to produce the final configuration, and the moments that they induce are distributed as usual. These moments added to those obtained with undeflected supports yield the final moments.

To apply this procedure, it is first necessary to know the fixed-end moments for a beam with supports at different levels. In Fig. 1-75a, the right end of a beam with span L is at a height d above the left end. To find the fixed-end moments, we first deflect the beam with both ends hinged, then fix the right end, leaving the left end hinged, as in Fig. 1-75b. Noting that a line connecting the two supports makes an angle approximately equal to d/L (its tangent) with the original position of the beam, we apply a moment at the hinged end to produce an end rotation there equal to d/L. By the definition of stiffness (Art. 1-62), this moment equals $K_L^F d/L$. The carry-over to the right end is C_R^F times this.

By the law of reciprocal deflections, the fixed-end moment at the right end of a beam due to a rotation of the other end equals the fixed-end moment at the left end of the beam due to the same rotation at the right end. Therefore, the carry-over moment for the right end also equals $C_L^F K_R^F d/L$ (see Fig. 1-75c). By adding the end moments for the loading conditions in Fig. 1-

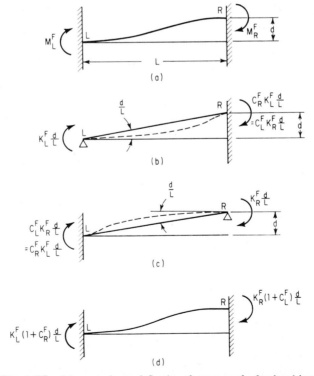

Fig. 1-75. Moments due to deflection of supports of a fixed-end beam.

$75b$ and c, we obtain the end moments in Fig. 1-75d, which is equivalent to the deflected beam in Fig. 1-75a:

$$M_L^F = K_L^F(1 + C_R^F)\frac{d}{L} \tag{1-68}$$

$$M_R^F = K_R^F(1 + C_L^F)\frac{d}{L} \tag{1-69}$$

In a similar manner, the fixed-end moment can be found for a beam with one end hinged and the supports at different levels:

$$M^F = K\frac{d}{l} \tag{1-70}$$

where K is the actual stiffness for the end of the beam that is fixed; for beams of variable moment of inertia K equals the fixed-end stiffness times $(1 - C_L^F C_R^F)$.

1-68. Procedure for Sidesway

The problem of computing sidesway moments in rigid frames is conveniently solved by the following method:

1. Apply forces to the structure to prevent sidesway while the fixed-end moments due to loads are distributed.
2. Compute the moments due to these forces.
3. Combine the moments obtained in steps 1 and 2 to eliminate the effect of the forces that prevented sidesway.

Example—Horizontal Axial Load ▪ Suppose the rigid frame in Fig. 1-76 is subjected to a 2000-lb horizontal load acting to the right at the level of beam BC. The first step is to compute the moment-influence factors by applying moments of $+1000$ at joints B and C (Art. 1-66), assuming sidesway is prevented, and enter the distributed moments in Table 1-3.

Since there are no intermediate loads on the beam and columns, the only fixed-end moments that need be considered are those in the columns due to lateral deflection of the frame.

This deflection, however, is not known initially. So we assume an arbitrary deflection, which produces a fixed-end moment of $-1000M$ at the top of column CD. M is an unknown constant to be determined from the fact that the sum of the shears in the deflected columns must equal

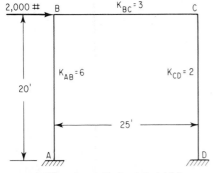

Fig. 1-76.   Laterally loaded rigid frame.

TABLE 1-3 Moment-Influence Factors for Fig. 1-76

Member	+1,000 at B	+1,000 at C
AB	351	−105
BA	702	−210
BC	298	210
CB	70	579
CD	−70	421
DC	−35	210

TABLE 1-4 Moment-Collection Table for Fig. 1-76

Remarks	AB	BA	BC	CB	CD	DC
1. Sidesway *FEM*	$-3,000M$	$-3,000M$	....	....	$-1,000M$	$-1,000M$
2. Distribution for *B*	$+1,053M$	$+2,106M$	$+894M$	$+210M$	$-210M$	$-105M$
3. Distribution for *C*	$-105M$	$-210M$	$+210M$	$+579M$	$+421M$	$+210M$
4. Final sidesway *M*	$-2,052M$	$-1,104M$	$+1,104M$	$+789M$	$-789M$	$-895M$
5. For 2,000-lb horizontal	$-17,000$	$-9,100$	$+9,100$	$+6,500$	$-6,500$	$-7,400$
6. 4,000-lb vertical *FEM*	....	....	$-12,800$	$+3,200$		
7. Distribution for *B*	$+4,490$	$+8,980$	$+3,820$	$+897$	-897	-448
8. Distribution for *C*	$+336$	$+672$	-672	$-1,853$	$-1,347$	-673
9. Moments with no sidesway	$+4,826$	$+9,652$	$-9,652$	$+2,244$	$-2,244$	$-1,121$
10. Sidesway *M*	$-4,710$	$-2,540$	$+2,540$	$+1,810$	$-1,810$	$-2,060$
11. For 4,000-lb vertical	$+116$	$+7,112$	$-7,112$	$+4,054$	$-4,054$	$-3,181$

the 2000-lb load. The deflection also produces a moment of $-1000M$ at the bottom of *CD* [see Eqs. (1-68) and (1-69)].

From the geometry of the structure, we furthermore note that the deflection of *B* relative to *A* equals the deflection of *C* relative to *D*. Then, according to Eqs. (1-68) and (1-69), the fixed-end moments of the columns of this frame are proportional to the stiffnesses of the columns and hence are equal in *AB* to $-1000M \times \frac{3}{2} = -3,000M$. The column fixed-end moments are entered in the first line of Table 1-4, the moment-collection table for Fig. 1-76.

In the deflected position of the frame, joints *B* and *C* are unlocked in succession. First, we apply a releasing moment of $+3000M$ at *B*. We distribute it by multiplying by 3 the entries in the columns marked "$+1000$ at *B*" in Table 1-3. Similarly, a releasing moment of $+1000M$ is applied at *C* and distributed with the aid of the moment-influence factors. The distributed moments are entered in the second and third lines of the moment-collection table. The final moments are the sum of the fixed-end moments and the distributed moments and are given in the fourth line of Table 1-4, in terms of *M*.

Isolating each column and taking moments about one end, we find that the overturning moment due to the shear equals the sum of the end moments. We have one such equation for each column. Adding these equations, noting that the sum of the shears equals 2000 lb, we obtain

$$-M(2052 + 1104 + 789 + 895) = -2000 \times 20$$

from which we find $M = 8.26$. This value is substituted in the sidesway totals (line 4) in the moment-collection table to yield the end moments for the 2000-lb horizontal load (line 5).

Example—Vertical Load on Beam ▪ Suppose a vertical load of 4000 lb is applied to *BC* of the rigid frame in Fig. 1-76, 5 ft from *B*. The same moment-influence factors and moment-collection table can again be used to determine the end moments with a minimum of labor.

The fixed-end moment at *B*, with sidesway prevented, is $-12,800$, and at *C* $+3200$ (Fig. 1-68a). With the joints still locked, the frame is permitted to move laterally an arbitrary amount, so that in addition to the fixed-end moments due to the 4000-lb load, column fixed-end moments of $-3000 M$ at *A* and *B* and $-1000M$ at *C* and *D* are induced. The moment-collection table already indicates in line 4 the effect of relieving these column moments by unlocking joints *B* and C. We now have to superimpose the effect of releasing joints B and C to relieve the fixed-end moments for the vertical load. This we can do with the aid of the moment-influence factors. The distribution is shown in lines 7 and 8 of Table 1-4, the moment-collection table. The sums of the fixed-end moments and distributed moments for the 4000-lb load are shown in line 9.

The unknown M can be evaluated from the fact that the sum of the horizontal forces acting on the columns must be zero. This is equivalent to requiring that the sum of the column end moments equal zero:

$$-M(2052 + 1104 + 789 + 895) + 4826 + 9652 - 2244 - 1121 = 0$$

from which $M = 2.30$. This value is substituted in line 4 of Table 1-4 to yield the sidesway moments for the 4000-lb load (line 10). Addition of these moments to the totals for no sidesway (line 9) gives the final moments (line 11).

Multistory Frames ▪ This procedure permits analysis of one-story bents with straight beams by solution of one equation with one unknown, regardless of the number of bays. If the frame is multistory, the procedure can be applied to each story. Since an arbitrary horizontal deflection is introduced at each floor or roof level, there are as many unknowns and equations as there are stories.

Arched Bents ▪ The procedure is more difficult to apply to bents with curved or polygonal members between the columns. The effect of the change in the horizontal projection of the curved or polygonal portion of the bent must be included in the calculations. In many cases, it may be easier to analyze the bent as a curved beam (arch).

1-69. Single-Cycle Moment Distribution

In the method of moment distribution by converging approximations (Art. 1-64), all joints but the one being unlocked are considered fixed. In distributing moments, the stiffnesses and carry-over factors used are based on this assumption (Arts 1-62 and 1-63). If, however, actual stiffnesses and carry-over factors are employed, moments can be distributed throughout continuous frames in a single cycle.

Formulas for actual stiffnesses and carry-over factors can be written in several simple forms. The equations given in this article were chosen to permit use of existing tables for beams of variable moment of inertia that are based on fixed-end stiffnesses and carry-over factors.

Considerable simplification of the formulas results if they are based on the simple-beam stiffness of members of continuous frames. This value can always be obtained from tables of fixed-end properties by multiplying the fixed-end stiffness by $(1 - C_L^F C_R^F)$, where C_L^F is the fixed-end carry-over factor to the left and C_R^F is the fixed-end carry-over factor to the right.

To derive the basic constants needed, apply a unit moment to one end of a member, isolated from the structure and simply supported (Fig. 1-77a). The end rotation at the support where the moment is applied is α, and at the far end, the rotation is β. By the dummy-load method (Art. 1-55), if x is measured from the β end,

$$\alpha = \frac{1}{L^2} \int_0^L \frac{x^2}{EI_x} \, dx \tag{1-71}$$

$$\beta = \frac{1}{L^2} \int_0^L \frac{x(L - x)}{EI_x} \, dx \tag{1-72}$$

in which I_x is the moment of inertia at a section a distance of x from the β end, and E is the modulus of elasticity.

Simple-beam stiffness K of the member is the moment required to produce a rotation of unity at the end where it is applied (Fig. 1-77b). Hence, at each end of a member, $K = 1/\alpha$.

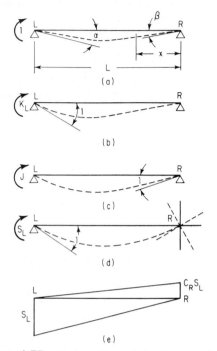

(a)

(b)

(c)

(d)

(e)

Fig. 1-77. End rotations of simple beams for determining stiffness.

For prismatic beams, K has the same value for both ends and equals $3EI/L$. For haunched beams, K for each end can be obtained from tables for fixed-end stiffnesses (Art. 1-62) or by numerical integration of Eq. (1-71).

Although the value of α, and consequently of K, is different at opposite ends of an unsymmetrical beam, the value of β is the same for both ends, in accordance with the law of reciprocal deflections. Now, apply a moment J at one end of the simply supported beam to produce a rotation of unity at the other end (Fig. 1-77c). This moment will be equal to $1/\beta$ and will have the same value regardless at which end it is applied. K/J equals the **fixed-end carry-over factor** for the beam.

J equals $6EI/L$ for prismatic beams. For haunched beams, it can be computed by numerical integration of Eq. (1-72).

Actual stiffness S of the end of an unloaded span is the moment producing a rotation of unity at the end where it is applied when the other end of the beam is restrained against rotation by the other members of the structure (Fig. 1-77d).

The bending-moment diagram for a moment S_L applied at the left end of a member of a continuous frame is shown in Fig. 1-77e. As indicated, the moment carried over to the far end is $C_R S_L$, where C_R is the carry-over factor to the right. At L, the rotation produced by S_L alone is S_L/K_L, and by $C_R S_L$ alone, $-C_R S_L/J$. By definition of stiffness, the sum of these angles must equal unity:

$$\frac{S_L}{K_L} - \frac{C_R S_L}{J} = 1$$

Solving for S_L and noting that $K_L/J = C_L^F$, the fixed-end carry-over factor to the left, we find the formula for the stiffness of the left end of a member:

$$S_L = \frac{K_L}{1 - C_L^F C_R} \qquad (1\text{-}73)$$

Similarly, the stiffness of the right end of a member is

$$S_R = \frac{K_R}{1 - C_R^F C_L} \qquad (1\text{-}74)$$

For prismatic beams, the stiffness formulas reduce to

$$S_L = \frac{K}{1 - C_R/2} \quad \text{and} \quad S_R = \frac{K}{1 - C_L/2} \qquad (1\text{-}75)$$

where $K = 3EI/L$.

Estimate of Stiffness ▪ When the far end of a beam is hinged, the carry-over factor is zero; the stiffness equals K. When the far end of a prismatic beam is fully fixed against rotation, the carry-over factor equals ½. Hence, the fixed-end stiffness equals $4K/3$. This indicates that the effect of partial restraint on prismatic beams is to vary the stiffness between K for no restraint and $1.33K$ for full restraint. Because of this small variation, in many cases, an estimate of the actual stiffness of a beam may be sufficiently accurate.

Restraint R at the end of an unloaded beam in a continuous frame is the moment applied at that end to produce a unit rotation in all the members of the joint. Since the sum of the moments at the joint must be zero, R must equal the sum of the stiffnesses of the adjacent ends of the members connected to the given beam at that joint.

Furthermore, the moment induced in or "distributed" to any of these other members bears the same ratio to the applied moment as the stiffness of the member does to the restraint. Consequently, *end moments are distributed at a joint in proportion to the stiffnesses of the members.*

Actual carry-over factors can be computed by modifying the fixed-end carry-over factors. In Fig. 1-77d and e, by definition of restraint, the rotation at joint R is $-C_R S_L/R_R$, which must equal the rotation of the beam at R due to the moments at L and R. The rotation of the beam due to $C_R S_L$ alone equals $C_R S_L/K_R$, and the rotation due to S_L alone, $-S_L/J$. Hence

$$- \frac{C_R S_L}{R_R} = \frac{C_R S_L}{K_R} - \frac{S_L}{J}$$

Solving for C_R and noting that $K_R/J = C_R^F$, the fixed-end carry-over factor to the right, we find the actual carry-over factor to the right:

$$C_R = \frac{C_R^F}{1 + K_R/R_R} \tag{1-76}$$

Similarly, the actual carry-over factor to the left is

$$C_L = \frac{C_L^F}{1 + K_L/R_L} \tag{1-77}$$

When analyzing a continuous beam, we generally know the carry-over factors toward the ends of the first and last spans. Starting with these values, we can calculate the rest of the carry-over factors and the stiffnesses of the members. But in many frames, there are no end conditions known in advance. To analyze these structures, we must assume several carry-over factors.

This will not complicate the analysis because in many cases it will be found unnecessary to correct the values of C based on assumed carry-over factors of preceding spans. The reason is that C is not very sensitive to the restraint at far ends of adjacent members. When carry-over factors are estimated, the greatest accuracy will be attained if the choice of assumed values is restricted to members subject to the greatest restraint.

Estimated Carry-Over Factors ▪ A very good approximation to the carry-over factor for prismatic beams may be obtained from the following formula, which is based on the assumption that far ends of adjacent members are subject to equal restraint:

$$C = \frac{\Sigma K - K}{2(\Sigma K - \delta K)} \tag{1-78}$$

where ΣK is the sum of the K values of all the members of the joint toward which the carry-over factor is acting, K is the simple-beam stiffness of the member for which the carry-over factor is being computed; and δ is a factor that varies from zero for no restraint to ¼ for full restraint at the far ends of the connecting members. Since δ varies within such narrow limits, it affects C very little.

To illustrate the estimating and calculation of carry-over factors, the carry-over factors and stiffnesses in the clockwise direction will be computed for the frame in Fig. 1-78a. Relative I/L, or K values, are shown in the circles.

A start will be made by estimating C_{AB}. Arbitrarily taking $\delta = \frac{1}{8}$, we apply Eq. (1-78) with $K = 3$ and $\Sigma K = 3 + 2 = 5$ and find $C_{AB} = 0.216$, as noted in Fig. 1-78a. The stiffness S_{AB}

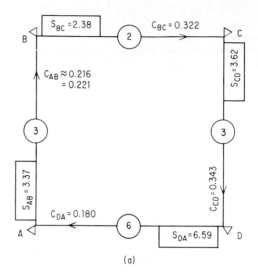

(a)

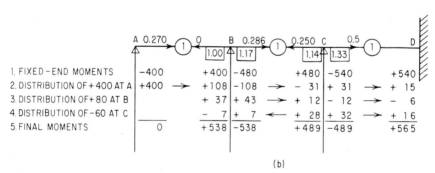

(b)

Fig. 1-78. Moments in (a) quadrangular frame, (b) continuous beam.

then equals $3/(1 - 0.216/2) = 3.37$. Because restraint $R_{AD} = S_{AB}$, we can now use the exact formula, Eq. (1-77), to obtain the carry-over factor from D to A: $C_{DA} = 0.5(1 + 6/3.37) = 0.180$. Continuing around the frame in this manner, we return to C_{AB} and recalculate it with Eq. (1-77), obtaining 0.221. This differs only slightly from the estimated value. The change in C_{DA} due to the new value of C_{AB} is negligible.

If a bending moment of 1000 ft-lb were introduced at A in AB, it would induce a moment of $1000C_{AB} = 221$ ft-lb at B; $221 \times 0.322 = 71$ ft-lb at C; $71 \times 0.343 = 24$ ft-lb at D; $24 \times 0.180 = 4$ ft-lb at A, and so on.

Example—Continuous Beam ▪ To demonstrate how the moments in a continuous beam would be computed, the end moments will be determined for the beam in Fig. 1-78b, which is identical with the one in Fig. 1-73, for which moments were obtained by converging approximations. Relative I/L, or K values, are shown in the circles on each span.

Since A is a hinged end, $C_{BA} = 0$. $S_{BA} = 1/(1 - 0) = 1$. Since there is only one member joined to BC at B, $R_{BC} = S_{BA}$, and $C_{CB} = 0.5(1 + 1/1) = 0.250$. With this value, we compute $S_{CB} = 1.14$. To obtain the carry-over factors in the opposite direction, we start with $C_{CD} = 0.5$, since we know D is a fixed end. This enables us to compute $S_{CD} = 1.33$ and the remainder of the beam constants.

The fixed-end moments are given on the first line of calculations in Fig. 1-78b. We start the distribution by unlocking A by applying a releasing moment of $+400$. Since A is a hinge, the full 400 is given to A and $400 \times 0.270 = 108$ is carried over to B. If several members had been connected to AB at B, this moment with sign changed would be distributed to them in proportion to their stiffnesses. But since only one member connects to B, the moment at BC is -108. Next, $-108 \times 0.286 = -31$ is carried over to C. Finally, a moment of $+15$ is carried to D.

Then, joint B is unlocked. The unbalanced moment of $400 - 480 = -80$ is counteracted with a moment of $+80$. This is distributed to BA and BC in proportion to their stiffnesses (shown in the boxes at the joint). BC, for example, gets $80 \times 1.17(1.17 + 1) = 43$. The carry-over to C is $43 \times 0.286 = 12$, and to D, -6.

Similarly, the unbalanced moment at C is counteracted and distributed to CB and CD in proportion to their stiffnesses, shown in the boxes at C, then carried over to B and D. The final moments are the sum of the fixed-end and distributed moments.

Shortcuts ▪ On occasion, advantage can be taken of certain properties of loads and structures to save work in distribution by using carry-over factors as the ratio of end moments in loaded members. For example, suppose it is obvious, from symmetry of loading and structure, that there will be no end rotation at an interior support. The part of the structure on one side of the support can be isolated and the moments distributed only in this part, with the carry-over factor toward the support taken as C^F.

Again, suppose it is evident that the final end moments at opposite ends of a span must be equal in magnitude and sign. Isolate the structure on each side of this beam and distribute moments only in each part, with the carry-over factor for this span taken as 1.

1-70. Method for Checking Moment Distribution

End moments computed for a continuous structure must satisfy both the laws of equilibrium and the requirements of continuity. At each joint, therefore, the sum of the moments must be zero (or equal to an external moment applied there). In addition, the end of every member connected there must rotate through the same angle. It is a simple matter to determine the sum of the moments, but further calculation is needed to prove that the moments yield the same rotation for the end of each member at a joint. The following method not only will indicate that the requirements of continuity are satisfied but will tend to correct automatically any mistakes that may have been made in computing the end moments.

Consider a joint O made of several members OA, OB, OC, and so forth. The members are assumed to be loaded and the calculation of end moments due to the loads to have started with fixed-end moments. For any one of the members, say OA, the end rotation at O for the fixed-end condition was zero; that is,

$$\frac{M^F_{OA}}{K_{OA}} - \frac{M^F_{AO}}{J_{OA}} - \phi_{OA} = 0 \qquad (1\text{-}79)$$

where M_{OA}^F = fixed-end moment at O
M_{AO}^F = fixed-end moment at A
K_{OA} = simple-beam stiffness at O (see Art. 1-69)
J_{OA} = moment required at A to produce unit rotation at O when span is considered simply supported
ϕ_{OA} = simple-beam end rotation at O due to loads
For the final end moments, the rotation at O is

$$\theta = \frac{M_{OA}}{K_{OA}} - \frac{M_{AO}}{J_{OA}} - \phi_{OA} \qquad (1\text{-}80)$$

Subtracting Eq. (1-79) from Eq. (1-80) and multiplying by K_{OA} yields

$$K_{OA}\theta = M_{OA} - M_{OA}^F - C_{AO}^F M_{OA}' \qquad (1\text{-}81)$$

In Eq. (1-81) the carry-over factor toward O, C_{AO}^F, has been substituted for K_{OA}/J_{OA}, and M_{OA}' for $M_{AO} - M_{AO}^F$. An analogous expression can be written for each of the other members at O. Summing these equations, we obtain

$$\theta\Sigma K_O = \Sigma M_O - \Sigma M_O^F - \Sigma C_O^F M_O' \qquad (1\text{-}82)$$

With this value of θ, we can solve for each of the final end moments at O and thus determine the equations that will check the joint for continuity. For example, using Eq. (1-82) with Eq. (1-81) yields

$$M_{OA} = M_{OA}^F + C_{AO}^F M_{OA}' - m_{OA} \qquad (1\text{-}83)$$

$$m_{OA} = \frac{K_{OA}}{\Sigma K_O}(-\Sigma M_O + \Sigma M_O^F + \Sigma C_O^F M_O') \qquad (1\text{-}84)$$

Similar equations can be written for the other members at O by substituting the proper letter for A in the subscripts.

If the calculations based on these equations are carried out in table form, the equations prove to be surprisingly simple (see Tables 1-5 and 1-6).

For prismatic beams, the terms $C^F M'$ become half the change in the fixed-end moment at the far end of each member at a joint.

Examples ▪ Suppose we want to check the end moments in the beam in Fig. 1-78b. Each joint and the ends of the members connected there are listed in Table 1-5, and a column is provided for the summation of the various terms for each joint. K values are given on line 1, the end moments to be checked on line 2, and the fixed-end moments on line 3. On line 4 is entered one-half the difference obtained when the fixed-end moment is subtracted from the final moment at the far end of each member. $-m$ is placed on line 5 and the corrected end moment on line 6. The $-m$ values are obtained from the summation columns by adding line 2 to the negative of the sum of lines 3 and 4 and distributing the result to the members of the joint in proportion to the K values. The corrected moment M is the sum of lines 3 to 5.

Assume that a mistake was made in computing the end moments for Fig. 1-78b giving the results shown in line 2 of the second part of Table 1-5 for the fixed-end moments on line 3. The correct moments can be obtained as follows:

At joint B, the sum of the incorrect moments is zero, as shown in the summation column on line 2. The sum of the fixed-end moments at B is -80, as indicated on line 3. For BA, ½M' is obtained from lines 2 and 3 of the column for AB: ½ × (0 + 400) = +200, which is entered on line 4. The line 4 entry for BC is obtained from CB: ½ × (450 − 480) = −15. The sum of

TABLE 1-5 Continuity Check for Fig. 1-78b

	A		B			C			D	
	AB	Σ	BA	BC	Σ	CB	CD	Σ	DC	Σ
1. K	1	1	1	1	2	1	1	2	1	∞
2. M	0	0	+538	−538	0	+489	−489	0	+565	
3. M^F	−400	−400	+400	−480	−80	+480	−540	−60	+540	
4. C^F M'			+200	+5	+205	−29	+13	−16	+26	
5. −m			−62	−63	−125	+38	+38	+76	0	
6. Check	0	0	+538	−538	0	+489	−489	0	+566	
			CHECK WHEN MOMENTS ARE INCORRECT							
2. Wrong M	0	0	+560	−560	0	+450	−450	0	+530	
3. M^F	−400	−400	+400	−480	−80	+480	−540	−60	+540	
4. C^F M'			+200	−15	+185	−40	−5	−45	+45	
5. −m			−53	−52	−105	+53	+52	+105	0	
6. New M	0	0	+547	−547	0	+493	−493	0	+585	
			SECOND CYCLE							
7. Trial M	0	0	+547	−547	0	+493	−493	0	+585	
8. M^F	−400	−400	+400	−480	−80	+480	−540	−60	+540	
9. C^F M'			+200	+3	+203	−34	+12	−22	+24	
10. +m			−62	−61	−123	+41	+41	+82	0	
11. New M	0	0	+538	−538	0	+487	−487	0	+564	

the line 4 values at B is therefore $200 - 15 = +185$. Entered in the summation column, this is then added to the summation value on line 3, the sign is changed, and the number on line 2 (in this case zero) is added to the sum, giving -105. This is inserted in the summation column on line 5. The values for BA and BC on line 5 are obtained by multiplying -105 by the ratio of the K value of each member to the sum of the K values at the joint; that is, for BA, $-m = -105 \times \frac{1}{2} = -53$. The corrected moment for BA, the sum of lines 3 to 5, is $+400 + 200 - 53 = +547$. Since this differs from the value on line 2, it indicates that one or more of the moments on that line were incorrect. The other corrected moments are found the same way and are shown on line 6.

A comparison with Fig. 1-78b shows that the new moments, although incorrect, are closer to the right answer than those with which we started. Even closer results can be obtained by repeating the calculations. Convergence, however, can be obtained much more quickly by starting first with fixed ends and joints that appear to have been most in error. Then, use the corrected values obtained for these in correcting adjacent joints.

For example, for the second cycle shown in the table, calculations were started with joint D, which is a fixed end. Using the values obtained at the end of the first cycle to compute M', we find the corrected value for DC to be $+564$, which is very close to the exact final moment. Then, we move to joint C. For M, we use the value obtained at the end of the first cycle. But M' for CD is based on the end moment just computed: $+564 - 540 = +24$, and half of this is placed on line 9 under CD. Continuing in this manner, we obtain moments that check closely the final moments in the first part of the table.

The procedure is useful also for estimating the effect of changing the stiffness of one or more members.

Check for Sidesway ▪ The checking equations can be generalized to include the effect of the movement d of a support in a direction normal to the initial position of a span of length L:

$$K_{OA}\theta = M_{OA} - M_{OA}^F - C_{AO}^F M_{OA}' + K_{OA}\frac{d}{L} \tag{1-85}$$

$$M_{OA} = M_{OA}^F + C_{AO}^F M_{OA}' - K_{OA}\frac{d}{L} - m_{OA} \tag{1-86a}$$

$$m_{OA} = \frac{K_{OA}}{\Sigma K_O}\left(-\Sigma M_O + \Sigma M_O^F + \Sigma C_O^F M_O' - \Sigma K_O\frac{d}{L}\right) \tag{1-86b}$$

For each span with a support movement, the term Kd/L can be obtained from the fixed-end moment due to this deflection alone; for OA, for example, Kd/L can be obtained by multiplying M_{OA}^F by $(1 - C_{AO}^F C_{OA}^F)/(1 + C_{OA}^F)$. [See Eqs. (1-68) and (1-73).] For prismatic beams, this factor reduces to ½. Equation (1-85) is called a **slope-deflection equation**.

In Table 1-6, the solution for the bent in Fig. 1-76 is checked for the condition in which a 4000-lb vertical load is placed 5 ft from B on span BC. The computations are similar to those

TABLE 1-6 Continuity Check for Fig. 1-76

	A		B			C			D	
	AB	Σ	BA	BC	Σ	CB	CD	Σ	DC	Σ
1. K	7	∞	6	3	9	3	2	5	2	∞
2. M	+116		+7,110	−7,110	0	+4,050	−4,050	0	−3,180	
3. M^F	0		0	−12,800	−12,800	+3,200	0	+3,200	0	
4. C^{F}M'	+3,555		+60	+430	+490	+2,840	−1,590	+1,250	−2,030	
5. −Kd/L	−3,450		−3,450	0	−3,450	0	−1,150	−1,150	−1,150	
6. −m	0		+10,510	+5,250	+15,760	−1,980	−1,320	−3,300	0	
7. M	+105		+7,120	−7,120	0	+4,060	−4,060	0	−3,180	

in the preceding table, except that the terms $-Kd/L$ are included for the columns to account for sidesway. These values are obtained from the sidesway fixed-end moments in the moment-collection table for Fig. 1-76. For BA, for example, $Kd/L = \frac{1}{2} \times 3000M$, with $M = 2.30$, as found in the solution. The check indicates that the original solution was sufficiently accurate for a slide-rule computation. If line 7 had contained a different set of moments, the shears would have had to be investigated again. A second cycle could be carried out by distributing the unbalance to the columns to obtain new Kd/L values.

1-71. Plane Grid Frameworks*

Grid framing comprises two or more sets of parallel girders or trusses interconnected at their intersections, or nodes. In plane grid framing, the girders lie in a single plane, usually horizontal. Grid framing, however, may also be composed of space trusses, in which case they usually are referred to as double-layer grid framing. Methods of analyzing single- and double-layer framing generally are similar. Our discussion here therefore illustrates the technique with the simpler plane framing and girders, instead of plane trusses. Loading is assumed vertical. Girders are assumed continuous at all nodes, except terminals.

Girders may be arranged in numerous ways for plane grid framing. Figure 1-79 shows some

*Reprinted with permission from F. S. Merritt, "Structural Steel Designers' Handbook," McGraw-Hill Book Company, New York.

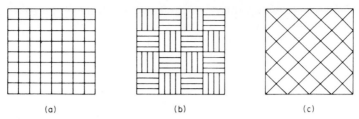

Fig. 1-79.   Orthogonal grids. (*a*) Girders on short spacing; (*b*) girders with wide spacing and beams between; (*c*) girders placed diagonally.

ways of placing two sets of girders. The grid in Fig. 1-79*a* consists of orthogonal sets laid perpendicular to boundary girders. Columns may be placed at the corners, along the boundaries, or at interior nodes. In the following analysis, for illustrative purposes, columns are assumed only at the corners, and interior girders are assumed simply supported on the boundary girders. With wider spacing of interior girders, the arrangement shown in Fig. 1-79*b* may be preferable. With beams in alternate bays spanning in perpendicular directions, loads are uniformly distributed to the girders. Alternatively, the interior girders may be set parallel to the main diagonals, as indicated in Fig. 1-79*c*. The method of analysis for this case is much the same as for girders perpendicular to boundary members. The structure, however, need not be rectangular or square, nor need the interior members be limited to two sets of girders.

Many methods have been used successfully to obtain exact or nearly exact solutions for grid framing, which may be highly indeterminate. These include consistent deflections, finite differences, moment distribution or slope deflection, flat plate analogy, and model analysis. This article is limited to illustrating the use of the method of consistent deflections.

In this method, each set of girders is temporarily separated from the other sets. Unknown loads satisfying equilibrium conditions then are applied to each set. Equations are obtained by expressing node deflections in terms of the loads and equating the deflection at each node of one set to the deflection of the same node in another set. Simultaneous solution of the equations yields the unknown loads on each set. With these now known, bending moments, shears, and deflections of all the girders can be computed by conventional methods.

For a simply supported grid, the unknowns generally can be selected and the equations formulated so that there are one unknown and one equation for each interior node. The number of equations required, however, can be drastically reduced if the framing is made symmetrical about perpendicular axes and the loading is symmetrical or antisymmetrical. For symmetrical grids subjected to unsymmetrical loading, the amount of work involved in analysis often can be decreased by resolving loads into symmetrical and antisymmetrical components. Figure 1-80 shows how this can be done for a single load unsymmetrically located on a grid. The analysis requires the solution of four sets of simultaneous equations and addition of the results, but there are fewer equations in each set than for unsymmetrical loading. The number of unknowns may be further decreased when the proportion of a load at a node to be assigned to a girder at that node can be determined by inspection or simple computation. For example, for a square orthog-

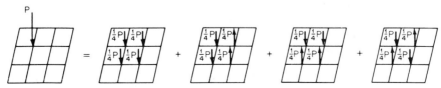

Fig. 1-80.   Resolution of load into symmetrical and antisymmetrical components.

onal grid, each girder at the central node carries half the load there when the grid loading is symmetrical or antisymmetrical.

For analysis of simply supported grid girders, influence coefficients for deflection at any point induced by a unit load are useful. They may be computed from the following formulas.

The deflection at a distance xL from one support of a girder produced by a concentrated load P at a distance kL from that support (Fig. 1-81) is given by

$$\delta = \frac{PL^3}{6EI} x(1 - k)(2k - k^2 - x^2) \qquad 0 \le x \le k \qquad (1\text{-}87)$$

$$\delta = \frac{PL^3}{6EI} k(1 - x)(2x - x^2 - k^2) \qquad k \le x \le 1 \qquad (1\text{-}88)$$

where L = span of simply supported girder
    E = modulus of elasticity of girder material
    I = moment of inertia of girder cross section
For deflections, the elastic curve is also the influence curve, when $P = 1$. Hence, the influence coefficient for any point of the girder may be written

$$\delta' = \frac{L^3}{EI} [x, k] \qquad (1\text{-}89)$$

where

$$[x, k] = \begin{array}{l} \dfrac{x}{6}(1 - k)(2k - k^2 - x^2) \qquad 0 \le x \le k \\[2mm] \dfrac{k}{6}(1 - x)(2x - x^2 - k^2) \qquad k \le x \le 1 \end{array} \qquad (1\text{-}90)$$

The deflection at a distance xL from one support of the girder produced by concentrated loads P at distances kL and $(1 - k)L$ from that support (Fig. 1-82) is given by

$$\delta = \frac{PL^3}{EI} (x, k) \qquad (1\text{-}91)$$

where

$$(x, k) = \begin{array}{l} \dfrac{x}{6}(3k - 3k^2 - x^2) \qquad 0 \le x \le k \\[2mm] \dfrac{k}{6}(3x - 3x^2 - k^2) \qquad k \le x \le \dfrac{1}{2} \end{array} \qquad (1\text{-}92)$$

The deflection at a distance xL from one support of the girder produced by a downward concentrated load P at distance kL from the support and an upward concentrated load P at a distance $(1 - k)L$ from the support (antisymmetric loading, Fig. 1-83) is given by

$$\delta = \frac{PL^3}{EI} \{x, k\} \qquad (1\text{-}93)$$

where

$$\{x, k\} = \begin{array}{l} \dfrac{x}{6}(1 - 2k)(k - k^2 - x^2) \qquad 0 \le x \le k \\[2mm] \dfrac{k}{6}(1 - 2x)(x - x^2 - k^2) \qquad k \le x \le \dfrac{1}{2} \end{array} \qquad (1\text{-}94)$$

For convenience in analysis, the loading carried by the grid framing is converted into concentrated loads at the nodes. Suppose, for example, that a grid consists of two sets of parallel

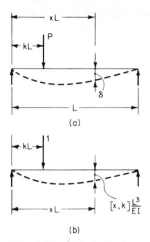

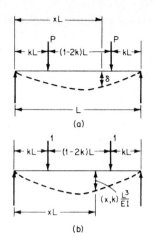

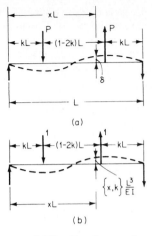

Fig. 1-81. Single load on a beam. (*a*) Deflection curve; (*b*) curve of influence coefficients for deflection at *xL* from support.

Fig. 1-82. Two equal downward loads symmetrically placed. (*a*) Deflection curve; (*b*) curve of influence coefficients.

Fig. 1-83. Equal upward and downward loads symmetrically placed on beam. (*a*) Deflection curve; (*b*) influence line.

girders as in Fig. 1-79, and the load at interior node r is P. Then, it is convenient to assume that one girder at the node is subjected to an unknown force X_r, there, and the other girder therefore carries a force $P_r - X_r$, at the node. With one set of girders detached from the other set, the deflections produced by these forces can be determined with the aid of Eqs. (1-87) to (1-94).

Square Grid ▪ A simple example illustrates the application of the method of consistent deflections. Assume an orthogonal grid within a square boundary (Fig. 1-84*a*). There are $n = 4$ equal spaces of width h between girders. Columns are located at the corners A, B, C, and D.

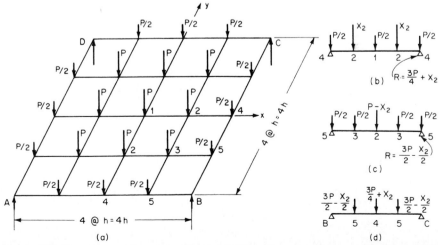

Fig. 1-84. Orthogonal grid. (*a*) Loads distributed to joints; (*b*) loads on midspan girder; (*c*) loads on quarter-point girder; (*d*) loads on boundary girder.

All girders have a span $nh = 4h$ and are simply supported at their terminals, though continuous at interior nodes. To simplify the example, all girders are assumed to have equal and constant moment of inertia I. Interior nodes carry a concentrated load P. Exterior nodes, except corners, are subjected to a load $P/2$.

Because of symmetry, only five different nodes need be considered. These are numbered from 1 to 5 in Fig. 1-84a, for identification. By inspection, loads P at nodes 1 and 3 can be distributed equally to the girders spanning in the x and y directions. Thus, when the two sets of parallel girders are considered separated, girder 4-4 in the x direction carries a load of $P/2$ at midspan (Fig. 1-84b). Similarly, girder 5-5 in the y direction carries loads of $P/2$ at the quarter points (Fig. 1-84c).

Let X_2 be the load acting on girder 4-4 (x direction) at node 2 (Fig. 1-84b). Then, $P - X_2$ acts on girder 5-5 (y direction) at midspan (Fig. 1-84c). The reaction R of girders 4-4 and 5-5 are loads on the boundary girders (Fig. 1-84d).

Because of symmetry, X_2 is the only unknown in this example. Only one equation is needed to determine it.

To obtain this equation, equate the vertical displacement of girder 4-4 (x direction) at node 2 to the vertical displacement of girder 5-5 (y direction) at node 2. The displacement of girder 4-4 equals its deflection plus the deflection of node 4 on BC. Similarly, the displacement of girder 5-5 equals its deflection plus the deflection of node 5 on AB or its equivalent BC.

When use is made of Eqs. (1-89) and (1-91), the deflection of girder 4-4 (x direction) at node 2 equals

$$\delta_2 = \frac{n^3 H^3}{EI}\left\{\left[\frac{1}{4},\frac{1}{2}\right]\frac{P}{2} + \left(\frac{1}{4},\frac{1}{4}\right)X_2\right\} + \delta_4 \tag{1-95 a}$$

where δ_4 is the deflection of BC at node 4. By Eq. (1-90), $[\frac{1}{4},\frac{1}{2}] = (\frac{1}{48})(\frac{11}{16})$. By Eq. (1-93), $(\frac{1}{4},\frac{1}{4}) = \frac{1}{48}$. Hence,

$$\delta_2 = \frac{n^3 h_3}{48 EI}\left(\frac{11}{32}P + X_2\right) + \delta_4 \tag{1-95 b}$$

For the loading shown in Fig. 1-84d,

$$\delta_4 = \frac{n^3 h^3}{EI}\left\{\left[\frac{1}{2},\frac{1}{2}\right]\left(\frac{3P}{4} + X_2\right) + \left(\frac{1}{2},\frac{1}{4}\right)\left(\frac{3P}{2} - \frac{X_2}{2}\right)\right\} \tag{1-96 a}$$

By Eq. (1-90), $[\frac{1}{2},\frac{1}{2}] = \frac{1}{48}$. By Eq. (1-93), $(\frac{1}{2},\frac{1}{4}) = (\frac{1}{48})(\frac{11}{8})$. Hence, Eq. (1-96$a$) becomes

$$\delta_4 = \frac{n^3 h^3}{48 EI}\left(\frac{45}{16}P + \frac{5}{16}X_2\right) \tag{1-96 b}$$

Similarly, the deflection of girder 5-5 (y direction) at node 2 equals

$$\delta_2 = \frac{n^3 h^3}{EI}\left\{\left[\frac{1}{2},\frac{1}{2}\right](P - X_2) + \left(\frac{1}{2},\frac{1}{4}\right)\frac{P}{2}\right\} + \delta_5$$

$$= \frac{n^3 h^3}{48 EI}\left(\frac{27}{16}P - X_2\right) + \delta_5 \tag{1-97}$$

For the loading shown in Fig. 1-84d,

$$\delta_5 = \frac{n^3 h^3}{EI}\left\{\left[\frac{1}{4},\frac{1}{2}\right]\left(\frac{3P}{4} + X_2\right) + \left(\frac{1}{4},\frac{1}{4}\right)\left(\frac{3P}{2} - \frac{X_2}{2}\right)\right\} \tag{1-98}$$

$$= \frac{n^3 h^3}{48 EI}\left(\frac{129}{64}P + \frac{3}{16}X_2\right)$$

The needed equation for determining X_2 is obtained by equating the right-hand side of Eqs. (1-95b) and (1-97) and substituting δ_4 and δ_5 given by Eqs. (1-96b) and (1-98). The result, after division of both sides of the equation by $n^3 h^3 / 48EI$, is

$$\frac{11}{32} P + X_2 + \frac{45}{16} P + \frac{5}{16} X_2 = \frac{27}{16} P - X_2 + \frac{129}{64} P + \frac{3}{16} X_2 \qquad (1\text{-}99)$$

Solution of the equation yields

$$X_2 = \frac{35P}{136} = 0.257P \quad \text{and} \quad P - X_2 = \frac{101P}{136} = 0.743P$$

With these forces known, the bending moments, shears, and deflections of the girders can be computed by conventional methods.

Rectangular Grid ▪ To examine a more general case of symmetrical framing, consider the orthogonal grid with rectangular boundaries in Fig. 1-85a. In the x direction, there are n spaces of width h. In the y direction, there are m spaces of width k. Only members symmetrically placed in the grid are the same size. Interior nodes carry a concentrated load P. Exterior nodes, except corners, carry $P/2$. Columns are located at the corners. For identification, nodes are numbered in one quadrant. Since the loading, as well as the framing, is symmetrical, corresponding nodes in the other quadrants may be given corresponding numbers.

At any interior node r, let X_r be the load carried by the girder spanning in the x direction. Then, $P - X_r$ is the load at that node applied to the girder spanning in the y direction. For this example, therefore, there are six unknowns X_r because r ranges from 1 to 6. Six equations are needed for determination of X_r; they may be obtained by the method of consistent deflections. At each interior node, the vertical displacement of the x-direction girder is equated to the vertical displacement of the y-direction girder, as in the case of the square grid.

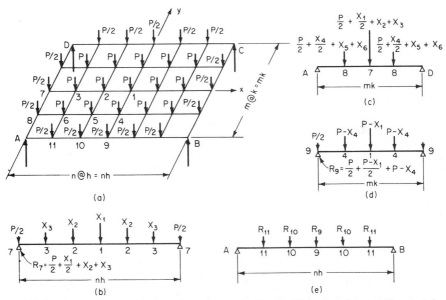

Fig. 1-85. Rectangular grid. (a) Loads distributed to joints; (b) loads on longer midspan girder; (c) loads on shorter boundary girder; (d) loads on shorter midspan girder; (e) loads on longer boundary girder.

To indicate the procedure for obtaining these equations, the equation for node 1 in Fig. 1-85a will be developed. When use is made of Eqs. (1-89) and (1-91), the deflection of girder 7-7 at node 1 (Fig. 1-85b) equals

$$\delta_1 = \frac{n^3 h^3}{EI_7}\left\{\left[\frac{1}{2},\frac{1}{2}\right]X_1 + \left(\frac{1}{2},\frac{1}{3}\right)X_2 + \left(\frac{1}{2},\frac{1}{6}\right)X_3\right\} + \delta_7 \qquad (1\text{-}100)$$

where I_7 = moment of inertia of girder 7-7
$\quad\;\delta_7$ = deflection of girder AD at node 7
Girder AD carries the reaction of the interior girders spanning in the x direction (Fig. 1-85c).

$$\delta_7 = \frac{m^3 k^3}{EI_{AD}}\left\{\left[\frac{1}{2},\frac{1}{2}\right]\left(\frac{P}{2} + \frac{X_1}{2} + X_2 + X_3\right)\right.$$
$$\left. + \left(\frac{1}{2},\frac{1}{4}\right)\left(\frac{P}{2} + \frac{X_4}{2} + X_5 + X_6\right)\right\} \qquad (1\text{-}101)$$

where I_{AD} = moment of inertia of girder AD. Similarly, the deflection of girder 9-9 at node 1 (Fig. 1-85d) equals

$$\delta_1 = \frac{m^3 k^3}{EI_9}\left\{\left[\frac{1}{2},\frac{1}{2}\right](P - X_1) + \left(\frac{1}{2},\frac{1}{4}\right)(P - X_4)\right\} + \delta_9 \qquad (1\text{-}102)$$

where I_9 = moment of inertia of girder 9-9
$\quad\;\delta_9$ = deflection of girder AB at node 9
Girder AB carries the reactions of the interior girders spanning in the y direction (Fig. 1-85c).

$$\delta_9 = \frac{n^3 h_3}{EI_{AB}}\left\{\left[\frac{1}{2},\frac{1}{2}\right]\left(\frac{P}{2} + \frac{P - X_1}{2} + P - X_4\right)\right.$$
$$+ \left(\frac{1}{2},\frac{1}{3}\right)\left(\frac{P}{2} + \frac{P - X_2}{2} + P - X_5\right) \qquad (1\text{-}103)$$
$$\left. + \left(\frac{1}{2},\frac{1}{6}\right)\left(\frac{P}{2} + \frac{P - X_3}{2} + P - X_6\right)\right\}$$

where I_{AB} = moment of inertia of girder AB. The equation for vertical displacement at node 1 is obtained by equating the right-hand side of Eqs. (1-100) and (1-102) and substituting δ_7 and δ, given by Eqs. (1-101) and (1-103).

After similar equations have been developed for the other five interior nodes, the six equations are solved simultaneously for the unknown forces X_r. When these have been determined, moments, shears, and deflections for the girders can be computed by conventional methods.

BUCKLING OF COLUMNS

Columns are compression members whose cross-sectional dimensions are small compared with their length in the direction of the compressive force. Failure of such members occurs because of instability when a certain load (called the critical or **Euler load**) is equaled or exceeded. The member may bend, or buckle, suddenly and collapse.

Hence, the strength of a column is determined not by the unit stress in Eq. (1-6) ($P = Af$) but by the maximum load it can carry without becoming unstable. The condition of instability is characterized by disproportionately large increases in lateral deformation with slight increase in load. It may occur in slender columns before the unit stress reaches the elastic limit.

1-72. Equilibrium of Columns

Figure 1-86 represents an axially loaded column with ends unrestrained against rotation. If the member is initially perfectly straight, it will remain straight as long as the load P is less than the critical load P_c (also called Euler load). If a small transverse force is applied, it will deflect, but it will return to the straight position when this force is removed. Thus, when P is less than P_c, internal and external forces are in stable equilibrium.

If $P = P_c$ and a small transverse force is applied, the column again will deflect, but this time, when the force is removed, the column will remain in the bent position (dashed line in Fig. 1-86).

The equation of this elastic curve can be obtained from Eq. (1-44):

$$EI \frac{d^2y}{dx^2} = -P_c y \qquad (1\text{-}104)$$

in which E = modulus of elasticity, psi
$\quad\quad I$ = least moment of inertia of cross section, in⁴
$\quad\quad y$ = deflection of bent member from straight position at distance x from one end, in

This assumes that the stresses are within the elastic limit.

Solution of Eq. (1-104) gives the smallest value of the Euler load as

$$P_c = \frac{\pi^2 EI}{L^2} \qquad (1\text{-}105)$$

Equation (1-105) indicates that there is a definite magnitude of an axial load that will hold a column in equilibrium in the bent position when the stresses are below the elastic limit.

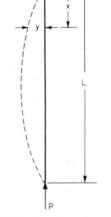

Fig. 1-86.   Buckling of a column.

Repeated application and removal of small transverse forces or small increases in axial load above this critical load will cause the member to fail by buckling. Internal and external forces are in a state of unstable equilibrium.

It is noteworthy that the Euler load, which determines the load-carrying capacity of a column, depends on the stiffness of the member, as expressed by the modulus of elasticity, rather than on the strength of the material of which it is made.

By dividing both sides of Eq. (1-105) by the cross-sectional area A, in², and substituting r^2 for I/A (r is the radius of gyration of the section), we can write the solution of Eq. (1-104) in terms of the average unit stress on the cross section:

$$\frac{P_c}{A} = \frac{\pi^2 E}{(L/r)^2} \qquad (1\text{-}106)$$

This holds only for the elastic range of buckling, that is, for values of the **slenderness ratio** L/r above a certain limiting value that depends on the properties of the material.

Effects of End Conditions ▪ Equation (1-106) was derived on the assumption that the ends of the columns are free to rotate. It can be generalized, however, to take into account the effect of end conditions:

$$\frac{P_c}{A} = \frac{\pi^2 E}{(kL/r)^2} \qquad (1\text{-}107)$$

where k is a factor that depends on the end conditions. For a pin-ended column, $k = 1$; for a column with both ends fixed, $k = \frac{1}{2}$; for a column with one end fixed and one end pinned, k is about 0.7; and for a column with one end fixed and one end free from all restraint, $k = 2$. When a column has different restraints or different radii of gyration about its principal axes, the largest value of kL/r for a principal axis should be used in Eq. (1-108 a).

Inelastic Buckling ▪ Equations (1-105) to (1-107), having been derived from Eq. (1-104), the differential equation for the elastic curve, are based on the assumption that the critical average stress is below the elastic limit when the state of unstable equilibrium is reached. In members with slenderness ratio L/r below a certain limiting value, however, the elastic limit is exceeded before the column buckles. As the axial load approaches the critical load the modulus of elasticity varies with the stress. Hence, Eqs. (1-105) to (1-107), based on the assumption that E is a constant, do not hold for these short columns.

After extensive testing and analysis, prevalent engineering opinion favors the Engesser equation for metals in the inelastic range:

$$\frac{P_t}{A} = \frac{\pi^2 E_t}{(kL/r)^2} \tag{1-108}$$

This differs from Eq. (1-107) only in that the tangent modulus E_t (the actual slope of the stress-strain curve for the stress P_t/A) replaces E, the modulus of elasticity in the elastic range. P_t is the smallest axial load for which two equilibrium positions are possible, the straight position and a deflected position.

Eccentric Loading ▪ Under eccentric loading, the maximum unit stress in short compression members is given by Eqs. (1-52) and (1-54), with the eccentricity e increased by the deflection given by Eq. (1-53). For columns, the stress within the elastic range is given by the **secant formula:**

$$f = \frac{P}{A} \left(1 + \frac{ec}{r^2} \sec \frac{kL}{2r} \sqrt{\frac{P}{AE}} \right) \tag{1-109}$$

When the slenderness ratio L/r is small, the formula approximates Eq. (1-52).

1-73. Column Curves

The result of plotting the critical stress in columns for various values of slenderness ratios (Art. 1-72) is called a column curve. For axially loaded, initially straight columns, it consists of two parts: the Euler critical values [Eq. (1-106)] and the Engresser, or tangent-modulus, critical values [Eq. (1-108)], with $k = 1$.

The second part of the curve is greatly affected by the shape of the stress-strain curve for the material of which the column is made, as indicated in Fig. 1-87. The stress-strain curve for a material, such as an aluminum alloy or high-strength steel, which does not have a sharply defined yield point, is shown in Fig. 1-87a. The corresponding column curve is plotted in Fig. 1-87b. In contrast, Fig. 1-87c presents the stress-strain curve for structural steel, with a sharply defined yield point, and Fig. 1-87d the related column curve. This curve becomes horizontal as the critical stress approaches the yield strength of the material and the tangent modulus becomes zero, whereas the column curve in Fig. 1-87b continues to rise with decreasing values of the slenderness ratio.

Examination of Fig. 1-87d also indicates that slender columns, which fall in the elastic range, where the column curve has a large slope, are very sensitive to variations in the factor k, which represents the effect of end conditions. On the other hand, in the inelastic range, where

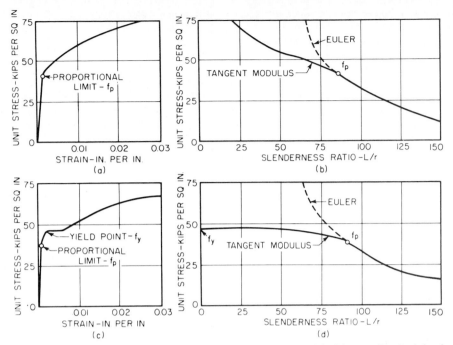

Fig. 1-87. Column curves. (*a*) Stress-strain curves for a material without a sharply defined yield point; (*b*) column curve for the material in (*a*); (*c*) stress-strain curve for a material with a sharply defined yield point; (*d*) column curve for the material in (*c*).

the column curve is relatively flat, the critical stress is relatively insensitive to changes in k. Hence, the effect of end conditions is of much greater significance for long columns than for short columns.

1-74. Behavior of Actual Columns

For many reasons, columns in structures behave differently from the ideal column assumed in deriving Eqs. (1-105) to (1-109). A major consideration is the effect of accidental imperfections, such as nonhomogeneity of materials, initial crookedness, and unintentional eccentricities of the axial load. These effects can be taken into account by a proper choice of safety factor.

There are, however, other significant conditions that must be considered in any design procedure: continuity in framed structures and eccentricity of the load. Continuity affects column action two ways: The restraint and sidesway at column ends determine the value of k, and bending moments are transmitted to the columns by adjoining structural members.

Because of the deviation of the behavior of actual columns from the ideal, columns generally are designed by empirical formulas. Separate equations usually are given for short columns, intermediate columns, and long columns, and still other equations for combinations of axial load and bending moment.

Furthermore, a column may fail not by buckling of the member as a whole but, as an alternative, by buckling of one of its components. Hence, when members like I beams, channels, and angles are used as columns, or when sections are built up of plates, the possibility that the critical load on a component (leg, half flange, web, lattice bar) will be less than the critical load on the column as a whole should be investigated.

Similarly, the possibility of buckling of the compression flange or the web of a beam should be investigated.

Local buckling, however, does not always result in a reduction in the load-carrying capacity of a column; sometimes it results in a redistribution of the stresses, which enables the member to carry additional load.

For more details on column action, see S. Timoshenko and J. M. Gere, "Theory of Elastic Stability," McGraw-Hill Book Company, New York; B. G. Johnston, "Guide to Stability Design Criteria for Metal Structures," John Wiley & Sons, Inc., New York; F. Bleich, "Buckling Strength of Metal Structures," McGraw-Hill Book Company, New York.

TORSION

Forces that cause a member to twist about a longitudinal axis are called torsional loads. Simple torsion is produced only by a couple, or moment, in a plane perpendicular to the axis.

If a couple lies in a nonperpendicular plane, it can be resolved into a torsional moment, in a plane perpendicular to the axis, and bending moments, in planes through the axis.

1-75. Shear Center

The point in each normal section of a member through which the axis passes and about which the section twists is called the shear center. (The location of the shear center in some commonly used shapes is given in Art. 1-37.) If the loads on a beam, for example, do not pass through the shear center, they cause the beam to twist.

1-76. Stresses Due to Torsion

Simple torsion is resisted by internal shearing stresses. These can be resolved into radial and tangential shearing stresses, which being normal to each other also are equal (see Art. 1-13). Furthermore, on planes that bisect the angles between the planes on which the shearing stresses act, there also occur compressive and tensile stresses. The magnitude of these normal stresses is equal to that of the shear. Therefore, when torsional loading is combined with other types of loading, the maximum stresses occur on inclined planes and can be computed by the methods of Arts. 1-14 and 1-17.

Circular Sections ▪ If a circular shaft (hollow or solid) is twisted, a section that is plane before twisting remains plane after twisting. Within the proportional limit, the shearing stress at any point in a transverse section varies with the distance from the center of the section. The maximum shear, psi, occurs at the circumference and is given by

$$v = \frac{Tr}{J} \tag{1-110}$$

where T = torsional moment, in-lb
$\quad r$ = radius of section, in
$\quad J$ = polar moment of inertia, in^4
Polar moment of inertia of a cross section is defined by

$$J = \int \rho^2 \, dA \tag{1-111}$$

where ρ = radius from shear center to any point in section
$\quad dA$ = differential area at point

In general, J equals the sum of the moments of inertia about any two perpendicular axes through the shear center. For a solid circular section, $J = \pi r^4/2$. For a hollow circular section with diameters D and d, $J = \pi(D^4 - d^4)/32$.

Within the proportional limit, the angular twist between two points L inches apart along the axis of a circular bar is, in radians (1 rad = 57.3°):

$$\theta = \frac{TL}{GJ} \qquad (1\text{-}112)$$

where G is the shearing modulus of elasticity (see Art. 1-5).

Noncircular Sections ▪ If a shaft is not circular, a plane transverse section before twisting does not remain plane after twisting. The resulting warping increases the shearing stresses in some parts of the section and decreases them in others, compared with the shearing stresses that would occur if the section remained plane. Consequently, shearing stresses in a noncircular section are not proportional to distances from the shear center. In elliptical and rectangular sections, for example, maximum shear occurs on the circumference at a point nearest the shear center.

For a solid rectangular section, this maximum may be expressed in the following form:

$$v = \frac{T}{kb^2d} \qquad (1\text{-}113)$$

where b = short side of rectangle, in
d = long side, in
k = constant depending on ratio of these sides:

$d/b =$	1.0	1.5	2.0	2.5	3	4	5	10	∞
$k =$	0.208	0.231	0.246	0.258	0.267	0.282	0.291	0.312	0.333

(S. Timoshenko and J. N. Goodier, "Theory of Elasticity," McGraw-Hill Book Company, New York.)

Hollow Tubes ▪ If a thin-shell hollow tube is twisted, the shearing force per unit of length on a cross section (**shear flow**) is given approximately by

$$H = \frac{T}{2A} \qquad (1\text{-}114)$$

where A is the area enclosed by the mean perimeter of the tube, in². And the unit shearing stress is given approximately by

$$v = \frac{H}{t} = \frac{T}{2At} \qquad (1\text{-}115)$$

where t is the thickness of the tube, in. For a rectangular tube with sides of unequal thickness, the total shear flow can be computed from Eq. (1-114) and the shearing stress along each side from Eq. (1-115), except at the corners, where there may be appreciable stress concentration.

Channels and I Beams ▪ For a narrow rectangular section, the maximum shear is very nearly equal to

$$v = \frac{T}{\frac{1}{3}b^2d} \qquad (1\text{-}116)$$

This formula also can be used to find the maximum shearing stress due to torsion in members, such as I beams and channels, made up of thin rectangular components. Let $J = \frac{1}{3}\Sigma b^3 d$, where b is the thickness of each rectangular component and d the corresponding length. Then, the maximum shear is given approximately by

$$v = \frac{Tb'}{J} \tag{1-117}$$

where b' is the thickness of the web or the flange of the member. Maximum shear will occur at the center of one of the long sides of the rectangular part that has the greatest thickness. (A. P. Boresi, O. Sidebottom, F. B. Seely, and J. O. Smith, "Advanced Mechanics of Materials," 3d ed., John Wiley & Sons, Inc., New York.)

ULTIMATE STRENGTH OF DUCTILE MEMBERS AND FRAMES

When an elastic material, such as structural steel, is loaded with a gradually increasing load, stresses are proportional to strains nearly to the yield point. If the material, like steel, also is ductile, then it continues to carry load beyond the yield point, although strains increase rapidly with little increase in load (Fig. 1-88a).

Similarly, a beam made of an elastic material continues to carry more load after the stresses in the outer fibers reach the yield point. However, the stresses will no longer vary with distance from the neutral axis; so the flexural formula [Eq. (1-36)] no longer holds. However, if simplifying assumptions are made, approximating the stress-strain relationship beyond the elastic limit, the load-carrying capacity of the beam can be computed with satisfactory accuracy.

1-77. Theory of Plastic Behavior

For a ductile material, the idealized stress-strain relationship in Fig. 1-88b may be assumed. Stress is proportional to strain until the yield-point stress f_y is reached, after which strain increases at a constant stress.

For a beam of this material, the following assumptions will also be made: Plane sections remain plane, strains thus being proportional to

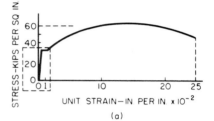

(a)

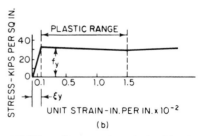

(b)

Fig. 1-88. Stress-strain relationship for a ductile material generally is similar to the curve in (a). To simplify plastic analysis, the portion of (a) enclosed by the dashed lines is approximated by the curve in (b), which extends to the range where strain hardening begins.

distance from the neutral axis; properties of the material in tension are the same as those in compression; its fibers behave the same in flexure as in tension; and deformations remain small.

Strain distribution across the cross section of a rectangular beam, based on these assumptions, is shown in Fig. 1-89a. At the yield point, the unit strain is ϵ_y and the curvature ϕ_y, as indicated in (1). In (2), the strain has increased several times, but the section still remains plane.

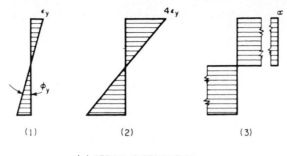

(a) STRAIN DISTRIBUTION

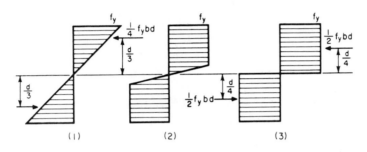

(b) STRESS DISTRIBUTION

Fig. 1-89.   Strain distribution is shown in (*a*) and stress distribution in (*b*) for a cross section of a beam as it is loaded beyond the yield point, assuming the idealized stress-strain relationship in Fig. 1-88*b*. Stage (1) shows the conditions at the yield point of the outer fibers; (2) after yielding starts; and (3) at ultimate load.

Finally, at failure (3), the strains are very large and nearly constant across upper and lower halves of the section.

Corresponding stress distributions are shown in Fig. 1-89*b*. At the yield point (1), stresses vary linearly and the maximum is f_y. With increase in load, more and more fibers reach the yield point, and the stress distribution becomes nearly constant, as indicated in (2). Finally, at failure (3), the stresses are constant across the top and bottom parts of the section and equal to the yield-point stress.

The resisting moment at failure for a rectangular beam can be computed from the stress diagram for stage 3. If *b* is the width of the member and *d* its depth, then the ultimate moment for a rectangular beam is

$$M_P = \frac{bd^2}{4} f_y \qquad (1\text{-}118)$$

Since the resisting moment at stage 1 is $M_y = f_y bd^2/6$, the beam carries 50% more moment before failure than when the yield-point stress is first reached in the outer fibers ($M_P/M_y = 1.5$).

A circular section has an M_P/M_y ratio of about 1.7, while a diamond section has a ratio of 2. The average wide-flange rolled-steel beam has a ratio of about 1.14.

The relationship between moment and curvature in a beam can be assumed to be similar to the stress-strain relationship in Fig. 1-88*b*. Curvature ϕ varies linearly with moment until M_y

$= M_P$ is reached, after which ϕ increases indefinitely at constant moment. That is, a **plastic hinge** forms.

This ability of a ductile beam to form plastic hinges enables a fixed-end or continuous beam to carry more load after M_P occurs at a section because a redistribution of moments takes place. Consider, for example, a uniformly loaded fixed-end beam. In the elastic range, the end moments are $M_L = M_R = WL/12$, while the midspan moment M_C is $WL/24$. The load when the yield point is reached in the outer fibers is $W_y = 12M_y/L$. Under this load, the moment capacity of the ends of the beam is nearly exhausted; plastic hinges form there when the moment equals M_P. As load is increased, the ends then rotate under constant moment and the beam deflects like a simply supported beam. The moment at midspan increases until the moment capacity at that section is exhausted and a plastic hinge forms. The load causing that condition is the ultimate load W_u since, with three hinges in the span, a link mechanism is formed and the member continues to deform at constant load. At the time the third hinge is formed, the moments at ends and center are all equal to M_P. Therefore, for equilibrium, $2M_P = W_uL/8$, from which $W_u = 16M_P/L$. Since, for the idealized moment-curvature relationship, M_P was assumed equal to M_y, the carrying capacity due to redistribution of moments is 33% greater.

1-78. Upper and Lower Bounds for Ultimate Loads

Methods for computing the ultimate strength of continuous beams and frames may be based on two theorems that fix upper and lower limits for load-carrying capacity.

Upper-Bound Theorem.   A load computed on the basis of an assumed link mechanism will always be greater than or at best equal to the ultimate load.

Lower-Bound Theorem.   The load corresponding to an equilibrium condition with arbitrarily assumed values for the redundants is smaller than or at best equal to the ultimate loading—provided that everywhere moments do not exceed M_P.

Equilibrium Method ▪ The equilibrium method, based on the lower-bound theorem, usually is easier for simple cases. The steps involved are:

1. Select redundants that if removed would leave the structure determinate.
2. Draw the moment diagram for the determinate structure.
3. Sketch the moment diagram for an arbitrary value of each redundant.
4. Combine the moment diagrams, forming enough peaks so that the structure will act as a link mechanism if plastic hinges are formed at those points.
5. Compute the value of the redundants from the equations of equilibrium, assuming that at the peaks $M = M_P$.
6. See that there are sufficient plastic hinges to form a mechanism and that M is everywhere less than or equal to M_P.

Consider, for example, the continuous beam $ABCD$ in Fig. 1-90a with three equal spans, uniformly loaded, the center span carrying double the load of the end spans. Assume that the ratio of the plastic moment for the end spans is k times that for the center span ($k < 1$). For what value of k will the ultimate strength be the same for all spans?

Figure 1-90b shows the moment diagram for the beam made determinate by ignoring the moments at B and C, and the moment diagram for end moments M_B and M_C applied to the determinate beam. Figure 1-90c gives the combined moment diagram. If plastic hinges form at all the peaks, a link mechanism will exist.

Since at B and C the joints can develop only the strength of the weakest beam, a plastic

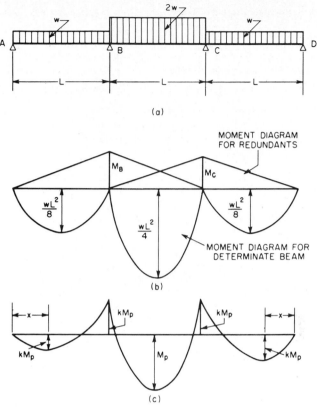

Fig. 1-90.   Continuous beam shown in (a) carries twice as much uniform load in the center span as in the side spans. (b) shows the moment diagrams for this loading condition with redundants removed and for the redundants. The two moment diagrams are combined in (c), producing peaks at which plastic hinges are assumed to form.

hinge will form when $M_B = M_C = kM_P$. A plastic hinge also will form at the center of span BC when the midspan moment is M_P. For equilibrium to be maintained, this will occur when

$$M_P = \frac{wL^2}{4} - \tfrac{1}{2}M_B - \tfrac{1}{2}M_C = \frac{wL^2}{4} - kM_P$$

from which

$$M_P = \frac{wL^2}{4(1 + k)} \qquad (1\text{-}119)$$

Maximum moment M will occur within spans AB and CD where $x = L/2 - M/wL$ —or if $M = kM_P$, where $x = L/2 - kM_P/wL$. A plastic hinge will form at this point when the moment equals kM_P. For equilibrium, therefore,

$$kM_P = \frac{w}{2}x(L - x) - \frac{x}{L}kM_P = \frac{w}{2}\left(\frac{L}{2} - \frac{kM_P}{wL}\right)\left(\frac{L}{2} + \frac{kM_P}{wL}\right) - \left(\frac{1}{2} - \frac{kM_P}{wL^2}\right)kM_P$$

This leads to the quadratic equation:

$$\frac{k^2 M_P^2}{wL^2} - 3kM_P + \frac{wL^2}{4} = 0 \tag{1-120}$$

Substitution of M_P from Eq. (1-119) in Eq. (1-120) yields

$$7k^2 + 4k = 4 \qquad \text{or} \qquad k(k + \tfrac{4}{7}) = \tfrac{4}{7} \tag{1-121}$$

from which $k = 0.523$. And the ultimate load is

$$wL = \frac{4M_P(1 + k)}{L} = 6.1 \frac{M_P}{L} \tag{1-122}$$

Mechanism Method ▪ The mechanism method, based on the upper-bound theorem, requires the following steps:

1. Determine the location of possible hinges (points of maximum moment).

2. Pick combinations of hinges to form possible link mechanisms. Certain types are classed as elementary, such as the mechanism formed when hinges are created at the ends and center of a fixed-end or continuous beam, or when hinges form at top and bottom of a column subjected to lateral load, or when hinges are formed at a joint in the members framing into it, permitting the joint to rotate freely. If there are n possible plastic hinges and x redundant forces or moments, then there are $n - x$ independent equilibrium equations and $n - x$ elementary mechanisms. Both elementary mechanisms and possible combinations of them must be investigated.

3. Apply a virtual displacement to each possible mechanism in turn and compute the internal and external work (Art. 1-52).

4. From the equality of internal and external work, compute the critical load for each mechanism. The mechanism with the lowest critical load is the most probable, and its load is the ultimate, or limit load.

5. Make an equilibrium check to ascertain that moments everywhere are less than or equal to M_P.

As an example of an application of the mechanism method, let us find the ultimate load for the rigid frame of constant section throughout in Fig. 1-91 a. Assume that the vertical load at midspan is equal to 1.5 times the lateral load.

Maximum moments can occur at five points—A, B, C, D, and E—and plastic hinges may form there. Since this structure has three redundants, the number of elementary mechanisms is $5 - 3 = 2$. These are shown in Fig. 1-91 b and c. A combination of these mechanisms is shown in in Fig. 1-91 d.

Let us first investigate the beam alone, with plastic hinges at B, E, and C. As indicated in Fig. 1-91 e, apply a virtual rotation θ to BE at B. The center deflection, then, is $\theta L/2$. Since C is at a distance of $L/2$ from E, the rotation at C must be θ and at E, 2θ. The external work, therefore, is $1.5P$ times the midspan deflection, $\theta L/2$, and equals $\tfrac{3}{4}\theta PL$. The internal work is the sum of the work at each hinge, or

$$M_P\theta + 2M_P\theta + M_P\theta = 4M_P\theta$$

Equating internal and external work:

$$\tfrac{3}{4}\theta PL = 4M_P\theta$$

from which $P = 5.3M_P/L$.

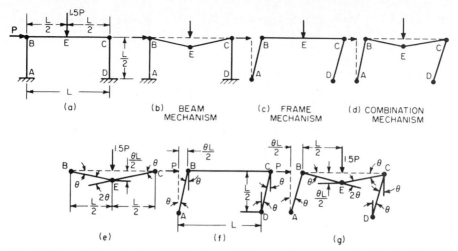

Fig. 1-91.   Ultimate-load possibilities for a rigid frame of constant section with fixed bases.

Next, let us investigate the frame mechanism, with plastic hinges at A, B, C, and D. As shown in Fig. 1-91f, apply a virtual rotation θ to AB. The point of application of the lateral load P will then move a distance $\theta L/2$, and the external work will be $\theta PL/2$. The internal work, being the sum of the work at the hinges, will be $4M_P\theta$. Equating internal and external work,

$$\frac{\theta PL}{2} = 4M_P\theta$$

from which $P = 8M_P/L$.

Finally, let us compute the critical load for the combination beam-frame mechanism in Fig. 1-91g.   Again, apply a virtual rotation θ to AB, moving B horizontally and E vertically a distance of $\theta L/2$. The external work, therefore, equals

$$\frac{\theta PL}{2} + \frac{3}{4}\theta PL = \frac{5\theta PL}{4}$$

The internal work is the sum of the work at hinges A, E, C, and D and is equal to

$$M_P\theta + 2M_P\theta + 2M_P\theta + M_P\theta = 6M_P\theta$$

Equating internal and external work,

$$\frac{5\theta PL}{4} = 6M_P\theta$$

from which $P = 4.8M_P/L$.

The combination mechanism has the lowest critical load. Now an equilibrium check must be made to insure that the moments everywhere in the frame are less than M_P. If they are, then the combination mechanism is the correct solution.

Consider first the length EC of the beam as a free body. Taking moments about E, noting that the moments at E and C equal M_P, the shear at C is computed to be $4M_P/L$. Therefore, the shear at B is $1.5 \times 4.8M_P/L - 4M_P/L = 3.2M_P/L$. By taking moments about B, with BE considered as a free body, the moment at B is found to be $0.6M_P$. Similar treatment of the columns as free bodies indicates that nowhere is the moment greater than M_P. Therefore, the combination mechanism is the correct solution, and the frame ultimate loads are $4.8M_P/L$ laterally and $7.2M_P/L$ vertically at midspan.

(R. O. Disque, "Applied Plastic Design in Steel," Van Nostrand Reinhold Company, New York; "Plastic Design in Steel—A Guide and Commentary," M & R no. 41, American Society of Civil Engineers, New York; B. G. Neal, "Plastic Methods of Structural Analysis," John Wiley & Sons, Inc., New York.)

STRESSES IN ARCHES

An arch is a curved beam, the radius of curvature of which is very large relative to the depth of section. It differs from a straight beam in that: (1) loads induce both bending and direct compressive stress in an arch; (2) arch reactions have horizontal components even though all loads are vertical, and (3) deflections have horizontal as well as vertical components.

The necessity of resisting the horizontal components of the reactions is an important consideration in arch design. Sometimes these forces are taken by tie rods between the supports, sometimes by heavy abutments or buttresses.

Arches may be built with fixed ends, as can straight beams, or with hinges at the supports. They may also be built with an internal hinge, usually located at the uppermost point, or crown.

1-79. Three-Hinged Arches

An arch with an internal hinge and hinges at both supports (Fig. 1-92) is statically determinate. There are four unknowns—two horizontal and two vertical components of the reactions—but four equations based on the laws of equilibrium are available: (1) The sum of the horizontal

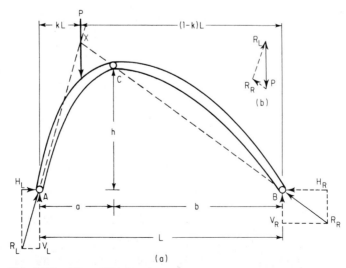

Fig. 1-92. Three-hinged arch.

forces must be zero. (In Fig. 1-92, $H_L = H_R = H$.) (2) The sum of the moments about the left support must be zero. ($V_R = Pk$). (3) The sum of the moments about the right support must be zero. [$V_L = P(1 - k)$.] (4) The bending moment at the crown hinge must be zero (not to be confused with the sum of the moments about the crown, which also must be equal to zero but which would not lead to an independent equation for the solution of the reactions). Hence, for the right half of the arch in Fig. 1-92, $Hh - V_Rb = 0$, and $H = V_Rb/h$. The influence line for H is a straight line, varying from zero for loads over the supports to the maximum of Pab/Lh for a load at C.

Reactions and stresses in three-hinged arches can be determined graphically by using the principles presented in Arts. 1-38 to 1-44 and 1-46, and by taking advantage of the fact that the bending moment at the crown hinge is zero. For example, in Fig. 1-92a, the load P is applied to segment AC of the arch. Then, since the bending moment at C must be zero, the line of action of the reaction R_R at B must pass through the crown hinge. It intersects the line of action of P at X. The line of action of the reaction R_L at A also must pass through X since P and the two reactions are in equilibrium. By constructing a force triangle with the load P and the lines of action of the reactions thus determined, you can obtain the magnitude of the reactions (Fig. 1-92b). After the reactions have been found, the stresses can be computed from the laws of statics or, in the case of a trussed arch, determined graphically.

1-80. Two-Hinged Arches

When an arch has hinges at the supports only (Fig. 1-93a), it is statically indeterminate; there is one more unknown reaction component than can be determined by the three equations of

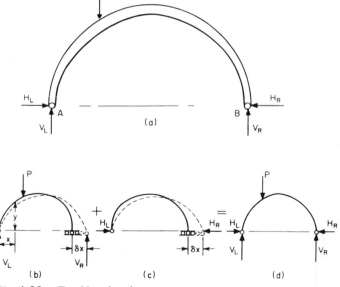

Fig. 1-93.   Two-hinged arch.

equilibrium. Another equation can be written from knowledge of the elastic behavior of the arch. One procedure is to assume that one of the supports is on rollers. The arch then is statically determinate, and the reactions and horizontal movement of the support can be computed for this condition (Fig. 1-93b), Next, the horizontal force required to return the movable support to its original position can be calculated (Fig. 1-93c). Finally, the reactions for the two-hinged arch (Fig. 1-93d) are obtained by superimposing the first set of reactions on the second.

For example, if δx is the horizontal movement of the support due to the loads on the arch, and if $\delta x'$ is the horizontal movement of the support due to a unit horizontal force applied to the support, then

$$\delta x + H \, \delta x' = 0 \tag{1-123a}$$

$$H = -\frac{\delta x}{\delta x'} \tag{1-123b}$$

where H is the unknown horizontal reaction. (When a tie rod is used to take the thrust, the right-hand side of the first equation is not zero but the elongation of the rod HL/A_sE_s, where L is the length of the rod, A_s its cross-sectional area, and E_s its modulus of elasticity. To account for the effect of an increase in temperature t, add to the left-hand side $EctL$, where E is the modulus of elasticity of the arch, c the coefficient of expansion.)

The dummy-unit-load method can be used to compute δx and $\delta x'$ (Art. 1-55):

$$\delta x = \int_A^B \frac{My \, ds}{EI} - \int_A^B \frac{N \, dx}{AE} \tag{1-124}$$

where M = bending moment at any section due to loads
y = ordinate of section measured from immovable end of arch
I = moment of inertia of arch cross section
A = cross-sectional area of arch
ds = differential length along arch axis
dx = differential length along the horizontal
N = normai thrust on cross section due to loads

$$\delta x' = -\int_A^B \frac{y^2 ds}{EI} - \int_A^B \frac{\cos \alpha \, dx}{AE} \tag{1-125}$$

where α = the angle the tangent to axis at the section makes with horizontal.

Equations (1-124) and (1-125) do not include the effects of shear deformation and curvature, which usually are negligible. Unless the thrust is very large, the second term on the right-hand side of Eq. (1-124) also can be dropped.

In most cases, integration is impracticable. The integrals generally must be evaluated by approximate methods. The arch axis is divided into a convenient number of elements of length Δs, and the functions under the integral sign are evaluated for each element. The sum of these terms is approximately equal to the integral. Thus, for the usual two-hinged arch

$$H = \frac{\sum_A^B (My \, \Delta s/EI)}{\sum_A^B (y^2 \, \Delta s/EI) + \sum_A^B (\cos \alpha \, \Delta x/AE)} \tag{1-126}$$

1-81. Fixed Arches

An arch is considered fixed when translation and rotation are prevented at the supports (Fig. 1-94a). Such an arch is statically indeterminate; there are six reaction components and only three equations are available from conditions of equilibrium. Three more equations must be obtained from a knowledge of the elastic behavior of the arch.

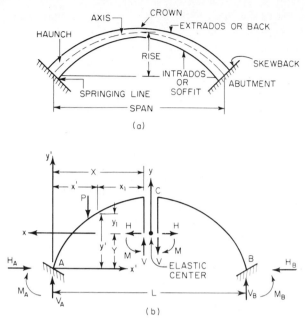

Fig. 1-94. Fixed-end arch.

One way to determine the reactions is to consider the arch cut at the crown, forming two cantilevers. First, the horizontal and vertical deflections and rotation produced at the end of each half arch by the loads are computed (Fig. 1-94*b*). Next, the deflection components and rotation at those ends are found for unit vertical force, unit horizontal force, and unit moment applied separately at the ends. These deformations, multiplied, respectively, by V, the unknown vertical shear; H, the unknown horizontal thrust at the crown; and M, the unknown moment there, yield the deformations caused by the unknown forces at the crown. Adding these deformations algebraically to the corresponding ones produced by the loads gives the net movement of the free end of each half arch. Since these ends must deflect and rotate the same amount to maintain continuity, three equations can be written for determination of V, H, and M. The various deflections can be computed by the dummy-unit-load method [Eq. (1-61)], as demonstrated in Art. 1-80 for two-hinged arches.

Elastic Center ▪ The solution of the equations, however, can be simplified considerably if the center of coordinates is shifted to the elastic center of the arch and the coordinate axes are properly oriented. If the unknown forces and moments V, H, and M are determined at the elastic center, each equation will contain only one unknown. When the unknowns at the elastic center are determined, the shears, thrusts, and moments at any point on the arch can be found by application of the laws of equilibrium.

Determination of the location of the elastic center of an arch is equivalent to finding the center of gravity of an area. Instead of an increment of area dA, however, an increment of length ds multiplied by a width $1/EI$ must be used. (E is the modulus of elasticity, I the moment of inertia.) Since, in general, integrals are difficult or impracticable to evaluate, the arch axis usually is divided into a convenient number of elements of length Δs and numerical integration is used, as described in Art. 1-80. Then, if the origin of coordinates is temporarily chosen at A, the left support of the arch in Fig. 1-94b, and if x' is the horizontal distance to a point on the arch and y' the vertical distance, the coordinates of the elastic center are

$$X = \frac{\sum\limits_{A}^{B} (x' \, \Delta s/EI)}{\sum\limits_{A}^{B} (\Delta s/EI)} \qquad Y = \frac{\sum\limits_{A}^{B} (y' \, \Delta s/EI)}{\sum\limits_{A}^{B} (\Delta s/EI)} \qquad (1\text{-}127)$$

If the arch is symmetrical about the crown, the elastic center lies on a normal to the tangent at the crown. In that case, there is a savings in calculations by taking the origin of the temporary coordinate system at the crown and measuring coordinates parallel to the tangent and the normal. To determine Y, the distance of the elastic center from the crown, Eq. (1-127) can be used with the summations limited to the half arch between crown and either support. For a symmetrical arch also, the final coordinate system should be chosen parallel to the tangent and normal to the crown.

After the elastic center has been located, the origin of a new coordinate system should be taken at the center. For convenience, the new coordinates, x_1, y_1 may be taken parallel to those in the temporary system. Then, for an unsymmetrical arch, the final coordinate axes should be chosen so that the x axis makes an angle α, measured clockwise, with the x_1 axis such that

$$\tan 2\alpha = \frac{2 \sum\limits_{A}^{B} (x_1 y_1 \, \Delta s/EI)}{\sum\limits_{A}^{B} (x_1^2 \, \Delta s/EI) - \sum\limits_{A}^{B} (y_1^2 \, \Delta s/EI)} \qquad (1\text{-}128)$$

The unknown forces H and V at the elastic center should be taken parallel, respectively, to the final x and y axes.

Forces at Elastic Center ▪ For a coordinate system with origin at the elastic center and axes oriented to satisfy Eq. (1-128),

$$H = \frac{\sum\limits_{A}^{B} (M'y \, \Delta s/EI)}{\sum\limits_{A}^{B} (y^2/\Delta s/EI)}$$

$$V = \frac{\sum\limits_{A}^{B} (M'x \, \Delta s/EI)}{\sum\limits_{A}^{B} (x^2 \, \Delta s/EI)} \qquad (1\text{-}129)$$

$$M = \frac{\sum\limits_{A}^{B} (M' \, \Delta s/EI)}{\sum\limits_{A}^{B} (\Delta s/EI)}$$

where M' is the average bending moment on each element due to loads. To account for the effect of an increase in temperature t, add $EctL$ to the numerator of H, where c is the coefficient of expansion and L the distance between abutments.

(S. Timoshenko and D. H. Young, "Theory of Structures," McGraw-Hill Book Company, New York; S. F. Borg and J. J. Gennaro, "Modern Structural Analysis," Van Nostrand Reinhold Company, New York.)

1-82.   Stresses in Arch Ribs

When the reactions have been found for an arch (Arts. 1-79 to 1-81), the principal forces acting on any cross section can be found by applying the equations of equilibrium. For example, consider the portion of a fixed arch in Fig. 1-95, where the forces at the elastic center H_e, V_e, and M_e are known and the forces acting at point X are to be found. The load P, H_e, and V_e may be resolved into components parallel to the axial thrust N and the shear S at X, as indicated in Fig. 1-95. Then, by equating the sum of the forces in each direction to zero, we get

$$N = V_e \sin \theta_x + H_e \cos \theta_x + P \sin (\theta_x - \theta) \tag{1-130}$$
$$S = V_e \cos \theta_x - H_e \sin \theta_x + P \cos (\theta_x - \theta)$$

And by taking moments about X and equating to zero, we obtain

$$M = V_e x + H_e y - M_e + Pa \cos \theta + Pb \sin \theta \tag{1-131}$$

The shearing unit stress on the arch cross section at X can be determined from S with the aid of Eq. (1-41). The normal unit stresses can be calculated from N and M with the aid of Eq. (1-49).

When designing an arch, it may be necessary to compute certain secondary stresses, in addition to those caused by live, dead, wind, and snow loads. Among the secondary stresses to be considered are those due to temperature changes, rib shortening due to thrust or shrinkage, deformation of tie rods, and unequal settlement of footings. The procedure is the same as for loads on the arch, with the deformations producing the secondary stresses substituted for or treated the same as the deformations due to loads.

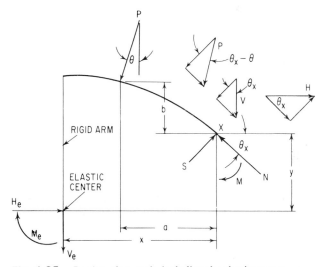

Fig. 1-95.   Portion of an arch, including the elastic center.

The many desirable characteristics of structural steels have brought the steels into widespread use in a large variety of applications. Structural steels are available in many product forms and offer an inherently high strength. They have a very high modulus of elasticity, so deformations under load are very small. Structural steels also possess high ductility. They have a linear or nearly linear stress-strain relationship up to relatively large stresses, and the modulus of elasticity is the same in tension and compression. Hence, structural steels' behavior under working loads can be accurately predicted by elastic theory. Structural steels are made under controlled conditions, so purchasers are assured of uniformly high quality.

Standardization of sections—shapes and plates—has facilitated design and kept down the cost of structural steels. For tables of properties of these sections, see "Steel Construction Manual," American Institute of Steel Construction, 400 North Michigan Ave., Chicago, Ill. 60611.

This section provides general information on structural-steel design and construction. Any use of this material for a specific application should be based on a determination of its suitability for the application by professionally qualified personnel.

2-1. Properties of Structural Steels

The term *structural steels* includes a large number of steels that, because of their economy, strength, ductility, and other properties, are suitable for load-carrying members in a wide variety of fabricated structures. Steel plates and shapes intended for use in bridges, buildings, transportation equipment, construction equipment, and similar applications are generally ordered to a specific specification of the American Society for Testing and Materials (ASTM) and furnished in "Structural Quality" according to the requirements (tolerances, frequency of testing, and so on) of ASTM A6. Plate steels

Section 2

Structural-Steel Design*

Roger L. Brockenbrough
Research Consultant
Research Laboratory
U.S. Steel Corporation
Monroeville, PA

*Revised from Sec. 9, "Structural-Steel Design and Construction," R.O. Disque and F.W. Stockwell, Jr., American Institute of Steel Construction, Chicago, IL, in "Standard Handbook for Civil Engineers," 2nd ed., F.S. Merritt (ed.), McGraw-Hill Book Co., NY.

for pressure vessels are furnished in "Pressure Vessel Quality" according to the requirements of ASTM A20.

Each structural steel is produced to specified minimum mechanical properties as required by the specific ASTM designation under which it is ordered. Generally, the structural steels include steels with yield points ranging from about 30 to 100 ksi. The various strength levels are obtained by varying the chemical composition and by heat treatment. Other factors that may affect mechanical properties include product thickness, finishing temperature, rate of cooling, and residual elements.

The following definitions aid in understanding the properties of steel.

Yield point F_y is that unit stress, ksi, at which the stress-strain curve exhibits a well-defined increase in strain without an increase in stress. Many design rules are based on yield point.

Tensile strength, or ultimate strength, is the largest unit stress the material can achieve in a tensile test.

Modulus of elasticity E is the slope of the stress-strain curve, computed by dividing the unit stress, ksi, by the unit strain, in/in, at yield stress. For all structural steels, it is usually taken as 29,000 ksi for design calculations.

Ductility is the ability of the material to undergo large inelastic deformations without fracture. It is generally measured by the percent elongation for a specified gage length (usually 2 or 8 in). Structural steel has considerable ductility, which is recognized in many design rules.

Weldability is the ability of steel to be welded without changing its basic mechanical properties. However, the welding materials, procedures, and techniques employed must be in accordance with the approved methods for each steel. Generally, weldability decreases with increase in carbon and manganese.

Notch toughness is an index of the propensity for brittle failure as measured by the impact energy necessary to fracture a notched specimen, such as a Charpy V-notch specimen.

Toughness reflects the ability of a smooth specimen to absorb energy as characterized by the area under a stress-strain curve.

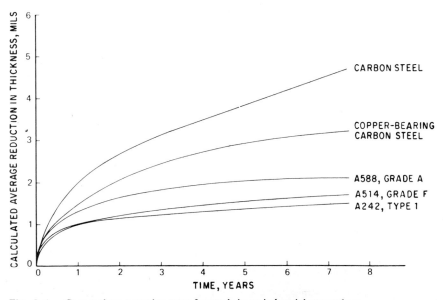

Fig. 2-1.   Curves show corrosion rates for steels in an industrial atmosphere.

Corrosion resistance has no specific index. However, relative corrosion-resistance ratings are based on the slopes of curves of corrosion loss (reduction in thickness) vs. time. The reference of comparison is usually the corrosion resistance of carbon steel without copper. Some high-strength structural steels are alloyed with copper and other elements to produce high resistance to atmospheric deterioration. These steels develop a tight oxide that inhibits further atmospheric corrosion. Figure 2-1 compares the rate of reduction of thickness of typical proprietary "corrosion-resistant" steels with that of ordinary structural steel.

(R. L. Brockenbrough and B. G. Johnston, "USS Steel Design Manual," ADUSS 27-3700-04, U.S. Steel Corp., Pittsburgh, Pa. 15222.)

2-2. Structural-Steel Shapes

Most structural steel used in building construction is fabricated from rolled shapes. In bridges, greater use is made of plates since girders spanning over about 90 ft are usually built-up sections.

Many different rolled shapes are available: W shapes (wide-flange shapes), M shapes (miscellaneous shapes), S shapes (standard I sections), angles, channels, and bars. The "Steel Construction Manual," American Institute of Steel Construction, lists properties of these shapes.

Wide-flange shapes normally range from a W4 $\times$ 13 (4 in deep weighing 13 lb/lin ft) to a W36 $\times$ 300 (36 in deep weighing 300 lb/lin ft). The heaviest shapes available are the "jumbo" column sections, which range up to W14 $\times$ 730.

In general, wide-flange shapes are the most efficient beam section. They have a high proportion of the cross-sectional area in the flanges and thus a high ratio of section modulus to weight. The 14-in W series includes shapes proportioned for use as column sections; the relatively thick web results in a large area-to-depth ratio.

Since the flange and web of a wide-flange beam do not have the same thickness, their yield points may differ slightly. In accordance with design rules for structural steel based on yield point, it is therefore necessary to establish a "design yield point" for each section. In practice, all beams that are rolled from A36 steel are considered to have a yield point of 36 ksi. Wide-flange shapes, plates, and bars rolled from higher-strength steels are required to have the minimum yields and tensile strength specified by ASTM.

Square, rectangular, and round structural tubular members are available with a variety of yield strengths. Suitable for columns because of their symmetry, these members are particularly useful in low buildings and where they are exposed for architectural effect. They may be used in tall buildings if closely spaced.

Connection Material ▪ Connections are normally made with A36 steel. If, however, higher-strength steels are used, the structural size groupings for angles and bars are:

Group 1: thicknesses of ½ in or less

Group 2: thicknesses exceeding ½ in but not more than ¾ in

Group 3: thicknesses exceeding ¾ in

Structural tees fall into the same group as the wide-flange or standard sections from which they are cut. (A WT7 $\times$ 13, for example, designates a tee formed by cutting in half a W14 $\times$ 26 and therefore is considered a Group 1 shape, as is a W14.)

2-3. Tolerances for Structural Shapes

ASTM Specification A6 lists mill tolerances for rolled-steel plates, shapes, sheet piles, and bars. Included are tolerances for rolling, cutting, section area, and weight, ends out of square, camber, and sweep. The "Steel Construction Manual," American Institute of Steel Construction, contains tables for applying these tolerances.

The AISC "Specification for the Design, Fabrication and Erection of Structural Steel for Buildings" and "Code of Standard Practice" gives fabrication and erection tolerances for structural steel for building work.

2-4. Structural-Steel Design Specifications

The basis for design of practically all structural steel for buildings in the United States is the AISC "Specification for the Design, Fabrication and Erection of Structural Steel for Buildings" (American Institute of Steel Construction, Inc., 400 N. Michigan Ave., Chicago, Ill. 60611). This specification relates allowable working stresses to yield point of material; the same relationships may be used for new steels when they are developed. Revisions may be made annually and are issued as supplements. Design criteria for buildings presented in this section derive from the basic 1978 Specification.

Design rules for bridges have been developed by the American Association of State Highway and Transportation Officials (AASHTO), 444 N. Capitol St., N.W., Washington, D.C. 20001 ("Standard Specifications for Highway Bridges"). They are slightly more conservative than AISC rules because bridges are subject to dynamic loading and fatigue (see Sec. 5).

In addition, there are other important specifications relating to steel structures:

American Iron and Steel Institute (AISI) publishes rules for the design of structural members cold-formed to shape from sheet or strip and used for load-carrying purposes in buildings.

American Welding Society (2501 N.W. 7th St., Miami, Fla. 33125) publishes the "Structural Welding Code," AWS D1.1, applicable to welding of bridges, buildings, and structures with tubular members.

American Railway Engineering Association publishers rules for design, fabrication, and erection of steel railway bridges (see Sec. 5).

AISC and the Steel Joist Institute (SJI) jointly publish specifications that cover design, manufacture, and use of open-web steel joists.

AISC also publishes a "Specification for the Design of Architecturally Exposed Structural Steel," which applies to steel structures subject to close inspection by the public.

2-5. Structural-Steel Design Theories

Structural steel for buildings may be designed by either the elastic or plastic theories. Design rules for both methods are included in the AISC "Specification for the Design, Fabrication and Erection of Structural Steel for Buildings" (Art. 2-4).

Elastic-design rules take into account the principles of plastic design, so that for many structures, neither method offers appreciable material saving compared with the other. Most engineers, however, consider plastic design simpler, more direct, and philosophically superior.

In elastic or "allowable-stress" design, a member is selected so that stresses will not exceed a specified value under working or service loads. Allowable stresses are listed in the AISC Specification and are determined by dividing yield stress by a safety factor.

Plastic design is based on the ability of structural steel to deform plastically at and above the yield point. If the load on a steel structure is increased sufficiently, the material will pass through the elastic range into the plastic range of stress.

The load-carrying capacity of a continuous beam is computed on the assumption that highly stressed areas may yield in bending. When these highly stressed zones yield, they no longer resist further bending. As a result, with further increases in load, the additional moment is transferred to less highly stressed areas. Thus, an overloaded continuous structure will readjust itself to carry the load.

Actually, these plastic hinges will not form under service loads. Plastic-design rules include a resistance to failure about the same as that for elastic design.

For buildings designed under the AISC Specification, plastic design may be applied to steels with yield points of 65 ksi and less. Braced and unbraced planar rigid frames, simple and continuous beams, and similar structures may be proportioned on the basis of plastic design (Art. 2-25).

AASHTO also specifies an alternative method of design for simple and continuous beam-and-girder structures of moderate length. Known as **load-factor design,** it is a method of proportioning structures for multiples of the design load. It does not depend on a complete internal redistribution of forces and moments as in plastic design, but it does recognize some of the higher strength indicated by plastic-design studies. Moments, shears, and other forces are determined by an elastic analysis, but the proportioning of members is accomplished on the basis of maximum strength, with additional checks on service behavior.

2-6. Allowable Tension in Steel

For buildings, AISC (Art. 2-4) specifies a basic allowable unit tensile stress, ksi, $F_t = 0.60F_y$, where F_y is the yield strength of the steel, ksi (Table 2-1). F_t is subjected to the further limitation that it should not exceed one-half the specified minimum tensile strength F_u of the material. This limitation, however, does not control for steels for which F_y does not exceed 65 ksi. On the net section through pinholes in eyebars, pin-connected plates, or built-up members, $F_t = 0.45F_y$.

TABLE 2-1 Allowable Tensile Stresses in Bridge and Building Steels, ksi

Yield strength	Buildings	Bridges
36	22	20
50	30	27

For bridges, AASHTO (Art. 2-4) specifies allowable tensile stresses as the smaller of $0.55F_y$ or $0.46F_u$ (Table 2-1).

Table 2-1 and subsequent tables apply to two strength levels, $F_y = 36$ ksi and $F_y = 50$ ksi, the ones generally used for construction.

These allowable stresses are applied to the net cross-sectional area of a member. The net section for a tension member with a chain of holes extending across a part in any diagonal or zigzag line is defined in the AISC Specification as follows: The net width of the part shall be obtained by deducting from the gross width the sum of the diameters of all the holes in the chain and adding, for each gage space in the chain, the quantity $s^2/4g$, where $s =$ longitudinal spacing (pitch), in, of any two consecutive holes and $g =$ transverse spacing (gage), in, of the same two holes. The critical net section of the part is obtained from the chain that gives the least net width. But the net section taken through a hole shall in no case be considered as more than 85% of the corresponding gross section.

2-7. Allowable Shear in Steel

The AISC Specification for buildings (Art. 2-4) specifies allowable unit shear, ksi, as $F_v = 0.40F_y$, where F_y is the yield point of the steel, ksi. The AASHTO Specification for bridges (Art. 2-4) gives $F_v = 0.33F_y$ (see Table 2-2).

TABLE 2-2 Allowable Shear
Stresses in Bridge and Building
Steels, ksi

Yield strength	Buildings	Bridges
36	14.5	12
50	20	17

The area used to compute shear stress in a rolled beam is defined as the product of the web thickness and the overall beam depth. The webs of all rolled structural shapes are of such thickness that shear is seldom the criterion for design.

At beam-end connections where the top flange is coped, and in similar situations in which failure might occur by shear along a plane through the fasteners or by a combination of shear along a plane through the fasteners and tension along a perpendicular plane, AISC limits the shear stress to $0.30F_u$ on the minimum net-failure surface bounded by the bolt holes. (F_u is the tensile strength of the beam.) This is required to prevent a web-tearing failure along the perimeter of the holes.

Within the boundaries of a rigid connection of two or more members whose webs lie in a common plane, the web shear stresses are generally high. For buildings, the Commentary on the AISC Specification states that such webs need no reinforcement if the web thickness, in, is greater than that calculated from

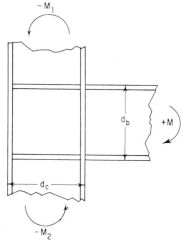

$$t_w = \frac{32M}{A_{bc}F_y} \qquad (2\text{-}1)$$

where M = algebraic sum of clockwise and counterclockwise moments applied on opposite sides of connection boundary, ft-kips

A_{bc} = planar area of connection web, in² (in Fig. 2-2, $d_b d_c$)

Additional requirements for column web stiffeners are presented in Art. 2-13.

2-8. Allowable Compression in Steel

(See also Art. 2-11.) The allowable compressive unit stress for a column is a function of its slenderness ratio. The slenderness ratio is defined in the AISC "Specification for the Design, Fabrication and Erection of Structural Steel for

Fig. 2-2. Rigid connection of two steel members with webs in a common plane.

Buildings" as Kl/r, where K = effective-length factor, which depends on restraints at top and bottom of the column; l = length of column between supports, in; and r = radius of gyration of column section, in.

The AISC Specification for buildings provides two formulas for computing allowable compressive stress F_a, ksi, for main members. The formula to use depends on the relationship of the largest effective slenderness ratio Kl/r of the cross section of any unbraced segment to a factor C_c defined by Eq. (2-2a). See Table 2-3a.

$$C_c = \sqrt{\frac{2\pi^2 E}{F_y}} = \frac{756.6}{\sqrt{F_y}} \qquad (2\text{-}2a)$$

TABLE 2-3a Values of C_c

F_y	C_c
36	126.1
50	107.0

TABLE 2-3b Allowable Stresses F_a, ksi, in Steel Building Columns for $Kl/r \leq 120$

Kl/r	Yield strength of steel F_y, ksi	
	36	50
10	21.16	29.26
20	20.60	28.30
30	19.94	27.15
40	19.19	25.83
50	18.35	24.35
60	17.43	22.72
70	16.43	20.94
80	15.36	19.01
90	14.20	16.94
100	12.98	14.71
110	11.67	12.34*
120	10.28	10.37*

*From Eq. (9-3c) because $Kl/r > C_c$.

TABLE 2-3c Allowable Stresses, ksi, in Steel Building Columns for $Kl/r > 120$

Kl/r	F_a	F_{as}
130	8.84	9.30
140	7.62	8.47
150	6.64	7.81
160	5.83	7.29
170	5.17	6.89
180	4.61	6.58
190	4.14	6.36
200	3.73	6.22

where E = modulus of elasticity of steel = 29,000 ksi
    F_y = yield stress of steel, ksi
When Kl/r is less than C_c,

$$F_a = \frac{\left[1 - \frac{(Kl/r)^2}{2C_c^2} \right] F_y}{F.S.}$$

(2-2b)

where $F.S.$ = safety factor = $\frac{5}{3} + \frac{3(Kl/r)}{8C_c} - \frac{(Kl/r)^3}{8C_c^3}$.

See Table 2-3b.

When Kl/r exceeds C_c,

$$F_a = \frac{12\pi^2 E}{23(Kl/r)^2} \tag{2-2c}$$

On the gross section of axially loaded bracing and secondary building members, when l/r exceeds 120 ($K = 1$), the allowable compressive stress, ksi, is

$$F_{as} = \frac{F_a}{1.6 - l/200r} \tag{2-2d}$$

In Eq. (2-2d), F_a is obtained from Eq. (2-2b) or (2-2c), depending on the value of C_c [see Table 2-3c and Eq. (2-2a)].

The effective-length factor K, equal to the ratio of effective column length to actual unbraced length, may be greater or less than 1.0. Theoretical K values for six idealized conditions, in which joint rotation and translation are either fully realized or nonexistent, are tabulated in Fig. 2-3.

An alternative and more precise method of calculating K for an unbraced column uses a nomograph given in the "Commentary on AISC Specification" ("Steel Construction Manual," American Institute of Steel Construction). This method requires calculation of "end-restraint factors" for the top and bottom of the column, to permit K to be determined from the chart.

In the AASHTO bridge-design Specifications, allowable stresses in concentrically loaded columns are determined from Eq. (2-3a) or (2-3b). When Kl/r is less than C_c,

$$F_a = \frac{F_y}{2.12} \left[1 - \frac{(Kl/r)^2}{2C_c^2} \right] \tag{2-3a}$$

BUCKLED SHAPE OF COLUMN IS SHOWN BY DASHED LINE	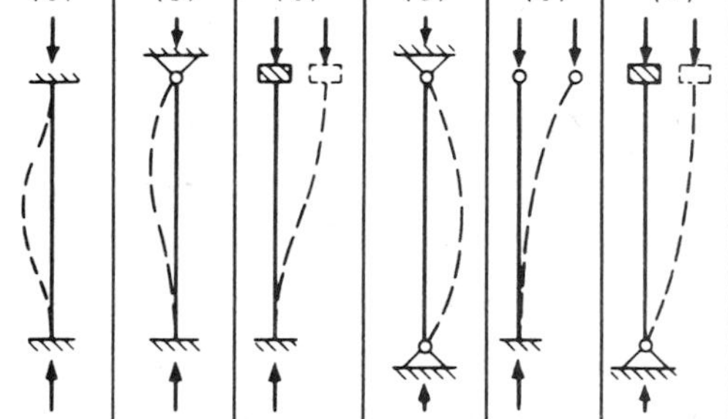 (a)	(b)	(c)	(d)	(e)	(f)
THEORETICAL K VALUE	0.5	0.7	1.0	1.0	2.0	2.0
RECOMMENDED DESIGN VALUE WHEN IDEAL CONDITIONS ARE APPROXIMATED	0.65	0.80	1.2	1.0	2.10	2.0
END CONDITION CODE	ROTATION FIXED AND TRANSLATION FIXED					
	ROTATION FREE AND TRANSLATION FIXED					
	ROTATION FIXED AND TRANSLATION FREE					
	ROTATION FREE AND TRANSLATION FREE					

Fig. 2-3. Values of effective-length factor K for columns.

TABLE 2-4 Column Formulas for Bridge Design

Yield strength, ksi	C_c	Allowable stress, ksi	
		$\dfrac{Kl}{r} \leq C_c$	$\dfrac{Kl}{r} \geq C_c$
36	126.1	$16.98 - 0.00053 \left(\dfrac{Kl}{r}\right)^2$	
50	107.0	$23.58 - 0.00103 \left(\dfrac{Kl}{r}\right)^2$	
90	79.8	$42.45 - 0.00333 \left(\dfrac{Kl}{r}\right)^2$	$\dfrac{135,000}{(Kl/r)^2}$
100	75.7	$47.17 - 0.00412 \left(\dfrac{Kl}{r}\right)^2$	

When Kl/r is equal to or greater than C_c,

$$F_a = \frac{\pi^2 E}{2.12(Kl/r)^2} \qquad (2\text{-}3b)$$

See Table 2-4.

2-9. Allowable Stresses in Bending

Bending stresses may be computed by elastic theory. The allowable stress in the compression flange usually governs the load-carrying capacity of steel beams or girders.

In building construction, the maximum fiber stress in bending for laterally supported beams and girders is $F_b = 0.66F_y$ for those sections classified as "compact" (Art. 2-23) and $F_b = 0.60F_y$ for those not conforming to the specified requirements. F_y is the yield strength of the steel, ksi. Table 2-5 lists values of F_b for two grades of steel.

TABLE 2-5 Allowable Bending Stresses in Braced Beams for Buildings, ksi

Yield strength, ksi	Compact $(0.66F_y)$	Noncompact $(0.60F_y)$
36	24	22
50	33	30

Because continuous steel beams have considerable reserve strength beyond the yield point, a redistribution of moments may be assumed when compact sections are continuous over supports or rigidly framed to columns. In that case, negative gravity-load moments over the supports may be reduced 10%. If this is done, the maximum positive moment in each span should be increased by 10% of the average negative moments at the span ends.

The allowable extreme-fiber stress of $0.60F_y$ applies to laterally supported, unsymmetrical members, except channels, and to noncompact-box sections. Compression on extreme fibers of channels should not exceed $0.60F_y$ or the value given by Eq. (2-6).

The allowable fiber stress of $0.66F_y$ for compact members should be reduced to $0.60F_y$ when the compression flange is unsupported for a length, in, exceeding

$$l_{max} = \frac{76.0b_f}{\sqrt{F_y}} \quad \text{or} \quad l_{max} = \frac{20,000}{F_y d / A_f}$$

where b_f = width of compression flange, in
d = beam depth, in
A_f = area, in^2, of compression flange

The allowable stress should be reduced even more when l/r_T exceeds certain limits, where l is the unbraced length, in, of the compression flange and r_T is the radius of gyration, in, of a portion of the beam consisting of the compression flange and one-third of the part of the web in compression.

For $\sqrt{102,000C_b/F_y} \leq l/r_T \leq \sqrt{510,000C_b/F_y}$, use

$$F_b = \left[\frac{2}{3} - \frac{F_y(l/r_T)^2}{1,530,000C_b} \right] F_y \tag{2-4a}$$

For $l/r_T > \sqrt{510,000C_b/F_y}$, use

$$F_b = \frac{170,000C_b}{(l/r_T)^2} \tag{2-5a}$$

where C_b = modifier for moment gradient (discussed later).

When, however, the compression flange is solid and nearly rectangular in cross section and its area is not less than that of the tension flange, the allowable stress may be taken as

$$F_b = \frac{12,000C_b}{ld/A_f} \tag{2-6}$$

See Table 2-6. When Eq. (2-6) applies (except for channels), F_b should be taken as the larger of the values computed from Eqs. (2-6) and (2-4a) or (2-5a) but not more than $0.60F_y$.

The moment-gradient factor C_b in Eqs. (2-4) to (2-6) may be computed from

$$C_b = 1.75 + 1.05 \frac{M_1}{M_2} + 0.3 \left(\frac{M_1}{M_2} \right)^2 \leq 2.3 \tag{2-7}$$

where M_1 = smaller beam end moment
M_2 = larger beam end moment

The algebraic sign of M_1/M_2 is positive for double-curvature bending and negative for single-curvature bending. When the bending moment at any point within an unbraced length is larger than that at both ends, the value of C_b should be taken as unity. For braced frames, C_b should be taken as unity for computation of F_{bx} and F_{by} with Eq. (2-12).

Equations (2-4a) and (2-5a) can be simplified by introduction of a new term:

$$Q = \frac{(l/r_T)^2 F_y}{510,000C_b} \tag{2-8}$$

Now, for $0.2 \leq Q \leq 1$,

$$F_b = \frac{(2 - Q)F_y}{3} \tag{2-4b}$$

For $Q > 1$,

$$F_b = \frac{F_y}{3Q} \tag{2-5b}$$

TABLE 2-6 Allowable Bending Stress F_b,* ksi, for Eq. (2-6)

ld/A_f	C_b							
	1.0	1.2	1.4	1.6	1.8	2.0	2.2	2.3
400	30.0	· · ·	· · ·	· · ·	· · ·	· · ·	· · ·	· · ·
450	26.7	32.0	· · ·	· · ·	· · ·	· · ·	· · ·	· · ·
500	24.0	28.8	· · ·	· · ·	· · ·	· · ·	· · ·	· · ·
550	21.8	26.2	30.5	· · ·	· · ·	· · ·	· · ·	· · ·
600	20.0	24.0	28.0	32.0	· · ·	· · ·	· · ·	· · ·
650	18.5	22.2	22.8	29.5	· · ·	· · ·	· · ·	· · ·
700	17.1	20.6	24.0	27.4	30.9	· · ·	· · ·	· · ·
750	16.0	19.2	22.4	25.6	28.8	· · ·	· · ·	· · ·
800	15.0	18.0	21.0	24.0	27.0	30.0	· · ·	· · ·
850	14.1	16.9	19.8	22.6	25.4	28.2	31.1	· · ·
900	13.3	16.0	18.7	21.3	24.0	26.7	29.3	30.7
950	12.6	15.2	17.7	20.2	22.7	25.3	27.8	29.1
1,000	12.0	14.4	16.8	21.6	21.6	24.0	26.4	27.6
1,100	10.9	13.1	15.3	17.5	19.6	21.8	24.0	25.1
1,200	10.0	12.0	14.0	16.0	18.0	20.0	22.0	23.0
1,300	9.23	11.1	12.9	14.8	16.6	18.5	20.3	21.2
1,400	8.57	10.3	12.0	13.7	15.4	17.1	18.9	19.7
1,500	8.00	9.60	11.2	12.8	14.4	16.0	17.6	18.4
1,600	7.50	9.00	10.5	12.0	13.5	15.0	16.5	17.3

*F_b may not exceed $0.60F_y$. Dots indicate that F_b is greater than 0.6×50 ksi; therefore, use 30 ksi. Values greater than 30 ksi are shown for interpolation purposes only and may not be used as F_b for $F_y = 50$ ksi material.

As for the preceding equations, when Eq. (2-6) applies (except for channels), F_b should be taken as the largest of the values given by Eqs. (2-6) and (2-4 b) or (2-5 b), but not more than $0.60F_y$.

For bridge design, AASHTO (Art. 2-4) gives the allowable unit (tensile) stress in bending as $F_b = 0.55F_y$ (Table 2-7). The same stress is permitted for compression when the compression flange is supported laterally for its full length by embedment in concrete or by other means.

TABLE 2-7   Allowable Bending
Stress in Braced Bridge Beams, ksi

F_y	F_b
36	20
50	27

When the compression flange is partly supported or unsupported in a bridge, the allowable bending stress, ksi, is

$$F_b = \left[1 - \frac{(l/r')^2}{2C_c^2} \right] 0.55F_y \tag{2-9}$$

where $r' = \sqrt{b^2/12}$ = radius of gyration, in
    b = flange width, in
    l = length of unsupported compression flange between lateral connections, knee braces, or other points of support, in, or twice length of cantilever if this is less than preceding

For $F_y = 36$ ksi, l/b should not exceed 36, and for $F_y = 50$ ksi, l/b should not be more than 30 (Table 2-8). For negative-moment regions of continuous girders, l may be taken as the distance from the interior support to point of dead-load contraflexure if this distance is less than

TABLE 2-8   Allowable
Compressive Stress in Flanges of
Bridge Beams, ksi

F_y	Max l/b	F_b
36	36	$20-0.0075(l/b)^2$
50	30	$27-0.0144(l/b)^2$

that between lateral connections, knee braces, or other supports. Continuous or cantilever bridge girders may be proportioned for negative moment at interior supports with an allowable unit stress 20% higher than permitted by Eq. (2-9) but in no case exceeding $0.55F_y$.

For each grade of steel, Eq. (2-9) reduces to the expressions given in Table 2-8.

2-10. Allowable Bearing Stress

In building construction, allowable bearing stress for milled surfaces, including bearing stiffeners, and pins in reamed, drilled, or bored holes, is $F_p = 0.90F_y$, where F_y is the yield strength of the steel, ksi.

For expansion rollers and rockers, the allowable bearing stress, kips/lin in, is

$$F_p = \frac{F_y - 13}{20} 0.66d \tag{2-10}$$

where d is the diameter, in, of the roller or rocker. When parts in contact have different yield strengths, F_y is the smaller value.

For highway design, AASHTO (Art. 2-4) limits the allowable bearing stress on milled stiffeners and other steel parts in contact to $F_p = 0.80F_y$ (Table 2-9). Allowable bearing stresses on pins are in Table 2-10.

The allowable bearing stress for expansion rollers and rockers used in bridges depends on the yield point in tension F_y of the steel in the roller or the base, whichever is smaller. For

TABLE 2-9   Allowable Bearing
Stress on Stiffeners of Bridge
Girders, ksi

F_y	F_p
36	29
50	40

TABLE 2-10 Allowable Bearing Stresses on
Pins, ksi

F_y	Buildings $F_p = 0.90F_y$	Bridges	
		Pins subject to rotation $F_p = 0.40F_y$	Pins not subject to rotation $F_p = 0.80F_y$
36	33	14	29
50	45	20	40

diameters up to 25 in, the allowable stress, kips/lin in, is

$$p = \frac{F_y - 13}{20} 0.6d \qquad (2\text{-}11\,a)$$

For diameters from 25 to 125 in,

$$p = \frac{F_y - 13}{20} 3\sqrt{d} \qquad (2\text{-}11\,b)$$

where d = diameter of roller or rocker, in.

2-11. Combined Axial Compression or Tension and Bending

The AISC Specification for buildings (Art. 2-4) includes three interaction formulas for combined axial compression and bending:

When the ratio of computed axial stress to allowable axial stress $f_a/F_a > 0.15$, both Eqs. (2-12) and (2-13) must be satisfied.

$$\frac{f_a}{F_a} + \frac{C_{mx}f_{bx}}{(1 - f_a/F'_{ex})F_{bx}} + \frac{C_{my}f_{by}}{(1 - f_a/F'_{ey})F_{by}} \leq 1 \qquad (2\text{-}12)$$

$$\frac{f_a}{0.60F_y} + \frac{f_{bx}}{F_{bx}} + \frac{f_{by}}{F_{by}} \leq 1 \qquad (2\text{-}13)$$

When $f_a/F_a \leq 0.15$, Eq. (2-14) may be used instead of Eqs. (2-12) and (2-13).

$$\frac{f_a}{F_a} + \frac{f_{bx}}{F_{bx}} + \frac{f_{by}}{F_{by}} \leq 1 \qquad (2\text{-}14)$$

In the preceding equations, subscripts x and y indicate the axis of bending about which the stress occurs, and

F_a = axial stress that would be permitted if axial force alone existed, ksi (see Arts. 2-6 and 2-8)

F_b = compressive bending stress that would be permitted if bending moment alone existed, ksi (see Art. 2-9)

F'_e = $149,000/(Kl_b/r_b)^2$, ksi; as for F_a, F_b, and $0.6F_y$, F'_e may be increased one-third for wind and seismic loads

l_b = actual unbraced length in plane of bending, in

r_b = radius of gyration about bending axis, in

K = effective-length factor in plane of bending

f_a = computed axial stress, ksi

f_b = computed compressive bending stress at point under consideration, ksi

C_m = adjustment coefficient

For compression members in frames subject to joint translation (sidesway), $C_m = 0.85$ in Eq. (2-12). For restrained compression members in frames braced against joint translation and not subject to transverse loading between supports in plane of bending, $C_m = 0.6 - 0.4M_1/M_2$, but not less than 0.4. M_1/M_2 is the ratio of the smaller to larger moment at the ends of that portion of the member unbraced in the plane of bending under consideration. M_1/M_2 is positive when the member is bent in reverse curvature and negative when it is bent in single curvature. For compression members in frames braced against joint translation in the plane of loading and subject to transverse loading between supports, the value of C_m may be determined by rational analysis. But in lieu of such analysis, the following values may be used: For members whose ends are restrained, $C_m = 0.85$. For members whose ends are unrestrained, $C_m = 1.0$.

Building members subject to combined axial tension and bending should satisfy Eq. (2-13), with f_b and F_b, respectively, as the computed and permitted bending tensile stress. But the computed bending compressive stress is limited by Eqs. (2-6) and (2-4 a) or (2-5 a).

Combined compression and bending stresses in bridge design are covered by equations similar to Eqs. (2-12) and (2-13) but adjusted to reflect the lower allowable stresses of AASHTO.

2-12. Web Crippling

The possibility of web crippling must be investigated in rolled beams and built-up girders at points of concentrated loads.

The AISC Specification for buildings (Art. 2-4) limits the compressive stress at the toe of the web of a rolled beam due to a concentrated load to $F_a = 0.75F_y$, where F_y is the yield point of the steel, ksi. In calculating the stress area, a 45° distribution should be assumed, as indicated in Fig. 2-4. Bearing stiffeners are required when F_a is exceeded.

For plate girders used in buildings, the sum of the compression stresses resulting from concentrated and distributed loads, bearing directly on or through a flange plate on the compression edge of the web plate, should not exceed the following:

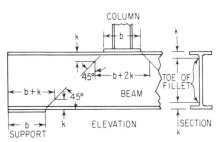

Fig. 2-4. Compressive stress is assumed to spread at 45° angles in the web of a rolled beam.

When the flange is restrained against rotation,

$$F_a = \left[5.5 + \frac{4}{(a/h)^2} \right] \frac{10,000}{(h/t)^2} \quad (2\text{-}15)$$

When the flange is not restrained against rotation,

$$F_a = \left[2 + \frac{4}{(a/h)^2} \right] \frac{10,000}{(h/t)^2} \quad (2\text{-}16)$$

where a = clear distance between transverse stiffeners, in

h = clear distance between flanges, in

t = web thickness, in

The load may be considered distributed over a web length equal to the panel length or girder depth, whichever is less (see also Art. 2-21).

Rolled beams used as bridge girders must be provided with suitable stiffeners at bearings when the unit shear in the web adjacent to the bearing exceeds 75% of the allowable shear for girder webs (Table 2-2).

With welded bridge girders, bearing stiffeners are always required over the end bearings and over the intermediate bearings of continuous girders. Design of these stiffeners is covered in Art. 2-22.

2-13. Restrained Members

AISC requires that fasteners or welds for end connections of beams, girders, and trusses be designed for the combined effect of forces resulting from moment and shear induced by the rigidity of the connection. When flanges or moment-connection plates for end connections of beams and girders are welded to the flange of an I- or H-shape column, a pair of column-web stiffeners having a combined cross-sectional area A_{st} not less than that calculated from Eq. (2-17) must be provided whenever the calculated value of A_{st} is positive.

$$A_{st} = \frac{P_{bf} - F_{yc}t(t_b + 5K)}{F_{yst}} \quad (2\text{-}17)$$

where F_{yc} = column yield stress, ksi
$\quad\ \ F_{yst}$ = stiffener yield stress, ksi
$\quad\ \ \ K$ = distance, in, between outer face of column flange and web toe of its fillet, if column is rolled shape, or equivalent distance if column is welded shape
$\quad\ \ P_{bf}$ = computed force, kips, delivered by flange of moment-connection plate multiplied by ⅗, when computed force is due to live and dead load only, or by ⅘, when computed force is due to live and dead load in conjunction with wind or earthquake forces
$\quad\ \ \ t$ = thickness of column web, in
$\quad\ \ t_b$ = thickness of flange or moment-connection plate delivering concentrated force, in

Notwithstanding the above requirements, a stiffener or a pair of stiffeners must be provided opposite the beam compression flange when the column-web depth clear of fillets d_c is greater than

$$d_c = \frac{4100t^3\sqrt{F_{yc}}}{P_{bf}} \tag{2-18}$$

and a pair of stiffeners should be provided opposite the tension flange when the thickness of the column flange t_f is less than

$$t_f = 0.4\ \sqrt{\frac{P_{bf}}{F_{yc}}} \tag{2-19}$$

Stiffeners required by the above provisions must comply with the following additional criteria:

1. The width of each stiffener plus half the thickness of the column web should not be less than one-third the width of the flange or moment-connection plate delivering the concentrated force.

2. The thickness of stiffeners should not be less than $t_b/2$.

3. When the concentrated force delivered occurs on only one column flange, the stiffener length need not exceed half the column depth.

4. The weld joining stiffeners to the column web must be sized to carry the force in the stiffener caused by unbalanced moments on opposite sides of the column.

Connections having high shear in the column web should be investigated as described by AISC. Equation (2-1) gives the condition for investigating high shear in the column web within the boundaries of the connection.

Stiffeners opposite the beam compression flange may be fitted to bear on the inside of the column flange. Stiffeners opposite the tension flange should be welded, and the welds must be designed for the forces involved.

2-14. Design of Beam Sections for Torsion

Torsional stresses may be induced in steel beams either by unsymmetrical loading or by symmetrical loading on unsymmetrical shapes, such as channels or angles. However, in most applications, they are much smaller than the concurrent axial or bending stresses.

2-15. Wind and Seismic Stresses

For buildings, allowable stresses may be increased one-third under wind or seismic forces acting alone or with gravity loads. The resulting design, however, should not be less than that required for dead and live loads without the increase in allowable stress. The increased stress is permitted because of the short duration of the load. Its validity has been justified by many years of satisfactory performance.

For allowable stresses, including wind and seismic effects on bridges, see Art. 5-4.

Successful wind or seismic design is dependent on close attention to connection details. It is good practice to provide as much ductility as practical in such connections so that the fasteners are not overstressed.

2-16. Fatigue Strength of Structural Components

Extensive research programs have been conducted to determine the fatigue strength of structural members and connections. These programs included large-scale beam specimens with various details such as flange-to-web fillet welds, flange cover plates, flange attachments, and web stiffeners. These studies showed that the stress range (algebraic difference between maximum and minimum stress) and the notch severity of the details were the dominant variables. For design purposes, the effects of steel yield point and stress ratio are not considered significant.

Allowable stress ranges based on this research have been adopted by AISC, AASHTO, and the American Welding Society (AWS) as indicated in Table 2-11. Plain material and various details have been grouped in categories of increasing severity, A through F. The allowable stress ranges are given for various numbers of cycles, from 20,000 to over 2 million. The over-2 million-cycles life corresponds to the fatigue limit; the detail is considered to have infinite life if the allowable stress range listed for over 2 million cycles is not exceeded. The allowable fatigue-stress range is applicable to any of the structural steels, but the maximum stress cannot exceed the maximum permitted under static loadings.

Note that the AISC, AASHTO, and AWS Specifications do not require fatigue checks in elements of members where calculated stresses are always in compression because although a crack may initiate in a region of tensile residual stress, the crack will generally not propagate beyond that region.

In design of a structural member to resist fatigue, each detail should be checked for the stress conditions that exist at that location. When a severe detail cannot be avoided, it is often

TABLE 2-11 Allowable Range of Stress under Fatique Loading, ksi*

Stress category†	Number of cycles			
	From 20,000–100,000	From 100,000–500,000	From 500,000–2,000,000	Over 2,000,000
A	60	36	24	24
B	45	27.5	18	16
C	32	19	13	10‡
D	27	16	10	7
E	21	12.5	8	5
E′	16	9.4	5.8	2.6
F	15	12	9	8

*Based on the requirements of AISC and AWS. Allowable ranges F_{sr} are for tension or reversal, except as noted. Values given represent 95% confidence limits for 95% survival. See AISC and AWS for description of stress categories. AASHTO requirements are similar, except as noted below, for structures with redundant load paths but are more severe for structures with nonredundant load paths.

†Typical details included in each category are as follows (see specifications for complete descriptions): A—base metal of plain material; B—base metal and weld metal at full-penetration groove welds, reinforcement ground off; C—base metal and weld metal at full-penetration groove welds, reinforcement not removed; D—base metal at certain attachment details; E—base metal at end of cover plate; E′—base metal at end of cover plate over 0.8 in thick (AASHTO requirement only); F—shear in weld metal of fillet welds.

‡Flexural stress range of 12 ksi permitted at toe of stiffener welds on webs or flanges.

advantageous to locate it in a region where the stress range is low so that the member can withstand the desired number of cycles.

Studies also have been conducted to determine the effect of weathering of bare steels on fatigue strength. Available information indicates that although the rougher surface of the weathered steel tends to reduce the fatigue strength, allowable stress ranges given in the above specifications may be used for bare steel.

2-17. Load Transfer and Stresses at Welds

Various coated stick electrodes for shielded metal-arc welding and various wire electrodes and flux or gas combinations for other processes may be selected to produce weld metals that provide a wide range of specified minimum-strength levels. AWS specifications give the electrode classes and welding processes that can be used to obtain *matching* weld metal, that is, weld metal that has a minimum tensile strength similar to that of various groups of steel. As indicated in Tables 2-12 and 2-13, however, matching weld metal is not always required, particularly in the case of fillet welds.

The differential cooling that accompanies welding causes residual stresses in the weld and the material joined. Although these stresses have an important effect on the strength of compression members, they do not usually have a significant effect on the strength of welded connections.

In groove welds, the loads are transferred directly across the weld by tensile or compressive stresses. For complete-penetration groove welds, the welding grade or electrode class is selected so that the resulting weld is as strong as the steel-joined one. Partial-penetration groove welds, in which only part of the metal thickness is welded, are sometimes used when stresses are low and there is no need to develop the complete strength of the material. The stress area of such a weld is the product of the length of the weld and an effective throat thickness. In single J- or U-type joints, the effective throat thickness is equal to the depth of the groove, and in bevel- or V-type joints, it is equal to the depth of the chamfer or the depth of the chamfer minus ⅛ in, depending on the included angle and the welding process. AWS does not permit partial-penetration groove welds to be used for cyclic tension normal to the weld axis; also, if the weld is made from one side only, it must be restrained from rotation. AISC permits such welds to be used for cyclic loading, but the allowable stress range is only one-third to one-half that of a complete-penetration groove weld. Details of recommended types of joints are given by AWS.

In fillet welds, the load is transferred between the connected plates by shear stresses in the welds. The shear stress in a fillet weld is computed on an area equal to the product of the length of the weld by the effective throat thickness.

The **effective throat thickness** is the shortest distance from the root to the face of the weld, a flat face being assumed, and is 0.707 times the nominal size or leg of a fillet weld with equal legs. The AISC specifies that the effective throat for submerged-arc fillet welds be taken equal to the leg size for welds ⅜ in or less and to the theoretical throat plus 0.11 in for larger welds.

Plug welds and slot welds are occasionally used to transfer shear stresses between plates. The shear area for the weld is the nominal cross-sectional area of the hole or slot. This type of connection should be avoided because of the difficulty in inspecting to ensure a satisfactory weld and the severe stress concentration created.

The basic allowable stresses for welds in buildings and bridges are shown in Tables 2-12 and 2-13. As indicated in the tables, complete-penetration groove welds in building or bridge construction and certain other welds in building construction have the same allowable stress as the steel that is joined. The allowable stresses shown for fillet welds provide a safety factor against ultimate weld shear failure of about 3 for building construction and about 10% higher for bridge construction.

TABLE 2-12 Allowable Stresses in Welds in Building Construction

Stress in weld*		Allowable stress	Required weld strength level†
		Complete-penetration groove welds	
Tension normal to the effective area		Same as base metal	Matching weld metal must be used.
Compression normal to the effective area		Same as base metal	Weld metal with a strength level equal to or one classification (10 ksi) less than matching weld metal may be used.
Tension or compression parallel to the axis of the weld		Same as base metal	
Shear on the effective area		0.30 nominal tensile strength of weld metal, ksi, except shear stress on base metal shall not exceed 0.40 yield strength of base metal	Weld metal with a strength level equal to or less than matching weld metal may be used.
		Partial-penetration groove welds	
Compression normal to effective area	Joint not designed to bear	0.50 nominal tensile strength of weld metal, ksi, except stress on base metal shall not exceed 0.60 yield strength of base metal	
	Joint designed to bear	Same as base metal	
Tension or compression parallel to the axis of the weld‡		Same as base metal	Weld metal with a strength level equal to or less than matching weld metal may be used.

		Weld metal with a strength level
Shear parallel to axis of weld	0.30 nominal tensile strength of weld metal, ksi, except shear stress on base metal shall not exceed 0.40 yield strength of base metal	
Tension normal to effective area	0.30 nominal tensile strength of weld metal, ksi, except tensile stress on base metal shall not exceed 0.60 yield strength of base metal	Weld metal with a strength level equal to or less than matching weld metal may be used.

Fillet welds

Shear on effective area	0.30 nominal tensile strength of weld metal, ksi, except shear stress on base metal shall not exceed 0.40 yield strength of base metal	
Tension or compression parallel to axis of weld‡	Same as base metal	

Plug and slot welds

Shear parallel to faying surfaces (on effective area)	0.30 nominal tensile strength of weld metal, ksi, except shear stress on base metal shall not exceed 0.40 yield strength of base metal	Weld metal with a strength level equal to or less than matching weld metal may be used.

*For definition of effective area, see AWS, D1.1.
†For matching weld metal, see AWS, D1.1.
‡Fillet welds and partial-penetration groove welds joining the component elements of built-up members, such as flange-to-web connections, may be designed without regard to the tensile or compressive stress in those elements parallel to the axis of the welds.

TABLE 2-13 Allowable Stresses in Welds in Bridge Construction

Stress in weld*		Allowable stress	Required weld strength level†
Complete-penetration groove welds			
Tension normal to the effective area		Same as base metal	Matching weld metal must be used.
Compression normal to the effective area		Same as base metal	Weld metal with a strength level equal to or one classification (10 ksi) less than matching weld metal may be used.
Tension or compression parallel to the axis of the weld		Same as base metal	
Shear on the effective area		0.27 nominal tensile strength of weld metal, ksi, except shear stress on base metal shall not exceed 0.36 yield strength of base metal.	Weld metal with a strength level equal to or less than matching weld metal may be used.
Partial-penetration groove welds			
Compression normal to effective area	Joint not designed to bear	0.45 nominal tensile strength of weld metal, ksi, except stress on base metal shall not exceed 0.55 yield strength of base metal	
	Joint designed to bear	Same as base metal	
Tension or compression parallel to the axis of the weld‡		Same as base metal	Weld metal with a strength level equal to or less than matching weld metal may be used.

Shear parallel to axis of weld	0.27 nominal tensile strength of weld metal, ksi, except shear stress on base metal shall not exceed 0.36 yield strength of base metal	Weld metal with a strength level equal to or less than matching weld metal may be used.
Tension normal to effective area	0.27 nominal tensile strength of weld metal, ksi, except tensile stress on base metal shall not exceed 0.55 yield strength of base metal	

Fillet welds

Shear on effective area	0.27 nominal tensile strength of weld metal, ksi, except shear stress on base metal shall not exceed 0.36 yield strength of base metal	
Tension or compression parallel to axis of weld‡	Same as base metal	

Plug and slot welds

Shear parallel to faying surfaces (on effective area)	0.27 nominal tensile strength of weld metal, ksi, except shear stress on base metal shall not exceed 0.36 yield strength of base metal	Weld metal with a strength level equal to or less than matching weld metal may be used.

*For definition of effective area, see AWS, D1.1.
†For matching weld metal, see AWS, D1.1.
‡Fillet welds and partial-penetration groove welds joining the component elements of built-up members, such as flange-to-web connections, may be designed without regard to the tensile or compressive stress in those elements parallel to the axis of the welds.

2-18. Stresses for Rivets and Bolts

For buildings, AISC (Art. 2-4) specifies allowable unit tension and shear stresses, ksi, on the area of rivets before being driven or on unthreaded body area of bolts and threaded parts, as given in Table 2-14. (Generally, rivets should not be used in direct tension.)

The allowable bearing stress, ksi, on the projected area of rivets and bolts in bearing-type connections is

$$F_p = 1.5F_u \tag{2-20}$$

where F_u is the tensile strength, ksi, of the connected part. (Bearing stresses are not restricted in friction-type connections assembled with A325 and A490 bolts.)

Friction-type connections are used when it is considered undesirable for the bolt to slip into bearing prior to failure. Bearing-type connections are usually specified in building construction because both types of connections have about the same ultimate capacity.

Combined Stresses ▪ In buildings, rivets and bolts subject to combined shear and tension should be so proportioned that the tension stress, ksi, produced by the resultant force does not exceed the following:
For A502, Grade 1, rivets:

$$F_t = 30 - 1.3f_v \leq 23 \tag{2-21a}$$

TABLE 2-14 Allowable Stresses for Bolts and Rivets in Buildings*

Type of fastener	Tensile stress, ksi	Shear stress, ksi	
		Friction-type connections	Bearing-type connections
A502 Grade 1 rivets, hot-driven	23	Not applicable	17.5
A502 Grade 2 rivets, hot-driven	29	Not applicable	22
A307 bolts	20	Not applicable	10
Threaded parts of suitable steels, threads not excluded from shear plane†	$0.33F_u$	Not applicable	$0.17F_u$
Threaded parts of suitable steels, threads excluded from shear plane	$0.33F_u$	Not applicable	$0.22F_u$
A325 bolts, threads not excluded from shear plane	44	17.5	21
A325 bolts, threads excluded from shear plane	44	17.5	30
A490 bolts, threads not excluded from shear plane	54	22	28
A490 bolts, threads excluded from shear plane	54	22	40

*Stresses are for nominal bolt and rivet areas, except as noted, and are based on AISC specifications. F_u is tensile strength, ksi. F_y is yield point, ksi.

Allowable stresses in tension are for static loads only, except for A325 and A490 bolts. For fatigue loadings, see AISC Specifications.

Allowable shear stresses for friction-type connections are for clean mill scale on faying surfaces. See AISC for allowable stresses for other surface conditions.

When bearing-type connections used to splice tension members have a fastener pattern whose length, measured parallel to the line of force, exceeds 50 in, allowable shear stresses must be reduced by 20%.

†In addition, the tensile capacity of the threaded portion of an upset rod, based on the cross-sectional area at its major thread diameter, must be larger than the nominal body area of the rod before upsetting times $0.60F_y$.

For A502, Grades 2 and 3, rivets:

$$F_t = 38 - 1.3f_v \le 29 \qquad (2\text{-}21\,b)$$

For A307 bolts on tension-stress area (Table 2-14):

$$F_t = 28 - 1.6f_v \le 20 \qquad (2\text{-}22\,a)$$

For A325 bolts in bearing-type joints, when threads are not excluded from shear planes:

$$F_t = 55 - 1.8f_v \le 44 \qquad (2\text{-}22\,b)$$

and when threads are excluded from shear planes:

$$F_t = 55 - 1.4f_v \le 44 \qquad (2\text{-}22\,c)$$

For A490 bolts in bearing-type joints, when threads are not excluded from shear planes:

$$F_t = 68 - 1.8f_v \le 54 \qquad (2\text{-}22\,d)$$

and when threads are excluded from shear planes:

$$F_t = 68 - 1.4f_v \le 54 \qquad (2\text{-}22\,e)$$

In the preceding equations, f_v is the shear stress, ksi, produced by the resultant force. This stress should not exceed the value for shear given in Table 2-14.

For bolts used in friction-type joints, the shear stress allowed in Table 2-14 should be reduced for combined shear and tension so that for A325 bolts:

$$F_v \le 17.5 \left(1 - \frac{f_t A_b}{T_b} \right) \qquad (2\text{-}22f)$$

and for A490 bolts:

$$F_v \le 22 \left(1 - \frac{f_t A_b}{T_b} \right) \qquad (2\text{-}22g)$$

where f_t = tensile stress, ksi, due to applied load
  A_b = area of bolt, in^2
  T_b = specified pretension load, kips, of bolt

When wind or seismic loading is combined with gravity loading, the allowable stresses in Table 2-14 may be increased one-third. For combined tension and shear in fasteners under such loadings, the constants of Eq. (2-22) may be increased one-third, but $1.4f_v$ and $1.8f_v$ should remain unchanged. For example, with wind or seismic loading, for A325 bolts in bearing-type joints, Eq. (2-22b) becomes

$$\begin{aligned} F_t &= \tfrac{4}{3} \times 55 - 1.8f_v \le \tfrac{4}{3} \times 44 \\ &= 73.3 - 1.8f_v \le 58.7 \end{aligned} \qquad (2\text{-}23)$$

Bridge Fasteners ▪ For bridges, AASHTO (Art. 2-4) specifies the working stresses for bolts and rivets listed in Table 2-15. Bearing-type connections with high-strength bolts are limited to members in compression and secondary members.

For combined shear and tension in friction-type joints, AASHTO requires that the shear stress be limited to a reduced allowable value F'_v, ksi.
For A325 bolts:

$$F'_v = (1 - 0.0159f_t) \qquad (2\text{-}24\,a)$$

TABLE 2-15 Allowable Stresses for Bolts and Rivets in Bridges*

Type of fastener	Tensile stress, ksi	Shear stress, ksi	
		Friction-type connections	Bearing-type connections
A307 bolts	13.5†	Not applicable	11.0
A502 Grade 1 rivets, hot-driven	Not applicable	Not applicable	13.5
A502 Grade 2 rivets, hot-driven	Not applicable	Not applicable	20.0
A325 bolts, threads not excluded from shear plane	39.5	16.0	19.0
A325 bolts, threads excluded from shear plane	39.5	16.0	27.0
A490 bolts, threads not excluded from shear plane	48.5	20.0	25.0
A490 bolts, threads excluded from shear plane	48.5	20.0	36.0

*Stresses are for nominal bolt and rivet area, except as noted, and are based on AASHTO Specifications. AASHTO specifies reduced shear stress values under certain conditions.

For fatigue loadings, see AASHTO Specifications.

Allowable shear stresses for friction-type connections are for clean mill scale on faying surfaces. See AASHTO for allowable stresses for other surface conditions.

In bearing-type connections whose length between extreme fasteners in each of the spliced parts measured parallel to the line of an axial force exceeds 50 in, allowable shear stresses must be reduced by 20%.

†Based on area at the root of thread.

For A490 bolts:

$$F_v' = (1 - 0.0127f_t) \qquad (2\text{-}24\,b)$$

where f_t is the tensile stress, ksi, due to applied loads, including any stress due to prying action.

For combined shear and tension in bearing-type joints, computed stresses in rivets and bolts should satisfy

$$f_v^2 + (kf_t)^2 \leq F_v^2 \qquad (2\text{-}24\,c)$$

where f_v = computed shear stress, ksi
f_t = computed tensile stress, ksi
F_v = allowable shear stress, ksi
k = 0.75 for rivets and 0.60 for A325 bolts with threads excluded from shear plane

2-19. Composite Construction

In composite construction, steel beams and a concrete slab are connected so that they act together to resist the load on the beam. The slab, in effect, serves as a cover plate. As a result, a lighter steel section may be used.

In building construction, there are two basic methods of composite construction.

Method 1. The steel beam is entirely encased in the concrete. Since the beam is completely braced laterally, the allowable stress in the flanges is $0.66F_y$, where F_y is the yield strength, ksi, of the steel. Assuming the steel to carry the full dead load and the composite section to carry the live load, the maximum unit stress, ksi, in the steel is

$$f_s = \frac{M_D}{S_s} + \frac{M_L}{S_{tr}} \leq 0.66F_y \qquad (2\text{-}25)$$

where M_D = dead-load moment, in-kips
M_L = live-load moment, in-kips

S_s = section modulus of steel beam, in³
S_{tr} = section modulus of transformed composite section, in³

An alternative, shortcut method is permitted by the AISC Specification (Art. 2-4). It assumes the steel beam will carry both live and dead loads and compensates for this by permitting a higher stress in the steel:

$$f_s = \frac{M_D + M_L}{S_s} \le 0.76F_y \qquad (2\text{-}26)$$

Method 2.   The steel beam is connected to the concrete slab by shear connectors: studs, wire spirals, or channels. Design is based on ultimate load and is independent of use of temporary shores to support the steel until the concrete hardens. The maximum stress in the bottom flange is

$$f_s = \frac{M_D + M_L}{S_{tr}} \le 0.66F_y \qquad (2\text{-}27)$$

To obtain the transformed composite section, treat the concrete above the neutral axis as an equivalent steel area by dividing the concrete area by n, the ratio of modulus of elasticity of steel to that of the concrete. When determining the transformed section, the width of the slab should not exceed one-fourth the beam span or, for an interior beam, eight times the slab thickness or half the clear distance to the adjacent beam. For an exterior beam, the effective projection should not be more than one-twelfth the beam span or six times the slab thickness or half the clear distance to the adjacent beam. (See Fig. 2-5.)

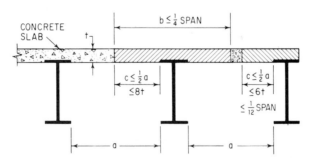

Fig. 2-5.   Limitations on effective width of concrete slab in composite steel-concrete beam.

For unshored construction, to limit the tension flange to considerably less than yield stress, the section modulus of the transformed section (referred to the tension flange) used in computation of the bottom-flange stress should not exceed

$$S_{tr} = \left(1.35 + 0.35\frac{M_L}{M_D} \right) S_s \qquad (2\text{-}28)$$

Shear on Connectors ▪ The total horizontal shear to be resisted by the shear connectors in building construction is taken as the smaller of the values given by Eqs. (2-29) and (2-30).

$$V_h = \frac{0.85f'_cA_c}{2} \qquad (2\text{-}29)$$

$$V_h = \frac{A_sF_y}{2} \qquad (2\text{-}30)$$

where V_h = total horizontal shear, kips, between maximum positive moment and each end of steel beams (or between point of maximum positive moment and point of contraflexure in continuous beam)

f'_c = specified compressive strength of concrete at 28 days, ksi

A_c = actual area of effective concrete flange, in^2

A_s = area of steel beam, in^2

In continuous composite construction, longitudinal reinforcing steel may be considered to act compositely with the steel beam in negative-moment regions. In this case, the total horizontal shear, kips, between an interior support and each adjacent point of contraflexure should be taken as

$$V_h = \frac{A_{sr}F_{yr}}{2} \tag{2-31}$$

where A_{sr} = area of longitudinal reinforcement at support within effective area, in^2

F_{yr} = specified minimum yield stress of longitudinal reinforcement, ksi

Number Required for Building Construction ▪ The total number of connectors to resist V_h is computed from V_h/q, where q is the allowable shear for one connector, kips. Values of q for connectors in buildings are in Table 2-16.

Table 2-16 is applicable only to composite construction with concrete made with stone aggregate conforming to ASTM C33. For lightweight concrete weighing at least 90 lb/ft^3 and made with rotary-kiln-produced aggregates conforming to ASTM C330, the allowable shears in Table 2-16 should be reduced by multiplying by the appropriate coefficient of Table 2-17.

The required number of shear connectors may be spaced uniformly between the sections of maximum and zero moment. Shear connectors should have at least 1 in of concrete cover in all directions, and unless studs are located directly over the web, stud diameters may not exceed 2.5 times the beam-flange thickness.

With heavy concentrated loads, the uniform spacing of shear connectors may not be sufficient between a concentrated load and the nearest point of zero moment. The number of shear connectors in this region should be at least

$$N_2 = \frac{N_1[(M\beta/M_{max}) - 1]}{\beta - 1} \tag{2-32}$$

TABLE 2-16 Allowable Shear Loads on Connectors for Composite Construction in Buildings

Type of connector	Allowable horizontal shear load q, kips (applicable only to concrete made with ASTM C33 aggregates)		
	f'_c, ksi		
	3.0	3.5	4.0
½-in dia × 2-in hooked or headed stud	5.1	5.5	5.9
⅝-in dia × 2½-in hooked or headed stud	8.0	8.6	9.2
¾-in dia × 3-in hooked or headed stud	11.5	12.5	13.3
⅞-in dia × 3½-in hooked or headed stud	15.6	16.8	18.0
3-in channel, 4.1 lb	4.3w	4.7w	5.0w
4-in channel, 5.4 lb	4.6w	5.0w	5.3w
5-in channel, 6.7 lb	4.9w	5.3w	5.6w

w = length of channel, in.

TABLE 2-17 Shear Coefficient for Lightweight Concrete with Aggregates Conforming to ASTM C330

	\multicolumn{7}{c}{Air dry unit weight, lb per ft³}						
	90	95	100	105	110	115	120
When $f'_c \leq 4$ ksi	0.73	0.76	0.78	0.81	0.83	0.86	0.88
When $f'_c \geq 5$ ksi	0.82	0.85	0.87	0.91	0.93	0.96	0.99

where M = moment at concentrated load, ft-kips

M_{max} = maximum moment in span, ft-kips

N_1 = number of shear connectors required between M_{max} and zero moment

β = S_{tr}/S_s or S_{eff}/S_s, as applicable

S_{eff} = effective section modulus for partial composite action, in³

Partial composite construction is used when the number N_1 of shear connectors required would provide a beam considerably stronger than necessary. In that case, the effective section modulus is used in stress computation instead of the transformed section modulus, and S_{eff} is calculated from Eq. (2-33).

$$S_{eff} = S_s + \sqrt{\frac{V'_h}{V_h}}(S_{tr} - S_s) \qquad (2\text{-}33)$$

where V'_h = number of shear connectors provided times allowable shear load q of Table 2-16 (times coefficient of Table 2-17, if applicable).

Composite construction of steel beams and concrete slab cast on a cold-formed-steel deck can also be designed with the information provided, but certain modifications are required as described in the AISC Specification. The total slab thickness, including the vertical ribs, is used in determining the effective slab width. Various dimensional requirements must be met. Also, the allowable shear loads for the stud connectors must be multiplied by a reduction factor. The ribs in the steel deck may be oriented perpendicular to or parallel to the steel beam or girder. The studs are typically welded directly through the deck by following procedures recommended by stud manufacturers.

Composite design for highway bridges is based on elastic rather than ultimate-load considerations. The total stress in the tension flange of the composite section must not exceed the allowable bending stress of the steel. When shores are used and kept in place until the concrete has attained 75% of its 28-day strength, the stress due to live and dead loads is computed for the composite section.

The effective width of the concrete flange is the same as for the section in building construction (Fig. 2-5), except that for interior bridge beams effective width may be increased from 8 to 12 times slab thickness.

Maximum spacing of shear connectors is 24 in, except over interior supports, where wider spacing may be used. Connectors must have a minimum of 2-in cover and must project a minimum of 2 in above the bottom of the slab (Fig. 2-6).

Bending stresses in composite beams in bridges depend on whether or not the members are shored; they are determined as for beams in buildings [see Eqs. (2-25) and (2-27)], except that the stresses in the steel may not exceed $0.55F_y$ [see Eqs. (2-34) and (2-35)]. Unshored:

$$f_s = \frac{M_D}{S_s} + \frac{M_L}{S_{tr}} \leq 0.55F_y \qquad (2\text{-}34)$$

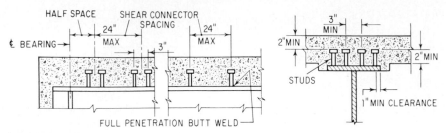

Fig. 2-6. Maximum pitch for stud shear connectors in composite beams.

Shored:

$$f_s = \frac{M_D + M_L}{S_{tr}} \le 0.55F_y \tag{2-35}$$

Shear Range ▪ Shear connectors in bridges are designed for fatigue and then are checked for ultimate strength. The horizontal-shear range for fatigue is computed from

$$S_r = \frac{V_r Q}{I} \tag{2-36}$$

where S_r = horizontal-shear range at juncture of slab and beam at point under consideration, kips/lin in

V_r = shear range (difference between minimum and maximum shears at the point) due to live load and impact, kips

Q = static moment of transformed compressive concrete area about neutral axis of transformed section, in^3

I = moment of inertia of transformed section, in^4

The transformed area is the actual concrete area divided by n (Table 2-18).

TABLE 2-18 Ratio of Moduli of Elasticity of Steel and Concrete for Bridges

f'_c for concrete, ksi	$n = \dfrac{E_s}{E_c}$
2.0–2.3	11
2.4–2.8	10
2.9–3.5	9
3.6–4.5	8
4.6–5.9	7
6.0 and over	6

The allowable range of horizontal shear Z_r, kips, for an individual connector is given by Eq. (2-37) or (2-38), depending on the connector used.

For channels:

$$Z_r = Bw \tag{2-37}$$

where w = channel length, in, in transverse direction on girder flange

B = cyclic variable = 4.0 for 100,000 cycles, 3.0 for 500,000 cycles, 2.4 for 2 million cycles, 2.1 for over 2 million cycles

For welded studs (with height-diameter ratio $H/d \ge 4$):

$$Z_r = \alpha d^2 \tag{2-38}$$

where d = stud diameter, in

α = cyclic variable = 13.0 for 100,000 cycles, 10.6 for 500,000 cycles, 7.85 for 2 million cycles, 5.5 for over 2 million cycles

Required pitch of shear connectors is determined by dividing the allowable range of horizontal shear of all connectors at one section Z_r, kips, by the horizontal range of shear S_r, kips/lin in.

Number of Connectors in Bridges ▪ The ultimate strength of the shear connectors is checked by computation of the number of connectors required from

$$N = \frac{P}{\phi S_u} \tag{2-39}$$

where N = number of shear connectors between maximum positive moment and end supports or dead-load points of contraflexure, or between maximum negative moment and points of contraflexure

S_u = ultimate shear connector strength, kips [see Eq. (2-41) and Table 2-19]

ϕ = reduction factor = 0.85

P = force in slab, kips

TABLE 2-19 Ultimate Horizontal-Shear Load for Connectors in Composite Beams in Bridges*

Connector	Ultimate shear, kips, for concrete compressive strength f'_c, ksi		
	$f'_c = 3.0$	$f'_c = 3.5$	$f'_c = 4.0$
Welded stud			
¾-in dia, 3 in min. high	21.8	24.4	27.0
⅞-in dia, 3.5 in min. high	29.6	33.3	36.8
Rolled channel†			
3 in deep, 4.1 lb/ft	10.78w	11.65w	12.45w
4 in deep, 5.4 lb/ft	11.69w	12.62w	13.50w
5 in deep, 6.7 lb/ft	12.50w	13.50w	14.43w

*The values are based on the requirements of AASHTO and include no safety factor. Values are for concrete with unit weight of 144 lb/ft.³

†w is channel length, in.

At points of maximum positive moments, P is the smaller of P_1 and P_2, computed from Eqs. (2-40a and b).

$$P_1 = A_s F_y \tag{2-40a}$$

$$P_2 = 0.85 f'_c A_c \tag{2-40b}$$

where A_c = effective concrete area, in²

f'_c = 28-day compressive strength of concrete, ksi

A_s = total area of steel section, in²

F_y = steel yield strength, ksi

At points of maximum negative moments, P is equal to P_3, computed from

$$P_3 = A_{sr} F_{yr} \tag{2-40c}$$

where A_{sr} = area of longitudinal reinforcing within effective flange, in²

F_{yr} = reinforcing steel yield strength, ksi

Ultimate Shear Strength of Connectors, Kips, in Bridges ▪ For channels:

$$S_u = 17.4(h + t/2)\sqrt{f'_c} \tag{2-41 a}$$

where h = average channel-flange thickness, in
t = channel-web thickness, in

For welded studs ($H/d \geq 4$ in):

$$S_u = 0.4d^2\sqrt{f'_c E_c} \tag{2-41 b}$$

Table 2-19 gives the ultimate shear for connectors as computed from Eqs. (2-41 a) and (2-41 b) for some commonly used concrete strengths.

Creep and Shrinkage ▪ AASHTO requires that the effects of creep be considered in the design of composite beams with dead loads acting on the composite section. For such beams, tension, compression, and horizontal shears produced by dead loads acting on the composite section should be computed for n or $3n$, whichever gives the higher stresses.

Shrinkage also should be considered. Resistance of a steel beam to longitudinal contraction of the concrete slab produces shear stresses along the contact surface. Associated with this shear are tensile stresses in the slab and compressive stresses in the steel top flange. These stresses also affect the beam deflection. The magnitude of the shrinkage effect varies within wide limits. It can be qualitatively reduced by appropriate casting sequences, for example, by placing concrete in a checkerboard pattern.

Span-Depth Ratios ▪ In bridges, for composite beams, preferably the ratio of span to steel-beam depth should not exceed 30 and the ratio of span to depth of steel beam plus slab should not exceed 25.

2-20. Plate-Girder Design Criteria for Buildings and Bridges

In computation of stresses in plate girders, the moment of inertia I, in^4, of the gross cross section generally is used. Bending stress f_b due to bending moment M is computed from $f_b = Mc/I$, where c is the distance, in, from the neutral axis to the extreme fiber. For determination of stresses in bolted or riveted girders for buildings or bridges, no deduction need be made for rivet or bolt holes unless the reduction in flange area, calculated as indicated in Art. 2-6, exceeds 15%; then the excess should be deducted.

In bolted or riveted plate girders, the flange angles should form as large a part of the area of the flange as practicable. Side plates should not be used unless the flange angles would have to exceed ⅞ in in thickness. The gross area of the compression flange should not be less than that of the tension flange. If several flange plates are used in a flange and the thicknesses are not the same, the plates should decrease in thickness from the flange angles outward. No plate should be thicker than the flange angles. At least one flange cover plate should extend the full length of the girder, unless the flange is covered with concrete. Any cover plate that is not full length should extend far enough to develop the capacity of the plate beyond the theoretical end, the section where flange stress without that cover plate equals the allowable stress.

In welded-plate girders, each flange should consist of a single plate. It may, however, comprise a series of shorter plates joined end to end by full-penetration groove welds. Flange thickness may be increased or decreased at a slope of not more than 1 in 2.5 as stress requirements permit. In bridges, the ratio of compression-flange width to thickness should not exceed 24 or $103/\sqrt{f_b}$, where f_b = computed maximum bending stress, ksi. In buildings, this ratio should not exceed $190/\sqrt{F_y}$, where F_y is the yield point of the flange, ksi.

The web depth-to-thickness ratio is defined as h/t, where h is the clear distance between flanges, in, and t is the web thickness, in. Several design rules for plate girders depend on this ratio.

See also Arts. 2-21 and 2-22.

2-21. Criteria for Plate Girders in Buildings

For greatest resistance to bending, as much of a plate girder cross section as practicable should be concentrated in the flanges, at the greatest distance from the neutral axis. (See also Art. 2-20.) This might require, however, a web so thin that the girder would fail by web buckling before it reached its bending capacity. To preclude this, the AISC Specification (Art. 2-4) limits the web clear-depth-to-thickness ratio h/t.

Critical Depth-Thickness Ratios ■ For an unstiffened web, this ratio should not exceed

$$\frac{h}{t} = \frac{14,000}{\sqrt{F_y(F_y + 16.5)}} \qquad (2\text{-}42\,a)$$

where F_y = yield strength of compression flange, ksi.

Larger values of h/t may be used, however, if the web is stiffened at appropriate intervals.

For this purpose, vertical angles may be fastened to the web or vertical plates welded to it. These transverse stiffeners are not required, though, when h/t is less than the value computed from Eq. (2-42 a) or given in Table 2-20.

TABLE 2-20 Critical h/t for Plate Girders in Buildings

F_y, ksi	$\dfrac{14,000}{\sqrt{F_y(F_y + 16.5)}}$	$\dfrac{2,000}{\sqrt{F_y}}$
36	322	333
50	243	283

With transverse stiffeners spaced not more than 1.5 times the girder depth apart, the web clear-depth-to-thickness ratio may be as large as

$$\frac{h}{t} = \frac{2000}{\sqrt{F_y}} \qquad (2\text{-}42\,b)$$

(See Table 2-20.) If, however, the web depth-to-thickness ratio h/t exceeds $760/\sqrt{F_b}$, where F_b, ksi, is the allowable bending stress in the compression flange that would ordinarily apply, this stress should be reduced to F'_b, given by Eq. (2-43 a).

$$F'_b = F_b \left[1 - 0.0005 \frac{A_w}{A_f} \left(\frac{h}{t} - \frac{760}{\sqrt{F_b}} \right) \right] \qquad (2\text{-}43\,a)$$

where A_w = web area, in^2

A_f = area of compression flange, in^2

In a hybrid girder, where the flange steel has a higher yield strength than the web, the allowable bending stress in the top or bottom flange should be taken as the smaller of the values given by Eqs. (2-43 a and b).

$$F'_b = F_b \left[\frac{12 + (A_w/A_f)(3\alpha - \alpha^3)}{12 + 2(A_w/A_f)} \right] \qquad (2\text{-}43\,b)$$

where α = ratio of web yield strength to flange yield strength. This protects against excessive yielding of the lower-strength web in the vicinity of the higher-strength flanges.

Stiffener Spacing ▪ The shear and allowable shear stress may determine required web area and stiffener spacing. Equations (2-44) and (2-45) give the allowable web shear F_v, ksi, for any panel of a building girder between transverse stiffeners.

When a factor C_v is less than 1.0, except for hybrid girders,

$$F_v = \frac{F_y}{2.89} \left[C_v + \frac{1 - C_v}{1.15\sqrt{1 + (a/h)^2}} \right] \qquad (2\text{-}44)$$

When C_v is more than 1.0 or when intermediate stiffeners are omitted, and for hybrid girders,

$$F_v = \frac{F_y C_v}{2.89} \leq 0.4F_y \qquad (2\text{-}45)$$

where a = clear distance between stiffeners, in
  h = clear distance between flanges, in
  $C_v = \dfrac{45,000k}{F_y(h/t)^2}$ when C_v is less than 0.8

  $= \dfrac{190}{h/t} \sqrt{\dfrac{k}{F_y}}$ when C_v is more than 0.8
  t = web thickness, in
  k = $5.34 + 4/(a/h)^2$ when $a/h > 1$
   = $4 + 5.34/(a/h)^2$ when $a/h < 1$

The average shear stress f_v, ksi, in a panel of a plate girder (web between successive stiffeners) is defined as the largest shear, kips, in the panel divided by the web cross-sectional area, in^2. As f_v approaches F_v given by Eq. (2-44), combined shear and tension become important. In that case, the tensile stress in the web due to bending in its plane should not exceed $0.6F_y$ or $(0.825 - 0.375f_v/F_v)F_y$, where F_v is given by Eq. (2-44).

The spacing between stiffeners at end panels, at panels containing large holes, and at panels adjacent to panels containing large holes, should be such that f_v does not exceed the value given by Eq. (2-45).

Intermediate stiffeners, when required, should be spaced so that a/h is less than 3 and less than $[260/(h/t)]^2$. Such stiffeners are not required when h/t is less than 260 and f_v is less than F_v computed from Eq. (2-45).

An infinite combination of web thicknesses and stiffener spacings is possible with a particular girder. Figure 2-7, developed for A36 steel, facilitates the trial-and-error process of selecting a suitable combination. Similar charts can be developed for other steels.

Stiffener Properties ▪ The required area of intermediate stiffeners is determined by

$$A_{st} = \frac{1 - C_v}{2} \left[\frac{a}{h} - \frac{(a/h)^2}{\sqrt{1 + (a/h)^2}} \right] YDht \qquad (2\text{-}46)$$

where A_{st} = gross stiffener area, in^2 (total area, if in pairs)
  Y = ratio of yield point of web steel to yield point of stiffener steel
  D = 1.0 for stiffeners in pairs
   = 1.8 for single-angle stiffeners
   = 2.4 for single-plate stiffeners

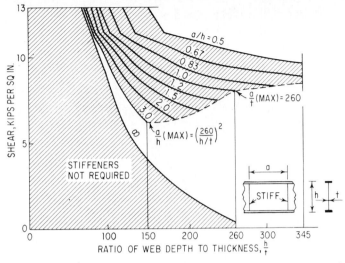

FIG. 2-7. Chart for determining spacing of girder stiffeners of A36 steel.

If the computed web-shear stress f_v is less than F_v computed from Eq. (2-44), A_{st} may be reduced by the ratio f_v/F_v.

The moment of inertia of a stiffener or pair of stiffeners should be at least $(h/50)^4$.

Stiffener Connections ▪ The stiffener-to-web connection should be designed for a shear, kips/lin in of single stiffener, or pair of stiffeners, of at least

$$f_{vs} = h \sqrt{\left(\frac{F_y}{340}\right)^3} \tag{2-47}$$

This shear may also be reduced by the ratio f_v/F_v.

Spacing of fasteners connecting stiffeners to the girder web should not exceed 12 in c to c. If intermittent fillet welds are used, the clear distance between welds should not exceed 10 in or 16 times the web thickness.

Bearing Stiffeners ▪ These are required on webs where ends of plate girders do not frame into columns or other girders. They may also be needed under concentrated loads and at reaction points. Bearing stiffeners should be designed as columns, assisted by a strip of web. The width of this strip may be taken as 25t at interior stiffeners and 12t at the end of the web. Effective length for l/r (slenderness ratio) should be at least 0.75 of the stiffener length. See Art. 2-12 for prevention of web crippling.

Splices ▪ Butt-welded splices should be complete-penetration groove welds and should develop the full strength of the smaller spliced section. Other types of splices in cross sections of plate girders should develop the strength required by the stresses at the point of splice but not less than 50% of the effective strength of the material spliced.

Flange Connections ▪ Rivets, high-strength bolts, or welds connecting flange to web, or cover plate to flange, should be proportioned to resist the total horizontal shear from bending.

The longitudinal spacing of the fasteners, in, may be determined from

$$p = \frac{R}{q} \qquad (2\text{-}48)$$

where R = allowable force, kips, on rivets, bolts, or welds that serve length p
q = horizontal shear, kips/in
For a rivet or bolt, $R = A_v F_v$, where A_v is the cross-sectional area, in², of the fastener and F_v the allowable shear stress, ksi. For a weld, R is the product of the length of weld, in, and allowable stress, kips/in. The horizontal shear may be computed from

$$q = \frac{VQ}{I} \qquad (2\text{-}49\,a)$$

where V = shear, kips, at point where pitch is to be determined
I = moment of inertia of section, in⁴
Q = static moment about neutral axis of flange cross-sectional area between outermost surface and surface at which horizontal shear is being computed, in³
Approximately,

$$q = \frac{V}{d} \frac{A}{A_f + A_w/6} \qquad (2\text{-}49\,b)$$

where d = depth of web, in, for welds between flange and web; distance between centers of gravity of tension and compression flanges, in, for bolts between flange and web; distance back to back of angles, in, for bolts between cover plates and angles
A = area of flange, in², for welds, rivets, and bolts between flange and web; area of cover plates only, in², for bolts and rivets between cover plates and angles
A_f = flange area, in²
A_w = web area, in²
If the girder supports a uniformly distributed load w, kips/in, on the top flange, the pitch should be determined from

$$p = \frac{R}{\sqrt{q^2 + w^2}} \qquad (2\text{-}50)$$

(See also Art. 2-12.)
   Maximum longitudinal spacing permitted in the compression-flange cover plates is 12 in or the thickness of the thinnest plate times $127/\sqrt{F_y}$ when fasteners are provided on all gage lines at each section or when intermittent welds are provided along the edges of the components. When rivets or bolts are staggered, the maximum spacing on each gage line should not exceed 18 in or the thickness of the thinnest plate times $190/\sqrt{F_y}$. Maximum spacing in tension-flange cover plates is 12 in or 24 times the thickness of the thinnest plate. Maximum spacing for connectors between flange angles and web is 24 in.

2-22. Design Criteria for Plate Girders in Bridges

For highway bridges, Table 2-21 gives critical web thicknesses t, in, for two grades of steel as a fraction of h, the clear distance, in, between flanges. When t is larger than the value in column 1, intermediate transverse (vertical) stiffeners are not required. If shear stress is less than the allowable, the web may be thinner. Thus, stiffeners may be omitted if $t \geq h\sqrt{f_v}/237$, where f_v = average unit shear, ksi (vertical shear at section, kips, divided by web cross-sectional area). But t should not be less than $h/150$.
   When t lies between the values in columns 1 and 2, transverse intermediate stiffeners are

TABLE 2-21 Minimum Web Thickness, in, for Highway-Bridge Plate Girders*

Yield strength, ksi	Without intermediate stiffeners (1)	Transverse stiffeners, no longitudinal stiffeners (2)	Longitudinal stiffener, transverse stiffeners (3)
36	$h/68$	$h/165$	$h/330$
50	$h/58$	$h/140$	$h/280$

*Standard Specifications for Highway Bridges, American Association of State Highway and Transportation Officials.

required. Webs thinner than the values in column 2 are permissible if they are reinforced by a longitudinal (horizontal) stiffener. If the computed maximum compressive bending stress f_b, ksi, at a section is less than the allowable bending stress, a longitudinal stiffener is not required if $t \geq h\sqrt{f_b}/727$; but t should not be less than $h/170$. When used, a plate longitudinal stiffener should be attached to the web at a distance $h/5$ below the inner surface of the compression flange. [See also Eqs. (2-65) and (2-66).]

Webs thinner than the values in column 3 are not permitted, even with transverse stiffeners and one longitudinal stiffener, unless the computed compressive bending stress is less than the allowable. When it is, t may be reduced to $h\sqrt{f_b}/1450$, but it should not be less than $h/340$.

Stiffener Spacing ▪ When transverse intermediate stiffeners are required, their spacing a_o must be such that the shear stress does not exceed the allowable shear stress F_v, ksi, computed from

$$F_v = \frac{F_y}{3}\left[C + \frac{0.87(1 - C)}{\sqrt{1 + (a_o/h)^2}} \right] \qquad (2\text{-}51\,a)$$

where F_y = yield strength, ksi, of the web steel.

$$C = \frac{2.2 \times 10^8[1 + (h/a_o)^2]}{F_y(h/t)^2} \leq 1 \qquad (2\text{-}51\,b)$$

The maximum spacing is limited to $1.5h$.

The spacing of the first intermediate stiffener at a simply supported end of a girder must be such that the end-panel shear stress does not exceed the value given by

$$F_v = \frac{70,000[1 + (h/a_o)^2]}{(h/t)^2} \leq \frac{F_y}{3} \qquad (2\text{-}51\,c)$$

In this case the maximum spacing is $0.5h$.

AASHTO also requires consideration of the interaction of shear and bending when the shear stress exceeds $0.6F_v$, by limiting the bending stress to $F_s = (0.754 - 0.34f_v/F_v)F_y$.

Stiffener Properties ▪ Intermediate stiffeners may be a single angle fastened to the web or a single plate welded to the web. But preferably they should be attached in pairs, one on each side of the web. Stiffeners on only one side of the web should be attached to the outstanding leg of the compression flange. At points of concentrated loading, stiffeners should be placed on both sides of the web and designed as bearing stiffeners.

The minimum moment of inertia, in⁴, of a transverse stiffener should be at least

$$I = a_o t^3 J \qquad (2\text{-}52\,a)$$

where $J = 2.5h^2/a_o^2 - 2 \geq 0.5$

    h = clear distance between flanges, in

    a = required stiffener spacing, in = h

    a_0 = actual stiffener spacing, in

    t = web thickness, in

For paired stiffeners, the moment of inertia should be taken about the center line of the web; for single stiffeners, about the face in contact with the web.

The width of an intermediate transverse stiffener, plate or outstanding leg of an angle, should be at least 2 in plus $\frac{1}{30}$ of the depth of the girder and preferably not less than one-fourth the width of the flange. Minimum thickness is $\frac{1}{16}$ of the width.

Transverse intermediate stiffeners should have a tight fit against the compression flange but need not be in bearing with the tension flange. The distance between the end of the stiffener weld and the near edge of the web-to-flange fillet weld should not be less than $4t$ or more than $6t$. However, if bracing or diaphragms are connected to an intermediate stiffener, care should be taken in design to avoid web flexing, which can cause premature fatigue failures.

Bearing Stiffeners ▪ These are required at all concentrated loads, including supports. Such stiffeners should be attached to the web in pairs, one on each side, and they should extend as nearly as practicable to the outer edges of the flanges. If angles are used, they should be proportioned for bearing on the outstanding legs of the flange angles or plates. (No allowance should be made for the portion of the legs fitted to the fillets of flange angles.) The stiffener angles should not be crimped.

Bearing stiffeners should be designed as columns. The allowable unit stress is given in Table 2-4, with $L = h$. For plate stiffeners, the column section should be assumed to consist of the plates and a strip of web. The width of the strip may be taken as 18 times the web thickness t for a pair of plates. For stiffeners consisting of four or more plates, the strip may be taken as the portion of the web enclosed by the plates plus a width of not more than $18t$. Minimum bearing-stiffener thickness is $(b'/12)\sqrt{F_y/33}$, where b' = stiffener width, in.

Bearing stiffeners must be ground to fit against the flange through which they receive their load or attached to the flange with full-penetration groove welds. But welding transversely across the tension flanges should be avoided to prevent creation of a severe fatigue condition.

Termination of Top Flange ▪ Upper corners of through-plate girders, where exposed, should be rounded to a radius consistent with the size of the flange plates and angles and the vertical height of the girder above the roadway. The first flange plate, or a plate of the same width, should be bent around the curve and continued to the bottom of the girder. In a bridge consisting of two or more spans, only the corners at the extreme ends of the bridge need to be rounded, unless the spans have girders of different heights. In such a case, the higher girders should have the top flanges curved down at the ends to meet the top corners of the girders in adjacent spans.

Seating at Supports ▪ Sole plates should be at least $\frac{3}{4}$ in thick. Ends of girders on masonry should be supported on pedestals so that the bottom flanges will be at least 6 in above the bridge seat.

Longitudinal Stiffeners ▪ These should be placed with the center of gravity of the fasteners $h/5$ from the toe, or inner face, of the compression flange. Moment of inertia, in^4, should be at least

$$I = ht^3 \left(2.4 \frac{a_0^2}{h^2} - 0.13 \right) \tag{2-52b}$$

where a_0 = actual distance between transverse stiffeners, in

t = web thickness, in

Thickness of stiffener, in, should be at least $b\sqrt{f_b}/71.2$, where b is the stiffener width, in, and f_b the flange compressive bending stress, ksi. The bending stress in the stiffener should not exceed the allowable for the material.

Longitudinal stiffeners usually are placed on one side of the web. They need not be continuous. They may be cut at their intersections with transverse stiffeners.

Splices ▪ These should develop the strength required by the stresses at the splices but not less than 75% of the effective strength of the material spliced. Splices in riveted flanges usually are avoided. In general, not more than one part of a girder should be spliced at the same cross section. Bolted web splices should have plates placed symmetrically on opposite sides of the web. Splice plates for shear should extend the full depth of the girder between flanges. At least two rows of bolts on each side of the joint should fasten the plates to the web.

Rivets, high-strength bolts, or welds connecting flange to web, or cover plate to flange, should be proportioned to resist the total horizontal shear from bending, as described for plate girders in buildings (*Flange Connections* in Art. 2-21). In riveted bridge girders, legs of angles 6 in or more wide connected to webs should have two lines of rivets. Cover plates over 14 in wide should have four lines of rivets.

See also Art. 2-20.

Hybrid Bridge Girders ▪ These may have flanges with larger yield strength than the web and may be composite or noncomposite with a concrete slab, or they may utilize an orthotropic-plate deck as the top flange. With composite or noncomposite girders, the web should have a yield strength at least 35% of the minimum yield strength of the tension flange. In noncomposite girders, both flanges should have the same yield strength. In composite girders, the compression flange may have a yield strength the same as that of the web. In girders with an orthotropic-plate deck, the yield strength of the web should be at least 35% of the bottom-flange yield strength in positive-moment regions and 50% in negative-moment regions.

Computation of bending stresses and allowable stresses is generally the same as for girders with uniform yield strength. Bending stress in the web, however, may exceed the allowable bending stress if the computed flange bending stress does not exceed the allowable stress multiplied by a factor R.

$$R = 1 - \frac{\beta\psi(1 - \alpha)^2(3 - \psi + \psi\alpha)}{6 + \beta\psi(3 - \psi)} \tag{2-53}$$

where α = ratio of web yield strength to flange yield strength

ψ = distance from outer edge of tension flange or bottom flange of orthotropic deck to neutral axis divided by depth of steel section

β = ratio of web area to area of tension flange or bottom flange of orthotropic-plate bridge

The rules for shear stresses are as previously described, except that for transversely stiffened girders, the allowable shear stress (throughout the length of the girder) is that given by Eq. (2-51c). Also, the limitation of the first stiffener spacing does not apply.

2-23. Dimensional Limitations on Steel Sections

Design specifications for structural steel include dimensional limitations on sections to prevent local buckling. AISC and AASHTO (Art. 2-4) stipulate maximum width-thickness (b/t) ratios for projecting elements under compression and depth-thickness ratios (d/t) for webs for various applications.

2-24. Deflection and Vibration Limitations

For buildings, beams and girders supporting plastered ceilings should not deflect under live load more than 1/360 of the span. To control deflection, fully stressed floor beams and girders should have a minimum depth of $F_y/800$ times the span, where F_y is the steel yield strength, ksi. Depth of fully stressed roof purlins should be at least $F_y/1000$ times the span, except for flat roofs, for which ponding conditions should be considered.

For bridges, simple-span or continuous girders should be designed so that deflection due to live load plus impact should not exceed 1/800 of the span. For bridges located in urban areas and used in part by pedestrians, however, deflection preferably should not exceed 1/1000 of the span. To control deflections, depth of noncomposite girders should be at least 1/25 of the span. For composite girders, overall depth, including slab thickness, should be at least 1/25 of the span, and depth of steel girder alone, at least 1/30 of the span. For continuous girders, the span for these ratios should be taken as the distance between inflection points.

In large open areas of buildings, where there are few partitions or other sources of damping, transient vibrations caused by pedestrian traffic may become annoying. The AISC suggests that perceptible vibration may be minimized by keeping depth of steel beams equal to or greater than 1/20 of the beam span.

2-25. Plastic-Design Criteria for Buildings

Plastic design (Art. 2-5) provides a method of selecting structural-steel members for load-carrying capacity rather than resistance to a particular stress. The AISC "Specification for Design, Fabrication, and Erection of Structural Steel for Buildings" (American Institute of Steel Construction) limits the method to steels with specified minimum yield strength up to 65 ksi.

The basic principles of plastic design are incorporated in the elastic-design provisions of the AISC Specification:

1. A 10% allowable stress increase is permitted for "compact" sections.
2. A 10% decrease in negative moment under gravity loads with a corresponding increase in positive moment is permitted in continuous construction.

Because of these adjustments in the elastic-design rules, some of the economies inherent in plastic design may be realized in elastic design. Plastic design, however, often requires less design time and is considered by most egineers to be more logical (Arts. 1-77 and 1-78).

Load Factors ▪ Using load factors by which service loads are multiplied, a uniform factor of safety (comparable with that of an elastically designed simple beam) is attained for plastic design for all conditions. These load factors are 1.7 for live and dead loads in combination with 1.3 times wind or earthquake loads.

Width-Thickness Ratios ▪ Plastic design is restricted to sections that meet certain dimensional limitations. These limitations were established to allow a member to yield sufficiently to form a plastic hinge without local buckling. The projecting-flange limits are listed in Table 2-22, where F_y is the steel yield strength, ksi; b_f is the flange width; and t_f the flange thickness.

Web Area ▪ Unless reinforced by diagonal stiffeners or a double plate, the webs of columns, beams, and girders must also be proportioned so that

$$V_u \leq 0.55 F_y dt \tag{2-54}$$

TABLE 2-22 Minimum Width-
Thickness Ratios for Plastic-Design
Beams

F_y, ksi	$b_f/2t_f$
36	8.5
42	8.0
45	7.4
50	7.0
55	6.6
60	6.3
65	6.0

where V_u = shear, kips
 d = depth of member, in
 t = web thickness, in
Web stiffeners are always required under concentrated loads if coincidental with a plastic hinge.

Connections ▪ In plastic design, connections also are designed on the ultimate-load basis. Rivets, welds, high-strength bolts, and A307 bolts are proportioned with a load factor of 1.7 applied to elastic unit stresses. When fasteners are used in tension, the minimum proof load is the design criterion.

Depth-Thickness Ratios ▪ Beam-depth limitation is given by Eq. (2-55) and is similar to that of elastic design of compact sections, except that the load factor of 1.7 has been incorporated.

$$\frac{d}{t} = \frac{412}{\sqrt{F_y}} \left(1 - 1.4 \frac{P}{P_y} \right) \le \frac{257}{\sqrt{F_y}} \qquad (2\text{-}55)$$

where P = factored axial load, kips
 $P_y = F_y A$, kips
 A = cross-sectional area of member, in^2

Slenderness Ratios ▪ The unbraced length of a beam's compression flange is critical and is given by Eqs. (2-56a) and (2-56b), which depend on the moment gradient applied to the beam. When $+1.0 > M/M_p > -0.5$,

$$\frac{l_c}{r_y} \le \frac{1375}{F_y} + 25 \qquad (2\text{-}56a)$$

When $-0.5 > M/M_p > -1.0$,

$$\frac{l_c}{r_y} \le \frac{1375}{F_y} \qquad (2\text{-}56b)$$

where l_c = length, in, of laterally unsupported compression flange
 r_y = radius of gyration, in, about weak axis
 M = lesser of end moments at ends of unbraced segment, ft-kips
 M_p = plastic moment, ft-kips = $F_y Z_x/12$
 Z_x = plastic section modulus, in^3 (See Art. 1-77 and tables for W shapes in AISC "Steel Construction Manual.")

The end moment ratio M/M_p is positive when the unbraced segment is bent in double curvature and negative when the segment is bent in single curvature.

Columns ▪ Under only axial load, columns are designed for a critical stress that is 1.7 times the allowable elastic axial stress F_a, computed as indicated in Art. 2-8. Thus, the factored axial load should not exceed

$$P_{cr} = 1.7AF_a \qquad (2\text{-}57)$$

For combined axial load and moment, however, columns should be designed to satisfy Eqs. (2-58a) and (2-58b).

$$\frac{P}{P_{cr}} + \frac{C_m M}{(1 - P/P_e)M_m} \leq 1 \qquad (2\text{-}58a)$$

$$\frac{P}{P_y} + \frac{M}{1.18M_p} \leq 1 \qquad M \leq M_p \qquad (2\text{-}58b)$$

where M = maximum applied moment, ft-kips
$P_e = (23/12)AF_e'$ (see Art. 2-11 for F_e')
C_m = coefficient defined in Art. 2-11
M_m = maximum moment that can be resisted by member without axial load, ft-kips
For columns braced about the weak axis,

$$M_m = M_p = F_y Z_x \qquad (2\text{-}59a)$$

For columns not braced about the weak axis,

$$M_m = \left[1.07 - \frac{(l/r_y)\sqrt{F_y}}{3160} \right] M_p \leq M_p \qquad (2\text{-}59b)$$

Bracing for Frames ▪ For plastic-design multistory frames under factored gravity loads (service loads times 1.7) or under factored gravity plus wind loads (service loads times 1.3), unbraced frames may be used if designed to preclude instability, including the effects of axial deformation of columns, without factored column axial loads exceeding $0.75AF_y$. Otherwise, frames should incorporate a vertical bracing system to maintain lateral stability. This vertical system may be used in selected braced bents that must carry not only horizontal loads directly applied to them but also the horizontal loads of unbraced bents. The latter loads may be transmitted through diaphragm action of the floor system.

Determination of Moments ▪ In multistory plastic design, the girders are designed as three-hinged mechanisms. The columns are designed for girder plastic moments distributed to the attached columns plus the moments due to girder shears at the column faces. Additional consideration should be given to moment-end rotation characteristics of the column above and the column below each joint.

("Plastic Design of Braced Multistory Steel Frames," American Institute of Steel Construction; R. Disque, "Applied Plastic Design in Steel," Van Nostrand Reinhold Company, New York.)

2-26. Load-Factor Design for Bridges

For bridges, AASHTO (Art. 2-4) permits load-factor design for simple and continuous beam-and-girder structures of moderate length. The design strength of the steel is taken as the specified minimum yield strength F_y, ksi, and the applied loads are factored to provide the appropriate safety factor.

Load Factors ▪ These are classified by groups. Group I is the normal design loading. For Group I:

$$\text{Load factor} = 1.30\left[D + \frac{5}{3}(L + \text{I})\right] \qquad (2\text{-}60\,a)$$

where D = dead-load effect (force, moment, stress, and so on)
L = live-load effect corresponding to D
I = impact effect corresponding to D
For bridges with less than H20 loading, Group IA is used to account for infrequent heavy loads. Live load is assumed to occur on only one lane.
For Group IA:

$$\text{Load factor} = 1.30[D + 2.2(L + I)] \qquad (2\text{-}60\,b)$$

Groups II through IX are load combinations that include the effects of wind, ice, earthquake, temperature, and so forth. The load factors and numerical load coefficients vary as given in the AASHTO Specification.

The preceding load factors are used with a variety of allowable stresses that depend on section used and design procedure employed.

Bending Strength ▪ For symmetrical beams and girders, there are three general types of members to consider: compact, braced noncompact, and unbraced sections. The maximum strength of each (moment, in-kips) depends on member dimensions and unbraced length as well as on applied shear and axial load (Table 2-23).

The maximum strengths given by the formulas in Table 2-23 apply only when the maximum

TABLE 2-23 Design Criteria for Symmetrical Flexural Sections for Load-Factor Design of Bridges

Type of section	Maximum bending strength M_u, in -kips	Flange minimum thickness t_f, in	Web minimum thickness t_w, in	Maximum unbraced length L_b, in	Maximum shear V, kips
Compact*	$F_y Z$	$\dfrac{b'\sqrt{F_y}}{50.6}$	$\dfrac{d\sqrt{F_y}}{421}$	$\dfrac{221 r_y}{\sqrt{F_y}}$ when $\dfrac{M_2}{M_1} \geq 0.7$ $\dfrac{379 r_y}{\sqrt{F_y}}$ when $\dfrac{M_2}{M_1} < 0.7$ and $\dfrac{20{,}000 A_f}{F_y d}$	$0.55 F_y d t_w$
Braced noncompact*	$F_y S$	$\dfrac{b'\sqrt{F_y}}{69.6}$	$\dfrac{h}{150}$	$\dfrac{20{,}000 A_f}{F_y d}$	$3.5 E t_w^3 /h$ but not more than $0.58 F_y h t_w$
Unbraced†	$F_y S\left[1 - R_u\left(\dfrac{L_b}{b'}\right)^2\right]$	$\dfrac{b'\sqrt{F_y}}{69.6}$	$\dfrac{h}{150}$		

*Straight-line interpolation between compact and braced noncompact moments may be used for intermediate criteria, except that $t_w \leq d\sqrt{F_y}/421$ should be maintained.
†If $M_2/M_1 < 0.7$, M_u may be increased by 20% but may not exceed $F_y S$.

axial stress does not exceed $0.15F_yA$, where A is the area of the member. Symbols used in **Table 2-23** are defined as follows:

F_y = steel yield strength, ksi
Z = plastic section modulus, in³ (See Art. 1-77.)
S = section modulus, in³
b' = width of projection of flange, in
d = depth of section, in
h = unsupported distance between flanges, in
M_2/M_1 = ratio of moments at braced points, where $M_1 \geq M_2$

$$R_u = \frac{3F_y}{4\pi^2 E} = \frac{1}{10,600} \text{ for } F_y = 36 \text{ ksi and } \frac{1}{7630} \text{ for } F_y = 50 \text{ ksi}$$

Transverse Stiffeners ▪ When the shear capacity exceeds the maximum shear given in Table 2-23, transverse stiffeners are required, and the minimum web thickness becomes $h\sqrt{F_y}/1154$. The shear capacity is given by

$$V_u = V_p\left[C + \frac{0.87(1 - C)}{\sqrt{1 + (d_o/h)^2}}\right] \tag{2-61}$$

where $V_p = 0.58F_yht_w$

$$C = \frac{569t_w}{h}\sqrt{\frac{1 + (h/d_o)^2}{F_y}} - 0.3 \leq 1$$

d_o = stiffener spacing, in
t_w = web thickness, in

If applied shear V and moment are both high and $V \geq 0.6V_u$, then the moment should not exceed

$$M = (1.375 - 0.625V/V_u)M_u \tag{2-62}$$

The stiffeners should have a moment of inertia of at least

$$I = d_o t_w^3 J \tag{2-63}$$

where $J = 2.5(h/d_o)^2 - 2 \geq 0.5$.
Stiffener area should be at least

$$A_{st} = \left[0.15Bht_w(1 - C)\frac{V}{V_u} - 18t_w^2\right]Y \tag{2-64}$$

where $Y = F_y \text{ (beam web)}/F_y \text{ (stiffener)}$
B = 1.0 for stiffener pairs
  = 1.8 for single angles
  = 2.4 for single plates

Longitudinal Stiffeners ▪ With longitudinal stiffeners at $h/5$ below the compression flange, the web may be as thin as $h\sqrt{F_y}/2308$, in. The stiffener moment of inertia should be at least

$$I_{st} = ht_w^3\left[2.4\left(\frac{d_o}{h}\right)^2 - 0.13\right] \tag{2-65}$$

and the stiffener radius of gyration, in, should be at least

$$r_{st} = \frac{d_o\sqrt{F_y}}{727} \tag{2-66}$$

where I_{st} and r_{st} refer to the stiffeners plus a centrally located strip of web not more than $18t_w$ in width.

Unsymmetrical and Other Types ▪ Unsymmetrical beams and girders as well as composite and hybrid sections used as bending members are also covered by AASHTO design rules in a similar manner to the preceding (see AASHTO "Standard Specifications for Highway Bridges").

Columns ▪ Compression members designed by load-factor design should have a maximum strength, kips,

$$P_u = 0.85 A_s F_{cr} \tag{2-67}$$

where A_s = gross effective area of column cross section, in².
For $KL_c/r \leq \sqrt{2\pi^2 E/F_y}$,

$$F_{cr} = F_y \left[1 - \frac{F_y}{4\pi^2 E} \left(\frac{KL_c}{r} \right)^2 \right] \tag{2-68 a}$$

For $KL_c/r > \sqrt{2\pi^2 E/F_y}$,

$$F_{cr} = \frac{\pi^2 E}{(KL_c/r)^2} \tag{2-69 a}$$

where F_{cr} = buckling stress, ksi
F_y = yield strength of the steel, ksi
K = effective length factor in plane of buckling
L_c = length of member between supports, in
r = radius of gyration in plane of buckling, in
E = modulus of elasticity of the steel, ksi

Equations (2-68 a) and (2-69 a) can be simplified by introducing a Q factor as was done in Art. 2-9. Define Q as

$$Q = \left(\frac{KL_c}{r} \right)^2 \frac{F_y}{2\pi^2 E} \tag{2-70}$$

Then, Eqs. (2-68 a) and (2-69 a) can be rewritten as follows:
For $Q \leq 1.0$:

$$F_{cr} = \left(1 - \frac{Q}{2} \right) F_y \tag{2-68 b}$$

For $Q > 1.0$:

$$F_{cr} = \frac{F_y}{2Q} \tag{2-69 b}$$

For combined axial load and bending, the interaction equations (2-71 a) and (2-71 b) must both be satisfied.

$$\frac{P}{0.85 A_s F_{cr}} + \frac{MC}{M_u[1 - (P/A_s F_e)]} \leq 1 \tag{2-71 a}$$

$$\frac{P}{0.85 A_s F_y} + \frac{M}{M_p} \leq 1 \tag{2-71 b}$$

where M_u = maximum strength of member in bending, in-kips

M = maximum moment, in-kips

P = maximum axial load, kips

$$F_e = \frac{\pi^2 E}{(KL_c/r)^2}$$

$M_p = F_y Z$

C = equivalent moment factor

  = $0.6 + 0.4(M_2/M_1) \geq 0.4$

M_1 and M_2 are the bending moments acting at the ends of the member and $M_1 \geq M_2$. M_2/M_1 is positive for single curvature and negative for double curvature.

2-27. Bracing

It usually is necessary to provide bracing for the main members or secondary members in most buildings and bridges.

In Buildings ▪ There are two general classifications of bracing for building construction: sway bracing for lateral loads and lateral bracing to increase the capacity of individual beams and columns.

Both low- and high-rise buildings require sway bracing to provide stability to the structure and to resist lateral loads from wind or seismic forces. This bracing can take the form of diagonal members or X bracing, knee braces, moment connections, and shear walls.

X bracing is probably the most efficient and economical bracing method. Fenestration or architectural considerations, however, often preclude it. This is especially true for high-rise structures.

Knee braces are often used in low-rise industrial buildings. They can provide local support to the column as well as stability for the overall structure.

Moment connections are frequently used in high-rise buildings. They can be welded, riveted, or bolted, or a combination of welds and bolts can be used. End-plate connections, with shop welding and field bolting, are an economical alternative. Figure 2-8 shows examples of various end-plate moment connections.

In many cases, moment connections may be used in steel frames to provide continuity and thus reduce the overall steel weight. This type of framing is especially suitable for welded construction; full moment connections made with bolts may be cumbersome and expensive.

In low buildings and the top stories of high buildings, moment connections may be designed to resist lateral forces alone. Although the overall steel weight is larger with this type of design, the connections are light and usually inexpensive.

Shear walls are also used to provide lateral bracing in steel-framed buildings. For this purpose, it often is convenient to reinforce the walls normally needed for fire walls, elevator shafts, divisional walls, and so on. Sometimes shear walls are used in conjunction with other forms of bracing.

Lateral bracing of columns, arches, beams, and trusses in building construction is used to

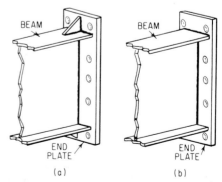

Fig. 2-8. End-plate connections for girders: (*a*) stiffened moment connection; (*b*) unstiffened moment connection.

reduce their critical or effective length, especially of those portions in compression. In floor or roof systems, for instance, it may be economical to provide a strut at midspan of long members to obtain an increase in the allowable stress for the load-carrying members. (See also Arts. 2-8 and 2-9 for effects on allowable stresses of locations of lateral supports.)

Usually, normal floor and roof decks can be relied on to provide sufficient lateral support to compression chords or flanges to warrant use of the full allowable compressive stress. Examples of cases where it might be prudent to provide supplementary support include purlins framed into beams well below the compression flange or precast-concrete planks inadequately secured to the beams.

In Bridges ▪ Bracing requirements for highway bridges are given in detail in "Standard Specifications for Highway Bridges," American Association of State Highway and Transportation Officials.

Through trusses require top and bottom lateral bracing. Top lateral bracing should be at least as deep as the top chord. Portal bracing of the two-plane or box type is required at the end posts and should take the full end reaction of the top-chord lateral system. In addition, sway bracing at least 5 ft deep is required at each intermediate panel point.

Deck-truss spans and spandrel arches also require top and bottom lateral bracing. Sway bracing, extending the full depth of the trusses, is required in the plane of the end posts and at all intermediate panel points. The end sway bracing carries the entire upper lateral stress to the supports through the end posts of the truss.

A special case arises with a half-through truss because top lateral bracing is not possible. The main truss and the floor beams should be designed for a lateral force of 300 lb/lin ft, applied at the top-chord panel points. The top chord should be treated as a column with elastic lateral supports at each panel point. The critical buckling force should be at least 50% greater than the maximum force from dead load, live load, and impact in any panel of the top chord.

Lateral bracing is not usually necessary for deck plate-girder or beam bridges. Most deck construction is adequate as top bracing, and substantial diaphragms (with depth preferably half the girder depth) or cross frames obviate the necessity of bottom lateral bracing. Cross frames are required at each end to resist lateral loads. In spans over 125 ft, lateral bracing should be placed near the bottom flange in at least one-third of the bays.

Through-plate girders should be stiffened against lateral deformation by gusset plates or knee braces attached to the floor beams. If the unsupported length of the inclined edge of a gusset plate exceeds 60 times the plate thickness, it should be stiffened with angles.

All highway bridges should be provided with cross frames or diaphragms spaced at a maximum of 25 ft.

("Structural Steel Detailing," American Institute of Steel Construction.)

2-28. Mechanical Fasteners

Rivets are produced from bar stock, and they come with one manufactured head. Riveted connections usually are made by driving the rivet hot (cherry-red) with pneumatic, hydraulic, or electric hammers, or guns. Large riveting machines are used for production-shop work. AASHTO (Art. 2-4) permits cold driving of rivets ⅜ in in diameter and less. Larger-size rivets should be heated to a light cherry-red for driving, and all rivets more than ⅞ in in diameter should be driven mechanically.

ASTM A502, Grade 1, rivets are usually used with A36 steel while A502, Grade 2, is most efficient with high-strength, low-alloy steels, such as A242 and A572. For A588 steel, special rivets with similar corrosion resistance, A502, Grade 3, are available and should be used. See also Art. 2-18.

For economic reasons, bolts or welds generally are preferred to rivets.

Unfinished bolts are used mainly in building construction where slip and vibration are not a factor. Characterized by a square head and nut, they also are known as machine, common, ordinary, or rough bolts. They are covered by ASTM A307 and are available in diameters over a wide range.

A325 bolts are identified by the notation A325. Additionally, Type 1 A325 bolts may optionally be marked with three radial lines 120° apart; Type 2 A325 bolts must be marked with three radial lines 60° apart; and Type 3 A325 bolts must have the A325 notation underlined. Heavy hexagonal nuts of the grades designated in A325 are manufactured and marked according to specification A563.

A490 bolts are identified by the notation A490. Additionally, Type 2 A490 bolts must be marked with six radial lines 30° apart, and Type 3 A490 bolts must have the A490 notation underlined. Heavy hexagonal nuts of the grades designated in A490 are manufactured and marked according to specification A563.

Bearing vs. Friction Connections ▪ Two different types of bolted connections are recognized for bridges and buildings: bearing and friction. Bearing-type connections are allowed higher shear stresses (Art. 2-18) and thus require fewer bolts. Friction-type connections offer greater resistance to repeated loads and therefore are used when connections are subjected to stress reversal or where slippage would be undesirable. To qualify for the higher allowable stresses in bearing-type connections, the bolt thread is excluded from the shear plane. Tests have demonstrated that the ultimate strength of both connections is about the same. Most building construction is done with bearing-type connections.

Bolt Tightening ▪ High-strength bolts are tightened by a calibrated wrench or by the "turn-of-the-nut" method. Calibrated wrenches are powered and have an automatic cutoff set for a predetermined torque. The turn-of-the-nut method requires snugging the plies together and then turning the nut a specified amount. From one-third to one turn is specified; increasing amounts of turn are required for long bolts or for bolts connecting parts with slightly sloped surfaces. Alternatively, a direct tension indicator, such as a load-indicating washer, may be used. This type of washer has on one side raised surfaces which when compressed to a predetermined height (0.005 in measured with a feeler gage) indicate attainment of required bolt tension. The "Specification for Structural Steel Joints Using A325 or A490 Bolts," adopted by

TABLE 2-24 Oversized-and Slotted-Hole Limitations for Structural Joints with A325 and A490 Bolts

Bolt diameter, in	Maximum hole size, in*		
	Oversize holes†	Short slotted holes‡	Long slotted holes‡
⅝	¹³⁄₁₆	¹¹⁄₁₆ × ⅞	¹¹⁄₁₆ × 1 ⁹⁄₁₆
¾	¹⁵⁄₁₆	¹³⁄₁₆ × 1	¹³⁄₁₆ × 1 ⅞
⅞	1¹⁄₁₆	¹⁵⁄₁₆ × 1 ⅛	¹⁵⁄₁₆ × 2 ³⁄₁₆
1	1¼	1 ¹⁄₁₆ × 1 ⁵⁄₁₆	1 ¹⁄₁₆ × 2 ½
1 ⅛	1⁷⁄₁₆	1 ³⁄₁₆ × 1 ½	1 ³⁄₁₆ × 2 ¹³⁄₁₆
1 ¼	1⁹⁄₁₆	1 ⁵⁄₁₆ × 1 ⅝	1 ⁵⁄₁₆ × 3 ⅛
1 ⅜	1¹¹⁄₁₆	1 ⁷⁄₁₆ × 1 ¾	1 ⁷⁄₁₆ × 3 ⁷⁄₁₆
1 ½	1¹³⁄₁₆	1 ⁹⁄₁₆ × 1 ⅞	1 ⁹⁄₁₆ × 3 ¾

*In friction-type connections, a lower allowable shear stress, as given by AISC, should be used for the bolts.
†Not allowed in bearing-type connections.
‡In bearing-type connections, slot must be perpendicular to direction of load application.

the American Institute of Steel Construction, gives detailed specifications for both tightening methods.

Overtightening a bolt is usually not a serious problem. The bolt works just as well in the plastic as in the elastic range. Excessive overtightening will cause failure, but the operator need only replace the bolt.

Undertightening will result in insufficient friction in a friction-type connection or eventual backing off of the nut and may lead to failure of the connection.

Holes ▪ These generally should be ¹⁄₁₆ in larger than the nominal fastener diameter. Oversize and slotted holes may be used subject to the limitations of Table 2-24.

Symbols used to represent fasteners on drawings are presented in Art. 2-31.

("Structural Steel Detailing," American Institute of Steel Construction.)

2-29. Welded Connections

Welding, a method of joining steel by fusion, is used extensively in both buildings and bridges. It usually requires less connection material than other methods. No general rules are possible regarding the economics of the various connection methods; each job must be individually analyzed.

Although there are many different welding processes, shielded-arc welding is used almost exclusively in construction. Shielding serves two purposes: It prevents the molten metal from oxidizing and it acts as a flux to cause impurities to float to the surface.

In manual arc welding, an operator maintains an electric arc between a coated electrode and the work. Its advantage lies in its versatility; a good operator can make almost any type of weld. It is used for fitting up as well as for finished work. The coating turns into a gaseous shield, protecting the weld and concentrating the arc for greater penetrative power.

Automatic welding, generally the submerged-arc process, is used in the shop, where long lengths of welds in the flat position are required. In this method, the electrode is a base wire (coiled) and the arc is protected by a mound of granular flux fed to the work area by a separate flux tube. Most welded bridge girders are fabricated by this method, including the welding of transverse stiffeners. Other processes, such as flux-cored arc welding, are also used.

There are basically two types of welds: fillet (Fig. 2-9) and groove (Fig. 2-10). AISC (Art. 2-4) permits partial-penetration groove welds (Fig. 2-11) with a reduction in allowable stress. AASHTO (Art. 2-4) does not recognize partial penetration groove welds for bridges. Allowable stresses for welds in buildings and bridges are presented in Art. 2-17.

("Structural Steel Detailing," American Institute of Steel Construction.)

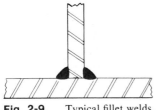

Fig. 2-9.   Typical fillet welds.

Fig. 2-10.   Typical complete-penetration groove weld.

Fig. 2-11.   Typical partial-penetration groove weld.

2-30. Combinations of Fasteners

In new construction, different types of fasteners (rivets, bolts, or welds) are generally not combined to share the same load because varying amounts of deformation are required to load the different fasteners properly. AISC (Art. 2-4) permits one exception to this rule: Friction-type high-strength bolt connections may be used with welds if the bolts are tightened prior to welding.

In an existing building, a connection may be reinforced for dead load by combining rivets and high-strength bolts, and for dead and live loads, welds may be used. This assumes that no slip is likely beyond that which has already occurred (see also Art. 2-28).

2-31. Fastener Symbols

These are used to denote the type and size of rivets, bolts, and welds on design drawings as well as on shop and erection drawings. The practice for buildings and bridges is similar.

Figure 2-12 shows the conventional signs for rivets and bolts. Figure 2-13 shows standard welding symbols, as recommended by the American Welding Society.

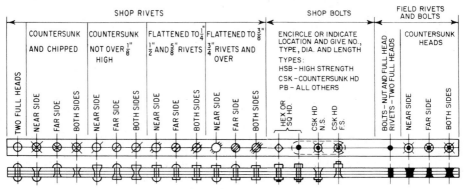

Fig. 2-12. Conventional symbols for rivets and bolts.

Fig. 2-13 (opposite). Symbols for welded joints recommended by the American Welding Society. Size, weld symbol, length of weld, and spacing should read in that order from left to right along the reference line, regardless of its orientation or arrow location. The perpendicular leg of symbols for fillet, bevel, J, and flare-bevel-groove welds should be at left. Arrow and Other Side welds should be the same size. Symbols apply between abrupt changes in direction of welding unless governed by the "all-around" symbol or otherwise dimensioned. When billing of detail material discloses the identity of the far side of a member (such as a stiffened web or a truss gusset) with the near side, welding shown for the near side should also be duplicated on the far side.

BACK	FILLET	PLUG OR SLOT	GROOVE OR BUTT						
			SQUARE	V	BEVEL	U	J	FLARE V	FLARE BEVEL

BASIC WELD SYMBOLS

WELD ALL AROUND	FIELD WELD	CONTOUR	
		FLUSH	CONVEX

SUPPLEMENTARY WELD SYMBOLS

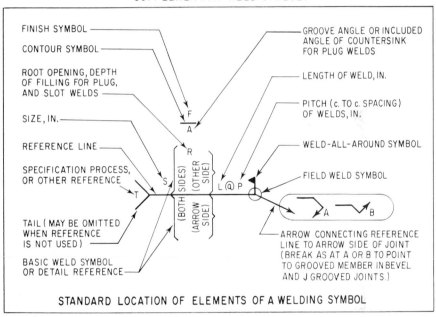

STANDARD LOCATION OF ELEMENTS OF A WELDING SYMBOL

Concrete made with portland cement is widely used as construction material because of its many favorable characteristics. One of the most important is a high strength-cost ratio in many applications. Another is that concrete, while plastic, may be cast in forms easily at ordinary temperatures to produce almost any desired shape. The exposed face may be developed into a smooth or rough hard surface, capable of withstanding the wear of truck or airplane traffic, or it may be treated to create desired architectural effects. In addition, concrete has high resistance to fire and penetration of water.

But concrete also has disadvantages. An important one is that quality control sometimes is not so good as for other construction materials because concrete often is manufactured in the field under conditions where responsibility for its production cannot be pinpointed. Another disadvantage is that concrete is a relatively brittle material—its tensile strength is small compared with its compressive strength. This disadvantage, however, can be offset by reinforcing or prestressing concrete with steel. The combination of the two materials, reinforced concrete, possesses many of the best properties of each and finds use in a wide variety of constructions, including building frames, floors, roofs, and walls; bridges; pavements; piles; dams; and tanks.

Section 3

Concrete Design and Construction*

Charles H. Thornton
Consulting Engineer; President
Lev Zetlin Associates, Inc.
New York, NY

I. Paul Lew
Consulting Engineer; Vice President
Lev Zetlin Associates, Inc.
New York, NY

3-1. Important Properties of Concrete

The characteristics of portland cement concrete can be varied to a considerable extent by controlling its ingredients. Thus, for a specific structure, it is economical to use a concrete that has exactly the characteristics needed, though weak in others. For example, concrete for a building frame should have high compressive strength, whereas concrete for a dam should be durable and watertight, and strength can be relatively small.

*Revised from Sec. 8, "Concrete Design and Construction," Lev Zetlin, President, Zetlin-Argo Guidance & Liaison Corp., New York, N.Y., and Donald Griff (deceased), Vice President, Lev Zetlin Associates, Inc., New York, N.Y., in "Standard Handbook for Civil Engineers," 2d ed., F.S. Merritt (ed.), McGraw-Hill Book Company, New York.

Workability is an important property for many applications of concrete. Difficult to evaluate, workability is essentially the ease with which the ingredients can be mixed and the resulting mix handled, transported, and placed with little loss in homogeneity. One characteristic of workability that engineers frequently try to measure is consistency, or fluidity. For this purpose, they often make a slump test.

In the slump test, a specimen of the mix is placed in a mold shaped as the frustum of a cone, 12 in high, with 8-in diameter base and 4-in diameter top (ASTM Specification C143). When the mold is removed, the change in height of the specimen is measured. When the test is made in accordance with the ASTM Specification, the change in height may be taken as the slump. (As measured by this test, slump decreases as temperature increases; thus the temperature of the mix at time of test should be specified, to avoid erroneous conclusions.)

Tapping the slumped specimen gently on one side with a tamping rod after completing the test may give additional information on the cohesiveness, workability, and placeability of the mix ("Concrete Manual," Bureau of Reclamation, Government Printing Office, Washington, D.C. 20402). A well-proportioned, workable mix settles slowly, retaining its original identity. A poor mix crumbles, segregates, and falls apart.

Slump of a given mix may be increased by adding water, increasing the percentage of fines (cement or aggregate), entraining air, or incorporating an admixture that reduces water requirements. But these changes affect other properties of the concrete, sometimes adversely. In general, the slump specified should yield the desired consistency with the least amount of water and cement.

Durability is another important property of concrete. Concrete should be capable of withstanding the weathering, chemical action, and wear to which it will be subjected in service. Much of the weather damage sustained by concrete is attributable to freezing and thawing cycles. Resistance of concrete to such damage can be improved by increasing the watertightness, by entraining 2 to 6% air, by using an air-entraining agent, or by applying a protective coating to the surface.

Chemical agents, such as inorganic acids, acetic and carbonic acids, and sulfates of calcium, sodium, magnesium, potassium, aluminum, and iron, disintegrate or damage concrete. When contact between these agents and concrete may occur, the concrete should be protected with a resistant coating; for resistance to sulfates, Type V portland cement may be used. (See "Concrete Manual," Bureau of Reclamation.) Resistance to wear usually is achieved by use of a high-strength, dense concrete made with hard aggregates.

Watertightness is an important property of concrete that can often be improved by reducing the amount of water in the mix. Excess water leaves voids and cavities after evaporation, and if they are interconnected, water can penetrate or pass through the concrete. Entrained air (minute bubbles) usually increases watertightness, as does prolonged thorough curing.

Volume change is another characteristic of concrete that should be taken into account. Expansion due to chemical reactions between the ingredients of concrete may cause buckling and drying shrinkage may cause cracking.

Expansion due to alkali-aggregate reaction can be avoided by selecting nonreactive aggregates. If reactive aggregates must be used, expansion may be reduced or eliminated by adding pozzolanic material, such as fly ash, to the mix. Expansion due to heat of hydration of cement can be reduced by keeping cement content as low as possible, using Type IV cement and chilling the aggregates, water, and concrete in the forms. Expansion due to increases in air temperature may be decreased by producing concrete with a lower coefficient of expansion, usually by using coarse aggregates with a lower coefficient of expansion.

Drying shrinkage can be reduced principally by cutting down on water in the mix. But less cement also will reduce shrinkage, as will adequate moist curing. Addition of pozzolans, however, unless enabling a reduction in water, will increase drying shrinkage.

Autogenous volume change, a result of chemical reaction and aging within the concrete and usually shrinkage rather than expansion, is relatively independent of water content. This type of shrinkage may be decreased by using less cement, and sometimes by using a different cement.

Whether volume change will damage the concrete often depends on the restraint present. For example, a highway slab that cannot slide on the subgrade while shrinking may crack; a building floor that cannot contract because it is anchored to relatively stiff girders also may crack. Hence, consideration should always be given to eliminating restraints or resisting the stresses they may cause.

Strength is a property of concrete that nearly always is of concern. Usually, it is determined by the ultimate strength of a specimen in compression, but sometimes flexural or tensile capacity is the criterion. Since concrete usually gains strength over a long period of time, the compressive strength at 28 days is commonly used as a measure of this property. In the United States, it is general practice to determine the compressive strength of concrete by testing specimens in the form of standard cylinders made in accordance with ASTM Specification C192 or C31. C192 is intended for research testing or for selecting a mix (laboratory specimens). C31 applies to work in progress (field specimens). The tests should be made as recommended in ASTM C39. Sometimes, however, it is necessary to determine the strength of concrete by taking drilled cores; in that case, ASTM C42 should be adopted. (See also American Concrete Institute Standard 214, "Recommended Practice for Evaluation of Strength Test Results of Concrete.")

The 28-day compressive strength of concrete can be estimated from the 7-day strength by a formula proposed by W. A. Slater (*Proceedings of the American Concrete Institute*, 1926):

$$S_{28} = S_7 + 30\sqrt{S_7} \tag{3-1}$$

where S_{28} = 28-day compressive strength, psi
S_7 = 7-day strength, psi

Concrete may increase significantly in strength after 28 days, particularly when cement is mixed with fly ash. Therefore, specification of strengths at 56 or 90 days is appropriate in design.

Concrete strength is influenced chiefly by the water-cement ratio; the higher this ratio, the lower the strength. In fact, the relationship is approximately linear when expressed in terms of the variable C/W, the ratio of cement to water by weight: For a workable mix,

$$S_{28} = 2700\frac{C}{W} - 760 \tag{3-2}$$

Strength may be increased by decreasing water-cement ratio, using higher-strength aggregates, grading the aggregates to produce a smaller percentage of voids in the concrete, moist curing the concrete after it has set, adding a pozzolan, such as fly ash, vibrating the concrete in the forms, and sucking out excess water with a vacuum from the concrete in the forms. The short-time strength may be increased by using Type III (high-early-strength) portland cement and accelerating admixtures, such as calcium chloride, and by increasing curing temperatures, but long-time strengths may not be affected. Strength-increasing admixtures generally accomplish their objective by reducing water requirements for the desired workability. Concrete strengths as high as 11,000 psi have been used in buildings.

The stress-strain diagram for concrete of a specified compressive strength is a curved line (Fig. 3-1). Maximum stress is reached at a strain of 0.002 in/in, after which the curve descends.

The modulus of elasticity generally used in design for concrete is a secant modulus. In ACI 318-7, "Building Code Requirements for Reinforced Concrete," it is determined by

$$E_c = w^{1.5}33\sqrt{f_c'} \tag{3-3a}$$

where w = weight of concrete, lb/ft^3
f_c' = specified compressive strength at 28 days, psi

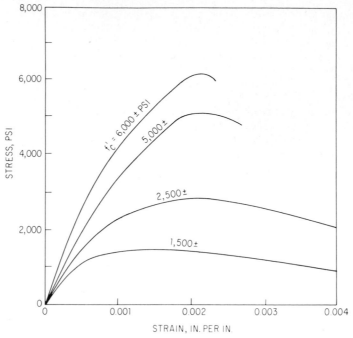

Fig. 3-1.   Stress-strain curves for concrete.

For ordinary concrete, with $w = 145$ lb/ft^3,

$$E_c = 57,000 \sqrt{f'_c} \qquad\qquad (3\text{-}3\,b)$$

The modulus increases with age, as does the strength.

Tensile strength of concrete is much lower than compressive strength and regardless of the types of test usually has poor correlation with f'_c. As determined in flexural tests, the tensile strength (modulus of rupture—not the true strength) is about $7\sqrt{f'_c}$ for the higher-strength concretes and $10\sqrt{f'_c}$ for the lower-strength concretes.

Creep is strain that occurs under a constant long-time load. The concrete continues to deform, but at a rate that diminishes with time. It is approximately proportional to the stress at working loads and increases with increasing water-cement ratio. It decreases with increase in relative humidity. In design of reinforced-concrete beams for allowable stress, the effects of creep are taken into account by reducing the modulus of elasticity of the concrete, usually by 50%. In design of prestressed-concrete beams, creep may be taken as 100% of the elastic strain for concrete in a very humid atmosphere to 300% for concrete in a very dry atmosphere. Part of the creep is recoverable on removal of the load.

Weight per cubic foot of ordinary sand-and-gravel concrete usually is about 145 lb. It may be slightly lower if the maximum size of coarse aggregate is less than 1½ in. It can be increased by using denser aggregate, and it can be decreased by using lightweight aggregate, increasing the air content, or incorporating a foaming, or expanding, admixture.

(G. E. Troxell, H. E. Davis, and J. W. Kelly, "Composition and Properties of Concrete," McGraw-Hill Book Company, New York; D. F. Orchard, "Concrete Technology," John Wiley & Sons, Inc., New York; J. J. Waddell, "Concrete Construction Handbook," McGraw-Hill Book Company, New York.)

3-2. Types of Reinforcing Steel

Because of the low tensile strength of concrete, steel is embedded in it to resist tensile stresses. Steel, however, also is used to take compression, in beams and columns, to permit use of smaller members. It serves other purposes too: It controls strains due to temperature and shrinkage and distributes load to the concrete and other reinforcing steel; it can be used to prestress the concrete; and it ties other reinforcing together for easy placement or to resist lateral stresses.

Most reinforcing is in the form of bars or wires whose surfaces may be smooth or deformed. The latter type is generally used because it produces better bond with the concrete because of the raised pattern on the steel.

Bars range in diameter from ¼ to 2¼ in. Sizes are designated by numbers, which are approximately eight times the nominal diameters. (See the latest edition of ASTM "Specifications for Steel Bars for Concrete Reinforcement." These also list the minimum yield points and tensile strengths for each type of steel.) Use of bars with yield points over 60,000 psi for flexural reinforcement is limited because special measures are required to control cracking and deflection.

Wires usually are used for reinforcing concrete pipe and, in the form of welded-wire fabric, for slab reinforcement. The latter consists of a rectangular grid of uniformly spaced wires, welded at all intersections, and meeting the minimum requirements of ASTM A185 and A497. Fabric offers the advantages of easy, fast placement of both longitudinal and transverse reinforcement and excellent crack control because of high mechanical bond with the concrete. (Deformed wires are designated by D followed by a number equal to the nominal area, in^2, times 100.) Bars and rods also may be prefabricated into grids, by clipping or welding (ASTM A184).

Sometimes, metal lath is used for reinforcing concrete, for example, in thin shells. It may serve as both form and reinforcing when concrete is applied by spray (gunite or shotcrete.)

3-3. Bending and Placing Reinforcing Steel

Bars are shipped by a mill to a fabricator in uniform long lengths and in bundles of 5 or more tons. The fabricator transports them to the job straight and cut to length or cut and bent.

Bends may be required for beam-and-girder reinforcing, longitudinal reinforcing of columns where they change size, stirrups, column ties and spirals, and slab reinforcing. Dimensions of standard hooks and typical bends and tolerances for cutting and bending are given in ACI 315, "Manual of Standard Practice for Detailing Reinforced Concrete Structures," American Concrete Institute.

Some preassembling of reinforcing steel is done in the fabricating shop or on the job. Beam, girder, and column steel often is wired into frames before placement in the forms. Slab reinforcing may be clipped or welded into grids, or mats, if not supplied as welded-wire fabric.

Some rust is permissible on reinforcing if it is not loose and there is no appreciable loss of cross-sectional area. In fact, rust, by creating a rough surface, will improve bond between the steel and concrete. But the bars should be free of loose rust, scale, grease, oil, or other coatings that would impair bond.

Bars should not be bent or straightened in any way that will damage them. If heat is necessary for bending, the temperature should not be higher than that indicated by a cherry-red color (1200°F), and the steel should be allowed to cool slowly, not quenched, to 600°F.

Reinforcing should be supported and tied in the locations and positions called for in the plans. The steel should be inspected before concrete is placed. Neither the reinforcing nor other parts to be embedded should be moved out of position before or during the casting of the concrete.

Bars and wire fabric should not be kinked or have unspecified curvatures when positioned. Kinked and curved bars, including those misshaped by workers walking on them, may cause the hardened concrete to crack when the bars are tensioned by service loads.

Usually, reinforcing is set on wire bar supports, preferably galvanized for exposed surfaces. Lower-layer bars in slabs usually are supported on bolsters consisting of a horizontal wire welded to two legs about 5 in apart. The upper layer generally is supported on bolsters with runner wires on the bottom so that they can rest on bars already in place. Or individual or continuous high chairs can be used to hold up a support bar, often a No. 5, at appropriate intervals, usually 5 ft. An individual high chair is a bar seat that looks roughly like an inverted U braced transversely by another inverted U in a perpendicular plane. A continuous high chair consists of a horizontal wire welded to two inverted-U legs 8 or 12 in apart. Beam and joist chairs have notches to receive the reinforcing. These chairs usually are placed at 5-ft intervals.

Although it is essential that reinforcement be placed exactly where called for in the plans, some tolerances are necessary. Reinforcement in beams and slabs should be within ±¼ in of the specified distance from the tension or compression face. Lengthwise, a cutting tolerance of ±1 in and a placement tolerance of ±2 in are normally acceptable. If length of embedment is critical, the designer should specify bars 3 in longer than the computed minimum to allow for accumulation of tolerances. Spacing of reinforcing in wide slabs and tall walls may be permitted to vary ±½ in or slightly more if necessary to clear obstructions, so long as the required number of bars are present.

Lateral spacing of bars in beams and columns, spacing between multiple reinforcement layers, and concrete cover over stirrups, ties, and spirals in beams and columns should never be less than that specified but may exceed it by ¼ in. A variation in setting of an individual stirrup or column hoop of 1 in may be acceptable, but the error should not be permitted to accumulate. ("CRSI Recommended Practice for Placing Reinforcing Bars," and "Manual of Standard Practice," Concrete Reinforcing Steel Institute, 933 N. Plum Grove Rd, Schaumburg, IL 60195.)

3-4. Spacing, Splicing, and Cover for Reinforcing

In buildings, the minimum clear distance between parallel bars should be 1 in for bars up to No. 8 and the nominal bar diameter for larger bars. For columns, however, the clear distance between longitudinal bars should be at least 1.5 in for bars up to No. 8 and 1.5 times the nominal bar diameter for larger bars. And the clear distance between multiple layers of reinforcement in building beams and girders should be at least 1 in; upper-layer bars should be directly above corresponding bars below. These minimum-distance requirements also apply to the clear distance between a contact splice and adjacent splices or bars.

A common requirement for minimum clear distance between parallel bars in highway bridges is 1.5 times the diameter of the bars, and spacing center to center should be at least 1.5 times the maximum size of coarse aggregate.

Many codes and specifications relate the minimum bar spacing to maximum size of coarse aggregate. This is done with the intention of providing enough space for all the concrete mix to pass between the reinforcing. But if there is a space to place concrete between layers of steel and between the layers and the forms, and the concrete is effectively vibrated, experience has shown that bar spacing or form clearance does not have to exceed the maximum size of coarse aggregate to insure good filling and consolidation. That portion of the mix which is molded by vibration around bars, and between bars and forms, is not inferior to that which would have filled those parts had a larger bar spacing been used. The remainder of the mix in the interior, if consolidated layer after layer, is superior because of its reduced mortar and water content ("Concrete Manual," U.S. Bureau of Reclamation, Government Printing Office, Washington, D.C. 20402).

Bundled Bars ▪ Groups of parallel reinforcing bars bundled in contact to act as a unit may be used only when they are enclosed by ties or stirrups. Four bars are the maximum permitted in a bundle, and all must be deformed bars. If full-length bars cannot be used between supports, then there should be a stagger of at least 40 bar diameters between any discontinuities. Also, the length of lap should be increased 20% for a three-bar bundle and 33% for a four-bar bundle. In determining minimum clear distance between a bundle and parallel reinforcing, the bundle should be treated as a single bar of equivalent area.

Maximum Spacing ▪ In walls and slabs in buildings, except for concrete-joist construction, maximum spacing, center to center, of principal reinforcement should be 18 in, or three times the wall or slab thickness, which ever is smaller.

Tension Development Lengths ▪ For bars in tension, the basic development length l_d, in for No. 11 and smaller bars is defined as

$$l_d = \frac{0.04 A_b f_y}{\sqrt{f'_c}} \tag{3-4 a}$$

where A_b = area of bar, in^2
f_y = yield strength of bar steel, psi
f'_c = 28-day compressive strength of concrete, psi
But l_d should not be less than $0.0004 d_b f_y$, where d_b = bar diameter, in, or less than 12 in, except in computation of lap splices or web anchorage.
For No. 14 bars,

$$l_d = 0.085 \frac{f_y}{\sqrt{f'_c}} \tag{3-4 b}$$

For No. 18 bars,

$$l_d = 0.11 \frac{f_y}{\sqrt{f'_c}} \tag{3-4 c}$$

and for deformed wire,

$$l_d = 0.03 d_b \frac{f_y}{\sqrt{f'_c}} \tag{3-4 d}$$

For top reinforcement, with horizontal bars so placed that more than 12 in of concrete is cast below the bars, the basic development length should be increased by 40%.
Development lengths for tension bars are listed in Table 3-1 a.
Added development length is also required where f_y exceeds 60 ksi or when lightweight concrete is used.
Where reinforcement bars being developed are spaced laterally at least 6 in on centers and at least 3 in from the side face of a member, the required development length l_d may be reduced by 20%.
Where bars are enclosed within a spiral formed by a bar at least ¼ in in diameter and with not more than a 4-in pitch, the required development length l_d may be reduced by 25%.

TABLE 3-1 Minimum Development Lengths l_d, in*

a. Tension Development in Normal-Weight Concrete†

Bar size no.	$f'_c = 3{,}000$ psi		$f'_c = 3{,}750$ psi		$f'_c = 4{,}000$ psi		$f'_c = 5{,}000$ psi		$f'_c = 6{,}000$ psi	
	Top‡ bars	Other bars	Top‡ bars	Other bars	Top‡ bars	Other bars	Top‡ bars	Other bars	Top‡ bars	Other bars
3	13	12	13	12	13	12	13	12	13	12
4	17	12	17	12	17	12	17	12	17	12
5	21	15	21	15	21	15	21	15	21	15
6	27	19	25	18	25	18	25	18	25	18
7	37	26	33	24	32	23	29	21	29	21
8	48	35	43	31	42	30	38	27	34	25
9	61	44	55	39	53	38	48	34	43	31
10	78	56	70	50	67	48	60	43	55	39
11	96	68	86	61	83	59	74	53	68	48
14	130	93	117	83	113	81	101	72	92	66
18	169	120	151	108	146	104	131	93	119	85

b. Compression Development in Normal-Weight Concrete§

Bar size no.	f'_c (Normal-weight concrete)			
	3,000 psi	3,750 psi	4,000 psi	Over 4,444 psi¶
3	8	8	8	8
4	11	10	10	9
5	14	12	12	11
6	17	15	14	14
7	19	17	17	16
8	22	20	19	18
9	25	22	22	20
10	28	25	24	23
11	31	28	27	25
14	37	33	32	31
18	50	44	43	41

*Courtesy Concrete Reinforcing Steel Institute.
†1. For bars enclosed in standard column spirals, use $0.75l_d$.
2. For bars, such as usual temperature bars, spaced 6 in or more, use $0.8l_d$.
3. Longer embedments for lightweight concrete are generally required, depending on the tensile splitting strength f_{ct}.
4. Standard 90 or 180° end hooks may be used to replace part of the required embedment.
‡ Horizontal bars with more than 12 in of concrete below.
§For embedments enclosed by spirals use 0.75 length given in *a*, above but not less than 8 in.
¶For $f'_c > 4444$ psi minimum embedment $= 18d_b$.

Compression Development Lengths ▪ For bars in compression, the basic development length l_d is defined as

$$l_d = \frac{0.02 f_y d_b}{\sqrt{f'_c}} \qquad (3\text{-}5)$$

but l_d should not be less than 8 in or $0.0003 f_y d_b$. See Table 3-1*b*.

For f_y greater than 60 ksi or concrete strengths less than 3000 psi, the required development length should be increased.

Bar Lap Splices ▪ Because of the difficulty of transporting very long bars, reinforcement cannot always be continuous. When splices are necessary, it is advisable that they should be made where the tensile stress is less than half the permissible stress.

Bars up to No. 11 in size may be spliced by overlapping them and wiring them together.

Bars spliced by noncontact lap splices in flexural members should not be spaced transversely farther apart than one-fifth the required lap length or 6 in.

Welded Splices ▪ These other positive connections should be used for bars larger than No. 11 and are an acceptable alternative for smaller bars. Welding should conform to AWS D12.1, "Reinforcing Steel Welding Code," American Welding Society, 2501 N.W. 7th St., Miami, Fla., 33125. Bars should be butted and welded so that the splice develops in tension at least 125% of their specified yield strength. Other positive connections should be equivalent in strength.

Tension Lap Splices ▪ The length of lap for bars in tension should conform to the following, with l_d taken as the tensile development length for the full yield strength f_y of the reinforcing steel [Eq. (3-4)] :

Class A splices (lap of l_d) are permitted where the maximum computed stress in the bar is always less than $0.5f_y$ and no more than three-fourths of the bars are lap spliced within a required lap length.

Class B splices (lap of $1.3l_d$) are required where either

1. The maximum computed stress in the bar is always less than $0.5f_y$ and more than three-fourths of the bars are lap spliced within a required lap length, or

2. Splices are used in regions of maximum moment and no more than one-half of the bars are spliced within a required lap length.

Class C splices (lap of $1.7l_d$) are required in regions of maximum moment where more than one-half of the bars are lap spliced within a lap length.

Splices for "tension tie members" should be fully welded or full mechanical connections and should be staggered at least $1.7l_d$. Where feasible, splices in regions of high stress also should be staggered and made with full welded connections.

Compression Lap Splices ▪ For a bar in compression, the minimum length of a lap splice should be the largest of the development length l_d given by Eq. (3-5), or 12 in, or $0.0005f_y d_b$, for f'_c of 3000 psi or larger and steel yield strength f_y of 60 ksi or less, where d_b is the bar diameter.

For tied compression members where the ties have an area, in^2, of at least $0.0015hs$ in the vicinity of the lap, the lap length may be reduced to 83% of the preceding requirements but not to less than 12 in (h is the overall thickness of the member, in, and s is the tie spacing, in).

For spirally reinforced compression members, the lap length may be reduced to 75% of the basic required lap but not to less than 12 in.

In columns where reinforcing bars are offset and one bar of a splice has to be bent to lap and contact the other one, the slope of the bent bar should not exceed 1 in 6. Portions of the bent bar above and below the offset should be parallel to the column axis. The design should account for a horizontal thrust at the bend taken equal to at least 1.5 times the horizontal component of the nominal stress in the inclined part of the bar. This thrust should be resisted

TABLE 3-2 Cast-in-Place Concrete Cover for Steel Reinforcement (Nonprestressed)

1. Concrete deposited against the ground, 3 in.
2. Concrete exposed to seawater, 4 in ; except precast-concrete piles, 3 in.
3. Concrete exposed to the weather or in contact with the ground after form removal, 2 in for bars larger than No. 5 and 1½ in for No. 5 or smaller.
4. Unexposed concrete slabs, walls, or joists, ¾ in for No. 11 and smaller, 1½ in for No. 14 and No. 18 bars. Beams, girders, and columns, 1½ in. Shells and folded-plate members, ¾ in for bars larger than No. 5, and ½ inch for No. 5 and smaller.

by steel ties, or spirals, or members framing into the column. This resistance should be provided within a distance of 6 in of the point of the bend.

Where column faces are offset 3 in or more, vertical bars should be lapped by separate dowels.

In columns, a minimum tensile strength at each face equal to one-fourth the area of vertical reinforcement multiplied by f_y should be provided at horizontal cross sections where splices are located. In columns with substantial bending, full tensile splices equal to double the factored tensile stress in the bar are required.

Welded-Wire Fabric ▪ Wire reinforcing normally is spliced by lapping. When the tensile stress at the splice is less than half the permissible stress, the overlap measured between outermost cross wires should be at least 2 in. Where stresses are higher, the overlap should equal the spacing of the cross wires plus 2 in.

Slab Reinforcement ▪ Structural floor and roof slabs with principal reinforcement in only one direction should be reinforced for shrinkage and temperature stresses in a perpendicular direction. The crossbars may be spaced at a maximum of 18 in or five times the slab thickness. The ratio of reinforcement area of these bars to gross concrete area should be at least 0.0025 for plain bars, 0.0020 for deformed bars with less than 60,000-psi yield strength, and 0.0018 for deformed bars with 60,000-psi yield strength and welded-wire fabric with welded intersections in the direction of stress not more than 12 in apart.

Concrete Cover ▪ To protect reinforcement against fire and corrosion, thickness of concrete cover over the outermost steel should be at least that given in Table 3-2.

(ACI 318, "Building Code Requirements for Reinforced Concrete," American Concrete Institute; "Standard Specifications for Highway Bridges," American Association of State Highway and Transportation Officials, 444 N. Capitol St., N.W., Washington, D.C. 20001).

3-5. Tendons

High-strength steel is required for prestressing concrete to make the stress loss due to creep and shrinkage of concrete and to other factors a small percentage of the applied stress. This type of loss does not increase as fast as increase in stress in the prestressing steel, or tendons.

Tendons should have specific characteristics in addition to high strength to meet the requirements of prestressed concrete. They should elongate uniformly up to initial tension for accuracy in applying the prestressing force. After the yield strength has been reached, the steel should continue to stretch as stress increases, before failure occurs. American Society for Testing and Materials Specifications for prestressing wire and strands, A421 and A416, set the yield

TABLE 3-3 Properties of Tendons

Diameter, in	Area, in^2	Weight per 1,000 ft, lb	Ultimate strength
	UNCOATED TYPE WA WIRE		
0.276	0.05983	203.2	235,000 psi
0.250	0.04909	166.7	240,000 psi
0.196	0.03017	102.5	250,000 psi
0.192	0.02895	98.3	250,000 psi
	UNCOATED TYPE BA WIRE		
0.250	0.04909	166.7	240,000 psi
0.196	0.03017	102.5	240,000 psi
	UNCOATED SEVEN-WIRE STRANDS 250 GRADE		
¼	0.04	122	9,000 lb
$^5/_{16}$	0.058	197	14,500 lb
⅜	0.080	272	20,000 lb
$^7/_{16}$	0.108	367	27,000 lb
½	0.144	490	36,000 lb
	270 GRADE		
⅜	0.085	290	23,000 lb
$^7/_{16}$	0.115	390	31,000 lb
½	0.153	520	41,300 lb

strength at 80 to 85% of the tensile strength. Furthermore, the tendons should exhibit little or no creep, or relaxation, at the high stresses used.

ASTM A421 covers two types of uncoated, stress-relieved, high-carbon-steel wire commonly used for linear prestressed-concrete construction. Type BA wire is used for applications in which cold-end deformation is used for end anchorages, such as buttonheads. Type WA wire is intended for end anchorages by wedges and where no cold-end deformation of the wire is involved. The wire is required to be stress-relieved by a continuous-strand heat treatment after it has been cold-drawn to size. Type BA usually is furnished 0.196 and 0.250 in in diameter, with an ultimate strength of 240,000 psi and yield strength (at 1% extension) of 192,000 psi. Type WA is available in those sizes and also 0.192 and 0.276 in in diameter, with ultimate strengths ranging from 250,000 for the smaller diameters to 235,000 psi for the largest. Yield strengths range from 200,000 for the smallest to 188,000 psi for the largest (Table 3-3).

For pretensioning, where the steel is tensioned before the concrete is cast, wires usually are used individually, as is common for reinforced concrete. For posttensioning, where the tendons are tensioned and anchored to the concrete after it has attained sufficient strength, the wires generally are placed parallel to each other in groups, or cables, sheathed or ducted to prevent bond with the concrete.

A seven-wire strand consists of a straight center wire and six wires of slightly smaller diameter winding helically around and gripping it. High friction between the center and outer wires is important where stress is transferred between the strand and concrete through bond. ASTM A416 covers strand with ultimate strengths of 250,000 and 270,000 psi (Table 3-3).

Galvanized strands sometimes are used for posttensioning, particularly when the tendons may not be embedded in grout. Sizes normally available range from a 0.5-in-diameter seven-wire strand, with 41,300-lb breaking strength, to $1\frac{11}{16}$-in-diameter strand, with 352,000-lb breaking strength. The cold-drawn wire comprising the strand is stress-relieved when galvanized, and stresses due to stranding are offset by prestretching the strand to about 70% of its ultimate strength. Tendons 0.5 and 0.6 in in diameter are typically used sheathed and unbonded.

Hot-rolled alloy-steel bars used for prestressing concrete generally are not so strong as wire or strands. The bars usually are stress-relieved, then cold-stretched to at least 90% of ultimate strength to raise the yield point. The cold stretching also serves as proof stressing, eliminating bars with defects.

(H. K. Preston and N. J. Sollenberger, "Modern Prestressed Concrete," McGraw-Hill Book Company, New York; J. R. Libby, "Modern Prestressed Concrete," Van Nostrand Reinhold Company, New York.)

3-6. Fabrication of Prestressed-Concrete Members

Prestressed concrete may be produced much like high-strength reinforced concrete, either cast in place or precast. Prestressing offers several advantages for precast members, which have to be transported from casting bed to final position and handled several times. Prestressed members are lighter than reinforced members of the same capacity, both because higher-strength concrete generally is used and the full cross section is effective. In addition, the prestressing normally counteracts handling stresses. And if a prestressed member survives the full prestress and handling, the probability of its failing under service loads usually is very small.

Two general methods of prestressing are commonly used—pretensioning and posttensioning—and both may be used for the same member. Pretensioning, where the tendons are tensioned before embedment in the concrete and stress transfer from steel to concrete usually is by bond, is especially useful for mass production of precast elements. Often, elements may be fabricated in long lines, by stretching the tendons (Art. 3-5) between abutments at the ends of the lines. By use of tiedowns and struts, the tendons may be draped in a vertical plane to develop upward and downward components on release. After the tendons have been jacked to their full stress, they are anchored to the abutments. The casting bed over which the tendons are stretched usually is made of a smooth-surface concrete slab with easily stripped side forms of steel. (Forms for pretensioned members must permit them to move on release of the tendons.) Separators are placed in the forms to divide the long line into members of required length and provide space for cutting the tendons. After the concrete has been cast and has attained its specified strength, generally after a preset period and steam curing, side forms are removed. Then, the tendons are detached from the anchorages at the ends of the line and relieved of their stress. Restrained from shortening by bond with the concrete, the tendons compress it. At this time, it is safe to cut the tendons between the members and remove the members from the forms.

Posttensioning frequently is used for cast-in-place members and long-span flexural members. Cables or bars (Art. 3-5) are placed in the forms in flexible ducts to prevent bond with the concrete. They may be draped in a vertical plane to develop upward and downward forces when tensioned. After the concrete has been placed and has attained sufficient strength, the tendons are tensioned by jacking against the member and then are anchored to it. Grout may be pumped into the duct to establish bond with the concrete and protect the tendons against corrosion. Applied at pressures of 75 to 100 psi, a typical grout consists of 1 part portland cement, 0.75 parts sand (capable of passing through a No. 30 sieve), and 0.75 parts water, by volume.

Concrete with higher strengths than ordinarily used for reinforced concrete offers economic advantages for prestressed concrete. In reinforced concrete, much of the concrete in a slab or beam is assumed to be ineffective because it is in tension and likely to crack under service loads. In prestressed concrete, the full section is effective because it is always under either compression or very low tension. Furthermore, high-strength concrete develops higher bond stresses with the tendons, greater bearing strength to withstand the pressure of anchorages, and a higher modulus of elasticity. The last indicates reductions in initial strain and camber when prestress is applied initially and in creep strain. The reduction in creep strain reduces the loss of prestress with time. Generally, concrete with a 28-day strength of 5000 psi or more is advantageous for prestressed concrete.

Concrete cover over prestressing steel, ducts, and nonprestressed steel should be at least 3 in for concrete surfaces in contact with the ground; 1½ in for prestressing steel and main reinforcing bars, and 1 in for stirrups and ties in beams and girders, 1 in in slabs and joists exposed to the weather; and ¾ in for unexposed slabs and joists. In extremely corrosive atmospheres or other severe exposures, the amount of protective cover should be increased.

Minimum clear spacing between pretensioning steel at the ends of a member should be four times the diameter of individual wires and three times the diameter of strands. Some codes also require that the spacing be at least 1⅓ times the maximum size of aggregate. (See also Art. 3-4.) Away from the ends of a member, prestressing steel or ducts may be bundled. Concentrations of steel or ducts, however, should be reinforced to control cracking.

Prestressing force may be determined by measuring tendon elongation, by checking jack pressure on a recently calibrated gage, or by using a recently calibrated dynamometer. If several wires or strands are stretched simultaneously, the method used should be such as to induce approximately equal stress in each.

Splices should not be used in parallel-wire cables, especially if a splice has to be made by welding, which would weaken the wire. Failure is likely to occur during tensioning.

Strands may be spliced, if necessary, when the coupling will develop the full strength of the tendon, not cause it to fail under fatigue loading, and does not replace sufficient concrete to weaken the member.

High-strength bars are generally spliced mechanically. The couplers should be capable of developing the full strength of the bars without decreasing resistance to fatigue and without replacing an excessive amount of concrete.

Anchor fittings are different for pretensioned and posttensioned members. For pretensioned members, the fittings hold the tendons temporarily against anchors outside the members and therefore can be reused. In posttensioning, the fittings usually anchor the tendons permanently to the members. In unbonded tendons, the sheathing is typically plastic or impregnated paper.

In pretensioning, the tendons may be tensioned one at a time to permit the use of relatively light jacks, in groups, or all simultaneously. A typical stressing arrangement consists of a stationary anchor post, against which jacks act, and a moving crosshead, which is pushed by the jacks and to which the tendons are attached. Usually, the tendons are anchored to a thick steel plate that serves as a combination anchor plate and template. It has holes through which the tendons pass to place them in the desired pattern. Various patented grips are available for anchoring the tendons to the plate. Generally, they are a wedge or chuck type capable of developing the full strength of the tendons.

A variety of patented fittings also are available for anchoring in posttensioned members. Such fittings should be capable of developing the full strength of the tendons under static and fatigue loadings. The fittings also should spread the prestressing force over the concrete or transmit it to a bearing plate. Sufficient space must be provided for the fittings in the anchor zone.

Generally, all the wires of a parallel-wire cable are anchored with a single fitting (Figs. 3-2 and 3-3). The type shown in Fig. 3-3 requires that the wires be cut to exact length and a buttonhead be cold-formed on the ends for anchoring.

The wedge type in Fig. 3-2 requires a double-acting jack. One piston, with the wires wedged to it, stresses them, and a second piston forces the male cone into the female cone to grip the

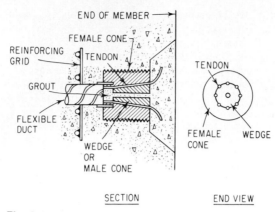

Fig. 3-2. Conical-wedge anchorage for prestressing wires.

tendons. Normally, a hole is provided in the male cone for grouting the wires. After final stress is applied, the anchorage may be embedded in concrete to prevent corrosion and improve appearance.

With the buttonhead type, a stressing rod may be screwed over threads on the circumference of a thick, steel stressing washer (Fig. 3-3b) or into a center hole in the washer (Fig. 3-3c). The rod then is bolted to a jack. When the tendons have been stressed, the washer is held in position by steel shims inserted between it and a bearing plate embedded in the member. The jack pressure then can be released and the jack and stressing rod removed. Finally, the anchorage is embedded in concrete.

Posttensioning bars may be anchored individually with steel wedges (Fig. 3-4a) or by tightening a nut against a bearing plate (Fig. 3-4b). The former has the advantage that the bars do not have to be threaded.

Posttensioning strands normally are shop-fabricated in complete assemblies, cut to length, anchor fittings attached, and sheathed in flexible duct. Swaged to the strands, the anchor fittings have a threaded steel stud projecting from the end. The threaded stud is used for jacking the

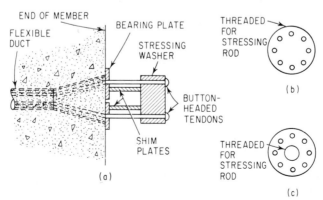

Fig. 3-3. (a) End anchorage for buttonheaded wires. The stressing head for tensioning the wires may be threaded externally (b) or internally (c) for attachment to jack.

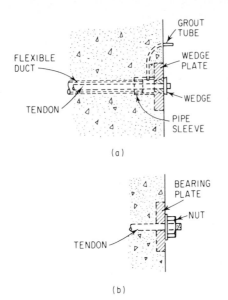

(a)

(b)

Fig. 3-4. End anchorages for bars. (*a*) Conical wedge. (*b*) Nut and washer at threaded end, acting against a bearing plate.

stress into the strand and for anchoring by tightening a nut against a bearing plate in the member (Fig. 3-5).

To avoid overstressing and failure in the anchorage zone, the anchorage assembly must be placed with care. Bearing plates should be placed perpendicular to the tendons to prevent eccentric loading. Jacks should be centered for the same reason and so as not to scrape the tendons against the plates. The entire area of the plates should bear against the concrete.

Prestress normally is applied with hydraulic jacks. The amount of prestressing force is determined by measuring tendon elongation and comparing with an average load-elongation curve for the steel used. In addition, the force thus determined should be checked against the jack pressure registered on a recently calibrated gage or by use of a recently calibrated dynamometer. Discrepancies of less than 5% may be ignored.

When prestressed-concrete beams do not have a solid rectangular cross section in the anchorage zone, an enlarged end section, called an end block, may be necessary to transmit the prestress from the tendons to the full concrete cross section a short distance from the anchor zone. End blocks also are desirable for transmitting vertical and lateral forces to supports and to provide adequate space for the anchor fittings for the tendons.

The transition from end block to main cross section should be gradual (Fig. 3-6). Length of end block, from beginning of anchorage area to the start of the main cross section, should be at least 24 in. The length normally ranges from three-fourths the depth of the member for deep beams to the full depth for shallow beams. The end block should be reinforced vertically and horizontally to resist tensile bursting and spalling forces induced by the concentrated loads of the tendons. In particular, a grid of reinforcing should be placed directly behind the anchorages to resist spalling.

Ends of pretensioned beams should be reinforced with vertical stirrups over a distance equal to one-fourth the beam depth. The stirrups should be capable of resisting in tension a force equal to at least 4% of the prestressing force.

Control of camber is important for prestressed members. Camber tends to increase with time because of creep. If a prestressed beam or slab has an upward camber under prestress and long-time loading, the camber will tend to increase upward. Excessive camber should be avoided, and for deck-type structures, such as highway bridges and building floors and roofs, the camber of all beams and girders of the same span should be the same.

Computation of camber with great accuracy is difficult, mainly because of the difficulty of ascertaining with accuracy the modulus of elasticity of the concrete, which varies with time. Other difficult-to-evaluate factors also influence

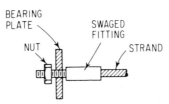

Fig. 3-5. Swaged fitting for anchoring strands. Prestress is maintained by tightening the nut against the bearing plate.

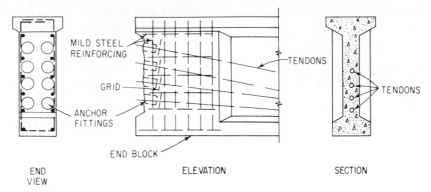

Fig. 3-6. Transition from end block of a prestressed-concrete beam to the main cross section.

camber: departure of the actual prestressing force from that calculated, effects of long-time loading, influence of length of time between prestressing and application of full service loads, methods of supporting members after removal from the forms, and influence of composite construction.

When camber is excessive, it may be necessary to use concrete with higher strength and modulus of elasticity (change from lightweight to ordinary concrete); increase the moment of inertia of the section; use partial prestressing; that is, decrease the prestressing force and add reinforcing steel to resist the tensile stresses; or use a larger prestressing force with less eccentricity.

To insure uniformity of camber, a combination of pretensioning and posttensioning is desirable for precast members. Sufficient prestress may be applied initially to permit removal of the member from the forms and transportation to a storage yard. After the member has increased in strength but before erection, additional prestress is applied by posttensioning to bring the camber to the desired value. During storage, the member should be supported in the same manner as it will be in the structure.

For handling prestressed precast members, inserts usually are embedded in the concrete. Stresses should be computed for use of these inserts as pickup points, and the member should not be handled by pickup at other points. Nor should it be supported upside down, sideways, or at points that would create greater stresses than it will sustain under service loads.

(H. K. Preston and N. J. Sollenberger, "Modern Prestressed Concrete," McGraw-Hill Book Company, New York; J. B. Libby, "Modern Prestressed Concrete," Van Nostrand Reinhold Company, New York.)

3-7. Precast Concrete

When concrete products are made in other than their final position, they are considered precast. They may be unreinforced, reinforced, or prestressed. They include in their number a wide range of products: block, brick, pipe, plank, slabs, conduit, joists, beams and girders, trusses and truss components, curbs, lintels, sills, piles, pile caps, and walls.

Precasting often is chosen because it permits efficient mass production of concrete units. With precasting, it usually is easier to maintain quality control and produce higher-strength concrete than with field concreting. Formwork is simpler, and a good deal of falsework can be eliminated. Also, since precasting normally is done at ground level, workers can move about more freely. But sometimes these advantages are more than offset by the cost of handling, transporting, and erecting the precast units. Also, joints may be troublesome and costly.

Design of precast products follows the same rules, in general, as for cast-in-place units. However, ACI 318, "Building Code Requirements for Reinforced Concrete" (American Concrete Institute), permits the concrete cover over reinforcing steel to be as low as ⅜ in for slabs, walls, or joists not exposed to weather. Also, ACI Standard 525, "Minimum Requirements for Thin-Section Precast Concrete Construction," permits the cover for units not exposed to weather to be only ⅜ in. Furthermore, for surfaces exposed to the weather, or in contact with the ground or water, cover for main reinforcement in beams, girders, and columns need only be ½ in, and reinforcement in slabs and secondary reinforcement in beams, girders, and columns need have only ⅜ in of cover. Hence, this standard permits units to be as thin as 1 in, reinforced with welded-wire fabric. But dense, watertight concrete must be used. For standard concrete, minimum cover should be that given in Table 3-2.

Concrete for precast elements not exposed to weather or in contact with the ground should have a minimum 28-day strength of 4000 psi. Exposed concrete should have a 5000-psi strength. Aggregate is restricted to a maximum of ¾ in or two-thirds the minimum clear distance between parallel reinforcing bars. In thin elements, spacing of wires in welded-wire fabric may not exceed 2 in.

Precast units must be designed for handling and erection stresses, which may be more severe than those they will be subjected to in service. Normally, inserts are embedded in the concrete for picking up the units. They should be picked up by these inserts, and when set down, they should be supported right side up, in such a manner as not to induce stresses higher than the units would have to resist in service. (See also Art. 3-6.)

For precast beams, girders, joists, columns, slabs, and walls, joints usually are made with cast-in-place concrete. Often, in addition, steel reinforcing projecting from the units to be joined is welded together. (ACI 512.1R, "Suggested Design of Joints and Connections in Precast Structural Concrete," American Concrete Institute.)

3-8.   Lift-Slab Construction

A type of precasting used in building construction involves casting floor and roof slabs at or near ground level and lifting them to their final position, hence the name lift-slab construction. It offers many of the advantages of precasting (Art. 3-7) and eliminates many of the storing, handling, and transporting disadvantages. It normally requires fewer joints than other types of precast building systems.

Typically, columns are erected first, but not necessarily for the full height of the building. Near the base of the columns, floor slabs are cast in succession, one atop another, with a parting compound between them to prevent bond. The roof slab is cast last, on top. Usually, the construction is flat plate, and the slabs have uniform thickness; waffle slabs or other types also can be used. Openings are left around the columns, and a steel collar is slid down each column for embedment in every slab. The collar is used for lifting the slab, connecting it to the column, and reinforcing the slab against shear.

To raise the slabs, jacks are set atop the columns and turn threaded rods that pass through the collars and do the lifting. As each slab reaches its final position, it is wedged in place and the collars are welded to the columns.

3-9.   Ultimate-Strength Theory for Reinforced
Concrete Beams

For consistent, safe, economical design of beams, their actual load-carrying capacity should be known. The safe load then can be determined by dividing this capacity by a safety factor. Or the design load can be multiplied by the safety factor to indicate what the capacity of the beams should be. It should be noted, however, that under design loads, stresses and deflections may be computed with good approximation on the assumptions of a linear stress-strain diagram and a cracked cross section.

ACI 318, "Building Code Requirements for Reinforced Concrete" (American Concrete Institute), provides for design by ultimate-strength theory. Bending moments in members are determined as if the structure were elastic. Ultimate-strength theory is used to design critical sections, those with the largest bending moments. The ultimate strength of each section is computed, and the section is designed for this capacity with an appropriate safety factor.

Stress Redistribution ▪ The ACI Code recognizes that, below ultimate load, a redistribution of stress occurs in continuous beams, frames, and arches, which allows the structure to carry loads higher than those indicated by elastic analysis. The code permits an increase or decrease of up to 10% in the negative moments calculated by elastic theory at the supports of continuous flexural members. But these modified moments must also be used for determining the moments at other sections for the same loading conditions. [The modifications, however, are permissible only for relatively small steel ratios at each support. The steel ratios ρ or $\rho - \rho'$ (see Arts. 3-12, 3-13, and 3-17 to 3-20) should be less than half ρ_b, the steel ratio for balanced conditions (concrete strength equal to steel strength) at ultimate load.] For example, suppose elastic analysis of a continuous beam indicates a maximum negative moment at a support of $wL^2/12$ and a maximum positive moment at midspan of $wL^2/8 - wL^2/12$, or $wL^2/24$. Then, the code permits the negative moment to be decreased to $0.9wL^2/12$, if the positive moment is increased to $WL^2/8 - 0.9wL^2/12$, or $1.2wL^2/24$.

Design Assumptions ▪ Ultimate strength of any section of a reinforced concrete beam may be computed assuming the following:

Strain in the concrete is directly proportional to the distance from the neutral axis (Fig. 3-7b).

Except in anchorage zones, strain in reinforcing steel equals strain in adjoining concrete.

At ultimate strength, maximum strain at the extreme compression surface equals 0.003 in/in.

When the reinforcing steel is not stressed to its yield strength, f_y, the steel stress is 29,000,000 psi times the steel strain, in/in. After the yield strength has been reached, the stress remains constant at f_y, though the strain increases.

Tensile strength of the concrete is negligible.

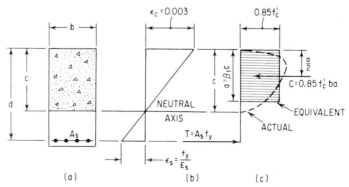

Fig. 3-7. Stresses and strains on a reinforced-concrete section (a) at ultimate load. (b) Strain diagram. (c) Actual and assumed equivalent-stress diagram.

At ultimate strength, concrete stress is not proportional to strain. The actual stress distribution may be represented by an equivalent rectangle that yields ultimate strengths in agreement with numerous, comprehensive tests (Fig. 3-7c).

The ACI Code recommends that the compressive stress for the equivalent rectangle be taken as $0.85f'_c$, where f'_c is the 28-day compressive strength of the concrete. The stress is assumed constant from the surface of maximum compressive strain over a depth $a = \beta_1 c$, where c is the distance to the neutral axis (Fig. 3-7c). For $f'_c \leq 4000$ psi, $\beta_1 = 0.85$; for greater concrete strengths, β_1 is reduced 0.05 for each 1000 psi in excess of 4000.

Formulas in the ACI Code based on these assumptions usually contain a factor ϕ to provide for the possibility that small adverse variations in materials, quality of work, and dimensions, while individually within acceptable tolerances, occasionally may combine, and actual capacity may be less than that computed. The coefficient ϕ may be taken as 0.90 for flexure, 0.85 for shear and torsion, 0.75 for spirally reinforced compression members, and 0.70 for tied compression members. Under certain conditions of load (as the value of the axial load approaches zero) and geometry, the ϕ value for compression members may increase linearly to a maximum value of 0.90.

Crack Control ▪ Because of the risk of large cracks opening up when reinforcement is subjected to high stresses, the ACI Code recommends that designs be based on a steel yield strength f_y no larger than 80 ksi. And when design is based on a yield strength f_y greater than 40 ksi, the cross sections of maximum positive and negative moment shall be proportioned for crack control so that specific limits are satisfied by

$$z = f_s \sqrt[3]{d_c A} \tag{3-6}$$

where f_s = calculated stress, ksi, in reinforcement at service loads
    d_c = thickness of concrete cover, in, measured from extreme tension surface to center of bar closest to that surface
    A = effective tension area of concrete, in² per bar. This area should be taken as that surrounding main tension reinforcement and having same centroid as that reinforcement multiplied by ratio of area of largest bar used to total area of tension reinforcement

These limits are $z \leq 175$ kips/in for interior exposures and $z \leq 145$ kips/in for exterior exposures. These correspond to limiting crack widths of 0.016 and 0.013 in, respectively, at the extreme tension edge under service loads. In Eq. (3-6), f_s should be computed by dividing the bending moment by the product of the steel area and the internal moment arm, but f_s may be taken as 60% of the steel yield strength without computation. [Use of d_c in Eq. (3-6) is questionable because it would lead to reduction of concrete cover, which is needed for corrosion protection.]

Required Strength ▪ For combinations of loads, the ACI Code requires that a structure and its members should have the following ultimate strengths (capacity to resist design loads and their related internal moments and forces):

Wind and earthquake loads not applied:

$$U = 1.4D + 1.7L \tag{3-7}$$

where D = effect of basic load consisting of dead load plus volume change (shrinkage, temperature)
    L = effect of live load plus impact

When wind loads are applied, the largest of Eq. (3-7) and Eq. (3-8a and b) determines the required strength.

$$U = 0.75 (1.4D + 1.7L + 1.7W) \tag{3-8a}$$
$$U = 0.9D + 1.3W \tag{3-8b}$$

where W = effect of wind load.

If the structure will be subjected to earthquake forces E, substitute $1.1E$ for W in Eq. (3-8).

Where the effects of differential settlement, creep, shrinkage, or temperature change may be critical to the structure, they should be included with the dead load D, and the strength should be at least equal to

$$U = 0.75 (1.4D + 1.7L) \tag{3-9}$$

For ultimate-strength loads (load-factor method) for bridges, see Art. 5-4.

Although structures may be designed by ultimate-strength theory, it is not anticipated that service loads will be substantially exceeded. Hence, deflections that will be of concern to the designer are those that occur under service loads. These deflections may be computed by working-stress theory. (See Art. 3-10.)

Deep Members ▪ Due to the nonlinearity of strain distribution and the possibility of lateral buckling, deep flexural members must be given special consideration. The ACI Code considers members with overall depth-to-span ratios greater than ⅖ for continuous spans (⅘ for simple spans) as deep members. The ACI Code provides special shear design requirements and minimum requirements for both horizontal and vertical reinforcement for such members. See Sec. 10.7, ACI 318-77.

(G. Winter and A. H. Nilson, "Design of Concrete Structures," McGraw-Hill Book Company, New York; P. F. Rice and E. S. Hoffman, "Structural Design Guide to the ACI Building Code," Van Nostrand Reinhold Company, New York.)

3-10. Working-Stress Theory for Reinforced-Concrete Beams

Stress distribution in a reinforced-concrete beam under service loads is different from that at ultimate strength (Art. 3-9). Knowledge of this stress distribution is desirable for many reasons, including the requirements of some design codes that specified working stresses in steel and concrete not be exceeded.

Working stresses in reinforced-concrete beams are computed from the following assumptions:

Longitudinal stresses and strains vary with distance from the neutral axis (Fig. 3-8c and d).

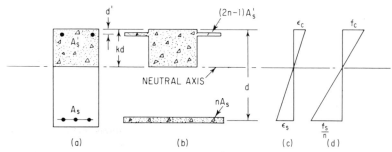

Fig. 3-8. Typical cross section (*a*) of a reinforced-concrete beam may be treated in design as an all-concrete (transformed) section (*b*). Strains (*c*) and stresses (*d*) are assumed to vary linearly.

The concrete develops no tensile stress.

Except in anchorage zones, strain in reinforcing steel equals strain in adjoining concrete.

But because of creep, strain in compressive steel in beams may be taken as half that in the adjoining concrete.

The modular ratio $n = E_s/E_c$ is constant. E_s is the modulus of elasticity of the reinforcing steel and E_c of the concrete.

The following allowable stresses may be used for flexure:

	Buildings	Bridges
Compression in extreme compression surface	$0.45f'_c$	$0.4f'_c$
Tension in reinforcement		
Grade 40 or 50 steel	20 ksi	20 ksi
Grade 60 or higher yield strength	24 ksi	24 ksi

where f'_c is the 28-day compressive strength of the concrete.

For other than such flexural stresses, allowable or maximum stresses to be used in design are stated as a percentage of the values given for ultimate-strength design. For example, for service loads:

Type of member and stress	Allowable stresses or capacity, % of ultimate (nominal)
Compression members, walls	40
Shear or tension in beams, joists, walls, one-way slabs	55
Shear or tension in two-way slabs, footings	50
Bearing in concrete	35

Allowable stresses may be increased one-third when wind or earthquake forces are combined with other loads, but the capacity of the resulting section should not be less than that required for dead plus live loads.

Other equivalency factors are also given in terms of ultimate strength values. Thus, the predominant design procedure is the ultimate-strength method, but for reasons of background and historical significance and because the working-stress design method is sometimes preferred for bridges and certain foundation and retaining-wall design, examples of working-stress design procedure are presented in Arts. 3-13, 3-18, and 3-20.

Transformed Section ▪ According to working-stress theory for reinforced-concrete beams (Art. 3-10), strains in reinforcing steel and adjoining concrete are equal. Hence f_s, the stress in the steel, is n times f_c, the stress in the concrete, where n is the ratio of modulus of elasticity of the steel E_s to that of the concrete E_c. The total force acting on the steel then equals $(nA_s)f_c$. This indicates that the steel area can be replaced in stress calculations by a concrete area n times as large.

The transformed section of a concrete beam is one in which the reinforcing has been replaced by an equivalent area of concrete (Fig. 3-8b). (In doubly reinforced beams and slabs, an effective modular ratio of $2n$ should be used to transform the compression reinforcement, to account for the effects of creep and nonlinearity of the stress-strain diagram for concrete. But the computed stress should not exceed the allowable tensile stress.) Since stresses and strains are assumed to vary with distance from the neutral axis, conventional elastic theory for homogeneous beams holds for the transformed section. Section properties, such as location of neutral axis, moment of inertia, and section modulus S, can be computed in the usual way, and stresses can be found from the flexure formula $f = M/S$, where M is the bending moment.

(G. Winter and A. H. Nilson, "Design of Concrete Structures," McGraw-Hill Book Company, New York; P. Rice and E. S. Hoffman, "Structural Design Guide to the ACI Building Code," Van Nostrand Reinhold Company, New York.)

3-11. Deflection Computations and Criteria

The assumptions of working-stress theory (Art. 3-10) may also be used for computing deflections under service loads; that is, elastic-theory deflection formulas may be used for reinforced concrete beams. In these formulas, the effective moment of inertia I_e should be taken as not greater than I_g, the moment of inertia of the gross concrete section, or as given by Eq. (3-10).

$$I_e = \left(\frac{M_{cr}}{M_a} \right)^3 I_g + \left[1 - \left(\frac{M_{cr}}{M_a} \right)^3 \right] I_{cr} \qquad (3\text{-}10).$$

where M_{cr} = cracking moment
M_a = moment for which deflection is being computed
I_{cr} = cracked concrete (transformed) section

If y_t is taken as the distance from the centroidal axis of the gross section, neglecting the reinforcement, to the extreme surface in tension, the cracking moment may be computed from

$$M_{cr} = \frac{f_r I_g}{y_t} \qquad (3\text{-}11)$$

with the modulus of rupture of the concrete $f_r = 7.5\sqrt{f'_c}$. Equation (3-10) takes into account the variation of the moment of inertia of a concrete section based on whether the section is cracked or uncracked. The modulus of elasticity of the concrete E_c may be computed from Eq. (3-2) in Art. 3-1.

The deflections thus calculated are those assumed to occur immediately on application of load. Additional long-time deflections can be estimated by multiplying the immediate deflection by 2 when there is no compression reinforcement or by $2 - 1.2A'_s/A_s \geq 0.6$, where A'_s is the area of compression reinforcement and A_s is the area of tension reinforcement.

Deflection Limitations ▪ The ACI Code recommends the following limits on deflections in buildings:

For roofs not supporting plastered ceilings or not attached to nonstructural elements, maximum immediate deflection under live load should not exceed $L/180$, where L is the span of beam or slab.

For floors not supporting partitions or not attached to nonstructural elements, the maximum immediate deflection under live load should not exceed $L/360$.

For a floor or roof construction intended to support or to be attached to partitions or other construction likely to be damaged by large deflections of the support, the allowable limit for the sum of immediate deflection due to live loads and the additional deflection due to shrinkage and creep under all sustained loads should not exceed $L/480$. If the construction is not likely to be damaged by large deflections, the deflection limit may be increased to $L/240$. But tolerances should be established and adequate measures should be taken to prevent damage to supported or nonstructural elements resulting from the deflections of structural members.

TABLE 3-4 Areas of Groups of Standard Bars. in^2

Bar No.	Diam, in	Weight, lb per ft	Number of bars								
			1	2	3	4	5	6	7	8	9
2	0.250	0.167	0.05	0.10	0.15	0.20	0.25	0.30	0.35	0.40	0.45
3	0.375	0.376	0.11	0.22	0.33	0.44	0.55	0.66	0.77	0.88	0.99
4	0.500	0.668	0.20	0.39	0.58	0.78	0.98	1.18	1.37	1.57	1.77
5	0.625	1.043	0.31	0.61	0.91	1.23	1.53	1.84	2.15	2.45	2.76
6	0.750	1.502	0.44	0.88	1.32	1.77	2.21	2.65	3.09	3.53	3.98
7	0.875	2.044	0.60	1.20	1.80	2.41	3.01	3.61	4.21	4.81	5.41
8	1.000	2.670	0.79	1.57	2.35	3.14	3.93	4.71	5.50	6.28	7.07
9	1.128	3.400	1.00	2.00	3.00	4.00	5.00	6.00	7.00	8.00	9.00
10	1.270	4.303	1.27	2.53	3.79	5.06	6.33	7.59	8.86	10.12	11.39
11	1.410	5.313	1.56	3.12	4.68	6.25	7.81	9.37	10.94	12.50	14.06
14	1.693	7.650	2.25	4.50	6.75	9.00	11.25	13.50	15.75	18.00	20.25
18	2.257	13.600	4.00	8.00	12.00	16.00	20.00	24.00	28.00	32.00	36.00

3-12. Ultimate-Strength Design of Rectangular Beams with Tension Reinforcement Only

Generally, the area A_s of tension reinforcement (Fig. 3-7) in a reinforced-concrete beam is represented by the ratio $\rho = A_s/bd$, where b is the beam width and d the distance from extreme compression surface to the centroid of tension reinforcement. At ultimate strength, the steel at a critical section of the beam will be at its yield strength f_y if the concrete does not fail in compression first (Art. 3-9). Total tension in the steel then will be $A_s f_y = \rho f_y bd$. It will be opposed, according to Fig. 3-7, by an equal compressive force, $0.85f'_c ba = 0.85f'_c b\beta_1 c$, where f'_c is the 28-day strength of the concrete, a the depth of the equivalent rectangular stress distribution, c the distance from the extreme compression surface to the neutral axis, and β_1 a constant (see Art. 3-9). Equating the compression and tension at the critical section yields

$$c = \frac{\rho f_y}{0.85\beta_1 f'_c} d \tag{3-12}$$

The criterion for compression failure is that the maximum strain in the concrete equals 0.003 in/in. In that case

$$c = \frac{0.003}{f_s/E_s + 0.003} d \tag{3-13}$$

where f_s = steel stress, psi
   E_s = modulus of elasticity of steel = 29,000,000 psi
Table 3-4 lists the nominal diameters, weights, and cross-sectional areas of standard steel reinforcing bars.

Balanced Reinforcing ▪ Under balanced conditions, the concrete will reach its maximum strain of 0.003 when the steel reaches its yield strength f_y. Then, c as given by Eq. (3-12) will equal c as given by Eq. (3-13) since c determines the location of the neutral axis. This determines the steel ratio for balanced conditions:

$$\rho_b = \frac{0.85\beta_1 f'_c}{f_y} \frac{87,000}{87,000 + f_y} \tag{3-14}$$

Reinforcing Limitations ▪ All structures are designed to collapse not suddenly but by gradual deformation when overloaded. To achieve this end in concrete, the reinforcement should yield before the concrete crushes. This will occur if the quantity of tensile reinforcement is less than the critical percentage determined by ultimate-strength theory [Eq. (3-14)]. The ACI Code, to avoid compression failures, limits the steel ratio ρ to a maximum of $0.75\rho_b$. The Code also requires that ρ for positive-moment reinforcement be at least $200/f_y$.

Moment Capacity ▪ For such underreinforced beams, the bending-moment capacity of ultimate strength is

$$M_u = 0.90[bd^2f'_c\omega(1 - 0.59\omega)] = 0.90\left[A_s f_y\left(d - \frac{a}{2}\right)\right] \tag{3-15}$$

where $\omega = \rho f_y/f'_c$
$a = A_s f_y/0.85 f'_c b$

Shear and Torsion Reinforcement ▪ The ultimate shear capacity V_n of a section of a beam equals the sum of the nominal shear strength of the concrete V_c and the nominal shear strength provided by the reinforcement V_s; that is, $V_n = V_c + V_s$. The factored shear force V_u on a section should not exceed

$$V_u = \phi V_n = \phi(V_c + V_s) \tag{3-16}$$

where ϕ = capacity reduction factor (0.85 for shear and torsion). Except for brackets and other short cantilevers, the section for maximum shear may be taken at a distance equal to d from the face of the support. V_c carried by the concrete alone should not exceed $2\sqrt{f'_c}b_w d$ where b_w is the width of the beam web and d the depth of the centroid of reinforcement. (As an alternative, the maximum for V_c may be taken as

$$V_c = \left(1.9\sqrt{f'_c} + 2500\,\rho_w\frac{V_u d}{M_u}\right)b_w d \le 3.5\sqrt{f'_c}b_w d \tag{3-17}$$

where $\rho_w = A_s/b_w d$ and V_u and M_u are the shear and bending moment, respectively, at the section considered, but M_u should not be less than $V_u d$.)

When V_u is larger than ϕV_c, the excess shear will have to be resisted by web reinforcement. In general, this reinforcement should be stirrups perpendicular to the axis of the member (Fig. 3-9) or welded-wire fabric with wires perpendicular to the axis of the member. In members without prestressing, however, the stirrups may be inclined, as long as the angle is at least 45° with the axis of the member. As an alternative, longitudinal bars may be bent up at an angle of 30° or more with the axis, or spirals may be used. Spacing should be such that every 45°

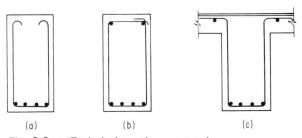

(a)　　　　　(b)　　　　　(c)

Fig. 3-9. Typical stirrups in a concrete beam.

line, representing a potential crack and extending from middepth $d/2$ to the longitudinal tension bars, should be crossed by at least one line of reinforcing.

The area of steel required in vertical stirrups, in^2 per stirrup, with a spacing s, in, is

$$A_v = \frac{V_s s}{f_y d} \qquad (3\text{-}18\,a)$$

where f_y = yield strength of the shear reinforcement. A_v is the area of the stirrups cut by a horizontal plane. V_s should not exceed $8\sqrt{f'_c}b_w\,d$ in sections with web reinforcement, nor should f_y exceed 60 ksi. Where shear reinforcement is required and is placed perpendicular to the axis of the member, it should not be spaced farther apart than $0.5d$, nor more than 24 in c to c. When V_s exceeds $4\sqrt{f'_c}b_w\,d$, however, the maximum spacing should be limited to $0.25d$.

Alternatively, for practical design, Eq. (3-18 a) can be transformed into Eq. (3-18 b) to indicate the stirrup spacing s for the design shear V_u, stirrup area A_v, and geometry of the member b_w and d:

$$s = \frac{A_v \phi f_y d}{V_u - 2\phi\sqrt{f'_c}b_w d} \qquad (3\text{-}18\,b)$$

The area required when a single bar or a single group of parallel bars are all bent up at the same distance from the support at angle α with the longitudinal axis of the member is

$$A_v = \frac{V_s}{f_y \sin \alpha} \qquad (3\text{-}18\,c)$$

in which V_s should not exceed $3\sqrt{f'_c}b_w d$. A_v is the area cut by a plane normal to the axis of the bars. The area required when a series of such bars are bent up at different distances from the support or when inclined stirrups are used is

$$A_v = \frac{V_s s}{(\sin \alpha + \cos \alpha)f_y d} \qquad (3\text{-}18\,d)$$

A minimum area of shear reinforcement is required in all members, except slabs, footings, and joists or where V_u is less than $0.5V_c$. When the ultimate torsion T_u, as discussed in the following, is less than the value calculated from Eq. (3-21), the area A_v of shear reinforcement should be at least

$$A_v = 50\frac{b_w s}{f_y} \qquad (3\text{-}19)$$

But when the ultimate torsion exceeds T_u calculated from Eq. (3-21) and where web reinforcement is required, either nominally or by calculation, the minimum area of closed stirrups required is

$$A_v + 2A_t = \frac{50 b_w s}{f_y} \qquad (3\text{-}20)$$

where A_t is the area of one leg of a closed stirrup resisting torsion within a distance s.

Shear or torsion reinforcement should extend the full depth d of the member and should be adequately anchored at both ends to develop the design yield strength of the reinforcement. While shear reinforcement may consist of stirrups (Fig. 3-9), bent-up longitudinal bars, spirals, or welded-wire fabric, torsion reinforcement should consist of closed ties, closed stirrups, or spirals—all combined with longitudinal bars. Closed ties or stirrups may be formed either in one piece by overlapping standard tie end hooks around a longitudinal bar (Fig. 3-9b) or in two pieces spliced as a Class C splice or adequately embedded. Pairs of U stirrups placed so as to form a closed unit should be lapped at least $1.7l_d$, where l_d is the development length (Art. 3-4).

Torsion effects should be considered whenever the ultimate torsion exceeds

$$T_u = \phi(0.5 \sqrt{f'_c} \Sigma x^2 y) \tag{3-21}$$

where ϕ = capacity reduction factor = 0.85
$\quad T_u$ = ultimate design torsional moment
$\quad \Sigma x^2 y$ = sum for component rectangles of section of product of square of shorter side and longer side of each rectangle (where T section applies, overhanging flange width used in design should not exceed three times flange thickness)
The torsion T_c carried by the concrete alone should not exceed

$$T_c = \frac{0.8 \sqrt{f'_c} \Sigma x^2 y}{1 + (0.4 V_u/C_t T_u)^2} \tag{3-22}$$

where $C_t = b_w d/\Sigma x^2 y$.

Torsion reinforcement should be provided in addition to that required for flexure, shear, and axial forces. The requirements for torsion reinforcement may be combined with those for other forces if the area provided equals or exceeds the sum of the individual required areas and if spacing of reinforcement meets the most restrictive of the spacing requirements.

Spacing of closed stirrups for torsion should be computed from

$$s = \frac{A_t \phi f_y \alpha_t x_1 y_1}{(T_u - \phi T_c)} \tag{3-23}$$

where A_t = area of one leg of closed stirrup
$\quad \alpha_t$ = $0.66 + 0.33 y_1/x_1$ but not more than 1.50
$\quad f_y$ = yield strength of torsion reinforcement
$\quad x_1$ = shorter dimension c to c of legs of closed stirrup
$\quad y_1$ = longer dimension c to c of legs of closed stirrup
The spacing of closed stirrups, however, should not exceed $(x_1 + y_1)/4$ nor 12 in. Torsion reinforcement should be provided over at least a distance of $d + b$ beyond the point where it is theoretically required, where b is the beam width.

At least one longitudinal bar should be placed in each corner of the stirrups. Size of longitudinal bars should be at least No. 3, and their spacing around the perimeters of the stirrups should not exceed 12 in. Longitudinal bars larger than No. 3 are required if indicated by the larger of the values of Al computed from Eqs. (3-24) and (3-25).

$$Al = 2A_t \frac{x_1 + y_1}{s} \tag{3-24}$$

$$Al = \left[\frac{400xs}{f_y} \left(\frac{T_u}{(T_u + V_u/3C_t)} \right) - 2A_t \right] \left(\frac{x_1 + y_1}{s} \right) \tag{3-25}$$

In Eq. (3-25), $50b_w s/f_y$ may be substituted for $2A_t$.
The maximum allowable torsion is $T_u = \phi 5 T_c$.

Development of Reinforcement ▪ To prevent bond failure or splitting, the calculated stress in any bar at any section must be developed on each side of the section by adequate embedment length, end anchorage, or hooks. The critical sections for development of reinforcement in flexural members are at points of maximum stress and at points within the span where adjacent reinforcement terminates. See Art. 3-14.

At least one-third of the positive-moment reinforcement in simple beams and one-fourth of the positive-moment reinforcement in continuous beams should extend along the same face of the member into the support, and for beams at least 6 in into the support. At simple supports and at points of inflection, the diameter of the reinforcement should be limited to a diameter such that the development length l_d defined in Art. 3-4 satisfies

$$l_d = \frac{M_n}{V_u} + l_a \qquad (3\text{-}26)$$

where M_n = computed flexural strength with all reinforcing steel at section stressed to f_y
   V_u = applied shear at section
   l_a = additional embedment length beyond inflection point or center of support
At an inflection point, l_a is limited to a maximum of d, the depth of the centroid of the reinforcement, or 12 times the reinforcement diameter.

Negative-moment reinforcement should have an embedment length into the span to develop the calculated tension in the bar, or a length equal to the effective depth of the member, or 12 bar diameters, whichever is greatest. At least one-third of the total negative reinforcement should have an embedment length beyond the point of inflection not less than the effective depth of the member, or 12 bar diameters, or one-sixteenth of the clear span, whichever is greatest.

Hooks on Bars ▪ Standard hooks are considered to develop a tensile stress in bar reinforcement of

$$f_h = \xi \sqrt{f_c'} \qquad (3\text{-}27)$$

where ξ has the values given in Table 3-5 for $f_y = 60$ ksi. For $f_y = 40$ ksi, $\xi = 360$ for No. 11 and smaller bars, 330 for No. 14 bars, and 220 for No. 18 bars. The given values may be increased 30% where enclosure is provided perpendicular to the plane of the hook.

An equivalent embedment length l_e for hooks may be computed from Eq. (3-4) by substituting l_e for l_d and f_h for f_y. Also, the development length may consist of a combination of embedment length plus the equivalent embedment length of a hook.

Hooks should not be considered effective in adding to the compressive resistance of reinforcement. Thus, hooks should not be used on footing dowels. Instead, when depth of footing is less than that required by large-size bars, the designer should substitute smaller-diameter bars with equivalent area and lesser embedment length. It may be possible sometimes to increase the footing depth where large-diameter dowel reinforcement is used so that footing dowels can have the proper embedment length. Footing dowels need only transfer the excess load above that transmitted in bearing and therefore may be bars with areas different from those required for compression design for the first column lift.

TABLE 3-5 ξ for $f_y = 60$ ksi

Bar size No.	Top bars	Other bars
3, 4, 5	540	540
6	450	540
7, 8, 9	360	540
10	360	480
11	360	420
14	330	330
18	220	220

(P. F. Rice and E. S. Hoffman, "Structural Design Guide to the ACI Building Code," Van Nostrand Reinhold Company, New York; "CRSI Handbook," Concrete Reinforcing Steel Institute, Schaumberg, Ill.; ACI SP-17, "Design Handbook in Accordance with the Strength Design Method of ACI 318-77," American Concrete Institute; G. Winter and A. H. Nilson, "Design of Concrete Structures," McGraw-Hill Book Company, New York.)

3-13. Working-Stress Design of Rectangular Beams with Tension Reinforcement Only

From the assumption that stress varies across a beam section with the distance from the neutral axis (Art. 3-10), it follows that (see Fig. 3-8d)

$$\frac{nf_c}{f_s} = \frac{k}{1-k} \tag{3-28}$$

where n = modular ratio E_s/E_c
  E_s = modulus of elasticity of steel reinforcement, psi
  E_c = modulus of elasticity of concrete, psi
  f_c = compressive stress in extreme surface of concrete, psi
  f_s = stress in steel, psi
  kd = distance from extreme compression surface to neutral axis, in
  d = distance from extreme compression to centroid of reinforcement, in
  When the steel ratio $\rho = A_s/bd$, where A_s = area of tension reinforcement, in^2, and b = beam width, in, is known, k can be computed from

$$k = \sqrt{2n\rho + (n\rho)^2} - n\rho \tag{3-29}$$

Wherever positive-moment steel is required, ρ should be at least $200/f_y$, where f_y is the steel yield stress. The distance jd between the centroid of compression and the centroid of tension, in, can be obtained from

$$j = 1 - \frac{k}{3} \tag{3-30}$$

The moment resistance of the concrete, in-lb, is

$$M_c = \tfrac{1}{2}f_c kjbd^2 = K_c bd^2 \tag{3-31}$$

where $K_c = \tfrac{1}{2}f_c kj$. The moment resistance of the steel is

$$M_s = f_s A_s jd = f_s \rho jbd^2 = K_s bd^2 \tag{3-32}$$

where $K_s = f_s \rho j$. Allowable stresses are given in Art. 3-10. Table 3-4 lists nominal diameters, weights, and cross-sectional areas of standard steel reinforcing bars.

Shear ▪ The nominal unit shear-stress acting on a section with shear V is

$$v = \frac{V}{bd} \tag{3-33}$$

Allowable shear stresses are 55% of those for ultimate-strength design (Art. 3-12). Otherwise, designs for shear by the working-stress and ultimate-strength methods are the same. Except for brackets and other short cantilevers, the section for maximum shear may be taken at a distance d from the face of the support. In working-stress design, the shear stress v_c carried by the concrete alone should not exceed $1.1\sqrt{f_c'}$. (As an alternative, the maximum for v_c may be taken as $\sqrt{f_c'} + 1300\rho Vd/M$, with a maximum of $1.9\sqrt{f_c'}$. M is the bending moment at the section but should not be less than Vd.)

At cross sections where the torsional stress v_t exceeds $0.825 \sqrt{f'_c}$, v_c should not exceed

$$v_c = \frac{1.1\sqrt{f'_c}}{\sqrt{1 + (v_t/1.2v)^2}} \tag{3-34}$$

The excess shear $v - v_c$ should not exceed $4.4\sqrt{f'_c}$ in sections with web reinforcement. Stirrups and bent bars should be capable of resisting the excess shear $V' = V - v_c bd$.

The area required in the legs of a vertical stirrup, in^2, is

$$A_v = \frac{V's}{f_v d} \tag{3-35}$$

where s = spacing of stirrups, in
  f_v = allowable stress in stirrup steel, psi (see Art. 3-10)
For a single bent bar or a single group of parallel bars all bent at an angle α with the longitudinal axis at the same distance from the support, the required area is

$$A_v = \frac{V'}{f_v \sin \alpha} \tag{3-36}$$

For inclined stirrups and groups of bars bent up at different distances from the support, the required area is

$$A_v = \frac{V's}{f_v d(\sin \alpha + \cos \alpha)} \tag{3-37}$$

Where shear reinforcing is required and the torsional moment T exceeds the value calculated from Eq. (3-38), the minimum area of shear reinforcement provided should be that given by Eq. (3-19).

Torsion ▪ Torsion effects should be considered whenever the torsion T due to service loads exceeds

$$T = 0.55(0.5f'_c \Sigma x^2 y) \tag{3-38}$$

where $\Sigma x^2 y$ = sum for the component rectangles of the section of the product of the square of the shorter side and the longer side of each rectangle. The allowable torsion stress on the concrete is 55% of that computed from Eq. (3-22). Spacing of closed stirrups for torsion should be computed from

$$s = \frac{3A_t \alpha_t x_1 y_1 f_v}{(v_t - v_{tc})\Sigma x^2 y} \tag{3-39}$$

where A_t = area of one leg of closed stirrup
  α_t = $0.66 + 0.33 y_1/x_1$ but not more than 1.50
  v_{tc} = allowable torsion stress on concrete
  x_1 = shorter dimension c to c of legs of closed stirrup
  y_1 = longer dimension c to c of legs of closed stirrup

Development of Reinforcement ▪ To prevent bond failure or splitting, the calculated stress in reinforcement at any section should be developed on each side of that section by adequate embedment length, end anchorage, or, for tension only, hooks. Requirements are the same as those given for ultimate-strength design in Art. 3-12. Embedment length required at simple supports and inflection points can be computed from Eq. (3-26) by substituting double the computed shears for V_u. In computation of M_t, the moment arm $d - a/2$ may be taken as $0.85d$ (Fig. 3-7). See also Art. 3-14.

3-14. Bar Cutoffs and Bend Points

It is common practice to stop or bend main reinforcement in beams and slabs where it is no longer required. But tensile steel should never be discontinued exactly at the theoretical cutoff or bend points. It is necessary to resist tensile forces in the reinforcement through embedment beyond those points.

All reinforcement should extend beyond the point at which it is no longer needed to resist flexure for a distance equal to the effective depth of the member or 12 bar diameters, whichever is greater. Lesser extensions, however, may be used at supports of a simple span and at the free end of a cantilever. See Art. 3-12 for embedment requirements at simple supports and inflection points and for termination of negative-moment bars. Continuing reinforcement should have an embedment length beyond the point where bent or terminated reinforcement is no longer required to resist flexure. The embedment should be at least as long as the development length l_d defined in Art. 3-4.

Flexural reinforcement should not be terminated in a tension zone unless one of the following conditions is satisfied:

1. Shear is less than two-thirds that normally permitted, including allowance for shear reinforcement, if any.

2. Continuing bars provide double the area required for flexure at the cutoff, and the shear does not exceed three-quarters of that permitted (No. 11 bar or smaller).

3. Stirrups in excess of those normally required are provided each way from the cutoff for a distance equal to 75% of the effective depth of the member. Area and spacing of the excess stirrups should be such that

$$A_v \geq 60 \frac{b_w s}{f_y} \tag{3-40}$$

where A_v = stirrup cross-sectional area, in^2
b_w = web width, in
s = stirrup spacing, in
f_y = yield strength of stirrup steel, psi

Stirrup spacing s should not exceed $d/8\beta_b$, where β_b is the ratio of the area of bars cut off to the total area of tension bars at the section and d is the effective depth of the member.

The location of theoretical cutoffs or bend points may usually be determined from bending moments since the steel stresses are approximately proportional to them. The bars generally are discontinued in groups or pairs. So, for example, if one-third the bars are to be bent up, the theoretical bend-up point lies at the section where the bending moment is two-thirds the maximum moment. The point may be found analytically or graphically.

(G. Winter and A. H. Nilson, "Design of Concrete Structures," McGraw-Hill Book Company, New York; P. F. Rice and E. S. Hoffman, "Structural Design Guide to the ACI Building Code," Van Nostrand Reinhold Company, New York, ACI 315, "Manual of Standard Practice for Detailing Reinforced Concrete Structures," American Concrete Institute.)

3-15. One-Way Slabs

If a slab supported on beams or walls spans a distance in one direction more than twice that in the perpendicular direction, so much of the load is carried on the short span that the slab may reasonably be assumed to be carrying all the load in that direction. Such a slab is called a one-way slab.

Generally, a one-way slab is designed by selecting a 12-in-wide strip parallel to the short direction and treating it as a rectangular beam. Reinforcing steel usually is spaced uniformly

TABLE 3-6   Areas of Bars in Slabs, in^2/ft of Slab

Spacing, in	Bar No.								
	3	4	5	6	7	8	9	10	11
3	0.44	0.78	1.23	1.77	2.40	3.14	4.00	5.06	6.25
3½	0.38	0.67	1.05	1.51	2.06	2.69	3.43	4.34	5.36
4	0.33	0.59	0.92	1.32	1.80	2.36	3.00	3.80	4.68
4½	0.29	0.52	0.82	1.18	1.60	2.09	2.67	3.37	4.17
5	0.26	0.47	0.74	1.06	1.44	1.88	2.40	3.04	3.75
5½	0.24	0.43	0.67	0.96	1.31	1.71	2.18	2.76	3.41
6	0.22	0.39	0.61	0.88	1.20	1.57	2.00	2.53	3.12
6½	0.20	0.36	0.57	0.82	1.11	1.45	1.85	2.34	2.89
7	0.19	0.34	0.53	0.76	1.03	1.35	1.71	2.17	2.68
7½	0.18	0.31	0.49	0.71	0.96	1.26	1.60	2.02	2.50
8	0.17	0.29	0.46	0.66	0.90	1.18	1.50	1.89	2.34
9	0.15	0.26	0.41	0.59	0.80	1.05	1.33	1.69	2.08
10	0.13	0.24	0.37	0.53	0.72	0.94	1.20	1.52	1.87
12	0.11	0.20	0.31	0.44	0.60	0.79	1.00	1.27	1.56

along both spans (Table 3-6). In addition to the main reinforcing in the short span, steel should be provided in the long direction to distribute concentrated loads and resist shrinkage and thermal stresses. The bars or wires should not be spaced farther apart than 18 in or five times the slab thickness.

For shrinkage and temperature stresses, ACI 318, "Building Code Requirements for Reinforced Concrete," requires the following minimum areas of steel, in^2/ft: deformed bars with yield strength less than 60,000 psi, 0.024; deformed bars with 60,000 psi yield strength or welded-wire fabric with wires not more than 12 in apart, 0.0216. For highway bridge slabs, "Standard Specifications for Highway Bridges" (American Association of State Highway and Transportation Officials) requires reinforcing steel in the bottoms of all slabs transverse to the main reinforcement for lateral distribution of wheel loads. The area of the distribution steel should be at least the following percentages of the main steel required for positive moment, where S is the effective span, ft (Art. 3-16): When main steel is parallel to traffic, $100/\sqrt{S}$, with a maximum of 50%; when the main steel is perpendicular to traffic, $200/\sqrt{S}$, with a maximum of 67%.

To control deflections, the ACI Code sets limitations on slab thickness unless deflections are computed and determined to be acceptable (Art. 3-11). Otherwise, thickness of one-way slabs must be at least $L/20$ for simply supported slabs; $L/24$ for slabs with one end continuous; $L/28$ for slabs with both ends continuous; and $L/10$ for cantilevers; where L is the span, in.

3-16. Design Spans for Beams and Slabs

ACI 318, "Building Code Requirements for Reinforced Concrete," specifies that the span of members not integral with supports should be taken as the clear span plus the depth of the member but not greater than the distance center to center of supports. For analysis of continuous frames, spans should be taken c to c of supports for determination of bending moments in beams and girders, but moments at the faces of supports may be used in the design of the members. Solid or ribbed slabs integral with supports and with clear spans up to 10 ft may be designed for the clear span.

"Standard Specifications for Highway Bridges" (American Association of State Highway and Transportation Officials) has the same requirements as the ACI Code for spans of simply supported beams and slabs. For slabs continuous over more than two supports, the effective span is the clear span for slabs monolithic with beams or walls (without haunches); the distance between stringer-flange edges plus half the stringer-flange width for slabs supported on steel stringers; clear span plus half the stringer thickness for slabs supported on timber stringers. For rigid frames, the span should be taken as the distance between centers of bearings at the top of the footings. The span of continuous beams should be the clear distance between faces of supports.

Where fillets or haunches make an angle of 45° or more with the axis of a continuous or restrained slab and are built integral with the slab and support, AASHTO requires that the span be measured from the section where the combined depth of the slab and fillet is at least 1.5 times the thickness of slab. The moments at the ends of this span should be used in the slab design, but no portion of the fillet should be considered as adding to the effective depth of the slab.

3-17. Rectangular Beams with Compression Bars— Ultimate-Strength Design

The steel ratio ρ_b for balanced conditions at ultimate strength of a rectangular beam is given by Eq. (3-14) in Art. 3-12. When the tensile steel ratio ρ exceeds $0.75\rho_b$, compression reinforcement should be used. When ρ is equal to or less than $0.75\rho_b$, the strength of the beam may be approximated by Eq. (3-15), disregarding any compression bars that may be present, since the strength of the beam will usually be controlled by yielding of the tensile steel.

The bending-moment capacity of a rectangular beam with both tension and compression steel is

$$M_u = 0.90 \left[(A_s - A_s')f_y\left(d - \frac{a}{2} \right) + A_s'f_y(d - d') \right] \tag{3-41}$$

where a = depth of equivalent rectangular compressive stress distribution
$\quad\quad = (A_s - A_s')f_y/f_c'b$
$\quad b$ = width of beam, in
$\quad d$ = distance from extreme compression surface to centroid of tensile steel, in
$\quad d'$ = distance from extreme compression surface to centroid of compressive steel, in
$\quad A_s$ = area of tensile steel, in^2
$\quad A_s'$ = area of compressive steel, in^2
$\quad f_y$ = yield strength of steel, psi
$\quad f_c'$ = 28-day strength of concrete, psi
Equation (3-41) is valid only when the compressive steel reaches f_y. This occurs when

$$(\rho - \rho') \geq 0.85\beta_1 \frac{f_c'd'}{f_yd} \frac{87,000}{87,000 - f_y} \tag{3-42}$$

where $\rho = A_s/bd$, $\rho' = A_s'/bd$, and β_1 is a constant defined in Art. 3-9. When $\rho - \rho'$ is less than the right-hand side of Eq. (3-42), calculate the moment capacity from Eq. (3-15) or from an analysis based on the assumptions of Art. 3-9. ACI 318, "Building Code Requirements for Reinforced Concrete," requires also that $\rho - \rho'$ not exceed $0.75\rho_b$ to avoid brittle failure of the concrete.

Compressive steel should be anchored by ties or stirrups at least ⅜ in in diameter and spaced no more than 16 bar diameters or 48 tie diameters apart. Tie reinforcement requirements are the same as those for columns.

Design for shear and development lengths of reinforcement is the same as for beams with tension reinforcement only (Art. 3-12).

3-18. Rectangular Beams with Compression Bars— Working-Stress Design

The following formulas, based on the linear variation of stress and strain with distance from the neutral axis (Fig. 3-8 in Art. 3-10) may be used in design:

$$k = \frac{1}{1 + f_s/nf_c} \tag{3-43}$$

where f_s = stress in tensile steel, psi
   f_c = stress in extreme compression surface, psi
   n = modular ratio, E_s/E_c

$$f'_s = \frac{kd - d'}{d - kd} 2f_s \tag{3-44}$$

where f'_s = stress in compressive steel, psi
   d = distance from extreme compression surface to centroid of tensile steel, in
   d' = distance from extreme compression surface to centroid of compressive steel, in

The factor 2 is incorporated into Eq. (3-44) in accordance with ACI 318, "Building Code Requirements for Reinforced Concrete," to account for the effects of creep and nonlinearity of the stress-strain diagram for concrete. But f'_s should not exceed the allowable tensile stress for the steel.

Since total compressive force equals total tensile force on a section,

$$C = C_c + C'_s = T \tag{3-45}$$

where C = total compression on beam cross section, lb
   C_c = total compression on concrete, lb, at section
   C'_s = force acting on compressive steel, lb
   T = force acting on tensile steel, lb

$$\frac{f_s}{f_c} = \frac{k}{2[\rho - \rho'(kd - d')/(d - kd)]} \tag{3-46}$$

where $\rho = A_s/bd$ and $\rho' = A'_s/bd$.

For reviewing a design, the following formulas may be used:

$$k = \sqrt{2n\left(\rho + \rho'\frac{d'}{d}\right) + n^2(\rho + \rho')^2} - n(\rho + \rho') \tag{3-47}$$

$$\bar{z} = \frac{(k^3d/3) + 4n\rho'd'[k - (d'/d)]}{k^2 + 4n\rho'[k - (d'/d)]} \tag{3-48}$$

$$jd = d - \bar{z} \tag{3-49}$$

where jd is the distance between the centroid of compression and the centroid of the tensile steel. The moment resistance of the tensile steel is

$$M_s = Tjd = A_s f_s jd \tag{3-50a}$$

$$f_s = \frac{M}{A_s jd} \tag{3-50b}$$

where M is the bending moment at the section of beam under consideration. The moment resistance in compression is

$$M_c = \frac{1}{2} f_c j b d^2 \left[k + 2n\rho' \left(1 - \frac{d'}{kd} \right) \right] \qquad (3\text{-}51\,a)$$

$$f_c = \frac{2M}{jbd^2\{k + 2n\rho'[1 - (d'/kd)]\}} \qquad (3\text{-}51\,b)$$

Solution of the preceding equations often is facilitated by tables and charts. Many designers, however, prefer the following approximate formulas:

$$M_1 = \frac{1}{2} f_c b k d \left(d - \frac{kd}{3} \right) \qquad (3\text{-}52)$$

$$M'_s = M - M_1 = 2f'_s A'_s(d - d') \qquad (3\text{-}53)$$

where M = bending moment
M'_s = moment-resisting capacity of compressive steel
M_1 = moment-resisting capacity of concrete

For determination of shear, see Art. 3-13. Compressive steel should be anchored by ties or stirrups at least No. 3 in size and spaced not more than 16 bar diameters or 48 tie diameters apart. At least one tie within the required spacing, throughout the length of the beam where compressive reinforcement is required, should extend completely around all longitudinal bars.

3-19. I and T Beams—Ultimate-Strength Design

A reinforced-concrete beam may be shaped in cross section like a T, or it may be composed of a slab and integral rectangular beam that, in effect, act as a T beam. According to ACI 318, "Building Code Requirements for Reinforced Concrete" (American Concrete Institute), and "Standard Specifications for Highway Bridges" (American Association of State Highway and Transportation Officials), when the slab forms the compression flange, its effective width b should not exceed one-fourth the beam span, and it should not be greater than the distance center to center of beams. In addition, the ACI Code requires that the overhanging width on either side of the beam web should not exceed eight times the slab thickness. The AASHTO Specifications more conservatively limit the effective width to 12 times the slab thickness plus the beam width. For beams with a flange on only one side, the effective overhanging flange width should not exceed one-twelfth the beam span, or six times the slab thickness, or half the clear distance to the next beam.

Two cases may occur in the design of T and I beams: The neutral axis lies in the compression flange (Fig. 3-10a and b) or in the web (Fig. 3-10c and d). For negative moment, a T beam should be designed as a rectangular beam with width b equal to that of the stem. (See Arts. 3-9 and 3-12.)

When the neutral axis lies in the flange, the member may be designed as a rectangular beam, with effective width b and depth d, by Eq. (3-15). For that condition, the flange thickness t will be greater than the distance from the extreme compression surface to the neutral axis,

$$c = \frac{1.18\omega d}{\beta_1} \qquad (3\text{-}54)$$

where β_1 = constant defined in Art. 3-9
$\omega = A_s f_y / b d f'_c$
A_s = area of tensile steel, in^2
f_y = yield strength of steel, psi
f'_c = 28-day strength of concrete

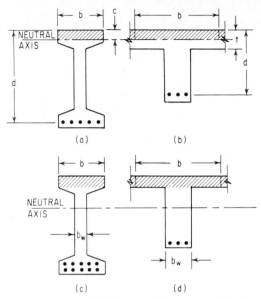

Fig. 3-10. I and T beams with neutral axis in the flange (*a*) and (*b*) are designed as rectangular beams. When the neutral axis lies in the web (*c*) and (*d*), the usual practice in design is to ignore the compression in the web.

When the neutral axis lies in the web, the ultimate moment should not exceed

$$M_u = 0.90 \left[(A_s - A_{sf}) f_y \left(d - \frac{a}{2} \right) + A_{sf} f_y \left(d - \frac{t}{2} \right) \right] \qquad (3\text{-}55)$$

where A_{sf} = area of tensile steel required to develop compressive strength of overhanging flange, in^2 = $0.85(b - b_w)tf'_c/f_y$

b_w = width of beam web or stem, in

a = depth of equivalent rectangular compressive stress distribution, in
= $(A_s - A_{sf})f_y/0.85f'_c b_w$

The quantity $\rho_w - \rho_f$ should not exceed $0.75\rho_b$, where ρ_b is the steel ratio for balanced conditions [Eq. (3-14)], $\rho_w = A_s/b_w d$, and $\rho_f = A_{sf}/b_w d$.

For determination of ultimate shear, see Art. 3-12. Note, however, that the web or stem width b_w should be used instead of b in these calculations.

3-20. I and T Beams—Working Stress Design

For T beams, effective width of compression flange is determined by the same rules as for ultimate-strength design (Art. 3-19). Also, for working-stress design, two cases may occur: The neutral axis may lie in the flange (Fig. 3-10*a* and *b*) or in the web (Fig. 3-10*c* and *d*). (For negative moment, a T beam should be designed as a rectangular beam with width b equal to that of the stem.) See Art. 3-13.

If the neutral axis lies in the flange, a T or I beam may be designed as a rectangular beam with effective width b. If the neutral axis lies in the web or stem, an I or T beam may be designed by the following formulas, which ignore the compression in the stem, as is customary:

$$k = \frac{I}{1 + f_s/nf_c} \tag{3-56}$$

where kd = distance from extreme compression surface to neutral axis, in
  d = distance from extreme compression surface to centroid of tensile steel, in
  f_s = stress in tensile steel, psi
  f_c = stress in concrete at extreme compression surface, psi
  n = modular ratio = E_s/E_c

Since the total compressive force C equals the total tension T,

$$C = \frac{1}{2}f_c(2kd - t)\frac{bt}{kd} = T = A_s f_s \tag{3-57}$$

$$kd = \frac{2ndA_s + bt^2}{2nA_s + 2bt} \tag{3-58}$$

where A_s = area of tensile steel, in^2
  t = flange thickness, in

The distance between the centroid of the area in compression and the centroid of the tensile steel is

$$jd = d - \bar{z} \tag{3-59}$$

$$\bar{z} = \frac{t(3kd - 2t)}{3(2kd - t)} \tag{3-60}$$

The moment resistance of the steel is

$$M_s = Tjd = A_s f_s jd \tag{3-61}$$

The moment resistance of the concrete is

$$M_c = Cjd = \frac{f_c btjd}{2kd}(2kd - t) \tag{3-62}$$

In design, M_s and M_c can be approximated by

$$M_s = A_s f_s \left(d - \frac{t}{2} \right) \tag{3-61 a}$$

$$M_c = \frac{1}{2}f_c bt \left(d - \frac{t}{2} \right) \tag{3-62 a}$$

derived by substituting $d - t/2$ for jd and $f_c/2$ for $f_c(1 - t/2kd)$, the average compressive stress on the section.

For determination of shear, see Art. 3-13. Note, however, that the web or stem width b_w should be used instead of b in these calculations.

3-21. Torsion in Reinforced-Concrete Members

Under twisting or torsional loads, a member develops normal (warping) and shear stresses. The warping, normal stresses help greatly in resisting torsion. But there are no accurate ways of computing this added resistance.

The maximum shears at any point are accompanied by equal tensile stresses on planes bisecting the angles between the planes of maximum shears.

As for ordinary shear, reinforcement should be incorporated to resist the diagonal tension in excess of the tensile capacity of the concrete. If web reinforcement is required for vertical shear in a horizontal beam subjected to both flexure and torsion, additional web reinforcement should be included to take care of the full torsional shear.

Design for torsion in combination with shear and requirements for torsion reinforcement are discussed in Art. 3-12.

3-22. Compression Members and Slenderness Effects

Very short compression members, piers or pedestals, may be unreinforced if the unit compressive stress on the cross-sectional area is less than the ultimate bearing stress $0.85\phi f'_c$, where f'_c is the 28-day compressive strength of the concrete and ϕ, the capacity reduction factor, is 0.70. The depth or width of an unreinforced pier or pedestal on soil shall be such that the flexural tensile stress in the concrete does not exceed $5\phi \sqrt{f'_c}$, where $\phi = 0.65$, when computed by the ultimate-strength method. The ratio of height to least dimension should not exceed 3 for unreinforced pedestals. In any event, pedestals must be designed as reinforced columns when loaded beyond the capacity of plain concrete.

In reinforced concrete columns, longitudinal steel bars help the concrete carry the load. Steel ties or spiral wrapping around those bars prevent the bars from buckling outward and spalling the outer concrete shell. Since spirals are more effective, columns with closely spaced spirals are allowed to carry greater loads than comparable columns with ties. Both types of columns may be designed by ultimate-load theory (Art. 3-23) or working stresses (Art. 3-24).

ACI 318, "Building Code Requirements for Reinforced Concrete," American Concrete Institute, sets limitations on columns geometry and reinforcement. Following are some of the more important.

Reinforcement Cover ▪ For cast-in-place columns, spirals and ties should be protected with a monolithic concrete cover of at least 1½ in. But for severe exposures, the amount of cover should be increased.

Spirals ▪ This type of transverse reinforcement should be at least ⅜ in in diameter. A spiral may be anchored at each of its ends by 1½ extra turns of the spiral. Splices may be made by welding or by a lap of 48 bar diameters (but at least 12 in). Spacing (pitch) of spirals should not exceed 3 in or be less than 1 in. Clear spacing should be at least 1⅓ times the maximum size of coarse aggregate.

A spiral should extend to the level of the lowest horizontal reinforcement in the slab, beam, or drop panel above. Where beams are of different depth or are not present on all sides of a column, ties should extend above the termination of the spiral to the bottom of the shallowest member. In a column with a capital, the spiral should extend to a plane at which the diameter or width of the capital is twice that of the column.

The ratio of the volume of spiral steel to volume of concrete core (out to out of spiral) should be at least

$$\rho_s = 0.45 \left(\frac{A_g}{A_c} - 1 \right) \frac{f'_c}{f_y} \tag{3-63}$$

where A_g = gross area of column
A_c = core area of column measured to outside of spiral
f_y = spiral steel yield strength
f'_c = 28-day compressive strength of concrete

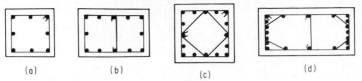

Fig. 3-11. Column ties provide lateral support at corners to alternate reinforcing bars.

Ties ■ Lateral ties should be at least ⅜ in in diameter for No. 10 or smaller bars and ½ in in diameter for No. 11 and larger bars. Spacing should not exceed 16 bar diameters, 48 tie diameters, or the least dimension of the column. The ties should be so arranged that every corner bar and alternate longitudinal bars will have lateral support provided by the corner of a tie having an included angle of not more than 135° (Fig. 3-11). No bar should be more than 6 in from such a laterally supported bar. Where bars are located around a circle, a complete circular tie may be used. (For more details, see ACI 315, "Manual of Standard Practice for Detailing Reinforced Concrete Structures," American Concrete Institute.)

Minimum Reinforcement ■ Columns should be reinforced with at least six longitudinal bars in a circular arrangement or with four longitudinal bars in a rectangular arrangement, at least No. 5 in bar size. Area of column reinforcement should not be less than 1% or more than 8% of the gross cross-sectional area of a column.

Excess Concrete ■ In a column that has a larger cross section than that required by load, the effective area A_g used to determine minimum reinforcement area and load capacity may be reduced proportionately, but not to less than half the total area.

Slenderness Ratios ■ Building columns generally are relatively short. Thus, an approximate evaluation of slenderness effects can usually be used in design. Slenderness, which is a function of column geometry and bracing, can reduce the load-carrying capacity of compression members by introducing bending stresses and can lead to a buckling failure.

Load-carrying capacity of a column decreases with increase in unsupported length l_u beyond a certain length. In buildings, l_u should be taken as the clear distance between floor slabs, girders, or other members capable of providing lateral support to the column or as the distance from a floor to a column capital or a haunch, if one is present.

In contrast, load-carrying capacity increases with increase in radius of gyration r of the column cross section. For rectangular columns, r may be taken as 30% of the overall dimension in the direction in which stability is being considered and for circular members as 25% of the diameter.

Also, the greater the resistance offered by a column to sidesway, or drift, because of lateral bracing or restraint against end rotations, the higher the load-carrying capacity. This resistance is represented by application of a factor k to the unsupported length of the column, and kl_u is referred to as the **effective length** of the column.

The combination of these factors, which is a measure of the slenderness of a column, kl_u/r, is called the slenderness ratio of the column.

The effective length factor k can be determined by analysis. If an analysis is not made, for compression members braced against sidesway, k should be taken as unity. For columns not braced against sidesway, k will be greater than unity; analysis should take into account the effects of cracking and reinforcement on relative stiffness.

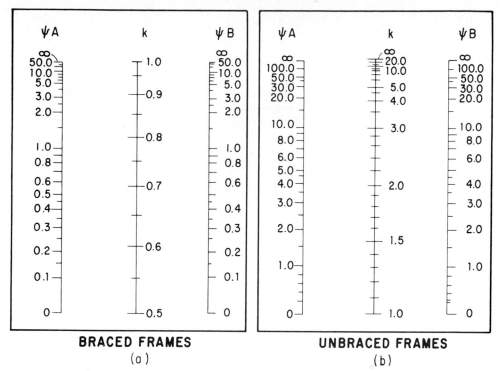

BRACED FRAMES

(a)

UNBRACED FRAMES

(b)

Fig. 3-12. Alignment charts for determination of effective-length factor k for columns. ψ is the ratio for each end of a column of $\Sigma EI/l_u$ for the compression members to $\Sigma EI/l$ for the girders.

ACI Committee 441 has proposed that k should be obtained from the Jackson and Moreland alignment chart, reproduced as Fig. 3-12 (Commentary on ACI 318-77, American Concrete Institute). For determination of k with this chart, a parameter ψ_A must be computed for end A of column AB, and a similar parameter ψ_B must be computed for end B. Each parameter equals the ratio at that end of the column of the sum of EI/l_u for the compression members meeting there to the sum of EI/l for the flexural members meeting there, where EI is the flexural stiffness of a member.

As a guide in judging whether a frame is braced or unbraced, note that the Commentary on ACI 318-77 indicates that a frame may be considered braced if the bracing elements, such as shear walls, shear trusses, or other means resisting lateral movement in a story, have a total stiffness at least six times the sum of the stiffnesses of all the columns resisting lateral movement in that story.

The slenderness effect may be neglected under the following conditions:

For columns braced against sidesway, when

$$\frac{kl_u}{r} < 34 - 12 \frac{M_1}{M_2} \tag{3-64}$$

where M_1 = smaller of two end moments on column as determined by conventional elastic frame analysis, with positive sign if column is bent in single curvature and negative sign if column is bent in double curvature

M_2 = absolute value of larger of the two end moments on column as determined by conventional elastic frame analysis

For columns not braced against sidesway, when

$$\frac{kl_u}{r} < 22 \tag{3-65}$$

Column Design Loads ▪ Analysis taking into account the influence of axial loads and variable moment of inertia on member stiffness and fixed-end moments, the effects of deflections on moments and forces, and the effects of duration of loads is required for all columns when

$$\frac{kl_u}{r} > 100 \tag{3-66}$$

For columns for which the slenderness ratio lies between 22 and 100, and therefore the slenderness effect on load-carrying capacity must be taken into account, either an elastic analysis can be performed to evaluate the effects of lateral deflections and other effects producing secondary stresses, or an approximate method presented in the ACI Code may be used. In the approximate method, the column is designed for the design axial load P_u and a magnified moment M_c defined by

$$M_c = \delta M_2 \tag{3-67}$$

where δ is the magnification factor, a function of the shape of the deflected column. δ may be determined from

$$\delta = \frac{C_m}{1 - P_u/\phi P_c} \geq 1 \tag{3-68}$$

where C_m = factor relating actual moment diagram to that for equivalent uniform moment
ϕ = capacity reduction factor = 0.75 for spiral-reinforced columns, otherwise 0.70
P_c = critical load for column
For members braced against sidesway and without transverse loads between supports,

$$C_m = 0.6 + 0.4 \frac{M_1}{M_2} \geq 0.4 \tag{3-69}$$

For other members, $C_m = 1$.
    The critical load is given by

$$P_c = \frac{\pi^2 EI}{(kl_u)^2} \tag{3-70}$$

where EI is the flexural stiffness of the column.
    The flexural stiffness EI may be computed approximately from

$$EI = \frac{E_c I_g/2.5}{1 + \beta_d} \tag{3-71}$$

where E_c = modulus of elasticity of concrete
    I_g = moment of inertia about centroidal axis of gross concrete section, neglecting load reinforcement
    β_d = ratio of maximum design dead load to total load moment (always taken positive)
    Because a column has different properties, such as stiffness, slenderness ratio, and δ, in different directions, it is necessary to check the strength of a column in each of its two principal directions.
    (G. Winter and A. H. Nilson, "Design of Concrete Structures," McGraw-Hill Book Com-

pany, New York; P. F. Rice and E. S. Hoffman, "Structural Design Guide to the ACI Building Code," Van Nostrand Reinhold Company, New York.)

3-23. Ultimate-Strength Design of Columns

At ultimate strength P_u, columns should be capable of sustaining loads as given by Eqs. (3-7) to (3-9) at actual eccentricities. P_u may not exceed ϕP_n, where ϕ is the capacity reduction factor and P_n is the column ultimate strength. If P_o is the column ultimate strength with zero eccentricity of load,

$$P_o = 0.85f'_c(A_g - A_{st}) + f_yA_{st} \qquad (3\text{-}72a)$$

where f_y = yield strength of reinforcing steel, psi
$\quad f'_c$ = 28-day compressive strength of concrete, psi
$\quad A_g$ = gross area of column, in^2
$\quad A_{st}$ = area of steel reinforcement, in^2
For members with spiral reinforcement then, for axial loads only,

$$P_u \leq 0.85\phi P_o \qquad (3\text{-}72b)$$

For members with tie reinforcement, for axial loads only,

$$P_u \leq 0.80\phi P_o \qquad (3\text{-}72c)$$

Eccentricities are measured from the plastic centroid. This is the centroid of the resistance to load computed for the assumptions that the concrete is stressed uniformly to $0.85f'_c$ and the steel is stressed uniformly to f_y.

Columns are designed under the ACI Code assumptions and requirements pertaining to members subject to combined flexure and axial load (Arts. 3-9 and 3-22). These assumptions are:

1. Loads and stresses are in equilibrium and strains are compatible.

2. Strains in both reinforcing steel and concrete are proportional to distance from the neutral axis.

3. The maximum strain at the extreme concrete compression surface does not exceed 0.003 in/in.

4. Stress in the reinforcement is E_s times the steel strain, where E_s is the modulus of elasticity of the steel, and the stress does not exceed f_y.

5. Tensile strength of concrete is negligible.

6. The concrete block may be taken as rectangular (Fig. 3-7c), with a concrete stress equal to $0.85f'_c$ extending from the extreme compression surface to a line parallel to the neutral axis and at a distance $a = \beta_1 c$ from the extreme compression surface. $\beta_1 = 0.85$ for concrete strengths up to $f'_c = 4000$ psi and decreases at a rate of 0.05 for each 1000 psi of strength above 4000 psi, and c is the distance from the extreme compression surface to the neutral axis.

Strength computed in accordance with these assumptions should be modified by a capacity reduction factor ϕ. It is equal to 0.75 for columns with spiral reinforcement and 0.70 for tied columns. A larger value of ϕ may be used for small design axial compression loads P_u. For symmetrically reinforced columns, ϕ generally may be increased when $P_u \leq 0.10f'_cA_g$, where A_g is the gross area of the section. For unsymmetrically reinforced columns, ϕ may be increased when P_u is less than the smaller of $0.10f'_cA_g$ and the design axial load for balanced conditions P_b. (In that case, the reinforcement ratio ρ should not exceed $0.75\rho_b$ for balanced conditions, as indicated for beams in Art. 3-12. Balanced conditions exist at a cross section when tension reinforcement reaches f_y just as the concrete in compression reaches its ultimate strain of 0.003 in/in.) When $P_u = 0$, $\phi = 0.90$, the capacity reduction factor for pure bending. ϕ may be assumed to increase linearly to 0.90 as P_u decreases from $0.10f'_cA_g$ or P_b to zero.

The axial-load capacity P_u, lb, of short, rectangular members subject to axial load and bending may be determined from

$$P_u = \phi(0.85f'_cba + A'_sf_y - A_sf_s) \qquad (3\text{-}73\,a)$$

$$P_ue' = \phi\left[0.85f'_cba\left(d - \frac{a}{2}\right) + A'_sf_y(d - d')\right] \qquad (3\text{-}73\,b)$$

where e' = eccentricity, in, of axial load at end of member with respect to centroid of tensile reinforcement, calculated by conventional methods of frame analysis
b = width of compression face, in
a = depth of equivalent rectangular compressive-stress distribution, in
A'_s = area of compressive reinforcement, in^2
A_s = area of tension reinforcement, in^2
d = distance from extreme compression surface to centroid of tensile reinforcement, in
d' = distance from extreme compression surface to centroid of compression reinforcement, in
f_s = tensile stress in steel, psi

Equations (3-73 a) and (3-73 b) assume that a does not exceed the column depth, that reinforcement is in one or two faces, parallel to axis of bending, and that reinforcement in any face is located at about the same distance from the axis of bending. Whether the compression steel will actually yield at ultimate strength, as assumed in these and the following equations, can be verified by strain compatibility calculations. That is, when the concrete crushes the strain in the compression steel, $0.003(c - d')/c$, must be larger than the strain when the steel starts to yield, f_y/E_s, where c is the distance, in, from the extreme compression surface to the neutral axis and E_s is the modulus of elasticity of the steel, psi.

The load P_b for balanced conditions can be computed from Eq. (3-73 a) with $f_s = f_y$ and

$$a = a_b = \beta_1c_b = \frac{87,000\beta_1d}{87,000 + f_y} \qquad (3\text{-}74)$$

The balanced moment can be obtained from

$$M_b = P_be_b = \phi\left[0.85f'_cba_b\left(d - d'' - \frac{a_b}{2}\right) + A'_sf_y(d - d' - d'') + A_sf_yd''\right] \qquad (3\text{-}75)$$

where e_b is the eccentricity, in, of the axial load with respect to the plastic centroid, and d'' is the distance, in, from plastic centroid to centroid of tension reinforcement.

When P_u is less than P_b or the eccentricity e is greater than e_b, tension governs. In that case, for unequal tension and compression reinforcement, the ultimate strength is

$$P_u = 0.85f'_cbd\phi\left\{\rho'm' - \rho m + \left(1 - \frac{e'}{d}\right)\right.$$

$$\left. + \sqrt{\left(1 - \frac{e'}{d}\right)^2 + 2\left[(\rho m - \rho'm')\frac{e'}{d} + \rho'm'\left(1 - \frac{d'}{d}\right)\right]}\right\} \qquad (3\text{-}76)$$

where $m = f_y/0.85f'_c$
$m' = m - 1$
$\rho = A_s/bd$
$\rho' = A'_s/bd$

For symmetrical reinforcement in two faces, Eq. (3-76) becomes

$$P_u = 0.85 f'_c bd\phi \left\{ -\rho + 1 - \frac{e'}{d} \right.$$

$$\left. + \sqrt{\left(1 - \frac{e'}{d}\right)^2 + 2\rho \left[m'\left(1 - \frac{d'}{d}\right) + \frac{e'}{d}\right]} \right\} \quad (3\text{-}77)$$

For no compression reinforcement, Eq. (3-76) becomes

$$P_u = 0.85 f'_c bd\phi \left[-\rho m + 1 - \frac{e'}{d} + \sqrt{\left(1 - \frac{e'}{d}\right)^2 + 2\frac{e'\rho m}{d}} \right] \quad (3\text{-}78)$$

When P_u is greater than P_b, or e is less than e_b, compression governs. In that case, the ultimate strength is approximately

$$P_u = P_o - (P_o - P_b)\frac{M_u}{M_b} \quad (3\text{-}79\,a)$$

$$P_u = \frac{P_o}{1 + (P_o/P_b - 1)(e/e_b)} \quad (3\text{-}79\,b)$$

where M_u = moment capacity under combined axial load and bending, in-lb
P_o = axial-load capacity of member when concentrically loaded, lb, as given by Eq. (3-72a)

For symmetrical reinforcement in single layers, the ultimate strength when compression governs in a column with depth h may be computed from

$$P_u = \phi \left(\frac{A'_s f_y}{\dfrac{e}{d - d'} + 0.5} + \frac{bhf'_c}{\dfrac{3he}{d^2} + 1.18} \right) \quad (3\text{-}80)$$

Ultimate strength of short, circular members with bars in a circle may be determined by the theory discussed in Art. 3-9 or from the following:
When tension controls:

$$P_u = 0.85 f'_c D^2 \phi \left[\sqrt{\left(\frac{0.85e}{D} - 0.38\right)^2 + \frac{\rho_t m D_s}{2.5D}} - \left(\frac{0.85e}{D} - 0.38\right) \right] \quad (3\text{-}81)$$

where D = overall diameter of section, in
D_s = diameter of circle through reinforcement, in
$\rho_t = A_{st}/A_g$
When compression governs:

$$P_u = \phi \left[\frac{A_{st} f_y}{\dfrac{3e}{D_s} + 1} + \frac{A_g f'_c}{\dfrac{9.6De}{(0.8D + 0.67D_s)^2} + 1.18} \right] \quad (3\text{-}82)$$

The eccentricity for the balanced condition is given approximately by

$$e_b = (0.24 + 0.39\rho_t m)D \quad (3\text{-}83)$$

Ultimate strength of short, square members with depth h and with bars in a circle may be computed from the following:

When tension controls:

$$P_u = 0.85 bh f_c' \phi \left[\sqrt{\left(\frac{e}{h} - 0.5 \right)^2 + 0.67 \frac{D_s}{h} \rho_t m} - \left(\frac{e}{h} - 0.5 \right) \right] \quad (3\text{-}84)$$

When compression governs:

$$P_u = \phi \left[\frac{A_{st} f_y}{\dfrac{3e}{D_s} + 1} + \frac{A_g f_c'}{\dfrac{12he}{(h + 0.67 D_s)^2} + 1.18} \right] \quad (3\text{-}85)$$

When the slenderness of a column has to be taken into account, the eccentricity should be determined from $e = M_c / P_u$, where M_c is the magnified moment given by Eq. (3-67).

As outside temperatures vary, exposed columns in tall buildings may undergo large changes in length relative to interior columns. The resulting floor warpage may crack partitions unless they are detailed to take the cracking.

(P. F. Rice and E. S. Hoffman, "Structural Design Guide to the ACI Building Code," Van Nostrand Reinhold Company, New York; "CRSI Handbook," Concrete Reinforcing Steel Institute, Schaumberg, Ill.; "Design Handbook in Accordance with the Strength Design Method of ACI 318-77," American Concrete Institute, Detroit, Mich.; "Strength Design of Reinforced Concrete Column Sections" (computer program), Portland Cement Association, Old Orchard Road, Skokie, Ill. 60076.)

3-24. Working-Stress Design of Columns

In working-stress design, the capacity of a column is taken as 40% of that determined by the ultimate-strength method (Art. 3-23) with $\phi = 1$. The capacity should be equal to or greater than the service loads on the column. (See also Art. 3-22.)

3-25. Walls

Concrete walls may be classified as non-load bearing, load-bearing, or shear walls. The last may be load-bearing or non-load-bearing.

Non-Load-Bearing Walls ▪ These are generally basement, retaining, or facade-type walls that support only their own weight and also resist lateral loads. Such walls are principally designed for flexure. By the ACI Code, design requirements include:

1. Ratio of vertical reinforcement to gross concrete area should be at least 0.0012 for deformed bars No. 5 or smaller, 0.0015 for deformed bars No. 6 and larger, and 0.0012 for welded-wire fabric not larger than ⅝ in in diameter.

2. Spacing of vertical bars should not exceed three times the wall thickness or 18 in.

3. Lateral or cross ties are not required if the vertical reinforcement is 1% or less of the concrete area, or where the vertical reinforcement is not required as compression reinforcement.

4. Ratio of horizontal reinforcement to gross concrete area should be at least 0.0020 for deformed bars No. 5 or smaller, 0.0025 for deformed bars No. 6 and larger, and 0.0020 for welded-wire fabric not larger than ⅝ in in diameter.

5. Spacing of horizontal bars should not exceed 3 times the wall thickness or 18 in.

Note that there is no requirement for the reinforcement to be placed at both faces of the wall, but it is good practice to provide nominal reinforcing to control shrinkage in the nonstressed face of foundation walls over 10 or 12 ft in height and also at the faces of walls exposed to view.

Load-Bearing Walls ▪ These are subject to axial compression loads in addition to their own weight and, where there is eccentricity of load or lateral loads, to flexure. Load-bearing walls may be designed in a manner similar to that for columns (Arts. 3-23 and 3-24), but including the preceding design requirements for non-load-bearing walls. As an alternative, load-bearing walls may be designed by an empirical procedure given in the ACI Code when the eccentricity of the resulting compressive load is equal to or less than one-sixth the thickness of the wall.

In the empirical method the axial capacity of the wall is

$$\phi P_n = 0.55 \phi f'_c A_g \left[1 - \left(\frac{l_c}{40h} \right)^2 \right] \tag{3-86}$$

where f'_c = 28-day compressive strength of concrete, psi
A_g = gross area of wall section, in^2
ϕ = capacity reduction factor = 0.70
l_c = vertical distance between supports, in
h = overall thickness of wall, in

The effective length of wall supporting a concentrated load should be taken as the smaller of the distance center to center between loads and the bearing width plus $4h$.

Reinforced bearing walls designed using Eq. (3-86) should have a thickness of at least $\frac{1}{25}$ of the unsupported height or width, whichever is shorter, but not less than 6 in. Also, walls more than 10 in thick, except for basement walls, should have two layers of reinforcement in each direction, with between one-half and two-thirds of the total steel area in the layer near the exterior face of the wall. This layer should be placed at least 2 in but not more than one-third the wall thickness from the face. Walls should be anchored to the floors, or to the columns, pilasters, or intersecting walls.

If structural analysis indicates adequate strength and stability with less thickness and reinforcement than the ACI Code requires, less may be used.

Walls designed as grade beams should have top and bottom reinforcement as required by the ACI Code for beam design.

Shear Walls ▪ Walls subject to horizontal shear forces in the plane of the wall should, in addition to satisfying flexural requirements, be capable of resisting the shear. The nominal shear stress can be computed from

$$v_u = \frac{V_u}{\phi h d} \tag{3-87}$$

where V_u = total design shear force
ϕ = capacity reduction factor = 0.85
$d = 0.8 l_w$
h = overall thickness of wall

The shear V_c carried by the concrete depends on whether N_u, the design axial load, lb, normal to the wall horizontal cross section and occurring simultaneously with V_u at the section, is a compression or tension force. When N_u is a compression force, V_c may be taken as $2\sqrt{f'_c}hd$. When N_u is a tension force, V_c should be taken as the smaller of the values calculated from Eq. (3-88 a) or (3-88 b).

$$V_c = 3.3 \sqrt{f'_c}hd - \frac{N_u d}{4 l_w} \tag{3-88 a}$$

where l_w = horizontal length of wall

$$V_c = hd \left[0.6 \sqrt{f_c'} + \frac{l_w (1.25 \sqrt{f_c'} - 0.2N_u/l_w h)}{M_u/V_u - l_w/2} \right] \tag{3-88b}$$

Equation $(3-88b)$ does not apply, however, when $M_u/V_u - l_w/2$ is negative.

When the factored shear V_u is less than $0.5\phi V_c$, reinforcement should be provided as required by the empirical method for bearing walls.

When V_u exceeds $0.5\phi V_c$, horizontal reinforcement should be provided in accordance with Eq. $(3-18a)$, with $V_s = A_v f_y d/s_2$, where s_2 = spacing of horizontal reinforcement and A_v = reinforcement area. Also, the ratio ρ_h of horizontal shear reinforcement to the gross concrete area of the vertical section of the wall should be at least 0.0025. Spacing of horizontal shear bars should not exceed $l_w/5$, $3h$, or 18 in. In addition, the ratio of vertical shear reinforcement area to gross concrete area of the horizontal section of wall need not be greater than that required for horizontal reinforcement but should not be less than

$$\rho_n = 0.0025 + 0.5 \left(2.5 - \frac{h_w}{l_w} \right) (\rho_h - 0.0025) \le 0.0025 \tag{3-89}$$

where h_w = total height of wall. Spacing of vertical shear reinforcement should not exceed $l_w/3$, $3h$, or 18 in.

In no case should the shear strength V_n be taken greater than $10 \sqrt{f_c'}hd$ at any section.

3-26. Composite Columns

A composite column consists of a structural steel shape, pipe, or tube compression member completely encased in concrete, with or without longitudinal reinforcement.

Composite compression members should be designed in accordance with the provisions applicable to ordinary reinforced concrete columns. Loads assigned to the concrete portion of a member must be transferred by direct bearing on the concrete through brackets, plates, reinforcing bars, or other structural shapes that have been welded to the central structural steel compression members prior to placement of the perimeter concrete. The balance of the load should be assigned to the structural steel shape and should be developed by direct connection to the structural shape.

Concrete-Filled Steel Columns ▪ When the composite member consists of a steel-encased concrete core, the required thickness of metal face of width b of a rectangular section is

$$t = b \sqrt{\frac{f_y}{3E_s}} \tag{3-90a}$$

and for circular sections of diameter h,

$$t = h \sqrt{\frac{f_y}{8E_s}} \tag{3-90b}$$

where f_y is the yield strength and E_s the modulus of elasticity of the steel.

Steel-Core Columns ▪ When the composite member consists of a spiral-bound concrete encasement around a structural-steel core, the concrete should have a minimum strength of 2500 psi, and spiral reinforcement should conform to the requirements of Art. 3-22.

When the composite member consists of a laterally tied concrete encasement around a steel core, the concrete should have a minimum strength of 2500 psi. The lateral ties should completely encase the core. Ties should be No. 3 bars or larger but should have a diameter of at least 1/50 the longest side of the cross section. Vertical spacing should not exceed one-half of the least width of the cross section, or 48 tie bar diameters, or 16 longitudinal bar diameters. The area of vertical reinforcing bars within the ties should not be less than 1% or more than 8% of the net concrete section. In rectangular sections, one longitudinal bar should be placed in each corner and other bars, if needed, spaced no farther apart than half the least side dimension of the section.

The design yield strength of the structural core should not be taken greater than 50 ksi, even though a larger yield strength may be specified.

3-27. Basic Principles of Prestressed Concrete

Prestressing is the application of permanent forces to a member or structure to counteract the effects of subsequent loading. Applied to concrete, prestressing takes the form of precompression, usually to eliminate disadvantages stemming from the weakness of concrete in tension. The usual procedure is to tension high-strength steel (Art. 3-5) and anchor it to the concrete, which resists the tendency of the stretched steel to shorten and thus is compressed. The amount of prestress used generally is sufficient to prevent cracking or sometimes to avoid tension entirely, under service loads. As a result, the whole concrete cross section is available to resist tension and bending, whereas in reinforced concrete construction, concrete in tension is considered ineffective. Hence, it is particularly advantageous with prestressed concrete to use high-strength concrete. (See also Art. 3-6.)

Prestressed-concrete pipe and tanks are made by wrapping steel wire under high tension around concrete cylinders. Domes are prestressed by wrapping tensioned steel wire around the ring girders. Beams and slabs are prestressed linearly with steel tendons anchored at their ends or bonded to the concrete (Art. 3-6). Piles also are prestressed linearly, usually to counteract handling stresses.

The final precompression of the concrete is not equal to the initial tension applied to the tendons. There are immediate losses, for example, due to elastic shortening of the concrete, frictional losses from curvature of the tendons, and slip at anchorages. And there are long-time losses, such as those due to shrinkage and creep of the concrete and possibly relaxation of the steel. These losses should be computed as accurately as possible, determined experimentally, or estimated. They should be deducted from the initial prestress to determine the effective prestress to be used in design. One reason high-tensioned tendons are used for prestressing is to maintain the sum of these losses at a small percentage of the applied prestress.

In determining stresses in prestressed members, the prestressing forces may be treated the same way as other external loads. If the prestress is large enough to prevent cracking under design loads, elastic theory may be applied to the entire concrete cross section.

For example, consider the simple beam in Fig. 3-13a. Prestress P is applied by a straight tendon at a distance e_1 below the neutral axis. The resulting prestress in the extreme surfaces throughout equals $P/A \pm Pe_1c/I$, where P/A is average stress on a cross section and Pe_1c/I, the bending stress ($+$ represents compression, $-$ represents tension), as indicated in Fig. 3-13c. If, now, stresses $\pm Mc/I$ due to downward-acting loads are superimposed at midspan, the net stresses in the extreme surfaces may become zero at the bottom and compressive at the top (Fig. 3-13c). Since the stresses due to loads at the beam ends are zero, however, the prestress is the final stress there. Hence, the top of the beam at the ends will be in tension.

If this is objectionable, the tendons may be draped, or harped, in a vertical curve, as shown in Fig. 3-13b. Stresses at midspan will be substantially the same as before (assuming the horizontal component of P approximately equal to P), and the stress at the ends will be a compression, P/A, since P passes through the centroid of the section there. Between midspan and the ends, the cross sections also are in compression (Fig. 3-13d).

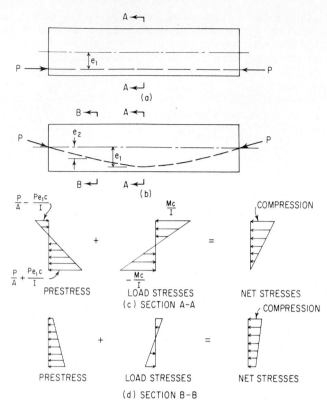

Fig. 3-13. Concrete beams may be prestressed with straight tendons (*a*) or curved (*b*). Stresses at midspan may be the same for both positions (*c*), but with the curved tendons, net stresses may remain compressive away from midspan (*d*), whereas they become tensile near the ends with straight tendons.

3-28. Losses in Prestress

As pointed out in Art. 3-27, the prestressing force acting on the concrete differs from the initial tension on the tendons by losses that occur immediately and over a long time.

Elastic Shortening of Concrete ▪ In pretensioned members (Art. 3-6), when the tendons are released from fixed abutments and the steel stress is transferred to the concrete by bond, the concrete shortens because of the compressive stress. For axial prestress, the decrease in inches per inch of length may be taken as P_i/AE_c, where P_i is the initial prestress, lb; A the concrete area, in^2; and E_c the modulus of elasticity of the concrete, psi. Hence, the decrease in unit stress in the tendons equals $P_iE_s/AE_c = nf_c$, where E_s is the modulus of elasticity of the steel, psi; n the modular ratio; and f_c the stress in the concrete, psi.

In posttensioned members, if tendons or cables are tensioned individually, the stress loss in each due to compression of the concrete depends on the order of tensioning. The loss will be greatest for the first tendon or cable tensioned and least for the last one. The total loss may be approximated by assigning half the loss in the first cable to all. As an alternative, the tendons may be brought to the final prestress in steps.

TABLE 3-7   Friction Coefficients for Posttensioned Tendons

Types of tendons and sheathing	Wobble coefficient, K	Curvature coefficient μ
Grouted tendons in metal sheathing:		
Wire tendons	0.0010–0.0015	0.15–0.25
High-strength bars	0.0001–0.0006	0.08–0.30
7-wire strand	0.0005–0.0020	0.15–0.25
Unbonded tendons		
Mastic-coated:		
Wire tendons	0.001–0.002	0.05–0.15
7-wire-strand	0.001–0.002	0.05–0.15
Pregreased:		
Wire tendons	0.0003–0.002	0.05–0.15
7-wire strand	0.0003–0.002	0.05–0.15

Frictional Losses ▪ In posttensioned members, there may be a loss of prestress where curved tendons rub against their enclosure. For harped tendons, the loss may be computed in terms of a curvature-friction coefficient μ. Losses due to unintentional misalignment may be calculated from a wobble-friction coefficient K (per lin ft). Since the coefficients vary considerably with duct material and construction methods, they should, if possible, be determined experimentally or obtained from the tendon manufacturer. Table 3-7 lists values of K and μ suggested in the Commentary of ACI 318-77, American Concrete Institute, for posttensioned tendons.

With K and μ known or estimated, the friction loss can be computed from

$$P_s = P_x e^{Kl+\mu\alpha} \tag{3-91}$$

where P_s = force in tendon at prestressing jack, lb
    P_x = force in tendon at any point x ft from jack, lb
    e = 2.718
    l = length of tendon from jacking point to point x, ft
    α = total angular change of tendon profile from jacking end to point x, rad
When $Kl + \mu\alpha$ does not exceed 0.3, P_s may be obtained from

$$P_s = P_x(1 + Kl + \mu\alpha) \tag{3-92}$$

Slip at Anchorages ▪ For posttensioned members, prestress loss may occur at the anchorages during the anchoring. For example, seating of wedges may permit some shortening of the tendons. If tests of a specific anchorage device indicate a shortening δl, the decrease in unit stress in the steel is $E_s \delta l / l$, where l is the length of the tendon.

Shrinkage of Concrete ▪ Change in length of a member due to concrete shrinkage results over time in prestress loss. This should be determined from test or experience. Generally, the loss is greater for pretensioned members than for posttensioned members, which are prestressed after much of the shrinkage has occurred. Assuming a shrinkage of 0.0002 in/in for a pretensioned member, the loss in tension in the tendons will be

$$0.0002E_s = 0.0002 \times 30 \times 10^6 = 6000 \text{ psi}$$

Creep of Concrete ▪ Change in length of concrete under sustained load induces a prestress loss over time. This loss may be several times the elastic shortening. An estimate of the loss may be made with a creep coefficient C_c, equal to the ratio of additional long-time deformation to initial elastic deformation, determined by test. Hence, for axial prestress, the loss in tension in the steel is $C_c n f_c$, where n is the modular ratio and f_c is the prestressing force divided by the concrete area. (Values ranging from 1.5 to 2.0 have been recommended for C_c.)

Relaxation of Steel ▪ Decrease in stress under constant high strain occurs with some steels. For example, for steel tensioned to 60% of ultimate strength, relaxation loss may be 3%. This type of loss may be reduced by temporary overtensioning, stabilizing the tendons by artificially accelerating relaxation and thus reducing the loss that will occur later at lower stresses.

The Commentary on ACI 318-77 and the "Standard Specifications for Highway Bridges" (American Association of State Highway and Transportation Officials) permit the sum of the prestress losses in the steel from all causes except friction to be assumed as 35,000 psi for pretensioned members and 25,000 psi for posttensioned members.

3-29. Allowable Stresses in Prestressed Concrete

In setting allowable stresses for prestressed concrete, design codes recognize two loading stages: application of initial stress and loading under service conditions. The codes permit higher stresses for the temporary loads during the initial stage.

Stresses due to the jacking force and those produced in the concrete and steel immediately after prestress transfer or tendon anchorage, before losses due to creep and shrinkage, are considered temporary. Permissible temporary stresses in the concrete are specified as a percentage of f'_{ci}, the compressive strength of the concrete, psi, at time of initial prestress, instead of the usual f'_c, 28-day strength of the concrete. This is done because prestress usually is applied only a few days after casting the concrete. With f_{pu} as the ultimate strength of tendons, the allowable stresses for prestressed concrete, in accordance with ACI 318, "Building Code Requirements for Reinforced Concrete" (American Concrete Institute), are in Table 3-8.

Bearing stress on the concrete from anchorages of posttensioned members with adequate

TABLE 3-8   Allowable Stresses in Prestressed Concrete Flexural Members

Stresses at transfer or anchoring:	
Compression in concrete	$0.60f'_{ci}$
Tension in concrete without auxiliary reinforcement in the tension zone*	$3\sqrt{f'_{ci}}$
Prestress in tendons due to jacking force†	$0.80f_{pu}$
Prestress in tendons immediately after transfer or anchoring	$0.70f_{pu}$
Stresses under service loads:	
Compression in concrete	$0.45f'_c$
Tension in concrete‡	$6\sqrt{f'_c}$
Tension in precompressed concrete where deflections based on transformed cracked section meet limits in Art 3-11§	$12\sqrt{f'_c}$

*Where the calculated tension stress exceeds this value, reinforcement should be provided to resist the total tension force on the concrete computed on the assumption of an uncracked section.

†But not greater than the maximum value recommended by the manufacturer of the steel or anchorages.

‡Allowable tension in members not exposed to freezing or to a corrosive environment. For members exposed to such conditions, more concrete cover should be provided around the steel than that normally required by the ACI Code (Art. 3-6), and crack-controlling reinforcement should be provided in the tension zone.

§Cover should be increased 50% over that normally required for prestressed concrete (Art. 3-6).

reinforcement in the end region should not exceed f_b calculated from Eq. (3-93a) and (3-93b).

$$f_b = 0.8 f'_{cc} \sqrt{A_b / A_b - 0.2} \leq 1.25 f'_{ci} \qquad (3\text{-}93\,a)$$

$$f_b = 0.6 \sqrt{f'_c} \sqrt{A_b / A'_b} \leq f'_c \qquad (3\text{-}93\,b)$$

where A_b = bearing area of anchor plate

A'_b = maximum area of portion of anchorage surface geometrically similar to and concentric with area of anchor plate

A more refined analysis may be applied in the design of the end-anchorage regions of prestressed members to develop the ultimate strength of the tendons. ϕ should be taken as 0.90 for the concrete. Adequate reinforcement should be provided to prevent bursting, horizontal splitting, and spalling. End blocks should be used to distribute end-bearing loads and those due to concentrated prestressing forces.

(J. R. Libby, "Modern Prestressed Concrete," Van Nostrand Reinhold Company, New York.)

3-30. Design of Prestressed-Concrete Beams

This involves selection of shape and dimensions of the concrete portion, type and positioning of tendons, and amount of prestress. After a concrete shape and dimensions have been assumed, determine geometric properties: cross-sectional area, center of gravity, distances of extreme surfaces from the centroid, section moduli, and dead load of member per unit of length. Treat prestressing forces as a system of external forces acting on the concrete (see Art. 3-27).

Compute bending stresses due to dead and live loads. From these, determine the magnitude and location of the prestressing force at points of maximum moment. This force must provide sufficient compression to offset the tensile stresses caused by the bending moments due to loads (Fig. 3-13). But at the same time, it must not create any allowable stresses exceeding those listed in Art. 3-29. Investigation of other sections will guide selection of tendons to be used and determine their position in the beam.

After establishing the tendon profile, prestressing forces, and tendon areas, check critical points along the beam under initial and final conditions, on removal from the forms, and during erection. Check ultimate strength in flexure and shear and the percentage of prestressing steel. Design anchorages, if required, and diagonal-tension steel. Finally, check camber.

The design may be based on the following assumptions. Strains vary linearly with depth. At cracked sections, the concrete cannot resist tension. Before cracking, stress is proportional to strain. The transformed area of bonded tendons may be included in pretensioned members and in posttensioned members after the tendons have been grouted. Areas of open ducts should be deducted in calculations of section properties before bonding of tendons. The modulus of rupture should be determined from tests, or the cracking stress may be assumed as $7.5 \sqrt{f'_c}$, where f'_c is the 28-day strength of the concrete, psi.

Prestressed beams should be checked by ultimate-strength theory (Art. 3-9). Beams for building should be capable of supporting the factored loads given by Eqs. (3-6) and (3-8). For bridge beams, the ultimate-load capacity should not be less than

$$U = \frac{1.30}{\phi} \left[D + \frac{5}{3} (L + I) \right] \qquad (3\text{-}94)$$

where D = effect of dead load

L = effect of design live load

I = effect of impact

ϕ = 1.0 for factory-produced precast, prestressed members

= 0.95 for posttensioned, cast-in-place members
= 0.90 for shear

Ultimate Strength in Bending ▪ After cracking, a prestressed beam behaves essentially as an ordinary reinforced concrete beam. For rectangular sections, or flanged sections in which the neutral axis lies within the flange, the design resisting moment, in-lb, can be computed from

$$M_t = 0.90[A_{ps}f_{ps}d(1 - 0.59\omega_p)] = 0.90\left[A_{ps}f_{ps}\left(d - \frac{a}{2}\right)\right] \tag{3-95}$$

where A_{ps} = area of tendons, in^2
  f_{ps} = calculated stress in tendons at design load, psi
  d = distance from extreme compression surface to centroid of prestressing force, in
  $\omega_p = \rho_p f_{ps}/f'_c$, with $\rho_p = A_{ps}/bd$
  b = width of compression, face, in
  a = depth of equivalent rectangular compressive-stress distribution, in
  = $A_{ps}f_{ps}/0.85f'_c b$

The neutral axis usually will lie within the flange when the flange thickness h_f, in, exceeds $1.4d\rho_p f_{ps}/f'_c$. For flanged sections with neutral axis outside the flange (usually when h_f is less than $1.4d\rho_p f_{ps}/f'_c$):

$$M_t = 0.90\left[A_{pw}f_{ps}\left(d - \frac{a}{2}\right) + 0.85f'_c(b - b_w)h_f\left(d - \frac{h_f}{2}\right)\right] \tag{3-96}$$

where A_{pf} = area of steel required to develop compressive strength of overhanging flanges, in^2
  = $0.85f'_c(b - b_w)h_f/f_{ps}$
  A_{pw} = area of tendons to develop web, in^2 = $A_{ps} - A_{pf}$
  b_w = minimum width of web, in

When information for determination of f_{ps} is not available, and if the effective prestress, after losses, is at least half the ultimate strength of the tendons, the following approximate value should be used for bonded tendons:

$$f_{ps} = f_{pu}\left(1 - \frac{0.5\rho_p f_{pu}}{f'_c}\right) \tag{3-97}$$

where f_{pu} = ultimate strength of the tendons, psi.
  For unbonded tendons:

$$f_{ps} = f_{se} + 10,000 + \frac{f'_c}{100\rho_p} \tag{3-98}$$

but not more than $f_{se} + 60,000$, where f_{se} is the effective prestress in tendons, psi, or more than the tendon yield strength.

Nonprestressed reinforcement, in combination with tendons, may be assumed equivalent at ultimate moment to its area times its yield point. But this is permitted only if

$$\frac{\rho_p f_{ps}}{f'_c} + \frac{\rho f_y}{f'_c} - \frac{\rho' f_y}{f'_c} \le 0.3 \tag{3-99}$$

where $\rho = A_s/bd$
  $\rho' = A'_s/bd$
  A_s = area of unprestressed reinforcement, in^2
  A'_s = area of compressive reinforcement, in^2
  f_y = yield strength of unprestressed reinforcement, psi

The total amount of prestressed and unprestressed reinforcement should be adequate to develop an ultimate load in flexure at least 1.2 times the cracking load calculated for a modulus of rupture of 7.5 $\sqrt{f_c'}$.

Limitations on Steel Ratios ▪ In calculating M_t from Eqs. (3-95) and (3-96), the amount of prestressing steel should be such that $\rho_p f_{ps}/f_c'$ is not more than 0.30. For flanged sections, ρ_p should be taken as A_{ps}/bd. For larger steel ratios, the ultimate moment should not exceed the following:

For rectangular sections, or flanged sections in which the neutral axis lies within the flange:

$$M_u = 0.225 f_c' b d^2 \tag{3-100}$$

For flanged sections with thickness h_f with the neutral axis outside the flange:

$$M_u = 0.90 \left[0.25 f_c' b_w d^2 + 0.85 f_c' (b - b_w) h_f (d - 0.5 h_f) \right] \tag{3-101}$$

Shear ▪ ACI 318, "Building Code Requirements for Reinforced Concrete" (American Concrete Institute), and "Standard Specifications for Highway Bridges" (American Association of State Highway and Transportation Officials) require that prestressed beams be designed to resist diagonal tension by ultimate-strength theory.

Shear reinforcement should consist of stirrups or welded-wire fabric. The area of shear reinforcement, in^2, set perpendicular to the beam axis, should not be less than

$$A_v = 50 \frac{b_w s}{f_y} \tag{3-102}$$

where s is the reinforcement spacing, in, except when the factored shear force V_u is less than one-half ϕV_c; or when the depth of the member h is less than 10 in or 2.5 times the thickness of the compression flange, or one-half the width of the web, whichever is largest. The capacity reduction factor ϕ should be taken as 0.85.

Alternatively, a minimum area

$$A_v = \frac{A_{ps} f_{pu} s}{80 f_y d} \sqrt{\frac{d}{b_w}} \tag{3-103}$$

may be used if the effective prestress force is at least equal to 40% of the tensile strength of the flexural reinforcement.

The yield strength of shear reinforcement, f_y used in design calculations should not exceed 60,000 psi.

Where shear reinforcement is required, it should be placed perpendicular to the axis of the member and should not be spaced farther apart than $0.75h$, where h is the overall depth of the member, or 24 in. Web reinforcement between the face of support and the section at a distance $h/2$ from it should be the same as the reinforcement required at that section.

When V_u exceeds the nominal shear strength ϕV_c of the concrete, shear reinforcement must be provided. V_c may be computed from Eq. (3-105) when the effective prestress force is 40% or more of the tensile strength of the flexural reinforcement, but this shear stress should not exceed 5 $\sqrt{f_c'} b_w d$.

$$V_c = \left(0.6 \sqrt{f_c'} + 700 \frac{V_u d}{M_u} \right) b_w d \geq 2 \sqrt{f_c'} b_w d \tag{3-104}$$

where M_u = design moment at section occurring simultaneously with shear V_u at section

b_w = web width

d = distance from extreme compression surface to centroid of tendons or $0.80h$, whichever is larger

$V_u d/M_u$ should not be taken greater than 1. For some sections, such as medium- and long-span I-shaped members, Eq. (3-104) may be overconservative, and the following more detailed analysis would be preferable.

The ACI Code requires a more detailed analysis when the prestress force is less than 40% of the tensile strength of the flexural reinforcement. The governing shear stress is the smaller of the values computed for inclined flexure-shear cracking V_{ci} from Eq. (3-105) and web-shear cracking V_{cw} from Eq. (3-106).

$$V_{ci} = 0.6\sqrt{f_c'}b_w d + V_d + V_i M_{cr}/M_{max} \geq 1.7\sqrt{f_c'}b_w d \qquad (3\text{-}105)$$

$$V_{cw} = (3.5\sqrt{f_c'}b_w d + 0.3f_{pc})b_w d + V_p \qquad (3\text{-}106)$$

where V_d = shear due to dead load

V_i = shear occurring simultaneously with M_{max} and produced by external loads

M_{cr} = cracking moment [see Eq. (3-107)]

M_{max} = maximum bending moment due to external design loads

b_w = web width or diameter of circular section

d = distance from extreme compression surface to centroid of prestressing force or 80% of overall depth of beam, whichever is larger

f_{pc} = compressive stress in concrete occurring, after all prestress losses have taken place, at center of cross section resisting applied loads or at junction of web and flange when centroid lies in flange

V_p = vertical component of effective prestress force at section considered

The cracking moment is given by

$$M_{cr} = \frac{I}{y_t}(6\sqrt{f_c'} + f_{pe} - f_d) \qquad (3\text{-}107)$$

where I = moment of inertia of section resisting external design loads, in

y_t = distance from centroidal axis of gross section, neglecting reinforcement, to extreme surface in tension, in

f_{pe} = compressive stress in concrete due to prestress only, after all losses, occurring at extreme surface of section at which tension is produced by applied loads, psi

f_d = stress due to dead loads in extreme surface of section at which tension is produced by applied loads, psi

Alternatively, V_{cw} may be taken as the shear force corresponding to the design load that induces a principal tensile force of $4\sqrt{f_c'}b_w d$ at the centroidal axis of the member or, when the centroidal axis is in the flange, induces this tensile stress at the intersection of flange and web.

The values of M_{max} and V_i used in Eq. (3-105) should be those resulting from the distribution of loads causing maximum moment to occur at the section.

In a pretensioned beam in which the section at a distance of half the overall beam depth $h/2$ from face of support is closer to the end of the beam than the transfer length of the tendon, the reduced prestress in the concrete at sections falling within the transfer length should be considered when calculating V_{cw}. The prestress may be assumed to vary linearly along the centroidal axis from zero at the beam end to a maximum at a distance from the beam end equal to the transfer length. This distance may be assumed to be 50 diameters for strand and 100 diameters for single wire.

When $V_u - \phi V_c$ exceeds $4\sqrt{f'_c}b_w d$, the maximum spacing of stirrups should be reduced to $0.375h$ but not to more than 12 in. But $V_u - \phi V_c$ should not exceed $8\sqrt{f'_c}b_w d$.

Bonded Reinforcement ▪ When prestressing steel is not bonded to the concrete, some bonded reinforcement should be provided in the precompressed tension zone of flexural members. The bonded reinforcement should be distributed uniformly over the tension zone near the extreme tension surface in beams and one-way slabs and should have an area of at least

$$A_s = 0.004A \tag{3-108}$$

where A = area, in^2, of that part of cross section between flexural tension face and center of gravity of cross section.

In positive-moment regions of two-way slabs where the tensile stress under service loads exceeds $2\sqrt{f'_c}$, the area of bonded reinforcement should be at least

$$A_s = \frac{N_c}{0.5f_y} \tag{3-109}$$

where N_c = tensile force on concrete under service (dead plus live) loads, lb
 f_y = yield strength, psi, of bonded reinforcement $\leq$ 60 ksi
At columns in negative-moment regions of two-way slabs, at least four bonded reinforcing bars should be placed in each direction and provide a minimum steel area

$$A_s = 0.00075hl \tag{3-110}$$

where l = span of slab in direction parallel to that of reinforcement being determined, in
 h = overall thickness of slab, in
The bonded reinforcement should be distributed, with a spacing not exceeding 12 in, over the slab width between lines that are $1.5h$ outside opposite faces of the columns.

Compression Members ▪ Members subject to axial compression and with an average prestress $f_{se}A_{ps}/A_c$ more than 225 psi or total axial load and bending should be designed by ultimate-strength methods (Art. 3-22 and 3-23), including effects of prestress, shrinkage, and creep. Reinforcement in columns with an average prestress less than 225 psi should have an area equal to at least 1% of the gross concrete area A_c. For walls subject to an average prestress greater than 225 psi and for which structural analysis shows adequate strength, the minimum reinforcement requirements given in Art. 3-25 may be waived.

Tendons in columns should be enclosed in spirals or closed lateral ties. The spiral should comply with the requirements given in Art. 3-22. Ties should be at least No. 3 bar size and spacing should not exceed 48 tie diameters or the least dimension of the column.

Ducts for Posttensioning ▪ Tendons for posttensioned members generally are sheathed in ducts before prestress is applied so that the tendons are free to move when tensioned. The tendons may be grouted in the ducts after transfer of prestress to the concrete and thus bonded to the concrete.

Ducts for grouting bonded tendons should be at least ¼ in larger than the diameter of the posttensioning tendons or large enough to produce an internal area at least twice the gross area of the tendons. The temperature of members at time of grouting should be above 50°F, and members must be maintained at this temperature for at least 48 h.

Unbonded tendons should be completely coated with suitable material to insure corrosion protection and protect the tendons against infiltration of cement during casting operations.

Deflections ▪ The immediate deflection of prestressed members may be computed by the usual formulas for elastic deflections. If cracking may occur, however, the effective moment of inertia (Art. 3-11) should be used. In these formulas, the moment of inertia used should be that of the gross uncracked concrete section. Long-time deflection computations should include effects of the sustained load and effects of creep and shrinkage and relaxation of the steel (Art. 3-11).

(P. F. Rice and E. S. Hoffman, "Structural Design Guide to the ACI Code," Van Nostrand Reinhold Company, New York; J. R. Libby, "Modern Prestressed Concrete," Van Nostrand Reinhold Company, New York; "PCI Design Handbook," Prestressed Concrete Institute, 20 North Wacker Drive, Chicago, Ill. 60606.)

3-31. Concrete Gravity Walls

Generally economical for walls up to about 15 ft high, gravity walls use their own weight to resist lateral forces from earth or other materials (Fig. 3-14a). Such walls usually are sufficiently massive to be unreinforced. In such cases, tensile stresses should not exceed $1.6\sqrt{f'_c}$, where f'_c is the 28-day strength of the concrete, as computed by the working-stress method.

Forces acting on gravity walls include the walls' own weight, the weight of the earth on the sloping back and heel, lateral earth pressure, and resultant soil pressure on the base. It is advis-

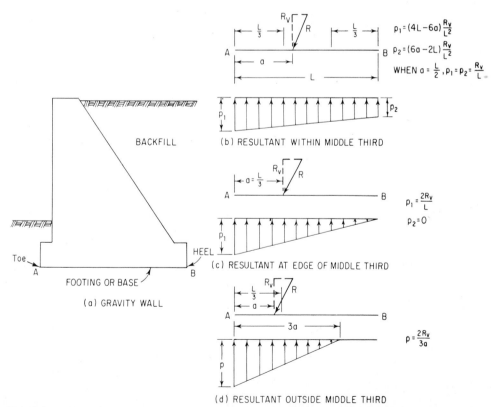

Fig. 3-14. Pressure diagram for base of concrete gravity wall (*a*) depends on whether the resultant of the forces acting on it lies within the middle third (*b*), at the edge of the middle third (*c*), or outside (*d*).

able to include a force at the top of the wall to account for frost action, perhaps 700 lb/lin ft. A wall, consequently, may fail by overturning or sliding, overstressing of the concrete, or settlement due to crushing of the soil.

Design usually starts with selection of a trial shape and dimensions, and this configuration is checked for stability. For convenience, when the wall is of constant height, a 1-ft-long section may be analyzed. Moments are taken about the toe. The sum of the righting moments should be at least 1.5 times the sum of the overturning moments. To prevent sliding

$$\mu R_v \geq 1.5 P_h \qquad (3\text{-}111)$$

where μ = coefficient of sliding friction
  R_v = total downward force on soil, lb
  P_h = horizontal component of earth thrust, lb

Next, the location of the vertical resultant R_v should be found at various sections of the wall by taking moments about the toe and dividing the sum by R_v. The resultant should act within the middle third of each section if there is to be no tension in the wall.

Finally, the pressure exerted by the base on the soil should be computed to insure that the allowable pressure will not be exceeded. When the resultant is within the middle third, the pressures, psf, under the ends of the base are given by

$$p = \frac{R_v}{A} \pm \frac{Mc}{I} = \frac{R_v}{A}\left(1 \pm \frac{6e}{L}\right) \qquad (3\text{-}112)$$

where A = area of base, ft^2
  L = width of base, ft
  e = distance, parallel to L, from centroid of base to R_v, ft

Figure 3-14b shows the pressure distribution under a 1-ft strip of wall for $e = L/2 - a$, where a is the distance of R_v from the toe. When R_v is exactly $L/3$ from the toe, the pressure at the heel becomes zero (Fig. 3-14c). When R_v falls outside the middle third, the pressure vanishes under a zone around the heel, and pressure at the toe is much larger than for the other cases (Fig. 3-14d). "Standard Specifications for Highway Bridges" (American Association of State Highway and Transportation Officials) requires that contraction joints be provided at intervals not exceeding 30 ft. Alternate horizontal bars should be cut at these joints for crack control. Expansion joints should be located at intervals of up to 90 ft.

3-32. Cantilever Retaining Walls

This type of wall resists the lateral thrust of earth pressure through cantilever action of a vertical stem and horizontal base (Fig. 3-15a). Cantilever walls generally are economical for heights from 10 to 20 ft. For lower walls, gravity walls may be less costly; for taller walls, counterforts may be less expensive.

Usually, the force acting on the stem is the lateral earth pressure, including the effect of frost action, perhaps 700 lb/lin ft. The base is loaded by the moment and shear from the stem, upward soil pressure, its own weight, and that of the earth above. The weight of the soil over the toe, however, may be ignored in computing stresses in the toe since the earth may not be in place when the wall is first loaded or may erode. For walls of constant height, it is convenient to design and analyze a 1-ft-long strip.

The stem is designed to resist the bending moments and shear due to the earth thrust. Then, the size of the base slab is selected to meet requirements for resisting overturning and sliding and to keep the pressure on the soil within the allowable. If the flat bottom of the slab does not provide sufficient friction [Eq. (3-111)], a key, or lengthwise projection, may be added on the bottom for that purpose. The key may be reinforced by extending and bending up the dowels between stem and base.

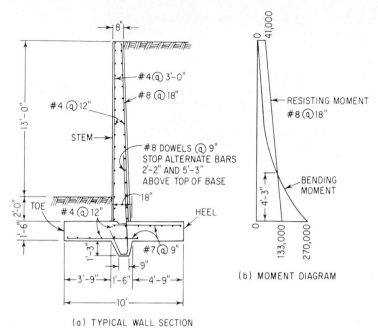

(a) TYPICAL WALL SECTION

Fig. 3-15. Cantilever retaining wall (*a*) has main reinforcing placed vertically in the stem. Reinforcing requirements may be determined from bending moment diagram (*b*).

To provide an adequate safety factor against overturning, the sum of the righting moments about the toe should be at least 1.5 times the sum of the overturning moments. The pressure under the base can be computed, as for gravity walls, from Eq. (3-112). (See also Fig. 3-14*b* to *d*.)

Generally, the stem is made thicker at the bottom than required for shear and balanced design for moment because of the saving in steel. Since the moment decreases from bottom to top, the earth side of the wall usually is tapered, and the top is made as thin as convenient concreting will permit (8 to 12 in). The main reinforcement is set, in vertical planes, parallel to the sloping face and 3 in away. The area of this steel at the bottom can be computed from Eq. (3-32). Some of the steel may be cut off where it no longer is needed. Cutoff points may be determined graphically (Fig. 3-15*b*). The bending-moment diagram is plotted and the resisting moment of steel not cut off is superimposed. The intersection of the two curves determines the theoretical cutoff point. The bars should extend upward beyond this point a distance equal to *d* or 12 bar diameters.

In addition to the main steel, vertical steel is set in the front face of the wall and horizontal steel in both faces to resist thermal and shrinkage stresses (Art. 3-15). "Standard Specifications for Highway Bridges" (American Association of State Highway and Transportation Officials) requires at least ⅛ in² of horizontal reinforcement per foot of height.

The heel and toe portions of the base are both designed as cantilevers supported by the stem. The weight of the earth tends to bend the heel down against relatively small resistance from soil pressure under the base. In contrast, the upward soil pressure tends to bend the toe up. So for the heel, main steel is placed near the top, and for the toe, near the bottom. Also, temperature steel is set lengthwise in the bottom. The area of the main steel may be computed from Eq. (3-32), but the bars should be checked for development length because of the relatively high shear.

To eliminate the need for diagonal-tension reinforcing, the thickness of the base should be sufficient to hold the shear stress, $v_c = V/bd$, below $1.1 \sqrt{f'_c}$, where f'_c is the 28-day strength of the concrete, as computed by the working-stress method. The critical section for shear is at a distance d from the face of the stem, where d is the distance from the extreme compression surface to the tensile steel.

The stem is constructed after the base. A key usually is formed at the top of the base to prevent the stem from sliding. Also, dowels are left projecting from the base to tie the stem to it, one dowel per stem bar. The dowels may be extended to serve also as stem reinforcing (Fig. 3-15a).

The AASHTO Specifications require that contraction joints be provided at intervals not exceeding 30 ft. Expansion joints should be located at intervals up to 90 ft.

To relieve the wall of water pressure, weep holes should be formed near the bottom of the stem. Also, porous pipe and backfill may be set behind the wall to conduct water to the weep holes.

(M. Fintel, "Handbook of Concrete Engineering," Van Nostrand Reinhold Company, New York; "CRSI Handbook," Concrete Reinforcing Steel Institute, 180 North La Salle St., Chicago, Ill. 60601.)

3-33. Counterfort Retaining Walls

Counterforts are ties between the vertical stem of a wall and its base (Fig. 3-16a). Placed on the earth side of the stem, they are essentially wedge-shaped cantilevers. (Walls with supports on the opposite side are called buttressed retaining walls.) Counterfort walls are economical for heights for which gravity and cantilever walls are not suitable.

Stability design is the same as for gravity walls (Art. 3-31) and cantilever walls (Art. 3-32). But the design is applied to a section of wall center to center of counterforts.

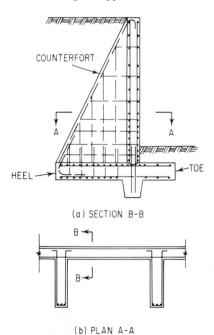

(a) SECTION B-B

(b) PLAN A-A

Fig. 3-16. Counterfort retaining wall.

The vertical face resists lateral earth pressure as a continuous slab supported by the counterforts. It also is supported by the base, but an exact analysis of the effects of the three-sided supports would not be worthwhile except for very long walls. Similarly, the heel portion of the base is designed as a continuous slab supported by the counterforts. In turn, the counterforts are subjected to lateral earth pressure on the sloping face and the pull of the vertical stem and base. The toe of the base acts as a cantilever, as in a cantilever wall.

Main reinforcing in the vertical face is horizontal. Since the earth pressure increases with depth, reinforcing area needed also varies with depth. It is customary to design a 1-ft-wide strip of slab spanning between counterforts at the bottom of the wall and at several higher levels. The steel area and spacing for each strip then are held constant between strips. Negative-moment steel should be placed near the earth face of the wall at the counterforts, and positive-moment steel near the opposite face between counterforts (Fig. 3-16b). Concrete cover should be 3 in over reinforcing throughout the wall. Design requirements are substantially the same as for rectangular beams and one-way slabs, except the thickness is made large enough to eliminate the need for shear reinforcing (Arts. 3-12 to 3-15). The vertical face also incorporates vertical steel, equal to about 0.3 to 1% of the concrete area, for placement purposes and to resist temperature and shrinkage stresses.

In the base, main reinforcing in the heel portion extends lengthwise, whereas that in the toe runs across the width. The heel is subjected to the downward weight of the earth above and its own weight and to the upward pressure of the soil below and the pull of the counterforts. So longitudinal steel should be placed in the top face at the counterforts and near the bottom between counterforts. Main transverse steel should be set near the bottom to resist the cantilever action of the toe.

The counterforts, resisting the lateral earth pressure on the sloping face and the pull of the vertical stem, are designed as T beams. Maximum moment occurs at the bottom. It is resisted by main reinforcing along the sloping face. (The effective depth should be taken as the distance from the outer face of the wall to the steel along a perpendicular to the steel.) At upper levels, main steel not required may be cut off. Some of the steel, however, should be extended and bent down into the vertical face. Also, dowels equal in area to the main steel at the bottom should be hooked into the base to provide anchorage.

Shear unit stress on a horizontal section of a counterfort may be computed from $v_c = V_1/bd$, where b is the thickness of the counterfort and d is the horizontal distance from face of wall to main steel.

$$V_1 = V - \frac{M}{d}(\tan \theta + \tan \phi) \qquad (3\text{-}113)$$

where V = shear on section
$\quad M$ = bending moment at section
$\quad \theta$ = angle earth face of counterfort makes with vertical
$\quad \phi$ = angle wall face makes with vertical

For a vertical wall face, $\phi = 0$ and $V_1 = V - (M/d) \tan \theta$. The critical section for shear may be taken conservatively at a distance up from the base equal to $d' \sin \theta \cos \theta$, where d' is the depth of counterfort along the top of the base.

Whether or not horizontal web reinforcing is needed to resist the shear, horizontal bars are required to dowel the counterfort to the vertical face (Fig. 3-16b). They should be designed for the full wall reaction. Also, vertical bars are needed in the counterfort to resist the pull of the base. They should be doweled to the base.

The base is concreted first. Vertical bars are left projecting from it to dowel the counterforts and the vertical face. Then, the counterforts and vertical stem are cast together.

3-34. Types of Footings

Footings should be designed to satisfy two objectives: limit total settlement to an acceptable small amount and eliminate differential settlement between parts of a structure as nearly as possible. To limit the amount of settlement, a footing should be constructed on soil with sufficient resistance to deformation, and the load should be spread over a large soil area. The load may be spread horizontally, as is done with spread footings, or vertically, as with friction-pile foundations.

There are a wide variety of spread footings. The most commonly used ones are illustrated in Fig. 3-17a to g. A simple pile footing is shown in Fig. 3-17h.

For walls, a spread footing is a slab wider than the wall and extending the length of the wall (Fig. 3-17a). Square or rectangular slabs are used under single columns (Fig. 3-17b to d). When two columns are so close that their footings would merge or nearly touch, a combined footing (Fig. 3-17e) extending under the two should be constructed. When a column footing

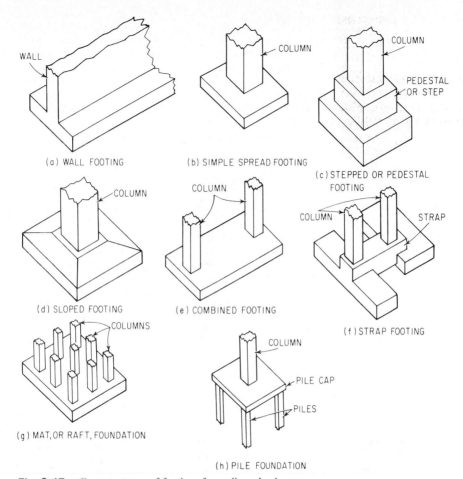

Fig. 3-17. Common types of footings for walls and columns.

cannot project in one direction, perhaps because of the proximity of a property line, the footing may be helped out by an adjacent footing with more space; either a combined footing or a strap (cantilever) footing (Fig. 3-17*f*) may be used under the two.

For structures with heavy loads relative to soil capacity, a mat or raft foundation (Fig. 3-17*g*) may prove economical. A simple form is a thick, two-way-reinforced-concrete slab extending under the entire structure. In effect, it enables the structure to float on the soil, and because of its rigidity, it permits negligible differential settlement. Even greater rigidity can be obtained by building the raft foundation as an inverted beam-and-girder floor, with the girders supporting the columns. Sometimes, also, inverted flat slabs are used as mat foundations.

In general, footings should be so located under walls or columns as to develop uniform pressure below. The pressure under adjacent footings should be as nearly equal as possible, to avoid differential settlement. In the computation of stresses in spread footings, the upward reaction of the soil may be assumed to vary linearly. For pile-cap stresses, the reaction from each pile may be assumed to act at the pile center.

Simple footings act as cantilevers under the downward column or wall loads and upward soil or pile reactions. Therefore, they can be designed as rectangular beams (Arts. 3-12 to 3-15) by working-stress or ultimate-strength theory.

3-35. Stress Transfer to Footings

For a footing to serve its purpose, column stresses must be distributed to it and spread over the soil or to piles, with a safety factor against failure of the footing. Stress in the longitudinal reinforcement of a column should be transferred to its pedestal or footing either by extending the longitudinal steel into the support or by dowels. At least four bars should be extended or four dowels used. In any case, a minimum steel area of 0.5% of the column area should be supplied for load transfer. The stress-transfer bars should project into the base a sufficient compression-embedment distance to transfer the stress in the column bars to the base concrete. Where dowels are used, their total area should be adequate to transfer the compression in excess of that transmitted by the column concrete to the footing in bearing, and the dowel diameter should not exceed the column-bar diameter by more than 0.15 in. If the required dowel length is larger than the footing depth less 3 in, either smaller-diameter bars with equivalent area should be used or a monolithic concrete cap should be added to increase the concrete depth. The dowels, in addition, should provide at least one-quarter of the tension capacity of the column bars on each column face. The dowels should extend into the column a distance equal to that required for compression lapping of column bars (Art. 3-4).

Stress in the column concrete should be considered transferred to the top of the pedestal or footing by bearing. ACI 318, "Building Code Requirements for Reinforced Concrete" (American Concrete Institute), specifies two bearing stresses:

For a fully loaded area, such as the base of a pedestal, allowable bearing stress is $0.85\phi f'_c$, where f'_c is the strength of the concrete and $\phi = 0.70$.

If the area A_1, the loaded portion at the top of a pedestal or footing, is less than the area of the top, the allowable pressure may be multiplied by $\sqrt{A_2/A_1}$, but not more than 2, where A_2 is the area of the top that is geometrically similar to and concentric with the loaded area A_1.

For working-stress design, the allowable bearing stress is $0.30f'_c$.

3-36. Wall Footings

The spread footing under a wall (Fig. 3-17a) distributes the wall load horizontally to preclude excessive settlement. (For retaining-wall footings, see Arts. 3-31 to 3-33.) The wall should be so located on the footing as to produce uniform bearing pressure on the soil (Fig. 3-18), ignoring the variation due to bending of the footing. The pressure, lb/ft^2, is determined by dividing the load per foot by the footing width, ft.

The footing acts as a cantilever on opposite sides of the wall under downward wall loads and upward soil pressure. For footings supporting concrete walls, the critical section for bending moment is at the face of the wall; for footings under masonry walls, halfway between the middle and edge of the wall. Hence, for a 1-ft-long strip of symmetrical concrete-wall footing, symmetrically loaded, the maximum moment, ft-lb, is

$$M = \frac{p}{8}(L - a)^2 \qquad (3\text{-}114)$$

where p = uniform pressure on soil, lb/ft^2
L = width of footing, ft
a = wall thickness, ft

If the footing is sufficiently deep that the tensile bending stress at the bottom, $6M/t^2$, where M is the factored moment and t is the footing depth, in, does not exeed $5\phi\sqrt{f'_c}$, where f'_c is the 28-day concrete strength, psi, and $\phi = 0.90$, the footing need not be reinforced. If the tensile stress is larger, the

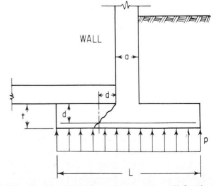

Fig. 3-18. Reinforced-concrete wall footing.

footing should be designed as a 12-in-wide, rectangular, reinforced beam. Bars should be placed across the width of the footing, 3 in from the bottom. Bar development length is measured from the point at which the critical section for moment occurs. Wall footings also may be designed by ultimate-strength theory.

ACI 318, "Building Code Requirements for Reinforced Concrete" (American Concrete Institute), requires at least 6 in of cover over the reinforcement at the edges. Hence, allowing about 1 in for the bar diameter, the minimum footing thickness is 10 in.

The critical section for shear is at a distance d from the face of the wall, where d is the distance from the top of the footing to the tensile reinforcement, in. Since diagonal-tension reinforcement is undersirable, d should be large enough to keep the shear unit stress, $V/12d$, below $1.1 \sqrt{f'_c}$, as computed by the working-stress method, or below $2 \sqrt{f'_c} b_w d$ for factored shear loads.

In addition to the main steel, some longitudinal steel also should be placed parallel to the wall to resist shrinkage stresses and facilitate placement of the main steel. (See also Art. 3-35.)

(G. Winter and A. H. Nilson, "Design of Concrete Structures," McGraw-Hill Book Company, New York.)

3-37. Single-Column Spread Footings

The spread footing under a column (Fig. 3-17b to d) distributes the column load horizontally to prevent excessive total and differential settlement. The column should be located on the footing so as to produce uniform bearing pressure on the soil (Fig. 3-19), ignoring the variation due to bending of the footing. The pressure equals the load divided by the footing area.

Single-column footings usually are square, but they may be made rectangular to satisfy space restrictions or to support elongated columns.

Under the downward load of the column and the upward soil pressure, a footing acts as a cantilever in two perpendicular directions. For rectangular concrete columns and pedestals, the critical section for bending moment is at the face of the loaded member (ab in Fig. 3-20a). (For round or octagonal columns or pedestals, the face may be taken as the side of a square with the same area.) For steel baseplates, the critical section for moment is halfway between the face of the column and the edge of the plate.

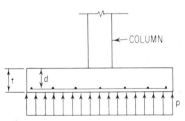

Fig. 3-19. Spread footing for column.

The bending moment on ab is produced by the upward pressure of the soil on the area abcd. That part of the footing is designed as a rectangular beam to resist the moment. Another critical section lies along a perpendicular column face and should be similarly designed. If the footing is sufficiently deep that the factored tensile bending stress at the bottom does not exceed $5\phi \sqrt{f'_c}$, where $\phi = 90$ and f'_c is the 28-day strength of the concrete, psi, the footing need not be reinforced. If the tensile stress is larger, reinforcement should be placed parallel to both sides of the footing, with the lower layer 3 in above the bottom of the footing and the upper layer a bar diameter higher. The critical section for anchorage (or bar embedment length) is the same as for moment.

In square footings, the steel should be uniformly spaced in each layer. Although the effective depth d is less for the upper layer, thus requiring more steel, it is general practice to compute the required area and spacing for the upper level and repeat them for the lower layer.

In rectangular footings, reinforcement parallel to the long side, with length A, ft, should be uniformly distributed over the width of the footing, B, ft. Bars parallel to the short side should be more closely spaced under the column than near the edges. ACI318, "Building Code Requirements for Reinforced Concrete" (American Concrete Institute), recommends that the

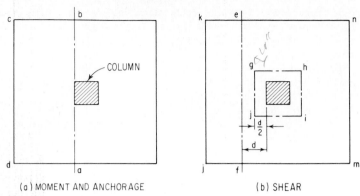

(a) MOMENT AND ANCHORAGE   (b) SHEAR

Fig. 3-20.   Critical sections in a column footing.

short bars should be given a constant but closer spacing over a width B centered under the column. The area of steel in this band should equal twice the total steel area required in the short direction divided by $A/B + 1$. The remainder of the reinforcement should be uniformly distributed on opposite sides of the band. (See also Art. 3-35.)

Two types of shear should be investigated: two-way action and beam-type shear. The critical section for beam-type shear lies at a distance d from the face of column or pedestal (ef in Fig. 3-20b). The shear equals the total upward pressure on area $efjk$. To eliminate the need for diagonal-tension reinforcing, d should be made large enough that the unit shear stress does not exceed $1.1 \sqrt{f_c'}$ ($2 \sqrt{f_c'}$ for ultimate-strength design).

The critical section for two-way action (punching shear) is concentric with the column or pedestal. It lies at a distance $d/2$ from the face of the loaded member ($ghij$ in Fig. 3-20b). The shear equals the column load less the upward soil pressure on area $ghij$. In this case, d should be large enough that the factored shear on the concrete does not exceed

$$V_c = \left(2 + \frac{4}{\beta_c} \right) \sqrt{f_c' b_o d} \qquad (3\text{-}115)$$

where β_c = ratio of long side to short side of critical shear section

b_o = perimeter of critical section, in

d = depth of centroid of reinforcement, in

Shearhead reinforcement (steel shapes), although generally uneconomical, may be used to obtain a shallow footing.

Footings for columns designed to take moment at the base should be designed against overturning and nonuniform soil pressures. When the moments are about only one axis, the footing may be made rectangular with the long direction perpendicular to that axis, for economy. Design for the long direction is similar to that for retaining-wall bases (Art. 3-31 to 3-33).

(G. Winter and A. H. Nilson, "Design of Concrete Structures," McGraw-Hill Book Company, New York; M. Fintel, "Handbook of Concrete Engineering," Van Nostrand Reinhold Company, New York; "CRSI Handbook," Concrete Reinforcing Steel Institute, Schaumberg, Ill.; ACI SP-17, "Design Handbook," American Concrete Institute, Detroit, Mich.)

3-38. Combined Footings

These are spread footings extended under more than one column (Fig. 3-17e). They may be necessary when two or more columns are so closely spaced that individual footings would inter-

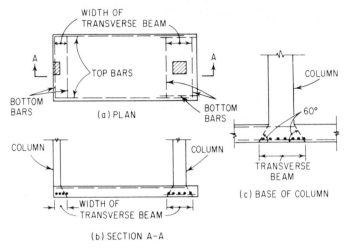

Fig. 3-21.   Combined footing.

fere with each other. Or they may be desirable when space is restricted for a column footing, such as an exterior member so close to a property line that an individual footing would be so short that it would have excessive eccentric loading. In that case, the footing may be extended under a rear column. If the footing can be continued past that column a sufficient distance, and the exterior column has a lighter load, the combined footing may be made rectangular (Fig. 3-21a). If not, it may be made trapezoidal.

If possible, the columns should be so placed on the combined footing as to produce a uniform pressure on the soil. Hence, the resultant of the column loads should coincide with the centroid of the footing in plan. This requirement usually determines the length of the footing. The width is computed from the area required to keep the pressure on the soil within the allowable.

In the longitudinal direction, the footing should be designed as a rectangular beam with overhangs. This beam is subjected to the upward pressure of the soil. Hence, the main steel consists of top bars between the columns and bottom bars at the columns where there are overhangs (Fig. 3-21b). Depth of footing may be determined by moment or shear (see Art. 3-37).

The column loads may be assumed distributed to the longitudinal beam by beams of the same depth as the footing but extending in the narrow, or transverse, direction. Centered, if possible, under each column, the transverse member should be designed as a rectangular beam subjected to the downward column load and upward soil pressure under the beam. The width of the beam may be estimated by assuming a 60° distribution of the column load, as indicated in Fig. 3-21c. Main steel in the transverse beam should be placed near the bottom.

Design procedure for a trapezoidal combined footing is similar. But the reinforcing steel in the longitudinal direction is placed fanwise, and alternate bars are cut off as the narrow end is approached. (See also Art. 3-35.)

(G. Winter and A. H. Nilson, "Design of Concrete Structures," McGraw-Hill Book Company, New York; M. Fintel, "Handbook of Concrete Engineering," Van Nostrand Reinhold Company, New York.)

3-39.   Strap or Cantilever Footings

In Art. 3-38, the design of a combined footing was explained for a column footing in restricted space, such as an exterior column at a property line. As the distance between such a column and a column with adequate space around it increases, the cost of a combined footing rises rapidly. For column spacing more than about 15 ft, a strap footing (Fig. 3-17f) may be more economical. It consists of a separate footing under each column connected by a beam or strap to distribute the column loads (Fig. 3-22a).

The footings are sized to produce the same, constant pressure under each (Fig. 3-22c). This requires that the centroid of their areas coincide with the resultant of the column loads. Usually, the strap is raised above the bottom of the footings so as not to bear on the soil. The sum of the footing areas, therefore, must be large enough for the allowable bearing capacity of the soil not to be exceeded. When these requirements are satisfied, the total net pressure under a footing does not necessarily equal the column design load on the footing.

The strap should be designed as a rectangular beam spanning between the columns. The loads on it include its own weight (when it does not rest on the soil) and the upward pressure from the footings. Width of the strap usually is selected arbitrarily as equal to that of the largest column plus 4 to 8 in so that column forms can be supported on top of the strap. Depth is determined by the maximum bending moment.

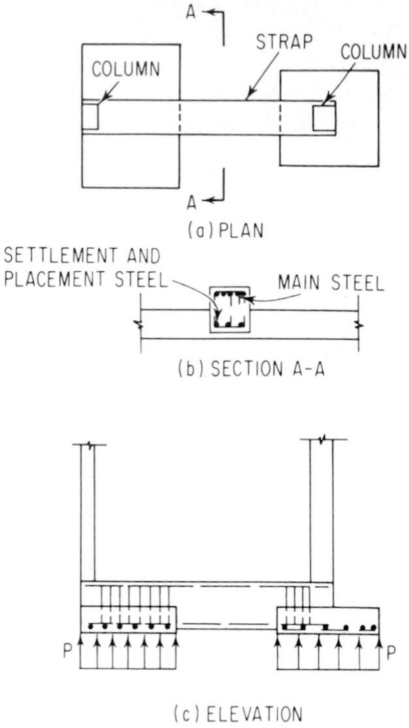

The main reinforcing in the strap is placed near the top. Some of the steel can be cut off where not needed. For diagonal tension, stirrups normally will be needed near the columns (Fig. 3-22b). In addition, longitudinal placement steel is set near the bottom of the strap, plus reinforcement to guard against settlement stresses.

(a) PLAN

(b) SECTION A-A

(c) ELEVATION

Fig. 3-22.   Strap (cantilever) footing.

The footing under the exterior column may be designed as a wall footing (Art. 3-36). The portions on opposite sides of the strap act as cantilevers under the constant upward pressure of the soil.

The interior footing should be designed as a single-column footing (Art. 3-37). The critical section for punching shear, however, differs from that for a conventional footing. This shear should be computed on a section parallel to the strap and at a distance $d/2$ from the sides and extending around the column at a distance $d/2$ from its faces; d is the effective depth of the footing, the distance from the bottom steel to the top of the footing.

(G. Winter and A. H. Nilson, "Design of Concrete Structures," McGraw-Hill Book Company, New York; M. Fintel, "Handbook of Concrete Engineering," Van Nostrand Reinhold Company, New York.)

3-40. Footings on Piles

When piles are required to support a structure, they are capped with a thick concrete slab, on which the structure rests. The pile cap should be reinforced. ACI 318, "Building Code Requirements for Reinforced Concrete" (American Concrete Institute), requires that the thickness above the tops of the piles be at least 12 in. The piles should be embedded from 6 to 9 in, preferably the larger amount, into the footing. They should be cut to required elevation before the footing is cast.

Like spread footings, pile footings for walls are continuous, the piles being driven in line under the wall. For a single column or pier, piles are driven in a cluster. "Standard Specifications for Highway Bridges" (American Association of State Highway and Transportation Officials) requires that piles be spaced at least 2 ft 6 in center to center. And the distance from the side of a pile to the nearest edge of the footing should be 9 in or more.

Whenever possible, the piles should be located so as to place their centroid under the resultant of the column load. If this is done, each pile will carry the same load. If the load is eccentric, then the load on a pile may be assumed to vary linearly with distance from an axis through the centroid.

The critical section for bending moment in the footing and embedment length of the reinforcing should be taken as follows:

At the face of the column, pedestal, or wall, for footings supporting a concrete column, pedestal, or wall

Halfway between the middle and edge of the wall, for footings under masonry walls

Halfway between the face of the column or pedestal and the edge of the metallic base, for footings under steel baseplates

The moment is produced at the critical section by the upward forces from all the piles lying between the section and the edge of the footing.

For diagonal tension, two types of shear should be investigated—punching shear and beam-like shear—as for single-column spread footings (Art. 3-37). The ACI Code requires that in computing the external shear on any section through a footing supported on piles, the entire reaction from any pile whose center is located half the pile diameter or more outside the section shall be assumed as producing shear on the section; the reaction from any pile whose center is located half the pile diameter or more inside the section shall be assumed as producing no shear on the section. For intermediate positions of the pile center, the portion of the pile reaction to be assumed as producing shear on the section shall be based on straight-line interpolation between the full value at half the pile diameter outside the section and zero value at that distance inside the section.

(G. Winter and A. H. Nilson, "Design of Concrete Structures," McGraw-Hill Book Company, New York; M. Fintel, "Handbook of Concrete Engineering," Van Nostrand Reinhold Company, New York).

Wood is remarkable for its beauty, versatility, strength, durability, and workability. It possesses a high-strength-to-weight ratio. It has flexibility. It performs well at low temperatures. It withstands substantial overloads for short periods. It has low electrical and thermal conductance. It resists the deteriorating action of many chemicals that are extremely corrosive to other building materials. There are few materials that cost less per pound than wood.

As a consequence of its origin, wood as a building material has inherent characteristics with which users should be familiar. For example, although cut simultaneously from trees growing side by side in a forest, two boards of the same species and size most likely do not have the same strength. The task of describing this nonhomogeneous material, with its variable biological nature, is not easy, but it can be described accurately, and much better than was possible in the past because research has provided much useful information on wood properties and behavior in structures.

Research has shown, for example, that a compression grade cannot be used, without modification, for the tension side of a deep bending member. Also, a bending grade cannot be used, unless modified, for the tension side of a deep bending member or for a tension member. Experience indicates that typical growth characteristics are more detrimental to tensile strength than to compressive strength. Furthermore, research has made possible better estimates of wood's engineering qualities. No longer is it necessary to use only visual inspection, keyed to averages, for estimating the engineering qualities of a piece of wood. With a better understanding of wood now possible, the availability of sound structural design criteria, and development of economical manufacturing processes, greater and more efficient use is being made of wood for structural purposes.

Improvements in adhesives also have contributed to the betterment of wood construction. In particular, the laminating process, employing adhesives to build up thin boards

Section 4

Wood Design and Construction

Maurice J. Rhude*

President
Sentinel Structures, Inc., Peshtigo, WI

*With revisions by Frederick S. Merritt, Consulting Engineer, West Palm Beach, FL.

into deep timbers, improves nature. Not only are stronger structural members thus made available, but also higher grades of lumber can be placed in regions of greatest stress and lower grades in regions of lower stress, for overall economy. Despite variations in strength of wood, lumber can be transformed into glued-laminated timbers of predictable strength and with very little variability in strength.

4-1. Basic Characteristics and How to Use Them

Wood differs in several significant ways from other building materials, mainly because of its cellular structure. Because of this structure, structural properties depend on orientation. Although most structural materials are essentially isotropic, with nearly equal properties in all directions, wood has three principal grain directions: longitudinal, radial, and tangential. (Loading in the longitudinal direction is referred to as parallel to the grain, whereas transverse loading is considered across the grain.) Parallel to the grain, wood possesses high strength and stiffness. Across the grain, strength is much lower. (In tension, wood stressed parallel to the grain is 25 to 40 times stronger than when stressed across the grain. In compression, wood loaded parallel to the grain is 6 to 10 times stronger than when loaded perpendicular to the grain.) Furthermore, a wood member has three moduli of elasticity, with a ratio of largest to smallest as large as 150:1.

Wood undergoes dimensional changes from causes different from those for dimensional changes in most other structural materials. For instance, thermal expansion of wood is so small as to be unimportant in ordinary usage. Significant dimensional changes, however, occur because of gain or loss in moisture. Swelling and shrinkage from this cause vary in the three grain directions; size changes about 6 to 16% tangentially, 3 to 7% radially, but only 0.1 to 0.3% longitudinally.

Wood offers numerous advantages nevertheless in construction applications—beauty, versatility, durability, workability, low cost per pound, high strength-to-weight ratio, good electrical insulation, low thermal conductance, and excellent strength at low temperatures. It is resistant to many chemicals that are highly corrosive to other materials. It has high shock-absorption capacity. It can withstand large overloads of short time duration. It has good wearing qualities, particularly on its end grain. It can be bent easily to sharp curvature. A wide range of finishes can be applied for decoration or protection. Wood can be used in both wet and dry applications. Preservative treatments are available for use when necessary, as are fire retardants. Also, there is a choice of a wide range of species with a wide range of properties.

In addition, many wood framing systems are available. The intended use of a structure, geographical location, configuration required, cost, and many other factors determine the framing system to be used for a particular project.

Design Recommendations ▪ The following recommendations aim at achieving economical designs with wood framing:

Use standard sizes and grades of lumber. Consider using standardized structural components, whether lumber, stock glued beams, or complex framing designed for structural adequacy, efficiency, and economy.

Use standard details wherever possible. Avoid specially designed and manufactured connecting hardware.

Use as simple and as few joints as possible. Place splices, when required, in areas of lowest stress. Do not locate splices where bending moments are large, and thus avoid design, erection, and fabrication difficulties.

Avoid unnecessary variations in cross section of members along their length.

Use identical member designs repeatedly throughout a structure, whenever practicable. Keep the number of different arrangements to a minimum.

Consider using roof profiles that favorably influence the type and amount of load on the structure.

Specify allowable design stresses rather than the lumber grade or combination of grades to be used.

Select an adhesive suitable for the service conditions, but do not overspecify. For example, waterproof resin adhesives need not be used where less expensive water-resistant adhesives will do the job.

Use lumber treated with preservatives where service conditions dictate. Such treatment need not be used where decay hazards do not exist. Fire-retardant treatments may be used to meet a specific flame-spread rating for interior finish but are not necessary for large-cross-sectional members that are widely spaced and already a low fire risk.

Instead of long, simple spans, consider using continuous or suspended spans or simple spans with overhangs.

Select an appearance grade best suited to the project. Do not specify premium appearance grade for all members if it is not required.

Table 4-1 is a guide to economical span ranges for roof and floor framing in buildings.

4-2. Standard Sizes of Lumber and Timber

Details regarding dressed sizes of various species of wood are given in the grading rules of agencies that formulate and maintain such rules. Dressed sizes in Table 4-2 are from the American Softwood Lumber Standard, "Voluntary Product Standard PS20-70." These sizes are generally available, but it is good practice to consult suppliers before specifying sizes not commonly used to find out what sizes are on hand or can be readily secured.

4-3. Sectional Properties of Lumber and Timber

Table 4-3 lists properties of sections of solid-sawn lumber and timber.

4-4. Standard Sizes of Glued-Laminated Timber

Standard finished sizes of structural glued-laminated timber should be used to the extent that conditions permit. These standard finished sizes are based on lumber sizes given in "Voluntary Product Standard PS20-70." Other finished sizes may be used to meet the size requirements of a design or other special requirements.

Nominal 2-in-thick lumber, surfaced to 1½ in before gluing, is used to laminate straight members and curved members with radii of curvature within the bending-radius limitations for the species. (Formerly, a net thickness of 1⅝ in was common for nominal 2-in lumber and may be used, depending on its availability.) Nominal 1-in-thick lumber, surfaced to ¾ in before gluing, may be used for laminating curved members when the bending radius is too short to permit use of nominal 2-in-thick laminations if the bending-radius limitations for the species are observed. Other limitation thicknesses may be used to meet special curving requirements.

4-5. Sectional Properties of Glued-Laminated Timber

Table 4-4 lists properties of sections of glued-laminated timber.

TABLE 4-1 Economical Span Range for Framing Members

Framing member	Economical span range, ft	Usual spacing, ft
Roof beams (generally used where a flat or low-pitched roof is desired):		
Simple span:		
Constant depth		
Solid-sawn	0–40	4–20
Glued-laminated	20–100	8–24
Tapered	25–100	8–24
Double tapered (pitched beams)	25–100	8–24
Curved beams	25–100	8–24
Simple beam with overhangs (usually more economical than simple span when span is over 40 ft):		
Solid-sawn	24	4–20
Glued-laminated	10–90	8–24
Continuous span:		
Solid-sawn	10–50	4–20
Glued-laminated	10–50	8–24
Arches (three-hinged for relatively high-rise applications and two-hinged for relatively low-rise applications):		
Three-hinged:		
Gothic	40–90	8–24
Tudor	30–120	8–24
A-frame	20–160	8–24
Three-centered	40–250	8–24
Parabolic	40–250	8–24
Radial	40–250	8–24
Two-hinged:		
Radial	50–200	8–24
Parabolic	50–200	8–24
Trusses (provide openings for passage of wires, piping, etc.):		
Flat or parallel chord	50–150	12–20
Triangular or pitched	50–90	12–20
Bowstring	50–200	14–24
Tied arches (where no ceiling is desired and where a long, clear span is desired with low rise):		
Tied segment	50–100	8–20
Buttressed segment	50–200	14–24
Domes	50–350	8–24
Simple-span floor beams:		
Solid-sawn	6–20	4–12
Glued-laminated	6–40	4–16
Continuous floor beams	25–40	4–16
Roof sheathing and decking:		
1-in sheathing	1–4	
2-in sheathing	6–10	
3-in roof deck	8–15	
4-in roof deck	12–20	
Plywood sheathing	1–4	
Sheathing on roof joists	1.33–2	
Plank floor decking (floor and ceiling in one):		
Edge to edge	4–16	
Wide face to wide face	4–16	

TABLE 4-2 Nominal and Minimum Dressed Sizes of Boards, Dimension, and Timbers

Item	Thickness, in			Face width, in		
	Nominal	Minimum dressed		Nominal	Minimum dressed	
		Dry*	Green†		Dry*	Green†
Boards	1	¾	$^{25}/_{32}$	2	1½	$1^9/_{16}$
	1¼	1	$1^1/_{32}$	3	2½	$2^9/_{16}$
	1½	1¼	$1^9/_{16}$	4	3½	$3^9/_{16}$
				5	4½	4⅝
				6	5½	5⅝
				7	6½	6⅝
				8	7¼	7½
				9	8¼	8½
				10	9¼	9½
				11	10¼	10½
				12	11¼	11½
				14	13¼	13½
				16	15¼	15½
Dimension	2	1½	$1^9/_{16}$	2	1½	$1^9/_{16}$
	2½	2	$2^1/_{16}$	3	2½	$2^9/_{16}$
	3	2½	$2^9/_{16}$	4	3½	$3^9/_{16}$
	3½	3	$3^1/_{16}$	5	4½	4⅝
				6	5½	5⅝
				8	7¼	7½
				10	9¼	9½
				12	11¼	11½
				14	13¼	13½
				16	15¼	15½
	4	3½	$3^9/_{16}$	2	1½	$1^9/_{16}$
	4½	4	$4^1/_{16}$	3	2½	$2^9/_{16}$
				4	3½	$3^9/_{16}$
				5	4½	4⅝
				6	5½	5⅝
				8	7¼	7½
				10	9¼	9½
				12	11¼	11½
				14		13½
				16		15½
Timbers	5 and thicker		½ in less	5 and wider		½ in less

*Dry lumber is defined as lumber seasoned to a moisture content of 19% or less.
†Green lumber is defined as lumber having a moisture content in excess of 19%.

TABLE 4-3 Properties of Sections of Solid-Sawn Wood

Nominal size, in	Standard dressed size, in. (S4S)	Area of section in²	Moment of inertia in⁴	Section modulus, in³	Weight, lb per lin ft, of piece when weight of wood, lb per ft³, equals					
					25	30	35	40	45	50
1 × 3	¾ × 2½	1.875	0.977	0.781	0.326	0.391	0.456	0.521	0.586	0.651
1 × 4	¾ × 3½	2.625	2.680	1.531	0.456	0.547	0.638	0.729	0.820	0.911
1 × 6	¾ × 5½	4.125	10.398	3.781	0.716	0.859	1.003	1.146	1.289	1.432
1 × 8	¾ × 7¼	5.438	23.817	6.570	0.944	1.133	1.322	1.510	1.699	1.888
1 × 10	¾ × 9¼	6.938	49.466	10.695	1.204	1.445	1.686	1.927	2.168	2.409
1 × 12	¾ × 11¼	8.438	88.989	15.820	1.465	1.758	2.051	2.344	2.637	2.930
2 × 3*	1½ × 2½	3.750	1.953	1.563	0.651	0.781	0.911	1.042	1.172	1.302
2 × 4	1½ × 3½	5.250	5.359	3.063	0.911	1.094	1.276	1.458	1.641	1.823
2 × 6	1½ × 5½	8.250	20.797	7.563	1.432	1.719	2.005	2.292	2.578	2.865
2 × 8	1½ × 7¼	10.875	47.635	13.141	1.888	2.266	2.643	3.021	3.398	3.776
2 × 10	1½ × 9¼	13.875	98.932	21.391	2.409	2.891	3.372	3.854	4.336	4.818
2 × 12	1½ × 11¼	16.875	177.979	31.641	2.930	3.516	4.102	4.688	5.273	5.859
2 × 14	1½ × 13¼	19.875	290.775	43.891	3.451	4.141	4.831	5.521	6.211	6.901
3 × 1	2½ × ¾	1.875	0.088	0.234	0.326	0.391	0.456	0.521	0.586	0.651
3 × 2	2½ × 1½	3.750	0.703	0.938	0.651	0.781	0.911	1.042	1.172	1.302
3 × 4	2½ × 3½	8.750	8.932	5.104	1.519	1.823	2.127	2.431	2.734	3.038
3 × 6	2½ × 5½	13.750	34.661	12.604	2.387	2.865	3.342	3.819	4.297	4.774
3 × 8	2½ × 7¼	18.125	79.391	21.901	3.147	3.776	4.405	5.035	5.664	6.293
3 × 10	2½ × 9¼	23.125	164.886	35.651	4.015	4.818	5.621	6.424	7.227	8.030
3 × 12	2½ × 11¼	28.125	296.631	52.734	4.883	5.859	6.836	7.813	8.789	9.766
3 × 14	2½ × 13¼	33.125	484.625	73.151	5.751	6.901	8.051	9.201	10.352	11.502
3 × 16	2½ × 15¼	38.125	738.870	96.901	6.619	7.943	9.266	10.590	11.914	13.238
4 × 1	3½ × ¾	2.625	0.123	0.328	0.456	0.547	0.638	0.729	0.820	0.911
4 × 2	3½ × 1½	5.250	0.984	1.313	0.911	1.094	1.276	1.458	1.641	1.823
4 × 3	3½ × 2½	8.750	4.557	3.646	1.519	1.823	2.127	2.431	2.734	3.038
4 × 4	3½ × 3½	12.250	12.505	7.146	2.127	2.552	2.977	3.403	3.828	4.253
4 × 6	3½ × 5½	19.250	48.526	17.646	3.342	4.010	4.679	5.347	6.016	6.684
4 × 8	3½ × 7¼	25.375	111.148	30.661	4.405	5.286	6.168	7.049	7.930	8.811
4 × 10	3½ × 9¼	32.375	230.840	49.911	5.621	6.745	7.869	8.933	10.117	11.241
4 × 12	3½ × 11¼	39.375	415.283	73.828	6.836	8.203	9.570	10.938	12.305	13.672
4 × 14	3½ × 13½	47.250	717.609	106.313	8.203	9.844	11.484	13.125	14.766	16.406
4 × 16	3½ × 15½	54.250	1,086.130	140.146	9.418	11.302	13.186	15.069	16.953	18.837
6 × 1	5½ × ¾	4.125	0.193	0.516	0.716	0.859	1.003	1.146	1.289	1.432
6 × 2	5½ × 1½	8.250	1.547	2.063	1.432	1.719	2.005	2.292	2.578	2.865
6 × 3	5½ × 2½	13.750	7.161	5.729	2.387	2.865	3.342	3.819	4.297	4.774
6 × 4	5½ × 3½	19.250	19.651	11.229	3.342	4.010	4.679	5.347	6.016	6.684
6 × 6	5½ × 5½	30.250	76.255	27.729	5.252	6.302	7.352	8.403	9.453	10.503
6 × 8	5½ × 7½	41.250	193.359	51.563	7.161	8.594	10.026	11.458	12.891	14.323
6 × 10	5½ × 9½	52.250	392.963	82.729	9.071	10.885	12.700	14.514	16.328	18.142
6 × 12	5½ × 11½	63.250	697.068	121.229	10.981	13.177	15.373	17.569	19.766	21.962
6 × 14	5½ × 13½	74.250	1,127.672	167.063	12.891	15.469	18.047	20.625	23.203	25.781
6 × 16	5½ × 15½	85.250	1,706.776	220.229	14.800	17.760	20.720	23.681	26.641	29.601
6 × 18	5½ × 17½	96.250	2,456.380	280.729	16.710	20.052	23.394	26.736	30.078	33.420
6 × 20	5½ × 19½	107.250	3,398.484	348.563	18.620	22.344	26.068	29.792	33.516	37.240
6 × 22	5½ × 21½	118.250	4,555.086	423.729	20.530	24.635	28.741	32.847	36.953	41.059
6 × 24	5½ × 23½	129.250	5,948.191	506.229	22.439	26.927	31.415	35.903	40.391	44.878
8 × 1	7¼ × ¾	5.438	0.255	0.680	0.944	1.133	1.322	1.510	1.699	1.888
8 × 2	7¼ × 1½	10.875	2.039	2.719	1.888	2.266	2.643	3.021	3.398	3.776
8 × 3	7¼ × 2½	18.125	9.440	7.552	3.147	3.776	4.405	5.035	5.664	6.293
8 × 4	7¼ × 3½	25.375	25.904	14.802	4.405	5.286	6.168	7.049	7.930	8.811
8 × 6	7½ × 5½	41.250	103.984	37.813	7.161	8.594	10.026	11.458	12.891	14.323
8 × 8	7½ × 7½	56.250	263.672	70.313	9.766	11.719	13.672	15.625	17.578	19.531
8 × 10	7½ × 9½	71.250	535.859	112.813	12.370	14.844	17.318	19.792	22.266	24.740
8 × 12	7½ × 11½	86.250	950.547	165.313	14.974	17.969	20.964	23.958	26.953	29.948
8 × 14	7½ × 13½	101.250	1,537.734	227.813	17.578	21.094	24.609	28.125	31.641	35.156
8 × 16	7½ × 15½	116.250	2.327.422	300.313	20.182	24.219	28.255	32.292	36.328	40.365
8 × 18	7½ × 17½	131.250	3,349.609	382.813	22.786	27.344	31.901	36.458	41.016	45.573
8 × 20	7½ × 19½	146.250	4,634.297	475.313	25.391	30.469	35.547	40.625	45.703	50.781
8 × 22	7½ × 21½	161.250	6,211.484	577.813	27.995	33.594	39.193	44.792	50.391	55.990
8 × 24	7½ × 23½	176.250	8,111.172	690.313	30.599	36.719	42.839	48.958	55.078	61.198
10 × 1	9¼ × ¾	6.938	0.325	0.867	1.204	1.445	1.686	1.927	2.168	2.409
10 × 2	9¼ × 1½	13.875	2.602	3.469	2.409	2.891	3.372	3.854	4.336	4.818
10 × 3	9¼ × 2½	23.125	12.044	9.635	4.015	4.818	5.621	6.424	7.227	8.030
10 × 4	9¼ × 3½	32.375	33.049	18.885	5.621	6.745	7.869	8.993	10.117	11.241
10 × 6	9½ × 5½	52.250	131.714	47.896	9.071	10.885	12.700	14.514	16.328	18.142
10 × 8	9½ × 7½	71.250	333.984	89.063	12.370	14.844	17.318	19.792	22.266	24.740
10 × 10	9½ × 9½	90.250	678.755	142.896	15.668	18.802	21.936	25.069	28.203	31.337
10 × 12	9½ × 11½	109.250	1,204.026	209.396	18.967	22.760	26.554	30.347	34.141	37.934
10 × 14	9½ × 13½	128.250	1,947.797	288.563	22.266	26.719	31.172	35.625	40.078	44.531
10 × 16	9½ × 15½	147.250	2,948.068	380.396	25.564	30.677	35.790	40.903	46.016	51.128
10 × 18	9½ × 17½	166.250	4,242.836	484.896	28.863	34.635	40.408	46.181	51.953	57.726
10 × 20	9½ × 19½	185.250	5,870.109	602.063	32.161	38.594	45.026	51.458	57.891	64.323
10 × 22	9½ × 21½	204.250	7,867.879	731.896	35.460	42.552	49.644	56.736	63.828	70.920
10 × 24	9½ × 23½	223.250	10,274.148	874.396	38.759	46.510	54.262	62.014	69.766	77.517
12 × 1	11¼ × ¾	8.438	0.396	1.055	1.465	1.758	2.051	2.344	2.637	2.930
12 × 2	11¼ × 1½	16.875	3.164	4.219	2.930	3.516	4.102	4.688	5.273	5.859
12 × 3	11¼ × 2½	28.125	14.648	11.719	4.883	5.859	6.836	7.813	8.789	9.766
12 × 4	11¼ × 3½	39.375	40.195	22.969	6.836	8.203	9.570	10.938	12.305	13.672
12 × 6	11½ × 5½	63.250	159.443	57.979	10.981	13.177	15.373	17.569	19.766	21.962
12 × 8	11½ × 7½	86.250	404.297	107.813	14.974	17.969	20.964	23.958	26.953	29.948
12 × 10	11½ × 9½	109.250	821.651	172.979	18.967	22.760	26.554	30.347	34.141	37.934
12 × 12	11½ × 11½	132.250	1,457.505	253.479	22.960	27.552	32.144	36.736	41.328	45.920
12 × 14	11½ × 13½	155.250	2,357.859	349.313	26.953	32.344	37.734	43.125	48.516	53.906

TABLE 4-3 Properties of Sections of Solid-Sawn Wood (*Continued*)

Nominal size, in	Standard dressed size, in. (S4S)	Area of section, in²	Moment of inertia in⁴	Section modulus, in³	Weight, lb per lin ft, of piece when weight of wood, lb per ft³, equals					
					25	30	35	40	45	50
12 × 16	11½ × 15½	178.250	3,568.713	460.479	30.946	37.135	43.325	49.514	55.703	61.892
12 × 18	11½ × 17½	201.250	5,136.066	586.979	34.939	41.927	48.915	55.903	62.891	69.878
12 × 20	11½ × 19½	224.250	7,105.922	728.813	38.932	46.719	54.505	62.292	70.078	77.865
12 × 22	11½ × 21½	247.250	9,524.273	885.979	42.925	51.510	60.095	68.681	77.266	85.851
12 × 24	11½ × 23½	270.250	12,437.129	1,058.479	46.918	56.302	65.686	75.069	84.453	93.837
14 × 2	13½ × 1½	19.875	3.727	4.969	3.451	4.141	4.831	5.521	6.211	6.901
14 × 3	13½ × 2½	33.125	17.253	13.802	5.751	6.901	8.051	9.201	10.352	11.502
14 × 4	13½ × 3½	46.375	47.34	27.052	8.047	9.657	11.266	12.877	14.485	16.094
14 × 6	13½ × 5½	74.250	187.172	68.063	12.891	15.469	18.047	20.625	23.203	25.781
14 × 8	13½ × 7½	101.250	474.609	126.563	17.578	21.094	24.609	28.125	31.641	35.156
14 × 10	13½ × 9½	128.250	964.547	203.063	22.266	26.719	31.172	35.625	40.078	44.531
14 × 12	13½ × 11½	155.250	1,710.984	297.563	26.953	32.344	37.734	43.125	48.516	53.906
14 × 16	13½ × 15½	209.250	4,189.359	540.563	36.328	43.594	50.859	58.125	65.391	72.656
14 × 18	13½ × 17½	236.250	6,029.297	689.063	41.016	49.219	57.422	65.625	73.828	82.031
14 × 20	13½ × 19½	263.250	8,341.734	855.563	45.703	54.844	63.984	73.125	82.266	91.406
14 × 22	13½ × 21½	290.250	11,180.672	1,040.063	50.391	60.469	70.547	80.625	90.703	100.781
14 × 24	13½ × 23½	317.250	14,600.109	1,242.563	55.078	66.094	77.109	88.125	99.141	110.156
16 × 3	15½ × 2½	38.125	19.857	15.885	6.619	7.944	9.267	10.592	11.915	13.240
16 × 4	15½ × 3½	53.375	54.487	31.135	9.267	11.121	12.975	14.828	16.682	18.536
16 × 6	15½ × 5½	85.250	214.901	78.146	14.800	17.760	20.720	23.681	26.641	29.601
16 × 8	15½ × 7½	116.250	544.922	145.313	20.182	24.219	28.255	32.292	36.328	40.365
16 × 10	15½ × 9½	147.250	1,107.443	233.146	25.564	30.677	35.790	40.903	46.016	51.128
16 × 12	15½ × 11½	178.250	1,964.463	341.646	30.946	37.135	43.325	49.514	55.703	61.892
16 × 14	15½ × 13½	209.250	3,177.984	470.813	36.328	43.594	50.859	58.125	65.391	72.656
16 × 16	15½ × 15½	240.250	4,810.004	620.646	41.710	50.052	58.394	66.736	75.078	83.420
16 × 18	15½ × 17½	271.250	6,922.523	791.146	47.092	56.510	65.929	75.347	84.766	94.184
16 × 20	15½ × 19½	302.250	9,577.547	982.313	52.474	62.969	73.464	83.958	94.453	104.948
16 × 22	15½ × 21½	333.250	12,837.066	1,194.146	57.856	69.427	80.998	92.569	104.141	115.712
16 × 24	15½ × 23½	364.250	16,763.086	1,426.646	63.238	75.885	88.533	101.181	113.828	126.476
18 × 6	17½ × 5½	96.250	242.630	88.229	16.710	20.052	23.394	26.736	30.078	33.420
18 × 8	17½ × 7½	131.250	615.234	164.063	22.786	27.344	31.901	36.458	41.016	45.573
18 × 10	17½ × 9½	166.250	1,250.338	263.229	28.863	34.635	40.408	46.181	51.953	57.726
18 × 12	17½ × 11½	201.250	2,217.943	385.729	34.939	41.927	48.915	55.903	62.891	69.878
18 × 14	17½ × 13½	236.250	3,588.047	531.563	41.016	49.219	57.422	65.625	73.828	82.031
18 × 16	17½ × 15½	271.250	5,430.648	700.729	47.092	56.510	65.929	75.347	84.766	94.184
18 × 18	17½ × 17½	306.250	7,815.754	893.229	53.168	63.802	74.436	85.069	95.703	106.337
18 × 20	17½ × 19½	341.250	10,813.359	1,109.063	59.245	71.094	82.943	94.792	106.641	118.490
18 × 22	17½ × 21½	376.250	14,493.461	1,348.229	65.321	78.385	91.450	104.514	117.578	130.642
18 × 24	17½ × 23½	411.250	18,926.066	1,610.729	71.398	85.677	99.957	114.236	128.516	142.795
20 × 6	19½ × 5½	107.250	270.359	98.313	18.620	22.344	26.068	29.792	33.516	37.240
20 × 8	19½ × 7½	146.250	685.547	182.813	25.391	30.469	35.547	40.625	45.703	50.781
20 × 10	19½ × 9½	185.250	1,393.234	293.313	32.161	38.594	45.026	51.458	57.891	64.323
20 × 12	19½ × 11½	224.250	2,471.422	429.813	38.932	46.719	54.505	62.292	70.078	77.865
20 × 14	19½ × 13½	263.250	3,998.109	592.313	45.703	54.844	63.984	73.125	82.266	91.406
20 × 16	19½ × 15½	302.250	6,051.297	780.813	52.474	62.969	73.464	83.958	94.453	104.948
20 × 18	19½ × 17½	341.250	8,708.984	995.313	59.245	71.094	82.943	94.792	106.641	118.490
20 × 20	19½ × 19½	380.250	12,049.172	1,235.813	66.016	79.219	92.422	105.625	118.828	132.031
20 × 22	19½ × 21½	419.250	16,149.859	1,502.313	72.786	87.344	101.901	116.458	131.016	145.573
20 × 24	19½ × 23½	458.250	21,089.047	1,794.813	79.557	95.469	111.380	127.292	243.203	159.115
22 × 6	21½ × 5½	118.250	298.088	108.396	20.530	24.635	28.741	32.847	36.953	41.059
22 × 8	21½ × 7½	161.250	755.859	201.563	27.995	33.594	39.193	44.792	50.391	55.990
22 × 10	21½ × 9½	204.250	1,536.130	323.396	35.460	42.552	49.644	56.736	63.828	70.920
22 × 12	21½ × 11½	247.250	2,724.901	473.896	42.925	51.510	60.095	68.681	77.266	85.851
22 × 14	21½ × 13½	290.250	4,408.172	653.063	50.391	60.469	70.547	80.625	90.703	100.781
22 × 16	21½ × 15½	333.250	6,671.941	860.896	57.856	69.427	80.998	92.569	104.141	115.712
22 × 18	21½ × 17½	376.250	9,602.211	1,097.396	65.321	78.385	91.450	104.514	117.578	130.642
22 × 20	21½ × 19½	419.250	13,284.984	1,362.563	72.786	87.344	101.901	116.458	131.016	145.573
22 × 22	21½ × 21½	462.250	17,806.254	1,656.396	80.252	96.302	112.352	128.403	144.453	160.503
22 × 24	21½ × 23½	505.250	23,252.023	1,978.896	87.717	105.260	122.804	140.347	157.891	175.434
24 × 6	23½ × 5½	129.250	325.818	118.479	22.439	26.927	31.415	35.903	40.391	44.878
24 × 8	23½ × 7½	176.250	826.172	220.313	30.599	36.719	42.839	48.958	55.078	61.198
24 × 10	23½ × 9½	223.250	1,679.026	353.479	38.759	46.510	54.262	62.014	69.766	77.517
24 × 12	23½ × 11½	270.250	2,978.380	517.979	46.918	56.302	65.686	75.069	84.453	93.837
24 × 14	23½ × 13½	317.250	4,818.234	713.813	55.078	66.094	77.109	88.125	99.141	110.156
24 × 16	23½ × 15½	364.250	7,292.586	940.979	63.238	75.885	88.533	101.181	113.828	126.476
24 × 18	23½ × 17½	411.250	10,495.441	1,199.479	71.398	85.677	99.957	114.236	128.516	142.795
24 × 20	23½ × 19½	458.250	14,520.797	1,489.313	79.557	95.469	111.380	127.292	143.203	159.115
24 × 22	23½ × 21½	505.250	19,462.648	1,810.479	87.717	105.260	122.804	140.347	157.891	175.434
24 × 24	23½ × 23½	552.250	25,415.004	2,162.979	95.877	115.052	134.227	153.403	172.578	191.753

*For lumber surfaced 1⅝ in thick, instead of 1½ in, the area, moment of inertia, and section modulus may be increased 8.33%.

TABLE 4-4 Properties of Sections of Glued-Laminated Timber*

2¼-in Width

No. of laminations 1½-in	No. of laminations ¾-in	d	C_F	A	S	I	Vol.
2	4	3.00	1.00	6.8	3.4	5.1	0.05
	5	3.75	1.00	8.4	5.3	9.9	0.06
3	6	4.50	1.00	10.1	7.6	17.1	0.07
	7	5.25	1.00	11.8	10.3	27.1	0.08
4	8	6.00	1.00	13.5	13.5	40.5	0.09
	9	6.75	1.00	15.2	17.1	57.7	0.11
5	10	7.50	1.00	16.9	21.1	79.1	0.12
	11	8.25	1.00	18.6	25.5	105.3	0.13
6	12	9.00	1.00	20.2	30.4	136.7	0.14
	13	9.75	1.00	21.9	35.6	173.8	0.15
7	14	10.50	1.00	23.6	41.3	217.0	0.16
	15	11.25	1.00	25.3	47.5	267.0	0.18
8	16	12.00	1.00	27.0	54.0	324.0	0.19
	17	12.75	1.00	28.7	61.0	388.6	0.20
9	18	13.50	1.00	30.4	68.5	461.3	0.21
	19	14.25	0.99	32.1	76.1	542.6	0.22
10	20	15.00	0.98	33.8	84.4	632.8	0.23

3⅛-in Width

No. of laminations 1½-in	No. of laminations ¾-in	d	C_F	A	S	I	Vol.
2	4	3.00	1.00	9.4	4.7	7.0	0.06
	5	3.75	1.00	11.7	7.3	13.7	0.08
3	6	4.50	1.00	14.1	10.5	23.7	0.10
	7	5.25	1.00	16.4	14.4	37.7	0.11
4	8	6.00	1.00	18.8	18.8	56.3	0.13
	9	6.75	1.00	21.1	23.7	80.1	0.15
5	10	7.50	1.00	23.4	29.3	109.9	0.16
	11	8.25	1.00	25.8	35.4	146.2	0.18
6	12	9.00	1.00	28.1	42.2	189.8	0.20
	13	9.75	1.00	30.5	49.5	241.4	0.21
7	14	10.50	1.00	32.8	57.4	301.5	0.23
	15	11.25	1.00	35.2	65.9	370.8	0.24
8	16	12.00	1.00	37.5	75.0	450.0	0.26
	17	12.75	0.99	39.8	84.7	539.8	0.28
9	18	13.50	0.99	42.2	94.9	640.7	0.29
	19	14.25	0.98	44.5	105.8	753.6	0.31
10	20	15.00	0.98	46.9	117.2	878.9	0.33
	21	15.75	0.97	49.2	129.2	1,017.4	0.34
11	22	16.50	0.97	51.6	141.8	1,169.8	0.36
	23	17.25	0.96	53.9	155.0	1,336.7	0.37
12	24	18.00	0.96	56.3	168.8	1,518.8	0.39
	25	18.75	0.95	58.6	183.1	1,716.6	0.41
13	26	19.50	0.95	60.9	198.0	1,931.0	0.42
	27	20.25	0.94	63.3	213.6	2,162.4	0.44

6¾-in Width

No. of laminations 1½-in	No. of laminations ¾-in	d	C_F	A	S	I	Vol.
4	8	6.00	1.00	40.5	40.5	121.5	0.28
	9	6.75	1.00	45.6	51.3	173.0	0.32
5	10	7.50	1.00	50.6	63.3	237.3	0.35
	11	8.25	1.00	55.7	76.6	315.9	0.39
6	12	9.00	1.00	60.8	91.1	410.1	0.42
	13	9.75	1.00	65.8	106.9	521.4	0.46
7	14	10.50	1.00	70.9	124.0	651.2	0.49
	15	11.25	1.00	75.9	142.4	800.9	0.53
8	16	12.00	1.00	81.0	162.0	972.0	0.56
	17	12.75	0.99	86.1	182.9	1,165.9	0.60
9	18	13.50	0.99	91.1	205.0	1,384.0	0.63
	19	14.25	0.98	96.2	228.4	1,627.7	0.67
10	20	15.00	0.98	101.3	253.1	1,898.4	0.70
	21	15.75	0.97	106.3	279.1	2,197.7	0.74
11	22	16.50	0.97	111.4	306.3	2,526.8	0.77
	23	17.25	0.96	116.4	334.8	2,887.5	0.81
12	24	18.00	0.96	121.5	364.5	3,280.5	0.84
	25	18.75	0.95	126.6	395.5	3,707.9	0.88
13	26	19.50	0.95	131.6	427.8	4,170.9	0.91
	27	20.25	0.94	136.7	461.3	4,670.9	0.95
14	28	21.00	0.94	141.8	496.1	5,209.3	0.98
	29	21.75	0.94	146.8	532.2	5,787.6	1.02
15	30	22.50	0.93	151.9	569.5	6,407.2	1.05
	31	23.25	0.93	156.9	608.1	7,069.5	1.09
16	32	24.00	0.92	162.0	648.0	7,776.0	1.12
	33	24.75	0.92	167.1	689.1	8,528.0	1.16
17	34	25.50	0.92	172.1	731.5	9,327.0	1.20
	35	26.25	0.91	177.2	775.2	10,174.4	1.23
18	36	27.00	0.91	182.3	820.1	11,071.7	1.27
	37	27.75	0.91	187.3	866.3	12,020.2	1.30
19	38	28.50	0.90	192.4	913.8	13,021.4	1.34
	39	29.25	0.90	197.4	962.5	14,076.7	1.37
20	40	30.00	0.90	202.5	1,012.5	15,187.5	1.41
	41	30.75	0.90	207.6	1,063.8	16,355.3	1.44
21	42	31.50	0.90	212.6	1,116.3	17,581.4	1.48
	43	32.25	0.89	217.7	1,170.1	18,867.4	1.51
22	44	33.00	0.89	222.8	1,225.1	20,214.6	1.55
	45	33.75	0.89	227.8	1,281.4	21,624.4	1.58
23	46	34.50	0.89	232.9	1,339.0	23,098.3	1.62
	47	35.25	0.88	237.9	1,397.9	24,637.7	1.65
24	48	36.00	0.88	243.0	1,458.0	26,244.0	1.69
	49	36.75	0.88	248.1	1,519.4	27,918.7	1.72
25	50	37.50	0.88	253.1	1,582.0	29,663.1	1.76
	51	38.25	0.88	258.2	1,645.9	31,478.7	1.79

8¾-in Width

No. of laminations 1½-in	No. of laminations ¾-in	d	C_F	A	S	I	Vol.
6	12	9.00	1.00	78.8	118.1	531.6	0.55
	13	9.75	1.00	85.3	138.6	675.8	0.59
7	14	10.50	1.00	91.9	160.8	844.1	0.64
	15	11.25	1.00	98.4	184.6	1,038.2	0.68
8	16	12.00	1.00	105.0	210.0	1,260.0	0.73
	17	12.75	0.99	111.6	237.1	1,511.3	0.77
9	18	13.50	0.99	118.1	265.8	1,794.0	0.82
	19	14.25	0.98	124.7	296.1	2,109.9	0.87
10	20	15.00	0.98	131.3	328.1	2,460.9	0.91
	21	15.75	0.97	137.8	361.8	2,848.8	0.96
11	22	16.50	0.97	144.4	397.0	3,275.5	1.00
	23	17.25	0.96	150.9	433.9	3,742.8	1.05
12	24	18.00	0.96	157.5	472.5	4,252.5	1.09
	25	18.75	0.95	164.1	512.7	4,806.5	1.14
13	26	19.50	0.95	170.6	554.5	5,406.7	1.18
	27	20.25	0.94	177.2	598.0	6,054.8	1.23
14	28	21.00	0.94	183.8	643.1	6,752.8	1.28
	29	21.75	0.94	190.3	689.9	7,502.5	1.32
15	30	22.50	0.93	196.9	738.3	8,305.7	1.37
	31	23.25	0.93	203.4	788.3	9,164.2	1.41
16	32	24.00	0.93	210.0	840.0	10,080.0	1.46
	33	24.75	0.92	216.6	893.3	11,054.8	1.50
17	34	25.50	0.92	223.1	948.3	12,090.6	1.55
	35	26.25	0.92	229.7	1,004.9	13,189.1	1.59
18	36	27.00	0.91	236.3	1,063.1	14,352.2	1.64
	37	27.75	0.91	242.8	1,123.0	15,581.7	1.69
19	38	28.50	0.91	249.4	1,184.5	16,879.6	1.73
	39	29.25	0.91	255.9	1,247.7	18,247.5	1.78
20	40	30.00	0.90	262.5	1,312.5	19,687.5	1.82
	41	30.75	0.90	269.1	1,378.9	21,201.3	1.87
21	42	31.50	0.90	275.6	1,447.0	22,790.7	1.91
	43	32.25	0.90	282.2	1,516.8	24,457.7	1.96
22	44	33.00	0.89	288.8	1,588.1	26,204.1	2.00
	45	33.75	0.89	295.3	1,661.1	28,031.6	2.05
23	46	34.50	0.89	301.9	1,735.8	29,942.2	2.10
	47	35.25	0.89	308.4	1,812.1	31,937.7	2.14
24	48	36.00	0.88	315.0	1,890.0	34,020.0	2.19
	49	36.75	0.88	321.6	1,969.6	36,190.9	2.23
25	50	37.50	0.88	328.1	2,050.8	38,452.2	2.28
	51	38.25	0.88	334.7	2,133.6	40,805.7	2.32
26	52	39.00	0.88	341.3	2,218.1	43,253.4	2.37
	53	39.75	0.88	347.8	2,304.3	45,797.1	2.42
27	54	40.50	0.87	354.4	2,392.2	48,438.6	2.46
	55	41.25	0.87	360.9	2,481.4	51,179.8	2.51

Beam cross-section (axis X–X; width b, depth d):

X ——— X
b (width) · d (depth)

Table — 8¾-in Width

		d	r	A	S	I	wt
28	56	42.00	0.87	367.5	2,572.5	54,022.5	2.60
	57	42.75	0.87	374.1	2,665.2	56,968.6	2.55
29	58	43.50	0.87	380.6	2,759.5	60,019.8	2.64
	59	44.25	0.86	387.2	2,855.5	63,178.1	2.69
30	60	45.00	0.86	393.8	2,953.1	66,445.3	2.73
	61	45.75	0.86	400.3	3,052.4	69,823.3	2.78
31	62	46.50	0.86	406.9	3,153.3	73,313.8	2.83
	63	47.25	0.86	413.4	3,255.8	76,918.8	2.87
32	64	48.00	0.86	420.0	3,360.0	80,640.0	2.92
	65	48.75	0.85	426.6	3,465.8	84,479.4	2.96
33	66	49.50	0.85	433.1	3,573.3	88,438.7	3.01
	67	50.25	0.85	439.7	3,682.4	92,519.9	3.05
34	68	51.00	0.85	446.3	3,793.1	96,724.7	3.10
	69	51.75	0.85	452.8	3,905.5	101,055.0	3.14
35	70	52.50	0.85	459.4	4,019.5	105,512.7	3.19
	71	53.25	0.85	465.9	4,135.2	110,099.6	3.24
36	72	54.00	0.85	472.5	4,252.5	114,817.5	3.28
	73	54.75	0.85	479.1	4,371.4	119,668.3	3.33
37	74	55.50	0.84	485.6	4,492.0	124,653.9	3.37
	75	56.25	0.84	492.2	4,614.3	129,776.0	3.42
38	76	57.00	0.84	498.8	4,738.1	135,036.6	3.46
	77	57.75	0.84	505.3	4,863.6	140,437.4	3.51
39	78	58.50	0.84	511.9	4,990.8	145,980.4	3.55
	79	59.25	0.84	518.4	5,119.6	151,667.3	3.60
40	80	60.00	0.84	525.0	5,250.0	157,500.0	3.65
	81	60.75	0.84	531.6	5,382.1	163,480.4	3.69
41	82	61.50	0.83	538.1	5,515.8	169,610.3	3.74
	83	62.25	0.83	544.7	5,651.1	175,891.5	3.78
42	84	63.00	0.83	551.3	5,788.1	182,326.0	3.83

Table — 6¾-in Width

		d	r	A	S	I	wt
26	52	39.00	0.88	263.3	1,711.1	33,366.9	1.83
	53	39.75	0.88	268.3	1,777.6	35,329.2	1.86
27	54	40.50	0.87	273.4	1,845.3	37,367.0	1.90
	55	41.25	0.87	278.4	1,914.3	39,481.6	1.93
28	56	42.00	0.87	283.5	1,984.5	41,674.5	1.97
	57	42.75	0.87	288.6	2,056.0	43,947.2	2.00
29	58	43.50	0.87	293.6	2,128.8	46,301.0	2.04
	59	44.25	0.86	298.7	2,202.8	48,737.4	2.07
30	60	45.00	0.86	303.8	2,278.1	51,257.8	2.11
	61	45.75	0.86	308.8	2,354.7	53,863.7	2.14
31	62	46.50	0.86	313.9	2,432.5	56,556.4	2.18
	63	47.25	0.86	318.9	2,511.6	59,337.3	2.21
32	64	48.00	0.86	324.0	2,592.0	62,208.0	2.25

Table — 5⅛-in Width

		d	r	A	S	I	wt
3	6	4.50	1.00	23.1	17.3	38.8	0.16
	7	5.25	1.00	26.9	23.5	61.8	0.19
4	8	6.00	1.00	30.8	30.8	92.3	0.21
	9	6.75	1.00	34.6	38.9	131.3	0.24
5	10	7.50	1.00	38.4	48.0	180.2	0.27
	11	8.25	1.00	42.3	58.1	239.8	0.29
6	12	9.00	1.00	46.1	69.2	311.3	0.32
	13	9.75	1.00	50.0	81.2	395.8	0.35
7	14	10.50	1.00	53.8	94.2	494.4	0.37
	15	11.25	1.00	57.7	108.1	608.1	0.40
8	16	12.00	1.00	61.5	123.0	738.0	0.43
	17	12.75	1.00	65.3	138.9	885.2	0.45
9	18	13.50	0.99	69.2	155.7	1,050.8	0.48
	19	14.25	0.99	73.0	173.4	1,235.8	0.51
10	20	15.00	0.98	76.9	192.2	1,441.4	0.53
	21	15.75	0.98	80.7	211.9	1,668.6	0.56
11	22	16.50	0.97	84.6	232.5	1,918.5	0.59
	23	17.25	0.97	88.4	254.2	2,192.2	0.61
12	24	18.00	0.96	92.3	276.8	2,490.8	0.64
	25	18.75	0.96	96.1	300.3	2,815.2	0.67
13	26	19.50	0.95	99.9	324.8	3,166.8	0.69
	27	20.25	0.95	103.8	350.3	3,546.4	0.72
14	28	21.00	0.94	107.6	376.7	3,955.2	0.75
	29	21.75	0.94	111.5	404.1	4,394.3	0.77
15	30	22.50	0.93	115.3	432.4	4,864.7	0.80
	31	23.25	0.93	119.2	461.7	5,367.6	0.83
16	32	24.00	0.93	123.0	492.0	5,904.0	0.85
	33	24.75	0.92	126.8	523.2	6,475.0	0.88
17	34	25.50	0.92	130.7	555.4	7,081.6	0.91
	35	26.25	0.92	134.5	588.6	7,725.0	0.93
18	36	27.00	0.91	138.4	622.7	8,406.3	0.96
	37	27.75	0.91	142.2	657.8	9,126.4	0.99
19	38	28.50	0.91	146.1	693.8	9,886.6	1.01
	39	29.25	0.91	149.9	730.8	10,687.8	1.04
20	40	30.00	0.90	153.8	768.8	11,531.3	1.07
	41	30.75	0.90	157.6	807.7	12,417.9	1.09
21	42	31.50	0.90	161.4	847.5	13,348.9	1.12
	43	32.25	0.90	165.3	888.4	14,325.2	1.15
22	44	33.00	0.89	169.1	930.2	15,348.1	1.17
	45	33.75	0.89	173.0	972.9	16,418.5	1.20
23	46	34.50	0.89	176.8	1,016.7	17,537.6	1.23
	47	35.25	0.89	180.7	1,061.4	18,706.4	1.25
24	48	36.00	0.88	184.5	1,107.0	19,926.0	1.28

Table — (3⅛-in Width, partial)

		d	r	A	S	I	wt
14	28	21.00	0.94	65.6	229.7	2,411.7	0.46
	29	21.75	0.94	68.0	246.4	2,679.5	0.47
15	30	22.50	0.93	70.3	263.7	2,966.3	0.49
	31	23.25	0.93	72.7	281.5	3,272.9	0.50
16	32	24.00	0.93	75.0	300.0	3,600.0	0.52

TABLE 4-4 Properties of Sections of Glued-Laminated Timber (*Continued*)

10¾-in Width

No. of laminations 1½-in	No. of laminations ¾-in	d	C_F	A	S	I	Vol.
7	14	10.50	1.00	112.9	197.5	1,037.0	0.78
	15	11.25	1.00	120.9	226.8	1,275.5	0.84
8	16	12.00	1.00	129.0	258.0	1,548.0	0.90
	17	12.75	0.99	137.1	291.3	1,856.8	0.95
9	18	13.50	0.99	145.1	326.5	2,204.1	1.01
	19	14.25	0.98	153.2	363.8	2,592.2	1.06
10	20	15.00	0.98	161.3	403.1	3,023.4	1.12
	21	15.75	0.97	169.3	444.4	3,500.0	1.18
11	22	16.50	0.97	177.4	487.8	4,024.2	1.23
	23	17.25	0.96	185.4	533.1	4,598.3	1.29
12	24	18.00	0.96	193.5	580.5	5,224.5	1.34
	25	18.75	0.95	201.6	629.9	5,905.2	1.40
13	26	19.50	0.95	209.6	681.3	6,642.5	1.46
	27	20.25	0.94	217.7	734.7	7,438.8	1.51
14	28	21.00	0.94	225.8	790.1	8,296.3	1.57
	29	21.75	0.94	233.8	847.6	9,217.3	1.62
15	30	22.50	0.93	241.9	907.0	10,204.1	1.68
	31	23.25	0.93	249.9	968.5	11,258.9	1.74
16	32	24.00	0.93	258.0	1,032.0	12,384.0	1.79
	33	24.75	0.92	266.1	1,097.5	13,581.7	1.85
17	34	25.50	0.92	274.1	1,165.0	14,854.1	1.90
	35	26.25	0.92	282.2	1,234.6	16,203.7	1.96
18	36	27.00	0.91	290.3	1,306.1	17,632.7	2.02
	37	27.75	0.91	298.3	1,379.7	19,143.3	2.07
19	38	28.50	0.91	306.4	1,455.3	20,737.8	2.13
	39	29.25	0.91	314.4	1,532.9	22,418.4	2.18
20	40	30.00	0.90	322.5	1,612.5	24,187.5	2.24
	41	30.75	0.90	330.6	1,694.1	26,047.3	2.30
21	42	31.50	0.90	338.6	1,777.8	28,000.1	2.35
	43	32.25	0.90	346.7	1,863.4	30,048.1	2.41
22	44	33.00	0.89	354.8	1,951.1	32,193.6	2.46
	45	33.75	0.89	362.8	2,040.8	34,438.8	2.52
23	46	34.50	0.89	370.9	2,132.5	36,786.2	2.58
	47	35.25	0.89	378.9	2,226.3	39,237.8	2.63
24	48	36.00	0.88	387.0	2,322.0	41,796.0	2.69
	49	36.75	0.88	395.1	2,419.8	44,463.1	2.74
25	50	37.50	0.88	403.1	2,519.5	47,241.2	2.80
	51	38.25	0.88	411.2	2,621.3	50,132.8	2.86
26	52	39.00	0.88	419.3	2,725.1	53,139.9	2.91
	53	39.75	0.88	427.3	2,830.9	56,265.0	2.97
27	54	40.50	0.87	435.4	2,938.8	59,510.3	3.02
	55	41.25	0.87	443.4	3,048.6	62,878.1	3.08
28	56	42.00	0.87	451.5	3,160.5	66,370.5	3.14
	57	42.75	0.87	459.6	3,274.4	69,989.9	3.19
29	58	43.50	0.87	467.6	3,390.3	73,738.6	3.25
	59	44.25	0.87	475.7	3,508.2	77,618.8	3.30
30	60	45.00	0.86	483.8	3,628.1	81,632.8	3.36
	61	45.75	0.86	491.8	3,750.1	85,782.9	3.42
31	62	46.50	0.86	499.9	3,874.0	90,071.2	3.47
	63	47.25	0.86	507.9	4,000.0	94,500.2	3.53
32	64	48.00	0.86	516.0	4,128.0	99,072.0	3.58
	65	48.75	0.86	524.1	4,258.0	103,789.0	3.64

12¼-in Width

No. of laminations 1½-in	No. of laminations ¾-in	d	C_F	A	S	I	Vol.
8	16	12.00	1.00	147.0	294.0	1,764.0	1.02
	17	12.75	0.99	156.2	331.9	2,115.8	1.08
9	18	13.50	0.99	165.4	372.1	2,511.6	1.15
	19	14.25	0.98	174.6	414.6	2,953.8	1.21
10	20	15.00	0.98	183.8	459.4	3,445.3	1.28
	21	15.75	0.97	192.9	506.4	3,988.6	1.34
11	22	16.50	0.97	202.1	555.8	4,585.7	1.40
	23	17.25	0.96	211.3	607.5	5,239.7	1.47
12	24	18.00	0.96	220.5	661.5	5,953.5	1.53
	25	18.75	0.95	229.7	717.7	6,728.9	1.60
13	26	19.50	0.95	238.9	776.3	7,569.4	1.66
	27	20.25	0.94	248.1	837.2	8,476.5	1.72
14	28	21.00	0.94	257.2	900.4	9,453.9	1.79
	29	21.75	0.94	266.4	965.8	10,503.1	1.85
15	30	22.50	0.93	275.6	1,033.6	11,627.9	1.91
	31	23.25	0.93	284.8	1,103.6	12,829.5	1.97
16	32	24.00	0.93	294.0	1,176.0	14,112.0	2.04
	33	24.75	0.92	303.2	1,250.6	15,476.3	2.10
17	34	25.50	0.92	312.4	1,327.6	16,926.8	2.17
	35	26.25	0.92	321.6	1,406.8	18,464.1	2.23
18	36	27.00	0.91	330.8	1,488.4	20,093.1	2.30
	37	27.75	0.91	339.9	1,572.2	21,813.7	2.36
19	38	28.50	0.91	349.1	1,658.3	23,631.4	2.42
	39	29.25	0.91	358.3	1,746.7	25,545.7	2.49
20	40	30.00	0.90	367.5	1,837.5	27,562.5	2.55
	41	30.75	0.90	376.7	1,930.5	29,680.8	2.62
21	42	31.50	0.90	385.9	2,025.8	31,907.0	2.68
	43	32.25	0.90	395.1	2,123.4	34,239.7	2.74
22	44	33.00	0.89	404.2	2,223.4	36,685.7	2.81
	45	33.75	0.89	413.4	2,325.6	39,244.1	2.87
23	46	34.50	0.89	422.6	2,430.1	41,919.1	2.93
	47	35.25	0.89	431.8	2,536.9	44,712.7	3.00
24	48	36.00	0.88	441.0	2,646.0	47,628.0	3.06
	49	36.75	0.88	450.2	2,757.4	50,667.0	3.13
25	50	37.50	0.88	459.4	2,871.1	53,833.0	3.19
	51	38.25	0.88	468.6	2,987.1	57,127.8	3.25
26	52	39.00	0.88	477.8	3,105.4	60,554.8	3.32
	53	39.75	0.88	486.9	3,226.0	64,115.8	3.38
27	54	40.50	0.87	496.1	3,348.8	67,814.1	3.45
	55	41.25	0.87	505.3	3,474.0	71,651.5	3.51
28	56	42.00	0.87	514.5	3,601.5	75,631.5	3.57
	57	42.75	0.87	523.7	3,731.3	79,755.7	3.64
29	58	43.50	0.87	532.9	3,863.3	84,027.7	3.70
	59	44.25	0.87	542.1	3,997.7	88,449.1	3.76
30	60	45.00	0.86	551.2	4,134.4	93,023.4	3.83
	61	45.75	0.86	560.4	4,273.3	97,752.2	3.89
31	62	46.50	0.86	569.6	4,414.6	102,639.3	3.96
	63	47.25	0.86	578.8	4,558.1	107,685.9	4.02
32	64	48.00	0.86	588.0	4,704.0	112,896.0	4.08
	65	48.75	0.86	597.2	4,852.1	118,270.7	4.15
33	66	49.50	0.85	606.4	5,002.6	123,814.2	4.21
	67	50.25	0.85	615.6	5,155.3	129,527.4	4.27

14¼-in Width

No. of laminations 1½-in	No. of laminations ¾-in	d	C_F	A	S	I	Vol.
9	18	13.50	0.99	192.4	432.8	2,921.7	1.34
	19	14.25	0.98	203.1	482.3	3,436.2	1.41
10	20	15.00	0.98	213.8	534.4	4,007.8	1.48
	21	15.75	0.97	224.4	589.1	4,639.5	1.56
11	22	16.50	0.97	235.1	646.6	5,334.4	1.63
	23	17.25	0.96	245.8	706.7	6,095.4	1.71
12	24	18.00	0.96	256.5	769.5	6,925.5	1.78
	25	18.75	0.95	267.2	835.0	7,827.8	1.86
13	26	19.50	0.95	277.9	903.1	8,805.2	1.93
	27	20.25	0.94	288.6	973.9	9,860.7	2.00
14	28	21.00	0.94	299.3	1,047.4	10,997.4	2.08
	29	21.75	0.94	309.9	1,123.5	12,218.3	2.15
15	30	22.50	0.93	320.6	1,202.3	13,526.4	2.23
	31	23.25	0.93	331.3	1,283.8	14,924.6	2.30
16	32	24.00	0.93	342.0	1,368.0	16,416.0	2.38
	33	24.75	0.92	352.7	1,454.8	18,003.6	2.45
17	34	25.50	0.92	363.4	1,544.3	19,690.4	2.52
	35	26.25	0.92	374.1	1,636.5	21,479.4	2.60
18	36	27.00	0.91	384.8	1,731.4	23,373.6	2.67
	37	27.75	0.91	395.4	1,828.9	25,376.0	2.75
19	38	28.50	0.91	406.1	1,929.1	27,489.6	2.82
	39	29.25	0.91	416.8	2,032.0	29,717.4	2.89
20	40	30.00	0.90	427.5	2,137.5	32,062.5	2.97
	41	30.75	0.90	438.2	2,245.7	34,527.8	3.04
21	42	31.50	0.90	448.9	2,356.6	37,116.4	3.12
	43	32.25	0.90	459.6	2,470.1	39,831.1	3.19
22	44	33.00	0.89	470.3	2,586.4	42,675.2	3.27
	45	33.75	0.89	480.9	2,705.3	45,651.5	3.34
23	46	34.50	0.89	491.6	2,826.8	48,763.1	3.41
	47	35.25	0.89	502.3	2,951.1	52,012.9	3.49
24	48	36.00	0.88	513.0	3,078.0	55,404.0	3.56
	49	36.75	0.88	523.7	3,207.6	58,939.4	3.64
25	50	37.50	0.88	534.4	3,339.8	62,622.1	3.71
	51	38.25	0.88	545.1	3,474.8	66,455.0	3.79
26	52	39.00	0.88	555.8	3,612.4	70,441.3	3.86
	53	39.75	0.88	566.4	3,752.6	74,583.9	3.93
27	54	40.50	0.87	577.4	3,895.6	78,885.8	4.01
	55	41.25	0.87	587.8	4,041.2	83,350.0	4.08
28	56	42.00	0.87	598.5	4,189.5	87,979.5	4.16
	57	42.75	0.87	609.2	4,340.5	92,777.4	4.23
29	58	43.50	0.87	619.9	4,494.1	97,746.5	4.30
	59	44.25	0.87	630.6	4,650.4	102,890.1	4.38
30	60	45.00	0.86	641.3	4,809.4	108,211.0	4.45
	61	45.75	0.86	651.9	4,971.0	113,712.2	4.53
31	62	46.50	0.86	662.6	5,135.3	119,396.7	4.60
	63	47.25	0.86	673.3	5,302.3	125,267.7	4.68
32	64	48.00	0.86	684.0	5,472.0	131,328.0	4.75
	65	48.75	0.86	694.7	5,644.3	137,580.7	4.82
33	66	49.50	0.85	705.4	5,819.3	144,028.8	4.90
	67	50.25	0.85	716.1	5,997.0	150,675.2	4.97
34	68	51.00	0.85	726.8	6,177.4	157,523.1	5.05
	69	51.75	0.85	737.4	6,360.4	164,575.3	5.12

Width = 10.75 in

No.	No.	d	C_F	I	S	A	Vol.
33	66	49.50	0.85	108,653.3	4,390.0	532.1	3.70
	67	50.25	0.85	113,667.3	4,524.1	540.2	3.75
34	68	51.00	0.85	118,833.2	4,660.1	548.3	3.81
	69	51.75	0.85	124,153.3	4,798.2	556.5	3.86
35	70	52.50	0.85	129,629.9	4,938.3	564.4	3.92
	71	53.25	0.85	135,265.2	5,080.4	572.4	3.98
36	72	54.00	0.85	141,061.5	5,224.5	580.5	4.03
	73	54.75	0.85	147,021.1	5,370.6	588.6	4.09
37	74	55.50	0.84	153,146.2	5,518.8	596.6	4.14
	75	56.25	0.84	159,439.1	5,668.9	604.7	4.20
38	76	57.00	0.84	165,902.1	5,821.1	612.8	4.26
	77	57.75	0.84	172,537.4	5,975.3	620.8	4.31
39	78	58.50	0.84	179,347.3	6,131.5	628.9	4.37
	79	59.25	0.84	186,334.1	6,289.8	636.9	4.42
40	80	60.00	0.84	193,500.0	6,450.0	645.0	4.48
	81	60.75	0.84	200,847.4	6,612.3	653.1	4.54
41	82	61.50	0.84	208,378.4	6,776.5	661.1	4.59
	83	62.25	0.84	216,095.3	6,942.9	669.2	4.65
42	84	63.00	0.83	224,000.5	7,111.1	677.3	4.70
	85	63.75	0.83	232,096.1	7,281.4	685.3	4.76
43	86	64.50	0.83	240,384.5	7,453.8	693.4	4.82
	87	65.25	0.83	248,867.9	7,628.1	701.4	4.87
44	88	66.00	0.83	257,548.5	7,804.5	709.5	4.93
	89	66.75	0.83	266,428.8	7,982.9	717.6	4.98
45	90	67.50	0.83	275,510.8	8,163.3	725.6	5.04
	91	68.25	0.83	284,796.9	8,345.7	733.7	5.10
46	92	69.00	0.82	294,289.3	8,530.1	741.8	5.15
	93	69.75	0.82	303,990.5	8,716.6	749.8	5.21
47	94	70.50	0.82	313,902.4	8,905.0	757.9	5.26
	95	71.25	0.82	324,027.5	9,095.5	765.9	5.32
48	96	72.00	0.82	334,368.0	9,288.0	774.0	5.38
	97	72.75	0.82	344,926.3	9,482.5	782.1	5.43
49	98	73.50	0.82	355,704.5	9,679.0	790.1	5.49
	99	74.25	0.82	366,704.8	9,877.6	798.2	5.54
50	100	75.00	0.82	377,929.7	10,078.1	806.3	5.60

Width = 12.25 in

No.	No.	d	C_F	I	S	A	Vol.
34	68	51.00	0.85	135,414.4	5,310.4	624.8	4.34
	69	51.75	0.85	141,476.6	5,467.1	633.9	4.40
35	70	52.50	0.85	147,717.8	5,627.3	643.1	4.47
	71	53.25	0.85	154,138.9	5,789.3	652.3	4.53
36	72	54.00	0.85	160,744.5	5,953.5	661.5	4.59
	73	54.75	0.84	167,535.1	6,120.0	670.7	4.66
37	74	55.50	0.84	174,515.4	6,288.8	679.9	4.72
	75	56.25	0.84	181,685.8	6,459.9	689.1	4.79
38	76	57.00	0.84	189,051.2	6,633.4	698.2	4.85
	77	57.75	0.84	196,611.7	6,809.1	707.4	4.91
39	78	58.50	0.84	204,372.5	6,987.1	716.6	4.98
	79	59.25	0.84	212,333.5	7,167.4	725.8	5.04
40	80	60.00	0.84	220,500.0	7,350.0	735.0	5.10
	81	60.75	0.84	228,871.8	7,534.9	744.2	5.17
41	82	61.50	0.83	237,454.4	7,722.1	753.4	5.23
	83	62.25	0.83	246,247.3	7,911.6	762.6	5.30
42	84	63.00	0.83	255,256.3	8,103.4	771.8	5.36
	85	63.75	0.83	264,460.7	8,297.4	780.9	5.42
43	86	64.50	0.83	273,926.5	8,493.8	790.1	5.49
	87	65.25	0.83	283,592.7	8,692.5	799.3	5.55
44	88	66.00	0.83	293,485.5	8,893.5	808.5	5.61
	89	66.75	0.83	303,603.8	9,096.7	817.7	5.68
45	90	67.50	0.83	313,954.1	9,302.3	826.9	5.74
	91	68.25	0.82	324,534.9	9,510.2	836.1	5.81
46	92	69.00	0.82	335,352.9	9,720.4	845.2	5.87
	93	69.75	0.82	346,406.5	9,932.8	854.4	5.93
47	94	70.50	0.82	357,702.7	10,147.6	863.6	6.00
	95	71.25	0.82	369,239.4	10,364.6	872.8	6.06
48	96	72.00	0.82	381,024.0	10,584.0	882.0	6.12
	97	72.75	0.82	393,054.2	10,805.6	891.2	6.19
49	98	73.50	0.82	405,337.6	11,029.6	900.4	6.25
	99	74.25	0.82	417,871.5	11,255.8	909.6	6.32
50	100	75.00	0.82	430,664.1	11,484.4	918.8	6.38
	101	75.75	0.82	443,712.2	11,715.2	927.9	6.44
51	102	76.50	0.82	457,024.1	11,948.3	937.1	6.51
	103	77.25	0.81	470,596.7	12,183.7	946.3	6.57
52	104	78.00	0.81	484,435.5	12,421.5	955.5	6.64
	105	78.75	0.81	498,545.9	12,661.5	964.7	6.70
53	106	79.50	0.81	512,927.8	12,903.8	973.9	6.76
	107	80.25	0.81	527,580.3	13,148.4	983.1	6.83
54	108	81.00	0.81	542,512.7	13,395.4	992.2	6.89
	109	81.75	0.81	557,720.6	13,644.5	1,001.4	6.95
55	110	82.50	0.81	573,213.9	13,896.1	1,010.6	7.02
	111	83.25	0.81	588,987.6	14,149.9	1,019.8	7.08
56	112	84.00	0.81	605,052.0	14,406.0	1,029.0	7.15

Width = 14.25 in

No.	No.	d	C_F	I	S	A	Vol.
35	70	52.50	0.85	171,835.5	6,546.1	748.1	5.20
	71	53.25	0.85	179,305.0	6,734.5	758.8	5.27
36	72	54.00	0.85	186,988.5	6,925.5	769.5	5.34
	73	54.75	0.85	194,888.4	7,119.2	780.2	5.42
37	74	55.50	0.84	203,007.7	7,315.6	790.9	5.49
	75	56.25	0.84	211,349.5	7,514.6	801.6	5.57
38	76	57.00	0.84	219,916.7	7,716.4	812.3	5.64
	77	57.75	0.84	228,712.4	7,920.8	822.9	5.71
39	78	58.50	0.84	237,739.5	8,127.8	833.6	5.79
	79	59.25	0.84	247,001.0	8,337.6	844.3	5.86
40	80	60.00	0.84	256,500.0	8,550.0	855.0	5.94
	81	60.75	0.84	266,239.5	8,765.1	865.7	6.01
41	82	61.50	0.84	276,222.5	8,982.8	876.4	6.09
	83	62.25	0.84	286,451.9	9,203.3	887.1	6.16
42	84	63.00	0.83	296,930.8	9,426.4	897.8	6.23
	85	63.75	0.83	307,662.3	9,652.1	908.4	6.31
43	86	64.50	0.83	318,649.2	9,880.6	919.1	6.38
	87	65.25	0.83	329,894.6	10,111.7	929.8	6.46
44	88	66.00	0.83	341,401.5	10,345.5	940.5	6.53
	89	66.75	0.83	353,173.0	10,582.0	951.2	6.61
45	90	67.50	0.83	365,212.0	10,821.1	961.9	6.68
	91	68.25	0.83	377,521.5	11,062.9	972.6	6.75
46	92	69.00	0.82	390,104.5	11,307.4	983.3	6.83
	93	69.75	0.82	402,964.0	11,554.5	993.9	6.90
47	94	70.50	0.82	416,103.1	11,804.3	1,004.6	6.98
	95	71.25	0.82	429,524.8	12,056.8	1,015.3	7.05
48	96	72.00	0.82	443,232.0	12,312.0	1,026.0	7.12
	97	72.75	0.82	457,227.8	12,569.8	1,036.7	7.20
49	98	73.50	0.82	471,515.1	12,830.3	1,047.4	7.27
	99	74.25	0.82	486,097.1	13,093.5	1,058.1	7.35
50	100	75.00	0.82	500,976.6	13,359.4	1,068.8	7.42
	101	75.75	0.82	516,156.7	13,627.9	1,079.4	7.50
51	102	76.50	0.82	531,640.5	13,899.1	1,090.1	7.57
	103	77.25	0.81	547,430.7	14,173.0	1,100.8	7.64
52	104	78.00	0.81	563,530.6	14,449.5	1,111.5	7.72
	105	78.75	0.81	579,943.1	14,728.7	1,122.2	7.79
53	106	79.50	0.81	596,671.2	15,010.6	1,132.9	7.87
	107	80.25	0.81	613,718.0	15,295.1	1,143.6	7.94
54	108	81.00	0.81	631,086.2	15,582.4	1,154.3	8.02
	109	81.75	0.81	648,779.2	15,872.8	1,164.9	8.09
55	110	82.50	0.81	666,799.8	16,164.8	1,175.6	8.16
	111	83.25	0.81	685,151.2	16,460.1	1,186.3	8.24
56	112	84.00	0.81	703,836.1	16,758.0	1,197.0	8.31
	113	84.75	0.81	722,857.7	17,058.6	1,207.7	8.39
57	114	85.50	0.81	742,218.8	17,361.8	1,218.4	8.46
	115	86.25	0.81	761,922.8	17,667.8	1,229.1	8.54
58	116	87.00	0.80	781,972.3	17,976.4	1,239.8	8.61
	117	87.75	0.80	802,370.6	18,287.6	1,250.4	8.68
59	118	88.50	0.80	823,120.6	18,601.6	1,261.1	8.76
	119	89.25	0.80	844,225.2	18,918.2	1,271.8	8.83
60	120	90.00	0.80	865,687.6	19,237.5	1,282.5	8.91
	121	90.75	0.80	887,510.6	19,559.5	1,293.2	8.98
61	122	91.50	0.80	909,697.3	19,884.1	1,303.9	9.05
	123	92.25	0.80	932,250.8	20,211.4	1,314.6	9.13
62	124	93.00	0.80	955,174.0	20,541.4	1,325.3	9.20
	125	93.75	0.80	978,470.0	20,874.0	1,335.9	9.28
63	126	94.50	0.79	1,002,141.6	21,209.3	1,346.6	9.35
	127	95.25	0.79	1,026,192.0	21,547.3	1,357.3	9.43
64	128	96.00	0.79	1,050,624.2	21,888.0	1,368.0	9.50

*d = section depth, in; C_F = size factor; A = cross-sectional area, in²; S = section modulus, in³; I = moment of inertia, in⁴; Vol. = volume, ft³/lin ft.

4-6. Basic and Allowable Stresses for Timber

Testing a species to determine average strength properties should be carried out from either of two viewpoints:

1. Tests should be made on specimens of large size containing defects. Practically all structural uses involve members of this character.
2. Tests should be made on small, clear specimens to provide fundamental data. Factors to account for the influence of various characteristics may be applied to establish the strength of structural members.

Tests made in accordance with the first viewpoint have the disadvantage that the results apply only to the particular combination of characteristics existing in the test specimens. To determine the strength corresponding to other combinations requires additional tests; thus, an endless testing program is necessary. The second viewpoint permits establishment of fundamental strength properties for each species and application of general rules to cover the specific conditions involved in a particular case.

This second viewpoint has been generally accepted. When a species has been adequately investigated under this concept, there should be no need for further tests on that species unless new conditions arise. ASTM Standard D143, "Standard Methods of Testing Small, Clear Specimens of Timber," gives the procedure for determining fundamental data on wood species.

Basic stresses are essentially unit stresses applicable to clear and straight-grained defect-free material These stresses, derived from the results of tests on small, clear specimens of green wood, include an adjustment for variability of material, length of loading period, and factor of safety. They are considerably less than the average for the species. They require only an adjustment for grade to become allowable unit stresses.

Allowable unit stresses are computed for a particular grade by reducing the basic stress according to the limitations on defects for that grade. The basic stress is multiplied by a strength ratio to obtain an allowable stress. This strength ratio represents that proportion of the strength of a defect-free piece that remains after taking into account the effect of strength-reducing features.

The principal factors entering into the establishment of allowable unit stress for each species include inherent strength of wood, reduction in strength due to natural growth characteristics permitted in the grade, effect of long-time loading, variability of individual species, possibility of some slight overloading, characteristics of the species, size of member and related influence of seasoning, and factor of safety. The effect of these factors is a strength value for practical-use conditions lower than the average value taken from tests on small, clear specimens.

Basic stresses for laminated timbers under wet service conditions are the same as for solid timbers; i.e., these stresses are based on the strength of wood in the green condition. When moisture content in a member will be low throughout its service, a second set of higher basic stresses, based on the higher strength of dry material, may be used. Technical Bulletin 479, U.S. Department of Agriculture, "Strength and Related Properties of Woods Grown in the United States," presents test results on small, clear, and straight-grained wood species in the green state and in the 12%-moisture-content, air-dry condition. Technical Bulletin 1069, U.S. Department of Agriculture, "Fabrication and Design of Glued-Laminated Structural Members," gives the basic stresses for clear, solid-sawn members and glued-laminated timbers in the wet condition and for clear, glued-laminated timbers in the dry condition.

Allowable unit stresses for commercial species of lumber, widely accepted for wood construction in economical and efficient designs, are given in "National Design Specification for Wood Construction," National Forest Products Association. Allowable unit stresses for glued-laminated timber are given in the AITC "Timber Construction Manual," John Wiley & Sons, Inc., and in the AITC "Standard Specifications for Structural Glued-Laminated Timber," AITC

117, American Institute of Timber Construction, 333 W. Hampden Ave., Englewood, Colo. 80110.

See also Arts. 4-9 to 4-12 and 4-14.

4-7. Structural Grading of Wood

Strength properties of wood are intimately related to moisture content and specific gravity. Therefore, data on strength properties unaccompanied by corresponding data on these physical properties are of little value.

The strength of wood is actually affected by many other factors, such as rate of loading, duration of load, temperature, direction of grain, and position of growth rings. Strength is also influenced by such inherent growth characteristics as knots, cross grain, shakes, and checks.

Analysis and integration of available data have yielded a comprehensive set of simple principles for grading structural timber ("Tentative Methods for Establishing Structural Grades of Lumber," ASTM D245).

The same characteristics, such as knots and cross grain, that reduce the strength of solid timber also affect the strength of laminated members. However, additional factors peculiar to laminated wood must be considered: Effect on strength of bending members is less from knots located at the neutral plane of the beam, a region of low stress. Strength of a bending member with low-grade laminations can be improved by substituting a few high-grade laminations at the top and bottom of the member. Dispersement of knots in laminated members has a beneficial effect on strength. With sufficient knowledge of the occurrence of knots within a grade, mathematical estimates of this effect may be established for members containing various numbers of laminations.

Allowable design stresses taking these factors into account are higher than for solid timbers of comparable grade. But cross-grain limitations must be more restrictive than for solid timbers, to justify these higher allowable stresses.

4-8. Weight and Specific Gravity of Commercial Lumber Species

Specific gravity is a reliable indicator of fiber content. Also, specific gravity and the strength and stiffness of solid wood or laminated products are interrelated. See Table 4-5 for weights and specific gravities of several commercial lumber species.

4-9. Moisture Content of Wood

Wood is unlike most structural materials in regard to the causes of its dimensional changes, which are primarily from gain or loss of moisture, not change in temperature. For this reason expansion joints are seldom required for wood structures to permit movement with temperature changes. It partly accounts for the fact that wood structures can withstand extreme temperatures without collapse.

A newly felled tree is green (contains moisture). When the greater part of this water is being removed, seasoning first allows free water to leave the cavities in the wood. A point is reached where these cavities contain only air, and the cell walls still are full of moisture. The moisture content at which this occurs, the fiber-saturation point, varies from 25 to 30% of the weight of the oven-dry wood.

During removal of the free water, the wood remains constant in size and in most properties (weight decreases). Once the fiber-saturation point has been passed, shrinkage of the wood begins as the cell walls lose water. Shrinkage continues nearly linearly down to zero moisture content (Table 4-6). (There are, however, complicating factors, such as the effects of timber

TABLE 4-5 Weights and Specific Gravities of Commercial Lumber Species

Species	Specific gravity based on oven-dry weight and volume at 12% moisture content	Weight, lb per ft³		Adjusting factor for each 1% change in moisture content	Moisture content when green (avg), %	Specific gravity based on oven-dry weight and volume when green	Weight when green, lb per ft³
		At 12% moisture content	At 20% moisture content				
Softwoods:							
Cedar:							
Alaska	0.44	31.1	32.4	0.170	38	0.42	35.5
Incense	0.37	25.0	26.4	0.183	108	0.35	42.5
Port Orford	0.42	29.6	31.0	0.175	43	0.40	35.0
Western red	0.33	23.0	24.1	0.137	37	0.31	26.4
Cypress, southern	0.46	32.1	33.4	0.167	91	0.42	45.3
Douglas fir:							
Coast region	0.48	33.8	35.2	0.170	38	0.45	38.2
Inland region	0.44	31.4	32.5	0.137	48	0.41	36.3
Rocky Mountain	0.43	30.0	31.4	0.179	38	0.40	34.6
Fir, white	0.37	26.3	27.3	0.129	115	0.35	39.6
Hemlock:							
Eastern	0.40	28.6	29.8	0.150	111	0.38	43.4
Western	0.42	29.2	30.2	0.129	74	0.38	37.2
Larch, western	0.55	38.9	40.2	0.170	58	0.51	46.7
Pine:							
Eastern white	0.35	24.9	26.2	0.167	73	0.34	35.1
Lodgepole	0.41	28.8	29.9	0.142	65	0.38	36.3
Norway	0.44	31.0	32.1	0.142	92	0.41	42.3
Ponderosa	0.40	28.1	29.4	0.162	91	0.38	40.9
Southern shortleaf	0.51	35.2	36.5	0.154	81	0.46	45.9
Southern longleaf	0.58	41.1	42.5	0.179	63	0.54	50.2
Sugar	0.36	25.5	26.8	0.162	137	0.35	45.8
Western white	0.38	27.6	28.6	0.129	54	9.36	33.0
Redwood	0.40	28.1	29.5	0.175	112	0.38	45.6
Spruce:							
Engelmann	0.34	23.7	24.7	0.129	80	0.32	32.5
Sitka	0.40	27.7	28.8	0.145	42	0.37	32.0
White	0.40	29.1	29.9	0.104	50	0.37	33.0
Hardwoods:							
Ash, white	0.60	42.2	43.6	0.175	42	0.55	47.4
Beech, American	0.64	43.8	45.1	0.162	54	0.56	50.6
Birch:							
Sweet	0.65	46.7	48.1	0.175	53	0.60	53.8
Yellow	0.62	43.0	44.1	0.142	67	0.55	50.8
Elm, rock	0.63	43.6	45.2	0.208	48	0.57	50.9
Gum	0.52	36.0	37.1	0.133	115	0.46	49.7
Hickory:							
Pecan	0.66	45.9	47.6	0.212	63	0.60	56.7
Shagbark	0.72	50.8	51.8	0.129	60	0.64	57.0
Maple, sugar	0.63	44.0	45.3	0.154	58	0.56	51.1
Oak:							
Red	0.63	43.2	44.7	0.187	80	0.56	56.0
White	0.68	46.3	47.6	0.167	68	0.60	55.6
Poplar, yellow	0.42	29.8	31.0	0.150	83	0.40	40.5

TABLE 4-6   Shrinkage Values of Wood Based on Dimensions When Green

Species	Dried to 20% MC*			Dried to 6% MC†			Dried to 0% MC		
	Ra-dial, %	Tan-gen-tial, %	Volu-met-ric, %	Ra-dial, %	Tan-gen-tial, %	Volu-met-ric, %	Ra-dial, %	Tan-gen-tial, %	Volu-met-ric, %
Softwoods:‡									
Cedar:									
Alaska	0.9	2.0	3.1	2.2	4.8	7.4	2.8	6.0	9.2
Incense	1.1	1.7	2.5	2.6	4.2	6.1	3.3	5.2	7.6
Port Orford	1.5	2.3	3.4	3.7	5.5	8.1	4.6	6.9	10.1
Western red	0.8	1.7	2.3	1.9	4.0	5.4	2.4	5.0	6.8
Cypress, southern	1.3	2.1	3.5	3.0	5.0	8.4	3.8	6.2	10.5
Douglas fir:									
Coast region	1.7	2.6	3.9	4.0	6.2	9.4	5.0	7.8	11.8
Inland region	1.4	2.5	3.6	3.3	6.1	8.7	4.1	7.6	10.9
Rocky Mountain	1.2	2.1	3.5	2.9	5.0	8.5	3.6	6.2	10.6
Fir, white	1.1	2.4	3.3	2.6	5.7	7.8	3.2	7.1	9.8
Hemlock:									
Eastern	1.0	2.3	3.2	2.4	5.4	7.8	3.0	6.8	9.7
Western	1.4	2.6	4.0	3.4	6.3	9.5	4.3	7.9	11.9
Larch, western	1.4	2.7	4.4	3.4	6.5	10.6	4.2	8.1	13.2
Pine:									
Eastern white	0.8	2.0	2.7	1.8	4.8	6.6	2.3	6.0	8.2
Lodgepole	1.5	2.2	3.8	3.6	5.4	9.2	4.5	6.7	11.5
Norway	1.5	2.4	3.8	3.7	5.8	9.2	4.6	7.2	11.5
Ponderosa	1.3	2.1	3.2	3.1	5.0	7.7	3.9	6.3	9.6
Southern (avg.)	1.6	2.6	4.1	4.0	6.1	9.8	5.0	7.6	12.2
Sugar	1.0	1.9	2.6	2.3	4.5	6.3	2.9	5.6	7.9
Western white	1.4	2.5	3.9	3.3	5.9	9.4	4.1	7.4	11.8
Redwood (old growth)	0.9	1.5	2.3	2.1	3.5	5.4	2.6	4.4	6.8
Spruce:									
Engelmann	1.1	2.2	3.5	2.7	5.3	8.3	3.4	6.6	10.4
Sitka	1.4	2.5	3.8	3.4	6.0	9.2	4.3	7.5	11.5
Hardwoods:‡									
Ash, white	1.6	2.6	4.5	3.8	6.2	10.7	4.8	7.8	13.4
Beech, American	1.7	3.7	5.4	4.1	8.8	13.0	5.1	11.0	16.3
Birch:									
Sweet	2.2	2.8	5.2	5.2	6.8	12.5	6.5	8.5	15.6
Yellow	2.4	3.1	5.6	5.8	7.4	13.4	7.2	9.2	16.7
Elm, rock	1.6	2.7	4.7	3.8	6.5	11.3	4.8	8.1	14.1
Gum, red	1.7	3.3	5.0	4.2	7.9	12.0	5.2	9.9	15.0
Hickory:									
Pecan§	1.6	3.0	4.5	3.9	7.1	10.9	4.9	8.9	13.6
True	2.5	3.8	6.0	6.0	9.0	14.3	7.5	11.3	17.9
Maple, hard	1.6	3.2	5.0	3.9	7.6	11.9	4.9	9.5	14.9
Oak:									
Red	1.3	2.7	4.5	3.2	6.6	10.8	4.0	8.2	13.5
White	1.8	3.0	5.3	4.2	7.2	12.6	5.3	9.0	15.8
Poplar, yellow	1.3	2.4	4.1	3.2	5.7	9.8	4.0	7.1	12.3

*MC = moisture content as a percent of weight of oven-dry wood. These shrinkage values have been taken as one-third the shrinkage to the oven-dry condition as given in the last three columns.

†These shrinkage values have been taken as four-fifths of the shrinkage to the oven-dry condition as given in the last three columns.

‡The total longitudinal shrinkage of normal species from fiber saturation to oven-dry condition is minor. It usually ranges from 0.17 to 0.3% of the green dimension.

§Average of butternut hickory, nutmeg hickory, water hickory, and pecan.

size and relative rates of moisture movement in three directions: longitudinal, radial, and tangential to the growth rings.) Eventually, the wood assumes a condition of equilibrium, with the final moisture-content dependent on the relative humidity and temperature of the ambient air. Wood swells when it absorbs moisture, up to the fiber-saturation point. The relationship of wood moisture content, temperature, and relative humidity can actually define an environment (Fig. 4-1).

This explanation has been simplified. Outdoors, rain, frost, wind, and sun can act directly on the wood. Within buildings, poor environmental conditions may be created for wood by localized heating, cooling, or ventilation. The conditions of service must be sufficiently well known to be specifiable. Then, the proper design stress can be assigned to wood and the most suitable adhesive selected.

Dry Condition of Use ▪ *Design Stresses.* Allowable unit stresses for dry conditions of use are applicable for normal loading when the wood moisture content in service is less than 16%, as in most covered structures.

Adhesive Selection. Dry-use adhesives are those that perform satisfactorily when the moisture content of wood does not exceed 16% for repeated or prolonged periods of service and are to be used only when these conditions exist.

Wet Condition of Use ▪ *Design Stresses.* Allowable unit stresses for wet condition of use are applicable for normal loading when the moisture content in service is 16% or more. This may occur in members not covered or in covered locations of high relative humidity.

Adhesive Selection. Wet-use adhesives will perform satisfactorily for all conditions, including exposure to weather, marine use, and where pressure treatments are used, whether before or after gluing. Such adhesives are required when the moisture content exceeds 16% for repeated or prolonged periods of service.

4-10. Checking in Timbers

Separation of grain, or checking, is the result of rapid lowering of surface moisture content combined with a difference in moisture content between inner and outer portions of the piece. As wood loses moisture to the surrounding atmosphere, the outer cells of the member lose at a more rapid rate than the inner cells. As the outer cells try to shrink, they are restrained by the inner portion of the member. The more rapid the drying, the greater the differential in shrinkage between outer and inner fibers and the greater the shrinkage stresses. Splits may develop.

Checks affect the horizontal shear strength of timber. A large reduction factor is applied to test values in establishing allowable unit stresses, in recognition of stress concentrations at the ends of checks. Allowable unit stresses for horizontal shear are adjusted for the amount of checking permissible in the various stress grades at the time of the grading. Since strength properties of wood increase with dryness, checks may enlarge with increasing dryness after shipment without appreciably reducing shear strength.

Cross-grain checks and splits that tend to run out the side of a piece, or excessive checks and splits that tend to enter connection areas, may be serious and may require servicing. Provisions for controlling the effects of checking in connection areas may be incorporated into design details.

To avoid excessive splitting between rows of bolts due to shrinkage during seasoning of solid-sawn timbers, the rows should not be spaced more than 5 in apart, or a saw kerf, terminating in a bored hole, should be provided between the lines of bolts. Whenever possible, maximum end distances for connections should be specified to minimize the effect of checks running into the joint area. Some designers require stitch bolts in members, with multiple connections loaded

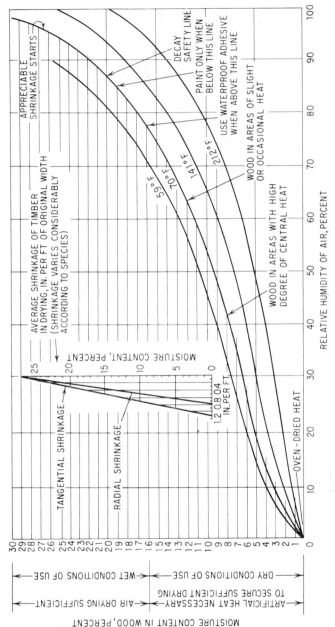

Fig. 4-1. Approximate relationship of wood equilibrium moisture content, temperature, and relative humidity is shown by the curves. The triangular diagram indicates the effect of wood moisture content on wood shrinkage.

at an angle to the grain. Stitch bolts, kept tight, will reinforce pieces where checking is excessive.

One principal advantage of glued-laminated timber construction is relative freedom from checking. Seasoning checks may, however, occur in laminated members for the same reasons that they exist in solid-sawn members. When laminated members are glued within the range of moisture contents set in AITC 103, "Standard for Structural Glued-Laminated Timber," they will approximate the moisture content in normal-use conditions, thereby minimizing checking. Moisture content of the lumber at the time of gluing is thus of great importance to the control of checking in service. However, rapid changes in moisture content of large wood sections after gluing will result in shrinkage or swelling of the wood, and during shrinking, checking may develop in both glued joints and wood.

Differentials in shrinkage rates of individual laminations tend to concentrate shrinkage stresses at or near the glue line. For this reason, when checking occurs, it is usually at or near glue lines. The presence of wood-fiber separation indicates glue bonds and not delamination.

In general, checks have very little effect on the strength of glued-laminated members. Laminations in such members are thin enough to season readily in kiln drying without developing checks. Since checks lie in a radial plane, and the majority of laminations are essentially flat grain, checks are so positioned in horizontally laminated members that they will not materially affect shear strength. When members are designed with laminations vertical (with wide face parallel to the direction of load application), and when checks may affect the shear strength, the effect of checks may be evaluated in the same manner as for checks in solid-sawn members.

Seasoning checks in bending members affect only the horizontal shear strength. They are usually not of structural importance unless the checks are significant in depth and occur in the midheight of the member near the support, and then only if shear governs the design of the members. The reduction in shear strength is nearly directly proportional to the ratio of depth of check to width of beam. Checks in columns are not of structural importance unless the check develops into a split, thereby increasing the slenderness ratio of the columns.

Minor checking may be disregarded since there is an ample factor of safety in allowable unit stresses. The final decision as to whether shrinkage checks are detrimental to the strength requirements of any particular design or structural member should be made by a competent engineer experienced in timber construction.

4-11.   Allowable Unit Stresses and Modifications for Stress-Grade Lumber

A wide range of allowable unit stresses for many species is given in the "National Design Specification for Wood Construction" (NDS). They apply to solid-sawn lumber under normal duration of load and under continuously dry service conditions, such as those in most covered structures. For continuously wet-use service conditions, apply the appropriate factor from Table 4-7. NDS may be obtained from the National Forest Products Association, 1961 Massachusetts Ave., N.W., Washington, D.C. 20036. (See also Arts. 4-6 to 4-10).

The modifications for duration of load temperature, treatment, and size are the same as for glued-laminated timber (Art. 4-12).

4-12.   Allowable Unit Stresses and Modifications for Structural Glued-Laminated Timber

The allowable unit stresses given in the AITC "Timber Construction Manual" and "Standard Specification for Structural Glued-Laminated Timber," AITC 117, are for normal conditions of loading, assuming uniform loading, horizontal laminating, and a 12-in-deep member with a

TABLE 4-7   Allowable-Stress Modification Factors for Moisture in Solid-Sawn Lumber

Service condition	Allowable unit stress				Modulus of elasticity E
	Extreme fiber in bending F_b or tension parallel to grain F_t	Compression parallel to grain F_c	Compression perpendicular to grain F_c	Horizontal shear F_v	
At or above fiber-saturation point, or continuously submerged	1.00	0.90	0.67	1.00	0.91

span-to-depth ratio of 21:1. Tables are given for dry- and wet-use conditions. AITC specifications may be obtained from the American Institute for Timber Construction, 333 W. Hampden Ave., Englewood, Colo. 80110. (See also Arts. 4-6 to 4-10.)

Allowable unit stresses for dry-use conditions are applicable when the moisture content in service is less than 16%, as in most covered structures. Allowable unit stresses for wet-use conditions are applicable when the moisture content in service is 16% or more, as may occur in exterior or submerged construction and in some structures housing wet processes or otherwise having constantly high relative humidities.

Allowable stresses for vertically laminated members made of combination grades of lumber are the weighted average of the lumber grades.

Requirements for limiting cross grain, type, and location of end joints, and certain manufacturing and other requirements, must be met for these allowable-unit-stress combinations to apply.

Species other than those referenced in these specifications may be used if allowable unit stresses are established for them in accordance with the provisions of U.S. Product Standard PS56-73.

For laminated bending members 16¼ in or more deep, the outermost tension-side laminations representing 5% of the total depth of the member should have additional grade restrictions, as described in AITC 117, "Standard Specification for Structural Glued-Laminated Timber of Douglas Fir, Southern Pine, Western Larch, and California Redwood."

Use of hardwoods in glued-laminated timbers is covered by AITC 119, "Standard Specification for Hardwood Glued-Laminated Timber."

Lumber that is E-rated, as well as visually graded and then positioned in laminated timber, is covered by AITC 120, "Standard Specification for Structural Glued-Laminated Timber Using E-Rated and Visually Graded Lumber of Douglas Fir, Southern Pine, Hemlock, and Lodgepole Pine." The lumber is sorted into E grades by measuring the modulus of elasticity of each piece and positioning the lumber in the member according to stiffness, but still grading visually so as to meet the visual grade requirements.

Modifications for Duration of Load ▪ Wood can absorb overloads of considerable magnitude for short periods; thus, allowable unit stresses are adjusted accordingly. The elastic limit and ultimate strength are higher under short-time loading. Wood members under continuous loading for years will fail at loads one-half to three-fourths as great as would be required to produce failure in a static-bending test when the maximum load is reached in a few minutes.

Normal load duration contemplates fully stressing a member to the allowable unit stress by the application of the full design load for a duration of about 10 years (either continuously or cumulatively).

When a member is fully stressed by maximum design loads for long-term loading conditions (greater than 10 years, either continuously or cumulatively), the allowable unit stresses are reduced to 90% of the tabulated values.

When duration of full design load (either continuously or cumulatively) does not exceed 10 years, tabulated allowable unit stresses can be increased as follows: 15% for 2-month duration, as for snow; 25% for 7-day duration; 33⅓% for wind or earthquake; and 100% for impact. These increases are not cumulative. The allowable unit stress for normal loading may be used without regard to impact if the stress induced by impact does not exceed the allowable unit stress for normal loading. These adjustments do not apply to modulus of elasticity, except when it is used to determine allowable unit loads for columns.

Modifications for Temperature ▪ Tests show that wood increases in strength as temperature is lowered below normal. Tests conducted at about −300°F indicate that the important strength properties of dry wood in bending and compression, including stiffness and shock resistance, are much higher at extremely low temperatures.

Some reduction of the allowable unit stresses may be necessary for members subjected to elevated temperatures for repeated or prolonged periods. This adjustment is especially desirable where high temperature is associated with high moisture content.

Temperature effect on strength is immediate. Its magnitude depends on the moisture content of the wood and, when temperature is raised, the duration of exposure.

Between 0 and 70°F, the static strength of dry wood (12% moisture content) roughly increases from its strength at 70°F about ⅛ to ½% for each 1°F decrease in temperature. Between 70 and 150°F, the strength decreases at about the same rate for each 1°F increase in temperature. The change is greater for higher wood moisture content.

After exposure to temperatures not much above normal for a short time under ordinary atmospheric conditions, the wood, when temperature is reduced to normal, may recover essentially all its original strength. Experiments indicate that air-dry wood can probably be exposed to temperatures up to nearly 150°F for a year or more without a significant permanent loss in most strength properties. But its strength while at such temperatures will be temporarily lower than at normal temperature.

When wood is exposed to temperatures of 150°F or more for extended periods of time, it will be permanently weakened. The nonrecoverable strength loss depends on a number of factors, including moisture content and temperature of the wood, heating medium, and time of exposure. To some extent, the loss depends on the species and size of the piece.

Glued-laminated members are normally cured at temperatures of less than 150°F. Therefore, no reduction in allowable unit stresses due to temperature effect is necessary for curing.

Adhesives used under standard specifications for structural glued-laminated members, for example, casein, resorcinol-resin, phenol-resin, and melamine-resin adhesives, are not affected substantially by temperatures up to those that char wood. Use of adhesives that deteriorate at high temperatures is not permitted by standard specifications for structural glued-laminated timber. Low temperatures appear to have no significant effect on the strength of glued joints.

Modifications for Pressure-Applied Treatments ▪ The allowable stresses given for wood also apply to wood treated with a preservative when this treatment is in accordance with American Wood Preservers Association (AWPA) standard specifications, which limit pressure and temperature. Investigations have indicated that, in general, any weakening of timber as a result of preservative treatment is caused almost entirely by subjecting the wood to temperatures and pressures above the AWPA limits.

Highly acidic salts, such as zinc chloride, tend to hydrolyze wood if they are present in appreciable concentrations. Fortunately, the concentrations used in wood preservative treatments are sufficiently small that strength properties other than impact resistance are not greatly affected under normal service conditions. A significant loss in impact strength may occur, however, if higher concentrations are used.

None of the other common preservatives is likely to form solutions as highly acidic as those of zinc chloride, so in most cases their effect on the strength of the wood can be disregarded.

In wood treated with highly acidic salts, such as zinc chloride, moisture is the controlling factor in corrosion of fastenings. Therefore, wood treated with highly acidic salts is not recommended for use under highly humid conditions.

The effects on strength of all treatments, preservative and fire-retardant, should be investigated, to ensure that adjustments in allowable unit stresses are made when required ("Manual of Recommended Practice," American Wood Preservers Association.)

Modifications for Size Factor ▪ When the depth of a rectangular beam exceeds 12 in, the tabulated unit stress in bending F_b should be reduced by multiplication by a size factor C_F.

$$C_F = \left(\frac{12}{d} \right)^{1/9} \tag{4-1}$$

where d = depth of member, in. Table 4-4 lists values of C_F for various cross-sectional sizes.

Equation (4-1) applies to bending members satisfying the following basic assumptions: simply supported beam, uniformly distributed load, and span-depth ratio $L/d = 21$. C_F may thus be applied with reasonable accuracy to beams usually used in buildings. Where greater accuracy is required for other loading conditions or span-depth ratios, the percentage changes given in Table 4-8 may be applied directly to the size factor given by Eq. (4-1). Straight-line interpolation may be used for L/d ratios other than those listed in the table.

TABLE 4-8   Change in Size Factor C_F

Span-depth ratio	Change, %	Loading condition for simply supported beams	Change, %
7	+6.2	Single concentrated load	+7.8
14	+2.3	Uniform load	0
21	0	Third-point load	−3.2
28	−1.6		
35	−2.8		

Modifications for Radial Tension or Compression ▪ The radial stress induced by a bending moment in a member of constant cross section may be computed from

$$f_r = \frac{3M}{2Rbd} \tag{4-2}$$

where M = bending moment, in-lb
  R = radius of curvature at center line of member, in
  b = width of cross section, in
  d = depth of cross section, in

Equation (4-2) can also be used to estimate the stresses in a member with varying cross section. Information on more exact procedures for calculating radial stresses in curved members with varying cross section can be obtained from the American Institute of Timber Construction.

When M is in the direction tending to decrease curvature (increase the radius), tensile stresses occur across the grain. For this condition, the allowable tensile stress across the grain is limited to one-third the allowable unit stress in horizontal shear for southern pine for all load conditions, and for Douglas fir and larch for wind or earthquake loadings. The limit is 15 psi for Douglas fir and larch for other types of loading. These values are subject to modification for duration of load. If these values are exceeded, mechanical reinforcement sufficient to resist all radial tensile stresses is required.

When M is in the direction tending to increase curvature (decrease the radius), the stress is compressive across the grain. For this condition, the allowable stress is limited to that for compression perpendicular to grain for all species covered by AITC 117.

Modifications for Curvature Factor ▪ For the curved portion of members, the allowable unit stress in bending should be modified by multiplication by the following curvature factor:

$$C_c = 1 - 2000 \left(\frac{t}{R} \right)^2 \tag{4-3}$$

where t = thickness of lamination, in
$\quad R$ = radius of curvature of lamination, in
t/R should not exceed $\frac{1}{100}$ for hardwoods and southern pine, or $\frac{1}{125}$ for softwoods other than southern pine. The curvature factor should not be applied to stress in the straight portion of an assembly, regardless of curvature elsewhere.

The recommended minimum radii of curvature for curved, structural glued-laminated members are 9 ft 4 in for ¾-in laminations, and 27 ft 6 in for 1½-in laminations. Other radii of curvature may be used with these thicknesses, and other radius-thickness combinations may be used.

Certain species can be bent to sharper radii, but the designer should determine the availability of such sharply curved members before specifying them.

Modifications for Lateral Stability ▪ The tabulated allowable bending unit stresses are applicable to members that are adequately braced. When deep and slender members not adequately braced are used, allowable bending unit stresses must be reduced. For the purpose, a slenderness factor should be applied, as indicated in the AITC "Timber Construction Manual." (See also Art. 4-13.)

The reduction in bending stresses determined by applying the slenderness factor should not be combined with a reduction in stress due to the application of the size factor. In no case may the allowable bending unit stress used in design exceed the stress determined by applying a size factor.

4-13. Lateral Support of Wood Framing

To prevent beams and compression members from buckling, they may have to be braced laterally. Need for such bracing and required spacing depend on the unsupported length and cross-sectional dimensions of members.

When buckling occurs, a member deflects in the direction of its least dimension b, unless prevented by bracing. (In a beam, b usually is taken as the width.) But if bracing precludes buckling in that direction, deflection can occur in the direction of the perpendicular dimension d. Thus, it is logical that unsupported length L, b, and d play important roles in rules for lateral support, or in formulas for reducing allowable stresses for buckling.

For glued-laminated beams, design for lateral stability is based on a function of Ld/b^2. For solid-sawn beams of rectangular cross section, maximum depth-width ratios should satisfy the approximate rules, based on nominal dimensions, summarized in Table 4-9. When the beams are adequately braced laterally, the depth of the member below the brace may be taken as the width. For glued-laminated arches, maximum depth-width ratios should satisfy the approximate rules, based on actual dimensions, in Table 4-10.

TABLE 4-9   Approximate Lateral-Support Rules for Sawn Beams*

Depth-width ratio (nominal dimensions)	Rule
2 or less	No lateral support required
3	Hold ends in position
4	Hold ends in position and member in line, e.g., with purlins or sag rods
5	Hold ends in position and compression edge in line, e.g., with direct connection of sheathing, decking, or joists
6	Hold ends in position and compression edge in line, as for 5 to 1, and provide adequate bridging or blocking at intervals not exceeding 6 times the depth
7	Hold ends in position and both edges firmly in line.

If a beam is subject to both flexure and compression parallel to grain, the ratio may be as much as 5:1 if one edge is held firmly in line, e.g., by rafters (or roof joists) and diagonal sheathing. If the dead load is sufficient to induce tension on the underside of the rafters, the ratio for the beam may be 6:1.

*From "National Specification for Wood Construction," National Forest Products Association.

TABLE 4-10   Approximate Lateral-Support Rules for Glued-Laminated Arches

Depth-width ratio (actual dimensions)	Rule
5 or less	No lateral support required
6	Brace one edge at frequent intervals

Where joists frame into arches or the compression chords of trusses and provide adequate lateral bracing, the depth, rather than the width, of arch or truss chord may be taken as b. The joists should be erected with upper edges at least ½ in above the supporting member, but low enough to provide adequate lateral support. The depth of arch or compression chord also may be taken as b where joists or planks are placed on top of the arch or chord and securely fastened to them and blocking is firmly attached between the joists.

For glued-laminated beams, no lateral support is required when the depth does not exceed the width. In that case also, the allowable bending stress does not have to be adjusted for lateral instability. Similarly, if continuous support prevents lateral movement of the compression flange, lateral buckling cannot occur and the allowable stress need not be reduced.

When the depth of a glued-laminated beam exceeds the width, bracing must be provided at supports. This bracing must be so placed as to prevent rotation of the beam in a plane perpendicular to its longitudinal axis. And unless the compression flange is braced at sufficiently close intervals between the supports, the allowable stress should be adjusted for lateral buckling. All other modifications of the stresses, except for size factor, should be made.

When the buckling factor

$$C = \frac{L_e d}{b^2} \tag{4-4}$$

where L_e is the effective length, in, and width b and depth d also are in inches, does not exceed 100, the allowable bending stress does not have to be adjusted for lateral buckling. Such beams are classified as short beams.

For intermediate beams, for which C exceeds 100 but is less than

$$K = \frac{3E}{5F_b} \tag{4-5}$$

the allowable bending stress should be determined from

$$F'_b = F_b \left[1 - \frac{1}{3} \left(\frac{C}{K} \right)^2 \right] \qquad (4\text{-}6)$$

where E = modulus of elasticity, psi

F_b = allowable bending stress, psi, adjusted, except for size factor

For long beams, for which C exceeds K but is less than 2500, the allowable bending stress should be determined from

$$F'_b = \frac{0.40E}{C} \qquad (4\text{-}7)$$

In no case should C exceed 2500.

The effective length L_e for Eq. (4-4) is given in terms of unsupported length of beam in Table 4-11. Unsupported length is the distance between supports or the length of a cantilever when the beam is laterally braced at the supports to prevent rotation and adequate bracing is not installed elsewhere in the span. When both rotational and lateral displacement are also prevented at intermediate points, the unsupported length may be taken as the distance between points of lateral support. If the compression edge is supported throughout the length of the beam and adequate bracing is installed at the supports, the unsupported length is zero.

TABLE 4-11   Ratio of Effective Length to
Unsupported Length of Glued-Laminated Beams

Simple beam, load concentrated at center	1.61
Simple beam, uniformly distributed load	1.92
Simple beam, equal end moments	1.84
Cantilever beam, load concentrated at unsupported end	1.69
Cantilever beam, uniformly distributed load	1.06
Simple or cantilever beam, any load (conservative value)	1.92

Acceptable methods of providing adequate bracing at supports include anchoring the bottom of a beam to a pilaster and the top of the beam to a parapet; for a wall-bearing roof beam, fastening the roof diaphragm to the supporting wall or installing a girt between beams at the top of the wall; for beams on wood columns, providing rod bracing.

For continuous lateral support of a compression flange, composite action is essential between deck elements, so that sheathing or deck acts as a diaphragm. One example is a plywood deck with edge nailing. With plank decking, nails attaching the plank to the beams must form couples, to resist rotation. In addition, the planks must be nailed to each other, for diaphragm action. Adequate lateral support is not provided when only one nail is used per plank and no nails are used between planks.

(American Institute of Timber Construction, "Timber Construction Manual," John Wiley & Sons, Inc., New York; "National Design Specification," National Forest Products Association; "Western Woods Use Book," Western Wood Products Association, 1500 Yeon Building, Portland, Ore. 97204.)

4-14. Combined Stresses in Timber

Allowable unit stresses given in the "National Design Specification for Wood Construction" apply directly to bending, horizontal shear, tension parallel or perpendicular to grain, and compression parallel or perpendicular to grain. For combined axial stress in short members ($L/d \leq 11$) P/A and bending stress M/S, stresses are limited by

$$\frac{P/A}{F} + \frac{M/S}{F_b} \leq 1 \qquad (4\text{-}8\,a)$$

where F = allowable tension stress parallel to grain or allowable compression stress parallel to grain adjusted for unsupported length, psi

F_b = allowable bending stress, psi

For combined axial compression and bending in members with $L/d > 11$, JP/A should be subtracted from F_b in Eq. (4-8 a), where

$$J = \frac{L/d - 11}{K - 11} \leq 1 \qquad (4\text{-}8\,b)$$

and K is a factor that depends on column cross-sectional properties [see Eqs. (4-12) and (4-15)].

When a bending stress f_x, shear stress f_{xy}, and compression or tension stress f_y perpendicular to the grain exist simultaneously, the stresses should satisfy the Norris interaction formula:

$$\frac{f_x^2}{F_x^2} + \frac{f_y^2}{F_y^2} + \frac{f_{xy}^2}{F_{xy}^2} \leq 1 \qquad (4\text{-}9)$$

where F_x = allowable bending stress F_b modified for duration of loading but not for size factor

F_y = allowable stress in compression or tension perpendicular to grain $F_{c\perp}$ or $F_{b\perp}$ modified for duration of loading

F_{xy} = allowable horizontal shear stress F_v modified for duration of loading

The usual design formulas do not apply to sharply curved glued-laminated beams. Nor do they hold where laminations of more than one species, each with markedly different modulus of elasticity, are comprised in the same member.

4-15. Effect of Shrinkage or Swelling on Shape of Curved Members

Wood shrinks or swells across the grain but has practically no dimensional change along the grain. Radial swelling causes a decrease in the angle between the ends of a curved member; radial shrinkage causes an increase in this angle.

Such effects may be of great importance in three-hinged arches that become horizontal, or nearly so, at the crest of a roof. Shrinkage, increasing the relative end rotations, may cause a depression at the crest and create drainage problems. For such arches, therefore, consideration must be given to moisture content of the member at time of fabrication and in service and to the change in end angles that results from change in moisture content and shrinkage across the grain.

4-16. Fabrication of Structural Timber

Fabrication consists of boring, cutting, sawing, trimming, dapping, routing, planing, and otherwise shaping, framing, and furnishing wood units, sawn or laminated, including plywood, to fit them for particular places in a final structure. Whether fabrictaion is performed in shop or field, the product must exhibit a high quality of work.

Jigs, patterns, templates, stops, or other suitable means should be used for all complicated and multiple assemblies to insure accuracy, uniformity, and control of all dimensions. All tolerances in cutting, drilling, and framing must comply with good practice in the industry and applicable specifications and controls. At the time of fabrication, tolerances must not exceed those listed below unless they are not critical and not required for proper performance. Specific jobs, however, may require closer tolerances.

Location of Fastenings ▪ Spacing and location of all fastenings within a joint should be in accordance with the shop drawings and specifications with a maximum permissible tolerance of ± ⅟₁₆ in. The fabrication of members assembled at any joint should be such that the fastenings are properly fitted.

Bolt-Hole Sizes ▪ Bolt holes in all fabricated structural timber, when loaded as a structural joint, should be ⅟₁₆ in larger in diameter than bolt diameter for ½-in and larger-diameter bolts, and ⅟₃₂ in larger for smaller-diameter bolts. Larger clearances may be required for other bolts, such as anchor bolts and tension rods.

Holes and Grooves ▪ Holes for stress-carrying bolts, connector grooves, and connector daps must be smooth and true within ⅟₁₆ in per 12 in of depth. The width of a split-ring connector groove should be within +0.02 in of and not less than the thickness of the corresponding cross section of the ring. The shape of ring grooves must conform generally to the cross-sectional shape of the ring. Departure from these requirements may be allowed when supported by test data. Drills and other cutting tools should be set to conform to the size, shape, and depth of holes, grooves, daps, and so on specified in the "National Design Specification for Stress-Grade Lumber and Its Fastenings," National Forest Products Association (formerly National Lumber Manufacturers Association).

Lengths ▪ Members should be cut within ± ⅟₁₆ in of the indicated dimension when they are up to 20 ft long and ± ⅟₁₆ in/20 ft of specified length when they are over 20 ft long. Where length dimensions are not specified or critical, these tolerances may be waived.

End Cuts ▪ Unless otherwise specified, all trimmed square ends should be square within ⅟₁₆ in/ft of depth and width. Square or sloped ends to be loaded in compression should be cut to provide contact over substantially the complete surface.

4-17. Fabrication of Glued-Laminated Timbers

Structural glued-laminated timber is made by bonding together layers of lumber with adhesive so that the grain direction of all laminations is essentially parallel. Narrow boards may be edge-glued; short boards, end-glued; and the resultant wide and long laminations then face-glued into large, shop-grown timbers.

Recommended practice calls for lumber of nominal 1- and 2-in thicknesses for laminating. The lumber is dressed to ¾- and 1½-in thicknesses, respectively, just prior to gluing. The thinner laminations are generally used in curved members.

Depth of constant-depth members normally is a multiple of the thickness of the lamination stock used. Depth of variable-depth members, due to tapering or special assembly techniques, may not be exact multiples of these lamination thicknesses.

Industry-standard finished widths correspond to the following nominal widths after allowance for drying and surfacing of nominal lumber widths:

Nominal width of stock, in	3	4	6	8	10	12	14	16
Standard member finished width, in	2¼	3⅛	5⅛	6¾	8¾	10¾	12¼	14¼

Standard widths are most economical since they represent the maximum width of board normally obtained from the lumber stock used in laminating.

When members wider than the stock available are required, laminations may consist of two boards side by side. These edge joints must be staggered, vertically in horizontally laminated beams (load acting normal to wide faces of laminations) and horizontally in vertically laminated beams (load acting normal to the edge of laminations). In horizontally laminated beams, edge joints need not be edge-glued. But edge gluing is required in vertically laminated beams.

Edge and face gluings are the simplest to make, end gluings the most difficult. Ends are also the most difficult surfaces to machine. Scarfs or finger joints generally are used to avoid end gluing.

A plane sloping scarf (Fig. 4-2), in which the tapered surfaces of laminations are glued together, can develop 85 to 90% of the strength of an unscarfed, clear, straight-grained control specimen. Finger joints (Fig. 4-3) are less waste-ful of lumber. Quality can be adequately controlled in machine cutting and in high-frequency gluing. A combination of thin tip, flat slope on the side of the individual fingers and a narrow pitch is desired. The length of fingers should be kept short for sav-ings of lumber but long for maximum strength.

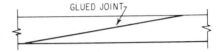

Fig. 4-2.   Plane sloping scarf.

The usefulness of structural glued-laminated timbers is determined by the lumber used and glue joint produced. Certain combinations of adhesive, treatment, and wood species do not produce the same quality of glue bond as other combinations, although the same gluing procedures are used. Thus, a combination must be supported by adequate experience with a laminator's gluing procedure (see also Art. 4-18).

The only adhesives currently recommended for wet-use and preservative-treated lumber, whether gluing is done before or after treatment, are the resorcinol and phenol-resorcinol resins. The prime adhesive for dry-use structural laminating is casein. Melamine and melamine-urea blends are used in smaller amounts for high-frequency curing of end gluings.

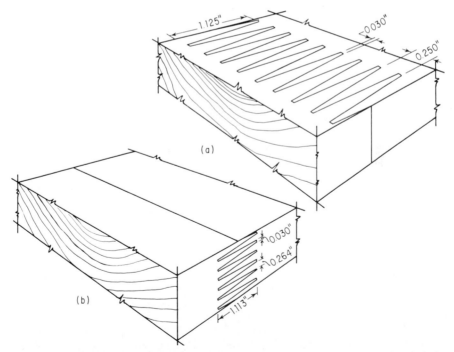

Fig. 4-3.   Finger joint: (*a*) fingers formed by cuts perpendicular to the wide face of the board; (*b*) fingers formed by cuts perpendicular to the edges.

Glued joints are cured with heat by several methods. R. F. (high-frequency) curing of glue lines is used for end joints and for limited-size members where there are repetitive gluings of the same cross section. Low-voltage resistance heating, where current is passed through a strip of metal to raise the temperature of a glue line, is used for attaching thin facing pieces. The metal may be left in the glue line as an integral part of the completed member. Printed electric circuits, in conjunction with adhesive films, and adhesive films, impregnated on paper or on each side of a metal conductor placed in the glue line, are other alternatives. Also, an aluminum-foil system, with the aluminum foil as the electrical conductor, faced with dry-film phenolic resin, may be used for resistance heating.

Preheating the wood to insure reactivity of the applied adhesive has limited application in structural laminating. The method requires adhesive application as a wet or dry film simultaneously to all laminations and then rapid handling of multiple laminations.

Curing the adhesive at room temperature has many advantages. Since wood is an excellent insulator, a long time is required for elevated ambient temperature to reach inner glue lines of a large assembly. With room-temperature curing, equipment needed to heat the glue line is not required, and the possibility of injury to the wood from high temperatures is avoided.

(AITC 103, "Standard for Structural Glued-Laminated Timber," American Institute of Timber Construction, Englewood, Colo.)

4-18. Preservative Treatments for Wood

Wood-destroying fungi must have air, suitable moisture, and favorable temperatures to develop and grow in wood. Submerge wood permanently and totally in water, keep the moisture content below 18 to 20%, or hold temperature below 40°F or above 110°F, and wood remains permanently sound. If wood moisture content is kept below the fiber-saturation point (25 to 30%) when the wood is untreated, decay is greatly retarded. Below 18 to 20% moisture content, decay is completely inhibited.

If wood cannot be kept dry, a wood preservative, properly applied, must be used. The following can be a guide to determine if treatment is necessary.

Wood members are permanent without treatment if located in enclosed buildings where good roof coverage, proper roof maintenance, good joint details, adequate flashing, good ventilation, and a well-drained site assure moisture content of the wood continuously below 20%. Also, in arid or semiarid regions, where climatic conditions are such that the equilibrium moisture content seldom exceeds 20%, and then only for short periods, wood members are permanent without treatment.

Where wood is in contact with the ground or water, where there is air and the wood may be alternately wet and dry, a preservative treatment, applied by a pressure process, is necessary to obtain an adequate service life. In enclosed buildings where moisture given off by wet-process operations maintains equilibrium moisture contents in the wood above 20%, wood structural members must be treated with a preservative, as must wood exposed outdoors without protective roof covering and where the wood moisture content can go above 18 to 20% for repeated or prolonged periods.

Where wood structural members are subject to condensation by being in contact with masonry, preservative treatment is necessary.

To obtain preservative-treated glued-laminated timber, lumber may be treated before gluing and the members then glued to the desired size and shape. Or the already glued and machined members may be treated with certain treatments. When laminated members do not lend themselves to treatment because of size and shape, gluing of treated laminations is the only method of producing adequately treated members.

There are problems in gluing some treated woods. Certain combinations of adhesive, treatment, and wood species are compatible; other combinations are not. All adhesives of the same type do not produce bonds of equal quality for a particular wood species and preservative. The bonding of treated wood depends on the concentration of preservative on the surface at the time of gluing and the chemical effects of the preservative on the adhesive. In general, longer curing times or higher curing temperatures, and modifications in assembly times, are needed for treated than for untreated wood to obtain comparable adhesive bonds (see also Art. 4-17).

Each type of preservative and method of treatment has certain advantages. The preservative to be used depends on the service expected of the member for the specific conditions of exposure. The minimum retentions shown in Table 4-12 may be increased where severe climatic or exposure conditions are involved.

Creosote and creosote solutions have low volatility. They are practically insoluble in water and thus are most suitable for severe exposure, contact with ground or water, and where painting is not a requirement or a creosote odor is not objectionable.

Oil-borne chemicals are organic compounds dissolved in a suitable petroleum carrier oil and are suitable for outdoor exposure or where leaching may be a factor, or where painting is not required. Depending on the type of oil used, they may result in relatively clean surfaces. There is a slight odor from such treatment, but it is usually not objectionable.

Water-borne inorganic salts are dissolved in water or aqua ammonia, which evaporates after treatment and leaves the chemicals in wood. The strength of solutions varies to provide net retention of dry salt required. These salts are suitable where clean and odorless surfaces are required. The surfaces are paintable after proper seasoning.

When treatment before gluing is required, water-borne salts, oil-borne chemicals in mineral spirits, or AWPA P9 volatile solvent are recommended. When treatment before gluing is not required or desired, creosote, creosote solutions, or oil-borne chemicals are recommended.

4-19. Resistance of Wood to Chemical Attack

Wood is superior to many building materials in resistance to mild acids, particularly at ordinary temperatures. It has excellent resistance to most organic acids, notably acetic. However, wood is seldom used in contact with solutions that are more than weakly alkaline. Oxidizing chemicals and solutions of iron salts, in combination with damp conditions, should be avoided.

Wood is composed of roughly 50 to 70% cellulose, 25 to 30% lignin, and 5% extractives with less than 2% protein. Acids such as acetic, formic, lactic, and boric do not ionize sufficiently at room temperature to attack cellulose and thus do not harm wood.

When the pH of aqueous solutions of weak acids is 2 or more, the rate of hydrolysis of cellulose is small and is dependent on the temperature. A rough approximation of this temperature effect is that for every 20°F increase, the rate of hydrolysis doubles. Acids with pH values above 2 or bases with pH below 10 have little weakening effect on wood at room temperature if the duration of exposure is moderate.

4-20. Designing for Fire Safety

Maximum protection of the occupants of a building and the property itself can be achieved in timber design by taking advantage of the fire-endurance properties of wood in large cross sections and by close attention to details that make a building fire-safe. Building materials alone, building features alone, or detection and fire-extinguishing equipment alone cannot provide maximum safety from fire in buildings. A proper combination of these three factors will provide the necessary degree of protection for the occupants and the property.

The following should be investigated:

Degree of protection needed, as dictated by occupancy or operations taking place.

Number, size, type (such as direct to the outside), and accessibility of exits (particularly stairways), and their distance from each other.

TABLE 4-12 Recommended Minimum Retentions of Preservatives, lb/ft*

Preservatives	Ground contact				Above ground			
	Sawn and laminated timbers		Laminations		Sawn and laminated timbers		Laminations	
	Western woods†	Southern pine	Western woods†	Southern pine	Western woods†	Southern pine	Western woods†	Southern pine
Creosote or creosote solutions:								
Creosote	12	12	12	12	6	6	6	6
Creosote—coal-tar solution	12	12	NR‡	12	6	6	NR‡	6
Creosote-petroleum solution	12	NR‡	12	NR‡	6	NR‡	6	NR‡
Oil-borne chemicals:								
Pentachlorophenol (5% in specified petroleum oil)	0.6	0.6	0.6	0.6	0.3§	0.3	0.3	0.3
Penta (water-repellent, moderate decay hazard)	NR‡	NR‡	NR‡	NR‡	0.2	0.2	0.25	0.25
Water-borne inorganic salts:								
Chromated zinc arsenate (Boliden salt)	NR‡	NR‡	NR‡	NR‡	0.50	0.50	0.45	0.45
Acid copper chromate (Celcure)	NR‡	NR‡	0.50	0.50	0.25	0.25	0.25	0.25
Ammoniacal copper arsenite (Chemonite)	0.40	0.40	0.40	0.40	0.25	0.25	0.25	0.25
Chromated zinc chloride	NR‡	NR‡	NR‡	NR‡	0.50	0.50	0.50	0.50
Copperized chromated zinc arsenate	NR‡	NR‡	NR‡	NR‡	0.45	0.45	0.45	0.45
Chromated copper arsenate (Greensalt, Erdalith)	0.60.	0.60	0.60	0.60	0.25	0.25	0.25	0.25
Fluor chrome arsenate phenol (Tanalith, Wolman salt, Osmossar, Osmosalt)	NR‡	NR‡	NR‡	NR‡	0.25	0.25	0.25	0.25

*See latest edition of AITC 109, "Treating Standard for Structural Timber Framing," American Institute of Timber Construction or AWPA Standards C2 and C28, American Wood Preservers Association.

†Douglas fir, western hemlock, western larch.

‡NR—not recommended.

§0.5 lb for timber less than 5 in thick.

Installation of automatic alarm and sprinkler systems.

Separation of areas in which hazardous processes or operations take place, such as boiler rooms and workshops.

Enclosure of stairwells and use of self-closing fire doors.

Fire stopping and elimination, or proper protection of concealed spaces.

Interior finishes to assure surfaces that will not spread flame at hazardous rates.

Roof venting equipment or provision of draft curtains where walls might interfere with production operations.

When exposed to fire, wood forms a self-insulating surface layer of char, which provides its own fire protection. Even though the surface chars, the undamaged wood beneath retains its strength and will support loads in accordance with the capacity of the uncharred section. Heavy-timber members have often retained their structural integrity through long periods of fire exposure and remained serviceable after the charred surfaces have been refinished. This fire endurance and excellent performance of heavy timber are attributable to the size of the wood members and to the slow rate at which the charring penetrates.

The structural framing of a building, which is the criterion for classifying a building as combustible or noncombustible, has little to do with the hazard from fire to the building occupants. Most fires start in the building contents and create conditions that render the inside of the structure uninhabitable long before the structural framing becomes involved in the fire. Thus, whether the building is classified as combustible or noncombustible has little bearing on the potential hazard to the occupants. However, once the fire starts in the contents, the material of which the building is constructed can significantly help facilitate evacuation, fire fighting, and property protection.

The most important protection factors for occupants, firefighters, and the property, as well as adjacent exposed property, are prompt detection of the fire, immediate alarm, and rapid extinguishment of the fire. Firefighters do not fear fires in buildings of heavy-timber construction as they do those in buildings of many other types of construction. They need not fear sudden collapse without warning; they usually have adequate time, because of the slow-burning characteristics of the timber, to ventilate the building and fight the fire from within the building or on top.

With size of member of particular importance to fire endurance of wood members, building codes specify minimum dimensions for structural members and classify buildings with wood framing as heavy-timber construction, ordinary construction, or wood-frame construction.

Heavy-timber construction is that type in which fire resistance is attained by placing limitations on the minimum size, thickness, or composition of all load-carrying wood members; by avoidance of concealed spaces under floors and roofs; by use of approved fastenings, construction details, and adhesives; and by providing the required degree of fire resistance in exterior and interior walls. (See AITC 108, "Heavy Timber Construction," American Institute of Timber Construction.)

Ordinary construction has exterior masonry walls and wood-framing members of sizes smaller than heavy-timber sizes.

Wood-frame construction has wood-framed walls and structural framing of sizes smaller than heavy-timber sizes.

Depending on the occupancy of a building or hazard of operations within it, a building of frame or ordinary construction may have its members covered with fire-resistive coverings. The interior finish on exposed surfaces of rooms, corridors, and stairways is important from the standpoint of its tendency to ignite, flame, and spread fire from one location to another. The fact that wood is combustible does not mean that it will spread flame at a hazardous rate. Most codes exclude the exposed wood surfaces of heavy-timber structural members from flame-spread requirements because such wood is difficult to ignite and, even with an external source of heat, such as burning contents, is resistant to spread of flame.

Fire-retardant chemicals may be impregnated in wood with recommended retentions to lower the rate of surface flame spread and make the wood self-extinguishing if the external source of heat is removed. After proper surface preparation, the surface is paintable. Such treatments are accepted under several specifications, including Federal and military. They are recommended only for interior or dry-use service conditions or locations protected against leaching. These treatments are sometimes used to meet a specific flame-spread rating for interior finish or as an alternate to noncombustible secondary members and decking meeting the requirements of Underwriters' Laboratories, Inc., NM 501 or NM 502, nonmetallic roof-deck assemblies in otherwise heavy-timber construction.

4-21. Mechanical Fastenings

Various kinds of mechanical fastenings are used in wood construction. The most common are nails, spikes, screws, lags, bolts, and timber connectors, such as shear plates and split rings. Joint-design data have been established by experience and tests because determination of stress distribution in wood and metal fasteners is complicated.

Allowable loads or stresses and methods of design for bolts, connectors, and other fasteners used in one-piece sawn members also are applicable to laminated members.

Problems can arise, however, if a deep-arch base section is bolted to the shoe attached to the foundation by widely separated bolts. A decrease in wood moisture content and shrinkage will set up considerable tensile stress perpendicular to grain, and splitting may occur. If the moisture content at erection is the same as that to be reached in service, or if the bolt holes in the shoe are slotted to permit bolt movement, the tendency to split will be reduced.

Fasteners subject to corrosion or chemical attack should be protected by painting, galvanizing, or plating. In highly corrosive atmospheres, such as in chemical plants, metal fasteners and connections should be galvanized or made of stainless steel. Consideration may be given to covering connections with hot tar or pitch. In such extreme conditions, lumber should be at or below equilibrium moisture content at fabrication, to reduce subsequent shrinkage, which could open avenues of attack for the corrosive atmosphere.

Iron salts are frequently very acidic and show hydrolytic action on wood in the presence of free water. This accounts for softening and discoloration of wood observed around corroded nails. This action is especially pronounced in acidic woods, such as oak, and in woods containing considerable tannin and related compounds, such as redwood. It can be eliminated, however, by using zinc-coated aluminum, or copper nails.

Nails and Spikes ▪ Common wire nails and spikes conform to the minimum sizes in Table 4-13.

Hardened deformed-shank nails and spikes are made of high-carbon-steel wire and are headed, pointed, annularly or helically threaded, and heat-treated and tempered, to provide greater strength than common wire nails and spikes. But the same loads are given for common wire nails and spikes or the corresponding lengths are used with a few exceptions.

Nails should not be driven closer together than half their length, unless driven in prebored holes. Nor should nails be closer to an edge than one-quarter their length. When one structural member is joined to another, penetration of nails into the second or farther timber should be at least half the length of the nails. Holes for nails, when necessary to prevent splitting, should be bored with a diameter less than that of the nail. If this is done, the same allowable load as for the same-size fastener with a bored hole applies in both withdrawal and lateral resistance.

Nails or spikes should not be loaded in withdrawal from the end grain of wood. Also, nails inserted parallel to the grain should not be used to resist tensile stresses parallel to the grain.

TABLE 4-13   Nail and Spike
Dimensions

Pennyweight	Length, in	Wire dia, in
Nails:		
6d	2	0.113
8d	2½	0.131
10d	3	0.148
12d	3¼	0.148
16d	3½	0.162
20d	4	0.192
30d	4½	0.207
40d	5	0.225
50d	5½	0.244
60d	6	0.263
Spikes:		
10d	3	0.192
12d	3¼	0.192
16d	3½	0.207
20d	4	0.225
30d	4½	0.244
40d	5	0.263
50d	5½	0.283
60d	6	0.283
$5/16$	7	0.312
⅜	8½	0.375

Safe lateral loads and safe resistance to withdrawal of common wire nails are in Table 4-14 and of spikes in Table 4-15. The tables provide a safety factor of about 6 against failure. Joint slippage would be objectionable long before ultimate load is reached.

For lateral loads, if a nail or spike penetrates the piece receiving the point for a distance less than 11 times the nail or spike diameter, the allowable load is determined by straight-line interpolation between full load at 11 diameters, as given in the tables, and zero load for zero penetration. But the minimum penetration in the piece receiving the point must be at least 3⅔ diameters. Allowable lateral loads for nails or spikes driven into end grain are two-thirds the tabulated values. If a nail or spike is driven into unseasoned wood that will remain wet or will be loaded before seasoning, the allowable load is 75% of the tabulated values, except that for hardened, deformed-shank nails the full load may be used. Where properly designed metal side plates are used, the tabulated allowable loads may be increased 25%.

Wood Screws ▪ The common types of wood screws have flat, oval, or round heads. The flat-head screw is commonly used if a flush surface is desired. Oval- and round-headed screws are used for appearance or when countersinking is objectionable.

Wood screws should not be loaded in withdrawal from end grain. They should be inserted perpendicular to the grain by turning into predrilled holes and should not be started or driven with a hammer. Spacings, end distances, and side distances must be such as to prevent splitting.

Table 4-16 gives the allowable loads for lateral resistance at any angle of load to grain, when the wood screw is inserted perpendicular to the grain (into the side grain of main member) and a wood side piece is used. Embedment must be seven times the shank diameter into the member holding the point. For less penetration, reduce loads in proportion; penetration, however, must not be less than four times the shank diameter.

Table 4-16 also gives allowable withdrawal loads in pounds per inch of penetration of the threaded portion of a screw into the member holding the point when the screw is inserted perpendicular to the grain of the wood.

When metal side plates rather than wood side pieces are used, the allowable lateral load, at

TABLE 4-14   Strength of Wire Nails

Size of nail, pennyweight	6d	8d	10d	12d	16d	20d	30d	40d	50d	60d
Length of nail, in	2	2 1/2	3	3 1/4	3 1/2	4	4 1/2	5	5 1/2	6
Safe lateral strength, lb (inserted perpendicular to the grain of the wood, penetrating 11 diameters), Douglas fir or southern pine	63	78	94	94	108	139	155	176	199	223
Safe resistance to withdrawal, lb/lin in of penetration into the member receiving the point (inserted perpendicular to the grain of the wood):										
Douglas fir (dense)	29	34	38	38	42	49	53	58	63	67
Southern pine (dense)	35	41	46	46	50	59	64	70	76	81

TABLE 4-15   Strength of Spikes

Size of spike, pennyweight	10d	12d	16d	20d	30d	40d	50d	60d	5/16 in	3/8 in
Length of spike, in	3	3 1/2	3 1/2	4	4 1/2	5	5 1/2	6	7	8 1/2
Safe lateral strength, lb (inserted perpendicular to the grain of the wood, penetrating 11 diameters), Douglas fir and southern pine	139	139	155	176	199	223	248	248	288	379
Safe resistance to withdrawal, lb/lin in of penetration into the member receiving the point (inserted perpendicular to the grain of the wood):										
Douglas fir (dense)	49	49	53	58	63	67	73	73	80	96
Southern pine (dense)	59	59	64	70	76	81	88	88	97	116

any angle of load to grain, may be increased 25%. Allowable lateral loads when loads act perpendicular to the grain and the screw is inserted into end grain are two-thirds of those shown.

For Douglas fir and southern pine, the lead hole for a screw loaded in withdrawal should have a diameter of about 70% of the root diameter of the screw. For lateral resistance, the part of the hole receiving the shank should be about seven-eighths the diameter of the screw at the root of the thread.

The loads in Table 4-16 are for wood screws in seasoned lumber, for joints used indoors or in a location always dry. When joints are exposed to the weather, use 75%, and where joints are always wet, 67% of the tabulated loads. For lumber pressure-impregnated with fire-retardant chemicals, use 75% of the tabulated loads.

TABLE 4-16 Strength of Wood Screws

Gage of screw	6	7	8	9	10	12	14	16	18	20	24
Diam of screw, in	0.138	0.151	0.164	0.177	0.190	0.216	0.242	0.268	0.294	0.320	0.372
Allowable lateral load, lb (normal duration): Douglas fir or southern pine	75	90	106	124	143	185	232	284	342	406	548
Allowable withdrawal load, lb/in of penetration (normal duration): Douglas fir	102	112	121	131	140	160	179	199	218	237	276
Southern pine	119	130	141	152	163	186	208	230	253	275	320

Lag Screws or Lag Bolts ▪ Lag screws are commonly used because of their convenience, particularly where it would be difficult to fasten a bolt or where a nut on the surface would be objectionable. They range, usually, from about 0.2 to 1.0 in in diameter and from 1 to 16 in in length. The threaded portion ranges from ¾ in for 1- and 1¼-in-long lag screws to half the length for all lengths greater than 10 in.

Lag screws, like wood screws, require prebored holes of the proper size. The lead hole for the shank should have the same diameter as the shank. The lead-hole diameter for the threaded portion varies with the density of the wood species. For Douglas fir and southern pine, the hole for the threaded portion should be 60 to 75% of the shank diameter. The smaller percentage applies to lag screws of smaller diameters. Lead holes slightly larger than those recommended for maximum efficiency should be used with lag screws of excessive length.

In determining withdrawal resistance, the allowable tensile strength of the lag screw at the net or root section should not be exceeded. Penetration of the threaded portion to a distance of about 7 times the shank diameter in the denser species and 10 to 12 times the shank diameter in the less dense species will develop approximately the ultimate tensile strength of a lag screw.

The resistance of a lag screw to withdrawal from end grain is about three-quarters that from side grain.

Table 4-17 gives the allowable normal-duration lateral and withdrawal loads for Douglas fir and southern pine. When lag screws are used with metal plates, the allowable lateral loads parallel to the grain are 25% higher than with wood side plates. No increase is allowed for load applied perpendicular to the grain.

Lag screws should preferably not be driven into end grain because splitting may develop under lateral load. If lag screws are so used, however, the allowable loads should be taken as two-thirds the lateral resistance of lag screws in side grain with loads acting perpendicular to the grain.

Spacings, edge and end distances, and net section for lag-screw joints should be the same as those for joints with bolts of a diameter equal to the shank diameter of the lag screw.

For more than one lag screw, the total allowable load equals the sum of the loads permitted for each lag screw, provided that spacings, end distances, and edge distances are sufficient to develop the full strength of each lag screw.

The allowable loads in Table 4-17 are for lag screws in lumber seasoned to a moisture content about equal to that which it will eventually have in service. For lumber installed unseasoned and seasoned in place, the full allowable lag-screw loads may be used for a joint having a single lag screw and loaded parallel or perpendicular to grain; or a single row of lag screws loaded parallel to grain; or multiple rows of lag screws loaded parallel to grain with separate splice plates for each row. For other types of lag-screw joints, the allowable loads are 40% of

TABLE 4-17 Strength of Lag Screws

Diam of lag screw, in	1/4	5/16	3/8	7/16	1/2	9/16	5/8	3/4	7/8	1	1 1/8	1 1/4
Allowable withdrawal load, lb/in, of penetration of the threaded part into the member holding the point (normal duration):												
Douglas fir	232	274	314	353	390	426	461	528	593	656	716	775
Southern pine	260	307	345	385	437	477	516	592	664	734	802	868

Allowable lateral load, lb per lag screw, in single shear with 1/2-in metal side plates (Douglas fir or southern pine) for normal duration

Length of lag screw in main members, in	1/4		5/16		3/8		7/16		1/2		9/16		5/8		3/4		7/8		1		1 1/8		1 1/4	
	Parallel to grain	Perpendicular to grain	Parallel to grain	Perpendicular to grain	Parallel to grain	Perpendicular to grain	Parallel to grain	Perpendicular to grain	Parallel to grain	Perpendicular to grain	Parallel to grain	Perpendicular to grain	Parallel to grain	Perpendicular to grain	Parallel to grain	Perpendicular to grain	Parallel to grain	Perpendicular to grain	Parallel to grain	Perpendicular to grain	Parallel to grain	Perpendicular to grain	Parallel to grain	Perpendicular to grain
3	210	160	265	180	320	195	370	210	415	215	455	225	490	235										
4	235	185	355	240	480	290	575	320	625	325	680	340	740	355										
5			375	255	535	325	710	405	850	440	930	460	1005	480	1190	525								
6			400	270	545	330	735	415	945	490	1095	540	1250	600	1480	650								
7					555	340	750	425	970	505	1210	600	1460	700	1840	810								
8							760	430	985	510	1240	615	1500	720	2130	935	2490	1035						
9									990	515	1250	620	1510	725	2160	950	2880	1200						
10													1540	740	2190	965	2960	1230	3710	1485				
11															2220	970	2990	1240	3880	1550				
12																	3000	1250	3900	1560	4900	1960		
13																	3030	1260	3930	1570	4920	1970		
14																			3950	1580	4950	1980	6060	2420
15																			3960		4980	1990	6110	2450
16																					5000	2000	6150	2460

the tabulated loads. For lumber partly seasoned when fabricated, proportionate intermediate loads between 100 and 40% may be used.

For lumber pressure-impregnated with fire-retardant chemicals, allowable loads for lag screws are the same as those for unseasoned lumber. Where joints are to be exposed to weather, use 75%, and where joints are always wet, 67% of the tabulated allowable loads.

Bolts ▪ Standard machine bolts with square heads and nuts are used extensively in wood construction. Spiral-shaped dowels are also used at times to hold two pieces of wood together; they are used to resist checking and splitting in railroad ties and other solid-sawn timbers.

Holes for bolts should always be prebored and have a diameter that permits the bolt to be driven easily (Art. 4-16). Careful centering of holes in main members and splice plates is necessary.

Center-to-center distance along the grain between bolts acting parallel to the grain should be at least four times the bolt diameter. For a tension joint, distance from end of wood to center of nearest bolt should be at least seven times the bolt diameter for softwoods and five times for hardwoods. For compression joints, the end distance should be at least four times the bolt diameter for both softwoods and hardwoods. If closer spacings are used, the loads allowed should be reduced proportionately.

Also for bolts acting parallel to the grain, the distance from the center of a bolt to the edge of the wood should be at least 1.5 times the bolt diameter. Usually, however, the edge distance is set at half the distance between bolt rows. In any event, the area at the critical section through the joint must be sufficient to keep unit stresses in the wood within the allowable.

The critical section is that section at right angles to the direction of the load that gives maximum stress in the member over the net area remaining after bolt holes at the section are deducted. For parallel-to-grain loads, the net area at a critical section should be at least 100% for hardwoods and 80% for softwoods of the total area in bearing under all the bolts in the joint.

For parallel- or perpendicular-to-grain loads, spacing between rows paralleling a member should not exceed 5 in unless separate splice plates are used for each row.

For bolts bearing perpendicular to the grain, center-to-center spacing across the grain should be at least four times the bolt diameter if wood side plates are used. But if the design load is less than the bolt bearing capacity of the plates, the spacing may be reduced proportionately. For metal side plates, the spacing only need be sufficient to permit tightening of the nuts. Distance from edge of wood to center of bolt should be at least four diameters, as should the distance between center of bolt and edge toward which the load is acting. The edge distance at the opposite edge is relatively unimportant.

A load applied to only one end of a bolt perpendicular to its axis may be taken as half the symmetrical two-end load.

Allowable bolt loads parallel and perpendicular to the grain are the same for dense Douglas fir and southern pine and are given in Table 4-18.

These basic loads are for permanent loads. For short-duration loads, basic loads may be modified as noted in Art. 4-12.

Groups of Bolts ▪ When bolts are properly spaced and aligned, the allowable load on a group of bolts may be taken as the sum of the individual load capacities.

Timber Connectors ▪ These are metal devices used with bolts for producing joints with fewer bolts without reduction in strength. Several types of connectors are available. Usually, they are either steel rings that are placed in grooves in adjoining members to prevent relative movement or metal plates embedded in the faces of adjoining timbers. The bolts used with these connectors prevent the timbers from separating. The load is transmitted across the joint through the connectors.

TABLE 4-18 Allowable Load, lb, on One Bolt Loaded in Double Shear and Bearing on Dense Douglas Fir or Southern Pine*

Diameter of bolt, in D	Length of bolt l, in, in main member											
	1 1/2		2		2 1/2		3		3 1/2		4	
	Parallel to grain P	Perpendicular to grain Q	Parallel to grain P	Perpendicular to grain Q	Parallel to grain P	Perpendicular to grain Q	Parallel to grain P	Perpendicular to grain Q	Parallel to grain P	Perpendicular to grain Q	Parallel to grain P	Perpendicular to grain Q
1/2	1100	500	1370	670	1480	840	1490	1010	1490	1140	1490	1180
5/8	1380	570	1820	760	2140	950	2290	1140	2320	1330	2330	1510
3/4	1660	630	2210	840	2710	1060	3080	1270	3280	1480	3340	1690
7/8	1940	700	2580	930	3210	1160	3770	1390	4190	1630	4450	1860
1	2220	760	2960	1010	3680	1270	4390	1520	5000	1770	5470	2030
1 1/4												
1 1/2												

Diameter of bolt, in D	Length of bolt l, in, in main member											
	4 1/2		5 1/2		7 1/2		9 1/2		11 1/2		13 1/2	
	Parallel to grain P	Perpendicular to grain Q	Parallel to grain P	Perpendicular to grain Q	Parallel to grain P	Perpendicular to grain Q	Parallel to grain P	Perpendicular to grain Q	Parallel to grain P	Perpendicular to grain Q	Parallel to grain P	Perpendicular to grain Q
1/2	2330	1640	2330	1650	2330	1480						
5/8	3350	1900	3350	2200	3350	2130	3350	1920				
3/4	4530	2090	4570	2550	4560	2840	4570	2660	4560	1980		
7/8	5770	2280	5930	2790	5950	3550	5950	3460	5950	3240	5960	2410
1	7980	2670	8940	3260	9310	4450	9300	5210	9300	5110	9300	4860
1 1/4							13410	6480	13410	7200	13400	7070
1 1/2												

*For three-member joints with side members at least half as thick as the main member. When steel plates are used for side members, increase tabulated values for parallel-to-grain loading 25%, but do not increase values for perpendicular-to-grain loading. Length of bolt l is measured in main member. For a two-member joint, use the smaller of the following; one-half the tabulated value for the thicker member or one-half the tabulated value for a piece twice the thickness of the thinner member. For a joint with four or more members of equal thickness, use for each shear plane one-half the tabulated value for the thickness of the corresponding member. If the members are not equally thick, resolve the joint into contiguous three-member joints. Determine the design load for each of these from the table and assign one-half to each shear plane. When a shear plane is assigned two different values, use the smaller.

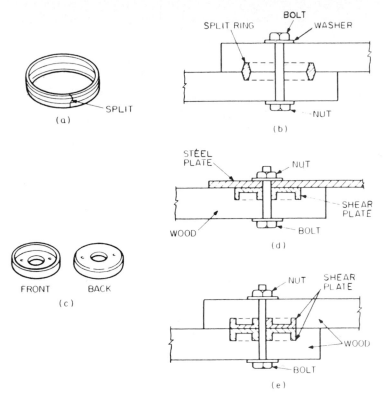

Fig. 4-4. Timber connectors: (*a*) split-ring; (*b*) wood members connected with split ring and bolt; (*c*) shear plate; (*d*) steel plate connected to wood member with shear plate and bolt; (*e*) wood members connected with a pair of shear plates and bolt. *(Reprinted with permission from F. S. Merritt, "Building Engineering and Systems Design," Van Nostrand Reinhold Company, New York.)*

Split rings (Fig. 4-4*a*) are the most efficient device for joining wood to wood. They are placed in circular grooves cut by a hand tool in the contact surfaces. About half the depth of each ring is in each of the two members in contact (Fig. 4-4*b*). A bolt hole is drilled through the center of the core encircled by the groove.

A split ring has a tongue-and-groove split to permit simultaneous bearing of the inner surface of the ring against the core and the outer surface of the ring against the outer wall of the groove. The ring is beveled for ease of assembly. Rings are manufactured in 2½-in and 4-in diameters.

Shear plates are intended for wood-to-steel connections (Fig. 4-4*c* and *d*). But when used in pairs, they may be used for wood-to-wood connections (Fig. 4-4*e*), replacing split rings. Set with one plate in each member at the contact surface, they enable the members to slide easily into position during fabrication of the joint, thus reducing the labor needed to make the connection. Shear plates are placed in precut daps and are completely embedded in the timber, flush with the surface. As with split rings, the role of the bolt through each plate is to prevent the components of the joint from separating; loads are transmitted across the joint through the plates. They come in 2⅝- and 4-in diameters.

Shear plates are useful in demountable structures. They may be installed in the members immediately after fabrication and held in position by nails.

Toothed rings and spike grids sometimes are used for special applications. But split rings and shear plates are the prime connectors for timber construction.

TABLE 4-19   Allowable Loads for One Split Ring and Bolt in Single Shear (Normal Loading)

Split-ring diam, in	Bolt diam, in	No. of faces of piece with connectors on same bolt	Thickness (net) of lumber, in	Loaded parallel to grain		Loaded perpendicular to grain		
				Min edge distance, in	Allowable load per connector and bolt, lb	Edge distance, in		Allowable load per connector and bolt, lb
						Unloaded edge, min	Loaded edge	
2½	½	1	1 min	1¾	2,270	1¾	1¾ min	1,350
							2¾ or more	1,620
			1½ or more	1¾	2,730	1¾	1¾ min	1,620
							2¾ or more	1,940
		2	1½ min	1¾	2,100	1¾	1¾ min	1,250
							2¾ or more	1,500
			2 or more	1¾	2,730	1¾	1¾ min	1,620
							2¾ or more	1,940
4	¾	1	1 min	2¾	3,510	2¾	2¾ min	2,030
							3¾ or more	2,440
			1½ or more	2¾	5,160	2¾	2¾ min	2,990
							3¾ or more	3,590
		2	1½ min	2¾	3,520	2¾	2¾ min	2,040
							3¾ or more	2,450
			2	2¾	4,250	2¾	2¾ min	2,470
							3¾ or more	2,960
			2½	2¾	5,000	2¾	2¾ min	2,900
							3¾ or more	3,480
			3 or more	2¾	5,260	2¾	2¾ min	3,050
							3¾ or more	3,660

Table 4-19 gives safe working loads for split rings and bolts. Table 4-20 gives safe working loads for shear plates and bolts.

When designing for wind forces acting alone or with dead and live loads, the safe working loads for connectors may be increased by one-third. But the number and size of connectors should not be less than those required for dead and live loads alone.

Impact may be disregarded up to the following percentages of the static effect of the live load producing the impact:

Connector	Impact allowance, %
Split ring, any size, bearing in any direction	100
Shear plate, any size, bearing parallel to grain	66⅔
Shear plate, any size, bearing perpendicular to grain	100

One-half of any impact load that remains after disregarding the percentages indicated should be included with the dead and live loads in designing the joint.

The above procedures for increasing the allowable loads on connectors for suddenly applied and short-duration loads do not reduce the actual factor of safety of the joint. They are recommended because of the favorable behavior of wood under such forces. Different values are allowed for different types and sizes of connectors and directions of bearing because of variations in the extent to which distortion of the metal, as well as the strength of the wood, affects the ultimate strength of the joint.

The tabulated loads apply to seasoned timbers used where they will remain dry. If the timbers will be more or less continuously damp or wet in use, two-thirds of the tabulated values should be used.

TABLE 4-20   Allowable Loads for One Shear Plate and Bolt in Single Shear (Normal Load Duration and Wood Side Plates*)

Shear-plate diam, in	Bolt diam, in	No. of faces of piece with connectors on same bolt	Thickness (net) of lumber, in	Loaded parallel to grain		Loaded perpendicular to grain		
				Min edge distance, in	Allowable load per connector and bolt, lb	Edge distance, in		Allowable load per connector and bolt, lb
						Unloaded edge, min	Loaded edge	
2 5/8	3/4	1	1 1/2 min	1 3/4	2670	1 3/4	1 3/4 min	1550
							2 3/4 or more	1860
		2	1 1/2 min	1 3/4	2080	1 3/4	1 3/4 min	1210
							2 3/4 or more	1450
			2	1 3/4	2730	1 3/4	1 3/4 min	1590
							2 3/4 or more	1910
			2 1/2 or more	1 3/4	2860	1 3/4	1 3/4 min	1660
							2 3/4 or more	1990
4	3/4 or 7/8	1	1 1/2 min	2 3/4	3750	2 3/4	2 3/4 min	2180
							3 3/4 or more	2620
			1 3/4 or more	2 3/4	4360	2 3/4	2 3/4 min	2530
							3 3/4 or more	3040
		2	1 3/4 min	2 3/4	2910	2 3/4	2 3/4 min	1680
							3 3/4 or more	2020
			2	2 3/4	3240	2 3/4	2 3/4 min	1880
							3 3/4 or more	2260
			2 1/2	2 3/4	3690	2 3/4	2 3/4 min	2140
							3 3/4 or more	2550
			3	2 3/4	4140	2 3/4	2 3/4 min	2400
							3 3/4 or more	2880
			3 1/2 or more	2 3/4	4320	2 3/4	2 3/4 min	2500
							3 3/4 or more	3000

*Tabulated loads also apply to metal side plates, except that for 4-in shear plates the parallel-to-grain (not perpendicular) loads for wood side plates shall be increased 11%; but loads shall not exceed: for all loadings, except wind, 2900 lb for 2⅝-in shear plates; 4970 and 6760 lb for 4-in shear plates with ¾- and ⅞-in bolts, respectively; or for wind loading, 3870, 6630, and 9020 lb, respectively. If bolt threads are in bearing on the shear plate, reduce the preceding values by one-ninth.

Metal side plates, when used, shall be designed in accordance with accepted metal practices. For A36 steel, the following unit stresses, psi, are suggested for all loadings except wind; net section in tension, 22,000; shear, 14,500; bearing, 33,000. For wind, these values may be increased one-third. If bolt threads are in bearing, reduce the preceding shear and bearing values by one-ninth.

Safe working loads for split rings and shear plates for angles between 0° (parallel to grain) and 90° (perpendicular to grain) may be obtained from the *Scholten nomograph* (Fig. 4-5). This determines the bearing strength of wood at various angles to the grain. The chart is a graphical solution of the Hankinson formula:

$$N = \frac{PQ}{P \sin^2 \theta + Q \cos^2 \theta} \qquad (4\text{-}10)$$

where N, P, and Q are, respectively, the allowable load, lb, or stress, psi, at inclination θ with the direction of grain, parallel to grain, and perpendicular to grain.

For example, given $P = 6000$ lb and $Q = 2000$ lb, find the allowable load at an angle of 40° with the grain. Connect 6 on the 0° line (point a) with the intersection of a vertical line through 2 and the 90° line (point b). $N = 3280$ lb is found directly below the intersection of line ab with the 40° line.

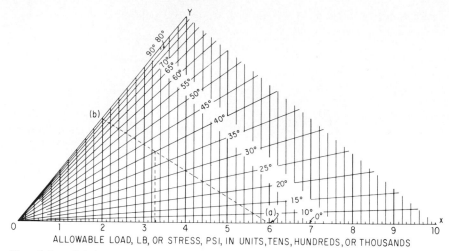

ALLOWABLE LOAD, LB, OR STRESS, PSI, IN UNITS,TENS, HUNDREDS, OR THOUSANDS

Fig. 4-5.   Scholten nomograph for determining allowable bolt load on or bearing stress in wood at various angles to the grain.

Safe working loads are based on the assumption that the wood at the joint is clear and relatively free from checks, shakes, and splits. If knots are present in the longitudinal projection of the net section within a distance from the critical section of half the diameter of the connector, the area of the knot should be subtracted from the area of the critical section. It is assumed that slope of the grain at the joint does not exceed 1 in 10.

The stress, whether tension or compression, in the net area, the area remaining at the critical section after subtracting the projected area of the connectors and the bolt from the full cross-sectional area of the member, should not exceed the safe stress of clear wood in compression parallel to the grain.

Minimum Thickness.   Tables 4-19 and 4-20 list the least thickness of member that should be used with the various sizes of connectors. The loads listed for the greatest thickness of member with each type and size of connector unit are the maximums to be used for all thicker material. Loads for members with thicknesses between those listed may be obtained by interpolation.

Rows of Fasteners ▪ For connectors in a multiple joint, there are several rules: Spacings, edge distances, and end distances should be measured as indicated in Fig. 4-6. The edge distances given in Tables 4-19 and 4-20 are the minimum permitted. End distances and appropriate percentages of tabulated allowable loads are listed in Table 4-21. For end distances

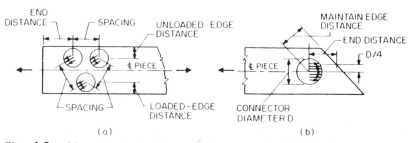

Fig. 4-6.   Measurement of spacing, edge distance, and end distance for connectors.

TABLE 4-21 Connector Spacings and End Distances (With Corresponding Percentages of Tabulated Loads)

Split ring diam, in	Shear plate diam, in	Spacing parallel to grain		Spacing perpendicular to grain		End distance		
		Spacing, in	Percent of tabulated load	Minimum, in	Percent of tabulated load	In tension members, in	In compression members, in	Percent of tabulated load*
		Parallel to grain loading						
2 1/2	2 5/8	6 3/4	100	3 1/2 min.	100	5 1/2	4	100
2 1/2	2 5/8	3 1/2 min	50	3 1/2 min.	100	2 3/4 min	2 1/2 min	62.5
4	4	9	100	5 min	100	7	5 1/2	100
4	4	5 min	50	5 min	100	3 1/2 min	3 1/4 min	62.5
		Perpendicular to grain loading						
2 1/2	2 5/8	3 1/2 min	100	4 1/4	100	5 1/2	5 1/2	100
2 1/2	2 5/8	3 1/2 min	50	3 1/2 min	50	2 3/4 min	2 3/4 min	62.5
4	4	5 min	100	6	100	7	7	100
4	4	5 min	50	5 min	50	3 1/2 min	3 1/2 min	62.5

*No reduction in end distance is permitted for compression members loaded parallel to grain.
Source: NFPA "National Design Specification for Wood Construction."

intermediate between the minimum and that required for full load, the allowable load may be determined by interpolation. Table 4-21 also lists spacings and appropriate percentages of tabulated design loads. For spacings intermediate between the minimum and that required for full load, the allowable load may be determined by interpolation.

The total allowable load for a joint with two or more connectors may be taken as the sum of the allowable loads for the individual connectors, but for any row of fasteners of the same size and type, the load should not exceed P_r computed from

$$P_r = P_s K \qquad (4\text{-}11)$$

where P_s = sum of allowable loads, lb, for individual fasteners in row
K = modification factor, as obtained from Table 4-22 or 4-23

Placement of connectors in joints with members at right angles to each other is subject to the limitations of either member. Since rules for alignment, spacing, and edge and end distance of connectors for all conceivable directions of applied load would be complicated, designers must rely on a sense of proportion and adequacy in applying the above rules to conditions of loading outside the specific limitations mentioned.

TABLE 4-22 Modification Factors K for Joints with Connectors, Bolts, and Laterally Loaded Lag Screws and Wood Side Plates

| A_1/A_2* | A_1, in^2*† | Number of fasteners in a row | | | | | | | | | | |
		2	3	4	5	6	7	8	9	10	11	12
	<12	1.00	0.92	0.84	0.76	0.68	0.61	0.55	0.49	0.43	0.38	0.34
	12–19	1.00	0.95	0.88	0.82	0.75	0.68	0.62	0.57	0.52	0.48	0.43
	>19–28	1.00	0.97	0.93	0.88	0.82	0.77	0.71	0.67	0.63	0.59	0.55
0.5‡	>28–40	1.00	0.98	0.96	0.92	0.87	0.83	0.79	0.75	0.71	0.69	0.66
	>40–64	1.00	1.00	0.97	0.94	0.90	0.86	0.83	0.79	0.76	0.74	0.72
	>64	1.00	1.00	0.98	0.95	0.91	0.88	0.85	0.82	0.80	0.78	0.76
	<12	1.00	0.97	0.92	0.85	0.78	0.71	0.65	0.59	0.54	0.49	0.44
	12–19	1.00	0.98	0.94	0.89	0.84	0.78	0.72	0.66	0.61	0.56	0.51
	>19–28	1.00	1.00	0.97	0.93	0.89	0.85	0.80	0.76	0.72	0.68	0.64
1.0‡	>28–40	1.00	1.00	0.99	0.96	0.92	0.89	0.86	0.83	0.80	0.78	0.75
	>40–64	1.00	1.00	1.00	0.97	0.94	0.91	0.88	0.85	0.84	0.82	0.80
	>64	1.00	1.00	1.00	0.99	0.96	0.93	0.91	0.88	0.87	0.86	0.85

*A_1 = cross-sectional area of main members before boring or grooving. A_2 = sum of cross-sectional areas of side members before boring or grooving.
†When A_1/A_2 exceeds 1.0 use A_2 instead of A_1.
‡When A_1/A_2 exceeds 1.0, use A_2/A_1.
For A_1/A_2 between 0 and 1.0, interpolate or extrapolate from the tabulated values.
Source: NFPA "National Design Specification for Wood Construction."

Anchor Bolts ▪ To attach columns or arch bases to concrete foundations, anchor bolts are embedded in the concrete, with sufficient projection to permit placement of angles or shores bolted to the wood. Sometimes, instead of anchor bolts, steel straps are embedded in the concrete with a portion projecting above for bolt attachment to the wood members.

Washers ▪ Bolt heads and nuts bearing on wood require metal washers to protect the wood and to distribute the pressure across the surface of the wood. Washers may be cast, malleable, cut, round-plate, or square-plate. When subjected to salt air or salt water, they should be galvanized or given some type of effective coating. Ordinarily, washers are dipped in red lead and oil prior to installation.

TABLE 4-23 Modification Factors K for Joints with Connectors, Bolts, and Laterally Loaded Lag Screws and Metal Side Plates

A_1/A_2*	A_1, in²*	Number of fasteners in a row										
		2	3	4	5	6	7	8	9	10	11	12
	25–39	1.00	0.94	0.87	0.80	0.73	0.67	0.61	0.56	0.51	0.46	0.42
2–12	40–64	1.00	0.96	0.92	0.87	0.81	0.75	0.70	0.66	0.62	0.58	0.55
	65–119	1.00	0.98	0.95	0.91	0.87	0.82	0.78	0.75	0.72	0.69	0.66
	120–199	1.00	0.99	0.97	0.95	0.92	0.89	0.86	0.84	0.81	0.79	0.78
	40–64	1.00	0.98	0.94	0.90	0.85	0.80	0.75	0.70	0.67	0.62	0.58
12–18	65–119	1.00	0.99	0.96	0.93	0.90	0.86	0.82	0.79	0.75	0.72	0.69
	120–199	1.00	1.00	0.98	0.96	0.94	0.92	0.89	0.96	0.83	0.80	0.78
	200	1.00	1.00	1.00	0.98	0.97	0.95	0.93	0.91	0.90	0.88	0.87
	40–64	1.00	1.00	0.96	0.93	0.89	0.84	0.79	0.74	0.69	0.64	0.59
18–24	65–119	1.00	1.00	0.97	0.94	0.92	0.89	0.86	0.83	0.80	0.76	0.73
	120–199	1.00	1.00	0.99	0.98	0.96	0.94	0.92	0.90	0.88	0.86	0.85
	200	1.00	1.00	1.00	1.00	0.98	0.96	0.95	0.93	0.92	0.92	0.91
	40–64	1.00	0.98	0.94	0.90	0.85	0.80	0.74	0.69	0.65	0.61	0.58
24–30	65–119	1.00	0.99	0.97	0.93	0.90	0.86	0.82	0.79	0.76	0.73	0.71
	120–199	1.00	1.00	0.98	0.96	0.94	0.92	0.89	0.87	0.85	0.83	0.81
	200	1.00	1.00	0.99	0.98	0.97	0.95	0.93	0.92	0.90	0.89	0.89
	40–64	1.00	0.96	0.92	0.86	0.80	0.74	0.68	0.64	0.60	0.57	0.55
30–35	65–119	1.00	0.98	0.95	0.90	0.86	0.81	0.76	0.72	0.68	0.65	0.62
	120–199	1.00	0.99	0.97	0.95	0.92	0.88	0.85	0.82	0.80	0.78	0.77
	200	1.00	1.00	0.98	0.97	0.95	0.93	0.90	0.89	0.87	0.86	0.85
	40–64	1.00	0.95	0.89	0.82	0.75	0.69	0.63	0.58	0.53	0.49	0.46
35–42	65–119	1.00	0.97	0.93	0.88	0.82	0.77	0.71	0.67	0.63	0.59	0.56
	120–199	1.00	0.98	0.96	0.93	0.89	0.85	0.81	0.78	0.76	0.73	0.71
	200	1.00	0.99	0.98	0.96	0.93	0.90	0.87	0.84	0.82	0.80	0.78

*A_1 = cross-sectional area of main member before boring or grooving.
A_2 = sum of cross-sectional areas of metal side plates before drilling.
Source: NFPA "National Design Specification for Wood Construction."

Setscrews should never be used against a wood surface. It may be possible, with the aid of proper washers, to spread the load of the setscrew over sufficient surface area of the wood that the compression strength perpendicular to grain is not exceeded.

Table 4-24 gives washer sizes capable of developing the capacity of A307 bolts.

Tie Rods ▪ To resist the horizontal thrust of arches not buttressed, tie rods are required. The tie rods may be installed at ceiling height or below the floor. Table 4-25 gives the allowable loads on bars used as tie rods.

Hangers ▪ Standard and special hangers are used extensively in timber construction. Stock hangers are available from a number of manufacturers. But by far the greater number of hangers are of special design. Where appearance is of prime importance, concealed hangers are frequently selected.

("National Design Specification for Wood Construction," National Forest Products Association, Washington; "Design Manual for TECO Timber Connector Construction," Timber Engineering Co., 1619 Massachusetts Ave., N.W., Washington, D.C. 20036; "Western Woods Use Book," Western Wood Products Association, 1500 Yeon Building, Portland, Ore. 97204.)

TABLE 4-24 Allowable Loads for Bolts and Washers

Rod or bolt		Plate washers†		Cut washers			
Dia, in	Tensile capacity,* lb	Side of square, in	Thickness, in	Outside dia, in	Hole dia, in	Thickness, in	Max load,‡ lb
3/8	1,550	1 3/4	3/16	1	7/16	5/64	290
7/16	2,100	2 1/8	1/4	1 1/4	1/2	5/64	460
1/2	2,750	2 3/8	1/4	1 3/8	9/16	7/64	550
5/8	4,300	3	5/16	1 3/4	11/16	9/64	910
3/4	6,190	3 3/4	3/8	2	13/16	5/32	1,100
7/8	8,420	4 3/8	1/2	2 1/4	15/16	11/64	1,100
1	11,000	5	1/2	2 1/2	1 1/16	11/64	1,800
1 1/8	13,920	5 5/8	5/8	2 3/4	1 1/4	11/64	2,100
1 1/4	17,180	6 3/8	3/4	3	1 3/8	11/64	2,500
1 3/8	20,800	7	3/4	3 1/4	1 1/2	3/16	2,900
1 1/2	24,700	7 3/4	7/8	3 1/2	1 5/8	3/16	3,400
1 5/8	29,000	8 3/8	7/8	3 3/4	1 3/4	3/16	3,900
1 3/4	33,700	9	1	4	1 7/8	3/16	4,400
1 7/8	38,700	9 3/4	1	4 1/4	2	3/16	5,000
2	44,000	10 1/4	1 1/8	4 1/2	2 1/8	3/16	5,500
2 1/8	49,700	11	1 1/8				
2 1/4	55,700	11 3/4	1 1/4	4 3/4	2 3/8	7/32	6,000
2 3/8	62,000	12 1/2	1 1/4				
2 1/2	68,700	13	1 3/8	5	2 5/8	15/64	6,500
2 5/8	75,800	13 3/4	1 1/2				
2 3/4	83,200	14 1/2	1 1/2	5 1/4	2 7/8	1/4	6,800
2 7/8	90,900	15 1/4	1 5/8				
3	99,000	15 3/4	1 5/8	5 1/2	3 1/8	9/32	7,200

*Based on ASTM A36 steel and A307 bolts.
†Size required to develop capacity of rod or bolt in accordance with note ‡.
‡Based on allowable strength of wood in compression perpendicular to grain of 450 psi.

4-22. Glued Fastenings

Glued joints are generally made between two pieces of wood where the grain directions are parallel (as between the laminations of a beam or arch). Or such joints may be between solid-sawn or laminated timber and plywood, where the face grain of the plywood is either parallel or at right angles to the grain direction of the timber.

Only in special cases may lumber be glued with the grain direction of adjacent pieces at an angle. When the angle is large, dimensional changes from changes in wood moisture content set up large stresses in the glued joint. Consequently, the strength of the joint may be considerably reduced over a period of time. Exact data are not available, however, on the magnitude of this expected strength reduction.

In joints connected with plywood gusset plates, this shrinkage differential is minimized because plywood swells and shrinks much less than does solid wood.

Glued joints can be made between end-grain surfaces, but they are seldom strong enough to meet the requirements of even ordinary service. Seldom is it possible to develop more than 25%

TABLE 4-25   Allowable Tension on Round and Square Upset Bars

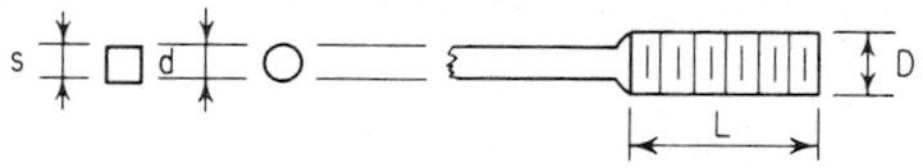

UNC and 4UN Class 2A Thread

Dia d or side s, in	Round bars			Square bars		
	Capacity,* lb	Upset		Capacity,* lb	Upset	
		Dia D, in	Length L, in		Dia. D, in	Length L, in
¾	9.700	1	4	12,400	1⅛	4
⅞	13,200	1⅛	4	16,900	1¼	4
1	17,300	1⅜	4	22,000	1½	4
1⅛	21,800	1½	4	27,800	1¾	4
1¼	27,000	1¾	4	34,400	2	4½
1⅜	32,700	1¾	4	41,600	2	4½
1½	38,900	2	4½	49,500	2¼	5
1⅝	45,600	2¼	5	58,100	2½	5½
1¾	53,000	2¼	5	67,400	2½	5½
1⅞	60,800	2⅛	5½	77.300	2¾	5½
2	69,100	2½	5½	88,000	2¾	5½
2⅛	78,000	2¾	5½	99,300	3	6
2¼	87,500	2¾	5½	111,400	3¼	6½
2⅜	97,500	3	6	124,100	3¼	6½
2½	108,000	3¼	6½	137,500	3½	7
2⅝	119,100	3¼	6½	151,600	3¾	7

*Based on ASTM A36 steel.

of the tensile strength of the wood in such butt joints. For this reason plane sloping scarfs of relatively flat slope (Fig. 4-2) or finger joints with thin tips and flat slope on the sides of the individual fingers (Fig. 4-3) are used to develop a high proportion of the strength of the wood.

Joints of end grain to side grain are also difficult to glue properly. When subjected to severe stresses as a result of unequal dimensional changes in the members due to changes in moisture content, joints suffer from severely reduced strength.

For the above reasons, joints between end-grain surfaces and between end-grain and side-grain surfaces should not be used if the joints are expected to carry load.

For joints made with wood of different species, the allowable shear stress for parallel-grain bonding is equal to the allowable shear stress parallel to the grain for the weaker species in the joint. This assumes uniform stress distribution in the joint. When grain direction is not parallel, the allowable shear stress on the glued area between the two pieces may be estimated from the Scholten nomograph (Fig. 4-5).

Adhesives used for fabricating structural glued-laminated timbers (Art. 4-12) also are satisfactory for other glued joints. In selecting an adhesive, consideration should be given to wood moisture content.

[AITC 103, "Standard for Structural Glued-Laminated Timber," American Institute of Timber Construction, Englewood, Colo.; Federal Specification MMM-A-125, "Adhesive, Casein-Type, Water- and Mold-Resistant," General Services Administration, Washington, D.C. 20405; Military Specification MIL-A-397B, "Adhesive, Room-Temperature and Intermediate-Temperature Setting Resin (Phenol, Resorcinol, and Melamine Base)," and Military Specification MIL A-5534A, "Adhesive, High-Temperature Setting Resin (Phenol, Melamine, and Resorcinol Base)," U.S. Naval Supply Depot, Philadelphia, Pa. 19120.]

4-23. Wood Columns

A wood compression member may be a solid piece of timber (Fig. 4-7*a*) or laminated (Fig. 4-7*b*) or built up of spaced members (Fig. 4-7*c*). The latter are comprised of two or more wood compression members with parallel longitudinal axes. The members are separated at ends and midpoints by blocking and joined to the end blocking with connectors with adequate shear resistance.

Short Columns ▪ Columns with a ratio of unsupported length L, in, to least dimension d, in, less than 11 fail by crushing. The allowable concentric load for such members equals the cross-sectional area, in², times F_c, the allowable compression stress, psi, parallel to grain for the species, adjusted for service conditions and duration of load.

Long and Intermediate Columns ▪ When the slenderness ratio L/d exceeds 11, wood columns generally fail by buckling, in which case the allowable stress is determined from formulas that yield values less than F_c. The computed allowable stress must be adjusted for duration of load.

For square or rectangular, solid-sawn or glued-laminated columns with L/d larger than 11 but smaller than K computed from

$$K = 0.671 \sqrt{\frac{E}{F_c}} \qquad (4\text{-}12)$$

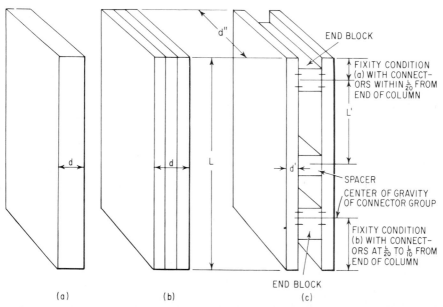

Fig. 4-7.   Behavior of wood columns depends on the ratio of length L or L' to least dimension d or d'. (*a*) Solid-sawn timber column; (*b*) glued-laminated column; (*c*) spaced column.

the allowable stress F_c' should not exceed

$$F_c' = \left[1 - \frac{1}{3} \left(\frac{L/d}{K} \right)^4 \right] F_c \leq F_c \qquad (4\text{-}13)$$

where F_c = allowable compression stress parallel to grain
   E = modulus of elasticity of wood, adjusted for duration of load
For long rectangular, solid-sawn or glued-laminated columns ($L/d \leq K$), the allowable unit stress may not exceed F_c or

$$F_c' = \frac{0.30E}{(L/d)^2} \qquad (4\text{-}14)$$

In no case may L/d exceed 50. The formula was derived for pin-end conditions but may also be used for square-cut ends.

For round columns, the allowable stress may not exceed that for a square column of the same cross-sectional area.

For tapered columns, d may be taken as the sum of the least dimension plus one-third the difference between this and the maximum thickness parallel to this dimension.

For spaced columns, L/d' should not exceed 80, L/d'' should not exceed 50, and L'/d' should not exceed 40 (see Fig. 4-7c). Allowable stresses depend on the fixity condition, a or b, as defined in Fig. 4-7c. The factor K for spaced columns is defined by

$$K = 0.671 \sqrt{\frac{C_x E}{F_c}} \qquad (4\text{-}15)$$

where C_x = 2.5 for fixity condition a
    = 3.0 for fixity condition b
For spaced columns with components having L'/d' larger than 11 but smaller than K, the allowable stress should not exceed F_c' computed from Eq. (4-13) with K from Eq. (4-15).

For long components of spaced columns ($L'/d' \geq K$), the allowable stress should not exceed

$$F_c' = \frac{0.30 C_x E}{(L'/d')^2} \qquad (4\text{-}16)$$

Each member of a spaced column should be designed separately on the basis of its L'/d'. The allowable load on each equals its cross-sectional area, in^2, times its allowable stress, psi, adjusted for duration of load. The sum of the allowable loads on the individual members should equal or exceed the total load on the column.

When a single spacer block is placed in the middle tenth of a spaced column, connectors are not required. They should be used for multiple spaced blocks. The distance between two adjacent blocks may not exceed half the distance between centers of connectors in the end blocks in opposite ends of the column.

For all types of columns, the distance between adequate bracing, including beams and struts, should be used as L in determining L/d. The largest L/d for the column or any component, whether it be computed for a major or minor axis, should be used in calculating the allowable unit stress.

Bending plus Compression ▪ For combined axial and bending stress, Eq. (4-8a) or (4-8b) governs. Bending caused by transverse loads, wind loads, or eccentric loads, or any combination of these, should be taken into account.

The critical section of columns supporting trusses frequently exists at the connection of knee

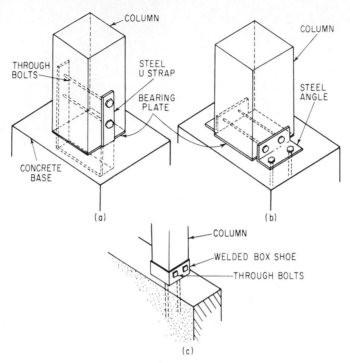

Fig. 4-8. Typical anchorages of wood column to base: (*a*) wood column anchored to concrete base with U strap; (*b*) anchorage with steel angles; (*c*) with a welded box shoe.

brace to column. Where no knee brace is used, or the column supports a beam, the critical section for moment usually occurs at the bottom of truss or beam. Then, a rigid connection must be provided to resist moment, or adequate diagonal bracing must be provided to carry wind loads into a support.

Figure 4-8 shows typical column base anchorages and Fig. 4-9 typical beam-to-column connections (AITC 104, "Typical Construction Details," American Institute of Timber Construction).

(American Institute of Timber Construction, "Timber Construction Manual," John Wiley & Sons, Inc., New York; "Western Woods Use Book," Western Wood Products Association, 1500 Yeon Building, Portland, Ore. 97204; "Wood Structural Design Data," National Forest Products Association, 1619 Massachusetts Ave., Washington, D.C. 20036.)

4-24. Design of Timber Joists

Joists are relatively narrow beams, usually spaced 12 to 24 in c to c. They generally are topped with sheathing and braced with diaphragms or cross bridging at intervals up to 10 ft. For joist spacings of 16 to 24 in c to c, 1-in sheathing usually is required. For spacings over 24 in, 2 in or more of wood decking is necessary.

Standard beam formulas for bending, shear, and deflection may be used to determine joist sizes. Connections shown in Figs. 4-9 and 4-10 may be used for joists.

See also Art. 4-25.

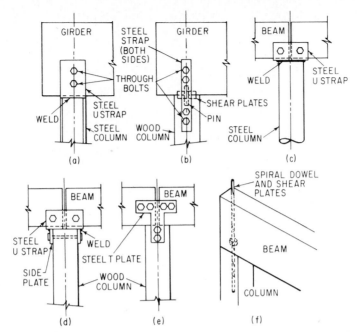

Fig. 4-9. Typical wood beam and girder connections to columns: (*a*) girder to steel column; (*b*) girder to wood column; (*c*) beam to pipe column; (*d*) beam to wood column, with steel strap welded to steel side plates; (*e*) beam to wood column, with a T plate; (*f*) with spiral dowel and shear plates.

4-25. Design of Timber Beams

Standard beam formulas for bending, shear, and deflection may be used to determine beam sizes. Ordinarily, deflection governs design, but for short, heavily loaded beams, shear is likely to control.

Figure 4-11 shows the types of beams commonly produced in timber. Straight and single- and double-tapered straight beams can be furnished solid-sawn or glued-laminated. The curved surfaces can be furnished only glued-laminated. Beam names describe the top and bottom surfaces of the beam: The first part describes the top surface, the word following the hyphen the bottom. Sawn surfaces on the tension side of a beam should be avoided.

Table 4-26 gives the load-carrying capacity for various cross-sectional sizes of glued laminated, simply supported beams.

Example ▪ Design a straight, glued-laminated beam, simply supported and uniformly loaded: span, 28 ft; spacing, 9 ft c to c; live load, 30 lb/ft²; dead load, 5 lb/ft² for deck and 7.5 lb/ft² for roofing. Allowable bending stress of combination grade is 2400 psi, with modulus of elasticity $E = 1,800,000$ psi. Deflection limitation is $L/180$, where L is the span, ft. Assume the beam is laterally supported by the deck throughout its length.

With a 15% increase for short-duration loading, the allowable bending stress F_b becomes 2760 psi and the allowable horizontal shear F_v, 230 psi.

Assume the beam will weigh 22.5 lb/lin ft, averaging 2.5 lb/ft². Then, the total uniform load comes to 45 lb/ft². So the beam carries $w = 45 \times 9 = 405$ lb/lin ft.

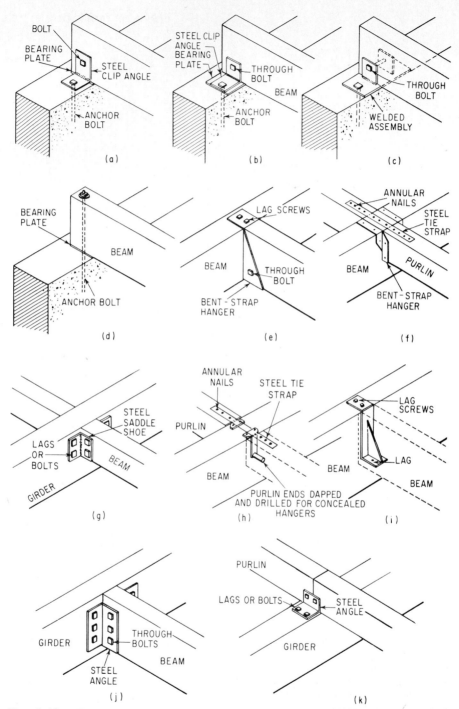

Fig. 4-10. Beam connections: (*a*) and (*b*) wood beam anchored on wall with steel angles; (*c*) with welded assembly; (*d*) beam anchored directly with bolt; (*e*) beam supported on girder with bent-strap hanger; (*f*) similar support for purlins; (*g*) saddle connects beam to girder (suitable for one-sided connection); (*h*) and (*i*) connections with concealed hangers; (*j*) and (*k*) connections with steel angles.

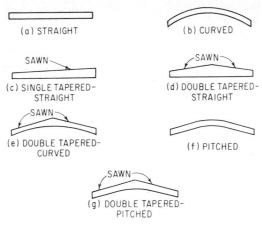

Fig. 4-11.   Types of glued-laminated beams.

The end shear $V = wL/2$ and the maximum shearing stress $= 3V/2 = 3wL/4$. Hence, the required area, in^2, for horizontal shear is

$$A = \frac{3wL}{4F_v} = \frac{wL}{306.7} = \frac{405 \times 28}{306.7} = 37.0$$

The required section modulus, in^3, is

$$S = \frac{1.5wL^2}{F_b} = \frac{1.5 \times 405 \times 28^2}{2760} = 172.6$$

If $D = 180$, the reciprocal of the deflection limitation, then the maximum deflection equals $5 \times 1728wL^4/384EI \leq 12L/D$, where I is the moment of inertia of the beam cross section, in^4. Hence, to control deflection, the moment of inertia must be at least

$$I = \frac{1.875DwL^3}{E} = \frac{1.875 \times 180 \times 405 \times 28^3}{1,800,000} = 1688 \text{ in}^4$$

Assume that the beam will be fabricated with 1½-in laminations. From Table 4-4, the most economical section satisfying all three criteria is 5⅛ × 16½, with $A = 84.6$, $S = 232.5$, and $I = 1918.5$. But it has a size factor of 0.97, so the allowable bending stress must be reduced to $2760 \times 0.97 = 2677$ psi. And the required section modulus must be increased accordingly to $172.6/0.97 = 178$. Nevertheless, the selected section still is adequate.

Suspended-Span Construction ▪ Cantilever systems may comprise any of the various types and combinations of beam illustrated in Fig. 4-12. Cantilever systems permit longer spans or larger loads for a given size member than do simple-span systems if member size is not controlled by compression perpendicular to grain at the supports or by horizontal shear. Substantial design economies can be effected by decreasing the depths of the members in the suspended portions of a cantilever system.

For economy, the negative bending moment at the supports of a cantilevered beam should be equal in magnitude to the positive moment.

TABLE 4-26 Load-Carrying Capacity of Simple-Span Laminated Beams*

Span, ft	Spacing, ft	Roof beam total load-carrying capacity						Floor beams total load
		30 lb/ft²	35 lb/ft²	40 lb/ft²	45 lb/ft²	50 lb/ft²	55 lb/ft²	50 lb/ft²
8	4	3⅛ × 4½	3⅛ × 4½	3⅛ × 6	3⅛ × 6	3⅛ × 6	3⅛ × 6	3⅛ × 6
	6	3⅛ × 4½	3⅛ × 4½	3⅛ × 6	3⅛ × 6	3⅛ × 6	3⅛ × 6	3⅛ × 6
	8	3⅛ × 4½	3⅛ × 4½	3⅛ × 6	3⅛ × 6	3⅛ × 6	3⅛ × 6	3⅛ × 7½
10	4	3⅛ × 4½	3⅛ × 4½	3⅛ × 6	3⅛ × 6	3⅛ × 6	3⅛ × 6	3⅛ × 7½
	6	3⅛ × 4½	3⅛ × 6	3⅛ × 6	3⅛ × 6	3⅛ × 6	3⅛ × 7½	3⅛ × 7½
	8	3⅛ × 6	3⅛ × 6	3⅛ × 7½	3⅛ × 7½	3⅛ × 7½	3⅛ × 7½	3⅛ × 9
	10	3⅛ × 6	3⅛ × 7½	3⅛ × 7½	3⅛ × 7½	3⅛ × 7½	3⅛ × 9	3⅛ × 9
12	6	3⅛ × 6	3⅛ × 6	3⅛ × 7½	3⅛ × 7½	3⅛ × 7½	3⅛ × 7½	3⅛ × 9
	8	3⅛ × 6	3⅛ × 7½	3⅛ × 9	3⅛ × 9	3⅛ × 9	3⅛ × 9	3⅛ × 10½
	10	3⅛ × 7½	3⅛ × 7½	3⅛ × 9	3⅛ × 9	3⅛ × 9	3⅛ × 10½	3⅛ × 10½
	12	3⅛ × 7½	3⅛ × 9	3⅛ × 9	3⅛ × 9	3⅛ × 10½	3⅛ × 10½	3⅛ × 12
14	8	3⅛ × 7½	3½ × 9	3⅛ × 9	3⅛ × 9	3⅛ × 10½	3⅛ × 10½	3⅛ × 12
	10	3⅛ × 9	3⅛ × 9	3⅛ × 10½	3⅛ × 10½	3⅛ × 10½	3⅛ × 12	3⅛ × 12
	12	3⅛ × 9	3⅛ × 10½	3⅛ × 10½	3⅛ × 10½	3⅛ × 12	3⅛ × 12	3⅛ × 13½
	14	3⅛ × 10½	3⅛ × 10½	3⅛ × 12	3⅛ × 12	3⅛ × 12	3⅛ × 13½	3⅛ × 13½
16	8	3⅛ × 9	3⅛ × 9	3⅛ × 10½	3⅛ × 10½	3⅛ × 12	3⅛ × 12	3⅛ × 13½
	12	3⅛ × 10½	3⅛ × 12	3⅛ × 12	3⅛ × 12	3⅛ × 13½	3⅛ × 13½	3⅛ × 15
	14	3⅛ × 12	3⅛ × 12	3⅛ × 13½	3⅛ × 13½	3⅛ × 15	3⅛ × 15	3⅛ × 15
	16	3⅛ × 12	3⅛ × 13½	3⅛ × 13½	3⅛ × 15	3⅛ × 15	3⅛ × 16½	3⅛ × 15
18	8	3⅛ × 9	3⅛ × 10½	3⅛ × 12	3⅛ × 12	3⅛ × 12	3⅛ × 13½	3⅛ × 15
	12	3⅛ × 12	3⅛ × 12	3⅛ × 13½	3⅛ × 13½	3⅛ × 15	3⅛ × 16½	3⅛ × 16½
	16	3⅛ × 13½	3⅛ × 15	3⅛ × 15	3⅛ × 16½	5⅛ × 13½	5⅛ × 13½	5⅛ × 15
	18	3⅛ × 15	3⅛ × 15	3⅛ × 16½	3⅛ × 18	5⅛ × 15	5⅛ × 15	5⅛ × 15
20	8	3⅛ × 12	3⅛ × 12	3⅛ × 13½	3⅛ × 13½	3⅛ × 13½	3⅛ × 15	3⅛ × 16½
	12	3⅛ × 13½	3⅛ × 13½	3⅛ × 15	3⅛ × 16½	3⅛ × 16½	5⅛ × 13½	5⅛ × 15
	16	3⅛ × 15	3⅛ × 16½	3⅛ × 18	3⅛ × 18	5⅛ × 15	5⅛ × 16½	5⅛ × 18
	18	3⅛ × 16½	3⅛ × 16½	3⅛ × 18	5⅛ × 15	5⅛ × 16½	5⅛ × 16½	5⅛ × 18
22	8	3⅛ × 13½	3⅛ × 13½	3⅛ × 13½	3⅛ × 15	3⅛ × 15	3⅛ × 16½	5⅛ × 15
	12	3⅛ × 15	3⅛ × 15	3⅛ × 16½	3⅛ × 18	3⅛ × 18	5⅛ × 15	5⅛ × 16½
	16	3⅛ × 16½	3⅛ × 18	5⅛ × 15	5⅛ × 16½	5⅛ × 16½	5⅛ × 18	5⅛ × 19½
	18	3⅛ × 18	5⅛ × 15	5⅛ × 16½	5⅛ × 16½	5⅛ × 18	5⅛ × 18	5⅛ × 19½
24	8	3⅛ × 13½	3⅛ × 15	3⅛ × 15	3⅛ × 16½	3⅛ × 16½	3⅛ × 18	5⅛ × 16½
	12	3⅛ × 16½	3⅛ × 16½	3⅛ × 18	5⅛ × 15	5⅛ × 16½	5⅛ × 16½	5⅛ × 18
	16	3⅛ × 18	5⅛ × 16½	5⅛ × 16½	5⅛ × 18	5⅛ × 18	5⅛ × 19½	5⅛ × 21
	18	5⅛ × 15	5⅛ × 16½	5⅛ × 18	5⅛ × 18	5⅛ × 19½	5⅛ × 21	5⅛ × 21
26	8	3⅛ × 15	3⅛ × 16½	3⅛ × 16½	3⅛ × 16½	3⅛ × 18	5⅛ × 16½	5⅛ × 18
	12	3⅛ × 18	3⅛ × 18	5⅛ × 16½	5⅛ × 16½	5⅛ × 18	5⅛ × 18	5⅛ × 19½
	16	5⅛ × 16½	5⅛ × 16½	5⅛ × 18	5⅛ × 18	5⅛ × 19½	5⅛ × 21	5⅛ × 22½
	18	5⅛ × 16½	5⅛ × 18	5⅛ × 18	5⅛ × 19½	5⅛ × 21	5⅛ × 21	5⅛ × 22½
28	8	3⅛ × 16½	3⅛ × 16½	3⅛ × 16½	3⅛ × 18	5⅛ × 16½	5⅛ × 16½	5⅛ × 19½
	12	3⅛ × 18	5⅛ × 16½	5⅛ × 18	5⅛ × 18	5⅛ × 18	5⅛ × 19½	5⅛ × 21
	16	5⅛ × 18	5⅛ × 18	5⅛ × 19½	5⅛ × 19½	5⅛ × 21	5⅛ × 22½	5⅛ × 24
	18	5⅛ × 18	5⅛ × 19½	5⅛ × 19½	5⅛ × 21	5⅛ × 22½	5⅛ × 24	5⅛ × 24
30	8	3⅛ × 18	3⅛ × 18	5⅛ × 16½	5⅛ × 16½	5⅛ × 18	5⅛ × 18	5⅛ × 21
	12	5⅛ × 16½	5⅛ × 18	5⅛ × 18	5⅛ × 19½	5⅛ × 19½	5⅛ × 21	5⅛ × 22½
	16	5⅛ × 18	5⅛ × 19½	5⅛ × 21	5⅛ × 21	5⅛ × 22½	5⅛ × 24	5⅛ × 25½
	18	5⅛ × 19½	5⅛ × 21	5⅛ × 21	5⅛ × 22½	5⅛ × 24	5⅛ × 25½	5⅛ × 27
32	8	3⅛ × 18	5⅛ × 16½	5⅛ × 18	5⅛ × 18	5⅛ × 18	5⅛ × 19½	5⅛ × 21
	12	5⅛ × 18	5⅛ × 19½	5⅛ × 19½	5⅛ × 21	5⅛ × 21	5⅛ × 22½	5⅛ × 24
	16	5⅛ × 19½	5⅛ × 21	5⅛ × 22½	5⅛ × 22½	5⅛ × 24	5⅛ × 25½	5⅛ × 27
	18	5⅛ × 21	5⅛ × 21	5⅛ × 22½	5⅛ × 24	5⅛ × 25½	5⅛ × 27	5⅛ × 28½
34	8	5⅛ × 16½	5⅛ × 18	5⅛ × 18	5⅛ × 19½	5⅛ × 19½	5⅛ × 21	5⅛ × 22½
	12	5⅛ × 19½	5⅛ × 19½	5⅛ × 21	5⅛ × 21	5⅛ × 22½	5⅛ × 24	5⅛ × 25½
	16	5⅛ × 21	5⅛ × 22½	5⅛ × 22½	5⅛ × 24	5⅛ × 25½	5⅛ × 27	5⅛ × 28½
	18	5⅛ × 22½	5⅛ × 22½	5⅛ × 24	5⅛ × 25½	5⅛ × 27	5⅛ × 28½	5⅛ × 28½
36	12	5⅛ × 19½	5⅛ × 21	5⅛ × 22½	5⅛ × 22½	5⅛ × 24	5⅛ × 25½	6¾ × 25½
	16	5⅛ × 22½	5⅛ × 24	5⅛ × 24	5⅛ × 25½	5⅛ × 27	5⅛ × 28½	6¾ × 27
	18	5⅛ × 22½	5⅛ × 24	5⅛ × 25½	5⅛ × 28½	5⅛ × 30	6¾ × 27	6¾ × 28½
	20	5⅛ × 24	5⅛ × 25½	5⅛ × 27	5⅛ × 30	6¾ × 27	6¾ × 28½	6¾ × 30
38	12	5⅛ × 21	5⅛ × 22½	5⅛ × 24	5⅛ × 24	5⅛ × 25½	5⅛ × 27	6¾ × 27
	16	5⅛ × 24	5⅛ × 24	5⅛ × 25½	5⅛ × 27	5⅛ × 28½	5⅛ × 30	6¾ × 28½
	18	5⅛ × 24	5⅛ × 25½	5⅛ × 27	5⅛ × 30	6¾ × 27	6¾ × 28½	6¾ × 30
	20	5⅛ × 25½	5⅛ × 27	5⅛ × 28½	6¾ × 27	6¾ × 28½	6¾ × 30	6¾ × 31½

TABLE 4-26 Load-Carrying Capacity of Simple-Span Laminated Beams *(Continued)*

Span, ft	Spacing, ft	Roof beams total load-carrying capacity						Floor beams total load
		30 lb/ft²	35 lb/ft²	40 lb/ft²	45 lb/ft²	50 lb/ft²	55 lb/ft²	50 lb/ft²
40	12	5⅛ × 22½	5⅛ × 24	5⅛ × 24	5⅛ × 25½	5⅛ × 27	6¾ × 25½	6¾ × 28½
	16	5⅛ × 24	5⅛ × 25½	5⅛ × 27	5⅛ × 28½	6¾ × 27	6¾ × 28½	6¾ × 31½
	18	5⅛ × 25½	5⅛ × 27	5⅛ × 28½	6¾ × 27	6¾ × 28½	6¾ × 30	6¾ × 31½
	20	5⅛ × 27	5⅛ × 28½	6¾ × 27	6¾ × 28½	6¾ × 30	6¾ × 31½	6¾ × 33
42	12	5⅛ × 24	5⅛ × 24	5⅛ × 25½	5⅛ × 27	6¾ × 25½	6¾ × 25½	6¾ × 30
	16	5⅛ × 25½	5⅛ × 27	5⅛ × 28½	5⅛ × 30	6¾ × 28½	6¾ × 30	6¾ × 33
	18	5⅛ × 27	5⅛ × 28½	5⅛ × 30	6¾ × 28½	6¾ × 30	6¾ × 31½	6¾ × 33
	20	5⅛ × 28½	5⅛ × 30	6¾ × 28½	6¾ × 30	6¾ × 31½	6¾ × 33	6¾ × 34½
44	12	5⅛ × 24	5⅛ × 25½	5⅛ × 27	5⅛ × 27	6¾ × 25½	6¾ × 27	6¾ × 31½
	16	5⅛ × 27	5⅛ × 28½	5⅛ × 30	6¾ × 28½	6¾ × 30	6⅜ × 31½	6¾ × 33
	18	5⅛ × 28½	5⅛ × 30	6¾ × 28½	6¾ × 30	6¾ × 31½	6¾ × 33	6¾ × 34½
	20	5⅛ × 30	6¾ × 27	6¾ × 30	6¾ × 30	6¾ × 33	6¾ × 34½	6¾ × 36
46	12	5⅛ × 25½	5⅛ × 27	5⅛ × 28½	6¾ × 25½	6¾ × 27	6¾ × 28½	6¾ × 31½
	16	5⅛ × 28½	5⅛ × 30	6¾ × 28½	6¾ × 28½	6¾ × 31½	6¾ × 33	6¾ × 36
	18	5⅛ × 28½	6¾ × 28½	6¾ × 30	6¾ × 31½	6¾ × 33	6¾ × 34½	6¾ × 36
	20	5⅛ × 30	6¾ × 28½	6¾ × 31½	6¾ × 33	6¾ × 34½	6¾ × 36	8¾ × 34½
48	12	5⅛ × 27	5⅛ × 28½	5⅛ × 30	5⅛ × 30	6¾ × 28½	6¾ × 30	6¾ × 33
	16	5⅛ × 30	6¾ × 28½	6¾ × 30	6¾ × 30	6¾ × 31½	6¾ × 34½	6¾ × 37½
	18	5⅛ × 30	6¾ × 30	6¾ × 30	6¾ × 33	6¾ × 34½	6¾ × 36	8¾ × 34½
	20	6¾ × 28½	6¾ × 30	6¾ × 31½	6¾ × 34½	6¾ × 36	6¾ × 37½	8¾ × 36
50	12	5⅛ × 28½	5⅛ × 28½	5⅛ × 30	6¾ × 28½	6¾ × 30	6¾ × 31½	6¾ × 34½
	16	5⅛ × 30	6¾ × 30	6¾ × 30	6¾ × 31½	6¾ × 33	6¾ × 36	8¾ × 34½
	18	6¾ × 28½	6¾ × 30	6¾ × 31½	6¾ × 34½	6¾ × 36	8¾ × 33	8¾ × 36
	20	6¾ × 30	6¾ × 31½	6¾ × 33	6¾ × 36	6¾ × 37½	8¾ × 34½	8¾ × 37½
52	12	5⅛ × 28½	5⅛ × 30	6¾ × 28½	6¾ × 30	6¾ × 31½	6¾ × 31½	6¾ × 36
	16	6¾ × 28½	6¾ × 30	6¾ × 31½	6¾ × 33	6¾ × 34½	6¾ × 37½	8¾ × 36
	18	6¾ × 30	6¾ × 31½	6¾ × 33	6¾ × 34½	6¾ × 37½	6¾ × 39	8¾ × 37½
	20	6¾ × 31½	6¾ × 33	6¾ × 34½	6¾ × 37½	6¾ × 39	8¾ × 36	8¾ × 39
54	12	5⅛ × 30	6¾ × 28½	6¾ × 30	6¾ × 31½	6¾ × 33	6¾ × 33	6¾ × 37½
	16	6¾ × 30	6¾ × 31½	6¾ × 33	6¾ × 34½	6¾ × 36	6¾ × 37½	8¾ × 37½
	18	6¾ × 31½	6¾ × 33	6¾ × 34½	6¾ × 36	6¾ × 39	8¾ × 36	8¾ × 39
	20	6¾ × 33	6¾ × 34½	6¾ × 36	6¾ × 39	8¾ × 36	8¾ × 37½	8¾ × 40½
56	12	6¾ × 28½	6¾ × 30	6¾ × 31½	6¾ × 33	6¾ × 33	6¾ × 34½	8¾ × 36
	16	6¾ × 31½	6¾ × 33	6¾ × 34½	6¾ × 36	6¾ × 37½	8¾ × 34½	8¾ × 39
	18	6¾ × 33	6¾ × 34½	6¾ × 36	6¾ × 37½	8¾ × 34½	8¾ × 37½	8¾ × 40½
	20	6¾ × 33	6¾ × 36	6¾ × 37½	8¾ × 34½	8¾ × 37½	8¾ × 39	8¾ × 42
58	12	6¾ × 30	6¾ × 31½	6¾ × 31½	6¾ × 33	6¾ × 34½	6¾ × 36	8¾ × 37½
	16	6¾ × 31½	6¾ × 34½	6¾ × 36	6¾ × 37½	6¾ × 39	8¾ × 36	8¾ × 40½
	18	6¾ × 33	6¾ × 34½	6¾ × 37½	6¾ × 39	8¾ × 36	8¾ × 39	8¾ × 42
	20	6¾ × 34½	6¾ × 36	6¾ × 39	8¾ × 36	8¾ × 39	8¾ × 40½	8¾ × 43½
60	12	6¾ × 30	6¾ × 31½	6¾ × 33	6¾ × 34½	6¾ × 36	6¾ × 37½	8¾ × 39
	16	6¾ × 33	6¾ × 34½	6¾ × 36	6¾ × 39	8¾ × 36	8¾ × 37½	8¾ × 42
	18	6¾ × 34½	6¾ × 36	6¾ × 39	8¾ × 36	8¾ × 37½	8¾ × 39	8¾ × 43½
	20	6¾ × 36	6¾ × 37½	8¾ × 36	8¾ × 37½	8¾ × 40½	8¾ × 42	8¾ × 45

*This table applies to straight, simply supported, laminated timber beams. Other beam support systems may be employed to meet varying design conditions.

1. Roofs should have a minimum slope of ¼ in/ft to eliminate water ponding.
2. Beam weight must be subtracted from total load carrying capacity. Floor beams are designed for uniform loads of 40-lb/ft² live load and 10-lb/ft² dead load.
3. Allowable stresses:

Bending stress, F_b = 2400 psi (reduced by size factor).
Shear stress F_v = 165 psi.
Modulus of elasticity, E = 1,800,000 psi.
For roof beams, F_b and F_v were increased 15% for short duration of loading.

4. Deflection limits:

Roof beams—1/180 span for total load. Floor beams—1/360 span for 40-lb/ft² live load only.
For preliminary design purposes only.
For more complete design information, see the AITC "Timber Construction Manual."

Fig. 4-12.   Cantilevered-beam systems. *A* is a single cantilever; *B* is a suspended beam; *C* has a double cantilever; and *D* is a beam with one end suspended.

Consideration must be given to deflection and camber in cantilevered multiple spans. When possible, roofs should be sloped the equivalent of ¼ in/ft of horizontal distance between the level of drains and the high point of the roof to eliminate water pockets, or provision should be made to insure that accumulation of water does not produce greater deflection and live loads than anticipated. Unbalanced loading conditions should be investigated for maximum bending moment, deflection, and stability.

(American Institute of Timber Construction, "Timber Construction Manual," John Wiley & Sons, Inc., New York; "Western Woods Use Book," Western Wood Products Association, 1500 Yeon Building, Portland, Ore. 97204; "Wood Structural Design Data," National Forest Products Association, 1619 Massachusetts Ave., Washington, D.C. 20036.)

4-26. Deflection and Camber of Timber Beams

The design of many structural systems, particularly those with long spans, is governed by deflection. Strength calculations based on allowable stresses alone may result in excessive deflection. Limitations on deflection increase member stiffness.

Table 4-27 gives recommended deflection limits, as a fraction of the beam span, for timber beams. The limitation applies to live load or total load, whichever governs.

Glued-laminated beams are cambered by fabricating them with a curvature opposite in direction to that corresponding to deflections under load. Camber does not, however, increase stiffness. Table 4-28 lists recommended minimum cambers for glued-laminated timber beams.

TABLE 4-27   Recommended Beam Deflection Limitations, in* (in Terms of Span *l*, in)

Use classification	Live load only	Dead load plus live load
Roof beams:		
Industrial	$l/180$	$l/120$
Commercial and institutional:		
Without plaster ceiling	$l/240$	$l/180$
With plaster ceiling	$l/360$	$l/240$
Floor beams:		
Ordinary usage †	$l/360$	$l/240$
Highway bridge stringers	$l/200$ to $l/300$	
Railway bridge stringers	$l/300$ to $l/400$	

*"Camber and Deflection," AITC 102, app. B, American Institute of Timber Construction.

†Ordinary usage classification is intended for construction in which walking comfort, minimized plaster cracking, and elimination of objectionable springiness are of prime importance. For special uses, such as beams supporting vibrating machinery or carrying moving loads, more severe limitations may be required.

TABLE 4-28   Recommended Minimum Camber for
Glued-Laminated Timber Beams*

Roof beams†	1½ times dead-load deflection
Floor beams‡	1½ times dead-load deflection
Bridge beams:§	
Long span	2 times dead-load deflection
Short span	2 times dead load + ½ applied-load deflection

*"Camber and Deflection," AITC 102, app. B, American Institute
of Timber Construction.
†The minimum camber of 1½ times dead-load deflection will produce a nearly level member under dead load alone after plastic deformation has occurred. Additional camber is usually provided to improve
appearance or provide necessary roof drainage (Art. 4-27).
‡The minimum camber of 1½ times dead-load deflection will produce a nearly level member under dead load alone after plastic deformation has occurred. On long spans, a level ceiling may not be desirable
because of the optical illusion that the ceiling sags. For warehouse or
similar floors where live load may remain for long periods, additional
camber should be provided to give a level floor under the permanently
applied load.
§Bridge members are normally cambered for dead load only on
multiple spans to obtain acceptable riding qualities.

4-27. Minimum Roof Slopes

Flat roofs have collapsed during rainstorms, although they were adequately designed on the
basis of allowable stresses and definite deflection limitations. The reason for these collapses was
the same, regardless of the structural framing used—the failures were caused by ponding of
water as increasing deflections permitted more and more water to collect.

Roof beams should have a continuous upward slope equivalent to ¼ in/ft between a drain
and the high point of a roof, in addition to minimum recommended camber (Table 4-28), to
avoid ponding. When flat roofs have insufficient slope for drainage (less than ¼ in/ft), the stiffness of supporting members should be such that a 5-lb/ft^2 load will cause no more than ½-in
deflection.

Because of ponding, snow loads or water trapped by gravel stops, parapet walls, or ice dams
magnify stresses and deflections from existing roof loads by

$$C_p = \frac{1}{1 - W'L^3/\pi^4 EI} \tag{4-17}$$

where C_p = factor for multiplying stresses and deflections under existing loads to determine
stresses and deflections under existing loads plus ponding
W' = weight of 1 in of water on roof area supported by beam, lb
L = span of beam, in
E = modulus of elasticity of beam material, psi
I = moment of inertia of beam, in^4

(Kuenzi and Bohannan, "Increases in Deflection and Stresses Caused by Ponding of Water
on Roofs," Forest Products Laboratory, Madison, Wis.)

4-28. Design of Wood Trusses

Type of truss and arrangement of members may be chosen to suit the shape of the structure,
the loads, and the stresses involved. The types most commonly built are bowstring, flat or parallel chord, pitched, triangular or A type, camelback, and scissors (Fig. 4-13). For most construction other than houses, trusses usually are spaced 12 to 20 ft apart. For houses, very light
trusses generally are erected 16 to 24 in c to c.

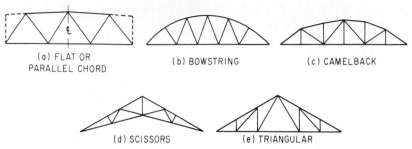

Fig. 4-13.   Types of wood trusses.

Joints are critical in the design of a truss. Use of a specific truss type is often governed by joint considerations.

Chords and webs may be single-leaf (or monochord), double-leaf, or multiple-leaf members. Monochord trusses and trusses with double-leaf chords and single-leaf web system are the most common arrangements. Web members may be attached to the sides of the chords, or the web members may be in the same plane as the chords and attached with straps or gussets.

Individual truss members may be solid-sawn, glued-laminated, or mechanically laminated. Glued-laminated chords and solid-sawn web members are usually used. Steel rods or other steel shapes may be used as members of timber trusses if they meet design and service requirements.

The bowstring truss is by far the most popular. In building construction, spans of 100 to 200 ft are common, with single or two-piece top and bottom chords of glued-laminated timber, webs of solid-sawn timber, and metal heel plates, chord splice plates, and web-to-chord connections. This system is light in weight for the loads that it can carry; it can be shop- or field-assembled. Attention to the top chord, bottom chord, and heel connections is of prime importance since they are the major stress-carrying components. Since the top chord is nearly the shape of an ideal arch, stresses in chords are almost uniform throughout a bowstring truss; web stresses are low under uniformly distributed loads.

Parallel-chord trusses, with slightly sloping top chords and level bottom chords, are used less often because chord stresses are not uniform along their length and web stresses are high. Hence, different cross sections are required for successive chords, and web members and web-to-chord connections are heavy. Eccentric joints and tension stresses across the grain should be avoided in truss construction whenever possible, but particularly in parallel-chord trusses.

Triangular trusses and the more ornamental camelback and scissors trusses are used for shorter spans. They usually have solid-sawn members for both chords and webs where degree of seasoning of timbers, hardware, and connections are of considerable importance.

Truss Joints ▪ For joints, split-ring and shear-plate connectors are generally most economical. Sometimes, when small trusses are field-fabricated, only bolted joints are used. However, grooving tools for connectors can also be used effectively in the field.

Framing between Trusses ▪ Longitudinal sway bracing perpendicular to the truss is usually provided by solid-sawn X bracing. Lateral wind bracing may be provided by end walls or intermediate walls, or both. The roof system and horizontal bracing should be capable of transferring the wind load to the walls. Knee braces between trusses and columns are often used to provide resistance to lateral loads.

Horizontal framing between trusses consists of struts between trusses at bottom-chord level and diagonal tie rods, often of steel with turnbuckles for adjustment.

TABLE 4-29   Bowstring-Truss Dimensions

Span range, ft	No. of panels	Avg truss height*	Avg arc length†
53–57	6	8 ft 7 in	60 ft 8 in
58–62		9 ft 3 in	65 ft 11 in
63–67		9 ft 11 in	71 ft 5 in
68–72	8	10 ft 7 in	76 ft 5 in
73–77		11 ft 3 in	81 ft 8 in
78–82		11 ft 11 in	86 ft 10 in
83–87		12 ft 7 in	92 ft 1 in
88–92		13 ft 3 in	97 ft 4 in
93–97		13 ft 11½ in	102 ft 7 in
98–102		14 ft 7½ in	107 ft 10 in
103–107	10	15 ft 5 in	113 ft 2 in
108–112		16 ft 1 in	118 ft 5 in
113–117		16 ft 9 in	123 ft 8 in
118–122		17 ft 5 in	128 ft 11 in
123–127		18 ft 1 in	134 ft 2 in
128–132	12	18 ft 11 in	139 ft 6 in
133–137		19 ft 7 in	144 ft 9 in

*The vertical distance from top of truss heel bearing plate to top of chord at midspan.

†Measured along the top of the top chord, wood to wood. Dimensions shown are for the longest of the span range. For smaller spans, lengths are proportionately shorter.

Bowstring-Truss Framing ▪ Table 4-29 gives typical bowstring-truss dimensions based on uniform loading conditions on the top chord, as normally imposed by roof joists.

For ordinary roof loads and spacing of 16 to 24 ft, vertical X bracing is required every 30 to 40 ft of chord length. This bracing is placed in alternate bays. Horizontal T-strut bracing should be placed from lower chord to lower chord in the same line as the vertical X bracing for the complete length of the building. If a ceiling is framed into the lower chords, these struts may be omitted.

Joists, spaced 12 to 24 in c to c, usually rest on the top chords of the trusses and secured there by toenailing. They may also be placed on ledgers attached to the sides of the upper chords or set in metal hangers, thus lowering the roof line.

Purlins, large-cross-sectional joists spaced 4 to 8 ft c to c, are often set on top of the top chords, where they are butted end to end and secured to the chords with clip angles. Purlins may also be set between the top chords on metal purlin hangers. Roof sheathing, 1-in D&M (dressed and matched) or ⅜- to ½-in-thick plywood, is laid directly on joists. For purlin construction, 2-in D&M roof sheathing is normally used.

("Design Manual for TECO Timber Connector Construction," Timber Engineering Co., 1619 Massachusetts Ave., Washington, D.C. 20036; AITC 102, app. A, "Trusses and Bracing," American Institute of Timber Construction, Englewood, Colo. 80110.)

Bridge engineering covers the planning, design, construction, and operation of structures that carry facilities for movement of humans, animals, or materials over natural or created obstacles.

Most of the diagrams used in this section were taken from the "Manual of Bridge Design Practice," State of California Dept. of Transportation. The authors express their appreciation for permission to use these illustrations from this comprehensive and authoritative publication.

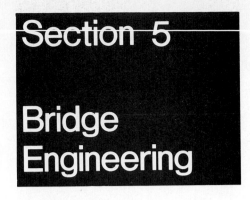

Section 5

Bridge Engineering

GENERAL DESIGN CONSIDERATIONS

5-1. Bridge Types

Bridges are of two general types: fixed and movable. They also can be grouped according to the following characteristics:

Supported facilities: Highway or railway bridges and viaducts, canal bridges and aqueducts, pedestrian or cattle crossings, material-handling bridges, pipeline bridges.

Bridge-over facilities or natural features: Bridges over highways and over railways; river bridges; bay, lake, and valley crossings.

Basic geometry: In plan—straight or curved, square or skewed bridges; in elevation—low-level bridges, including causeways and trestles, or high-level bridges.

Structural systems: Single-span or continuous-beam bridges, single- or multiple-arch bridges, suspension bridges, frame-type bridges.

Construction materials: Timber, masonry, concrete, and steel bridges.

5-2. Design Specifications

Designs of highway and railway bridges of concrete or steel often are based on the Standard Specifications for Highway Bridges of the American Association of State Highway and Transportation Officials (AASHTO) and the "Manual for Railway Engineering" of the American Railway Engineering Association (AREA). Also useful are the "Standard Plans for Highway Bridges," Federal Highway Administration (FHWA), and standard plans issued by various highway administrations and railway companies.

John J. Kozak
Formerly Chief, Division of Structures
California Department of Transportation
Sacramento, CA

James E. Roberts
Project Director
Sacramento Transit Development Agency
Sacramento, CA

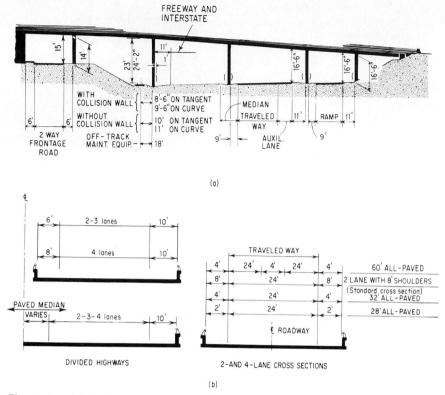

Fig. 5-1. (*a*) Minimum clearances for highway structures. (*b*) Typical bridge cross sections. Major long-span bridges may vary from these.

Length, width, elevation, alignment, and angle of intersection of a bridge must satisfy the functional requirements of the supported facilities and the geometric or hydraulic requirements of the bridged-over facilities or natural features. Figures 5-1 and 5-2 show typical highway and railway clearance diagrams.

Selection of the structural system and of the construction material and detail dimensions is governed by requirements of structural safety; economy of fabrication, erection, operation and maintenance; and esthetic considerations.

Highway bridge decks should offer comfortable, well-drained riding surfaces. Longitudinal grades and cross sections are subject to standards similar to those for open highways. Provisions for roadway lighting and emergency services should be made on long bridges.

Barrier railings should keep vehicles within the roadways and, if necessary, separate vehicular lanes from pedestrians. Utilities carried on or under bridges should be adequately protected and equipped to accommodate expansion or contraction of the structures.

Most railroads require that the ballast bed be continuous across bridges to facilitate vertical track adjustments. Long bridges should be equipped with service walkways.

5-3. Design Loads

Bridges must support the following loads without exceeding permissible stresses and deflections:

Dead load D, including permanent utilities

Live load L and impact I

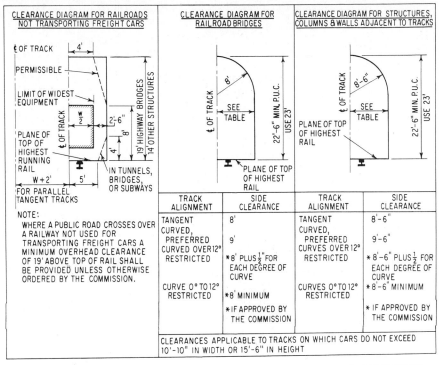

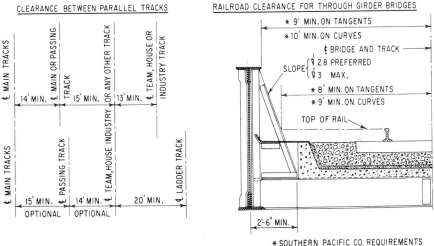

Fig. 5-2.   Minimum clearances for railroad structures.

Longitudinal forces due to acceleration or deceleration *LF* and friction *F*

Centrifugal forces *CF*

Wind pressure acting on the structure *W* and the moving load *WL*

Earthquake forces *EQ*

Earth *E*, water and ice pressure *ICE*, stream flow *SF*, and uplift *B* acting on the substructure

Forces resulting from elastic deformations, including rib shortening R

Forces resulting from thermal deformations T, including shrinkage S

Highway Bridge Loads ▪ *Vehicular live load* of highway bridges is expressed in terms of design lanes and lane loadings. The number of design lanes depends on the width of the roadway.

Each lane load is represented by a standard truck with trailer (Fig. 5-3) or, alternatively,

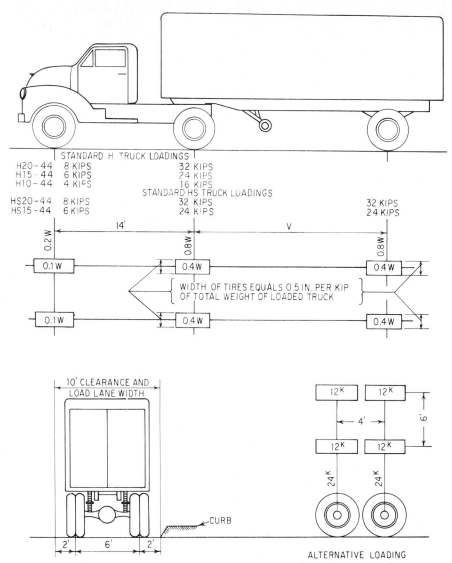

Fig. 5-3. Standard truck loading. For H trucks, W = total weight of truck and load, and for HS trucks, W = combined weight on the first two axles, which is the same weight as for H trucks. V indicates a variable spacing from 14 to 30 ft that should be selected to produce maximum stress.

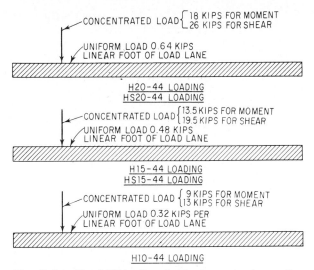

Fig. 5-4. H and HS loadings for simply supported spans. For maximum negative moment in continuous spans, an additional concentrated load of equal weight should be placed in one other span for maximum effect. For maximum positive moment, only one concentrated load should be used per lane, but combined with as many spans loaded uniformly as required for maximum effect.

as a 10-ft-wide uniform load in combination with a concentrated load (Fig. 5-4). As indicated in Fig. 5-3, there are five classes of loading: H20, H15, and H10, which represent a truck with two loaded axles, and HS20 and HS15, which represent a truck and trailer with three loaded axles. These loading designations are followed by a 44, which indicates that the loading standard was adopted in 1944.

When proportioning any member, all lane loads should be assumed to occupy, within their respective lanes, the positions that produce maximum stress in that member. Table 5-1 gives maximum moments, shears, and reactions for one loaded lane. Effects resulting from the simultaneous loading of more than two lanes may be reduced by a loading factor, which is 0.90 for three lanes and 0.75 for four lanes.

In design of steel grid and timber floors for H20 or HS20 loading, one axle load of 24 kips or two axle loads of 16 kips each, spaced 4 ft apart, may be used, whichever produces the greater stress, instead of the 32-kip axle shown in Fig. 5-3. For slab design, the center line of the wheel should be assumed to be 1 ft from the face of the curb.

Sidewalks and their direct supports should be designed for a uniform live load of 85 lb/ft². The effect of sidewalk live loading on main bridge members should be computed from

$$P = \left(30 + \frac{3000}{l} \right) \frac{55 - b}{50} \leq 60 \text{ lb/ft}^2 \tag{5-1}$$

where P = sidewalk live load, lb/ft²
l = loaded length of sidewalk, ft
b = sidewalk width, ft

Curbs should resist a force of 500 lb/lin ft acting 10 in above the floor. For design loads for *railings,* see Fig. 5-5.

TABLE 5-1 Maximum Moments, Shears, and Reactions for Truck Loads on One Lane, Simple Spans*

Span, ft	H15		H20		HS15		HS20	
	Moment †	End shear and end reaction ‡	Moment †	End shear and end reaction ‡	Moment †	End shear and end reaction ‡	Moment †	End shear and end reaction ‡
10	60.0 §	24.0 §	80.0 §	32.0 §	60.0 §	24.0 §	80.0 §	32.0 §
20	120.0 §	25.8 §	160.0 §	34.4 §	120.0 §	32.2 §	160.0 §	41.6 §
30	185.0 §	27.2 §	246.6 §	36.3 §	211.6 §	37.2 §	282.1 §	49.6 §
40	259.5 §	29.1	346.0 §	38.8	337.4 §	41.4 §	449.8 §	55.2 §
50	334.2 §	31.5	445.6 §	42.0	470.9 §	43.9 §	627.9 §	58.5 §
60	418.5	33.9	558.0	45.2	604.9 §	45.6 §	806.5 §	60.8 §
70	530.3	36.3	707.0	48.4	739.2 §	46.8 §	985.6 §	62.4 §
80	654.0	38.7	872.0	51.6	873.7 §	47.7 §	1,164.9 §	63.6 §
90	789.8	41.1	1,053.0	54.8	1,008.3 §	48.4 §	1,344.4 §	64.5 §
100	937.5	43.5	1,250.0	58.0	1,143.0 §	49.0 §	1,524.0 §	65.3 §
110	1,097.3	45.9	1,463.0	61.2	1,277.7 §	49.4 §	1,703.6 §	65.9 §
120	1,269.0	48.3	1,692.0	64.4	1,412.5 §	49.8 §	1,883.3 §	66.4 §
130	1,452.8	50.7	1,937.0	67.6	1,547.3 §	50.7	2,063.1 §	67.6
140	1,648.5	53.1	2,198.0	70.8	1,682.1 §	53.1	2,242.8 §	70.8
150	1,856.3	55.5	2,475.0	74.0	1,856.3	55.5	2,475.1	74.0
160	2,076.0	57.9	2,768.0	77.2	2,076.0	57.9	2,768.0	77.2
170	2,307.8	60.3	3,077.0	80.4	2,307.8	60.3	3,077.0	80.4
180	2,551.5	62.7	3,402.0	83.6	2,551.5	62.7	3,402.0	83.6
190	2,807.3	65.1	3,743.0	86.8	2,807.3	65.1	3,743.0	86.8
200	3,075.0	67.5	4,100.0	90.0	3,075.0	67.5	4,100.0	90.0
220	3,646.5	72.3	4,862.0	96.4	3,646.5	72.3	4,862.0	96.4
240	4,266.0	77.1	5,688.0	102.8	4,266.0	77.1	5,688.0	102.8
260	4,933.5	81.9	6,578.0	109.2	4,933.5	81.9	6,578.0	109.2
280	5,649.0	86.7	7,532.0	115.6	5,649.0	86.7	7,532.0	115.6
300	6,412.5	91.5	8,550.0	122.0	6,412.5	91.5	8,550.0	122.0

*Based on "Standard Specifications for Highway Bridges," American Association of State Highway and Transportation Officials. Impact not included.

†Moments in thousands of ft-lb (ft-kips).

‡Shear and reaction in kips. Concentrated load is considered placed at the support. Loads used are those stipulated for shear.

§Maximum value determined by standard truck loading. Otherwise, standard lane loading governs.

Impact is expressed as a fraction of live-load stress and determined by the formula:

$$I = \frac{50}{125 + l} \qquad 30\% \text{ maximum} \qquad (5\text{-}2)$$

where l = span, ft; or for truck loads on cantilevers, length from moment center to farthermost axle; or for shear due to truck load, length of loaded portion of span. For negative moments in continuous spans, use the average of two adjacent loaded spans. For cantilever shear, use I = 30%. Impact is not figured for abutments, retaining walls, piers, piles (except for steel and concrete piles above ground rigidly framed into the superstructure), foundation pressures and footings, and sidewalk loads.

Longitudinal forces on highway bridges should be assumed at 5% of the live load headed in one direction, plus forces resulting from friction in bridge expansion bearings.

Centrifugal forces should be computed as a percentage of design live load

$$C = \frac{6.68S^2}{R} \qquad (5\text{-}3)$$

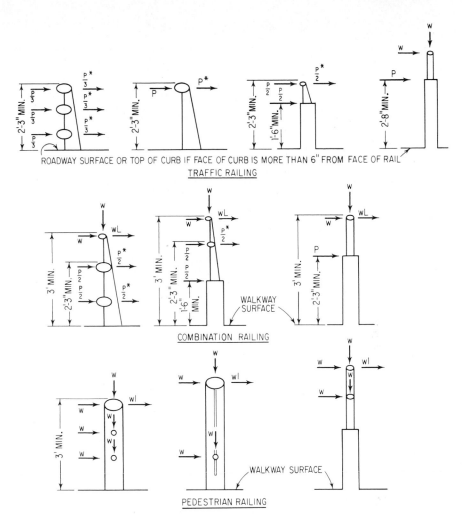

ROADWAY SURFACE OR TOP OF CURB IF FACE OF CURB IS MORE THAN 6" FROM FACE OF RAIL
TRAFFIC RAILING

COMBINATION RAILING

WALKWAY SURFACE

PEDESTRIAN RAILING

* WITH SIMULTANEOUS LONGITUDINAL LOAD OF $\frac{1}{2}$ THIS AMOUNT,
DIVIDED AMONG POSTS IN A CONTINUOUS RAIL LENGTH

Fig. 5-5. Design loads for railings. $P = 10$ kips; $L =$ post spacing for traffic railing; $w = 50$ lb/lin ft; $l =$ post spacing for pedestrian railing. Rail loads are shown on the left, post loads on the right. (The shapes of rail members are for illustrative purposes only.)

where $S =$ design speed, mi/h
$\qquad R =$ radius of curvature, ft
These forces are assumed to act horizontally 6 ft above deck level and perpendicular to the bridge center line.

Wind forces generally are considered as moving loads that may act horizontally in any direction. They apply pressure to the exposed area of the superstructure, as seen in side elevation; to traffic on the bridge, with the center of gravity 6 ft above the deck; and to the exposed areas of the substructure, as seen in lateral or front elevation. Wind loads in Tables 5-2 and 5-3 were

TABLE 5-2   Wind Loads for Superstructure Design

	Trusses and arches	Beams and girders	Live load
Wind load	75 lb/ft²	50 lb/ft²	100 lb/lin ft
Minimums:			
On loaded chord	300 lb/lin ft		
On unloaded chord	150 lb/lin ft		
On girders		300 lb/lin ft	

TABLE 5-3   Wind Loads for Substructure Design

a. Loads transmitted by superstructure to substructure slab and girder bridges (up to 125-ft span)

	Transverse	Longitudinal
Wind on superstructure when not carrying live load, lb/ft²	50	12
Wind on superstructure when carrying live load, lb/ft²	15	4
Wind on live load, lb/lin ft*	100	40

Major and unusual structures

Skew angle or wind, deg	No live load on bridge				Live load on bridge				Wind on live load, lb/lin ft*	
	Wind on trusses, lb/ft²		Wind on girders, lb/ft²		Wind on trusses, lb/ft²		Wind on girders, lb/ft²			
	Lateral load	Longitudinal load	Lateral load	Longitudinal load	Lateral load	Longitudinal load	Lateral load	Longitudinal load	Lateral load	Longitudinal load
0	75	0	50	0	22.5	0	15	0	100	0
15	70	12	44	6	21	3.6	13.2	1.8	88	12
30	65	28	41	12	19.5	8.4	12.3	3.6	82	24
45	47	41	33	16	14.1	12.3	9.9	4.8	66	32
60	25	50	17	19	7.5	15	5.1	5.7	34	38

b. Loads from wind acting directly on the substructure†

Horizontal wind—no live load on bridge, lb/ft²	40
Horizontal wind—live load on bridge, lb/ft²	12

*Acting 6 ft of above deck.
†Resolve wind forces acting at a skew into components perpendicular to side and front elevations of the substructure and apply at centers of gravity of exposed areas. These loads act simultaneously with wind loads from superstructure.

derived from "Standard Specifications for Highway Bridges," American Association of State Highway and Transportation Officials. They are based on 100-mi/h wind velocity. They should be multiplied by $(V/100)^2$ for other design velocities except for Group III loading (Art. 5-4).

When investigating overturning, add to horizontal wind forces acting normal to the longitudinal bridge axis an upward force of 20 lb/ft² for the structure without live load or 6 lb/ft² when the structure carries live load. This force should be applied to the deck and sidewalk area in plan at the windward quarter point of the transverse superstructure width.

Thermal Forces.   Provision should be made for expansion and contraction due to temperature variations, and on concrete structures, also for shrinkage. For the continental United States, Table 5-4 covers temperature ranges of most locations and includes the effect of shrinkage on ordinary beam-type concrete structures. The coefficient of thermal expansion for both

TABLE 5-4   Expansion and Contraction of Structures*

Air temp range	Steel		Concrete†	
	Temp rise and fall, °F	Movement per unit length	Temp rise and fall, °F	Movement per unit length
Extreme: 120 °F, certain mountain and desert locations	60	0.00039	40	0.00024
Moderate: 100 °F, interior valleys and most mountain locations	50	0.00033	35	0.00021
Mild: 80 °F, coastal areas, Los Angeles. and San Francisco Bay area	40	0.00026	30	0.00018

*This table was developed for California. For other parts of the United States, the temperature limits given by AASHTO "Standard Specifications for Highway Bridges" should be used.
†Includes shrinkage.

concrete and steel per °Fahrenheit is 0.0000065 (approximately 1/150,000). The shrinkage coefficient for concrete arches and rigid frames should be assumed as 0.002, equivalent to a temperature drop of 31°F.

Restraint forces, generated by preventing deformations, must be considered in design.

Earthquake (seismic) forces should be assumed to act horizontally at the center of gravity of the structure in the direction that produces maximum stresses in the member or part of the structure under consideration. These forces may be computed from

$$EQ = \phi KCD \qquad (5\text{-}4a)$$

where EQ = total earthquake force, kips, affecting member or part
$\quad D$ = total dead load supported by member or part, kips
$\quad K$ = coefficient indicating energy-absorption characteristics of structure
$\quad$ = 1.33 for bridges in which a wall with ratio of height to length of 2.5 or less resists horizontal forces acting along the wall
$\quad$ = 1.00 for bridges in which single columns or piers with ratio of height to length exceeding 2.5 resist horizontal forces
$\quad$ = 0.67 for bridges in which continuous frames resist horizontal forces acting along frames
$\quad C$ = $0.05/\sqrt[3]{T}$ = coefficient representing stiffness of structure
$\quad T$ = natural period of vibration of structure, s = $0.32\sqrt{D_c/P}$
$\quad P$ = force, kips, required to produce lateral deflection of structure equal to 1 in
$\quad D_c$ = "contributory" dead load, kips
$\quad \phi$ = coefficient related to regional earthquake probability (California Division of Highways requires ϕ = 2.0 for structures with spread footings and ϕ = 2.5 for structures on piles)

The EQ forces calculated from Eq. (5-4a) should not be less than 0.40D. In the structural analysis, special consideration should be given to bridges with large skews or with columns having large differences in stiffness. Careful investigation and accurate analysis also are necessary for structures founded on very poor material, structures adjacent to active faults, and large and high structures.

Restraining features should be provided to limit the displacement of superstructures, such as hinge ties and shear blocks. These should be designed for the force computed from Eq. (5-4b):

$$EQ = 0.25D_c - V \tag{5-4b}$$

where V = column shear due to EQ. D_c should be determined from an examination of the entire frame. For example, a simple span with fixed bearings at one end and sliding bearings at the other end will have the weight of the entire superstructure as the contributing dead load for longitudinal forces at the fixed abutment, while one-half of the superstructure dead load will act at each abutment for transverse forces.

For a frame, such as a two-span continuous structure, the full length of the bridge should be used as the contributory length in the longitudinal direction. The resulting force can be reduced by deduction of V. For hinge restrainers, use 0.25 times D_c of the smaller of the two frames and deduct V.

For complex structures, a more rigorous dynamic analysis should be made.

Stream-flow pressure on a pier should be computed from

$$P = KV^2 \tag{5-5}$$

where P = pressure, lb/ft^2
      V = velocity of water, ft/s
      K = ⅔ for square ends, ½ for angle ends when angle is 30° or less, and ⅔ for circular piers

Ice pressure should be assumed as 400 psi. The design thickness should be determined locally.

Earth pressure on piers and abutments should be computed by recognized soil-mechanics formulas, but the equivalent fluid pressure should be at least 36 lb/ft^3 when it increases stresses and not more than 27 lb/ft^3 when it decreases stresses.

Railway Bridge Loads ▪ *Live load* is specified by axle-load diagrams or by the E number of a "Cooper's train," consisting of two locomotives and an indefinite number of freight cars. Table 5-5 shows the typical axle spacing and axle loads for E10 loading and corresponding simple-beam moments, shears, and reactions for spans from 7 to 250 ft. Values in the table should be increased proportionally for specified loading other than E10, for example, multiplied by 7.2 for E72 loading. (The American Railway Engineering Association recommends E80 loading for main-line structures in its "Manual for Railway Engineering," 1981.)

Members receiving load from more than one track should be assumed to be carrying the following proportions of live load: For two tracks, full live load; for three tracks, full live load from two tracks and half from the third track; for four tracks, full live load from two, half from one, and one-fourth from the remaining one.

Impact of railway loads on steel structures is composed of two components that act vertically on the rails. One is the rolling effect, which increases the load on one rail and decreases the load on the other, each by 10% of the axle load. The second is the vertical effect due to track irregularities, speed, and car impact. This effect acts equally on both rails. Its magnitude depends on type of equipment (steam, electric, or diesel locomotive), type of bridge (rolled beam, girder, truss), and loaded length of structure. For equipment other than steam engines, the impact, expressed as a percentage I of live load for steel bridges (without the rolling effect), may be computed from either of the following:

For l less than 80 ft,

$$I = 40 - \frac{3l^2}{1600} \tag{5-6a}$$

For l greater than 80 ft,

$$I = 16 + \frac{600}{l - 30} \qquad (5\text{-}6b)$$

where l = span, ft, center to center of supports for stringers, transverse floor beams without stringers, longitudinal girders, and main members of trusses; or length, ft, of the longer adjacent supported stringer, longitudinal beam, girder, or truss for impact in floor beams, floor-beam hangers, subdiagonals of trusses, transverse girders, supports for longitudinal and transverse girders, and viaduct columns.

Members receiving load from more than two tracks should be assumed to take full impact from any two tracks. Members receiving load from two tracks should be assumed to take the full impact from the two tracks when l is less than 175 ft and from only one track when l is greater than 225 ft. For l between 175 and 225 ft, the members should take full impact from one track and a percentage of full impact from the other as given by $450 - 2l$.

Impact on concrete structures, as a percentage of live load, may be computed from

$$I = \frac{100L}{L + D} \qquad (5\text{-}7)$$

where L is the total live load and D the dead load on the member for which stresses are being computed.

Longitudinal forces should be computed for one track only. They should total either 15% of the entire moving load on the bridge or 25% of the load on the driving axles, whichever is greater.

TABLE 5-5   Maximum Moments, Shears, and Reductions for Class E10 Engine Loading*

One Track of Two Rails

Axle loads, kips

5.0 10.0 10.0 10.0 10.0 6.5 6.5 6.5 6.5   5.0 10.0 10.0 10.0 10.0 10.0 6.5 6.5 6.5 6.5   1 kip per lin ft uniform load   12.5 12.5

|—8 ft—|—5 ft—|—5 ft—|—9 ft—|—5 ft—|—6 ft—|—5 ft—|— 8 ft —|—8 ft—|—5 ft—|—5 ft—|—5 ft—|—9 ft—|—5 ft—|—6 ft—|—5 ft—|—5 ft—| or |—7 ft—|

Span, ft	Max moment, ft-kips	Max shear, kips	Max floor beam reaction, kips†	Equivalent uniform load, kips per ft		
				Moment	Shear	Reaction
10	31.2	16.2	20.0	2.50	3.25	2.00
15	62.5	20.0	27.3	2.22	2.67	1.82
20	103.1	25.0	32.8	2.06	2.50	1.64
25	152.5	28.4	37.8	1.95	2.27	1.51
30	205.2	31.5	43.1	1.82	2.10	1.44
35	261.5	34.6	48.8	1.71	1.98	1.39
40	327.8	37.7	54.0	1.64	1.88	1.35
50	475.5	43.5	64.3	1.52	1.74	1.29
60	649.5	48.8	76.6	1.44	1.63	1.28
70	853.7	55.3	88.5	1.39	1.58	1.26
80	1,080.0	62.1	99.4	1.35	1.55	1.24
90	1,334.7	68.6	109.3	1.32	1.53	1.22
100	1,609.7	75.0	118.6	1.29	1.50	1.19
125	2,497.7	89.7	140.5	1.28	1.44	1.12
150	3,531.0	103.7	162.7	1.25	1.38	1.08
175	4,676.3	117.3	185.8	1.22	1.34	1.06
200	5,939.0	130.5	209.5	1.19	1.31	1.05
250	8,796.3	156.6	257.6	1.13	1.25	1.03

*The standard Class E10 load train consists of two Class E10 engines, coupled front to rear, followed by an indefinite, uniform load of 1 kip per lin ft of track. To obtain the actual design moments, shears, and reactions, the tabulated figures must be multiplied by 8.0 for E80 loading.
†From two spans.

TABLE 5-6   Moments at Midspan of Highway and Railway Bridges, ft-kips

Span, ft	Highway One lane of HS20			Railway One track of E60		
	Live load	Impact	Total	Live load	Impact	Total
50	628	180	808	2,853	1,007	3,860
100	1,524	339	1,863	9,660	2,380	12,040
200	4,100	632	4,732	35,634	6,960	42,594

Centrifugal forces should be computed from Eq. (5-3).

Table 5-6 compares bending moments due to live load plus impact for highway and railway loadings on single-span bridges.

5-4.   Proportioning of Members and Sections

The forces—axial forces N, bending moments M, shears V, and torques M_T—generated by each type of loading that a structure may be subjected to should be computed for all members and relevant sections of the structure in accordance with recognized methods of static analysis (see Sec. 1).

Members and sections then should be proportioned to fulfill either or both of the following conditions:

Working-Stress Design ▪ The sum of the stresses induced by various combinations of loadings should not exceed the percentage of the basic working stresses for the given materials indicated in Table 5-7.

TABLE 5-7   Allowable Stresses for Loading Combinations*

Group	Combination	% of basic unit stresses
I	$D + L + I + E + B + SF$	100
II	$D + E + B + SF + W$	125
III	Group I + $LF + F$ + 30% $W + WL + CF$	125
IV	Group I + $R + S + T$	125
V	Group II + $R + S + T$	140
VI	Group III + $R + S + T$	140
VII	$D + E + B + SF + EQ$	133⅓
VIII	Group I + ICE	140
IX	Group II + ICE	150

*In frame-type structures, temperature stresses should be included in Group I.

D = dead load	CF = centrifugal force
L = live load	F = longitudinal force due
I = live-load impact	to friction
E = earth pressure	R = rib shortening
B = buoyancy	S = shrinkage
W = wind load on structure	T = temperature
WL = wind load on live load	EQ = earthquake
LF = longitudinal force from	SF = stream flow pressure
live load	ICE = ice pressure

See also Art. 5-3.

Groupings other than those listed in Table 5-7 may be specified for spans over 500 ft. Higher working stresses are also permitted for the operational rating of existing bridges. In anticipation of such future relaxations of the safety factor, some major bridges have been designed initially with up to 33% higher dead-load stresses than current specifications would permit.

Load-Factor Design ▪ The totals of the effects (shears, moments, stresses) of the following load groups should not exceed the capacity of the structure, member, or connection.

$$\text{Group I} = 1.3D + \frac{5}{3}(L + I) \tag{5-8a}$$

For design loadings less than H20, capacity also must be adequate for

$$\text{Groups IA} = 1.3D + 2.2(L + I) \tag{5-8b}$$

$$\text{Group II} = 1.3(D + W + F + SF + B + S + T) \tag{5-9}$$

(If a structure may be subjected to earthquake forces, use EQ instead of W; and if to ice pressure, use ICE instead of SF.)

$$\text{Group III} = 1.3(D + L + I + CF + 0.3W + WL + F + LF) \tag{5-10}$$

Allowable working stresses and design capacities depend on the quality of the materials of the particular member or connection, on shape or geometry for components or members in compression, and on frequency of loading (fatigue) for connections. For steel structures, capacities are functions of the ultimate strength F_u, yield strength F_y, or modulus of elasticity E. For concrete, capacities are functions of the 28-day compressive strength f'_c.

See also Secs. 2 and 3.

STEEL BRIDGES

Steel is competitive as a construction material for medium spans and favored for long-span bridges for the following reasons: It has high strength in tension and compression. It behaves as a nearly perfect elastic material within the usual working ranges. It has strength reserves beyond the yield point. The high standards of the fabricating industry guarantee users uniformity of the controlling properties within narrow tolerances. Connection methods are reliable, and workers skilled in their application are available.

The principal disadvantage of steel in bridge construction, its susceptibility to corrosion, is being increasingly overcome by chemical additives or improved protective coatings.

5-5.   Systems Used for Steel Bridges

The following are typical components of steel bridges. Each may be applied to any of the functional types and structural systems listed in Art. 5-1.

Main support: Rolled beams, plate girders, box girders, or trusses.

Connections: (See also Art. 5-8.) Riveted, high-strength bolted, welded, or combinations.

Materials for traffic-carrying deck: Timber stringers and planking, reinforced concrete slab or prestressed-concrete slab, stiffened steel plate (orthotropic deck), or steel grid.

Timber decks are restricted to bridges on roads of minor importance. Plates of corrosion-resistant steel should be used as ballast supports on through plate-girder bridges for railways. For roadway decks of stiffened steel plates, see Art. 5-13.

Deck framing: Deck resting directly on main members or supported by grids of stringers and floor beams.

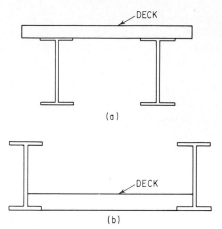

Fig. 5-6. (*a*) Deck-type bridge; (*b*) through bridge.

Location of deck: On top of main members: deck spans (Fig. 5-6*a*); between main members, the underside of the deck framing being flush with that of the main members: through spans (Fig. 5-6*b*).

5-6. Grades and Permissible Stresses of Steel for Bridges

Preferred steel grades, permissible stresses, and standards of details, materials, and quality of work for steel bridges are covered in the "Manual for Railway Engineering," American Railway Engineering Association, and "Standard Specifications for Highway Bridges," American Association of State Highway and Transportation Officials. Properties of the various grades of steel and the testing methods to be used to control them are regulated by specifications of the American Society for Testing and Materials (ASTM). Properties of the structural steels presently preferred in bridge construction are tabulated in Table 5-8.

Dimensions and geometric properties of commercially available rolled plates and shapes are tabulated in the "Steel Construction Manual," American Institute of Steel Construction (AISC), and in manuals issued by the major steel producers.

All members and parts of steel bridges should be designed in accordance with the recognized rules of elastic analysis. They should be so proportioned that all stresses remain within permissible limits.

The permissible tensile stresses are obtained by applying to the yield strength of the given grade of steel a safety factor of 1.8. Compression stresses are subject to further reduction to compensate for slenderness of member of element, as well as unintentional and calculated eccentricities. The basic stresses are compiled in Sec. 2 and summarized for convenience in Table 5-9. (See also Art. 5-4.)

Members, connection material, and fasteners subject to repeated variations or reversals of stress should be proportioned for the allowable stress ranges in Art. 2-16.

5-7. Other Design Limitations

Depth Ratios, Slenderness Ratios, Deflections. AASHTO and AREA specifications restrict the depth-to-span ratios D/L of bridge structures and the slenderness ratios l/r of individual truss or bracing members to the values in Table 5-10,

TABLE 5-8   Mechanical Properties of Structural Steel for Bridges

Characteristic	ASTM designation	Plates thicknesses, in	Shape groups*	Min tensile strength, ksi	Yield strength, ksi
Structural carbon steel	A36	To 8 incl.	All	58	36
High-strength low-alloy steel	A242 A440 A441 A588 A572		See Table 2 of AISC Manual		
High-yield-strength, quenched and tempered alloy steel	A514	To 2½ incl. Over 2½ to 4 incl.	Not applicable	110 100	100 90
	A517	To 2½ incl.	Not applicable	115–135	100
Structural steel for bridges	A709 Grade 36	To 8 incl.	To 426 lb per ft	58–80	36
			Over 426 per ft	58 min	36
		Over 8	. . . .	58 min	32
	Grade 50	To 2 incl.	1, 2, 3, 4	65 min	50
	Grade 50W	To 4 incl.	All	70 min	50
	Grades 100 and 100W	To 2½ incl.	Not applicable	100–130	100
		Over 2½ to 4 incl.	Not applicable	100–130	90

*See Table 2 of AISC "Manual of Steel Construction."

where D = depth of construction, ft

L = span, ft, c to c bearings for simple spans or distance between points of contraflexure for continuous spans

l = unsupported length of member, in

r = radius of gyration, in

These are minimum values; preferred values are higher.

For plate-girder design criteria, see Arts. 2-20 and 2-22.

Both specifications limit the elastic deflections of bridges under design live load plus impact to ⅟₈₀₀ of the span, measured c to c bearings, except that ⅟₁₀₀₀ may be used for bridges used by pedestrians; ⅟₃₀₀ of the length of cantilever arms. Deflection calculations should be based on the gross sections of girders or truss members. Anticipated dead-load deflections must be compensated by adequate camber in the fabrication of steel structures.

Splices. Shop, assembly yard, or erection splices must be provided for units whose overall length exceeds available rolled lengths of plates and shapes or the clearances of available shipping facilities. Splices also must be provided when total weight exceeds the capacity of available erection equipment.

Accessibility. All parts should be accessible and adequately spaced for fabrication, assembly, and maintenance. Closed box girders and box-type sections should be equipped with handholes or manholes.

On long and high bridges, installation of permanent maintenance travelers may be justified.

5-8.   Steel Connections in Bridges

Riveted Connections. Rivets function as shear connectors and clamping devices, the heads preventing the parts connected from falling apart. Rivet diameters most commonly used on

TABLE 5-9 Permissible Stresses for Structural Steel for Highway Bridges, ksi

Axial tension on net section of members with holes: The smaller of

$$F_a = 0.55F_y$$
$$F_a = 0.46F_u$$

where F_y = minimum yield strength, ksi
 F_u = minimum tensile strength, ksi

Axial tension in members without holes
Tension in extreme surfaces of rolled shapes, girders, and built-up sections subject to bending, on net section
Axial compression in stiffeners of plate girders, on gross section
Compression on gross section of splice material

$$F = 0.55F_y$$

Compression in extreme surfaces of rolled shapes, girders, and built-up sections subject to bending (on gross section) when compression flange is
(a) supported laterally for its full length by embedment in concrete

$$F_b = 0.55F_y$$

(b) partly supported or unsupported with ratio of unsupported length L_u to flange width b_f not more than about $215/\sqrt{F_y}$

$$F_b = 0.55F_y\left[1 - \frac{(L_u/b_f)^2 F_y}{95,400}\right]$$

Compression in concentrically loaded columns with slenderness ratios not exceeding

Max L'/r	130	125	120	115	110	90	85
For F_y	36	42–50	55	60	65	90	100

where L' = length of member, in
 r = least radius of gyration, in
(a) with riveted or bolted ends

$$F_a = 0.44f_y\left[1 - \frac{(L'/r)^2 F_y}{2,000,000}\right]$$

(b) with pinned ends

$$F_a = 0.44f_y\left[1 - \frac{(L'/r)^2 F_y}{1,500,000}\right]$$

Shear in girder webs

$$F_v = 0.33F_y$$

Bearing on milled stiffeners and other steel parts in contact except rivets and bolts
Stress in extreme surface of pins
Bearing on pins not subject to rotation

$$F = 0.80F_y$$

Shear in pins

$$F_v = 0.40F_y$$

Bearing on pins subject to rotation (as for rockers and hinges)

$$F_p = 0.40F_y$$

Bearing on power-driven rivets and high-strength bolts

$$F_p = 1.22F_y$$

but not more than the allowable bearing on the fasteners

bridges are ¾, ⅞, 1, and occasionally 1⅛ in. Rivet holes are either punched or subpunched and reamed, or drilled full size. They are usually made ⅟₁₆ in larger than the nominal rivet diameter.

The most commonly used rivet steel is ASTM A502, Grade 1. Less common are A502, Grade 2 high-strength structural rivets. Allowable stresses are given in Art. 2-18.

TABLE 5-10   Dimensional Limitations for Bridge Members

	AASHTO	AREA
Min depth-span ratios:		
For rolled beams	$1/25$	$1/15$
For girders	$1/25$	$1/12$
For trusses	$1/10$	$1/10$
Max slenderness ratios:		
For main members in compression	120	100
For bracing members in compression	140	120
For main members in tension	200	200
For bracing members in tension	240	200

Limitations on rivet spacing and edge distances are spelled out in design specifications. Minimum pitch is controlled by fabrication requirements. Edge distances are controlled by the size of the rivet head, curvature of rolled edge, and rolling tolerances. The maximum pitch of stitch rivets in compression members is limited to prevent buckling of the individual plates, whereas spacing of rivets along free edges of the material is restricted to assure sealing of the joint.

Connections with High-Strength Bolts.   The parts may be clamped together by bolts of quenched and tempered steel, ASTM A325. The nuts are tightened to specified amounts. Setting of the bolts requires less preparation of the faying surfaces and more labor than riveting.

Details and quality of work are covered by the "Specifications for Structural Joints Using ASTM A325 and A490 Bolts," approved by the Research Council on Riveted and Bolted Structural Joints of the Engineering Foundation. Permissible working stresses are given in Art. 2-18.

Welded Connections.   In welding, the parts to be connected are fused at high temperatures, usually with addition of suitable metallic material. The "Structural Welding Code," AWS D1.1, American Welding Society, regulates application of the various types and sizes of welds, permissible stresses in the weld and parent metal, permissible edge configurations, kinds and sizes of electrodes, details of quality of work, and qualification of welding procedures and welders. (For allowable welding stresses, see Art. 2-17.)

Welded construction has the following advantages over riveted construction that make it attractive to bridge designers: savings of steel due to elimination of holes that weaken the effective section in tension, and to omission of additional splice material; smoother appearance, easier maintenance, easier repairs; less noise during erection. These advantages are partly offset by some disadvantages in fabrication: restriction in choice of steel to weldable steels; greater shop-space requirements; and necessity of extensive, often costly, inspection. To these must be added some structural shortcomings: distortions that result from differential cooling of the weld and its surrounding metal, and, consequently, formation of hidden, or "locked-up," stresses, which are quantitatively difficult to assess; brittleness of the weld or its surroundings at low temperatures.

For the last reason, welded construction requires special controls for bridges in cold climates. It is still not often used in railway bridges. When welding is used for field splices of bridges under conditions that are less favorable for effective controls than sheltered shops, the splices should receive thorough inspection.

Many designers favor the combination of shop welding with high-strength bolted field connections.

Pin Connections.   Hinges between members subject to relative rotation are usually formed with pins, machined steel cylinders. They are held in either semicircular machined recesses or smoothly fitting holes in the connected members.

For fixation of the direction of the pin axis, pins up to 10-in diameter have threaded ends for recessed nuts, which bear against the connected members. Pins over 10-in diameter are held by recessed caps. These in turn are held either by tap bolts or a rod that runs axially through a hole in the pin itself and is threaded and secured by nuts at its ends.

Pins are designed for bending and shear and for bearing against the connected members. (For stresses, see Table 5-9 and Art. 2-10.)

5-9. Bridge Bearings

Bearings are structural assemblies installed to secure the safe transfer of all reactions from the superstructure to the substructure. They must fulfill two basic requirements: They must spread the reactions over adequate areas of the substructure, and they must be capable of adapting to elastic, thermal, and other deformations of the superstructure without generating harmful restraining forces.

Generally, bearings are classified as fixed or expansion, and as metal or elastomeric.

Fixed bearings adapt only to angular deflections. They must be designed to resist both vertical and horizontal components of reactions.

Expansion bearings adapt to both angular deflections and longitudinal movements of the superstructure. Except for friction, they resist only those components of the superstructure reactions perpendicular to these movements.

In both types of bearings, provision must be made for the safe transfer of all forces transverse to the direction of the span.

Metal bearings are preferably of structural steel, cast steel, or cast iron. Their basic components are an upper unit, which is bolted to the superstructure, and a lower unit (shoe or pedestal), which is anchored to the substructure. Inserted between these, if required, are elements for centering and for adaptation to angular deflections and, in the case of expansion bearings, to longitudinal movements of the superstructure.

According to AASHTO "Standard Specifications for Highway Bridges," no provision for angular deflections need be made when spans are less than 50 ft. Bearings then may consist of two plane steel plates in contact with each other. However, for better centering and maintenance, a *bearing bar* of rectangular cross section with chamfered or rounded top may be welded to the lower plate, and a *keeper plate,* cut out to fit over the bearing bar, may be welded to the upper plate.

For steel bearings of spans over 50 ft, AASHTO "Standard Specifications for Highway Bridges" requires curved plates, hinges, or pins (Fig. 5-7). For expansion bearings in addition, sliding plates, rockers (Fig. 5-7a) or rollers (Fig. 5-8b), or elastomeric pads are required. The top and bottom of each bearing must be held together effectively without obstructing the required movements.

Heavier bearings are made of cast steel or built up from structural steel by riveting, bolting, or welding (Figs. 5-7 and 5-8). Such bearings have machined contact surfaces without or with centering pins. Weldments must be stress-relieved before machining.

The plan dimensions of bearings are governed by the permissible bearing pressures on the bridge seat. The allowable concrete stresses are:

Under properly hinged bearings, not subject to high pressures	1000 psi
Under bearing plates and nonhinged shoes	700 psi

These stresses should be reduced by 25% if the concrete edges project less than 3 in beyond the edge of the base plate.

Permissible stresses for proportioning of pins and expansion rollers and rockers are given in Table 5-9 and Art. 2-10.

Plate thicknesses and other vertical dimensions should be proportioned to resist the flexural

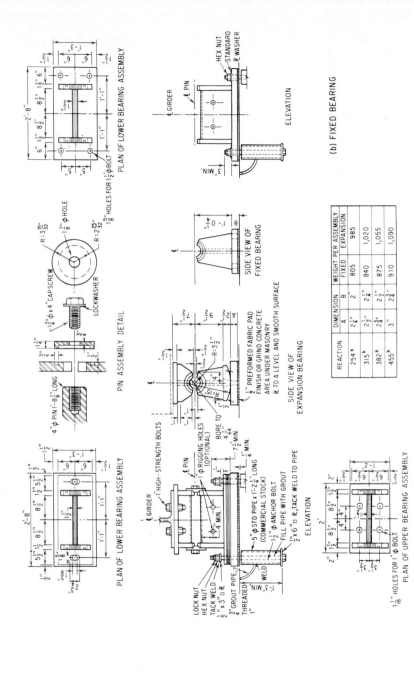

Fig. 5-7. Steel bearings for spans over 50 ft. Details for the fixed bearings are similar to those for the expansion bearing, except as shown. All joints should be welded all around with fillet welds. Weldments should be stress-relieved by heat treatment after welding has been completed.

PLAN OF LOWER BEARING ASSEMBLY

ELEVATION

(b) FIXED BEARING

PIN ASSEMBLY DETAIL

LOCKWASHER

SIDE VIEW OF
FIXED BEARING

REACTION	DIMENSION		WEIGHT PER ASSEMBLY	
	A	B	FIXED	EXPANSION
254 K	$2\frac{1}{4}"$	$2"$	805	985
315 K	$2\frac{1}{2}"$	$2\frac{1}{4}"$	840	1,020
382 K	$2\frac{3}{4}"$	$2\frac{1}{2}"$	875	1,055
455 K	$3"$	$2\frac{3}{4}"$	910	1,090

PLAN OF LOWER BEARING ASSEMBLY

SIDE VIEW OF
EXPANSION BEARING

ELEVATION

PLAN OF UPPER BEARING ASSEMBLY

(a) EXPANSION BEARING

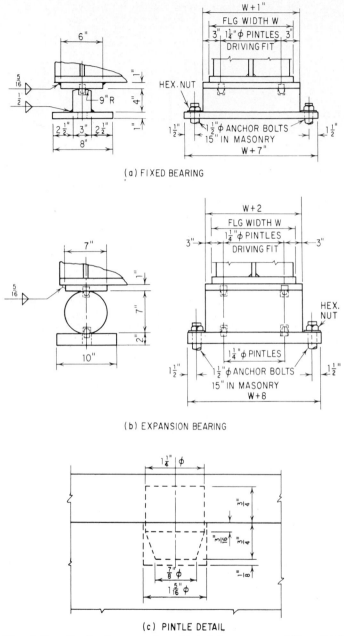

(a) FIXED BEARING

(b) EXPANSION BEARING

(c) PINTLE DETAIL

Fig. 5-8.   Bearing for composite welded girders on spans of 90 to 180 ft.

stresses that result from the spreading of the load from the upper lines of contact to the bearing areas. For anchorage of base plates to the substructure, AASHTO specifications give the minimum requirements listed in Table 5-11. The mass of masonry engaged by anchor bolts in tension should be sufficient to resist at least 1.5 times the calculated uplift.

TABLE 5-11   Requirements for Anchorage of Base Plates

Type and spans of superstructure	Bolts		Embedment in concrete, in
	Diameter, in	Number	
Outer beam of I-beam spans	1	2	10
Truss and girder spans:			
To 50 ft	1	2	10
51–100 ft	1¼	2	12
101–150 ft	1½	2	15
Over 150 ft	1½	4	15

Elastomeric bearings are either plain bearings, consisting of elastomer only, or laminated bearings, consisting of layers of elastomer, restrained at their interfaces by bonded laminas. The elastomer may be either natural rubber (polyisoprene) or synthetic rubber (chloroprene, Neoprene). Laminas should be rolled, mild-steel sheets.

The bearing pressure of elastomeric bearings should not exceed the following values: For dead load alone: 500 psi; for dead load plus live load: 800 psi. If the pressure from dead load minus uplift drops below 200 psi, restraints are required to prevent lateral crawling.

The capacity of an elastomeric bearing to absorb angular deflections and longitudinal movements of the superstructure is a function of its thickness (or of the sum of the thicknesses of its rubber elements between steel laminas), its shape factor (area of the loaded face divided by the sum of the side areas free to bulge), and the properties of the elastomer.

AASHTO Specifications limit the overall thickness of a plain elastomer bearing to one-fifth its length or width, of a laminated bearing to one-third its length (in the direction of the bridge) or one-half its width, in both cases whichever is smaller. The thickness should be at least twice the sum of the positive and negative movements expected because of temperature changes.

5-10. Rolled-Beam Bridges

The simplest steel bridges consist of rolled I beams or wide-flange beams, which either support the traffic-carrying deck or are fully embedded in it. Rolled beams serve also as floor beams and stringers for decks of plate-girder and truss bridges.

Reductions in steel weight may be obtained, but with greater labor costs, by adding cover plates in the area of maximum moments, by providing continuity over several spans, by utilizing the deck in composite action, or by a combination of these measures. The principles of design and details are essentially identical with those of plate girders (Art. 5-11).

Standard designs by the Federal Highway Administration uses shapes, up to W36 × 245, up to the following maximum spans:

Simple spans, without cover plates . . . . 70 ft
Simple spans, composite and with welded cover plates . . . . 90 ft
Continuous spans with welded cover plates and high-strength bolted splices . . . . 80–100–80 ft

For design of a concrete deck slab, see Art. 5-20.

5-11. Plate-Girder Bridges

The term plate girder applies to structural elements of I-shaped cross section that are either riveted together from plates and angles or welded from plates alone. Plate girders are used as primary supporting elements in many structural systems: as simple beams on abutments or, with overhanging ends, on piers; as continuous or hinged multispan beams; as stiffening girders of arches and suspension bridges, and in frame-type bridges. They also serve as floor beams and stringers on these and other bridge systems.

Their prevalent application on highway and railway bridges is in the form of *deck-plate girders* in combination with concrete decks (Fig. 5-9). (For design of concrete deck slabs, see Art. 5-20.) For girders with steel decks (ortho-tropic decks), see Art. 5-13. Girders with track ties mounted directly on the top flanges, *open-deck girders,* are used on branch railways and industrial spurs. *Through plate girders* (Fig. 5-10) are now practically restricted to railway bridges where allowable structure depth is limited.

The two or more girders supporting each span must be braced against each other to provide stability against overturning and flange buckling, to resist transverse forces (wind, earthquake, centrifugal), and to distribute concentrated heavy loads. On *deck girders,* this is done by systems of diagonals in the planes of the top and bottom flanges (Fig. 5-9c) and by transverse bracing in vertical planes. The top lateral system can be omitted if the deck and its connections to the girder are designed to take its place. Bottom lateral systems are required for deck-plate girders with spans greater than 125 ft. Transverse bracing should be installed over each bearing and, on bridges of over 40-ft span, at intermediate locations not over 25 ft apart. This bracing may consist either of full-depth cross frames (Fig. 5-9b) or of solid diaphragms of not less than one-third and preferably half the girder depth. On *through girder spans,* since there can be no provision for top lateral and transverse bracing systems, the top flanges must be braced against the floor system by heavy gusset plates or knee braces (Fig. 5-10b).

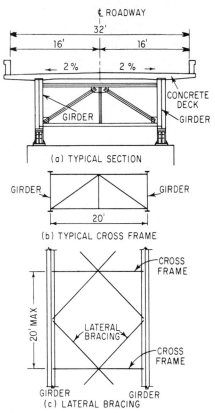

Fig. 5-9.   Two-lane, deck-girder highway bridge.

Standard designs for two-lane welded deck-girder bridges have been prepared by the Federal Highway Administration for spans from 90 to 200 ft. Figure 5-11 shows typical welded-girder details, and Figs. 5-12 and 5-13, typical riveted or bolted girder details.

Flexural Analysis ▪ The extreme fiber stresses in flexure of rolled beams and girders are computed under the assumption of linear stress distribution across each section from the equations

$$f_c = \frac{Me_c}{I_{\text{gross}}} \tag{5-11a}$$

$$f_t = \frac{Me_t}{I_{\text{net}}} \tag{5-11b}$$

where M = maximum bending moment, in-kips

e_c and e_t = distances of extreme fibers in compression and tension from neutral axis, in

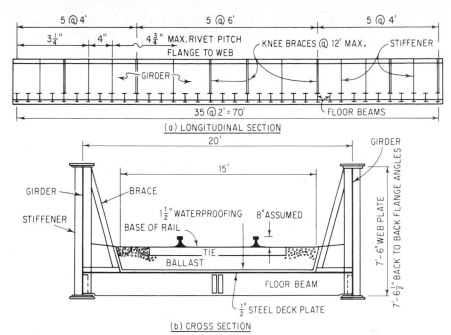

Fig. 5-10. Single-track, through-girder railway bridge.

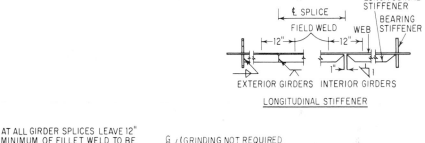

Fig. 5-11. Typical welded-girder details. Top and bottom flange plates may be offset. Web splices should be full-penetration groove welds. Where weld areas on exterior girders are visible from a traveled way, the areas should be ground flush.

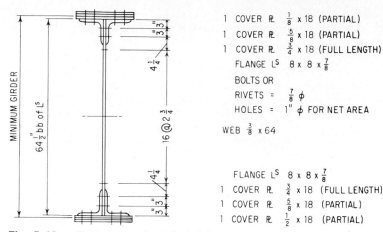

1 COVER ℞ $\frac{1}{8}$ x 18 (PARTIAL)
1 COVER ℞ $\frac{5}{8}$ x 18 (PARTIAL)
1 COVER ℞ $\frac{3}{4}$ x 18 (FULL LENGTH)
FLANGE L^S 8 x 8 x $\frac{7}{8}$
BOLTS OR
RIVETS = $\frac{7}{8}$ ϕ
HOLES = 1" ϕ FOR NET AREA

WEB $\frac{3}{8}$ x 64

FLANGE L^S 8 x 8 x $\frac{7}{8}$
1 COVER ℞ $\frac{3}{4}$ x 18 (FULL LENGTH)
1 COVER ℞ $\frac{5}{8}$ x 18 (PARTIAL)
1 COVER ℞ $\frac{1}{2}$ x 18 (PARTIAL)

Fig. 5-12.   Typical riveted or bolted girder cross section.

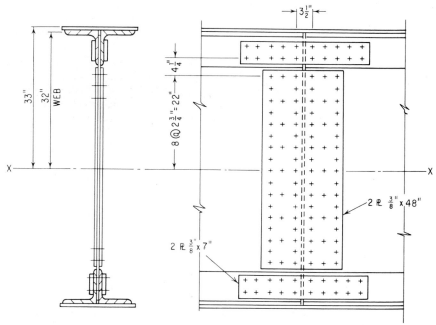

Fig. 5-13.   Full splice in riveted or bolted plate girder.

I_{gross} = moment of inertia of gross section, in^4
I_{net} = moment of inertia of net section, in^4, referred to neutral axis of gross section
f_c and f_t = extreme fiber stresses in compression and tension, ksi
The web shear stresses are computed from

$$v_s = \frac{V}{D_t} \qquad (5\text{-}12)$$

where V = maximum static shear, kips
D = depth of web, in

t = thickness of web plate, in

v_s = average web shear on gross section, ksi

For permissible stresses, see Sec. 2.

In riveted or bolted design, the web thickness and flange angle sizes are preferably constant along the girder. The section modulus is adapted to the moment variations by successive addition of flange cover plates to a maximum of three per flange, with thicknesses not exceeding that of the angles. Field connections and splices usually are made with high-strength bolts.

In welded design, variations in moment resistance are obtained by using flange plates of different thicknesses, widths, or steel grades, butt-welded to each other in succession. Stacking of plates, as in riveted or bolted design, is not recommended. Web thickness too may be varied.

Girder webs should be protected against buckling by transverse and, in the case of deep webs, longitudinal stiffeners. Transverse bearing stiffeners are required to transfer end reactions from the web into the bearings and to introduce concentrated loads into the web. Intermediate and longitudinal stiffeners are required if the girder depth-to-thickness ratios exceed critical values (see Art. 2-22).

Stiffeners may be plain plates, angles, or T sections. Transverse stiffeners should preferably be in pairs, although single stiffeners are allowed. The AASHTO Specifications contain restrictions on width-to-thickness ratios and minimum widths of plate stiffeners (Art. 2-22).

In riveted or bolted design, intermediate transverse stiffeners may be crimped at the flange angles, but bearing stiffeners must be straight. Hence, filler plates of the same thickness as the flange angles are required. No such problem exists with welded girders. Transverse intermediate stiffeners on only one side of the web should be riveted or welded to the compression flange.

Web-to-flange connections should be capable of carrying the stress flow from web to flange at every section of the girder. At an unloaded point, the stress flow equals the horizontal shear per linear inch. Where a wheel load may act, for example, at upper flange-to-web connections of deck girders, the stress flow is the vectorial sum of the horizontal shear per inch and the wheel load (assumed distributed over a web length equal to twice the deck thickness).

Similarly, connection rivets or welds between flange cover plates and between cover plates and flange angles should be capable of carrying the stress flow (see Art. 2-22).

Rivets or welds connecting bearing stiffeners to the web must be designed for the full bearing reaction.

Space restrictions in the shop, clearance restrictions in transportation, and erection considerations may require dividing long girders into shorter sections, which are then joined (spliced) in the field. Individual segments, plates or angles, must be spliced either in the shop or in the field if they exceed in length the sizes produced by the rolling mills or if shapes are changed in thickness to meet stress requirements.

Specifications require splices to be designed for the average between the stress due to design loads and the capacity of the unspliced segment, but for not less than 75% of the latter. In riveted or bolted design, material may have to be added at each splice to satisfy this requirement. Each splice element must be connected by a sufficient number of rivets or bolts to develop its full strength. Whenever it is possible to do so, splices of individual segments should be staggered. No splices should be located in the vicinity of the highest-stressed parts of the girder, for example, at midspan of simple-beam spans, or over the bearings on continuous beams.

(F. S. Merritt, "Structural Steel Designers' Handbook," McGraw-Hill Book Company, New York.)

5-12. Composite-Girder Bridges

Installation of appropriately designed shear connectors between the top flange of girders or beams and the concrete deck allows use of the deck as part of the top flange (equivalent cover plate). The resulting increase in effective depth of the total section and possible reductions of the top-flange steel usually allow some savings in steel compared with the noncomposite steel section. The overall economy depends on the cost of the shear connectors and any other additions to the girder or the deck that may be required and on possible limitations in effectiveness of the composite section as such.

In areas of negative moment, composite effect may be assumed only if the calculated tensile stresses in the deck are either taken up fully by reinforcing steel or compensated by prestressing. The latter method requires special precautions to assure slipping of the deck on the girder during the prestressing operation but rigidity of connection after completion.

If the steel girder is not shored up while the deck concrete is placed, computation of dead-load stresses must be based on the steel section alone.

Shear connectors should be capable of resisting all forces tending to separate the abutting concrete and steel surfaces, both horizontally and vertically. Connectors should not obstruct placement and thorough compaction of the concrete. Their installation should not harm the structural steel.

The types of shear connectors presently preferred are channels, or welded studs. Channels should be placed on beam flanges normal to the web and with the channel flanges pointing toward the girder bearings.

For stress calculations, see Arts. 2-19 and 5-20.

(F. S. Merritt, "Structural Steel Designers' Handbook," McGraw-Hill Book Company, New York.)

5-13. Orthotropic-Deck Bridges

An orthotropic deck is, essentially, a continuous, flat steel plate, with stiffeners (ribs) welded to its underside in a parallel or rectangular pattern. The term *orthotropic* is shortened from *orthogonal anisotropic,* referring to the mathematical theory used for the flexural analysis of such decks.

When used on steel bridges, orthotropic decks are usually joined quasi-monolithically, by welding or high-strength bolting, to the main girders and floor beams. They then have a dual function as roadway and as structural top flange.

The combination of plate or box girders with orthotropic decks allows the design of bridges of considerable slenderness and of nearly twice the span reached by girders with concrete decks. The most widespread application of orthotropic decks is on continuous, two- to five-span girders on low-level river crossings in metropolitan areas, where approaches must be kept short and grades low. This construction has been used for main spans up to 1100 feet with bracing by overhead cables and up to 856 ft without such bracing. There also are some spectacular high-level orthotropic girder bridges and some arch and suspension bridges with orthotropic stiffening girders. On some of the latter, girders and deck have been combined in a single lens-shaped box section that has great stiffness and low aerodynamic resistance.

Box Girders ▪ Single-web or box girders are used for orthotropic bridges. Box girders are preferred if structure depth is restricted. Their inherent stiffness makes it possible to reduce, or to omit, unsightly transverse bracing systems. In cross section, they usually are rectangular, occasionally trapezoidal. Minimum dimensions of box girders are controlled by considerations of accessibility and ease of fabrication.

Wide decks are supported by either single box girders or twin boxes. Wide single boxes have been built with multiple webs or secondary interior trusses. Overhanging floor beams sometimes are supported by diagonal struts.

Depth-Span Ratios ▪ Girder soffits are parallel to the deck, tapered, or curved. Parallel flanges, sometimes with tapered side spans, generally are used on unbraced girders with depth-to-main-span ratios as low as 1:70. Parallel-flange unbraced girders are practically restricted to high-level structures with unrestricted clearance. Unbraced low-level girders usually are designed with curved soffits with minimum depth-to-main-span ratios of about 1:25 over the main piers and 1:50 at the shallowest section.

Cable Systems ▪ Cable-braced girder systems have either two main girders or one, with two towers or one for each girder. The cables are curved if the girders are suspended at each floor beam; otherwise they are straight. In the latter (more usual) case, the cables either are arranged in parallel tiers or converge toward the tower or towers.

Each cable adds one degree of statical indeterminancy to a system. To make the actual conditions conform to design assumptions, the cable length must be adjustable either at the anchorages to the girders or at the saddles on the towers. (See also Art. 5-17.)

Steel Grades ▪ The steel most commonly used for orthotropic plates is weldable high-strength, low-alloy structural steel A441. The minimum plate thickness is seldom less than $\frac{7}{16}$ in (10 mm), to avoid excessive deflections under heavy wheel loads. The maximum thickness seldom exceeds $\frac{3}{4}$ in because of the decrease in permissible working stresses of high-strength low-alloy steel and the increase of fillet- and butt-weld sizes for plates of greater thickness.

Floor Beams ▪ If, as in most practical cases, the deck spans transversely between main girders, transverse ribs are replaced by the floor beams, which are then built up of inverted T sections, with the deck plate acting as top flange. Floor-beam spacings are preferably kept constant on any given structure. They range from less than 5 ft to over 15 ft. Longer spacings have been suggested for greater economy.

Ribs ▪ These are either open (Fig. 5-14a) or closed (Fig. 5-14b). The spacing of open ribs is seldom less than 12 in or more than 15 in. The lower limit is determined by accessibility for fabrication and maintenance, the upper by considerations of deck-plate stiffness. To reduce deformations of the surfacing material under concentrated traffic loads, some specifications require the plate thickness to be not less than $\frac{1}{25}$ of the spacing between open ribs or between the weld lines of closed ribs.

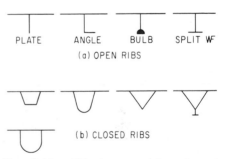

Usually, the longitudinal ribs are made continuous through slots or cutouts in the floor-beam webs to avoid a multitude of butt welds. Rib splices can then be coordinated with the transverse deck splices.

Closed ribs, because of their greater tor-

Fig. 5-14.   Rib shapes used in orthotropic-plate decks.

sional rigidity, give better load distribution and, other things being equal, require less steel and less welding than open ribs. Disadvantages of closed ribs are their inaccessibility for inspection and maintenance and more complicated splicing details. There have also been some difficulties in defining the weld between closed ribs and deck plate.

Fabrication ▪ Orthotropic decks are fabricated in the shop in panels as large as transportation and erection facilities permit. Deck-plate panels are fabricated by butt-welding available rolled plates. Ribs and floor beams are fillet-welded to the deck plate in upside-down position. Then, the deck is welded to the girder webs.

It is important to schedule all welding sequences to minimize distortion and locked-up stresses. The most effective method has been to fit up all components of a panel—deck plate, ribs, and floor beams—before starting any welding, then to place the fillet welds from rib to rib and from floor beam to floor beam, starting from the panel center and uniformly proceeding toward the edges. Since this sequence practically requires manual welding throughout, American fabricators prefer to join the ribs to the deck by automatic fillet welding before assembly with the floor beams. After slipping the floor-beam webs over the ribs, the fabricators weld manually only the beam webs to the deck. This method requires careful preevaluation of rib distortions, wider floor-beam slots, and consequently more substantial or only one-sided rib-to-floor-beam welds.

Analysis ▪ Stresses in orthotropic decks are considered as resulting from a superposition of four static systems:

System I consists of the deck plate considered as an isotropic plate elastically supported by the ribs (Fig. 5-15a). The deck is subject to bending from wheel loads between the ribs.

System II combines the deck plate, as transverse element, and the ribs, as longitudinal elements. The ribs are continuous over, and elastically supported by, the floor beams (Fig. 5-15b). The orthotropic analysis furnishes the distribution of concentrated (wheel) loads to the ribs, their flexural and torsional stresses, and thereby the axial and torsional stresses of the deck plate as their top flange.

System III combines the ribs with the floor beams and is treated either as an orthotropic

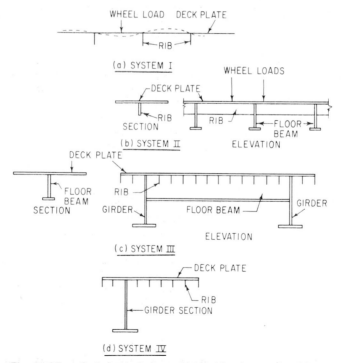

Fig. 5-15. Orthotropic deck may be considered to consist of four systems. (*a*) Deck plate supported on ribs; (*b*) rib-deck T beams spanning between floor beams; (*c*) floor beam with deck plate as top flange, supported on girders; (*d*) girder with deck plate as top flange.

system or as a grid (Fig. 5-15c). Analysis of this system furnishes the flexural stresses of the floor beams, including the stresses the deck plate receives as their top flange.

System IV comprises the main girders with the orthotropic deck as top flange (Fig. 5-15d). Axial stresses in the deck plate and ribs and shear stresses in the deck plate are obtained from the flexural and torsional analysis of the main girders by conventional methods.

Theoretically, the deck plate should be designed for the maximum principal stresses that may result from the simultaneous effect of all four systems. Practically, because of the rare coincidence of the maxima from all systems and in view of the great inherent strength reserve of the deck as a membrane (second-order stresses), a design is generally satisfactory if the stresses from any one system do not exceed 100% of the ordinarily permissible working stresses and 125% from a combination of any two systems.

In the design of long-span girder bridges, special attention must be given to buckling stability of deep webs and of the deck. Also, consideration should be given to conditions that may arise at intermediate stages of construction.

Steel-Deck Surfacing ▪ All traffic-carrying steel decks require a covering of some nonmetallic material to protect them from accidental damage, distribute wheel loads, compensate for surface irregularities, and provide a nonskid, plane riding surface. To be effective, the surfacing must adhere firmly to the base and resist wear and distortion from traffic under all conditions. Problems arise because of the elastic and thermal properties of the steel plate, its sensitivity to corrosion, the presence of bolted deck splices, and the difficulties of replacement or repair under traffic.

The surfacing material usually is asphaltic. Strength is provided by the asphalt itself (mastic-type pavements) or by mineral aggregate (asphalt-concrete pavement). The usefulness of mastic-type pavements is restricted to a limited temperature range, below which they become brittle and above which they may flow. The effectiveness of the mineral aggregate of asphalt concrete depends on careful grading and adequate compaction, which on steel decks sometimes is difficult to obtain. Asphalt properties may be improved by admixtures of highly adhesive or ductile chemicals of various plastics families.

("Design Manual for Orthotropic Steel Plate Deck Bridges," American Institute of Steel Construction, Chicago, Ill.; F. S. Merritt, "Structural Steel Designers' Handbook," McGraw-Hill Book Company, New York.)

5-14. Truss Bridges

Trusses are lattices formed of straight members in triangular patterns. Although truss-type construction is applicable to practically every static system, the term is restricted here to beam-type structures: simple spans and continuous and hinged (cantilever) structures. For other applications see Arts. 5-15 and 5-16. For typical single-span bridge truss configurations, see Fig. 1-48. For the stress analysis of bridge trusses, see Arts. 1-45 through 1-51.

Truss bridges require more field labor than comparable plate girders. Also, trusses are more costly to maintain because of the more complicated makeup of members and poor accessibility of the exposed steel surfaces. For these reasons, and as a result of changing esthetic preferences, use of trusses is increasingly restricted to long-span bridges on which the relatively low weight and consequent easier handling of the individual members are decisive advantages.

The superstructure of a typical truss bridge is composed of two main trusses, the floor system, a top lateral system, a bottom lateral system, cross frames, and bearing assemblies.

Decks for highway truss bridges are usually concrete slabs on steel framing. On long-span railway bridges, the tracks are sometimes mounted directly on steel stringers, although continuity of the track ballast across the deck is usually preferred. Orthotropic decks are rarely used on truss bridges.

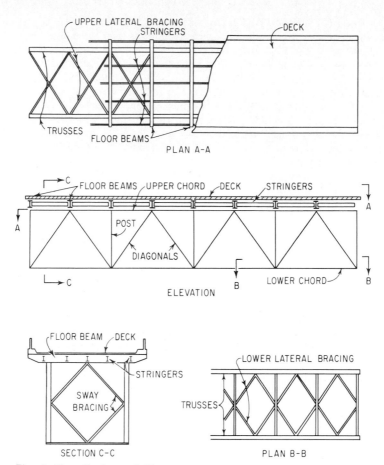

Fig. 5-16.   Deck truss bridge.

Most truss bridges have the deck located between the main trusses, with the floor beams framed into the truss posts. As an alternative, the deck framing may be stacked on top of the top chord. *Deck trusses* have the deck at or above top-chord level (Fig. 5-16); *through trusses,* near the bottom chord (Fig. 5-17). Through trusses whose depth is insufficient for the installation of a top lateral system are referred to as *half through trusses* or *pony trusses.*

Figure 5-17 illustrates a typical cantilever truss bridge. The main span comprises a suspended span and two cantilever arms. The side, or anchor, arms counterbalance the cantilever arms.

Sections of truss members are selected to insure effective use of material, simple details for connections, and accessibility in fabrication, erection, and maintenance. Preferably, they should be symmetrical.

In bolted design, the members are formed of channels or angles and plates, which are combined into open or half-open sections. Open sides are braced by lacing bars, stay plates, or perforated cover plates. Welded truss members are formed of plates. Figure 5-18 shows typical truss-member sections.

For slenderness restrictions of truss members, see Art. 5-7. For permissible stresses, see Table 5-9. The design strength of tensile members is controlled by their net section, that is,

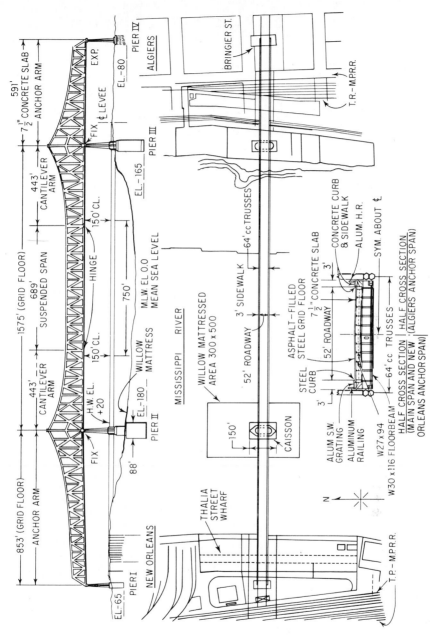

Fig. 5-17. Typical cantilever truss bridge.

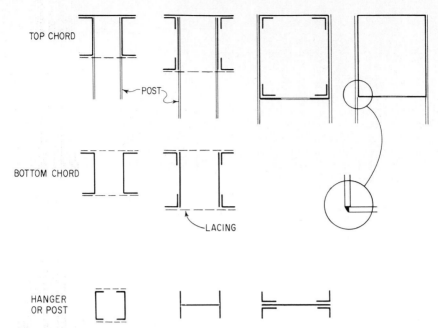

Fig. 5-18. Typical sections used in steel bridge trusses.

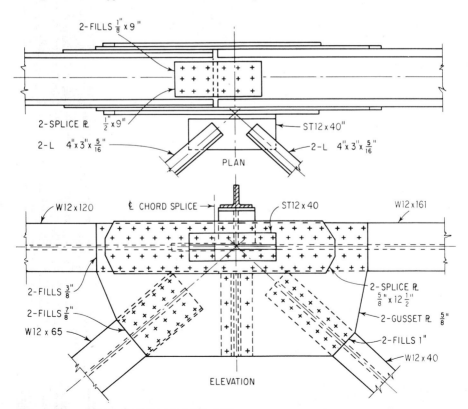

Fig. 5-19. Upper-chord joint in bridge truss.

by the section area that remains after deduction of rivet or bolt holes. In shop-welded field-bolted construction, it is sometimes economical to build up tensile members by butt-welding three sections of different thickness or steel grades. Thicker plates or higher-strength steel is used for the end sections to compensate for the section loss at the holes.

The permissible stress of compression members depends on the slenderness ratio (see Art. 2-8). Design specifications also impose restrictions on the width-to-thickness ratios of webs and cover plates to prevent local buckling.

The magnitude of stress variation is restricted for members subject to stress reversal during passage of a moving load (Art. 2-16).

All built-up members must be stiffened by diaphragms in strategic locations to secure their squareness. Accessibility of all members and connections for fabrication and maintenance should be a primary design consideration.

Whenever possible, each web member should be fabricated in one piece reaching from the top to the bottom chord. The shop length of chord members may extend over several panels. Chord splices should be located near joints and may be incorporated into the gusset plates of a joint.

In most trusses, members are joined by bolting or welding with gusset plates. Pin connections, which were used frequently in earlier truss bridges, are now the exception. As a rule, the center lines or center-of-gravity lines of all members converging at a joint intersect in a single point (Figs. 5-19 and 5-20).

Stresses in truss members and connections are divided into primary and secondary stresses. Primary stresses are the axial stresses in the members of an idealized truss, all of whose joints are made with frictionless pins and all of whose loads are applied at pin centers (Arts. 1-45 to 1-51). Secondary stresses are the stresses resulting from the incorrectness of these assumptions. Somewhat higher stresses are allowed when secondary stresses are considered. (Some specifications require computation of the flexural stresses in compression members caused by their own weight as primary stresses.) Under ordinary conditions, secondary stresses must be computed only for members whose depth is more than one-tenth of their length.

(F. S. Merritt, "Structural Steel Designers' Handbook," McGraw-Hill Book Company, New York.)

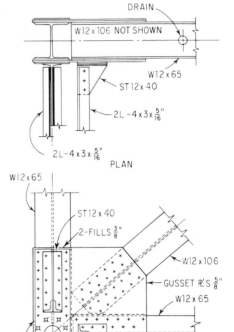

Fig. 5-20.   Bridge-truss lower-chord pin joint at support.

5-15.   Suspension Bridges

These are presently the exclusive bridge type for spans over 1800 ft, and they compete with other systems on shorter spans.

The basic structural system consists of flexible main cables (or, occasionally, eyebar chains) and, suspended from them, stiffening girders or trusses (collectively referred to as "stiffening beams"), which carry the deck framing. The vehicular traffic lanes are as a rule accommodated between the main supporting systems. Sidewalks may lie between the main systems or cantilever out on both sides.

Stiffening Beams ▪ Stiffening beams distribute concentrated loads, reduce local deflections, act as chords for the lateral system, and secure the aerodynamic stability of the structure. Spacing of the stiffening beams is controlled by the roadway width but is seldom less than $\frac{1}{50}$ the span.

Stiffening beams may be either plate girders, box girders, or trusses, the last being preferred because of their smaller air resistance. On major bridges, their depth is at least $\frac{1}{180}$ of the main span. Stiffening trusses of variable depth have been built. Part of the top chord may be formed by the main supporting eyebar chain. Panel lengths may be equal to, twice, or one-half the floor-beam spacing, so the truss diagonals will be close to 45°.

Anchorages ▪ The main cables are anchored in massive concrete blocks or, where rock subgrade is capable of resisting cable tension, in concrete-filled tunnels. Or the main cables are connected to the ends of the stiffening girders, which then are subjected to longitudinal compression equal to the horizontal component of the cable tension.

Continuity ▪ Single-span suspension bridges are rare in engineering projects. They may occur in crossings of narrow gorges where the rock on both sides provides a reliable foundation for high-level cable anchorages.

The overwhelming majority of suspension bridges have main cables draped over two towers. Such bridges consist, thus, of a main span and two side spans. Preferred ratios of side span to main span are 1:4 to 1:2. Ratios of cable sag to main span are preferably in the range of 1:9 to 1:11, seldom less than 1:12.

If the side spans are short enough, the main cables may drop directly from the tower tops to the anchorages, in which case the deck is carried to the abutments on independent, single-span plate girders or trusses. Otherwise, the suspension system is extended over both side spans to the next piers. There, the cables are deflected to the anchorages. The first system allows the designer some latitude in alignment, for example, curved roadways. The second requires straight side spans, in line with the main span. It is the common system for all those suspension bridges that are links in a chain of multiple-span crossings.

When side spans are not suspended, the stiffening beam is of course restricted to the main span. When side spans are suspended, the stiffening beams of the three spans may be continuous or discontinuous at the towers. Continuity of stiffening beams is required in self-anchored suspension bridges, where the cable ends are anchored to the stiffening beams.

Cable Systems ▪ The suspenders between main cables and stiffening beams are usually equally spaced and vertical. Sometimes, for greater aerodynamic stability, the suspenders are interwoven with diagonals that originate at the towers. Zigzag suspender systems have also been used.

Main cables, suspenders, and stiffening beams (girders or trusses) are usually arranged in vertical planes, symmetrical with the longitudinal bridge axis. Bridges with inward- or outward-sloping cables and suspenders and with offset stiffening beams are rare.

Three-dimensional stability is provided by top and bottom lateral systems and transverse frames, similar to those in ordinary girder and truss bridges. Rigid roadway decks may take the place of either or both lateral systems.

In the United States, the main cables are usually made up of 6-gage galvanized bridge wire of 220 to 225 ksi ultimate and 82 to maximum 90 ksi working stress. The wires are placed either parallel or in strands and compacted and wrapped with No. 9 wire. In Europe, strands containing elaborately shaped heat-treated cast-steel wires are sometimes used. Strands must be prestretched. They have a lower and less reliable modulus of elasticity than parallel wires. The heaviest cables, those of the Golden Gate Bridge, are about 36 in in diameter. Twin cables are used if larger sections are required.

Suspenders may be eyebars, rods, single steel ropes, or pairs of ropes slung over the main cable. Connections to the main cable are made with cable bands. These are cast steel whose inner faces are molded to fit the main cable. The bands are clamped together with high-strength bolts.

Floor System ▪ In the design of the floor system, reduction of dead load and resistance to vertical air currents should be the governing considerations. The deck is usually lightweight concrete or steel grating partly filled with concrete. Expansion joints should be provided every 100 to 120 ft to prevent mutual interference of deck and main structure. Stringers should be made composite with the deck for greater strength and stiffness. Floor beams may be plate girders or trusses, depending on available clearance. With trusses, wind resistance is less.

As an alternative to conventional floor framing, orthotropic decks may be used advantageously (Art. 5-13).

Towers ▪ The towers may be portal type, multistory, or diagonally braced frames (Fig. 5-21). They may be of cellular construction, made of steel plates and shapes, or steel lattices, or of reinforced concrete. The substructure below the "spray" line is concrete. The base of steel towers is usually fixed, but it may be hinged. (Hinged towers, however, offer some erection

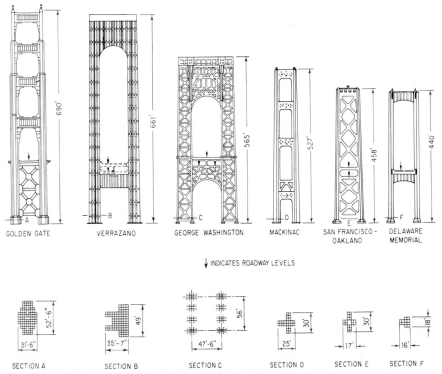

Fig. 5-21.   Types of tower used for large suspension bridges.

difficulties.) The cable saddles at the top of fixed towers are sometimes placed on rollers to reduce the effect on the towers of unbalanced cable deflections. Cable bents can be considered as short towers, either fixed or hinged, whose axis coincides with the bisector of the angle formed by the cable.

Analysis ▪ For gravity loads, the three elements of a suspension bridge in a vertical plane—the main cable or chain, the suspenders, and the stiffening beam—are considered as a single system. The system of discrete suspenders often is idealized as one of continuous suspension.

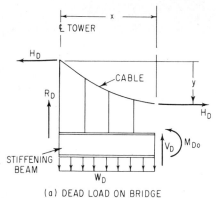

(a) DEAD LOAD ON BRIDGE

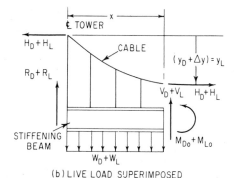

(b) LIVE LOAD SUPERIMPOSED

Fig. 5-22. Stresses in cable and stiffening beam of a suspension bridge.

The stiffening beam is assumed stressless under dead load, a condition approximated by appropriate methods of erection. Moments and shears are produced by that part of the live load not taken up by the main cable through the suspenders. Also, moments and shears result from changes in cable length and sag due to temperature variations or unbalanced loadings of adjacent spans. Deflections of the stiffening beam are strictly elastic; that is, neglecting the effect of shear, the curvature at any section of the elastic line of the loaded beam is proportional to the bending moment divided by the moment of inertia of that section.

The suspenders are subject to tension only. Their elongation under live load is usually neglected in the analysis.

The main cable too is assumed to have no flexural stiffness and to be subject to axial tension only. Its shape is that of a funicular polygon of the applied forces (which include the dead weight of the cable). The pole distance H, lb, which is the horizontal component of the cable tension, is constant for a given loading and a given sag. The shape of the cable under given loads, that is, its ordinate y, ft, and slope $\tan \alpha$ at any point with abscissa x, ft, can be expressed in terms of the moment M_o, ft-kips, and shear V, kips, that a simple beam of the same span L, ft, as the cable would have under the same load (Fig. 5-22).

$$y = \frac{M_o}{H} \qquad \tan \alpha = \frac{V}{H} \qquad (5\text{-}13)$$

In the special case of a uniform load w, kips/lin ft,

$$H = \frac{wL^2}{8f} \qquad (5\text{-}14)$$

$$y = \frac{wx(L - x)}{2H} \qquad \text{or} \qquad y = \frac{4fx(L - x)}{L^2} \qquad (5\text{-}15)$$

where f = cable sag, ft.

The shape of the cable under its own weight without suspended load would be a catenary; under full dead load, the cable shape is usually closer to a parabola. The difference is small.

Concentrated or sectionally uniform live load superimposed on the dead load subjects the cable to additional strain and causes it to adjust its shape to the changed load configuration. The resulting deformations are not exactly proportional to the additional loading; their magnitude is influenced by the already existing dead-load stresses.

If M_o is the bending moment of the stiffening beam under the applied load but without cooperation of the cable, the beam moment M with cooperation of the cable will be

$$M = M_o - Hy \qquad (5\text{-}16)$$

More specifically, using subscripts D and L, respectively, for dead and live load and considering that

$$y_L = y_D + \Delta y \qquad (5\text{-}17)$$

one gets the following expression for the dead- plus live-load bending moment of the beam (see Fig. 5-22b):

$$M = M_D + M_L = M_{D0} + M_{L0} - (H_D + H_L)(y_D + \Delta y)$$

But, since $M_D = M_{D0} - H_D y_D = 0$, because the stiffening beam has no bending moment under dead load (ideally),

$$M = M_{L0} - (H_D + H_L)\,\Delta y - H_L y_D \qquad (5\text{-}18)$$

This is the basic equation of the cable-beam system.

In this equation, M_{L0}, H_D, and y_D are given. H_L and Δy must be so determined that the conditions of static equilibrium of all forces and geometric compatibility of all deformations are satisfied throughout the system.

The mathematically exact solution of the problem is known as the deflection theory. A less exact, older theory is known as the elastic theory. Besides these, there are several approximate methods based on observed regularities in the behavior of suspension bridges, which are sufficiently accurate to serve for preliminary design.

Wind Resistance ▪ Wind acting on the main cables and on part of the suspenders is carried to the towers by the cables. Wind acting on the deck, stiffening beams, and live load is resisted mainly by the lateral bracing system and slightly by the cables because of the gravity component resulting from any elastic lateral deflection of the main supporting system.

Oscillations of the structure may be generated by live load, earthquake, or wind. Live-load vibrations are insignificant in major bridges. Earthquake load seldom governs the design (N. C. Raab and H. C. Wood, "Earthquake Stresses in the San Francisco–Oakland Bay Bridge," *Transactions of the American Society of Civil Engineers,* vol. 106, 1941). Oscillations due to wind, however, can become dangerous if excessive amplitudes build up; that is, if the exciting impulses approach the natural frequency of the structure. Oscillating wind forces are caused by eddies, which may be generated outside the structure or by the structure itself, especially on the lee side of large plates. Oscillations of the structure may be purely flexural, purely torsional, or coupled (flutter), the last two being the more dangerous.

Methods used to predict the aerodynamic behavior of suspension bridges include:

Mathematical analysis of the natural frequency of the structure in flexure and torsion [F. Bleich, C. B. McCullogh, R. Rosecrans, and G. S. Vincent, "Mathematical Theory of Vibration in Suspension Bridges," Government Printing Office, Washington, D.C.: A. G. Pugsley, "Theory of Suspension Bridges," Edward Arnold (Publishers) Ltd., London].

Wind-tunnel tests on scale models of the entire structure or of typical sections ("Aerodynamic Stability of Suspension Bridges with Special Reference to the Tacoma Narrows Bridge," *University of Washington Engineering Experiment Station Bulletin* 116).

Application of Steinman's criteria (these are controversial) (D. B. Steinman, "Rigidity and Aerodynamic Stability of Suspension Bridges," with discussion, *Transactions of the American Society of Civil Engineers,* vol. 110, 1945).

To prevent large or annoying oscillations, reduce air resistance, use trusses instead of girders for stiffening beams, and provide openings in the deck (for example, use a steel grid). Also, increase the rigidity of the structure, make the stiffening beams continuous past the towers, and use diagonal stays. ("Aerodynamic Stability of Suspension Bridges," 1952 Report of Advisory Board, *Transactions of the American Society of Civil Engineers,* vol. 120, 1955).

Tower Stresses ▪ The towers must resist the forces imposed on them by the main cables in addition to the gravity and wind loads acting directly.

The following forces must be considered: The vertical components of the main cables in main and side spans under dead load, live load, temperature change, and wind; wind acting on the main cables, both parallel and transverse to the bridge axis; reactions to longitudinal cable movements due to unbalanced loading. These reactions will develop unless the movements are taken up by hinges or friction-free roller nests. Theoretically, the magnitude of these movements will be affected by the flexural resistance Q of the towers, but this effect, being comparatively small, is usually neglected.

Movement of the tower top generates bending moments. These increase from the top to the bottom at the rate of

$$M_x = Vy + Qx \tag{5-19}$$

where V = vertical cable reaction
x = distance below top
y = horizontal deflection at x
Q = horizontal resistance at top

The magnitude of Q is such that the total deflection equals the longitudinal cable movement. It is found by solving the differential equation for the elastic curve of the tower axis. Thus,

$$y = A \sin cx + B \cos cx - \frac{Q}{V}x = \frac{Q}{V}\left(\frac{\sin cx}{c \cos cL} - x\right) \tag{5-20}$$

in which $c = \sqrt{V/EI}$, I = moment of inertia, and E = modulus of elasticity of tower, if the towers have constant cross sections. The bending moment at x is

$$M_x = \frac{Q}{c}\frac{\sin cx}{\cos cL} \tag{5-21}$$

where L = height of tower.

If, as is usual, the tower cross section varies in several steps, the coefficients A and B in Eq. (5-20) differ from section to section. They are found from the continuity conditions at each step.

Anchorages and footings should be designed for adequate safety against uplift, tipping, and sliding under any possible combination of acting forces.

(S. Hardesty and H. E. Wessman, "Preliminary Design of Suspension Bridges," *Transactions of the American Society of Civil Engineers,* vol. 104, 1939; R. J. Atkinson and R. V. Southard, "On the Problem of Stiffened Suspension Bridges and Its Treatment by Relaxation Methods," *Proceedings of the Institute of Civil Engineers,* 1939; C. D. Crosthwaite, "The Corrected Theory of the Stiffened Suspension Bridge," *Proceedings of the Institute of Civil Engineers,* 1946; Ling-Hi Tsien, "A Simplified Method of Analyzing Suspension Bridges," *Transactions of the American Society of Civil Engineers,* vol. 114, 1947; F. S. Merritt, "Structural Steel Designers' Handbook," McGraw-Hill Book Company, New York.)

5-16. Steel Arch Bridges

A typical arch bridge consists of two or (rarely) more parallel arches or series of arches, plus necessary lateral bracing and end bearings, and columns or hangers for supporting the deck framing. Types of arches correspond roughly to positions of the deck relative to the arch ribs.

Bridges with decks above the arches and clear space underneath (Fig. 5-23*a*) are designed as open spandrel arches on thrust-resisting abutments. Given enough underclearance and adequate foundations, this type is usually the most economical. Often, it is competitive in cost with other bridge systems.

Bridges with decks near the level of the arch bearings (Fig. 5-23*b*) are usually designed as tied arches; that is, tie bars take the arch thrust. End bearings and abutments are similar to those for girder or truss bridges. Tied arches compete in cost with through trusses in locations where underclearances are restricted. Arches sometimes are preferred for esthetic reasons. Unsightly overhead laterals can be avoided by using arches with sufficiently high moment of inertia to resist buckling.

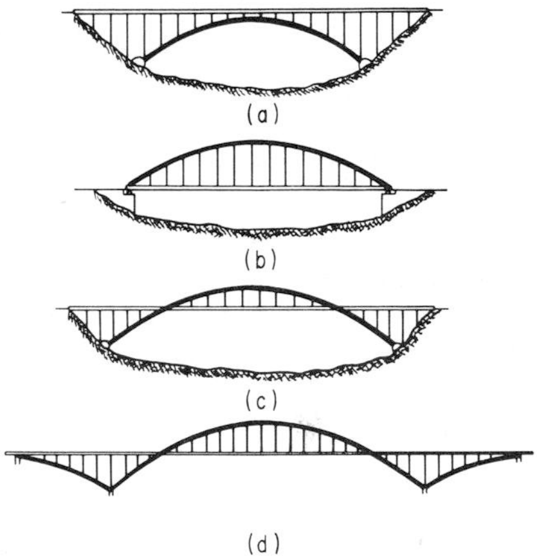

Bridges with decks at an intermediate level

Fig. 5-23. Basic types of steel arch bridges. (*a*) Open spandrel arch; (*b*) tied arch; (*c*) arch with deck at intermediate level; (*d*) multiple-arch bridge.

(Fig. 5-23*c*) may be tied, may rest on thrust-resisting abutments, or may be combined structurally with side spans that alleviate the thrust of the main span on the main piers (Fig. 5-23*d*). Intermediate deck positions are used for long, high-rising spans on low piers.

Spans of multiple-arch bridges are usually structurally separated at the piers. But such bridges may also be designed as continuous structures.

Hinges ▪ Whether or not hinges are required for arch bridges depends on foundation conditions. Abutment movements may sharply increase rib stresses. Fully restrained arches are more sensitive to small abutment movements (and temperature variations) than hinged arches. Flat arches are more sensitive than high arches. If foundations are not fully reliable, hinged bearings should be used.

Complete independence from small abutment movements is achieved by installing a third hinge, usually at the crown. This hinge may be either permanent or temporary during erection, to be locked after all dead-load deformations have been accounted for.

Arch Analysis ▪ The elementary analysis of steel arches is based on the elastic, or first-order, theory, which assumes that the geometric shape of the center line remains constant, irrespective of the imposed load. This assumption is never mathematically correct. The effects of deviations caused by overall flattening of the arch due to the elastic rib shortening, elastic or inelastic displacements of the abutment, and local deformation due to live-load concentrations increase with initial flatness of the arch. An effort is usually made to eliminate the dead-load part of the effect of rib shortening and abutment yielding during erection by jacking the legs of an arch toward each other or the crown section apart before final closure. Arches subject to substantial deformation must be checked by the second-order, or deflection, theory.

For heavy moving loads, it is sometimes advantageous to assign the flexural resistance of the system to special stiffening girders or trusses, analogous to those of suspension bridges (Art. 5-15). The arches themselves are then subject, essentially, to axial stresses only and can be designed as slender as buckling considerations permit.

Arch Design ▪ In general, steel arches must be designed for combined stresses due to axial loads and bending.

The height-to-span ratio used for steel arches varies within wide limits. Minimum values are around 1:10 for tied arches, 1:16 for open spandrel arches.

In cross section, steel arches may be I-shaped, box-shaped, or tubular. Or they may be designed as space trusses.

Deck Construction ▪ The roadway deck of steel arch bridges is usually of reinforced concrete, often of lightweight concrete, on a framing of steel floor beams and stringers. To avoid undesirable cooperation with the primary steel structure, concrete decks either are provided with appropriately spaced expansion joints or prestressed. Orthotropic decks that combine the functions of traffic deck, tie bar, stiffening girder, and lateral diaphragm have been used on some major arch bridges.

(F. S. Merritt, "Structural Steel Designers' Handbook," McGraw-Hill Book Company, New York.)

5-17. Cable-Stayed Bridges*

The cable-stayed bridge, also called the stayed-girder (or truss), has come into wide use since about 1950 for medium- and long-span bridges because of its economy, stiffness, esthetic qualities, and ease of erection without falsework. Design of such bridges utilizes taut cables connecting pylons to a span to provide intermediate support. This principle has been understood by bridge engineers for at least the last two centuries, as indicated by the bridge in Fig. 5-24. The

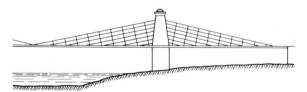

Fig. 5-24.   Cable-stayed chain bridge (Hatley system, 1840).

Roeblings used cable stays as supplementary stiffening elements in the famous Brooklyn Bridge (1883). Many recently built and proposed suspension bridges also incorporate taut cable stays when dynamic (railroad) and long-span effects have to be contended with, such as in the Salazar Bridge.

*Extracted with permission from F. S. Merritt, "Structural Steel Designers' Handbook," McGraw-Hill Book Company, New York.

Characteristics of Cable-Stayed Bridges ■ The cable-stayed bridge offers a proper and economical solution for bridge spans intermediate between those suited for deck girders (usually up to 600 to 800 ft but requiring extreme depths, up to 33 ft) and the longer-span suspension bridges (over 1000 ft). The cable-stayed bridge thus finds application in the general range of 600- to 1600-ft spans but may be competitive in cost for spans as long as 2600 ft.

A cable-stayed bridge has the advantage of greater stiffness over a suspension bridge. Use of single or multiple box girders gains large torsional and lateral rigidity. These factors make the structure stable against wind and aerodynamic effects.

The true action of a cable-stayed bridge (Fig. 5-25) is considerably different from that of a suspension bridge. As contrasted with the relatively flexible cable of the latter, the inclined, taut cables of the cable-stayed structure furnish relatively stable point supports in the main span. Deflections are thus reduced. The structure, in effect, becomes a continuous girder over the piers, with additional intermediate, elastic (yet relatively stiff) supports in the span. As a result, the girder may be shallow. Depths usually range from $\frac{1}{60}$ to $\frac{1}{80}$ the main span, sometimes even as small as $\frac{1}{100}$ the span.

Cable forces are usually balanced between the main and flanking spans, and the structure is internally anchored; that is, it requires no massive masonry anchorages. Analogous to the self-anchored suspension bridge, second-order effects of the type requiring analysis by a deflection theory are of relatively minor importance. Thus, static analysis is simpler, and the structural behavior may be more clearly understood.

The above remarks apply to the common, self-anchored type of cable-stayed bridges, characterized by compression in the main bridge girders (Fig. 5-25a). It is possible to conceive of the opposite extreme of a fully anchored (earth-anchored) cable bridge in which the main girders are in tension. This could be achieved by pinning the girders to the abutments and providing sliding joints in the side-span girders adjacent to the pylons (Fig. 5-25b). The fully anchored system is stiffer than the self-anchored system and may be advantageously analyzed

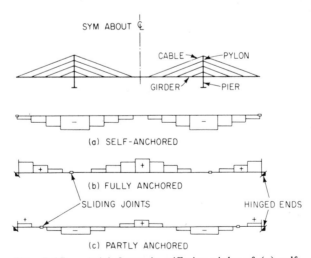

Fig. 5-25. Axial forces in stiffening girder of (*a*) self-anchored, (*b*) fully anchored, (*c*) partly anchored cable-stayed bridges. *(From N. J. Gimsing, "Anchored and Partially Anchored Stayed Bridges," International Symposium on Suspension Bridges, Laboratorio Nacional de Engenharia Civil, Lisbon, 1966.)*

by second-order deflection theory because (analogous to suspension bridges) bending moments are reduced by the deformations.

A further increase in stiffness of the fully anchored system is possible by providing piers in the side spans at the cable attachments (Fig. 5-26). This is advantageous if the side spans are not used for boat traffic below, and if, as is often the case, the side spans cross over low water or land (Kniebrücke at Düsseldorf, Fig. 5-29i).

A partly anchored cable-stayed system (Fig. 5-25c) has been proposed wherein some of the cables are self-anchored and some fully anchored. The axial forces in the girders are then partly compression and partly tension, but their magnitudes are considerably reduced.

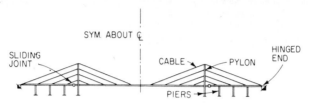

Fig. 5-26.   Anchorage of side-span cables at piers and abutments increases stiffness of center span.

Classification of Cable-Stayed Bridges ▪ The relatively small diameter of the cables and the absolute minimum amount of overhead structure required are the principal features contributing to the excellent architectural appearance of cable-stayed bridges. The functional character of the structural design produces, as a by-product, a graceful and elegant solution for a bridge crossing. This is encouraged by the wide variety of possible types, using single or multiple cables, including the bundle, harp, fan, and star configurations, as seen in elevation (Fig. 5-27). These may be symmetrical or asymmetrical.

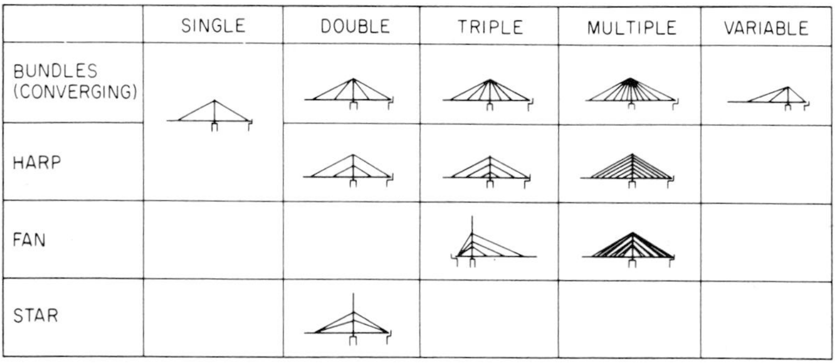

	SINGLE	DOUBLE	TRIPLE	MULTIPLE	VARIABLE
BUNDLES (CONVERGING)					
HARP					
FAN					
STAR					

Fig. 5-27.   Classification of cable-stayed bridges by form. *(From A. Feige, "Evolution of German Cable-Stayed Bridges: An Over-All Survey," Acier-Stahl-Steel, vol. 12, 1966.)*

A wide latitude of choice of cross section of the bridge at the pylons is also possible (Fig. 5-28). The most significant distinction occurs between those with twin pylons (individual, portal, or A frame) and those with single pylons in the center of the roadway. The single pylons usually require a large box girder to resist the torsion of eccentric loadings, and the box is most frequently of steel with an anisotropic steel deck. The single-pylon type is advantageous in allowing a clear unobstructed view from cars passing over the bridge. The pylons may (as with

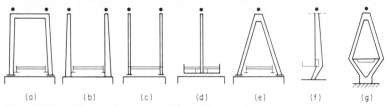

Fig. 5-28. Shapes of towers used for cable-stayed bridges. (*a*) Portal-frame type with top cross member; (*b*) tower fixed to pier and without top cross member; (*c*) tower fixed to superstructure and without top cross member; (*d*) axial tower fixed to the superstructure; (*e*) A-shaped tower; (*f*) lateral tower fixed to the pier; (*g*) diamond-shaped tower. *(From A. Feige, "Evolution of German Cable-Stayed Bridges: An Over-All Survey," Acier-Stahl-Steel, vol. 12, 1966.)*

suspension-bridge towers) be either fixed or pinned at their bases. In the case of fixity, this may be either with the girders or directly with the pier.

Some details of cable-stayed bridges are shown in the elevations and cross sections in Fig. 5-29.

Cable-Stayed Bridge Analysis ▪ The static behavior of a cable-stayed girder can best be gaged from the simple, two-span example of Fig. 5-30. The girder is supported by one stay cable in each span, at *E* and *F*, and the pylon is fixed to the girder at the center support *B*. The static system has two internal cable redundants and one external support redundant.

If the cable and pylon were infinitely rigid, the structure would behave as a continuous four-span beam *AC* on five rigid supports *A*, *E*, *B*, *F*, and *C*. The cables are elastic, however, and correspond to springs. The pylon also is elastic but much stiffer because of its large cross section. If cable stiffness is reduced to zero, the girder assumes the shape of a deflected two-span beam *ABC*.

Cable-stayed bridges of the nineteenth century differed from those of the 1960s in that their tendons constituted relatively soft spring supports. Heavy and long, the tendons could not be stressed highly. Usually, the cables were installed with significant slack or sag. Consequently, large deflections occurred under live load as the sag was diminished. Modern cables have high-strength steel, are relatively short and taut, and their weight is low. Their elastic action may therefore be considered linear, and an equivalent modulus of elasticity may be used. The action of such cables then produces something more nearly like the four-span beam for a structure like the one in Fig. 5-30.

If the pylon were hinged at its base connection with the girder at *B*, the pylon would act as a pendulum column. This would have an important effect on the stiffness of the system, for the spring support at *E* would become more flexible. In magnitude, the effect might exceed that due to the elastic stretch of the cables. In contrast, the elastic shortening of the tower has no appreciable effect.

Relative girder stiffness plays a dominant role in the structural action. The girder tends to approach a beam on rigid supports *A*, *E*, *B*, *F*, *C* as girder stiffness decreases toward zero. With increasing girder stiffness, however, the action of the cables becomes minor and the bridge approaches a girder supported on its piers and abutments *A*, *B*, *C*.

In a three-span bridge, a side-span cable connected to the abutment furnishes more rigid support to the main span than does a cable attached to some point in the side span. In Fig. 5-30, for example, the support of the load *P* in the position shown would be improved if the cable attachment at *F* were shifted to *C*. This explains why cables from the pylon top to the abutment are structurally more efficient, although not as esthetically pleasing as other arrangements.

The stiffness of the system is also affected by whether the cables are fixed at the towers (at

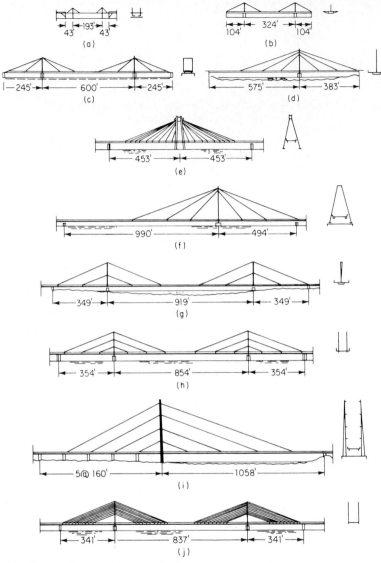

Fig. 5-29. Some cable-stayed bridges, with cross sections taken at towers. (*a*) Büchenauer crossing at Bruchsal, 1956; (*b*) Jülicherstrasse crossing at Düsseldorf, 1964; (*c*) bridge over the Strömsund, Sweden, 1955; (*d*) bridge over the Rhine near Maxau, 1966; (*e*) bridge on the elevated highway at Ludwigshafen, 1969; (*f*) Severin Bridge, Cologne, 1959; (*g*) bridge over the Rhine near Levenkusen, 1965; (*h*) North Bridge at Düsseldorf, 1958; (*i*) Kniebrücke at Düsseldorf, 1969; (*j*) bridge over the Rhine at Rees, 1967.

D, for example, in Fig. 5-30) or whether they run continuously over (or through) the towers. Most designs with more than one cable to a pylon from the main span require one of the cables to be fixed to the pylon and the others to be on movable saddle supports.

The curves of maximum-minimum girder moments for all load variations usually show a large range of stress. Designs providing for the corresponding normal forces in the girder may

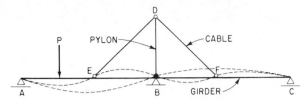

Fig. 5-30. Deflected positions (dash lines) of a cable-stayed bridge.

require large variation in cross sections. By prestressing the cables or by raising or lowering the support points, it is possible to achieve a more uniform and economical moment capacity. The amount of prestressing to use for this purpose may be calculated by successively applying a unit force in each cable and drawing the respective moment diagrams. Then, by trial, the proper multiples of each force are determined so that when their moments are superimposed on the maximum-minimum moment diagrams, an optimum balance results.

Static Analysis—Elastic Theory ▪ Cable-stayed bridges may be analyzed by the general method of indeterminate analysis with the Maxwell-Mohr equations of virtual work.

The degree of internal redundancy of the system depends on the number of cables, types of connections (fixed or movable) of cables with the pylons, and the nature of the pylon connection at its base with the girder or pier. The girder is usually made continuous over three spans. Figure 5-31 shows the order of redundancy for various single-plane systems of cables.

If the bridge has two planes of cables, two girders, and double pylons, it usually also must

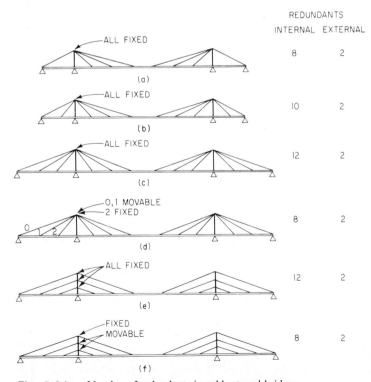

Fig. 5-31. Number of redundants in cable-stayed bridges.

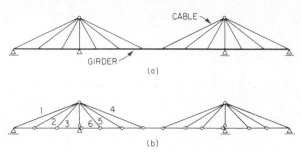

Fig. 5-32.   Cable-stayed bridge. (*a*) Girder actually is continuous; (*b*) insertion of hinges in girder at cable-attachment points makes system statically determinate.

be provided with a number of intermediate cross diaphragms in the floor system, each of which is capable of transmitting moment and shear. The bridge may also have cross girders across the top of the pylons. Each cross member adds two redundants, to which must be added twice the internal redundancy of the single plane structure, and any additional reactions in excess of those needed for external equilibrium as a space structure. The redundancy of the space structure is very high, usually of the order of 40 to 60. Therefore, the methods of plane statics are normally used, except for larger structures.

It is convenient to select as redundants the bending moments in the girder at those points where the cables and pylons join the girder. When these redundants are set equal to zero, an articulated, statically determinate truss base system is obtained (Fig. 5-32*b*). When the loads are applied to this choice of base system, the stresses in the cables do not differ greatly from their final values, so the cables may be dimensioned in a preliminary way.

Other approaches are also possible. One is to use the continuous girder itself as a statically indeterminate base system, with the cable forces as redundants. But computation is generally increased.

A third method involves imposition of hinges, for example at *a* and *b* (Fig. 5-33), so placed as to form two coupled symmetrical base systems, each statically indeterminate to the fourth

Fig. 5-33.   Insertion of hinges at *a* and *b* reduces indeterminacy of cable-stayed girder.

degree. The influence lines for the four indeterminate cable forces of each partial base system are at the same time also the influence lines of the cable forces in the real system. The two redundant moments X_a and X_b are treated as symmetrical and antisymmetrical group loads, $Y = X_a + X_b$, and $Z = X_a - X_b$ to calculate influence lines for the 10-degree-indeterminate structure shown. Kern moments are plotted to determine maximum effects of combined bending and axial forces.

Note that the bundle system in Fig. 5-31*c* and *d* generally has more favorable bending moments for long spans than does the harp system of Fig. 5-31*e* and *f*. Cable stresses also are somewhat lower for the bundle system because the steeper cables are more effective. But the concentration of cable forces at the top of the pylon introduces detailing and construction difficulties. When viewed at an angle, the bundle system presents esthetic problems because of the different intersection angles when the cables are in two planes. Furthermore, fixity of the cables at pylons with the bundle system in Fig. 5-31c and d produces a wider range of stress than does a movable arrangement. This can influence design for fatigue.

The secondary effect of creep of cables can be incorporated into the analysis. The analogy of a beam on elastic supports is changed thereby to a beam on linear viscoelastic supports.

Static Analysis—Deflection Theory ▪ Distortion of the structural geometry of a cable-stayed bridge under action of loads is considerably less than in comparable suspension bridges. The influence on stresses of distortion is relatively small for cable-stayed bridges. In any case, the effect of distortion is to increase stresses, as in arches, rather than the reverse, as in suspension bridges. This effect for the Severin Bridge is 6% for the girder and less than 1% for the cables. Similarly for the Düsseldorf North Bridge, stress increase due to distortion amounts to 12% for the girders.

The calculations, therefore, most expeditiously take the form of a series of successive corrections to results from first-order theory. The magnitude of vertical and horizontal displacements of the girder and pylons can be calculated from the first-order theory results. If the cable stress is assumed constant, the vertical and horizontal cable components V and H change by magnitudes ΔV and ΔH by virtue of the new deformed geometry. The first approximate correction determines the effects of these ΔV and ΔH forces on the deformed system, as well as the effect of V and H due to the changed geometry. This process is repeated until convergence, which is fairly rapid.

Dynamic Analysis—Aerodynamic Stability ▪ The aerodynamic action of cable-stayed bridges is less severe than that of suspension bridges because of increased stiffness due to the taut cables and the widespread use of torsion box decks.

Preliminary Design of Cable-Stayed Bridges ▪ In general, height of a pylon in a cable-stayed bridge is about ⅙ to ⅛ the span. Depth of girder ranges from ⅟₆₀ to ⅟₈₀ the span and is usually 8 to 14 ft, averaging 11 ft.

Wide box girders are mandatory for single-plane systems to resist the torsion of eccentric loads. Box girders, even narrow ones, are also desirable for double-plane systems to enable cable connections to be made without eccentricity. Single-web girders, however, are occasionally used.

To achieve symmetry of cables at pylons the ratio of side to main spans should be about 3:7 where three cables are used on each side of the pylons, and about 2:5 where two cables are used. A proper balance of side-span length to main-span length must be established if uplift at the abutments is to be avoided. Otherwise, movable (pendulum-type) tiedowns must be provided at the abutments.

The usual range of live-load deflections is from ⅟₄₀₀ to ⅟₅₀₀ the span.

Since elastic-theory calculations are relatively simple to program for a computer, a formal set is usually made for preliminary design after the general structure and components have been proportioned.

Design Details for Cable-Stayed Bridges ▪ These structures differ from usual long-span girder bridges in only a few details.

Towers and Floor System. The towers are composed basically of two parts: the pier (below the deck) and the pylon (above the deck). The pylons are frequently of steel box cross section, although concrete may also be used.

Deck Girders. Although cable-stiffened bridges usually incorporate an orthotropic steel deck with steel box girders, to reduce the dead load, other types of construction also are in use. For the Lower Yarra River Bridge in Australia, a concrete deck was specified to avoid site welding and to reduce the amount of shop fabrication. The Maracaibo Bridge likewise incorporates a concrete deck, and the Bridge of the Isles (Canada) has a concrete-slab deck supported on longitudinal and transverse steel box girders and steel floor beams. The Büchenauer Bridge also has a concrete deck. Use of a concrete deck in place of orthotropic-plate construction is largely a matter of local economics. The cost of structure to carry the added dead load should be compared with the lower cost per square foot of the concrete deck and other possible advantages, such as better durability and increased stability against wind.

5-18.   Horizontally Curved Steel Girders

For bridges with curved steel girders, the effects of torsion must be taken into consideration by the designer. Also, careful attention should be given to cross frames—spacing, design, and connection details. The effects of torsion decrease the stresses in the inside girders (those nearer the center of curvature). But there is a corresponding increase in stresses in the outside girders. Although the differences are not large for multiple-girder systems, the differences in stress for two-girder systems with short-radius curves and long spans can be as high as 50%. The torsional forces translate into vertical and horizontal forces, which must be transferred from the outside to inside girders through the cross frames.

An approximate method for analysis of curved girder stresses is in the U.S. Steel "Highway

Fig. 5-34.   Curved girders of Tuolumne River Bridge, California, were erected in pairs with their cross frames between them. *(State of California, Department of Public Works.)*

Structures Design Handbook." This approximate method has proven satisfactory for many structures, but for complex structures (those with long spans, short-radius curves, or with only two girders), it is recommended that a rigorous analysis using a computer program such as STRUDL be used. For the structure in Fig. 5-34, the stress differentials in the two girders are 50% and the cross frames transfer up to 70 kips of vertical and horizontal forces between girders. The center of the main span rotated 4 in when the deck was placed. Such rotations should be anticipated and the girders erected "out of plumb" so that the final web position will be vertical.

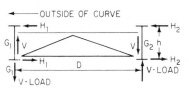

Fig. 5-35. Cross section of curved-girder bridge at cross frame, showing forces resulting from curvature.

Design of curved-girder bridges should consider the following:

1. Full-depth cross frames should be used to transfer the lateral forces from the flanges. (See Fig. 5-35.)

2. The cross frames should be designed as primary stress-carrying members to transfer the loads.

3. Flange-plate width should be increased above the normal design minimums to provide stability during handling and erection.

4. Cross-frame connections at the web plates are critical. The web plate should be thickened to provide bending resistance, as shown in Fig. 5-36.

Approximate Analysis of a Curved, Two-Girder System ▪ The approximate analysis of a curved, two-girder system is illustrated with reference to the framing shown in Figs. 5-35 to 5-37. The system consists of two lines of girders continuous over several spans. Cross frames that brace the girders and provide the primary resistance to torsion are spaced at intervals not more than ⅛ the span length. Since the torsional stiffness of the individual girders is small, it is neglected in the analysis. When computing the moments and shears produced by applied loads, the girders are assumed to be straight with spans equal to their developed lengths.

The girders are curved, however, and there are radial components F_r of the resultant forces F_n in the girders. Since the resultant forces in the upper and lower portions of the girder are in opposite directions (one compression and one tension), the radial components also are in opposite directions. The resulting twisting of the girders is resisted by the cross frames. They provide reactive moments equal and opposite to the sum of the torsional moments over a distance along the girder equal to the cross-frame spacing. Between the cross frames, the torsional moments are resisted internally by the girders.

The cross frames, which span between adjacent curved girders, connect the side of one girder nearer the center of curvature to the side of the other girder farther from the center. Hence, the torsional moments resisted by and at each end of the cross frames are in the same rotational direction. To balance these moments and provide equilibrium, vertical shears also exist at each end of the cross frames. These shears increase bending moments and shears on the outside girder and decrease moments and shears on the inside girder (Fig. 5-35).

Final stresses in a girder are the sum of the following:

1. Stresses due to the applied loads on the assumed equivalent straight girder of length equal to the developed length of the curved girder.

2. Stresses due to the reactive vertical shears at each end of the cross frames.

3. Stresses due to the torsional moments in the girders between the cross frames.

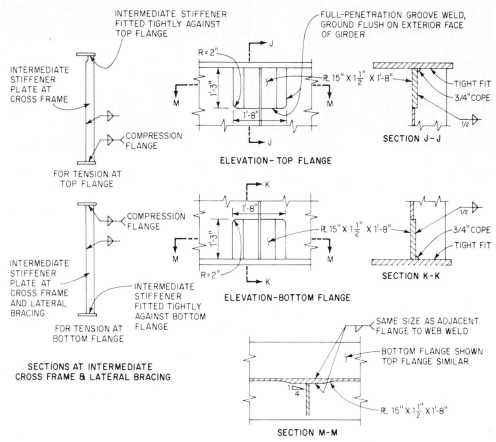

Fig. 5-36.   Connection of a cross frame and lateral bracing at curved-girder tension flanges.

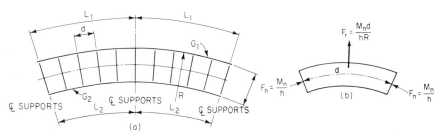

Fig. 5-37.   (*a*) Plan view of a bridge with two curved girders. (*b*) Plan view of a segment of the bridge included between two cross frames with spacing *d*.

Development of Basic Equations ▪ The stresses due to applied loading are determined on the basis of the assumption that the girders are straight and have spans equal to their developed lengths. The formula for determining shear loads V on the cross frames is derived as follows:

Bending produces at point n in outer girder G_1 (Fig. 5-37*a*) flange forces, kips, given, as indicated in Fig. 5-37*b*, approximately by

$$F_n = \frac{M_1}{h} \tag{5-22}$$

where M_1 = final moment, ft-kips, at any point n on G_1

$\qquad h$ = depth, ft, of girder between centers of gravity of flanges

For a unit length of flange, the radial component of F_n is M_1/hR_1, where R_1 is the radius of curvature of G_1, ft. Hence, for a length of flange d, ft, between cross frames, measured along the axis of G_1,

$$F_n = \frac{M_1 d}{hR_1} \tag{5-23}$$

As shown in Fig. 5-37b, the radial components of the flange forces act in opposite directions on top and bottom flanges, as indicated by H_1 and H_2 in Fig. 5-35. For the outer girder G_1,

$$H_1 = \frac{M_1 d}{hR_1} \tag{5-24a}$$

For the final moment M_2 in the inner girder G_2,

$$H_2 = \frac{M_2 d}{hR_2} \tag{5-24b}$$

These forces comprise couples $H_1 h$ and $H_2 h$, respectively, which are resisted by vertical shears V acting on the girders at the cross frame (Fig. 5-35), with moment arm D, the distance, ft, between the girders.

Summation of the moments on the cross frame gives

$$VD = (H_1 + H_2)h \tag{5-25}$$

Substitution of H_1 and H_2 from Eq. (5-24) in Eq. (5-25) and using R_2 as an approximation for R_1 gives for the shear, kips,

$$V = \left(\frac{M_1}{R_1} + \frac{M_2}{R_2} \right) \frac{d}{D} \approx \frac{M_1 + M_2}{R_2 D/d} \approx \frac{M_1 + M_2}{K} \tag{5-26}$$

where $K = R_2 D/d$.

The final moment in the outside girder G_1 is given by

$$M_1 = M_1' + M_{v1} \tag{5-27}$$

where M_1' = moment, ft-kips, at n in G_1 for specific loading applied to equivalent girder (straight with span equal to developed length of G_1)

$\qquad M_{v1}$ = bending moment at n in G_1 due to shears V (V loads), which act downward on G_1

Similarly, the final moment in the inside girder G_2 is given by

$$M_2 = M_2' + M_{v2} \tag{5-28}$$

Because the magnitudes of M_{v1} and M_{v2} are proportional to the girder spans L_1 and L_2, and M_{v2} and M_{v1} act in opposite directions,

$$M_{v2} = - \frac{M_{v1} L_2}{L_1} \tag{5-29}$$

Substitution in Eq. (5-28) yields

$$M_2 = M_2' - \frac{M_{v1} L_2}{L_1} \tag{5-30}$$

Then, from Eqs. (5-26), (5-27), and (5-30),

$$VK = M_1 + M_2 = M_1' + M_2' + M_{v1}\left(1 - \frac{L_2}{L_1}\right) \tag{5-31}$$

But M_{v1} is small compared with $M_1' + M_2'$, and $1 - L_2/L_1$ also is small. Therefore, the last term in Eq. (5-31) has little effect on the value of V. Hence, V is given closely by

$$V = \frac{M_1' + M_2'}{K} \tag{5-32}$$

With the V loads determined for a particular loading condition, bending moments and resulting flange stresses can be readily calculated by application of the V loads to the equivlent straight girders.

For lateral bending, the flange acts as a continuous beam supported by the cross frames. Horizontal reactions at a cross frame are nearly equal to H as given by Eq. (5-24). An approximation of the maximum lateral bending moment that can occur at any cross frame is

$$M_L = \frac{Hd}{10} \tag{5-33}$$

For the outside girder, therefore, from Eq. (5-24),

$$M_L = \left(\frac{M_1 d}{hR_1}\right)\left(\frac{d}{10}\right) = \frac{M_1 d^2}{10hR_1} \tag{5-34}$$

The moment M_1 to be used in Eq. (5-34) should include both the moment due to applied loads and the secondary moment M_{v1} due to the V loads.

Approximate Analysis of a Curved, Three-Girder System ▪ Analysis of a curved three-girder system, or of any curved multiple arrangement beyond two girders, is complicated by another structural action, independent of the curvature. The diaphragms connecting multiple girders (beyond two) tend to equalize deflections of all girders under any condition of loading, true whether the girders are straight or curved. The degree of deflection equalization, however, depends on the relative stiffness of the girders and cross frames, the spacing of cross frames, the stiffness of connections, and other similar factors. Accordingly, any approximate method of analysis for three or more girders framed with curved alignment involves some means of determining not only the effect of curvature but, in addition, the effect of the cross-frame equalizing action. The latter effect is the more difficult to approximate accurately.

Procedures may be simplified by applying to curved girders the rules in the AASHTO Specifications for multiple straight-girder systems. With these rules, moments and shears are determined as for straight girders but with spans equal to the developed lengths of the curved girders. The effect of curvature is then added to obtain the total moments and shears. This effect in a three-girder system may be found by a process similar to that developed for the two-girder arrangement.

Cross-frame shears between the outer girder and center girder, and between the center girder and inner girder, are assumed equal. It may then be shown, by a derivation similar to that used for two girders, that

$$V = \frac{M_1 + M_2 + M_3}{K} \tag{5-35}$$

where M_1, M_2, and M_3 are moments due to the applied loads on the outer girder, center girder, and inner girder, respectively. Since the cross-frame shears on either side of the center girder are opposite in direction and are assumed equal, the curvature effect on the center girder is zero for the V loads. However, the effect of lateral bending of the flanges supported at the cross frames should still be taken into account.

Extension of Approximate Analysis for Three-Girder System to Multiple-Girder Systems
Approximate analysis for determining stresses in a curved, multiple-girder system involves the following two considerations:

1. Distribution of loading to the system depends on the relative stiffness of longitudinal girders and connecting cross frames. Other elements, such as ridigity of connections, also have some effect. This grid action is present whether the girders are straight or curved.
2. Curvature of the girders affects distribution of loading. In general, loading on girders outside the center line of the system is increased and loading on girders inside the center line is decreased.

CONCRETE BRIDGES

Reinforced concrete is used extensively in highway bridges because of its economy in short and medium spans, durability, low maintenance costs, and easy adaptability to horizontal and vertical curvature. The principal types of cast-in-place supporting elements are the longitudinally reinforced slab, T beam or girder, and cellular or box girder. Precast construction, usually prestressed, often employs an I-beam or box-girder cross section. In long-span construction, posttensioned box girders often are used.

5-19. Slab Bridges

Concrete slab bridges, longitudinally reinforced, may be simply supported on piers or abutments, monolithic with wall supports, or continuous over supports.

Design Span ▪ For simple spans, the span is the distance center to center of supports but need not exceed the clear span plus slab thickness. For slabs monolithic with walls (without haunches), use the clear span. For slabs on steel or timber stringers, use the clear span plus half the stringer width.

Load Distribution ▪ In design, usually a 1-ft-wide longitudinal, typical strip is selected and its thickness and reinforcing determined for the appropriate HS loading. Wheel loads may be assumed distributed over a width, ft,

$$E = 4 + 0.06S \leq 7 \tag{5-36}$$

where S = span, ft. Lane loads should be distributed over a width of $2E$.

For simple spans, the maximum live-load moment, ft-kips, per foot width of slab, without impact, for HS20 loading is closely approximated by

$$M = 0.9S \qquad S \leq 50 \text{ ft} \tag{5-37a}$$
$$M = 1.30S - 20 \qquad 50 > S < 100 \tag{5-37b}$$

For HS15 loading, use three-quarters of the value given by Eqs. (5-37).

For longitudinally reinforced cantilever slabs, wheel loads should be distributed over a width, ft,

$$E = 0.35X + 3.2 \leq 7 \text{ ft} \tag{5-38}$$

where X = distance from load to point of support, ft.

The moment, ft-kips per foot width of slab, is

$$M = \frac{P}{E}X \qquad (5\text{-}39)$$

where $P = 16$ kips for H20 loading and 12 kips for H15.

Reinforcement ▪ Slabs should also be reinforced transversely to distribute the live loads laterally. The amount should be at least the following percentage of the main reinforcing steel required for positive moment: $100/\sqrt{S}$, but it need not exceed 50%.

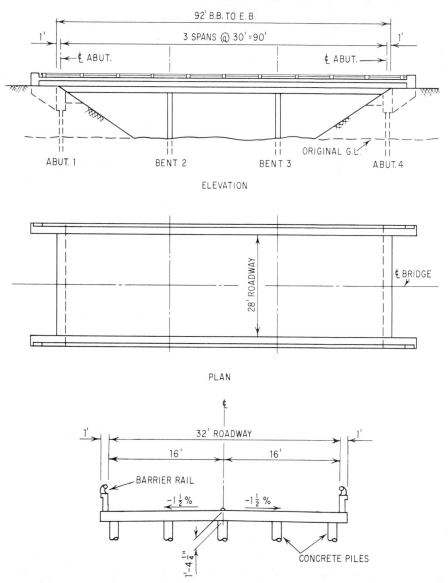

Fig. 5-38.   Three-span, concrete-slab bridge.

The slab should be strengthened at all unsupported edges. In the longitudinal direction, strengthening may consist of a slab section additionally reinforced, a beam integral with and deeper than the slab, or an integral reinforced section of slab and curb. These should be designed to resist a live-load moment, ft-kips, of $1.6S$ for H20 loading and $1.2S$ for H15 loading on simply supported spans. Values for continuous spans may be reduced 20%. Greater reductions are permissible if justified by more exact analysis.

At bridge ends and intermediate points where continuity of the slabs is broken, the edges should be supported by diaphragms or other suitable means. The diaphragms should be designed to resist the full moment and shear produced by wheel loads that can pass over them.

Design Procedure ▪ The following procedure may be used for design of a typical longitudinally reinforced concrete slab bridge (Fig. 5-38).

Step 1. Determine the live-load distribution (effective width). For the three-span, 90-ft-long bridge in Fig. 5-38, $S = 30$ ft and

$$E = 4 + 0.06 \times 30 = 5.8 \text{ ft}$$

The distributed load for a 4-kip front wheel then is 4/5.8, or 0.69 kips, and for a 16-kip rear or trailer wheel load, 16/5.8, or 2.76 kips, per foot of slab width. For an alternative 12-kip wheel load, the distributed load is 12/5.8, or 2.07 kips per foot of slab width (see Fig. 5-39).

Step 2. Determine moment-distribution constants.

Step 3. Make up a table of distributed unit moments.

Step 4. Distribute the uniform-load fixed-end moment coefficients.

Step 5. Assume a slab depth.

Step 6. Distribute dead-load moments for the assumed slab depth.

Step 7. Determine live-load moment at point of maximum moment. (This is done at this stage to get a check on the assumed slab depth.)

Step 8. Combine dead-load, live-load, and impact moments at point of maximum moment. Compare the required slab depth with the assumed depth.

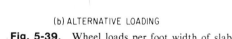

(a) HS 20-44 TRUCK

(b) ALTERNATIVE LOADING

Fig. 5-39. Wheel loads per foot width of slab for bridge of Fig. 5-38.

Step 9. Adjust the slab depth, if necessary.

If the required depth differs from the assumed depth of step 5, the dead-load moments should be revised and step 8 repeated. Usually, the second assumption is sufficient to yield the proper slab depth. Steps 2 through 9 follow conventional structural theory.

Step 10. Place live loads for maximum moments at other points on the structure to obtain intermediate values for drawing envelope curves of maximum moment.

Step 11. Draw the envelope curves (Fig. 5-40). Determine the sizes and points of cutoff for reinforcing bars.

Step 12. Determine distribution steel. Figure 5-41 shows the reinforcing-bar layout.

Step 13. Determine the number of piles required at each bent.

Table 5-12 lists slab thicknesses and reinforcing requirements for typical three-span slab bridges designed for HS20 loading. In Fig. 5-41 and Table 5-12, an A span is an end span with the same length as interior span B; a D end span has three-fourths the length of B, which gives better moment balance. Data for B spans also can be used for interior spans of slabs continuous over more than three spans, while the data for A and D can be used for the end spans.

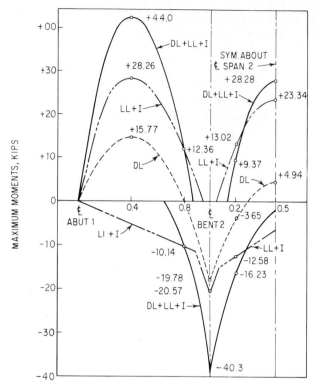

Fig. 5-40.   Maximum moments per foot width of slab in bridge of Fig. 5-38 for dead load plus live load plus impact.

("Bridge Planning and Design Manual," Bridge Department, Division of Highways, State of California Department of Public Works.)

5-20.   T-Beam Bridges

Widely used in highway construction, this type of bridge consists of a concrete slab supported on, and integral with, girders. It is especially economical in the 50- to 80-ft range. Where falsework is prohibited, because of traffic conditions or clearance limitations, precast construction of reinforced or prestressed concrete may be used. But adequate bond and shear resistance must be provided at the junction of slab and girders to justify the assumption that they are integral.

Since the girders parallel traffic, main reinforcing in the slab is perpendicular to traffic. For simply supported slabs, the span should be the distance center to center of supports but need not exceed the clear distance plus thickness of slabs. For slabs continuous over more than two girders, the clear distance may be taken as the span.

Bending Moments ▪ The live-load moment, ft-kips, for HS20 loading on simply supported slab spans is given by

$$M = 0.5(S + 2) \qquad (5\text{-}40)$$

where S = span, ft.

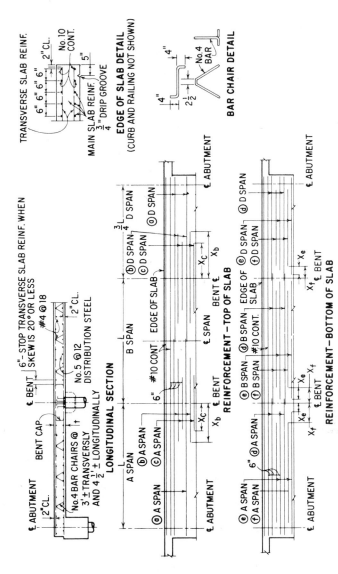

Fig. 5-41. Reinforcing bar layout for typical three-span concrete bridge.

TABLE 5-12 Slab Thicknesses and Reinforcing for Continuous Slab Bridges*

Length of span, ft	16			18			20			22			24		
Type of span	A	B	D	A	B	D	A	B	D	A	B	D	A	B	D
Reinforcement:†															
Top of slab															
a bars															
Size	No. 7		No. 6	No. 7		No. 6	No. 7		No. 6	No. 7		No. 6	No. 8		No. 6
Length	27'0"		22'0"	30'0"		25'0"	33'0"		27'0"	36'0"		30'0"	39'0"		32'0"
b bars															
Size	No. 6		No. 6	No. 7		No. 6	No. 7		No. 6	No. 7		No. 7	No. 8		No. 7
Length	12'0"		12'0"	12'0"		12'0"	13'0"		12'0"	15'0"		12'0"	14'0"		13'0"
X_b	6'0"		6'0"	6'0"		6'0"	6'6"		6'0"	7'6"		6'0"	7'0"		6'6"
c bars															
Size	No. 6		No. 6	No. 6		No. 6	No. 7		No. 6	No. 8		No. 6	No. 7		No. 7
Length	6'0"		6'0"	6'0"		6'0"	7'0"		6'0"	8'0"		7'0"	10'0"		7'0"
X_c	3'0"		3'0"	3'0"		3'0"	3'6"		3'0"	4'0"		3'6"	5'0"		3'6"
Bottom of slab															
d bars															
Size	No. 8	No. 6	No. 6	No. 8	No. 6	No. 6	No. 8	No. 7	No. 7	No. 9	No. 6	No. 7	No. 9	No. 6	No. 7
Length	20'0"	18'0"	14'0"	22'0"	20'0"	16'0"	24'0"	22'0"	17'0"	27'0"	24'0"	19'0"	29'0"	26'0"	20'0"
e bars															
Size	No. 6	No. 6	No. 6	No. 8	No. 6	No. 6	No. 9	No. 7	No. 6	No. 9	No. 8	No. 7	No. 10	No. 8	No. 7
Length	14'0"	13'0"	11'0"	15'0"	15'0"	12'0"	18'0"	16'0"	11'0"	18'0"	17'0"	14'0"	21'0"	18'0"	15'0"
X_e	1'6"	1'6"	1'0"	2'6"	1'6"	1'6"	2'0"	2'0"	3'6"	4'0"	2'6"	2'6"	3'0"	3'0"	3'0"
f bars															
Size	No. 6	No. 6	No. 6	No. 7	No. 7	No. 6	No. 7	No. 7	No. 6	No. 7	No. 7	No. 6	No. 7	No. 6	No. 6
Length	10'0"	8'0"	8'0"	11'0"	11'0"	9'0"	11'0"	13'0"	9'0"	10'0"	14'0"	9'0"	12'0"	16'0"	10'0"
X_f	4'0"	3'0"	3'0"	4'6"	3'6"	3'6"	5'6"	3'6"	4'6"	8'0"	4'0"	5'6"	8'0"	4'0"	6'0"
Quantities per ft of slab width:															
Concrete, ft³	14.9	14.1	11.4	18.2	17.3	15.1	21.0	20.0	16.0	24.0	23.0	18.2	28.1	27.0	21.4
Steel, lb	135	115	85	155	135	100	180	160	115	200	130	125	260	215	140
Deflection at midspan Δ, ft:															
ABA spans							0.02'			0.03'			0.03'		
DBD spans							Negligible deflection						0.02		
t = thickness of slab, in *t*	10½"			11½"			12"			12½"			13½"		

*For HS20 loading on any width roadway. *A* span is end span with same length as interior span *B* (see Fig. 5-41). *D* spans are three-fourths the length of span indicated in the heading. Design stresses: $f'_c = 0.4f'_c = 1.3$ ksi; Grade 60 reinforcing bars, $f_s = 24$ ksi.

†Splices in top main bars to be located near center of span.
 Splices in bottom main bars to be located near bent.
 Spacing of all transverse bars is measured along center line of roadway.

Bar Splice Length, In

Skew 0° to 20°: Place all transverse bars parallel to bent.
Skew over 20°: Place transverse slab bars perpendicular to center line of bridge.
‡Add 1/2 in for "marine environment" and adjust concrete quantity.

TABLE 5-12 (Continued)

Bar Size No.	4	5	6	7	8	9	10	11
All bars, except top bars in spans over 26 ft	18	23	27	32	36	41	51	57
Top bars in spans over 26 ft	26	32	38	45	51	58	72	80

Length of span, ft	26			28			30			32			34		
Type of span	A	B	D	A	B	D	A	B	D	A	B	D	A	B	D
Reinforcement:‡															
Top of slab															
a bars															
Size	No. 7		No. 7	No. 9		No. 7	No. 10		No. 7	No. 10		No. 7	No. 10		No. 8
Length	42'0"		35'0"	46'0"		37'0"	49'0"		40'0"	52'0"		42'0"	55'0"		45'0"
b bars															
Size	No. 9		No. 7	No. 8		No. 7	No. 8		No. 7	No. 9		No. 8	No. 9		No. 8
Length	17'0"		14'0"	13'0"		14'0"	11'0"		14'0"	13'0"		14'0"	15'0"		14'0"
X_b	8'6"		7'0"	6'6"		7'0"	5'6"		7'0"	6'6"		7'0"	7'6"		7'0"
c bars															
Size	No. 9		No. 7	No. 7		No. 7	No. 7		No. 8	No. 7		No. 8	No. 9		No. 8
Length	7'0"		7'0"	8'0"		7'0"	10'0"		8'0"	9'0"		8'0"	10'0"		7'0"
X_c	3'6"		3'6"	4'0"		3'6"	5'0"		4'0"	4'6"		4'0"	5'6"		3'6"
Bottom of slab															
d bars															
Size	No. 9	No. 7	No. 7	No. 9	No. 7	No. 7	No. 9	No. 7	No. 7	No. 9	No. 7	No. 7	No. 10	No. 8	No. 8
Length	31'0"	28'0"	22'0"	33'6"	30'0"	23'0"	35'0"	32'0"	25'0"	37'0"	34'0"	26'0"	39'0"	37'0"	28'0"
e bars															
Size	No. 9	No. 8	No. 7	No. 9	No. 8	No. 7	No. 9	No. 8	No. 7	No. 11	No. 8	No. 7	No. 10	No. 8	No. 7
Length	22'0"	19'0"	16'0"	23'0"	20'0"	17'0"	25'0"	21'0"	18'0"	27'0"	24'0"	19'0"	26'0"	22'0"	19'0"
X_e	4'0"	3'6"	3'6"	4'6"	4'0"	4'0"	4'6"	4'6"	4'6"	4'6"	4'0"	5'0"	7'0"	6'0"	6'0"
f bars															
Size	No. 9	No. 7	No. 7	No. 9	No. 7	No. 7	No. 9	No. 7	No. 7	No. 7	No. 8	No. 7	No. 8	No. 7	No. 7
Length	17'0"	15'0"	13'0"	18'0"	16'0"	13'0"	19'0"	17'0"	14'0"	15'0"	19'0"	15'0"	17'0"	17'0"	15'0"
X_f	7'0"	4'6"	5'0"	8'0"	6'0"	6'0"	8'6"	6'6"	6'0"	11'6"	6'6"	7'0"	11'6"	8'6"	8'0"
Quantities per ft of slab width:															
Concrete, ft³	31.5	30.3	23.9	36.2	35.0	27.5	41.3	40.0	31.4	45.4	44.0	33.0	49.5	48.1	37.5
Steel, lb	290	225	155	310	255	175	340	285	190	370	315	210	400	345	220
Deflection at midspan Δ, ft:															
ABA spans	0.04'	0.02'		0.05'	0.02'		0.06'	0.03'		0.07'	0.03'		0.08'	0.04'	
DBA spans						0.02'			0.01'			0.01'			0.02'
t = thickness of slab, in ‡	14"			15"			16"			16½"			17"		

TABLE 5-12 (*Continued*)

Length of span, ft	36			38			40			42			44		
Type of span	A	B	D	A	B	D	A	B	D	A	B	D	A	B	D
Reinforcement:															
Top of slab															
a bars															
Size	No. 10		No. 8	No. 10		No. 8	No. 10		No. 8	No. 10		No. 8	No. 11		No. 8
Length	58'0"		48'0"	67'0"		50'0"	70'0"		53'0"	73'0"		55'0"	76'0"		58'0"
b bars															
Size	No. 10		No. 8	No. 8		No. 9	No. 10		No. 8	No. 11		No. 8	No. 11		No. 10
Length	17'0"		15'0"	17'0"		16'0"	19'0"		17'0"	19'0"		19'0"	21'0"		21'0"
X_b	8'6"		7'6"	8'6"		8'0"	9'6"		8'6"	9'6"		9'6"	10'6"		10'6"
c bars															
Size	No. 9		No. 9	No. 11		No. 9	No. 10		No. 10	No. 10		No. 11	No. 11		No. 10
Length	8'0"		11'0"	14'0"		9'0"	11'0"		12'0"	10'0"		15'0"	12'0"		11'0"
X_c	4'0"		5'6"	7'0"		4'6"	5'6"		6'0"	5'0"		7'6"	6'0"		5'6"
Bottom of slab															
d bars															
Size	No. 11	No. 8	No. 8	No. 10	No. 8	No. 8	No. 10	No. 8	No. 8	No. 11	No. 9	No. 8	No. 11	No. 10	No. 8
Length	41'0"	39'0"	30'0"	43'0"	41'0"	31'0"	45'0"	43'0"	33'0"	47'0"	46'0"	34'0"	49'0"	48'0"	36'0"
e bars															
Size	No. 11	No. 8	No. 7	No. 11	No. 8	No. 8	No. 10	No. 9	No. 8	No. 11	No. 8	No. 8	No. 11	No. 8	No. 8
Length	29'0"	26'0"	20'0"	30'0"	27'0"	22'0"	32'0"	27'0"	22'0"	33'0"	28'0"	25'0"	34'0"	26'0"	25'0"
X_e	6'6"	5'0"	6'0"	7'0"	5'6"	6'0"	8'0"	6'6"	5'6"	8'6"	7'0"	6'0"	9'0"	9'0"	7'0"
f bars															
Size	No. 8	No. 8	No. 7	No. 8	No. 8	No. 7	No. 10	No. 8	No. 7	No. 10	No. 8	No. 8	No. 10	No. 7	No. 8
Length	18'0"	19'0"	15'0"	19'0"	20'0"	15'0"	24'0"	22'0"	15'0"	23'0"	20'0"	18'0"	24'0"	17'0"	19'0"
X_f	12'6"	8'6"	9'0"	14'6"	9'0"	9'6"	12'0"	9'0"	10'6"	13'6"	11'0"	10'0"	14'0"	13'6"	10'6"
Quantities per ft of slab width:															
Concrete, ft³	54.0	52.5	40.8	58.5	57.0	44.3	64.9	63.3	49.0	69.8	68.2	52.8	75.0	73.3	56.7
Steel, lb	430	375	250	480	415	280	550	435	290	590	485	320	635	535	340
Deflection at midspan Δ, ft:															
ABA spans	0.10'	0.05'		0.11'	0.01'		0.13'	0.01'		0.15'	0.01'		0.16'	0.01'	
DBD spans			0.02'		0.05'	0.02'		0.06'	0.03'		0.07'	0.03'		0.08'	0.03'

t = thickness of slab, in †

	36	38	40	42
	17½"	18"	19"	19½"

For slabs continuous over three or more supports, multiply M in Eq. (5-40) by 0.8 for both positive and negative moment. For HS15 loading, multiply M by ¾.

Reinforcement also should be placed in the slab parallel to traffic to distribute concentrated live loads. The amount should be the following percentage of the main reinforcing steel required for positive moment: $220/\sqrt{S}$, but need not exceed 67%.

Where a slab cantilevers over a girder, the wheel load should be distributed over a distance, ft, parallel to the girder of

$$E = 0.8X + 3.75 \qquad (5\text{-}41)$$

where X = distance, ft, from load to point of support. The moment, fit-kips per foot of slab parallel to girder, is

$$M = \frac{P}{E} X \qquad (5\text{-}42)$$

where P = 16 kips for H20 loading and 12 kips for H15.

Equations (5-40) to (5-42) apply also to concrete slabs supported on steel girders, including composite construction.

Slab Design ▪ In design of the slabs, a 1-ft-wide strip is selected and its thickness and reinforcing determined. The dead-load moments, ft-kips, positive and negative, can be assumed to be $wS^2/10$, where w is the dead load, kips/ft^2. Live-load moments are given by Eq. (5-40) with a 20% reduction for continuity. Impact is a maximum of 30%. With these values, standard charts can be developed for design of slabs on steel and concrete girders. An example is illustrated by Table 5-13 and Fig. 5-42. A typical slab-reinforcement layout is shown in Fig. 5-43.

T-Beam Design ▪ The structure shown in Fig. 5-44 is a typical four-span grade-separation structure. The structural frame assumed for analysis is shown in Fig. 5-45. The hinged conditions at the base of the columns is probably close to the condition existing in most structures of this type. Completely fixed column bases are difficult to attain because of footing rotation. In addition, economy in footing design results from the uniform footing pressures under a hinged column base.

Ratios of beam depths to spans used in continuous T-beam bridges generally range from 0.065 to 0.075. An economical depth usually results when a small amount of compressive reinforcement is required at the interior supports.

Girder spacing ranges from about 7 to 9 ft. Usually, a deck slab overhang of about 2 ft 6 in is economical.

When the slab is made integral with the girder, its effective width in design may not exceed distance center to center of girders, one-fourth the girder span, or 12 times the least thickness of slab, plus girder web width. For exterior girders, however, effective overhang width may not exceed half the clear distance to the next girder, one-twelfth the girder span, or six times the slab thickness.

For concrete girder design, curves of maximum moments for dead load plus live load plus impact may be developed to determine reinforcement. For live-load moments, truck loadings are moved across the bridge. As they move, they generate changing moments, shears, and reactions. It is necessary to accumulate maximum combinations of moments to provide an adequate design. For heavy moving loads, extensive investigation is necessary to find the maximum stresses in continuous structures.

Figure 5-46 shows curves of maximum moments consisting of dead load plus live load plus impact combinations that are maximum along the span. From these curves, reinforcing steel amounts and lengths may be determined by plotting the moments developed by bars. Maximum-shear requirements are derived theoretically by a point-to-point study of variations. Usu-

TABLE 5-13 Design Data for Concrete Slabs on Girders with H20 Loading*

Effective span S	t slab thickness, in	F	A, B, C bars Size No.	A, B, C bars Spacing, in	D bars, No. of No. 5 bars	Effective span S	t slab thickness, in	F	A, B, C bars Size No.	A, B, C bars Spacing, in	D bars, No. of No. 5 bars
4'0"	6	6"	5	13	4	12'0"	8⅝	1'7"	6	11	16
4'3"	6	7"	5	13	4	12'3"	8⅝	1'7"	6	11	16
4'6"	6	7"	5	13	4	12'6"	8¾	1'7"	6	11	16
4'9"	6	7"	5	12	5	12'9"	8⅞	1'8"	6	11	17
5'0"	6	8"	5	12	5	13'0"	8⅞	1'8"	6	10	17
5'3"	6	8"	5	12	5	13'3"	9	1'9"	6	10	17
5'6"	6	9"	5	11	5	13'6"	9	1'9"	6	10	18
5'9"	6	9"	5	11	6	13'9"	9⅛	1'9"	6	10	18
6'0"	6⅛	9"	5	11	6	14'0"	9¼	1'10"	6	10	18
6'3"	6¼	10"	5	11	6	14'3"	9⅜	1'10"	6	10	19
6'6"	6¼	10"	5	11	7	14'6"	9½	1'11"	7	14	19
6'9"	6¼	11"	5	10	7	14'9"	9⅝	1'11"	7	14	20
7'0"	6⅜	11"	5	10	7	15'0"	9¾	1'11"	7	14	20
7'3"	6½	11"	5	10	8	15'3"	9¾	2'0"	7	13	20
7'6"	6⅝	1'0"	5	10	8	15'6"	9¾	2'0"	7	13	21
7'9"	6¾	1'0"	5	10	8	15'9"	9⅞	2'1"	7	13	21
8'0"	7	1'1"	5	10	9	16'0"	10	2'1"	7	13	22
8'3"	7⅜	1'1"	6	13	9	16'3"	10	2'1"	7	13	22
8'6"	7½	1'1"	6	13	10	16'6"	10⅛	2'2"	7	13	22
8'9"	7½	1'2"	6	13	10	16'9"	10¼	2'2"	7	12	23
9'0"	7⅝	1'2"	6	13	11	17'0"	10¼	2'3"	7	12	23
9'3"	7¾	1'2"	6	13	11	17'3"	10⅜	2'3"	7	12	24
9'6"	7¾	1'3"	6	13	11	17'6"	10½	2'3"	7	12	24
9'9"	7⅞	1'3"	6	13	12	17'9"	10½	2'4"	7	12	24
10'0"	8	1'4"	6	12	12	18'0"	10⅝	2'4"	7	12	25
10'3"	8	1'4"	6	12	13	18'3"	10¾	2'4"	7	11	25
10'6"	8⅛	1'4"	6	12	13	18'6"	10¾	2'5"	7	11	26
10'9"	8⅛	1'5"	6	12	14	18'9"	10⅞	2'5"	7	11	26
11'0"	8¼	1'5"	6	12	14	19'0"	11	2'6"	7	11	26
11'3"	8⅜	1'6"	6	12	14	19'3"	11⅛	2'6"	7	11	27
11'6"	8½	1'6"	6	11	15	19'6"	11¼	2'6"	7	11	27
11'9"	8½	1'6"	6	11	15	19'9"	11¼	2'7"	7	11	28
						20'0"	11⅜	2'7"	7	11	28

*Design based on decks having three or more girders. (Add ½ in to slab thickness above 4000-ft elevation.) Design stresses: f_c = 1200 psi; f_s = 20,000 psi; n = 10. Dead-load moment = $wS^2/10$. Impact factor = 30% for all spans. Moment for live load plus impact, with 80% continuity factor, ft-kips = $0.52(S + 2)$.

ally, a straight line between center line and end maximums is adequate. Figure 5-47 shows curves of maximum shears. Figure 5-48 shows the girder reinforcement layout.

Design of intermediate supports or bents varies widely, according to the designer's preference. A simple two-column bent is shown in Fig. 5-44. But considerable shape variations in column cross section and elevation are possible.

Abutments are usually of the L type or a monolithic end diaphragm supported on piles.

("Bridge Planning and Design Manual," Bridge Department, Division of Highways, State of California Department of Public Works.)

5-21. Box-Girder Bridges

Box or hollow concrete girders are favored by many designers because of the smooth plane of the bottom surface, uncluttered by lines of individual girders. Provision of space in the open cells for utilities is both a structural and an esthetic advantage. Utilities are supported by the bottom slab, and access can be made available for inspection and repair of utilities.

For sites where structure depth is not severely limited, box girders and T beams have been

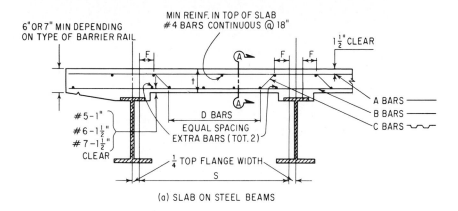

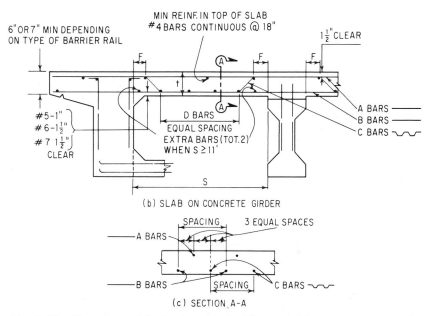

Fig. 5-42. Transverse reinforcing for concrete slabs on steel beams or concrete girders. (See also Table 5-13.) For bridges skewed 20° or less, place the reinforcing parallel to the abutments and bents. Space it along the center line of the structure. For skews greater than 20°, place the reinforcing perpendicular to the girders. Use *A* and *B* bars at acute corners at half the given spacing. Add ½ in to slab thickness above the 4000-ft elevation.

about equal in price in the 80-ft span range. For shorter spans, T beams usually are cheaper, and for longer spans, box girders. These cost relations hold in general, but box girders have, in some instances, been economical for spans as short as 50 ft when structure depth was restricted.

Girder Design ▪ Structural analysis is usually based on two typical segments, interior and exterior girders (Fig. 5-49). An argument could be made for analyzing the entire cross section as a unit because of its inherent transverse stiffness. Requirements in "Standard Specifications for Highway Bridges," American Association of State Highway and Transportation Officials,

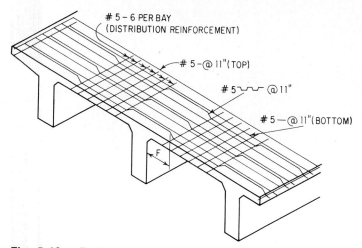

Fig. 5-43. Typical reinforcement layout for slab supported on girders.

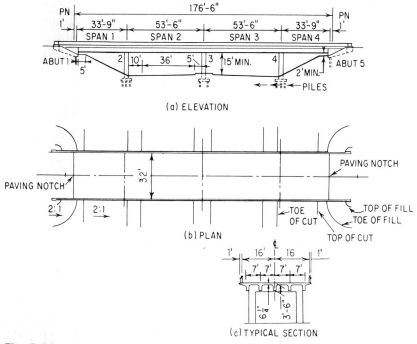

Fig. 5-44. Four span reinforced concrete T beam.

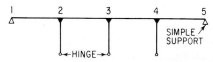

Fig. 5-45. Assumed support conditions for bridge of Fig. 5-44.

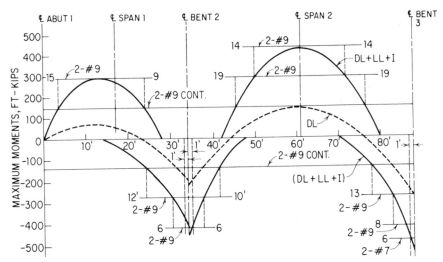

Fig. 5-46. Reinforcing for T beams of Fig. 5-44 is determined from curves of maximum moment. Numbers at ends of bars are distances, ft, from center line of span or bent.

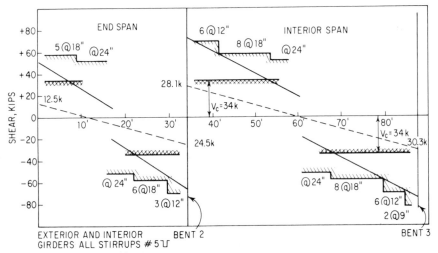

Fig. 5-47. Curves of maximum shear for T beams of Fig. 5-44.

however, are based on live-load distributions for individual girders, and so design usually is based on the assumption that a box-girder bridge is composed of separate girders.

Effective width of slab as top flange of an interior girder may be taken as the smallest of the distance center to center of girders, one-fourth the girder span, and 12 times the least thickness of slab, plus girder web width. Effective overhang width for an exterior girder may be taken as the smallest of half the clear distance to the next girder, one-twelfth the girder span, and six times the least thickness of the slab.

Usual depth-to-span ratio for continuous spans is 0.055. This may be reduced to about 0.048 with balanced spans, at some sacrifice in economy and increase in deflections. Simple spans usually require a minimum depth-to-span ratio of 0.065.

A typical concrete box-girder highway bridge is illustrated in Fig. 5-50. Spacing of webs could be either 7 ft 4 in or 9 ft 4 in. The wider spacing is chosen to eliminate one web. Minimum

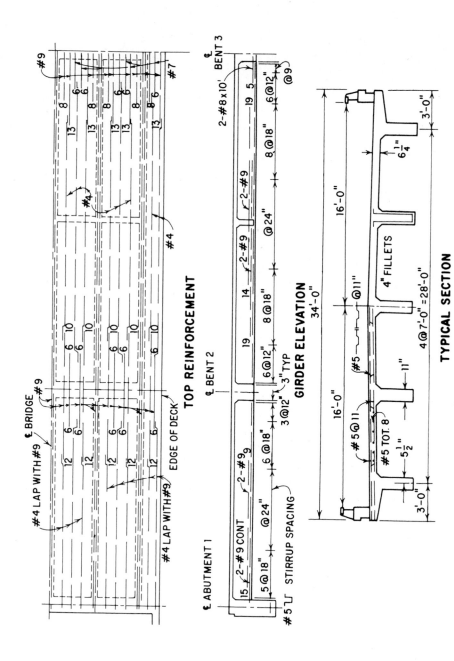

TOP REINFORCEMENT

GIRDER ELEVATION

TYPICAL SECTION

Fig. 5-48. Reinforcement layout for T beams of Fig. 5-44. Reinforcement is symmetrical about center lines of bridge and Bent 3. Numbers at ends of bars indicate distances, ft, from center line of bent or span.

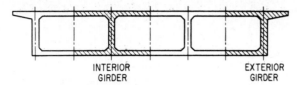

Fig. 5-49. Typical design sections (crosshatched) for box-girder bridges.

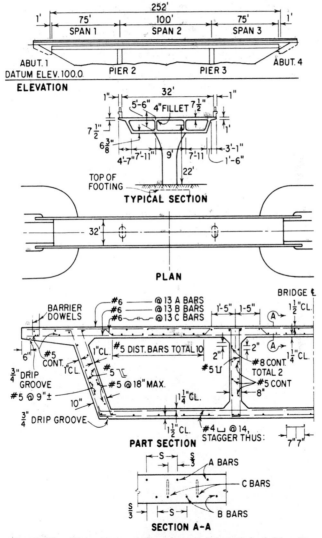

Fig. 5-50. Three-span, reinforced concrete box-girder bridge. For more reinforcing details, see Fig. 5-53.

web thickness is determined by shear but generally is at least 8 in. Changes should be gradual, spread over a distance at least 12 times the difference in web thickness.

Top-slab design follows the procedure described for T-beam bridges in Art. 5-20. Bottom-slab thickness and secondary reinforcement are usually controlled by specification minimums. AASHTO Specifications require that slab thickness be at least one-sixteenth the clear distance between girders but not less than 6 in for the top slab and 5½ in for the bottom slab. Fillets should be provided at the intersections of all surfaces within the cells.

Minimum flange reinforcement parallel to the girder should be 0.6% of the flange area. This steel may be distributed at top and bottom or placed in a single layer at the center of the slab. Spacing should not exceed 18 in. Minimum flange reinforcing normal to the girder should be 0.5% and similarly distributed. Bottom-flange bars should be bent up into the exterior-girder webs at least 10 diameters. The top-flange bars should extend to the exterior face of all exterior girders. At least one-third of these bars should be anchored with 90° bends or, where the flange projects beyond the girder sufficiently, extended far enough to develop bar strength in bond.

When the top slab is placed after the web walls have set, at least 10% of the negative-moment reinforcing should be placed in the web. The bars should extend a distance of at least one-fourth the span each side of the intermediate supports of continuous spans, one-fifth the span from the restrained ends of continuous spans, and the entire length of cantilevers. In any event, the web should have reinforcing placed horizontally in both faces, to prevent temperature and shrinkage cracks. The bars should be spaced not more than 2 ft c to c. Total area of this steel should be at least ⅛ in²/ft of web height.

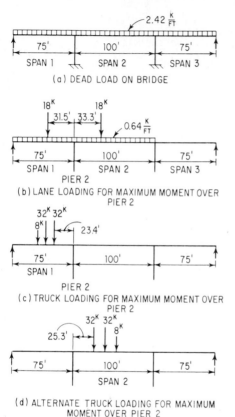

Fig. 5-51. Loading patterns for maximum stresses in a box-girder bridge.

Analysis of the structure in Fig. 5-50 for dead loads follows conventional moment-distribution procedure. Assumed end conditions are shown in Fig. 5-51a.

Live loads, positioned to produce maximum negative moments in the girders over Pier 2, are shown in Fig. 5-51b to d. Similar loadings should be applied to find maximum positive and negative moments at other critical points. Moments should be distributed and points plotted on a maximum-moment diagram (for dead load plus live load plus impact), as shown in Fig. 5-52. Layout of main girder reinforcement follows directly from this diagram. Figure 5-53 shows a typical layout.

("Bridge Planning and Design Manual," Bridge Department, Division of Highways, State of California Department of Public Works.)

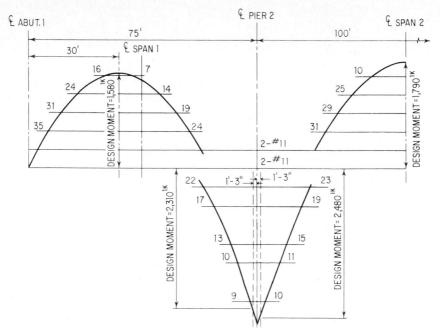

Fig. 5-52. Reinforcing for box girder is determined from curves of maximum moment. Numbers at ends of bars indicate distances, ft, from center line of piers or span.

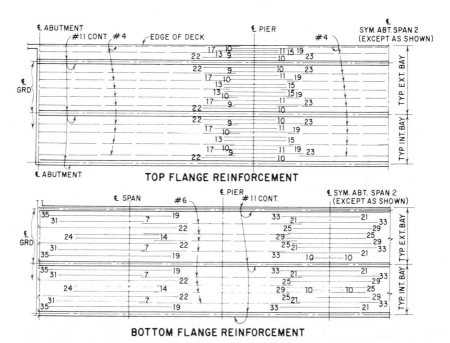

TOP FLANGE REINFORCEMENT

BOTTOM FLANGE REINFORCEMENT

Fig. 5-53. Reinforcing layout for the box-girder bridge of Fig. 5-50. Numbers at ends of bars indicate distances, ft, from center line of pier for top reinforcement and center line of span for bottom bars. Design stresses for HS20 loading: $f_c = 1.3$ ksi, except 1.2 ksi for transverse deck slabs; $f_s = 24$ ksi, except 20 ksi for transverse deck slabs and stirrups.

Geotechnical engineering can be defined as the application of the principles of soil and engineering mechanics to evaluation of the behavior of earth materials, usually in the context of engineering investigation, design, and construction. Applications typically include dams, transportation systems, structural foundations, and site development.

Foundation engineering deals with the support of structures on or within the earth and the associated earth-structure interactions. Design-related engineering activities include formulation of criteria for:

1. Design of foundation systems and related earthworks
2. Earth pressures
3. Permanent groundwater control
4. Subgrade stabilization or improvement techniques

Construction-related activities include:

1. Formulation of provisions and technical specifications for load-bearing fills
2. Temporary earth-retaining structures and excavations
3. Temporary groundwater control
4. Protection of existing facilities from construction-induced damages

Associated quality-assurance or quality-control activities during construction may also be required.

Related site engineering applications are usually associated with the formulation of design and construction criteria for site grades, earthworks, groundwater control, roadways, parking areas, and surface drainage over relatively large development areas. In some instances, this work involves siting facilities to minimize foundation and earthwork costs or risks associated with potential natural or constructed hazards, such as sinkholes, landslides, and subsurface mines.

Geotechnical engineers should have a thorough knowledge of soil and engineering mechanics, subsurface exploration, and laboratory investigation techniques as well as of earthworks and foundation construction. A good understanding of relevant geological and geophysical applications is also essen-

Section 6

Geotechnical Engineering

Frederick S. Merritt
Consulting Engineer
West Palm Beach, FL.

William S. Gardner
Consulting Engineer
Executive Vice President,
Woodward-Clyde Consultants
Plymouth Meeting, PA.

tial. Because the practice of geotechnical engineering is more of an art than a science, the practitioner should have a broad base of practical experience. This requirement was clearly expressed by Karl Terzaghi as "The magnitude of the difference between the performance of real soils under field conditions and the performance predicted on the basis of theory can only be ascertained by field experience."

6-1. Lessons from Construction Claims and Failures

Unanticipated subsurface conditions encountered during construction are by far the largest source of construction-related claims for additional payment by contractors and cost overruns. Failures of structures as a result of foundation deficiencies can entail even greater costs and moreover jeopardize public safety. A large body of experience has identified consistently recurring factors contributing to these occurrences. It is important for the engineer to be aware of the causes of cost overruns, claims, and failures and to use these lessons to help minimize similar future occurrences.

Unanticipated conditions (changed conditions) are the result of a variety of factors. The most frequent cause is the lack of definition of the constituents of rock and soil deposits and their variation throughout the construction site. Related claims are for unanticipated or excessive quantities of soil and rock excavation, misrepresentation of the quality and depth of bearing levels, unsuitable or insufficient on-site borrow materials, and unanticipated obstructions to pile driving or shaft drilling. Misrepresentation of groundwater conditions is another common contributor to work *extras* as well as to costly construction delays and emergency redesigns. Significant claims have also been generated by the failure of geotechnical investigations to identify natural hazards, such as swelling soils and rock minerals, unstable natural and cut slopes, and old fill deposits.

Failures of structures during construction are usually related to undesirable subsurface conditions not detected before or during construction, faulty design, or poor quality of work. Examples are foundations supported on expansive or collapsing soils, on solutioned rock, or over undetected weak or compressible subsoils; foundation designs too difficult to construct properly; foundations that do not perform as anticipated; and deficient construction techniques or materials. Another important design-related cause of failure is underestimation or lack of recognition of extreme loads associated with natural events, such as earthquakes, hurricanes, floods, and prolonged precipitation. Related failures include soil liquefaction during earthquakes, hydrostatic uplift or water damage to structures because of a rise in groundwater level, undermining of foundations by scour and overtopping, or wave erosion of earth dikes and dams.

It is unlikely that conditions leading to construction claims and failures can ever be completely precluded, inasmuch as discontinuities and extreme variation in subsurface conditions occur frequently in many types of soil deposits and rock formations. An equally important constraint that must be appreciated by both engineers and clients is the limitations of the current state of geotechnical engineering practice.

Mitigation of claims and failures, however, can be achieved by fully integrated geotechnical investigation, design, and construction quality assurance conducted by especially qualified professionals. Integration, rather than departmentalization of these services, ensures a continuity of purpose and philosophy that effectively reduces the risks associated with unanticipated subsurface conditions and design and construction deficiencies. It is also extremely important that owners and prime design professionals recognize that cost savings which reduce the quality of geotechnical services may purchase liabilities several orders of magnitude greater than their initial "savings."

SOIL AND ROCK CLASSIFICATION

All soils are initially the products of chemical alteration or mechanical disintegration of bedrock that has been exposed to weathering processes. Soil constituents may have been subsequently modified by transportation processes such as water, wind, and ice and by inclusion and decomposition of organic matter. Consequently, soil deposits may be given a geologic as well as a constitutive classification.

Rock types are broadly classified by their mode of formation into igneous, metamorphic, and sedimentary deposits. The supporting ability (quality) assigned to rock for design or analysis should reflect the degree of alteration of the rock minerals due to weathering, the frequency of discontinuities within the rock mass, and the susceptibility of the rock to deterioration upon exposure.

6-2. Geologic Classification of Soils

The classification of a soil deposit with respect to its mode of deposition and geologic history is an important step in understanding the variation in soil type and the maximum stresses imposed on the deposit since deposition. (A geologic classification that identifies the mode of deposition of soil deposits is shown in Table 6-1.) The geologic history of a soil deposit may also provide valuable information on the rate of deposition, the amount of erosion, and the tectonic forces that may have acted on the deposit subsequent to deposition.

Geological and agronomic soil maps and detailed reports are issued by the U.S. Department of Agriculture, U.S. Geological Survey, and corresponding state offices. Old surveys are useful for locating original shore lines, stream courses, and surface-grade changes.

TABLE 6-1 Geologic Classification of Soil Deposits

Classification	Mode of formation
Aeolian	
Dune	Wind deposition (coastal and desert)
Loess	Deposition during glacial periods
Alluvial	
Alluvium	River and stream deposition
Lacustrine	Lake waters, including glacial lakes
Flood plain	Flood waters
Colluvial	
Colluvium	Downslope soil movement
Talus	Downslope movement of rock debris
Glacial	
Ground moraine	Deposited and consolidated by glaciers
Terminal moraine	Scour and transport at ice front
Outwash	Glacier melt waters
Marine	
Beach or bar	Wave deposition
Estuarine	River estuary deposition
Lagoonal	Deposition in lagoons
Salt marsh	Deposition in sheltered tidal zones
Residual	
Residual soil	Complete alteration by in situ weathering
Saprolite	Incomplete but intense alteration and leaching
Laterite	Complex alteration in tropical environment
Decomposed rock	Advanced alteration within parent rock

6-3. Unified Soil Classification System

This is the most widely used of the various constitutive classification systems and correlates soil type with generalized soil behavior. All soils are classified as coarse-grained (50% of the particles > 0.074 mm), fine-grained (50% of the particles < 0.074 mm), or predominantly organic (see Table 6-2).

Coarse-grained soils are categorized by their particle size into boulders (particles larger than 8 in), cobbles (3 to 8 in), gravel, and sand. For sands (S) and gravels (G), grain-size distribution is identified as either poorly graded (P) or well-graded (W), as indicated by the group symbol in Table 6-2. The presence of fine-grained soil fractions (under 50%), such as silt and clay, is indicated by the symbols M and C, respectively. Sands may also be classified as coarse (larger than No. 10 sieve), medium (smaller than No. 10 but larger than No. 40), or fine (smaller than No. 40). Because properties of these soils are usually significantly influenced by relative density D_r, rating of the in situ density and D_r is an important consideration (see Art. 6-6).

Fine-grained soils are classified by their liquid limit and plasticity index as organic clays OH or silts OL, inorganic clays CH or CL, or silts or sandy silts MH or ML, as shown in Table 6-2. For the silts and organic soils, the symbols H and L denote a high and low potential compressibility rating; for clays, they denote a high and low plasticity. Typically, the consistency of cohesive soils is classified from pocket penetrometer or Torvane tests on soil samples. The consistency ratings are expressed as follows:

Soft—under 0.25 tons/ft^2

Firm—0.25 to 0.50 tons/ft^2

Stiff—0.50 to 1.0 ton/ft^2

Very stiff—1.0 to 2.0 tons/ft^2

Hard—more than 2.0 tons/ft^2

6-4. Rock Classification

Rock, obtained from core samples, is commonly characterized by its type, degree of alteration (weathering), and continuity of the core. (Where outcrop observations are possible, rock structure may be mapped.) Rock-quality classifications are typically based on the results of compressive strength tests or the condition of the core samples, or both. Rock types typical of igneous deposits include basalt, granite, diorite, rhyolite, and andesite. Typical metamorphic rocks include schist, gneiss, quartzite, slate, and marble. Rocks typical of sedimentary deposits include shale, sandstone, conglomerate, and limestone.

Rock structure and degree of fracturing usually control the behavior of a rock mass that has not been significantly altered by weathering processes. It is necessary to characterize both regional and local structural features that may influence design of foundations, excavations, and underground openings in rock. Information from geologic publications and maps are useful for defining regional trends relative to the orientation of bedding, major joint systems, faults, and so on.

Rock-quality indices determined from inspection of rock cores include the fracture frequency *(FF)* and rock-quality designation *(RQD)*. *FF* is the number of naturally occurring fractures per foot of core run, whereas *RQD* is the cumulative length of naturally separated core pieces, 4 in or more in dimension, expressed as a percentage of the length of core run. The rock-quality rating also may be based on the velocity index obtained from laboratory and in situ seismic-wave-propagation tests. The velocity index is given by $(V_s/V_l)^2$, where V_s and V_l represent seismic-wave velocities from in situ and laboratory core measurements, respectively. Proposed *RQD* and velocity index rock-quality classifications and in situ deformability correlations

TABLE 6-2 Unified Soil Classification Including Identification and Description[a]

Major division	Group symbol	Typical name	Field identification procedures[b]	Laboratory classification criteria[c]
A. Coarse-grained soils (more than half of material larger than No. 200 sieve)[d]				
1. Gravels (more than half of coarse fraction larger than No. 4 sieve)[e]				
Clean gravels (little or no fines)	GW	Well-graded gravels, gravel-sand mixtures, little or no fines	Wide range in grain sizes and substantial amounts of all intermediate particle sizes	$D_{60}/D_{10} > 4$ $1 < D_{30}{}^2/D_{10}D_{60} < 3$ $D_{10}, D_{30}, D_{60} = $ sizes corresponding to 10, 30, and 60% on grain-size curve
	GP	Poorly graded gravels or gravel-sand mixtures, little or no fines	Predominantly one size, or a range of sizes with some intermediate sizes missing	Not meeting all gradation requirements for GW
Gravels with fines (appreciable amount of fines)	GM	Silty gravels, gravel-sand-silt mixtures	Nonplastic fines or fines with low plasticity (see ML soils)	Atterberg limits below A line or PI < 4
	GC	Clayey gravels, gravel-sand-clay mixtures	Plastic fines (see CL soils)	Atterberg limits above A line with PI > 7
				Soils above A line with 4 < PI < 7 are borderline cases, require use of dual symbols
2. Sands (more than half of coarse fraction smaller than No. 4 sieve)[e]				
Clean sands (little or no fines)	SW	Well-graded sands, gravelly sands, little or no fines	Wide range in grain sizes and substantial amounts of all intermediate particle sizes	$D_{60}/D_{10} > 6$ $1 < D_{30}{}^2/D_{10}D_{60} < 3$
	SP	Poorly graded sands or gravelly sands, little or no fines	Predominantly one size, or a range of sizes with some intermediate sizes missing	Not meeting all gradation requirements for SW
Sands with fines (appreciable amount of fines)	SM	Silty sands, sand-silt mixtures	Nonplastic fines or fines with low plasticity (see ML soils)	Atterberg limits above A line or PI < 4
	SC	Clayey sands, sand-clay mixtures	Plastic fines (see CL soils)	Atterberg limits above A line with PI > 7
				Soils with Atterberg limits above A line while 4 < PI < 7 are borderline cases; require use of dual symbols

TABLE 6-2 (Continued)

Information required for describing coarse-grained soils:

For undisturbed soils, add information on stratification, degree of compactness, cementation, moisture conditions, and drainage characteristics. Give typical name; indicate approximate percentage of sand and gravel; maximum size, angularity, surface condition, and hardness of the coarse grains; local or geological name and other pertinent descriptive information; and symbol in parentheses. Example: *Silty sand, gravelly; about 20% hard, angular gravel particles, ½ in. maximum size; rounded and subangular sand grains, coarse to fine; about 15% nonplastic fines with low dry strength; well compacted and moist in place; alluvial sand; (SM)*.

Major division	Group symbol	Typical names	Identification procedure[f]			Laboratory classification criteria[c]
			Dry strength (crushing characteristics)	Dilatancy (reaction to shaking)	Toughness (consistency near PL)	

B. Fine-grained soils (more than half of material smaller than No. 200 sieve)[d]

Major division	Group symbol	Typical names	Dry strength (crushing characteristics)	Dilatancy (reaction to shaking)	Toughness (consistency near PL)
Silts and clays with liquid limit less than 50	ML	Inorganic silts and very fine sands, rock flour, silty or clayey fine sands, or clayey silts with slight plasticity	None to slight	Quick to slow	None
	CL	Inorganic clays of low to medium plasticity, gravelly clays, sandy clays, silty clays, lean clays	Medium to high	None to very slow	Medium
	OL	Organic silts and organic silty clays of low plasticity	Slight to medium	Slow	Slight
Silts and clays with liquid limit more than 50	MH	Inorganic silts, micaceous or diatomaceous fine sandy or silty soils, elastic silts	Slight to medium	Slow to none	Slight to medium
	CH	Inorganic clays of high plasticity, fat clays	None to very high	None	High
	OH	Organic clays of medium to high plasticity	Medium to high	None to very slow	Slight to medium

C. Highly organic soils

Major division	Group symbol	Typical names	
	Pt	Peat and other highly organic soils	Readily identified by color, odor, spongy feel, and often by fibrous texture

Plasticity chart for laboratory classification of fine-grained soils compares them at equal liquid limit. Toughness and dry strength increase with increasing plasticity index (PI)

Field identification procedures for fine-grained soils or fractions[g]:

Dilatancy (reaction to shaking)

After removing particles larger than No. 40 sieve, prepare a pat of moist soil with a volume of about ½ cu in. Add enough water if necessary to make the soil soft but not sticky.

Place the pat in the open palm of one hand and shake horizontally, striking vigorously against the other hand several times. A positive reaction consists of the appearance of water on the surface of the pat, which changes to a livery consistency and becomes glossy. When the sample is squeezed between the fingers, the water and gloss disappear from the surface, the pat stiffens, and finally it cracks or crumbles. The rapidity of appearance of water during shaking and of its disappearance during squeezing assist in identifying the character of the fines in a soil.

Very fine clean sands give the quickest and most distinct reaction, whereas a plastic clay has no reaction. Inorganic silts, such as a typical rock flour, show a moderately quick reaction.

Dry strength (crushing characteristics)

After removing particles larger than No. 40 sieve, mold a pat of soil to the consistency of putty, adding water if necessary. Allow the pat to dry completely by oven, sun, or air drying, then test its strength by breaking and crumbling between the fingers. This strength is a measure of character and quantity of the colloidal fraction contained in the soil. The dry strength increases with increasing plasticity.

High dry strength is characteristic of clays of the CH group. A typical inorganic silt possesses only very slight dry strength. Silty fine sands and silts have about the same slight dry strength but can be distinguished by the feel when powdering the dried specimen. Fine sand feels gritty, whereas a typical silt has the smooth feel of flour.

Toughness (consistency near PL)

After particles larger than the No. 40 sieve are removed, a specimen of soil about ½ cu in. in size is molded to the consistency of putty. If it is too dry, water must be added. If it is too sticky, the specimen should be spread out in a thin layer and allowed to lose some moisture by evaporation. Then, the specimen is rolled out by hand on a smooth surface or between the palms into a thread about ⅛ in. in diameter. The thread is then folded and rerolled repeatedly. During this manipulation, the moisture content is gradually reduced and the specimen stiffens, finally loses its plasticity, and crumbles when the plastic limit (PL) is reached.

After the thread crumbles, the pieces should be lumped together and a slight kneading action continued until the lump crumbles.

The tougher the thread near the PL and the stiffer the lump when it finally crumbles, the more potent is the colloidal clay fraction in the soil. Weakness of the thread at the PL and quick loss of coherence of the lump below the PL indicate either organic clay of low plasticity or materials such as kaolin-type clays and organic clays that occur below the A line.

Highly organic clays have a very weak and spongy feel at PL.

TABLE 6-2 (*Continued*)

Information required for describing fine-grained soils:

For undisturbed soils, add information on structure, stratification, consistency in undisturbed and remolded states, moisture, and drainage conditions. Give typical name; indicate degree and character of plasticity; amount and maximum size of coarse grains; color in wet condition; odor, if any; local or geological name and other pertinent descriptive information; and symbol in parentheses. Example: *Clayey silt*, brown; slightly plastic; small percentage of fine sand; numerous vertical root holes; firm and dry in place; loess; (ML).

---

[a] Adapted from recommendations of Corps of Engineers and U.S. Bureau of Reclamation. All sieve sizes United States standard.

[b] Excluding particles larger than 3 in. and basing fractions on estimated weights.

[c] Use grain-size curve in identifying the fractions as given under field identification.

For coarse-grained soils, determine percentage of gravel and sand from grain-size curve. Depending on percentage of fines (fractions smaller than No. 200 sieve), coarse-grained soils are classified as follows:

    Less than 5% fines . . . . GW, GP, SW, SP

    More than 12% fines . . . . GM, GC, SM, SC

    5% to 12% fines . . . . Borderline cases requiring use of dual symbols

Soils possessing characteristics of two groups are designated by combinations of group symbols; for example, GW-GC indicates a well-graded, gravel-sand mixture with clay binder.

[d] The No. 200 sieve size is about the smallest particle visible to the naked eye.

[e] For visual classification, the ¼-in. size may be used as equivalent to the No. 4 sieve size.

[f] Applicable to fractions smaller than No. 40 sieve.

[g] These procedures are to be performed on the minus 40-sieve-size particles (about $1/_{64}$ in.).

For field classification purposes, screening is not intended. Simply remove by hand the coarse particles that interfere with the tests.

TABLE 6-3 Rock-Quality Classification and Deformability Correlation

Classification	RQD	Velocity index	Deformability E_d/E_t*
Very poor	0–25	0–0.20	Under 0.20
Poor	25–50	0.20–0.40	Under 0.20
Fair	50–75	0.40–0.60	0.20–0.50
Good	75–90	0.60–0.80	0.50–0.80
Excellent	90–100	0.80–1.00	0.80–1.00

*E_d = in situ deformation modulus of rock mass; E_t = tangent modulus at 50% of UC strength of core specimens.
Source: Deere, Patton and Cording, "Breakage of Rock," *Proceedings*, 8th Symposium on Rock Mechanics, American Institute of Mining and Metallurgical Engineers, Minneapolis, MN.

are in Table 6-3. A relative-strength rating of the quality of rock cores representative of the intact elements of the rock mass, proposed by Deere and Miller, is based on the uniaxial compressive *(UC)* strength of the core and its tangent modulus at one-half of the *UC*.

(D. U. Deere and R. P. Miller, "Classification and Index Properties for Intact Rock," Technical Report AFWL-TR-65-116, Airforce Special Weapons Center, Kirtland Airforce Base, New Mexico, 1966.)

Inasmuch as some rocks tend to disintegrate rapidly (slake) upon exposure to the atmosphere, the potential for slaking should be rated from laboratory tests. These tests include emersion in water, Los Angeles abrasion, repeated wetting and drying, and other special tests, such as a slaking-durability test. Alteration of rock minerals due to weathering processes is often associated with reduction in rock hardness and increase in porosity and discoloration. In an advanced stage of weathering, the rock may contain soil-like seams, be easily abraded (friable), readily broken, and may (but not necessarily) exhibit a reduced *RQD* or *FF*. Rating of the degree of rock alteration when logging core specimens is a valuable aid in assessing rock quality.

SOIL PROPERTIES AND PARAMETERS

Basic soil properties and parameters can be subdivided into physical, index, and engineering categories. Physical soil properties include density, particle size and distribution, specific gravity, and water content. Index parameters of cohesive soils include liquid limit, plastic limit, shrinkage limits, and activity. Such parameters are useful for classifying cohesive soils and providing correlations with engineering soil properties.

Engineering soil properties and parameters describe the behavior of soil under induced stress and environmental changes. Of interest to most geotechnical applications are the strength, deformability, and permeability of in situ and compacted soils. The American Society for Testing and Materials (ASTM) test procedure designations for the more common soil properties and parameters are in Table 6-4.

6-5. Physical Properties of Soils

The **water content** w of a soil sample represents the weight of free water contained in the sample expressed as a percentage of its dry weight.

The **degree of saturation** S of the sample is the ratio, expressed as percentage, of the volume of free water contained in a sample to its total volume of voids V_v.

Porosity n, which is a measure of the relative amount of voids, is the ratio of void volume to the total volume V of soil:

$$n = \frac{V_v}{V} \tag{6-1}$$

TABLE 6-4 Selected ASTM Standards for Soil and Rock*

D1140–54 (1971)	Amount of Material in Soils Finer Than the No. 200 (75-μm) Sieve
D2166–66 (1979)	Compressive Strength, Unconfined, of Cohesive Soil
D2922–78	Density of Soil and Soil-Aggregate in Place by Nuclear Methods (Shallow Depth)
D2937–71 (1976)	Density of Soil in Place by the Drive-Cylinder Method
D2167–66 (1977)	Density of Soil in Place by the Rubber-Balloon Method
D1556–64 (1974)*e*	Density of Soil in Place by the Sand-Cone Method
D3080–72 (1979)	Direct Shear Test of Soils under Consolidated Drained Conditions
D2573–72 (1978)	Field Vane Shear Test in Cohesive Soil
D3689–78	Individual Piles under Static Axial Tensile Load, Testing
D 422–63 (1972)	Particle-Size Analysis of Soils
D3155–73 (1978)	Lime Content of Uncured Soil-Lime Mixtures
D 423–66 (1972)	Liquid Limit of Soils
D3017–78	Moisture Content of Soil and Soil-Aggregate in Place by Nuclear Methods (Shallow Depth)
D 698–78	Moisture-Density Relations of Soils and Soil-Aggregate Mixtures Using 5.5-lb (2.49-kg) Rammer and 12-in (305-mm) Drop
D1557–78	Moisture-Density Relations of Soils and Soil-Aggregate Mixtures Using 10-lb (4.54-kg) Rammer and 18-in (457-mm) Drop
D2435–81	One-Dimensional Consolidation Properties of Soils
D3877–80	One-Dimensional Expansion, Shrinkage, and Uplift Pressure of Soil-Lime Mixtures
D2434–68 (1974)	Permeability of Granular Soils (Constant Head)
D1143–74	Piles under Axial Compressive Load, Testing
D 424–59 (1971)	Plastic Limit and Plasticity Index of Soils
D2049–69*e*	Relative Density of Cohesionless Soils
D2844–69 (1975)	Resistance R Value and Expansion Pressure of Compacted Soils
D3550–77*e*	Ring-Lined Barrel Sampling of Soils
D 427–61 (1974)	Shrinkage Factors of Soils
D 854–58 (1979)	Specific Gravity of Soils
D2850–70	Unconsolidated, Undrained Strength of Cohesive Soils in Triaxial Compression
D2216–80	Water (Moisture) Content of Soil, Rock, and Soil-Aggregate Mixtures, Laboratory Determination of
D1411–69 (1975)	Water-Soluble Chlorides Present as Admixes in Graded Aggregate Road Mixes
D2487–69 (1975)	Classification of Soils for Engineering Purposes
D3441–79	Deep, Quasi-Static, Cone and Friction-Cone Penetration Tests of Soil
D2113–70 (1976)*e*	Diamond Core Drilling for Site Investigation
D 421–58 (1978)	Dry Preparation of Soil Samples for Particle-Size Analysis and Determination of Soil Constants
D1586–67 (1974)	Penetration Test and Split-Barrel Sampling of Soils
D1587–74	Thin-Walled Tube Sampling of Soils
D2217–66 (1978)	Wet Preparation of Soil Samples for Particle-Size Analysis and Determination of Soil Constants
D2488–69 (1975)	Description of Soils (Visual-Manual Procedure)
D2936–78	Direct Tensile Strength of Intact Rock Core Specimens
D3148–80	Elastic Moduli of Intact Rock Core Specimens in Uniaxial Compression
D2845–69 (1976)	Laboratory Determination of Pulse Velocities and Ultrasonic Elastic Constants of Rock
D2664–80	Triaxial Compressive Strength of Undrained Rock Core Specimens without Pore Pressure Measurements
D2938–79	Unconfined Compressive Strength of Intact Rock Core Specimens

*In the specification designations, the number following the dash indicates the year of original issue or year of last revision. Number in parentheses indicates the year of reapproval without change. Superscript *e* indicates that editorial changes have been made.

The ratio of V_v to the volume occupied by the soil particles V_s defines the **void ratio** e. Given e, the degree of saturation may be computed from

$$S = \frac{wG_s}{e} \tag{6-2}$$

where G_s represents the specific gravity of the soil particles. For most inorganic soils, G_s is usually in the range of 2.67 ± 0.05.

The dry unit weight γ_d of a soil specimen with any degree of saturation may be calculated from

$$\gamma_d = \frac{\gamma_w G_s S}{1 + wG_s} \tag{6-3}$$

where γ_w is the unit weight of water and is usually taken as 62.4 lb/ft³ for fresh water and 64.0 lb/ft³ for seawater.

The particle-size distribution (gradation) of soils can be determined by mechanical (sieve) analysis and combined with hydrometer analysis if the sample contains a significant amount of particles finer than 0.074 mm (No. 200 sieve). The soil particle gradation in combination with the maximum, minimum, and in situ density of cohesionless soils can provide useful correlations with engineering properties (see Arts. 6-6 and 6-8).

6-6. Index Parameters for Soils

The **liquid limit** of cohesive soils represents a near liquid state, that is, an undrained shear strength about 0.01 lb/ft². The water content at which the soil ceases to exhibit plastic behavior is termed the **plastic limit**. The **shrinkage limit** represents the water content at which no further volume change occurs with a reduction in water content. The most useful classification and correlation parameters are the plasticity index I_p, the liquidity index I_l, the shrinkage index I_s, and the activity A_c. These parameters are defined in Table 6-5.

Relative density D_r of cohesionless soils may be expressed in terms of void ratio e or unit dry weight γ_d:

$$D_r = \frac{e_{max} - e_o}{e_{max} - e_{min}} \tag{6-4a}$$

$$D_r = \frac{1/\gamma_{min} - 1/\gamma_d}{1/\gamma_{min} - 1/\gamma_{max}} \tag{6-4b}$$

D_r provides cohesionless soil property and parameter correlations, including friction angle, permeability, compressibility, small-strain shear modulus, cyclic shear strength, and so on. (H.

TABLE 6-5 Soil Indices

Index	Definition*	Correlation
Plasticity	$I_p = W_l - W_p$	Strength, compressibility, compactibility, and so forth
Liquidity	$I_l = \dfrac{W_n - W_p}{I_p}$	Compressibility and stress state
Shrinkage	$I_s = W_p - W_s$	Shrinkage potential
Activity	$A_c = \dfrac{I_p}{\mu}$	Swell potential, and so on

*W_l = liquid limit; W_p = plastic limit; W_n = moisture content, %; W_s = shrinkage limit; μ = percent of soil finer than 0.002 mm (clay size).

B. Seed and I. M. Idriss, "Report no. EERC 70-10," Earthquake Engineering Research Center, University of California, Berkeley, 1970.)

6-7. Shear Strength of Cohesive Soils

The undrained shear strength c_u of cohesive soils under static loading can be determined by several types of laboratory tests, including uniaxial compression, triaxial compression *(TC)* or extension *(TE)*, simple shear, direct shear, and torsion shear. The *TC* test is the most versatile and best understood laboratory test. Triaxial tests involve application of a controlled confining pressure σ_3 and axial stress σ_1 to a soil specimen. σ_3 may be held constant and σ_1 increased to failure (*TC* tests), or σ_1 may be held constant while σ_3 is decreased to failure (*TE* tests). Specimens may be sheared in a drained or undrained condition.

The unconsolidated-undrained *(UU)* triaxial compression test is appropriate and commonly used for determining the c_u of relatively good-quality samples. For soils that do not exhibit changes in soil structure under elevated consolidation pressures, consolidated-undrained *(CU)* tests following the SHANSEP testing approach mitigate the effects of sample disturbance.

(C. C. Ladd and R. Foott, "New Design Procedures for Stability of Soft Clays," *ASCE Journal of Geotechnical Engineering Division,* vol. 99, no. GT7, 1974.)

For cohesive soils exhibiting a normal clay behavior, a relationship between the normalized undrained shear strength c_u/σ'_{vo} and the overconsolidation ratio *OCR* can be defined independently of the water content of the test specimen by

$$\frac{c_u}{\sigma'_{vo}} = K(OCR)^n \tag{6-5}$$

where c_u is normalized by the preshear vertical effective stress, the effective overburden pressure σ'_{vo}, or the consolidation pressure σ'_{lc} triaxial test conditions. *OCR* is the ratio of preconsolidation pressure to overburden pressure. The parameter K represents the c_u/σ'_{vo} of the soil in a normally consolidated state, and n primarily depends on the type of shear test. For *CU* triaxial compression tests, K is approximately 0.32 ± 0.02 and is lowest for low plasticity soils and is a maximum for soils with plasticity index I_p over 40%. The exponent n is usually within the range of 0.70 ± 0.05 and tends to be highest for *OCR* less than about 4.

In situ vane shear tests also are often used to provide c_u measurements in soft to firm clays. Tests are commonly made on both the undisturbed and remolded soil to investigate the **sensitivity,** the ratio of the undisturbed to remolded soil strength. This test is not applicable in sand or silts or where hard inclusions (nodules, shell, gravel, and so forth) may be present. (See also Art. 6-15.)

Drained shear strength of cohesive soils is important in design and control of construction embankments on soft ground as well as in other evaluations involving effective-stress analyses. Conventionally, drained shear strength τ_f is expressed by the Mohr-Coulomb failure criteria as:

$$\tau_f = c' + \sigma'_n \tan \phi' \tag{6-6}$$

The c' and ϕ' parameters represent the effective cohesion and effective friction angle of the soil, respectively. σ'_n is the effective stress normal to the plane of shear failure and can be expressed in terms of total stress σ_n as $(\sigma_n - u_e)$, where u_e is the excess pore-water pressure at failure. u_e is induced by changes in the principal stresses $(\Delta\sigma_1, \Delta\sigma_3)$. For saturated soils, it is expressed in terms of the pore-water-pressure parameter A_f at failure as:

$$u = \Delta\sigma_3 + A_f(\Delta\sigma_1 - \Delta\sigma_3)_f \tag{6-7}$$

The effective-stress parameters c', ϕ', and A_f are readily determined by *CU* triaxial shear tests employing pore-water-pressure measurements or, excepting A_f, by consolidated-drained *(CD)* tests. A correlation between the plasticity index I_p and ϕ' for normally consolidated soils is shown in Fig. 6-1.

After large movements along preformed failure planes, cohesive soils exhibit a significantly reduced (residual) shear strength. The corresponding effective friction angle ϕ'_r is dependent on I_p. For many cohesive soils, ϕ'_r is also a function of σ'_n. The ϕ'_r parameter is applied in analysis of the stability of soils where prior movements (slides) have occurred.

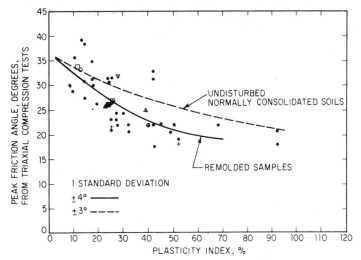

Fig. 6-1. Curves indicate the variation of effective friction angle with plasticity index for remolded and normally consolidated undisturbed soils.

Cyclic loading with complete stress reversals decreases the shearing resistance of saturated cohesive soils by inducing a progressive buildup in pore-water pressure. The amount of degradation depends primarily on the intensity of the cyclic shear stress, the number of load cycles, the stress history of the soil, and the type of cyclic test used. The strength degradation potential can be determined by postcyclic, *UU* tests.

6-8. Strength of Cohesionless Soils

The shear strength of cohesionless soils under static loading is conventionally interpreted from results of drained or undrained *TC* tests incorporating pore-pressure measurements. The effective angle of internal friction ϕ' can also be expressed by Eq. (6-6), except that c' is usually interpreted as zero. For cohesionless soils, ϕ' is dependent on density or void ratio, gradation, grain shape, and grain mineralogy. Markedly stress-dependent, ϕ' decreases with increasing σ'_n, the effective stress normal to the plane of shear failure. (M. M. Baligh, "Cavity Expansion in Sands with Curved Envelopes," *ASCE Journal of Geotechnical Engineering Division*, vol. 102, no. GT3, 1976.)

In situ cone penetration tests in sands may be used to estimate ϕ' from cone resistance q_c records. One approach relates the limiting q_c values directly to ϕ'. Where q_c increases approximately linearly with depth, ϕ' can also be interpreted from the slope of the curve for $q_c - \sigma_{vo}$ vs. σ'_{vo}, where σ_{vo} = total vertical stress, $\sigma'_{vo} = \sigma_{vo} - u$, and u = pore-water pressure. The third approach is to interpret the relative density D_r from q_c and then relate ϕ' to D_r as a function of the gradation and grain shape of the sand. (J. H. Schmertmann, "Measurement of In-Situ

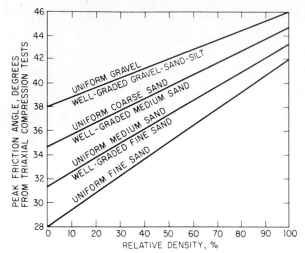

Fig. 6-2. Chart determines friction angles for sands of various relative densities. *(After J. H. Schmertmann, 1978.)*

Shear Strength," ASCE Specialty Conference on In-Situ Measurement of Soil Properties, Raleigh, N.C., 1975.)

Relative density provides good correlation with ϕ' for a given gradation, grain shape, and normal stress range. A widely used correlation is shown in Fig. 6-2. D_r can be interpreted from standard penetration resistance tests (Fig. 6-16) and cone penetration resistance tests (see Arts. 6-14 and 6-15) or calculated from the results of in situ or maximum and minimum density tests.

Dense sands typically exhibit a reduction in shearing resistance at strains greater than those required to develop the peak resistance. At relatively large strains, the stress-strain curves of loose and dense sands converge. The void ratio at which there is no volume change during shear

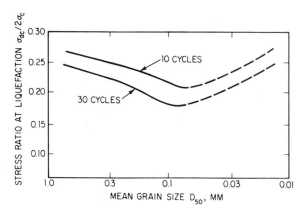

Fig. 6-3. Curves indicate the variation, with mean grain size, of the stress ratio causing liquefaction in sands of medium grain size in 10 and 30 cycles (for cyclical triaxial compression tests on soils with 50% relative density). $\sigma_{dc} = \sigma_1 - \sigma_3$ under cyclic loading.

is called the critical void ratio. A volume increase during shear (dilatancy) of saturated, dense, cohesionless soils produces negative pore-water pressures and a temporary increase in shearing resistance. Subsequent dissipation of negative pore-water pressure accounts for the "relaxation effect" sometimes observed after piles have been driven into dense, fine sands.

Saturated, cohesionless soils subject to cyclic loads exhibit a significant reduction in strength if cyclic loading is applied at periods smaller than the time required to achieve significant dissipation of pore pressure. Should the number of load cycles N_c be sufficient to generate pore pressures that approach the confining pressure within a soil zone, excessive deformations and eventually failure (liquefaction) is induced. For a given confining pressure and cyclic-stress level, the number of cycles required to induce initial liquefaction N_{cl} increases with an increase in relative density D_r. The variation of the ratio of cyclic shear stress σ_{dc} to effective confining pressure σ_c with median grain size is demonstrated by Fig. 6-3. Cyclic shear strength is commonly investigated by cyclic triaxial tests and occasionally by cyclic, direct, simple-shear tests.

6-9. Deformability of Fine-Grained Soils

Deformations of fine-grained soils can be classified as those that result from volume change, (elastic) distortion without volume change, or a combination of these causes. Volume change may be a one-dimensional or, in the presence of imposed shear stresses, a three-dimensional mechanism and may occur immediately or be time-dependent. **Immediate deformations** are realized without volume change during undrained loading of saturated soils and as a reduction of air voids (volume change) within unsaturated soils.

The rate of volume change of saturated, fine-grained soils during loading or unloading is controlled by the rate of pore-fluid drainage from or into the stressed soil zone. The compression phase of delayed volume change associated with pore-pressure dissipation under a constant load is termed **primary consolidation.** Upon completion of primary consolidation, some soils (particularly those with a significant organic content) continue to decrease in volume at a decreasing rate. This response is usually approximated as a straight line for a plot of log time vs. compression and is termed **secondary compression.**

As the imposed shear stresses become a substantial fraction of the undrained shear strength of the soil, time-dependent deformations may occur under constant load and volume conditions. This phenomenon is termed **creep deformation.** Failure by creep may occur if safety factors are insufficient to maintain imposed shear stresses below the creep threshold of the soil. (Also see Art. 6-19.)

One-dimensional volume-change parameters are conveniently interpreted from consolidation (oedometer) tests. A typical curve for log consolidation pressure vs. volumetric strain ϵ_v (Fig. 6-4a) demonstrates interpretation of the strain-referenced compression index C'_c, recompression index C'_r, and swelling index C'_s. The secondary compression index C'_α represents the slope of the near-linear portion of the volumetric strain vs. log-time curve following primary consolidation (Fig. 6-4b). The parameters C'_c, C'_r, and C'_α can be roughly estimated from soil-index properties. ("Design Manual—Soil Mechanics, Foundations, and Earth Structures," NAVDOCKS DM-7, U.S. Navy.)

Deformation moduli representing three-dimensional deformation can be interpreted from the stress-strain curves of laboratory shear tests for application to either volume change or elastic deformation problems. Moduli may be interpreted to represent a secant modulus E_s or a tangent modulus E_t for any given stress level. Moduli representing combined drained and undrained deformation can be derived from stress-strain curves obtained from stress-path triaxial tests. The stress paths followed are chosen to represent the sequence of loading and drainage conditions that best simulates the field conditions. (T. W. Lambe and R. V. Whitman, "Soil Mechanics," John Wiley & Sons, Inc., New York.)

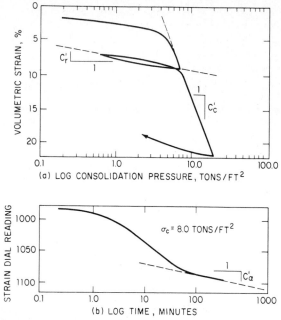

Fig. 6-4. Typical curves obtained from consolidation tests for load increment ratios of unity.

6-10. Deformability of Coarse-Grained Soils

Deformation of most coarse-grained soils occurs almost exclusively by volume change at a rate essentially equivalent to the rate of stress change. Deformation moduli are markedly nonlinear with respect to stress change and dependent on the initial state of soil stress. Some coarse-grained soils exhibit a delayed volume-change phenomenon known as **friction lag.** This response is analogous to the secondary compression of fine-grained soils and can account for a significant amount of the compression of coarse-grained soils composed of weak or sharp-grained particles. (J. H. Schmertmann, "Static Cone to Compute Static Settlement over Sand," *ASCE Journal of Soil Mechanics and Foundation Engineering Division,* vol. 96, no. SM3, 1970.)

The laboratory approach previously described for derivation of drained deformation parameters for fine-grained soils has a limited application for coarse-grained soils because of the difficulty in obtaining reasonably undisturbed samples. Tests may be carried out on reconstituted samples but should be used with caution since the soil fabric, aging, and stress history cannot be adequately simulated in the laboratory. As a consequence, in situ testing techniques currently appear to be a more promising approach to the characterization of cohesionless soil properties (see Art. 6-15) and are gaining increasing acceptance.

The Quasi-Static Cone Penetration Test (CPT) ▪ The CPT is one of the most useful in situ tests for investigating the deformability of cohesionless soils. The secant modulus E'_s, tons/ft^2, of sands has been related to cone resistance q_c by correlations of small-scale plate load tests and load tests on footings. The relationship is given by Eq. (6-8). The empirical correlation coefficient α in Eq. (6-8) is influenced by the relative density, grain characteristics, and stress history of the soil (see Art. 6-15).

$$E'_s = \alpha q_c \qquad (6-8)$$

The α parameter has been reported to range between 1.5 and 3 for sands and can be expressed in terms of relative density D_r as $2(1 + D_r^2)$. α may also be derived from correlations between q_c and the standard penetration resistance N by assuming q_c/N for mechanical (Delft) cones or $q_c/N + 1$ for electronic (Fugro)-type cone tips is about 6 for sandy gravel, 5 for gravelly sand, 4 for clean sand, and 3 for sandy silt. Bear in mind, however, that E_s' characterizations from q_c or N are strictly empirical and can provide erroneous characterizations under some circumstances. (See also Art. 6-22 and J. H. Schmertmann, "Static Cone to Compute Static Settlement over Sand," *ASCE Journal of Soil Mechanics and Foundation Engineering Division,* vol. 96, no. SM3, 1970.)

One of the earliest methods for evaluating the in situ deformability of coarse-grained soils is the small-scale *load-bearing test.* Data developed from these tests have been used to provide a scaling factor to express the settlement ρ of a full-size footing from the settlement ρ_1 of a 1-ft square plate. This factor ρ/ρ_1 is given as a function of the width B of the full-size bearing plate as:

$$\frac{\rho}{\rho_1} = \left(\frac{2B}{1 + B} \right)^2 \tag{6-9}$$

From an elastic half-space solution, E_s' can be expressed from results of a plate load test in terms of the ratio of bearing pressure to plate settlement k_v as:

$$E_s' = \frac{k_v(1 - \mu^2)\pi/4}{4B/(1 + B)^2} \tag{6-10}$$

μ represents Poisson's ratio, usually considered to range between 0.30 and 0.40. Equation (6-10) assumes that ρ_1 is derived from a rigid, 1-ft-diameter circular plate and that B is the equivalent diameter of the bearing area of a full-scale footing. Empirical formulations such as Eq. (6-9) may be significantly in error because of the limited footing-size range used and the large scatter of the data base. Furthermore, consideration is not given to variations in the characteristics and stress history of the bearing soils.

(J. K. Mitchell and W. S. Gardner, "In-Situ Measurement of Volume Change Characteristics," ASCE Specialty Conference on In-Situ Measurement of Soil Properties, Raleigh, N.C., 1975; K. Terzaghi and R. B. Peck, "Soil Mechanics and Engineering Practice," John Wiley & Sons, Inc., New York.)

Pressuremeter tests (PMTs) in soils and soft rocks have been used to characterize E_s' from radial pressure vs. volume-change data developed by expanding a cylindrical probe in a drill hole (see Art. 6-15). Because cohesionless soils are sensitive to comparatively small degrees of soil disturbance, proper access-hole preparation is critical.

6-11. State of Stress of Soils

Assessment of the vertical σ_{vo}' and horizontal σ_{ho}' effective stresses within a soil deposit and the maximum effective stresses imposed on the deposit since deposition σ_{vm}' is a general requirement for characterization of soil behavior. The ratio $\sigma_{vm}'/\sigma_{vo}'$ is termed the **overconsolidation ratio** *(OCR)*. Another useful parameter is the ratio of $\sigma_{ho}'/\sigma_{vo}'$, which is called the **coefficient of earth pressure at rest** (K_o).

For a simple gravitation piezometric profile, the effective overburden stress σ_{vo}' is directly related to the depth of groundwater below the surface and the effective unit weight of the soil strata. Groundwater conditions, however, may be characterized by irregular piezometric profiles that cannot be modeled by a simple gravitational system. For these conditions, **sealed piezometer measurements** are required to assess σ_{vo}'.

The Maximum Past Consolidation Stress ▪ σ'_{vm} of a soil deposit may reflect stresses imposed prior to geologic erosion or during periods of significantly lower groundwater, as well as desiccation effects and man-made excavations. The maximum past consolidation stress is conventionally interpreted from consolidation (oedometer) tests on undisturbed samples using a procedure proposed by A. Casagrande. An alternate procedure proposed by D. M. Burmister requires that the consolidation test incorporate an unload-reload cycle. J. M. Schmertmann proposed an interpretation technique that is recommended for obviously disturbed test samples. (D. M. Burmister, "Application of Controlled Test Methods in Consolidation Testing," ASTM Spec. Tech. Publ. 126, 1951; J. M. Schmertmann, "Undisturbed Consolidation of Clay," *ASCE Transactions,* vol. 120, 1955.)

Normalized-shear-strength concepts provide an alternative method for estimating *OCR* from good-quality *UU* compression tests. In the absence of site-specific data relating c_u/σ'_{vo} and *OCR*, a form of Eq. (6-5) may be applied to estimate *OCR*. In this interpretation, σ'_{vo} represents the effective overburden pressure at the depth of the *UU*-test sample. A very approximate estimate of σ'_{vm} can also be obtained for cohesive soils from relationships proposed between liquidity index and effective vertical stress ("Design Manual—Soil Mechanics, Foundations, and Earth Structures," NAVDOCKS DM-7, U.S. Navy). In addition to laboratory techniques, pressuremeter test (PMT) results have been used to interpret σ'_{vm} by empirically equating it to yield pressure P_y (see Fig. 6-5). (See also W. S. Gardner, "Soil Property Characterization in Geotechnical Engineering Practice," *Geotechnical/Environmental Bulletin,* vol. X, no. 7, Woodward-Clyde Consultants, San Francisco, Cal.)

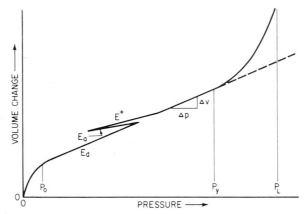

Fig. 6-5. Idealized PMT curve. E_d = deformation modulus; E_a = alternated or rebound modulus; E^+ = recompression modulus; P_o = horizontal earth pressure at rest; P_y = creep limit; P_L = limit pressure.

For coarse-grained soil deposits, it is difficult to characterize σ'_{vm} reliably from either in situ or laboratory tests because of an extreme sensitivity to disturbance. Attempts have been made to interpret σ'_{vm} or *OCR* from the results of in situ cone penetration and screw-plate load tests. (See J. K. Mitchell and W. S. Gardner, "In-Situ Measurement of Volume-Change Characteristics," ASCE Specialty Conference on In-Situ Measurement of Soil Properties, Raleigh, N.C., 1975.) A procedure for estimating *OCR* from CPTs in sands of known relative density D_r has also been proposed. (See W. S. Gardner and S. V. Nathan, "The Suitcase Cone System," *Symposium on Cone Penetration Testing and Experience,* Proceedings of ASCE National Convention, St. Louis, Mo., 1981.)

The coefficient of earth pressure at rest K_o can be determined in the laboratory from "no-lateral strain" TC tests on undisturbed soil samples or from consolidation tests conducted in specially constructed oedometers. Interpretation of K_o from in situ CPT, PMT, and dilatometer tests has also been proposed. In view of the significant impact of sample disturbance on laboratory results and the empirical nature of in situ test interpretations, the following correlations of K_o with friction angle ϕ' and OCR are useful. For both coarse- and fine-grained soils:

$$K_o = (1 - \sin \phi')OCR^m \qquad (6\text{-}11)$$

A value for m of 0.5 has been proposed for overconsolidated cohesionless soils, whereas for cohesive soils it is proposed that m be estimated in terms of the plasticity index I_p as $0.58I_p^{-0.12}$.

6-12. Soil Permeability

The coefficient of permeability k is a measure of the rate of flow of water through saturated soil under a given hydraulic gradient i, cm/cm and is defined in accordance with Darcy's law as:

$$V = kiA \qquad (6\text{-}12)$$

where V = rate of flow, cm^3/s.

A = cross-sectional area of soil conveying flow, cm^2

k is dependent on the grain-size distribution, void ratio, and soil fabric and typically may vary from as much as 10 cm/s for gravel to less than 10^{-7} cm/s for clays. For typical soil deposits, k for horizontal flow is greater than k for vertical flow, often by an order of magnitude.

Soil-permeability measurements can be conducted in tests under falling or constant head, either in the laboratory or the field. Large-scale pumping (drawdown) tests also may be conducted in the field to provide a significantly larger-scale measurement of formation permeability. Correlations of k with soil gradation and relative density or void ratio have been developed for a variety of coarse-grained materials. General correlations of k with soil index and physical properties are less reliable for fine-grained soils because other factors than porosity may control.

(R. E. Olson and D. E. Daniel, "Field and Laboratory Measurement of the Permeability of Saturated and Partially Saturated Fine-Grained Soils," *Geotechnical Engineering Report GE79-1,* Department of Civil Engineering, University of Texas, Austin, 1979; D. M. Burmister, "Physical, Stress-Strain and Strength Responses of Granular Soils," ASTM Spec. Tech. Publ. no. 322, 1962; T. W. Lambe and R. V. Whitman, "Soil Mechanics," John Wiley & Sons, Inc., New York.)

SITE INVESTIGATIONS

The objective of most geotechnical site investigations is to obtain information on the site and subsurface conditions that is required for design and construction of engineered facilities and for evaluation and mitigation of geologic hazards, such as landslides, subsidence, and liquefaction. The site investigation is part of a fully integrated process that includes:

1. Synthesis of available data
2. Field and laboratory investigations
3. Characterization of site stratigraphy and soil properties
4. Engineering analyses
5. Formulation of design and construction criteria or engineering evaluations

6-13. Planning and Scope

In the planning stage of a site investigation, all pertinent topographical, geologic, and geotechnical information available should be reviewed and assessed. In urban areas, the development history of the site should be studied and evaluated. It is particularly important to provide or require that a qualified engineer direct and witness all field operations.

The scope of the geotechnical site investigation varies with the type of project but typically includes topographic and location surveys, exploratory borings, and groundwater monitoring. Frequently, the borings are supplemented by soundings and test pits. Occasionally, air photo interpretations, in situ testing, and geophysical investigations are conducted.

6-14. Exploratory Borings

Typical boring methods employed for geotechnical exploration consist of rotary drilling, auger drilling, percussion drilling, or any combination of these. Deep soil borings (greater than about 100 ft) are usually conducted by rotary-drilling techniques recirculating a weighted drilling fluid to maintain borehole stability. Auger drilling, with hollow-stem augers to facilitate sampling, is a widely used and economical method for conducting short- to intermediate-length borings. Most of the drill rigs are truck-mounted and have a rock-coring capability.

With percussion drilling, a casing is usually driven to advance the boring. Water circulation or driven, clean-out spoons are often used to remove the soil (cuttings) in the casing. This method is employed for difficult-access locations where relatively light and portable drilling equipment is required. A rotary drill designed for rock coring is often included.

Soil Samples ▪ These are usually obtained by driving a split-barrel sampler or by hydraulically or mechanically advancing a thin-wall (Shelby) tube sampler. Driven samplers, usually 2 in outside diameter (OD), are advanced 18 in by a 140-lb hammer dropped 30 in (ASTM D1586). The number of blows required to drive the last 12 in of penetration constitutes the **standard penetration resistance** (SPT) value. The **Shelby tube sampler,** used for undisturbed sampling, is typically a 12- to 16-gage seamless steel tube and is nominally 3 in OD (ASTM D1587). In soils that are soft or otherwise difficult to sample, a **stationary piston sampler** is used to advance a Shelby tube either hydraulically (pump pressure) or by the down-crowd system of the drill.

Rotary core drilling is typically used to obtain core samples of rock and hard, cohesive soils that cannot be penetrated by conventional sampling techniques. Typically, rock cores are obtained with diamond bits that yield core-sample diameters from ⅞ in (AX) to 2⅛ in (NX). For hard clays and soft rocks, a 3- to 6-in OD undisturbed sample can also be obtained by rotary drilling with a **Dennison** or **Pitcher sampler.**

Test Boring Records (Logs) ▪ These typically identify the depths and material classification of the various strata encountered, the sample location and penetration resistance, rock-core interval and recovery, groundwater levels encountered during and after drilling. Special subsurface conditions should be noted on the log, for example, changes in drilling resistance, hole caving, voids, and obstructions. General information required includes the location of the boring, surface elevation, drilling procedures, sampler and core barrel types, and other information relevant to interpretation of the boring log.

Groundwater Monitoring ▪ Monitoring groundwater levels is an integral part of boring and sampling operations. Groundwater measurements during and at least 12 h after drilling are usually required. Standpipes are often installed in test borings to provide longer-term observations; they are typically small-diameter pipes perforated in the bottom few feet of casing.

If irregular piezometric profiles are suspected, piezometers may be set and sealed so as to measure hydrostatic heads within selected strata. Piezometers may consist of watertight ½- to ¾-in OD standpipes or plastic tubing attached to porous ceramic or plastic tips. Piezometers with electronic or pneumatic pressure sensors have the advantage of quick response and automated data acquisition. However, it is not possible to conduct in situ permeability tests with these closed-system piezometers.

6-15. In Situ Testing of Soils

In situ tests can be used under a variety of circumstances to enhance profile definition, to provide data on soil properties, and to obtain parameters for empirical analysis and design applications.

Quasi-static and dynamic cone penetration tests (CPTs) quite effectively enhance profile definition by providing a continuous record of penetration resistance. Quasi-static cone penetration resistance is also correlated with the relative density, *OCR,* friction angle, and compressibility of coarse-grained soils and the undrained shear strength of cohesive soils. Empirical foundation design parameters are also provided by the CPT.

The standard CPT in the United States consists of advancing a 10-cm², 60° cone at a rate between 1.5 and 2.5 cm/s and recording the resistance to cone penetration (ASTM D3441). A friction sleeve may also be incorporated to measure frictional resistance during penetration. The cone may be incrementally (mechanical penetrometer) or continuously (electronic penetrometer) advanced.

Dynamic cones are available in a variety of sizes, but in the United States, they typically have a 2-in upset diameter with a 60° apex. They are driven by blows of a 140-lb hammer dropped 30 in. Automatically driven cone penetrometers are widely used in western Europe and are portable and easy to operate. The self-contained ram-sounding method has the broadest application and has been standardized.

Pressuremeter tests (PMTs) provide an in situ interpretation of soil compressibility and undrained shear strength. Pressuremeters have also been used to provide parameters for foundation design.

The PMT is conducted by inserting a probe containing an expandable membrane into a drill hole and then applying a hydraulic pressure to radially expand the membrane against the soil, to measure its volume change under pressure. The resulting curve for volume change vs. pressure is the basis for interpretation of soil properties. Figure 6-5 in Art. 6-11 presents a typical PMT curve and demonstrates conventional interpretations.

Vane shear tests provide in situ measurements of the undrained shear strength of soft to firm clays, usually by rotating a four-bladed vane and measuring the torsional resistance T. Undrained shear strength is then calculated by dividing T by the cylindrical side and end areas inscribing the vane. Account must be taken of torque rod friction (if unsleeved), which can be determined by calibration tests (ASTM D2573). Vane tests are typically run in conjunction with borings, but in soft clays the vane may be advanced without a predrilled hole.

Other in situ tests occasionally used to provide soil-property data include plate load tests (PLTs), borehole shear (BHS) tests, and dilatometer tests. The PLT technique may be useful for providing data on the compressibility of soils and rocks. The BHS may be useful for characterizing effective-shear-strength parameters for relatively free-draining soils as well as total-stress (undrained) shear-strength parameters for fine-grained soils. Dilatometer tests provide a technique for investigating the horizontal effective stress σ'_{ho} and soil compressibility. Some of the newer types of tests use small-diameter probes to measure pore-pressure response, acoustical emissions, bulk density, and moisture content during penetration.

Prototype load testing as part of the geotechnical investigation represents a variation of in situ testing. It may include pile load tests, earth load tests to investigate settlement and stability, and tests on small-scale or full-size shallow foundation elements. Feasibility of construction can also be evaluated at this time by test excavations, indicator pile driving, drilled shaft excavation, rock rippability trials, dewatering tests, and so on.

(G. Sanglerat, "The Penetrometer and Soil Exploration," Elsevier Publishing Company, Amsterdam; F. Baguelin, J. F. Jezequel, and D. H. Shields, "The Pressuremeter and Foundation Engineering," Trans Tech Publications, Clausthal, Germany; *Proceedings of the ASCE Specialty Conference on In-Situ Measurement of Soil Properties,* vol. 1, 1975, and in particular, J. W. Wineland, "Borehole Shear Device," and S. Marchetti, "A New In-Situ Test for the Measurement of Horizontal Soil Deformability," Raleigh, N.C..)

6-16. Geophysical Investigations

Geophysical measurements are often valuable when evaluating the continuity of soil and rock strata between boring locations, and under some circumstances they can reduce the number of borings required. Certain of these measurements can also provide data for interpreting soil and rock properties. The techniques most common to engineering applications are now described.

Seismic-wave-propagation techniques include seismic refraction, seismic reflection, and direct wave-transmission measurements. Refraction techniques measure the travel time of seismic waves generated from a single-pulse energy source to detectors (geophones) located at various distances from the source. The principle of seismic refraction surveying is based on refraction of the seismic waves at boundaries of layers with different acoustical impedances. This technique is illustrated in Fig. 6-6.

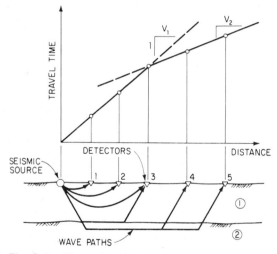

Fig. 6-6.   Illustration of seismic refraction concept.

Compression P wave velocities are interpreted to define velocity profiles that may be correlated with stratigraphy and the depth to rock. The P wave velocity may also help identify type of soil. However, in saturated soils, the velocity measured represents wave transmission through water-filled voids. This velocity is about 4800 ft/s regardless of the soil type. Low-cost single- and dual-channel seismographs have been developed for routine engineering applications.

Seismic reflection involves measuring the times required for a seismic wave induced at the surface to return to the surface after reflection from the interfaces of strata that have different acoustical impedances. Unlike refraction techniques, which usually record only the first arrivals of the seismic waves, wave trains are concurrently recorded by several detectors at different positions so as to provide a pictorial representation of formation structure. This type of survey can be conducted in both marine and terrestrial environments and usually incorporates comparatively expensive multiple-channel recording systems.

Direct seismic-wave-transmission techniques include measurements of the arrival times of shear S and P waves after they have traveled between a seismic source and geophones placed at similar elevations in adjacent drill holes. By measuring the precise distances between source and detectors, both S and P wave velocities can be measured for a given soil or rock interval if the hole spacing is chosen to ensure a direct wave-transmission path.

Alternatively, geophones can be placed at different depths in a drill hole to measure seismic waves propagated down from a surface source near the drill hole. The detectors and source locations can also be reversed to provide up-hole instead of down-hole wave propagation. Although this method does not provide as precise a measure of interval velocity as the cross-hole technique, it is substantially less costly.

Direct wave-transmission techniques are usually conducted so as to maximize S wave energy generation and recognition by polarization of the energy input. S wave interpretations allow calculation of the small-strain shear modulus G_{max} required for dynamic response analysis. Poisson's ratio can also be determined if both P and S wave velocities can be recorded.

Resistivity and conductance investigation techniques relate to the proposition that stratigraphic details can be derived from differences in the electrical resistance or conductivity of individual strata. Resistivity techniques for engineering purposes usually apply the Wenner method of investigation, which involves four equally spaced steel electrodes (pins). The current is introduced through the two end pins, and the associated potential drop is measured across the two center pins. The apparent resistivity ρ is then calculated as a function of current I, potential difference V, and pin spacing a as:

$$\rho = \frac{2\pi a V}{I} \tag{6-13}$$

To investigate stratigraphic changes, tests are run at successively greater pin spacings. Interpretations are made by analyzing accumulative or discrete-interval resistivity profiles or by theoretical curve-matching procedures.

A conductivity technique for identifying subsurface anomalies and stratigraphy involves measuring the transient decay of a magnetic field with the source (dipole transmitter) in contact with the surface. The depth of apparent conductivity measurement depends on the spacing and orientation of the transmitter and receiver loops.

Both resistivity and conductivity interpretations are influenced by groundwater chemistry. This characteristic has been utilized to map the extent of some groundwater pollutant plumes by conductivity techniques.

Other geophysical methods with more limited engineering applications include gravity and magnetic field measurements. Surveys using these techniques can be airborne, shipborne, or ground-based. Microgravity surveys have been useful in detecting subsurface solution features in carbonate rocks.

(M. B. Dobrin, "Introduction to Geophysical Prospecting," McGraw-Hill Book Company, New York.)

SHALLOW FOUNDATIONS

Shallow foundation systems can be classified as spread footings, wall and continuous (strip) footings, and mat (raft) foundations. Variations are combined footings, cantilevered (strapped) footings, two-way strip (grid) footings, and discontinuous (punched) mat foundations.

6-17. Types of Footings

Spread (individual) footings (Fig. 6-7) are the most economical shallow foundation types but are more susceptible to differential settlement. They usually support single concentrated loads, such as those imposed by columns.

Combined footings (Fig. 6-8) are used where the bearing areas of closely spaced columns overlap. Cantilever footings (Fig. 6-9) are designed to accommodate eccentric loads.

Continuous wall and strip footings (Fig. 6-10) can be designed to redistribute bearing-stress concentrations and associated differential settlements in the event of variable bearing conditions or localized ground loss beneath footings.

Mat foundations have the greatest facility for load distribution and for redistribution of subgrade stress concentrations caused by localized anomalous bearing conditions. Mats may be constant section, ribbed, waffled, or arched. Buoyancy mats are used on compressible soil sites in combination with basements or subbasements, to create a permanent unloading effect, thereby reducing the net stress change in the foundation soils.

(M. J. Tomlinson, "Foundation Design and Construction," John Wiley & Sons, Inc., New York; J. E. Bowles, "Foundation Analysis and Design," McGraw-Hill Book Company, New York.)

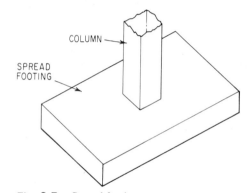

Fig. 6-7. Spread footing.

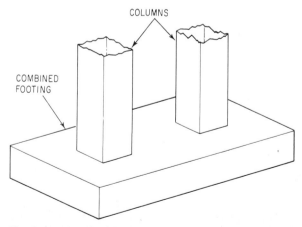

Fig. 6-8. Combined footing.

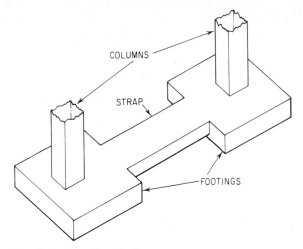

Fig. 6-9.   Cantilever footing.

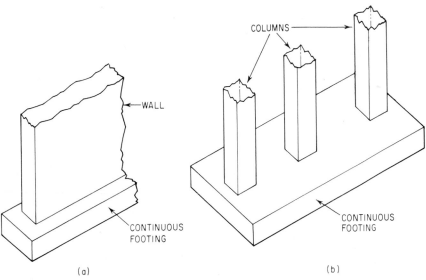

Fig. 6-10.   Continuous footings for: (*a*) a wall; (*b*) several columns.

6-18. Approach to Foundation Analysis

Shallow-foundation analysis and formulation of geotechnical design provisions are generally approached in the following steps:

1. Establish project objectives and design or evaluation conditions.
2. Characterize site stratigraphy and soil properties.
3. Evaluate load-bearing fill support or subsoil-improvement techniques, if applicable.
4. Identify bearing levels; select and proportion candidate foundation systems.

5. Conduct performance, constructibility, and economic feasibility analyses.

6. Repeat steps 3 through 5 as required to satisfy the design objectives and conditions.

The scope and detail of the analyses vary according to the project objectives.

Project objectives to be quantified are essentially the intent of the project assignment and the specific scope of associated work. The conditions that control geotechnical evaluation or design work include criteria for loads and grades, facility operating requirements and tolerances, construction schedules, and economic and environmental constraints. Failure to provide a clear definition of relevant objectives and design conditions can result in significant delays, extra costs, and, under some circumstances, unsafe designs.

During development of design conditions for structural foundations, tolerances for total and differential settlements are commonly established as a function of the ability of a structure to tolerate movement. Suggested structure tolerances in terms of angular distortion are in Table 6-6. Angular distortion represents the differential vertical movement between two points divided by the horizontal distance between the points. (See G. A. Leonards, "Foundation Engineering," McGraw-Hill Book Company, New York; R. Grant, J. T. Christian, and E. K. Van Marcke, "Differential Settlement of Buildings," *ASCE Journal of Geotechnical Engineering Division,* vol. 100, no. GT9, 1974.)

Development of *design profiles* for foundation analysis ideally involves a synthesis of geologic and geotechnical data relevant to site stratigraphy and soil and rock properties. This usually requires site investigations (see Arts. 6-13 to 6-16) and in situ or laboratory testing, or both, of representative soil and rock samples (see Arts. 6-5 to 6-12).

To establish and proportion candidate foundation systems, consideration must first be given to identification of feasible bearing levels. The *depth of the foundation* must also be sufficient to protect exposed elements against frost heave and to provide sufficient confinement to produce a factor of safety not less than 2.5 (preferably 3.0) against shear failure of the bearing soils. Frost penetration has been correlated with a **freezing index,** which equals the number of days with temperature below $32°F$ multiplied by $T - 32$, where T = average daily temperature. Such correlations can be applied in the absence of local codes or experience. Generally, footing depths below final grade should be a minimum of 2.0 to 2.5 ft. (H. F. Winterkorn and H. Y. Fang, "Foundation Engineering Handbook," Van Nostrand Reinhold Company, New York.)

For marginal bearing conditions, consideration should be given to improvement of the quality of potential bearing strata. Soil-improvement techniques include excavation and replacement or overlaying of unsuitable subsoils by *load-bearing fills, preloading* of compressible subsoils, *soil densification, and grout injection.* Densification methods include high-energy surface

TABLE 6-6 Limiting Angular Distortions*

Structural response	Angular distortion
Cracking of panel and brick walls	1/100
Structural damage to columns and	
beams	1/150
Impaired overhead crane operation	1/300
First cracking of panel walls	1/300
Limit for reinforced concrete frame	1/400
Limit for wall cracking	1/500
Limit for diagonally braced frames	1/600
Limit for settlement-sensitive machines	1/750

*Limits represent the maximum distortions that can be safely accommodated.
Source: After L. Bjerrum, European Conference on Soil Mechanics and Foundation Engineering, Wiesbaden, Germany, vol.2, 1963.

impact (dynamic consolidation), on-grade vibratory compaction, and subsurface vibratory compaction by Vibroflotation or Terra-Probe techniques. ("Soil Improvement, History, Capabilities, and Outlook," Report of Committee on Placement and Improvement of Soils, *ASCE Geotechnical Engineering Division*, 1978.)

Another approach to improving bearing conditions involves in situ reinforcement. Associated techniques are *stone columns, lime columns, geofabric reinforcing,* and *reinforced earth.* The choice of an appropriate soil-improvement technique is highly dependent on the settlement tolerance of the structure as well as the magnitude and nature of the applied loads.

Assessing the suitability of candidate foundation systems includes evaluating the factor of safety against both catastrophic failure and excessive deformation under sustained and transient design loads. Catastrophic-failure assessment must consider overstressing and creep of the bearing soils as well as lateral displacement of the foundation. Evaluating the probable settlement behavior involves analyzing the stresses imposed within the soil and predicting, with the use of appropriate soil parameters, foundation settlements. Typically, *settlement analyses* provide estimates of total and differential settlement at strategic locations within the foundation area and may include time-rate predictions of settlement. Under most circumstances, the suitability of shallow foundation systems is governed by the systems' load-settlement response rather than bearing capacity.

6-19. Foundation-Stability Analysis

The maximum load that can be sustained by shallow foundation elements at incipient failure *(bearing capacity)* is a function of the cohesion and friction angle of bearing soils as well as the width B and shape of the foundation. The **net bearing capacity** q_u is conventionally expressed as:

$$q_u = \alpha_f c_u N_c + \sigma'_{vo} N_q + \beta_f \gamma B N_\gamma \tag{6-14}$$

where α_f = 1.0 for strip footings and 1.3 for circular and square footings
   c_u = undrained shear strength of soil
   σ'_{vo} = effective vertical shear stress in soil at level of bottom of footing
   β_f = 0.5 for strip footings, 0.4 for square footings, and 0.6 for circular footings
   γ = unit weight of soil
   B = width of footing for square and rectangular footings and radius of footing for circular footings
N_c, N_q, N_γ = bearing-capacity factors, functions of angle of internal friction ϕ
   For undrained (rapid) loading of cohesive soils, $\phi = 0$ and Eq. (6-14) reduces to

$$q_u = N'_c c_u \tag{6-15}$$

where $N'_c = \alpha_f N_c$. Solutions for N'_c are given in Fig. 6-11 as a function of the ratio of foundation depth to footing width. For drained (slow) loading of cohesive soils, ϕ and c_u are defined in terms of effective friction angle ϕ' and effective stress c'_u.

Modifications of Eq. (6-14) are also available to predict the bearing capacity of layered soils and for eccentric loading.

Rarely, however, does q_u control foundation design when the safety factor is within the range of 2.5 to 3. (Should creep or local yield be induced, excessive settlements may occur. This consideration is particularly important when selecting a safety factor for foundations on soft to firm clays with medium to high plasticity. Note that the PMT may be used to identify the yield threshold of the soil, as indicated in Fig. 6-5.)

(K. Terzaghi and R. B. Peck, "Soil Mechanics and Engineering Practice," John Wiley & Sons, Inc., New York; W. H. Perloff and W. Baron, "Soil Mechanics, Principles and Applica-

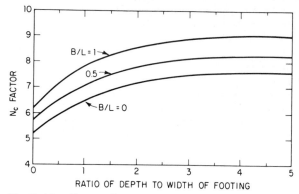

Fig. 6-11. Curves relate N_c to the ratio of depth of footing to its width for various ratios of footing width B to length L. *(After A. W. Skempton, "The Bearing Capacity of Clays," Building Research Congress, Institute of Civil Engineers, London, 1951.)*

tions," The Ronald Press Company, New York; D. J. D'Appolonia, H. G. Poulos, and C. C. Ladd, "Initial Settlement of Structures on Clay," *ASCE Journal of Soil Mechanics and Foundation Engineering Division,* vol. 97, no. SM10, 1971.)

Eccentric loading can have a significant impact on selection of the bearing value for foundation design. The conventional approach is to proportion the foundation to maintain the resultant force within its middle third. The footing is assumed to be rigid and the bearing pressure is assumed to vary linearly as shown by Fig. 6-12b. If the resultant lies outside the middle third of the footing, it is assumed that there is bearing over only a portion of the footing, as shown in Fig. 6-12d. For the conventional case, the maximum and minimum bearing pressures are:

$$q_m = \frac{P}{BL}\left(1 \pm \frac{6e}{B}\right) \tag{6-16}$$

where B = width of rectangular footing
L = length of rectangular footing
e = eccentricity of loading
For the other case (Fig. 6-12d), the soil pressure ranges from 0 to a maximum of:

$$q_m = \frac{2P}{3L(B/2 - e)} \tag{6-17}$$

For square or rectangular footings subject to overturning about two principal axes and for unsymmetrical footings, the loading eccentricities e_1 and e_2 are determined about the two principal axes. For the case where the full bearing area of the footings is engaged, q_m is given in terms of the distances from the principal axes c_1 and c_2, the radius of gyration of the footing area about the principal axes r_1 and r_2, and the area of the footing A as:

$$q_m = \frac{P}{A}\left(1 + \frac{e_1 c_1}{r_1^2} + \frac{e_2 c_2}{r_2^2}\right) \tag{6-18}$$

For the case where only a portion of the footing is bearing, the maximum pressure may be approximated by trial and error.

For all cases of *sustained eccentric loading,* the maximum (edge) pressures should not exceed the shear strength of the soil, and also the factor of safety should be at least 1.5 (preferably 2.0) against overturning.

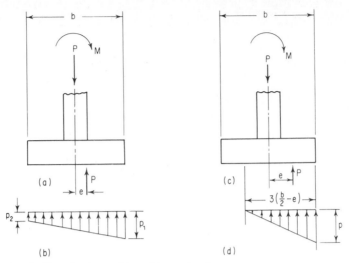

Fig. 6-12. Footings subjected to overturning.

The foregoing analyses, except for completely rigid foundation elements, are a very conservative approximation. Because most mat foundations and large footings are not completely rigid, their deformation during eccentric loading acts to produce a more uniform distribution of bearing pressures than would occur under a rigid footing and to reduce maximum contact stresses.

In the event of *transient eccentric loading,* experience has shown that footings can sustain maximum edge pressures significantly greater than the shear strength of the soil. Consequently, some building codes conservatively allow increases in sustained-load bearing values of 30% for transient loads. Reduced safety factors have also been used in conjunction with transient loading. For cases where significant cost savings can be realized, finite-element analyses that model soil-structure interaction can provide a more realistic evaluation of an eccentrically loaded foundation.

Resistance to horizontal forces is also of interest in foundation-stability analysis. The horizontal resistance of shallow foundations is mobilized by a combination of the passive soil resistance on the vertical projection of the embedded foundation and the friction between the foundation base and the subgrade. The soil pressure mobilized at full passive resistance, however, requires lateral movements greater than can be sustained by some foundations. Consequently, a soil resistance between the at-rest and passive-pressure cases should be determined on the basis of the allowable lateral deformations of the foundation.

The frictional resistance f to horizontal translation is conventionally estimated as a function of the sustained, real, load-bearing stresses q_d from

$$f = q_d \tan \delta \qquad (6\text{-}19)$$

where δ is the friction angle between the foundation and bearing soils and may be taken as equivalent to the internal-friction angle ϕ' of the subgrade soils. In the case of cohesive soils, $f = c_u$. Again, some relative movement must be realized to develop f, but this movement is less than that required for passive-pressure development.

If a factor of safety against translation of at least 1.5 is not realized with friction and passive soil pressure, footings should be keyed to increase soil resistance or tied to engage additional resistance. Building basement and shear walls are also commonly used to sustain horizontal loading.

6-20. Stress Analysis of Soils

Stress changes imposed in bearing soils by earth and foundation loads or by excavations are conventionally predicted from elastic half-space theory as a function of the foundation shape and the position of the desired stress profile. Elastic solutions available may take into account foundation rigidity, depth of the compressible zone, superposition of stress from adjacent loads, layered profiles, and moduli that increase linearly with depth. An elastic half space for flexible footings is shown in Fig. 6-13.

An approximate estimate of the average stress $\Delta\sigma_{av}$ at any depth z below rectangular footings of width B and length L can be derived by assuming that the vertical load spreads at the rate of 1 ft horizontally to 2 ft vertically beneath the footing; that is,

$$\Delta\sigma_{av} = P/(z + B)(z + L) \tag{6-20}$$

where P = footing load. This approximation, however, is not useful where it is necessary to characterize deviatoric stresses at a point within the stressed soil zone.

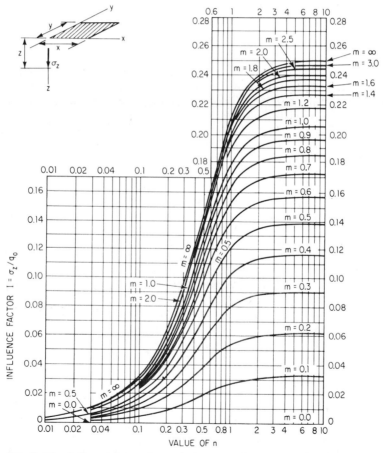

Fig. 6-13. Influence factors for vertical stress under the corner of a rectangular area. $I = \sigma_z/q_o$, where σ_z = vertical unit stress and q_o = load per unit area. $m = x/z$ and $n = y/z$ are interchangeable. (See sketch.) *(After "Design Manual— Soil Mechanics, Foundations, and Earth Structures," NAVDOCKS DM-7, U.S. Navy.)*

6-21. Settlement Analyses of Cohesive Soils

Settlement of foundations supported on cohesive soils is usually represented as the sum of the primary one-dimensional consolidation ρ_c, immediate ρ_i, and secondary ρ_s settlement components. Settlement due to primary consolidation is conventionally predicted for n soil layers by Eq. (6-21). For normally consolidated soils,

$$\rho_c = \sum_{i=1}^{n} H_i \left(C_c' \log \frac{\sigma_v}{\sigma_{vo}'} \right) \tag{6-21 a}$$

where H_i = depth below grade of ith soil layer
$\quad C_c'$ = strain referenced compression index for ith soil layer (Art. 6-9)
$\quad \sigma_v$ = sum of average σ_{vo}' and average imposed vertical-stress change $\Delta\sigma_v$ in ith soil layer
$\quad \sigma_{vo}'$ = initial effective overburden pressure at middle of ith layer (Art. 6-11)
For overconsolidated soils with $\sigma_v > \sigma_{vm}'$,

$$\rho_c = \sum_{i=1}^{n} H_i \left(C_r' \log \frac{\sigma_{vm}'}{\sigma_{vo}'} + C_c' \log \frac{\sigma_v}{\sigma_{vm}'} \right) \tag{6-21 b}$$

where C_r' = strain referenced recompression index of ith soil layer (Art. 6-9)
$\quad \sigma_{vm}'$ = preconsolidation (maximum past consolidation) pressure at middle of ith layer (Art. 6-11)
The maximum thickness of the compressible soil zone contributing significant settlement can be taken to be equivalent to the depth where $\Delta\sigma_v = 0.1\sigma_{vo}'$.

Equation (6-21 a) can also be applied to overconsolidated soils if σ_v is less than σ_{vm}' and C_r' is substituted for C_c'.

Inasmuch as Eq. (6-21) represents one-dimensional compression, it may provide rather poor predictions for cases of three-dimensional loading. Consequently, corrections to ρ_c have been derived for cases of three-dimensional loading. These corrections are approximate but represent an improved approach when loading conditions deviate significantly from the one-dimensional case. (A. W. Skempton and L. Bjerrum, "A Contribution to Settlement Analysis of Foundations on Clay," *Geotechnique,* vol. 7, 1957.)

The stress-path method of settlement analysis attempts to simulate field loading conditions by conducting triaxial tests so as to track the sequential stress changes of an average point or points beneath the foundation. The strains associated with each drained and undrained load increment are summed and directly applied to the settlement calculation. Deformation moduli can also be derived from stress-path tests and used in three-dimensional deformation analysis.

Three-dimensional settlement analyses using elastic solutions have been applied to both drained and undrained conditions. Immediate (elastic) foundation settlements ρ_i, representing the undrained deformation of saturated cohesive soils, can be calculated with elastic theory from

$$\rho_i = \frac{C_1 C_2 (1 - \mu^2) Bq}{E} \tag{6-22 a}$$

where q = net bearing value
$\quad B$ = width of rectangular footing or diameter of circular footing
$\quad E$ = undrained deformation modulus
$\quad \mu$ = Poisson's ratio
$\quad C_1, C_2$ = correction coefficients for footing shape and thickness of compressible layer
(H. G. Poulos and E. H. Davis, "Elastic Solutions for Soil and Rock Mechanics," John Wiley & Sons, Inc., New York; T. W. Lambe, "Methods of Estimating Settlements," *Proceedings,* ASCE Settlement Conference, Northwestern Univerity, 1964.)

Alternatively, it is often more convenient to calculate ρ_i by discrete layer analysis [Eq. (6-22b)].

$$\rho_i = \sum_{i=1}^{n} H_i \frac{\sigma_1 - \sigma_3}{E_i} \tag{6-22 b}$$

where $\sigma_1 - \sigma_3$ = change in average deviator stress within each layer influenced by applied load. Note that Eq. (6-22b) is strictly applicable only for axisymmetrical loading. Drained three-dimensional deformation can be estimated from Eq. (6-22) by substituting the secant modulus E'_s for E (see Art. 6-9).

The rate of one-dimensional consolidation can be evaluated with Eq. (6-24) in terms of the degree of consolidation U and the nondimensional time factor T_v. U is defined by

$$U = \frac{\rho_t}{\rho_c} = 1 - \frac{u_t}{u_i} \tag{6-23}$$

where ρ_t = settlement at time t after instantaneous loading
ρ_c = ultimate consolidation settlement
u_t = excess pore-water pressure at time t
u_i = initial pore-water pressure ($t = 0$)

To correct approximately for the assumed instantaneous load application, ρ_t at the end of the loading period can be taken as the settlement calculated for one-half of the load application time. The time t required to achieve U is evaluated as a function of the shortest drainage path within the compressible zone h, the coefficient of consolidation C_v, and the dimensionless time factor T_v from

$$t = T_v \frac{h^2}{C_v} \tag{6-24}$$

Closed-form solutions for T_v vs. U are available for a variety of initial pore-pressure distributions. (H. F. Winterkorn and H. Y. Fang, "Foundation Engineering Handbook," Van Nostrand Reinhold Company, New York.)

Solutions for constant and linearly increasing u_i are shown in Fig. 6-14. Equation (6-25) presents an approximate solution that can be applied to the constant initial u_i distribution case for $T_v > 0.2$.

$$U = 1 - \frac{8}{\pi^2} e^{-\pi^2 T_v/4} \tag{6-25}$$

where $e = 2.71828$. Numerical solutions for any u_i configuration in a single compressible layer as well as solutions for contiguous clay layers may be derived with finite-difference techniques. (R. F. Scott, "Principles of Soil Mechanics," Addison-Wesley Publishing Company, Inc., Reading, Mass.)

The coefficient of consolidation C_v is usually derived from conventional consolidation tests by fitting the curve for time vs. deformation (for an appropriate load increment) to the theoretical solution for constant u_i. For tests of samples drained at top and bottom, C_v may be interpreted from the curve for log time or square root of time vs. strain (or dial reading) as

$$C_v = \frac{T_v H^2}{4t} \tag{6-26}$$

where H = height of sample, in
t = time for 90% consolidation ($\sqrt{t}$ curve) or 50% consolidation (log t curve), days
T_v = 0.197 for 90% consolidation or 0.848 for 50% consolidation
(See T. W. Lambe and R. V. Whitman, "Soil Mechanics," John Wiley & Sons, Inc., New

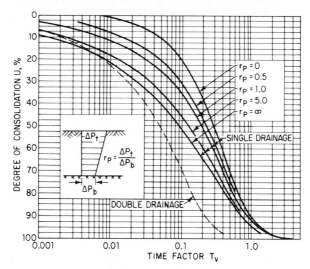

Fig. 6-14. Curves relate degree of consolidation and time factor T_v. *(After Janbu et al., 1956.)*

York, for curve-fitting procedures.) Larger values of C_v are usually obtained with the $\sqrt{t}$ method and appear to be more representative of field conditions.

Secondary compression settlement ρ_s is assumed, for simplicity, to begin on completion of primary consolidation, at time t_{100} corresponding to 100% primary consolidation. ρ_s is then calculated from Eq. (6-27) for a given period t after t_{100}.

$$\rho_s = \sum_{i=1}^{n} H_i C_\alpha \log \frac{t}{t_{100}} \tag{6-27}$$

H_i represents the thickness of compressible layers and C_α is the coefficient of secondary compression given in terms of volumetric strain (Art. 6-9).

The ratio C_α to compression index C_c is nearly constant for a given soil type and is generally within the range of 0.045 ± 0.015. C_α, as determined from consolidation tests (Fig. 6-4b), is extremely sensitive to pressure-increment ratios of less than about 0.5 (standard is 1.0). The effect of overconsolidation, either from natural or construction preload sources, is to significantly reduce C_α. This is an important consideration in the application of preloading for soil improvement. (R. D. Holtz and W. D. Kovacs, "An Introduction to Geotechnical Engineering," Prentice-Hall, Inc., Englewood Cliffs, N.J.)

The rate of *consolidation due to radial drainage* is important for the design of vertical *sand or wick drains*. As a rule, drains are installed in compressible soils to reduce the time required for consolidation and to accelerate the associated gain in soil strength. Vertical drains are typically used in conjunction with preloading as a means of improving the supporting ability and stability of the subsoils.

(S. J. Johnson, "Precompression for Improving Foundation Soils," *ASCE Journal of Soil Mechanics and Foundation Engineering Division,* vol. 96, no. SM1, 1970.)

6-22. Settlement Analysis of Sands

The methods most frequency used to estimate the settlement of foundations supported by relatively free-draining cohesionless soils generally employ empirical correlations between field observations and in situ tests. The primary correlative tests are plate bearing (PLT), cone penetration resistance (CPT), and standard penetration resistance (SPT) (see Art. 6-15). These methods, however, are developed from data bases that contain a number of variables not considered in the correlations and, therefore, should be applied with caution.

Plate Bearing Tests ▪ The most common approach is to scale the results of PLTs to full-size footings in accordance with Eq. (6-9). A less conservative modification of this equation proposed by A. R. S. S. Barazaa is

$$\rho = \left[\frac{2.5B}{1.5 + B} \right]^2 \rho_1 \tag{6-28}$$

where B = footing width, ft
ρ = settlement of full-size bearing plate
ρ_1 = settlement of 1-ft square bearing plate
These equations are not sensitive to the relative density, gradation, and OCR of the soil or to the effects of depth and shape of the footing.

Use of *large-scale load tests* or, ideally, full-scale load tests mitigates many of the difficulties of the preceding approach but is often precluded by costs and schedule considerations. Unless relatively uniform soil deposits are encountered, this approach also requires a number of tests, significantly increasing the cost and time requirements. (See J. K. Mitchell and W. S. Gardner, "In-Situ Measurement of Volume-Change Characteristics," ASCE Specialty Conference on In-Situ Measurement of Soil Properties, Raleigh, N.C., 1975.)

Cone Penetrometer Methods ▪ Correlations between quasi-static penetration resistance q_c and observation of the settlement of bearing plates and small footings form the basis of foundation settlement estimates using CPT data. The Buisman-DeBeer method utilizes a one-dimensional compression formulation. A recommended modification of this approach that considers the influence of the relative density of the soil D_r and increased secant modulus E'_s is

$$\rho = \sum_{i=1}^{n} H_i \frac{1.15\sigma'_{vo}}{(1 + D_r^2)q_c} \log \frac{\sigma'_{vo} + \Delta\sigma_v}{\sigma'_{vo}} \tag{6-29}$$

where ρ = estimated footing settlement. The σ'_{vo} and $\Delta\sigma_v$ parameters represent the average effective overburden pressure and vertical stress change for each layer considered below the base of the foundation (see Art. 6-21). Equation (6-29) has limitations because no consideration is given to: (1) soil stress history, (2) soil gradation, and (3) three-dimensional compression. Also, Eq. (6-29) incorporates an empirical representation of E'_s, given by Eq. (6-30), and has all the limitations thereof (see Art. 6-10).

$$E'_s = 2(1 + D_r^2)q_c \tag{6-30}$$

An alternate approach employs a simplied distribution of vertical strain beneath footings together with q_c data to predict settlements. The settlement ρ [Eq. (6-31)] is given as a function of the strain-influence factor I_z (Art. 6-20), net bearing pressure q_o, deformation modulus E'_s, and correction factors for footing embedment C_1 and time-dependent settlement C_2.

$$\rho = C_1 C_2 q_o \sum_{i=1}^{n} H_i \frac{I_z}{E'_s} \tag{6-31}$$

I_z and E'_s should be determined for each subdivision of the bearing soil zone extending to a depth below the footing base up to four times the footing width. C_1 is given in terms of the overburden pressure at foundation level p_o and the net increase in bearing pressure q_o as:

$$C_1 = 1 - 0.5 \frac{p_o}{q_o} \tag{6-32}$$

The time-dependent settlement (friction lag) component is given by

$$C_2 = 1 + 0.2 \log \left(\frac{t}{0.1} \right) \tag{6-33}$$

where t = time in years after construction.

Figure 6-15 presents a chart for interpreting I_z for rectangular footings. The peak-strain-influence factor I_{zp} required is a function of q_o and the effective overburden pressure p_{vp} at the depth where $I_z = 0$:

$$I_{zp} = 0.5 + 0.1 \left(\frac{q_o}{p_{vp}} \right) \tag{6-34}$$

E_s' in Eq. (6-31) may be taken as $2.5q_c$ for footings for which $L/B \leq 2$ and as $3.5q_c$ for footings with $L/B \geq 10$, where L is the footing length and B the least width of the footing. For intermediate cases, linear interpolation is recommended. It has been suggested that a 50% reduction in the settlement prediction be made for highly overconsolidated soils (say, for OCR > 5).

The foregoing procedures are not applicable to large footings and foundation mats. From field observations relating foundation width B, meters, to ρ/B, the upper limit for ρ/B, for $B > 13.5$ m, is given, in percent, approximately by

$$\frac{\rho}{B} = 0.194 - 0.115 \log \frac{B}{10} \tag{6-35}$$

For the same data base, the best fit of the average ρ/B measurements ranges from about 0.09% $(B = 20$ m$)$ to 0.06% $(B = 80$ m$)$.

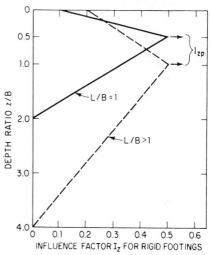

Fig. 6-15. Influence factors for analysis of settlement of footings on sands. *(After J. H. Schmertmann, "Guidelines for a Cone Penetration Test Performance and Design," Report FHWA-TS-78-209, Office of Research and Development, Federal Highway Administration, U.S. Department of Transportation, Washington, D.C., 1978.)*

Standard Penetration Resistance Methods ▪ A variety of methods have been proposed to relate foundation settlement to standard penetration resistance N. An approach proposed by I. Alpan and G. G. Meyerhof appears reasonable and has the advantage of simplicity. Settlement S, in, is computed for $B \leq 4$ ft from

$$S = \frac{8q}{N'} \tag{6-36a}$$

and for $B \geq 4$ ft from

$$S = \frac{12q}{N'} \left(\frac{2B}{1 + B} \right)^2 \tag{6-36b}$$

where q = bearing capacity of soil, tons/ft^2
   B = footing width, ft
N' is given approximately by Eq. (6-37) for $\sigma_{vo}' \leq 40$ psi.

$$N' = \frac{50N}{\sigma_{vo}' + 10} \tag{6-37}$$

and represents N (blows per foot) normalized for $\sigma_{vo}' = 40$ psi (see Fig. 6-16).

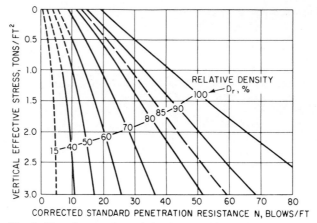

Fig. 6-16. Curves relate relative density with standard penetration resistance and effective vertical stress. *(After W. G. Holtz and H. J. Gibbs, 1956.)*

(G. G. Meyerhof, "Shallow Foundations," *ASCE Journal of Soil Mechanics and Foundation Engineering Division,* vol. 91, no. SM92, 1965; W. G. Holtz and H. J. Gibbs, "Shear Strength of Pervious Gravelly Soils," *Proceedings ASCE,* paper 867, 1956; R. B. Peck, W. E. Hanson, and T. H. Thornburn, "Foundation Engineering," John Wiley & Sons, Inc., New York.)

Laboratory Test Methods ▪ The limitations in developing representative deformation parameters from reconstituted samples were described in Art. 6-10. A possible exception may be for the settlement analyses of foundations supported by compacted fill. Under these circumstances, consolidation tests and stress-path, triaxial shear tests on the fill materials may be appropriate for providing the parameters for application of settlement analyses described for cohesive soils. (D. J. D'Appolonia, E. D'Appolonia, and R. F. Brisette, "Settlement of Spread Footings on Sand," *ASCE Journal of Soil Mechanics and Foundation Engineering Division,* vol. 94, no. SM3, 1968.)

HAZARDOUS SITE AND FOUNDATION CONDITIONS

There are a variety of natural hazards of potential concern in site development and foundation design. Frequently, these hazards are overlooked or not given proper attention, particularly in areas where associated failures have been infrequent.

6-23. Solution-Prone Formations

Significant areas in the eastern and midwestern United States are underlain by formations (carbonate and evaporate rocks) that have been susceptible to dissolution during geologic time. Subsurface voids created by this process range from open jointing to huge caverns. These features have caused catastrophic failures and detrimental settlements of structures as a result of ground loss or surface subsidence.

Special investigations designed to identify rock-solution hazards include geologic reconnaissance, air photo interpretation, and geophysical (resistivity, microgravity, and so on) surveys. To mitigate these hazards, careful attention should be given to:

1. Site drainage to minimize infiltration of surface waters near structures
2. Limitation of excavations to maximize the thickness of soil overburden
3. Continuous foundation systems designed to accommodate a partial loss of support beneath the foundation system
4. Deep foundations socketed into rock and designed solely for socket bond resistance

It is prudent to conduct special **proof testing** of the bearing materials during construction in solution-prone formations. Proof testing often consists of soundings continuously recording the penetration resistance through the overburden and the rate of percussion drilling in the rock. Suspect zones are thus identified and can be improved by excavation and replacement or by in situ grouting.

6-24. Expansive Soils

Soils with a medium to high potential for causing structural damage on expansion or shrinkage are found primarily throughout the Great Plains and Gulf Coastal Plain Physiographic Provinces. Heave or settlement of *active* soils occurs because of a change in soil moisture in response to climatic changes, construction conditions, changes in surface cover, and other conditions that influence the groundwater and evapotransportation regimes. Differential foundation movements are brought about by differential moisture changes in the bearing soils. Figure 6-17 presents a method for classifying the volume-change potential of clay as a function of activity.

Investigations in areas containing potentially expansive soils typically include laboratory swell tests. Infrequently, soil suction measurements are made to provide quantitative evaluations of volume-change potential. Special attention during the field investigation should be given to evaluation of the groundwater regime and to delineation of the depth of active moisture changes. (L. D. Johnson, "Predicting Potential Heave and Heave with Time and Swelling Foundation Soils," Technical Report S-78-7, U.S. Army Engineers Waterways Experiment Station, Vicksburg, Miss., 1978.)

Common design procedures for preventing structural damage include mitigation of moisture changes, removal or modification of expansive materials, and deep foundation support.

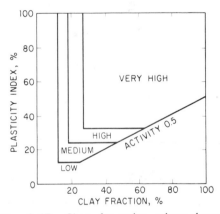

Fig. 6-17. Chart for rating volume-change potential of expansive soils. *(Adapted from Williams, 1958.)*

Horizontal and vertical moisture barriers have been utilized to minimize moisture losses due to evaporation or infiltration and to cut off subsurface groundwater flow into the area of construction. Excavation of potentially active materials and replacement with inert material or with excavated soil modified by the addition of lime have proved feasible where excavation quantities are not excessive. ("Engineering and Design of Foundations on Expansive Soils," U.S. Department of the Army, 1981.)

Deep foundations (typically drilled shafts) have been used to bypass the active zone and to resist or minimize uplift forces that may develop on the shaft. Associated grade beams are usually constructed to prevent development of uplift forces.

6-25. Landslide Hazards

Landslides are usually associated with areas of significant topographic relief, which are characterized by relatively weak sedimentary rocks (shales, siltstones, and so forth) or by relatively impervious soil deposits containing interbedded water-bearing strata. Under these circumstances, slides that have occurred in the geologic past, whether or not currently active, represent a significant risk for hillside site development. In general, hillside development in potential landslide areas is a most hazardous undertaking. If there are alternatives, one of those should be adopted.

Detailed geologic studies are required to evaluate slide potential and should emphasize detection of old slide areas. Procedures that tend to stabilize an active slide or to provide for the continued stability of an old slide include excavation at the head of the sliding mass to reduce the driving force, subsurface drainage to depress piezometric levels along potential sliding surface, and buttressing at the toe of the potential sliding mass to provide a force resistant to slide movement.

(P. B. Schnabel, J. Lysmer, and H. B. Seed, "A Computer Program for Earthquake Response Analysis of Horizontally Layered Sites," Report EERC 72-12, Earthquake Engineering Research Center, University of California, Berkeley, 1972; H. B. Seed, P. P. Martin, and J. Lysmer, "Pore-Water Pressure Changes During Soil Liquefaction," *ASCE Journal of Geotechnical Engineering Division,* vol. 102, no. GT4, 1975; K. L. Lee and A. Albaisa, "Earthquake-Induced Settlements in Saturated Sands," *ASCE Journal of Geotechnical Engineering Division,* vol. 100, no. GT4, 1974.)

EARTH AND WATER PRESSURE ON FOUNDATIONS

6-26. Lateral Active Pressure

Water exerts against a vertical surface a horizontal pressure equal to the vertical pressure. At any level, the vertical pressure equals the weight of a 1-ft^2 column of water above that level. Hence, the horizontal pressure p, lb/ft^3, at any level is

$$p = wh \tag{6-38}$$

where w = unit weight of water, lb/ft^3
h = depth of water, ft
The pressure diagram is triangular (Fig. 6-18). Equation (6-38) also can be written

$$p = Kwh \tag{6-39}$$

where K = pressure coefficient = 1.00.

The resultant, or total, pressure, lb/lin ft, represented by the area of the hydrostatic-pressure diagram, is

$$P = K \frac{wh^2}{2} \tag{6-40}$$

It acts at a distance $h/3$ above the base of the triangle.

Soil also exerts lateral pressure. But the amount of this pressure depends on the type of soil, its compaction or consistency, and its degree of saturation, and on the resistance of the structure to the pressure. Also, the magnitude of passive pressure differs from that of active pressure.

Active pressure tends to move a structure in the direction in which the pressure acts. **Passive pressure** opposes motion of a structure.

Free-standing walls retaining cuts in sand tend to rotate slightly around the base. Behind such a wall, a wedge of sand ABC (Fig. 6-19a) tends to shear along plane AC. C. A. Coulomb determined that the ratio of sliding resistance to sliding force is a minimum when AC makes an angle of $45° + \phi/2$ with the horizontal, where ϕ is the angle of internal friction of the soil, deg.

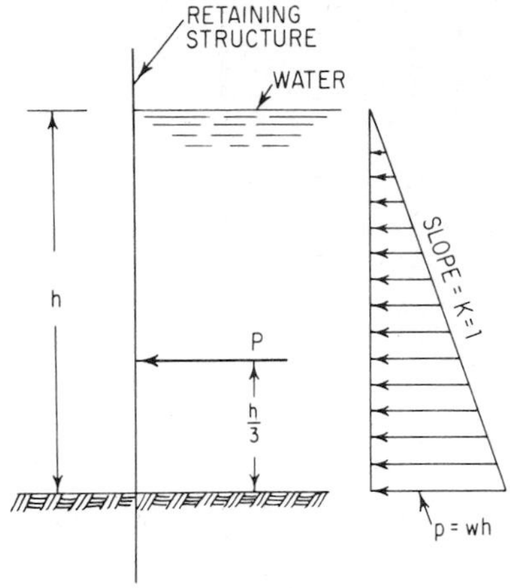

For triangular pressure distribution (Fig. 6-19b), the active lateral pressure of a cohesionless soil at a depth h, ft, is

$$p = K_a wh \qquad (6-41)$$

where K_a = coefficient of active earth pressure
w = unit weight of soil, lb/ft³
The total active pressure, lb/lin ft, is

$$E_a = K_a \frac{wh^2}{2} \qquad (6-42)$$

Because of frictional resistance to sliding at the face of the wall, E_a is inclined at an angle δ with the normal to the wall, where δ is the angle of wall friction, deg (Fig. 6-19a). If the face of the wall is vertical, the horizontal active pressure equals $E_a \cos \delta$. If the face makes an angle β with the vertical (Fig. 6-19a), the active pressure equals $E_a \cos (\delta + \beta)$. The resultant acts at a distance of $h/3$ above the base of the wall.

Fig. 6-18. Pressure diagram for water.

If the ground slopes upward from the top of the wall at an angle α, deg, with the horizontal, then for cohesionless soils

$$K_a = \frac{\cos^2 (\phi - \beta)}{\cos^2 \beta \cos (\delta + \beta) \left[1 + \sqrt{\dfrac{\sin (\phi + \delta) \sin (\phi - \alpha)}{\cos (\delta + \beta) \cos (\alpha - \beta)}} \right]^2} \qquad (6-43)$$

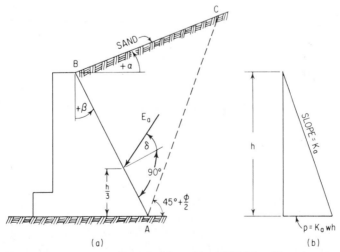

Fig. 6-19. Free-standing wall with sand backfill (a) is subjected to triangular pressure distribution (b).

The effect of wall friction on K_a is small and usually is neglected. For $\delta = 0$,

$$K_a = \frac{\cos^2 (\phi - \beta)}{\cos^3 \beta \left[1 + \sqrt{\dfrac{\sin \phi \sin (\phi - \alpha)}{\cos \beta \cos (\alpha - \beta)}} \right]^2} \tag{6-44}$$

Table 6-7 lists values of K_a determined from Eq. (6-44). Approximate values of ϕ and unit weights for various soils are given in Table 6-8.

TABLE 6-7   Active-Lateral-Pressure Coefficients K_a

$\phi =$		10°	15°	20°	25°	30°	35°	40°
	$\alpha = 0$	0.70	0.59	0.49	0.41	0.33	0.27	0.22
	$\alpha = 10°$	0.97	0.70	0.57	0.47	0.37	0.30	0.24
$\beta = 0$	$\alpha = 20°$	. . .	. . .	0.88	0.57	0.44	0.34	0.27
	$\alpha = 30°$	. . .	. . .	. . .	. . .	0.75	0.43	0.32
	$\alpha = \phi$	0.97	0.93	0.88	0.82	0.75	0.67	0.59
	$\alpha = 0$	0.76	0.65	0.55	0.48	0.41	0.43	0.29
	$\alpha = 10°$	1.05	0.78	0.64	0.55	0.47	0.38	0.32
$\beta = 10°$	$\alpha = 20°$	. . .	. . .	1.02	0.69	0.55	0.45	0.36
	$\alpha = 30°$	. . .	. . .	. . .	. . .	0.92	0.56	0.43
	$\alpha = \phi$	1.05	1.04	1.02	0.98	0.92	0.86	0.79
	$\alpha = 0$	0.83	0.74	0.65	0.57	0.50	0.43	0.38
	$\alpha = 10°$	1.17	0.90	0.77	0.66	0.57	0.49	0.43
$\beta = 20°$	$\alpha = 20°$	. . .	. . .	1.21	0.83	0.69	0.57	0.49
	$\alpha = 30°$	. . .	. . .	. . .	. . .	1.17	0.73	0.59
	$\alpha = \phi$	1.17	1.20	1.21	1.20	1.17	1.12	1.06
	$\alpha = 0$	0.94	0.86	0.78	0.70	0.62	0.56	0.49
	$\alpha = 10°$	1.37	1.06	0.94	0.83	0.74	0.65	0.56
$\beta = 30°$	$\alpha = 20°$	. . .	. . .	1.51	1.06	0.89	0.77	0.66
	$\alpha = 30°$	. . .	. . .	. . .	. . .	1.55	0.99	0.79
	$\alpha = \phi$	1.37	1.45	1.51	1.54	1.55	1.54	1.51

TABLE 6-8   Angles of Internal Friction and Unit Weights of Soils

Type of soil	Density or consistency	Angle of internal friction ϕ, deg	Unit weight w, lb per ft^3
Coarse sand or sand and gravel	Compact	40	140
	Loose	35	90
Medium sand	Compact	40	130
	Loose	30	90
Fine silty sand or sandy silt	Compact	30	130
	Loose	25	85
Uniform silt	Compact	30	135
	Loose	25	85
Clay-silt	Soft to medium	20	90–120
Silty clay	Soft to medium	15	90–120
Clay	Soft to medium	0–10	90–120

For level ground at the top of the wall ($\alpha = 0$):

$$K_a = \frac{\cos^2(\phi - \beta)}{\cos^3 \beta \left(1 + \dfrac{\sin \phi}{\cos \beta} \right)^2}$$

If in addition, the back face of the wall is vertical ($\beta = 0$), Rankine's equation is obtained:

$$K_a = \frac{1 - \sin \phi}{1 + \sin \phi} \qquad\qquad (6\text{-}45\,a)$$

Coulomb derived the trigonometric equivalent:

$$K_a = \tan^2 \left(45° - \frac{\phi}{2} \right) \qquad\qquad (6\text{-}45\,b)$$

When information on the value of the angle of wall friction is not available, δ may be taken equal to $\phi/2$ for determining the horizontal component of E_a.

Note: Even light compaction may permanently increase the earth pressure into the passive range. This may be compensated for in wall design by use of a safety factor of at least 2.5.

Unyielding walls retaining cuts in sand, such as the abutment walls of a rigid-frame concrete bridge or foundation walls braced by floors, do not allow shearing resistance to develop in the sand along planes that can be determined analytically. For such walls, triangular pressure diagrams may be assumed, and K_a may be taken equal to 0.5. But only sand or gravel should be permitted for the backfill, and compaction should be light within 5 to 10 ft of the walls. See preceding Note for free-standing walls.

Braced walls retaining cuts in sand (Fig. 6-20a) are subjected to earth pressure gradually and develop resistance in increments as excavation proceeds and braces are installed. Such walls tend to rotate about a point in the upper portion. Hence, the active pressures do not vary linearly with depth. Field measurements have yielded a variety of curves for the pressure diagram, of

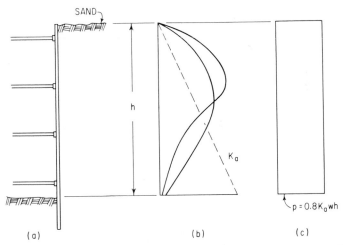

Fig. 6-20. Braced wall retaining sand (a) may have to resist pressure of the type shown in (b). Uniform pressure distribution (c) may be assumed for design.

which two types are shown in Fig. 6-20b. Consequently, some authorities have recommended a trapezoidal pressure diagram, with a maximum ordinate

$$p = 0.8K_a wh \tag{6-46}$$

K_a may be obtained from Table 6-7. The total pressure exceeds that for a triangular distribution.

Figure 6-21 shows earth-pressure diagrams developed for a sandy soil and a clayey soil. In both cases, the braced wall is subjected to a 3-ft-deep surcharge, and height of wall is 34 ft. For the sandy soil (Fig. 6-21a), Fig. 6-21b shows the pressure diagram assumed. The maximum pressure can be obtained from Eq. (6-46), with $h = 34 + 3 = 37$ ft and K_a assumed as 0.30 and w as 110 lb/ft^3.

$$p_1 = 0.8 \times 0.3 \times 110 \times 37 = 975 \text{ lb/ft}^2$$

The total pressure is estimated as

$$P = 0.8 \times 975 \times 37 = 28,900 \text{ lb/lin ft}$$

The equivalent maximum pressure for a trapezoidal diagram for the 34-ft height of the wall then is

$$p = \frac{28,900}{0.8 \times 34} = 1060 \text{ lb/ft}^2$$

Assumption of a uniform distribution (Fig. 6-20c), however, simplifies the calculations and has little or no effect on the design of the sheeting and braces, which should be substantial to withstand construction abuses. Furthermore, trapezoidal loading terminating at the level of the excavation may not apply if piles are driven inside the completed excavation. The shocks may temporarily decrease the passive resistance of the sand in which the wall is embedded and lower the inflection point. This would increase the span between the inflection point and the lowest brace and increase the pressure on that brace. Hence, uniform pressure distribution may be more applicable than trapezoidal for such conditions.

See Note for free-standing walls.

Flexible bulkheads retaining sand cuts are subjected to active pressures that depend on the fixity of the anchorage. If the anchor moves sufficiently or the tie from the anchor to the upper portion of the bulkhead stretches enough, the bulkhead may rotate slightly about a point near the bottom. In that case, the sliding-wedge theory may apply. The pressure distribution may be taken as triangular, and Eqs. (6-41) to (6-45) may be used. But if the anchor does not yield, then pressure distributions much like those in Fig. 6-20b for a braced cut may occur. Either a trapezoidal or uniform pressure distribution may be assumed, with maximum pressure given by Eq. (6-46). Stresses in the tie should be kept low because it may have to resist unanticipated pressures, especially those resulting from a redistribution of forces from soil arching. The safety factor for design of ties and anchorages should be at least twice that used in conventional design.

Free-standing walls retaining plastic-clay cuts (Fig. 6-22a) may have to resist two types of active lateral pressure, both with triangular distribution. If the shearing resistance is due to cohesion only, a clay bank may be expected to stand with a vertical face without support for a height, ft, of

$$h' = \frac{2c}{w} \tag{6-47}$$

where $2c$ = unconfined compressive strength of clay, lb/ft^2
w = unit weight of clay, lb/ft^3

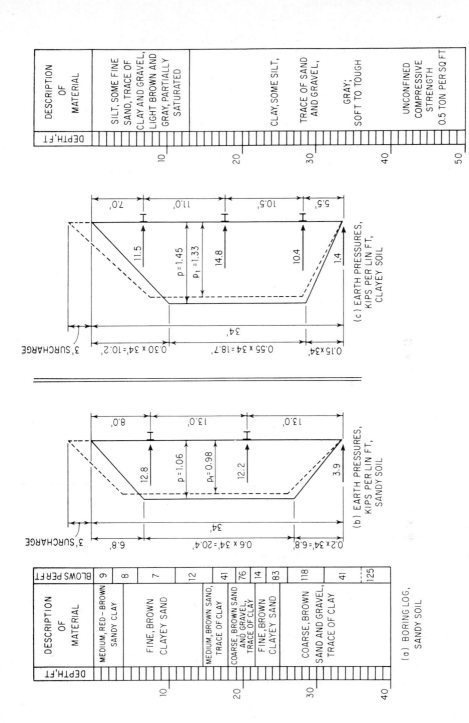

Fig 6-21. Assumed trapezoidal pressure diagrams for a braced wall.

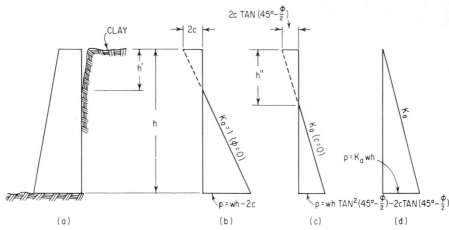

Fig. 6-22. Free-standing wall retaining clay (a) may have to resist the pressure distribution shown in (b) and (d). For mixed soils, the distribution may approximate that shown in (c).

So if there is a slight rotation of the wall about its base, the upper portion of the clay cut will stand vertically without support for a depth h'. Below that, the pressure will increase linearly with depth as if the clay were a heavy liquid (Fig. 6-22b):

$$p = wh - 2c$$

The total pressure, lb/lin ft, then is

$$E_a = \frac{w}{2}\left(h - \frac{2c}{w}\right)^2 \tag{6-48}$$

It acts at a distance $(h - 2c/w)/3$ above the base of the wall. These equations assume wall friction is zero, the back face of the wall is vertical, and the ground is level.

This condition is likely to be temporary. In time, the clay will consolidate. The pressure distribution may become approximately triangular (Fig. 6-22d) from the top of the wall to the base. The pressures then may be calculated from Eqs. (6-41) to (6-45) with an apparent angle of internal friction for the soil (for example, see the values of ϕ in Table 6-8). The wall should be designed for the pressures producing the highest stresses and overturning moments.

Note: The finer the backfill material, the more likely it is that pressures greater than active will develop, because of plastic deformations, water-level fluctuation, temperature changes, and other effects. As a result, it would be advisable to use in design at least the coefficient for earth pressure at rest:

$$K_o = 1 - \sin\phi \tag{6-49}$$

The safety factor should be at least 2.5.

Clay should not be used behind retaining walls, where other economical alternatives are open. The swelling type especially should be avoided because it can cause high pressures and progressive shifting or rotation of the wall.

For a mixture of cohesive and cohesionless soils, the pressure distribution may temporarily be as shown in Fig. 6-22c. The height, ft, of the unsupported vertical face of the clay is

$$h'' = \frac{2c}{w\tan(45° - \phi/2)} \tag{6-50}$$

The pressure at the base is

$$p = wh \tan^2 \left(45° - \frac{\phi}{2} \right) - 2c \tan \left(45° - \frac{\phi}{2} \right) \tag{6-51}$$

The total pressure, lb/lin ft, is

$$E_a = \frac{w}{2} \left[h \tan \left(45° - \frac{\phi}{2} \right) - \frac{2c}{w} \right]^2 \tag{6-52}$$

It acts at a distance $(h - h'')/3$ above the base of the wall.

Braced walls retaining clay cuts (Fig. 6-23a) also may have to resist two types of active lateral pressure. As for sand, the pressure distribution may temporarily be approximated by a

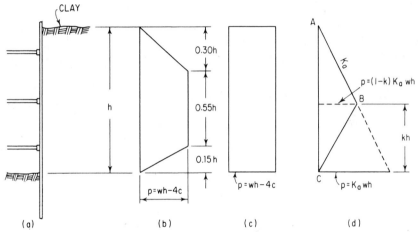

Fig. 6-23. Braced wall retaining clay (a) may have to resist pressures approximated by the pressure distributions in (b) and (d). Uniform pressure distribution (c) may be assumed in design.

trapezoidal diagram (Fig. 6-23b). On the basis of field observations, R. B. Peck has recommended a maximum pressure of

$$p = wh - 4c \tag{6-53}$$

and a total pressure, lb/lin ft, of

$$E_a = \frac{1.55h}{2} (wh - 4c) \tag{6-54}$$

[R. B. Peck, "Earth Pressure Measurements in Open Cuts, Chicago (Ill.) Subway," *Transactions, American Society of Civil Engineers*, 1943, pp. 1008–1036.]

Figure 6-21e shows a trapezoidal earth-pressure diagram determined for the clayey-soil condition of Fig. 6-21d. The weight of the soil is taken as 120 lb/ft³; c is assumed as zero and the active-lateral-pressure coefficient as 0.3. Height of the wall is 34 ft, surcharge 3 ft. Then, the maximum pressure, obtained from Eq. (6-41) since the soil is clayey, not pure clay, is

$$p_1 = 0.3 \times 120 \times 37 = 1330 \text{ lb/ft}^2$$

From Eq. (6-54) with the above assumptions, the total pressure is

$$P = \frac{1.55}{2} \times 37 \times 1330 = 38,100 \text{ lb/lin ft}$$

The equivalent maximum pressure for a trapezoidal diagram for the 34-ft height of wall is

$$p = \frac{38,100}{34} \times \frac{2}{1.55} = 1450 \text{ lb/ft}^2$$

To simplify calculations, a uniform pressure distribution may be used instead (Fig. 6-23c).

If after a time the clay should attain a consolidated equilibrium state, the pressure distribution may be better represented by a triangular diagram ABC (Fig. 6-23d), as suggested by G. P. Tschebotarioff. The peak pressure may be assumed at a distance of $kh = 0.4h$ above the excavation level for a stiff clay; that is, $k = 0.4$. For a medium clay, k may be taken as 0.25, and for a soft clay, as 0. For computing the pressures, K_a may be estimated from Table 6-7 with an apparent angle of friction obtained from laboratory tests or approximated from Table 6-8. The wall should be designed for the pressures producing the highest stresses and overturning moments.

See also Note for free-standing walls.

Flexible bulkheads retaining clay cuts and anchored near the top similarly should be checked for two types of pressures. When the anchor is likely to yield slightly or the tie to stretch, the pressure distribution in Fig. 6-23d with $k = 0$ may be applicable. For an unyielding anchor, any of the pressure distributions in Fig. 6-23 may be assumed, as for a braced wall. The safety factor for design of ties and anchorages should be at least twice that used in conventional design. See also Note for free-standing walls.

Backfill placed against a retaining wall should preferably be gravel to facilitate drainage. Also, weepholes should be provided through the wall near the bottom and a drain installed along the footing, to conduct water from the back of the wall and prevent buildup of hydrostatic pressures.

Saturated or submerged soil imposes substantially greater pressure on a retaining wall than dry or moist soil. The active lateral pressure for a soil-fluid backfill is the sum of the hydrostatic pressure and the lateral soil pressure based on the buoyed unit weight of the soil. This weight roughly may be 60% of the dry weight.

Surcharge, or loading imposed on a backfill, increases the active lateral pressure on a wall and raises the line of action of the total, or resultant, pressure. A surcharge w_s, lb/ft^2, uniformly distributed over the entire ground surface may be taken as equivalent to a layer of soil of the same unit weight w as the backfill and with a thickness of w_s/w. The active lateral pressure, lb/ft^2, due to the surcharge, from the backfill surface down, then will be $K_a w_s$. This should be added to the lateral pressures that would exist without the surcharge. K_a may be obtained from Table 6-7.

(A. Caquot and J. Kérisel, "Tables for Calculation of Passive Pressure, Active Pressure, and Bearing Capacity of Foundations," Gauthier-Villars, Paris.)

6-27. Passive Lateral Pressure

As defined in Art. 6-26, active pressure tends to move a structure in the direction in which pressure acts, whereas passive pressure opposes motion of a structure.

Passive pressures of cohesionless soils, resisting movement of a wall or anchor, develop because of internal friction in the soils. Because of friction between soil and wall, the failure surface is curved, not plane as assumed in the Coulomb sliding-wedge theory (Art. 6-26). Use of the Coulomb theory yields unsafe values of passive pressure when the effects of wall friction are included.

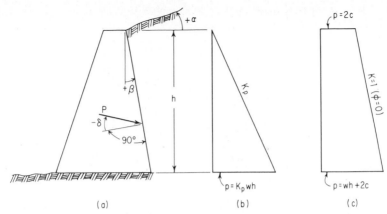

Fig. 6-24. Passive pressure on a wall (*a*) may vary as shown in (*b*) for sand or in (*c*) for clay.

Total passive pressure, lb/lin ft, on a wall or anchor extending to the ground surface (Fig. 6-24*a*) may be expressed for sand in the form

$$P = K_p \frac{wh^2}{2} \tag{6-55}$$

where K_p = coefficient of passive lateral pressure
w = unit weight of soil, lb/ft^3
h = height of wall or anchor to ground surface, ft
The pressure distribution usually assumed for sand is shown in Fig. 6-24*b*. Table 6-9 lists values of K_p for a vertical wall face ($\beta = 0$) and horizontal ground surface ($\alpha = 0$), for curved surfaces of failure. (Many tables and diagrams for determining passive pressures are given in A. Caquot and J. Kérisel, "Tables for Calculation of Passive Pressure, Active Pressure, and Bearing Capacity of Foundations," Gauthier-Villars, Paris.).

TABLE 6-9   Passive Lateral-Pressure Coefficients K_p*

ϕ =	10°	15°	20°	25°	30°	35°	40°
$\delta = 0$	1.42	1.70	2.04	2.56	3.00	3.70	4.60
$\delta = -\phi/2$	1.56	1.98	2.59	3.46	4.78	6.88	10.38
$\delta = -\phi$	1.65	2.19	3.01	4.29	6.42	10.20	17.50

*For vertical wall face ($\beta = 0$) and horizontal ground surface ($\alpha = 0$).

Since a wall usually transmits a downward shearing force to the soil, the angle of wall friction δ correspondingly is negative (Fig. 6-24*a*). For embedded portions of structures, such as anchored sheetpile bulkheads, δ and the angle of internal friction ϕ of the soil reach their peak values simultaneously in dense sand. For those conditions, if specific information is not available, δ may be assumed as $- \frac{2}{3}\phi$ (for $\phi > 30°$). For such structures as a heavy anchor block subjected to a horizontal pull or thrust, δ may be taken as $-\phi/2$ for dense sand. For those cases, the wall friction develops as the sand is pushed upward by the anchor and is unlikely to reach its maximum value before the internal resistance of the sand is exceeded.

When wall friction is zero ($\delta = 0$), the failure surface is a plane inclined at an angle of $45° - \phi/2$ with the horizontal. The sliding-wedge theory then yields

$$K_p = \frac{\cos^2(\phi + \beta)}{\cos^3 \beta \left[1 - \sqrt{\dfrac{\sin \phi \sin(\phi + \alpha)}{\cos \beta \cos(\alpha - \beta)}}\,\right]^2} \tag{6-56}$$

When the ground is horizontal ($\alpha = 0$):

$$K_p = \frac{\cos^2(\phi + \beta)}{\cos^3 \beta(1 - \sin \phi/\cos \beta)^2} \tag{6-57}$$

If, in addition, the back face of the wall is vertical ($\beta = 0$):

$$K_p = \frac{1 + \sin \phi}{1 - \sin \phi} = \tan^2\left(45° + \frac{\phi}{2}\right) = \frac{1}{K_a} \tag{6-58}$$

The first line of Table 6-9 lists values obtained from Eq. (6-58).

Continuous anchors in sand ($\phi = 33°$), when subjected to horizontal pull or thrust, develop passive pressures, lb/lin ft, of about

$$P = 1.5wh^2 \tag{6-59}$$

where h = distance from bottom of anchor to the surface, ft.

This holds for ratios of h to height d, ft, of anchor of 1.5 to 5.5, and assumes a horizontal ground surface and vertical anchor face. For a square anchor within the same range of h/d, approximately

$$P = \left(2.50 + \frac{h}{8d}\right)^2 d\,\frac{wh^2}{2} \tag{6-60}$$

where P = passive lateral pressure, lb
d = length and height of anchor, ft

Passive pressures of cohesive soils, resisting movement of a wall or anchor extending to the ground surface, depend on the unit weight of the soil w and its unconfined compressive strength $2c$, psf. At a distance h, ft, below the surface, the passive lateral pressure, psf, is

$$p = wh + 2c \tag{6-61}$$

The total pressure, lb/lin ft, is

$$P = \frac{wh^2}{2} + 2ch \tag{6-62}$$

and acts at a distance, ft, above the bottom of the wall or anchor of

$$\frac{h(wh + 6c)}{3(wh + 4c)}$$

The pressure distribution for plastic clay is shown in Fig. 6-24c.

Continuous anchors in plastic clay, when subjected to horizontal pull or thrust, develop passive pressures, lb/lin ft, of about

$$P = cd\left[8.7 - \frac{11,600}{(h/d + 11)^3}\right] \tag{6-63}$$

where h = distance from bottom of anchor to surface, ft
d = height of anchor, ft

Equation (6-63) is based on tests made with horizontal ground surface and vertical anchor face.

Safety factors should be applied to the passive pressures computed from Eqs. (6-55) to (6-63) for design use. Experience indicates that a safety factor of 2 is satisfactory for clean sands and gravels. For clay, a safety factor of 3 may be desirable because of uncertainties as to effective shearing strength.

(G. P. Tschebotarioff, "Soil Mechanics, Foundations, and Earth Structures," Mc-Graw-Hill Book Company, New York; K. Terzaghi and R. B. Peck, "Soil Mechanics in Engineering Practice," John Wiley & Sons, Inc., New York; Leo Casagrande, "Comments on Conventional Design of Retaining Structures," *ASCE Journal of Soil Mechanics and Foundations Engineering Division,* 1973, pp. 181–198; H. F. Winterkorn and H. Fang, "Foundation Engineering Handbook," Van Nostrand Reinhold Company, New York.)

6-28. Vertical Earth Pressure on Conduit

The vertical load on an underground conduit depends principally on the weight of the prism of soil directly above it. But the load also is affected by vertical shearing forces along the sides of this prism. Caused by differential settlement of the prism and adjoining soil, the shearing forces may be directed up or down. Hence, the load on the conduit may be less or greater than the weight of the soil prism directly above it.

Conduits are classified as ditch or projecting, depending on installation conditions that affect the shears. A ditch conduit is a pipe set in a relatively narrow trench dug in undisturbed soil (Fig. 6-25). Backfill then is placed in the trench up to the original ground surface. A projecting conduit is a pipe over which an embankment is placed.

A projecting conduit may be positive or negative, depending on the extent of the embankment vertically. A positive projecting conduit is installed in a shallow bed with the pipe top above the surface of the ground. Then, the embankment is placed over the pipe (Fig. 6-26a). A negative projecting conduit is set in a narrow, shallow trench with the pipe top below the original ground surface (Fig. 6-26b). Then, the ditch is backfilled, after which the embankment is placed. The load on the conduit is less when the backfill is not compacted.

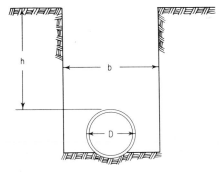

Fig. 6-25. Ditch conduit.

Load on underground pipe also may be reduced by the imperfect-ditch method of construction. This starts out as for a positive projecting conduit, with the pipe at the original ground surface. The embankment is placed and compacted for a few feet above the pipe. But then, a trench as wide as the conduit is dug down to it through the compacted soil. The trench is backfilled with a loose, compressible soil (Fig. 6-26c). After that, the embankment is completed.

The load, lb/lin ft, on a rigid ditch conduit may be computed from

$$W = C_D whb \qquad (6\text{-}64)$$

and on a flexible ditch conduit from

$$W = C_D whD \qquad (6\text{-}65)$$

where C_D = load coefficient for ditch conduit
w = unit weight of fill, lb/ft^3
h = height of fill above top of conduit, ft
b = width of ditch at top of conduit, ft
D = outside diameter of conduit, ft

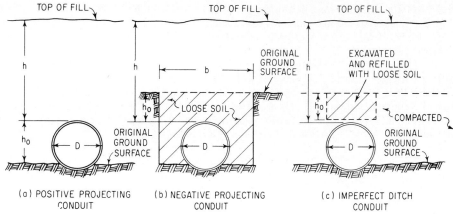

Fig. 6-26.   Projecting conduit.

From the equilibrium of vertical forces, including shears, acting on the backfill above the conduit, C_D may be determined:

$$C_D = \frac{1 - e^{-kh/b}}{k} \frac{b}{h} \qquad (6\text{-}66)$$

where $e = 2.718$
$\quad k = 2K_a \tan \theta$
$\quad K_a =$ coefficient of active earth pressure [Eq. (6-45) and Table 6-7]
$\quad \theta =$ angle of friction between fill and adjacent soil ($\theta \le \phi$, angle of internal friction of fill)

Table 6-10 gives values of C_D for $k = 0.33$ for cohesionless soils, $k = 0.30$ for saturated topsoil, and $k = 0.26$ and 0.22 for clay (usual maximum and saturated).

Vertical load, lb/lin ft, on conduit installed by tunneling may be estimated from

$$W = C_D b(wh - 2c) \qquad (6\text{-}67)$$

where $c =$ cohesion of the soil, or half the unconfined compressive strength of the soil, psf. The load coefficient C_D may be computed from Eq. (6-66) or obtained from Table 6-10 with $b =$

TABLE 6-10 Load Coefficients C_D for Ditch Conduit

h/b	Cohesionless soils	Saturated topsoil	Clay $k = 0.26$	$k = 0.22$
1	0.85	0.86	0.88	0.89
2	0.75	0.75	0.78	0.80
3	0.63	0.67	0.69	0.73
4	0.55	0.58	0.62	0.67
5	0.50	0.52	0.56	0.60
6	0.44	0.47	0.51	0.55
7	0.39	0.42	0.46	0.51
8	0.35	0.38	0.42	0.47
9	0.32	0.34	0.39	0.43
10	0.30	0.32	0.36	0.40
11	0.27	0.29	0.33	0.37
12	0.25	0.27	0.31	0.35
Over 12	$3.0b/h$	$3.3b/h$	$3.9b/h$	$4.5b/h$

maximum width of tunnel excavation, ft, and h = distance from tunnel top to ground surface, ft.

For a ditch conduit, shearing forces extend from the pipe top to the ground surface. For a projecting conduit, however, if the embankment is sufficiently high, the shear may become zero at a horizontal plane below grade, the plane of equal settlement. Load on a projecting conduit is affected by the location of this plane.

Vertical load, lb/lin ft, on a positive projecting conduit may be computed from

$$W = C_P whD \tag{6-68}$$

where C_P = load coefficient for positive projecting conduit. Formulas have been derived for C_P and the depth of the plane of equal settlement. These formulas, however, are too lengthy for practical application, and the computation does not appear to be justified by the uncertainties in actual relative settlement of the soil above the conduit. Tests may be made in the field to determine C_P. If so, the possibility of an increase in earth pressure with time should be considered. For a rough estimate, C_P may be assumed as 1 for flexible conduit and 1.5 for rigid conduit.

The vertical load, lb/lin ft, on negative projecting conduit may be computed from

$$W = C_N whb \tag{6-69}$$

where C_N = load coefficient for negative projecting conduit
h = height of fill above top of conduit, ft
b = horizontal width of trench at top of conduit, ft
The load on an imperfect ditch conduit may be obtained from

$$W = C_N whD \tag{6-70}$$

where D = outside diameter of conduit, ft.

Formulas have been derived for C_N, but they are complex, and insufficient values are available for the parameters involved. As a rough guide, C_N may be taken as 0.9 when depth of cover exceeds conduit diameter.

Superimposed surface loads increase the load on an underground conduit. The magnitude of the increase depends on the depth of the pipe below grade and the type of soil. For moving loads, an impact factor of about 2 should be applied. A superimposed uniform load w', lb/ft^2, of large extent may be treated for projecting conduit as an equivalent layer of embankment with a thickness, ft, of w'/w. For ditch conduit, the load due to the soil should be increased by $bw'e^{-kh/b}$, where $k = 2K_a \tan \theta$, as in Eq. (6-66). The increase caused by concentrated loads can be estimated by assuming the loads to spread out linearly with depth, at an angle of about 30° with the vertical (see Art. 6-20).

(G. A. Leonards, "Foundations Engineering," McGraw-Hill Book Company, New York; M. G. Spangler, "Soil Engineering," International Textbook Company, Scranton, Pa.; "Handbook of Steel Drainage and Highway Construction Products," American Iron and Steel Institute, New York.)

SOIL IMPROVEMENT

Soil for foundations can be altered to conform to desired characteristics. Whether this should be done depends on the cost of alternatives.

Investigations of soil and groundwater conditions on a site should indicate whether soil improvement, or stabilization, is needed. Tests may be necessary to determine which of several applicable techniques may be feasible and economical. Table 6-11 lists some conditions for which soil improvement should be considered and the methods that may be used.

TABLE 6-11   Where Soil Improvement May Be Economical

Soil deficiency	Probable type of failure	Probable cause	Possible remedies
Slope instability	Slides on slope	Pore-water pressure	Drain; flatten slope; freeze
		Loose granular soil	Compact
		Weak soil	Mix or replace with select material
	Mud flow	Excessive water	Exclude water
	Slides—movement at toe	Toe instability	Place toe fill, and drain
Low bearing capacity	Excessive settlement	Saturated clay	Consolidate with surcharge, and drain
		Loose granular soil	Compact; drain; increase footing depth; mix with chemicals
		Weak soil	Superimpose thick fill; mix or replace with select material; inject or mix with chemicals; freeze (if saturated); fuse with heat (if unsaturated)
Heave	Excessive rise	Frost	For buildings: place foundations below frost line; insulate refrigeration-room floors; refrigerate to keep ground frozen
			For roads: Remove fines from gravel; replace with nonsusceptible soil
		Expansion of clay	Exclude water; replace with granular soil
Excessive permeability	Seepage	Pervious soil or fissured rock	Mix or replace soil with select material; inject or mix soil with chemicals; construct cutoff wall with grout; enclose with sheetpiles and drain
"Quick" bottom	Loss of strength	Flow under cofferdam	Add berm against cofferdam inner face; increase width of cofferdam between lines of sheeting; drain with wellpoints outside the cofferdam

As indicated in the table, soil stabilization may increase strength, increase or decrease permeability, reduce compressibility, improve stability, or decrease heave due to frost or swelling. The main techniques used are constructed fills, replacement of unsuitable soils, surcharges, reinforcement, mechanical stabilization, thermal stabilization, and chemical stabilization.

6-29. Mechanical Stabilization of Soils

This comprises a variety of techniques for rearranging, adding, or removing soil particles. The objective usually is to increase soil density, decrease water content, or improve gradation. Particles may be rearranged by blending the layers of a statified soil, remolding an undisturbed soil, or densifying a soil. Sometimes, the desired improvement can be obtained by drainage alone. Often, however, compactive effort plus water control is needed.

Embankments ▪ Earth often has to be placed over the existing ground surface to level or raise it. Such constructed fills may create undesirable conditions because of improper compaction, volume changes, and unexpected settlement under the weight of the fill. To prevent such conditions, fill materials and their gradation, placement, degree of compaction, and thickness should be suitable for properly supporting the expected loads.

Fills may be either placed dry with conventional earthmoving equipment and techniques or wet by hydraulic dredges. Wet fills are used mainly for filling behind bulkheads or for large fills.

A variety of soils and grain sizes are suitable for fills for most purposes. Inclusion of organic matter or refuse should, however, be prohibited. Economics usually require that the source of fill material be as close as possible to the site. For most fills, soil particles in the 18 in below foundations, slabs, or the ground surface should not be larger than 3 in in any dimension.

For determining the suitability of a soil as fill and for providing a standard for compaction, the moisture-density relationship test, or Proctor test (ASTM D698 and D1557), is often used. Several of these laboratory tests should be performed on the borrowed material, to establish moisture-density curves. The peak of a curve indicates the maximum density achievable in the laboratory by the test method as well as the optimum moisture content. ASTM D1557 should be used as the standard when high bearing capacity and low compressibility are required; ASTM D698 should be used when requirements are lower, for example, for fills under parking lots.

The two ASTM tests represent different levels of compactive effort. But much higher compactive effort may be employed in the field than that used in the laboratory. Thus, a different moisture-density relationship may be produced on the site. Proctor test results, therefore, should not be considered an inherent property of the soil. Nevertheless, the test results indicate the proposed fill material's sensitivity to moisture content and the degree of field control that may be required to obtain the specified density.

Fill Compaction ▪ The degree of compaction required for a fill is usually specified as a minimum percentage of the maximum dry density obtained in the laboratory tests. This compaction is required to be accomplished within a specific moisture range. Minimum densities of 90 to 95% of the maximum density are suitable for most fills. Under roadways, footings, or other highly loaded areas, however, 100% compaction is often required. In addition, moisture content within 2 to 4% of the optimum moisture content usually is specified.

Field densities can be greater than 100% of the maximum density obtained in the laboratory test. Also, with greater compactive effort, such densities can be achieved with moisture contents that do not lie on the curves plotted from laboratory results. (Fine-grained soils should not be overcompacted on the dry side of optimum because when they get wet, they may swell and soften significantly.)

For most projects, lift thickness should be restricted to 8 to 12 in, each lift being compacted before the next lift is placed. On large projects where heavy compaction equipment is used, a lift thickness of 18 to 24 in is appropriate.

Compaction achieved in the field should be determined by performing field density tests on each lift. For that purpose, wet density and moisture content should be measured and the dry density computed. Field densities may be ascertained by the sand-cone (ASTM D1556) or balloon volume-meter (ASTM D2167) method, from an undisturbed sample, or with a nuclear moisture-density meter. Generally, one field density test for each 4000 to 10,000 ft^2 of lift surface is adequate.

Hydraulically placed fills composed of dredged soils normally need not be compacted during placement. Although segregation of the silt and clay fractions of the soils may occur, it usually is not harmful. But accumulation of the fine-grained material in pockets at bulkheads or under structures should be prevented. For the purpose, internal dikes, weirs, or decanting techniques may be used.

Soil Replacement or Blending ▪ When materials at or near grade are unsuitable, it may be economical to remove them and substitute a fill of suitable soil, as described in Embankments.

When this is not economical, consideration should be given to improving the soil by other methods, such as densification or addition or removal of soil particles.

Mixing an existing soil with select materials or removing selected sizes of particles from an existing soil can change its properties considerably. Adding clay to a cohesionless soil in a nonfrost region, for example, may make the soil suitable as a base course for a road (if drainage is not too greatly impaired). Adding clay to a pervious soil may reduce its permeability sufficiently to permit its use as a reservoir bottom. Washing particles finer than 0.02 mm from gravel makes the soil less susceptible to frost heave (desirable upper limit for this fraction is 3%).

Surcharges ▪ Where good soils are underlain by soft, compressible clays that would permit unacceptable settlement, the site often can be made usable by surcharging, or preloading, the surface. The objective is to use the weight of the surcharge to consolidate the underlying clays. This offsets the settlement of the completed structure that would otherwise occur. A concurrent objective may be to increase the strength of the underlying clays.

If the soft clay is overlain by soils with adequate bearing capacity, the area to be improved may be loaded with loose, dumped earth, until the weight of the surcharge is equivalent to the load that later will be imposed by the completed structure. (If highly plastic clays or thick layers with little internal drainage are present, it may be necessary to insert sand drains to achieve consolidation within a reasonable time.) During and after placement of the surcharge, settlement of the original ground surface and the clay layer should be closely monitored. The surcharge may be removed after little or no settlement is observed. If surcharging has been properly executed, the completed structure should experience no further settlement due to primary consolidation. Potential settlement due to secondary compression, however, should be evaluated, especially if the soft soils are highly organic.

Densification ▪ Any of a variety of techniques, most involving some form of vibration, may be used for soil densification. The density achieved with a specific technique, however, depends on the grain size of the soil. Consequently, grain-size distribution must be taken into account when selecting a densification method.

Compaction of clean sands to depths of about 6 ft usually can be achieved by rolling the surface with a heavy, vibratory, steel-drum roller. Although the vibration frequency is to some extent adjustable, the frequencies most effective are in the range of 25 to 30 Hz. Bear in mind, however, that little densification will be achieved below a depth of 6 ft, and the soil within about 1 ft of the surface may actually be loosened. Compactive effort in the field may be measured by the number of passes made with a specific machine of given weight and at a given speed. For a given compactive effort, density varies with moisture content. For a given moisture content, increasing the compactive effort increases the soil density and reduces the permeability.

Compaction piles also may be used to densify sands. For that purpose, the piles usually are made of wood or are a sand replacement (sand pile). To create a sand pile, a wood pile or a steel shell is driven and the resulting hole is filled with sand. Densification of the surrounding soils results from soil displacement during driving of the pile or shell and from the vibration produced during pile driving. The foundations to be constructed need not bear directly on the compaction piles but may be seated anywhere on the densified mass.

Vibroflotation and Terra-Probe are alternative methods that increase sand density by multiple insertions of vibrating probes. These form cylindrical voids, which are then filled with off-site sand, stone, or blast-furnace slag. The probes usually are inserted in clusters, with typical spacing of about 4½ ft, where footings will be placed. Relative densities of 85% or more can be achieved throughout the depth of insertion, which may exceed 40 ft. Use of vibrating probes may not be effective, however, if the fines content of the soil exceeds about 15% or if organic matter is present in colloidal form in quantities exceeding about 5% by weight.

Another technique for densification is dynamic compaction, which in effect subjects the site to numerous miniearthquakes. In saturated soils, densification by this method also results from partial liquefaction, and the elevated pore pressures produced must be dissipated between each application of compactive energy if the following application is to be effective. As developed by Techniques Louis Menard, dynamic compaction is achieved by dropping weights ranging from 10 to 40 tons from heights up to 100 ft onto the ground surface. Impact spacings range up to 60 ft. Multiple drops are made at each location to be densified. This technique is applicable to densification of large areas and a wide range of grain sizes and materials.

Drainage ▪ This is effective in soil stabilization because strength of a soil generally decreases with an increase in amount and pressure of pore water. Drainage may be accomplished by gravity, pumping, compression with an external load on the soil, electroosmosis, heating, or freezing.

Pumping often is used for draining excavations (Art. 6-32). For permanent stabilization of slopes, however, advantage must be taken of gravity flow. Usually, intercepting drains, laid approximately along contours, suffice. Vertical wells may be used to relieve artesian pressures. Where mud flows may occur, water must be excluded from the area. Surface and subsurface flow must be intercepted and conducted away at the top of the area. Also, cover, such as heavy mulching and planting, should be placed over the entire surface to prevent water from percolating downward into the soil.

Electrical drainage adapts the principle that water flows to the cathode when a direct current passes through saturated soil. The water may be pumped out at the cathode. Electroosmosis is relatively expensive and therefore usually is limited to special conditions, such as drainage of silts, which ordinarily are hard to drain by other methods.

Vertical sand drains, or piles, may be used to compact loose, saturated cohesionless soils or to consolidate saturated cohesive soils. They provide an escape channel for water squeezed out of the soil by an external load. A surcharge of pervious material placed over the ground surface also serves as part of the drainage system as well as part of the fill, or external load. Usually, the surcharge is placed before the sand piles are formed, to support equipment, such as pile drivers, over the soft soil. Fill should be placed in thin layers to avoid formation of mud flows, which might shear the sand drains and cause mud waves. Analyses should be made of embankment stability at various stages of construction.

Geotextiles ▪ Also known as geofabrics, filter cloths, support fabrics, or civil engineering fabrics, geotextiles are membranes used to stabilize soils. For that purpose, permeable fabrics made of synthetic fibers with high tensile strength, even when wet, large modulus of elasticity, high ductility, and negligible creep when under load are usually used. These fabrics also are required to be biologically inert so that they will not decompose in the soil; resistant to wear, tear, puncture, and abrasion; and unaffected by untraviolet light before installation, acids, alkalis, oils, and a wide variety of chemical solvents.

Primary functions of geotextiles are reinforcement of soils, giving them tensile resistance they would not otherwise have, and separation of different types of soils, to prevent intermixing and resulting undesirable changes in soil properties. Other functions of geotextiles include erosion control, filtration, and drainage through and along the plane of the fabric.

In the past, many different materials have been used for soil separation or reinforcement, including grasses, rushes, wood logs, wood boards, metal mats, cotton, and jute. Because they deteriorated in a relatively short time, required maintenance frequently, or were costly, however, use of more efficient, more permanent materials was desirable. As a result, permeable synthetic fabrics are now used as an alternative. These fabrics may be woven or nonwoven. Made of such synthetics as polypropylene, the nonwoven fabrics are needle-punched for permeability to water. Specific properties of synthetic fabrics for use in selection of appropriate materials and thicknesses for specific applications may be obtained from the manufacturers.

Typical applications of geotextiles include the following:

1. **Retaining Earth Walls.** The fabric may be sandwiched between successive layers of soil to prevent slippage and improve compaction.
2. **Constructed Fills of Surcharges over Soft Soils.** The fabric may be laid over the surface and covered with about 1 ft of fill, to support light soil-spreading equipment. A second layer of fabric and fill may be placed to permit transit of dump trucks for building up the fill.
3. **Paved Surfaces.** The fabric may be placed under an aggregate subbase to separate it from the subsoil and to strengthen the subsoil by permitting drainage of free water.
4. **Railroads.** The fabric may be placed between ballast and subgrade as a separator and to facilitate drainage.
5. **Drains.** The fabric may be installed around drains to serve as a filter and prevent movement of soil into the drains.
6. **Erosion Control.** The fabric may be inserted under riprap and armor rock to prevent erosion of underlying soil and provide a rough surface for adhesion of facing materials.
7. **Storage Ponds.** The fabric may be placed under an impermeable liner to protect it from damage by sharp irregularities in the soil surface.

6-30. Thermal Stabilization of Soils

Still in the experimental stage, thermal stabilization generally is costly and is restricted to conditions for which other methods are not suitable. Heat has been used to strengthen nonsaturated loess and to decrease the compressibility of cohesive soils. One technique is to burn liquid or gas fuel in a borehole. Another is to inject into the soil through pipes spaced about 10 ft apart a mixture of liquid fuel and air under pressure. Then, the mixture is burned for about 10 days, producing a solidified soil.

Freezing a wet soil converts it into a rigid material with considerable strength, but it must be kept frozen. The method is excellent for a limited excavation area, for example, freezing the ground to sink a shaft. For the purpose, a network of pipes must be placed in the ground and a liquid, usually brine, at low temperature circulated through the pipes. Care must be taken that the freezing does not spread beyond the area to be stabilized and cause heaving damage.

6-31. Chemical Stabilization of Soils

Including use of portland cement and bitumens, chemical stabilization meets many needs. In surface treatments, it supplements mechanical stabilization to make the effects more lasting. In subsurface treatments, chemicals may be used to improve bearing capacity or decrease permeability.

Soil-cement, a mixture of portland cement and soil, is suitable for subgrades, base courses, and pavements of roads not carrying heavy traffic ("Essentials of Soil-Cement Construction," Portland Cement Association). Bitumen-soil mixtures are extensively used in road and airfield construction and sometimes as a seal for earth dikes (Guide Specifications for Highway Construction, American Association of State Highway and Transportation Officials, 444 North Capitol St., N.W., 20001). Hydrated, or slaked, lime may be used alone as a soil stabilizer, or with fly ash, portland cement, or bitumen ("Lime Stabilization of Roads," National Lime Association, 925 15th St., N.W., Washington D.C. 20006). Calcium or sodium chloride is used as a dust palliative and an additive in construction of granular base and wearing courses for roads ("Calcium Chloride for Stabilization of Bases and Wearing Courses," Calcium Chloride Institute, Ring Building, Washington, D.C. 20036).

Grouting, with portland cement or other chemicals, often is used to fill rock fissures, decrease soil permeability, form underground cutoff walls to eliminate seepage, and stabilize soils at considerable depth. The chemicals may be used to fill the voids in the soil, to cement the particles, or to form a rocklike material. The process, however, generally is suitable only in pervious soils. Also, rapid setting of chemicals may interfere with thorough injection of the underground region. Chemicals used include sodium silicate and salts or acids; chrome-lignin; and low-viscosity organics ("Chemical Grouting," *ASCE Journal of Soil Mechanics and Foundation Engineering Division,* vol. 83, November 1957).

(R. A. Barron, "Consolidation of Fine-Grained Soils by Drain Wells," *ASCE Transactions,* vol. 113, p. 718, 1948; L. Casagrande, "Electro-Osmotic Stabilization of Soils," *Journal, Boston Society of Civil Engineers,* vol. 39, p. 51, 1952; K. Terzaghi and R. B. Peck, "Soil Mechanics in Engineering Practice," John Wiley & Sons, Inc., New York; G. P. Tschebotarioff, "Soil Mechanics, Foundations, and Earth Structures," McGraw-Hill Book Company, New York; H. F. Winterkorn and H. Fang, "Foundation Engineering Handbook," Van Nostrand Reinhold Company, New York.)

6-32. Dewatering Methods for Excavations

The main purpose of dewatering is to enable construction to be carried out under relatively dry conditions. But proper drainage also stabilizes excavated slopes, reduces lateral loads on sheeting and bracing, reduces required air pressure in tunneling, makes excavated material lighter and easier to handle, prevents loss of soil below slopes or from the bottom of the excavation, and prevents a "quick" or "boiling" bottom. In addition, permanent lowering of the groundwater table or relief of artesian pressure may allow a less expensive design for the structure, especially when the soil consolidates or becomes compact. If lowering of the water level or pressure relief is temporary, however, the improvement of the soil should not be considered in design. Increases in strength and bearing capacity may be lost when the soil again becomes saturated.

To keep an excavation reasonably dry, the groundwater table should be kept at least 2 ft, and preferably 5 ft, below the bottom in most soils.

Site investigations should yield information useful for deciding on the most suitable and economical dewatering method. Important is a knowledge of the types of soil in and below the site, probable groundwater levels during construction, permeability of the soils, and quantities of water to be handled. A pumping test may be desirable for estimating capacity of pumps needed and drainage characteristics of the ground.

Many methods have been used for dewatering excavations. Those used most often are listed in Table 6-12 with conditions for which they generally are most suitable. (See also Art. 6-29.)

In many small excavations, or where there are dense or cemented soils, water may be collected in ditches or sumps at the bottom and pumped out. This is the most economical method of dewatering, and the sumps do not interfere with future construction as does a comprehensive wellpoint system. But the seepage may slough the slopes, unless they are stabilized with gravel, and may hold up excavation while the soil drains. Also, springs may develop in fine sand or silt and cause underground erosion and subsidence of the ground surface.

For sheetpile-enclosed excavations in pervious soils, it is advisable to intercept water before it enters the enclosure. Otherwise, the water will put high pressures on the sheeting. Seepage also can cause the excavation bottom to become quick, overloading the bottom bracing, or create piping, undermining the sheeting. Furthermore, pumping from the inside of the cofferdam is likely to leave the soil to be excavated wet and tough to handle.

Wellpoints often are used for lowering the water table in pervious soils. They are not suitable, however, in soils that are so fine that they will flow with the water or in soils with low perme-

TABLE 6-12 Methods for Dewatering Excavations

Saturated-soil conditions	Dewatering method probably suitable
Surface water	Ditches; dikes; sheetpiles and pumps or underwater excavation and concrete tremie seal
Gravel	Underwater excavation, grout curtain; gravity drainage with large sumps with gravel filters
Sand (except very fine sand)	Gravity drainage
Waterbearing strata near surface; water table does not have to be lowered more than 15 ft	Wellpoints with vacuum and centrifugal pumps
Waterbearing strata near surface; water table to be lowered more than 15 ft, low pumping rate	Wellpoints with jet-eductor pumps
Excavations 30 ft or more below water table; artesian pressure; high pumping rate; large lowering of water table—all where adequate depth of pervious soil is available for submergence of well screen and pump	Deep wells, plus, if necessary, wellpoints
Sand underlain by rock near excavation bottom	Wellpoints to rock, plus ditches, drains, automatic "mops"
Sand underlain by clay	Wellpoints in holes 3 or 4 into the clay, backfilled with sand
Silt; very find sand (permeability coefficient between 0.01 and 0.0001 mm/s)	For lifts up to 15 ft, wellpoints with vacuum; for greater lifts, wells with vacuum; sumps
Silt or silty sand underlain by pervious soil	At top of excavation, and extending to the pervious soil, vertical sand drains plus wellpoints or wells
Clay-silts, silts	Electroosmosis
Clay underlain by pervious soil	At top of excavation, wellpoints or deep wells extending to pervious soil
Dense or cemented soils; small excavations	Ditches and sumps

ability. Also, other methods may be more economical for deep excavations, very heavy flows, or considerable lowering of the water table (Table 6-12).

Wellpoints are metal well screens about 2 to 3 in in diameter and up to about 4 ft long. A pipe connects each wellpoint to a header, from which water is pumped to discharge (Fig. 6-27). Each pump usually is a combined vacuum and centrifugal pump. Spacing of wellpoints generally ranges from 3 to 12 ft c to c.

A wellpoint may be jetted into position or set in a hole made with a hole puncher or heavy steel casing. Accordingly, wellpoints may be self-jetting or plain-tip. To insure good drainage in fine and dirty sands or in layers of silt or clay, the wellpoint and riser should be surrounded by sand to just below the water table. The space above the filter should be sealed with silt or clay to keep air from getting into the wellpoint through the filter.

Wellpoints generally are relied on to lower the water table 15 to 20 ft. Deep excavations may be dewatered with multistage wellpoints, with one row of wellpoints for every 15 ft of depth. Or when the flow is less than about 15 gal/min per wellpoint, a single-stage system of wellpoints may be installed above the water table and operated with jet-eductor pumps atop the wellpoints. These pumps can lower the water table up to about 100 ft, but they have an efficiency of only about 30%.

Deep wells may be used in pervious soils for deep excavations, large lowering of the water table, and heavy water flows. They may be placed along the top of an excavation to drain it, to intercept seepage before pressure makes slopes unstable, and to relieve artesian pressure before it heaves the excavation bottom.

Usual spacing of wells ranges from 20 to 250 ft. Diameter generally ranges from 6 to 20 in.

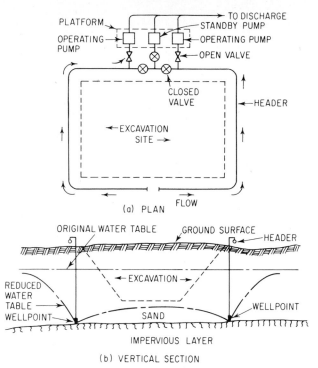

Fig. 6-27. Wellpoint system for dewatering an excavation.

Well screens may be 20 to 75 ft long, and they are surrounded with a sand-gravel filter. Generally, pumping is done with a submersible or vertical turbine pump installed near the bottom of each well.

Figure 6-28 shows a deep-well installation used for a 300-ft-wide by 600-ft-long excavation for a building of the Smithsonian Institution, Washington, D.C. Two deep-well pumps lowered the general water level in the excavation 20 ft. The well installation proceeded as follows: (1) Excavation to water level (elevation 0.0). (2) Driving of sheetpiles around the well area (Fig. 6-28a). (3) Excavation underwater inside the sheetpile enclosure to elevation -37.0 ft (Fig. 6-28b). Bracing installed as digging progressed. (4) Installation of a wire-mesh-wrapped timber frame, extending from elevation 0.0 to -37.0 (Fig. 6-28c). Weights added to sink the frame. (5) Backfilling of space between sheetpiles and mesh with $\frac{3}{16}$- to $\frac{3}{8}$-in gravel. (6) Removal of sheetpiles. (7) Installation of pump and start of pumping.

Vacuum well or wellpoint systems may be used to drain silts with low permeability (coefficient between 0.01 and 0.0001 mm/s). In these systems, wells or wellpoints are closely spaced, and a vacuum is held with vacuum pumps in the well screens and sand filters. At the top, the filter, well, and risers should be sealed to a depth of 5 ft with bentonite or an impervious soil to prevent loss of the vacuum. Water drawn to the well screens is pumped out with submersible or centrifugal pumps.

Where a pervious soil underlies silts or silty sands, vertical sand drains and deep wells can team up to dewater an excavation. Installed at the top, and extending to the pervious soil, the sand piles intercept seepage and allow it to drain down to the pervious soil. Pumping from the deep wells relieves the pressure in that deep soil layer.

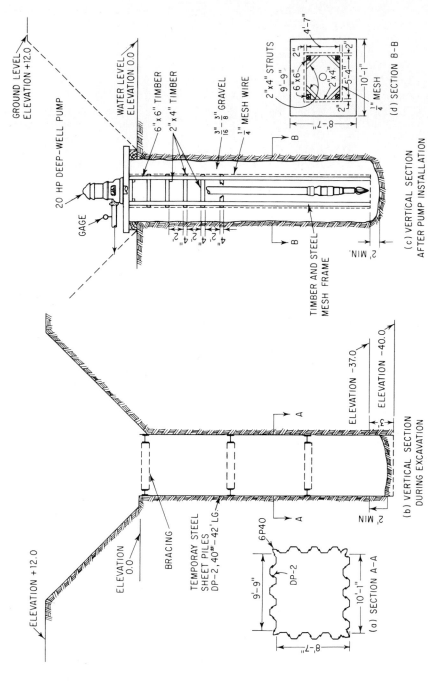

Fig. 6-28. Deep-well installation used at Smithsonian Institution, Washington, D.C. (*Spencer, White & Prentis, Inc.*)

GROUND LEVEL ELEVATION +12.0

20 HP DEEP-WELL PUMP

WATER LEVEL ELEVATION 0.0

GAGE

6"×6" TIMBER

2"×4" TIMBER

$\frac{3}{16}$"—$\frac{3}{8}$" GRAVEL

$\frac{1}{4}$" MESH WIRE

B

B

2"×4" STRUTS

$\frac{1}{4}$" MESH

4'-7"

2" 2"

$\frac{1}{2}$"

6"×6"

9'-9"

2"×4"

5'-4"

2"

2"

10'-1"

8'-7"

(d) SECTION B-B

2"

2"

4"

2"

4"

2" MIN.

TIMBER AND STEEL MESH FRAME

(c) VERTICAL SECTION AFTER PUMP INSTALLATION

ELEVATION +12.0

ELEVATION 0.0

BRACING

TEMPORARY STEEL SHEET PILES DP-2, 40#-42' LG.

A

A

ELEVATION -37.0

ELEVATION -40.0

2' MIN.

2' MIN.

(b) VERTICAL SECTION DURING EXCAVATION

6P40

DP-2

9'-9"

10'-1"

8'-7"

(a) SECTION A-A

For some silts and clay-silts, electrical drainage with wells or wellpoints may work, whereas gravity methods may not (Art. 6-29). In saturated clays, thermal or chemical stabilization may be necessary (Arts. 6-30 and 6-31).

Small amounts of surface water may be removed from excavations with "mops." Surrounded with gravel to prevent clogging, these drains are connected to a header with suction hose or pipe. For automatic operation, each mop should be opened and closed by a float and float valve.

When structures on silt or soft material are located near an excavation to be dewatered, care should be taken that lowering of the water table does not cause them to settle. It may be necessary to underpin the structures or to pump discharge water into recharge wells near the structures to maintain the water table around them.

(G. A. Leonards, "Foundation Engineering," McGraw-Hill Book Company, New York; L. Zeevaert, "Foundation Engineering for Difficult Subsoil Conditions," Van Nostrand Reinhold Company, New York; H. F. Winterkorn and H. Fang, "Foundation Engineering Handbook," Van Nostrand Reinhold Company, New York.)

Water engineering includes the application of fluid mechanics, hydraulics, hydrology, and water-supply theories. Fluid mechanics describes the behavior of water under various static and dynamic conditions. This theory, in general, has been developed for an ideal liquid, a frictionless, inelastic liquid whose particles follow smooth flow paths. Since water only approaches an ideal liquid, empirical coefficients and formulas are used to describe more accurately the behavior of water. These empiricisms are intended to compensate for all neglected and unknown factors.

The relatively high degree of dependence on empiricism, however, does not minimize the importance of an understanding of the basic theory. Since major hydraulic problems are seldom identical to the experiments from which the empirical coefficients were derived, the application of fundamentals is frequently the only means available for attacking problems.

Section 7

Water Engineering

7-1. Dimensions and Units

A list of symbols and their dimensions used in this section is given in Table 7-1. Table 7-2 lists conversion factors for commonly used quantities, including the basic equivalents between the English and metric systems.

Samuel B. Nelson

Former Director of Public Works, State of California; Retired General Manager and Chief Engineer, Department of Water and Power, City of Los Angeles; Former General Manager, Southern California Rapid Transit District; Vice President, Daniel, Mann, Johnson & Mendenhall; Vice Chairman, Board of Directors, Metropolitan Water District of Southern California; and Member, California Water Commission

FLUID MECHANICS

7-2. Properties of Fluids

Specific weight or **unit weight** w is defined as weight per unit volume. The specific weight of water varies from 62.42 lb/ft³ at 32°F to 62.22 lb/ft³ at 80°F but is commonly taken as 62.4 lb/ft³ for the majority of engineering problems. The specific weight of sea water is about 64.0 lb/ft³.

Density, ρ is defined as mass per unit volume and is significant in all flow problems where acceleration is important. It is obtained by dividing the specific weight w by the acceleration due to gravity g. The variation of g with latitude and altitude is small

TABLE 7-1 Symbols, Dimensions, and Units Used in Water Engineering

Symbol	Terminology	Dimensions	Units
A	Area	L^2	ft^2
C	Chezy roughness coefficient	$L^{1/2}/T$	$ft^{1/2}/s$
C_1	Hazen-Williams roughness coefficient	$L^{0.37}/T$	$ft^{0.37}/s$
d	Depth	L	ft
d_c	Critical depth	L	ft
D	Diameter	L	ft
E	Modulus of elasticity	F/L^2	psi
F	Force	F	lb
g	Acceleration due to gravity	L/T^2	ft/s^2
H	Total head, head on weir	L	ft
h	Head or height	L	ft
h_f	Head loss due to friction	L	ft
L	Length	L	ft
M	Mass	FT^2/L	$lb \cdot s^2/ft$
n	Manning's roughness coefficient	$T/L^{1/3}$	$s/ft^{1/3}$
P	Perimeter, weir height	L	ft
P	Force due to pressure	F	lb
p	Pressure	F/L^2	lb/ft^2
Q	Flow rate	L^3/T	ft^3/s
q	Unit flow rate	$L^3/T \cdot L$	$ft^3/(s \cdot ft)$
r	Radius	L	ft
R	Hydraulic radius	L	ft
T	Time	T	s
t	Time, thickness	T,L	s, ft
V	Velocity	L/T	ft/s
W	Weight	F	lb
w	Specific weight	F/L^3	lb/ft^3
y	Depth in open channel, distance from solid boundary	L	ft
Z	Height above datum	L	ft
ϵ	Size of roughness	L	ft
μ	Viscosity	FT/L^2	$lb \cdot s/ft^2$
ν	Kinematic viscosity	L^2/T	ft^2/s
ρ	Density	FT^2/L^4	$lb \cdot s^2/ft^4$
σ	Surface tension	F/L	lb/ft
τ	Shear stress	F/L^2	psi

Symbols for dimensionless quantities

Symbol	Quantity
C	Weir coefficient, coefficient of discharge
C_c	Coefficient of contraction
C_v	Coefficient of velocity
$\mathbf{F}$	Froude number
f	Darcy-Weisbach friction factor
K	Head-loss coefficient
$\mathbf{R}$	Reynolds number
S	Friction slope—slope of energy grade line
S_c	Critical slope
η	Efficiency
Sp. gr.	Specific gravity

TABLE 7-2   Conversion Table for Commonly Used Quantities

Area	Discharge
1 acre = 43,560 ft^2	1 ft^3/s = 449 gal/min = 646,000 gpd
1 mi^2 = 640 acres	1 ft^3/s = 1.98 acre-ft/day = 724 acre-ft/year

Volume	1 ft^3/s = 50 miner's inches in Idaho, Kansas, Nebraska, New Mexico, North Dakota, and South Dakota
1 ft^3 ≒ 7.4805 gal	1 ft^3/s = 4C miner's inches in Arizona, California, Montana, and Oregon
1 acre-ft = 325,850 gal	1 mgd = 3.07 acre-ft/day = 1120 acre-ft/year
1 mg = 3.0689 acre-ft	

Power	1 mgd = 3.07 acre-ft/day = 1120 acre-ft/year
1 hp = 550 ft · lb/s	1 mgd = 1.55 ft^3/s = 694 gpm
1 hp = 0.746 kW	1 million acre-ft/year = 1380 ft^3/s
1 hp = 6535 kWh/year	

Weight of water	Pressure
1 ft^3 weighs 62.4 lb	1 psi = 2.31 ft of water
1 gal weighs 8.34 lb	= 51.7 mm of mercury
	1 in of mercury = 1.13 ft of water
	1 ft of water = 0.433 psi

Metric equivalents
Length: 1 ft = 0.3048 m
Area: 1 acre = 4046.9 m^2
Volume: 1 gal = 3.7854 L
            1 m^3 = 264.17 gal
Weight: 1 lb = 0.4536 kg

enough to warrant the assumption that its value is constant at 32.2 ft/s^2 in hydraulics computations.

The **specific gravity** of water is the ratio of its density at some temperature to that of pure water at 68.2°F (20°C).

Modulus of elasticity E of a fluid is defined as the change in pressure intensity divided by the corresponding change in volume per unit volume. Its value for water is about 300,000 psi, varying slightly with temperature. The modulus of elasticity of water is large enough to permit the assumption that it is incompressible for all hydraulics problems except those involving water hammer (Art. 7-13).

Surface tension and **capillarity** are a result of the molecular forces of liquid molecules. **Surface tension** σ is due to the cohesive forces between liquid molecules. It shows up as the apparent skin that forms when a free liquid surface is in contact with another fluid. It is expressed as the force in the liquid surface normal to a line of unit length drawn in the surface. Surface tension decreases with increasing temperature and is also dependent on the fluid with which the liquid surface is in contact. The surface tension of water at 70°F in contact with air is 0.00498 lb/ft.

Capillarity is due to both the cohesive forces between liquid molecules and adhesive forces of liquid molecules. It shows up as the difference in liquid surface elevations between the inside and outside of a small tube that has one end submerged in the liquid. Since the adhesive forces of water molecules are greater than the cohesive forces between water molecules, water wets a surface and rises in a small tube, as shown in Fig. 7-1. Capillarity is commonly expressed as the height of this rise. In equation form,

$$h = \frac{2\sigma \cos \theta}{(w_1 - w_2)r} \tag{7-1}$$

where h = capillary rise, ft
        σ = surface tension, lb/ft

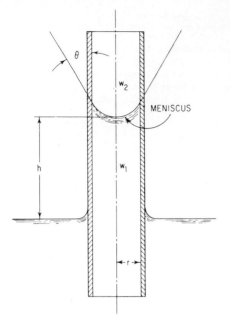

Fig. 7-1. Capillary action raises water in a small tube. Meniscus, or liquid surface, is concave upward.

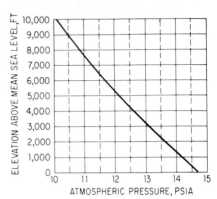

Fig. 7-2. Atmospheric pressure changes with elevation. The curve is based on the ICAO standard atmosphere.

w_1 and w_2 = specific weights of fluids below and above meniscus, respectively, lb/ft
θ = angle of contact
r = radius of capillary tube, ft

Capillarity, like surface tension, decreases with increasing temperature. Its temperature variation, however, is small and insignificant in most problems.

Surface tension and capillarity, although negligible in many water engineering problems, are significant in others, such as capillary rise and flow of liquids in narrow spaces, formation of spray from water jets, interpretation of the results obtained on small models, and freezing damage to concrete.

Atmospheric pressure is the pressure due to the weight of the air above the earth's surface. Its value at sea level is 2116 lb/ft² or 14.7 psi. The variation in atmospheric pressure with elevation from sea level to 10,000 ft is shown in Fig. 7-2. **Gage pressure,** *psi,* is pressure above or below atmospheric. **Absolute pressure,** *psia,* is the total pressure including atmospheric pressure. Thus, at sea level, a gage pressure of 10 psi is equivalent to 24.7 psia. Gage pressure is positive when pressure is greater than atmospheric and is negative when pressure is less than atmospheric.

Vapor pressure is the partial pressure caused by the formation of vapor at the free surface of a liquid. When the liquid is in a closed container, the partial pressure due to the molecules leaving the surface increases until the rates at which the molecules leave and reenter the liquid are equal. The vapor pressure at this equilibrium condition is called the saturation pressure. Vapor pressure increases with increasing temperature, as shown in Fig. 7-3.

Cavitation occurs in flowing liquids at pressures below the vapor pressure of the liquid. Cavitation is a major problem in the design of pumps and turbines since it causes mechanical vibra-

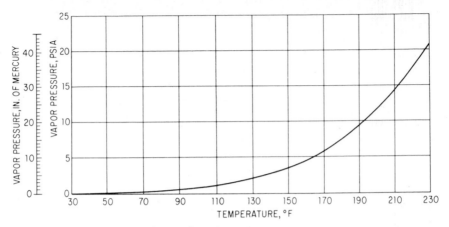

Fig. 7-3.   Vapor pressure of water depends on temperature.

tions, pitting, and loss of efficiency through gradual destruction of the impeller. The cavitation phenomenon may be described as follows:

Because of low pressures, portions of the liquid vaporize, with subsequent formation of vapor cavities. As these cavities are carried a short distance downstream, abrupt pressure increases force them to collapse, or implode. The implosion and ensuing inrush of liquid produce regions of very high pressure, which extend into the pores of the metal. (Pressures as high as 350,000 psi have been measured in the collapse of vapor cavities by the Fluid Mechanics Laboratory at Stanford University.) Since these vapor cavities form and collapse at very high frequencies, weakening of the metal results as fatigue develops, and pitting appears.

Cavitation may be prevented by designing pumps and turbines so that the pressure in the liquid at all points is always above its vapor pressure.

Viscosity μ of a fluid, also called the **coefficient of viscosity, absolute viscosity,** or **dynamic viscosity,** is a measure of its resistance to flow. It is expressed as the ratio of the tangential shearing stresses between flow layers to the rate of change of velocity with depth:

$$\mu = \frac{\tau}{dV/dy} \tag{7-2}$$

where τ = shearing stress, lb/ft^2
$\quad\;\; V$ = velocity, ft/s
$\quad\;\; y$ = depth, ft

Viscosity decreases as temperature increases but may be assumed independent of changes in pressure for the majority of engineering problems. Water at 70°F has a viscosity of 0.00002050 lb · s/ft².

Kinematic viscosity ν is defined as viscosity μ divided by density ρ. It is so named because its units, ft²/s, are a combination of the kinematic units of length and time. Water at 70°F has a kinematic viscosity of 0.00001059 ft²/s.

In hydraulics, viscosity is most frequently encountered in the calculation of Reynolds number (Art. 7-8) to determine whether laminar, transitional, or completely turbulent flow exists.

7-3. Fluid Pressure

Pressure or **intensity of pressure** p is the force per unit area acting on any real or imaginary surface within a fluid. Fluid pressure acts normal to the surface at all points. At any depth, the pressure acts equally in all directions. This results from the inability of a fluid to transmit shear when at rest. Liquid and gas pressures differ in that the variation of pressure with depth is linear for a liquid and nonlinear for a gas.

Hydrostatic pressure is the pressure due to depth. It may be derived by considering a submerged rectangular prism of water of height Δh, ft, and cross-sectional area A, ft², as shown in Fig. 7-4. The boundaries of this prism are imaginary. Since the prism is at rest, the summation of all forces in both the vertical and horizontal directions must be zero. Let w equal the specific weight of the liquid, lb/ft³. Then, the forces acting in the vertical direction are the weight of the prism $wA\,\Delta h$, the force due to pressure p_1, lb/ft², on the top surface, and the force due to pressure p_2, lb/ft², on the bottom surface. Summing these vertical forces and setting the total equal to zero yields

$$p_2 A - wA\,\Delta h - p_1 A = 0 \qquad (7\text{-}3\,a)$$

$$p_2 = w\,\Delta h + p_1 \qquad (7\text{-}3\,b)$$

For the special case where the top of the prism coincides with the water surface, p_1 is atmospheric pressure. Since most hydraulics problems involve gage pressure, p_1 is zero (gage pressure is zero at atmospheric pressure). Taking Δh to be h, the depth below the water surface, ft, then p_2 is p, the pressure, lb/ft², at depth h. Equation (7-3 b) then becomes

$$p = wh \qquad h = \frac{p}{w} \qquad (7\text{-}4)$$

Equation (7-4) gives the depth of water h of specific weight w required to produce a gage pressure p. By adding atmospheric pressure p_a to Eq. (7-4), absolute pressure p_{ab} is obtained as shown in Fig. 7-4. Thus,

$$p_{ab} = wh + p_a \qquad (7\text{-}5)$$

Pressure on submerged plane surfaces is important in the design of dams, tanks, and outlet works, such as sluice gates. For horizontal surfaces, the pressure-force determination is a simple

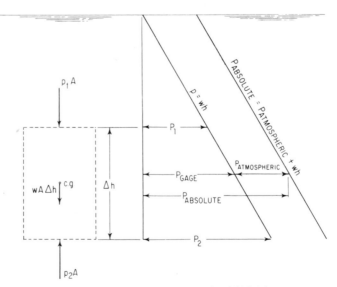

Fig. 7-4. Hydrostatic pressure varies linearly with depth.

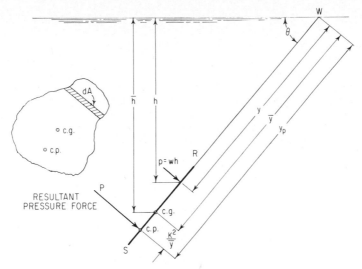

Fig. 7-5.   Total pressure on a submerged plane surface depends on pressure at center of gravity (c.g.) but acts at a point (c.p.) that is below the c.g.

matter since the pressure is constant. For determination of the pressure force on inclined or vertical surfaces, however, the summation concepts of integral calculus must be used.

Figure 7-5 represents any submerged plane surface of negligible thickness inclined at an angle θ with the horizontal. The resultant pressure force P, lb, acting on the surface is equal to $\int p \, dA$. Since $p = wh$ and $h = y \sin \theta$, where w is the specific weight of water, lb/ft³,

$$P = w \int y \sin \theta \, dA \tag{7-6}$$

Equation (7-6) can be simplified by setting $\int y \, dA = \bar{y} A$, where A is the area of the submerged surface, ft²; and $\bar{y} \sin \theta = \bar{h}$, the depth of the centroid, ft. Therefore,

$$P = w \bar{h} A = p_{cg} A \tag{7-7}$$

where p_{cg} is the pressure at the centroid, lb/ft².

The point on the submerged surface at which the resultant pressure force acts is called the **center of pressure** (c.p.). It is below the center of gravity because the pressure intensity increases with depth. The location of the center of pressure, represented by the length y_p, is calculated by summing the moments of the incremental forces about an axis in the water surface through point W (Fig. 7-5). Thus, $P y_p = \int y \, dP$. Since $dP = wy \sin \theta \, dA$ and $P = w \int y \sin \theta \, dA$,

$$y_p = \frac{\int y^2 \, dA}{\int y \, dA} \tag{7-8}$$

The quantity $\int y^2 \, dA$ is the moment of inertia of the area about the axis through W. It also equals $AK^2 + A\bar{y}^2$, where K is the radius of gyration, ft, of the surface about its centroidal axis. The denominator of Eq. (7-8) equals $\bar{y} A$. Hence

$$y_p = \bar{y} + \frac{K^2}{\bar{y}} \tag{7-9}$$

and $K^2/\bar{y}$ is the distance between the centroid and center of pressure.

Values of K^2 for some common shapes are given in Fig. 7-6 (see also Fig. 1-29). For areas

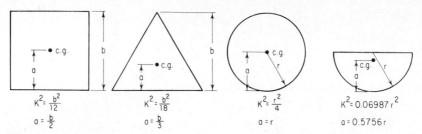

Fig. 7-6.   Radius of gyration and location of center of gravity (c.g.) of common shapes.

for which radius of gyration has not been determined, y_p may be calculated directly from Eq. (7-8).

The horizontal location of the center of pressure may be determined as follows: It lies on the vertical axis of symmetry for surfaces symmetrical about the vertical. It lies on the locus of the midpoints of horizontal lines located on the submerged surface, if that locus is a straight line. Otherwise, the horizontal location may be found by taking moments about an axis perpendicular to the one through W in Fig. 7-5 and lying in the plane of the submerged surface.

Example 7-1 ▪ Determine the magnitude and point of action of the resultant pressure force on a 5-ft-square sluice gate inclined at an angle θ of 53.2° to the horizontal (Fig. 7-7).

From Eq. (7-7), the total force $P = w\bar{h}A$, with

$$\bar{h} = [2.5 + \tfrac{1}{2}(5)]\ \sin 53.2° = 5.0 \times 0.8 = 4.0 \text{ ft}$$
$$A = 5 \times 5 = 25 \text{ ft}^2$$

Thus, $P = 62.4 \times 4 \times 25 = 6240$ lb. From Eq. (7-9), its point of action is a distance $y_p = \bar{y} + K^2/\bar{y}$ from point G, and $\bar{y} = 2.5 + \tfrac{1}{2}(5.0) = 5.0$ ft. Also, $K^2 = b^2/12 = 5^2/12 = 2.08$. Therefore, $y_p = 5.0 + 2.08/5 = 5.0 + 0.42 = 5.42$ ft.

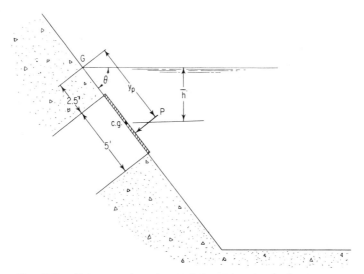

Fig. 7-7.   Sluice gate (crosshatched) is subjected to hydrostatic pressure. (See Example 7-1.)

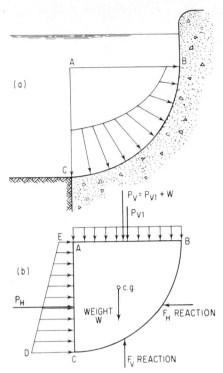

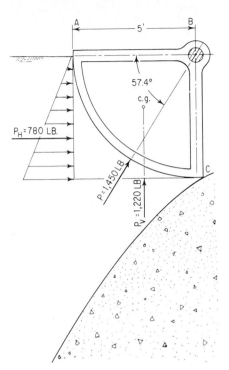

Fig. 7-8. Hydrostatic pressure on a submerged curved surface. (*a*) General configuration showing pressure variation. (*b*) Free-body diagram.

Fig. 7-9. Taintor gate has submerged curved surface under pressure. Vertical component of pressure acts upward. (See Example 7-2.)

Pressure on Submerged Curved Surfaces ▪ The resultant pressure force on submerged curved surfaces cannot be calculated from the equations developed for the pressure force on submerged plane surfaces because of the variation in direction of the pressure force. The resultant pressure force can be calculated, however, by determining its horizontal and vertical components and combining them vectorially.

A typical configuration of pressure on a submerged curved surface is shown in Fig. 7-8. Consider *ABC* a 1-ft-thick prism and analyze it as a free body by the principles of statics. Note:

1. The horizontal component P_H of the resultant pressure force has a magnitude equal to the pressure force on the vertical projection *AC* of the curved surface and acts at the centroid of pressure diagram *ACDE*.

2. The vertical component P_V of the resultant pressure force has a magnitude equal to the sum of the pressure force on the horizontal projection *AB* of the curved surface and the weight of the water vertically above *ABC*. The horizontal location of the vertical component is calculated by taking moments of the two vertical forces about point *C*.

When water is below the curved surface, such as for a taintor gate (Fig. 7-9), the vertical component P_V of the resultant pressure force has a magnitude equal to the weight of the imaginary volume of water vertically above the surface. P_V acts upward through the center of gravity of this imaginary volume.

Example 7-2 ▪ Calculate the magnitude and direction of the resultant pressure on a 1-ft-wide strip of the semicircular taintor gate in Fig. 7-9.

The magnitude of the horizontal component P_H of the resultant pressure force equals the pressure force on the vertical projection of the taintor gate. From Eq. (7-7), $P_H = w\bar{h}A = 62.4 \times 2.5 \times 5 = 780$ lb.

The magnitude of the vertical component of the resultant pressure force equals the weight of the imaginary volume of water in the prism ABC above the curved surface. The volume of this prism is $\pi R^2/4 = 3.14 \times 25/4 = 19.6$ ft^3, so the weight of the water is $19.6w = 19.6 \times 62.4 = 1220$ lb $= P_V$.

The magnitude of the resultant pressure force equals

$$P = \sqrt{P_H^2 + P_V^2} = \sqrt{780^2 + 1220^2} = 1450 \text{ lb}$$

The tangent of the angle the resultant pressure force makes with the horizontal $= P_V/P_H = 1220/780 = 1.564$. The corresponding angle is 57.4°.

The positions of the horizontal and vertical components of the resultant pressure force are not required to find the point of action of the resultant. Its angle with the horizontal is known, and for a constant-radius surface, the resultant must act perpendicular to the surface.

7-4. Submerged and Floating Bodies

The principles of buoyancy govern the behavior of submerged and floating bodies and are important in determining the stability and draft of cargo vessels.

The buoyant force acting on a submerged body equals the weight of the volume of liquid displaced.

A floating body displaces a volume of liquid equal to its weight.

The buoyant force acts vertically through the center of buoyancy c.b., which is located at the center of gravity of the volume of liquid displaced.

For a body to be in equilibrium, whether floating or submerged, the center of buoyancy and center of gravity must be on the same vertical line AB (Fig. 7-10a). The stability of a ship, its tendency not to overturn when it is in a nonequilibrium position, is indicated by the *metacenter.* It is the point at which a vertical line through the center of buoyancy intersects the rotated position of the line through the centers of gravity and buoyancy for the equilibrium condition

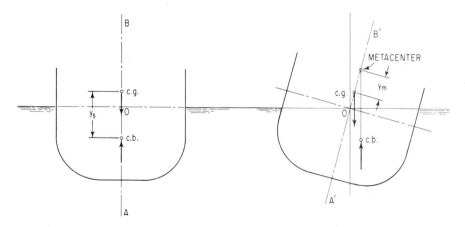

Fig. 7-10.   Stability of a ship depends on the location of its metacenter relative to its center of gravity (c.g.).

$A'B'$ (Fig. 7-10b). The ship is stable only if the metacenter is above the center of gravity since the resulting moment for this condition tends to right the ship.

The distance between the ship's metacenter and center of gravity is called the *metacentric height* and is designated by y_m in Fig. 7-10b. Given in feet by Eq. (7-10), y_m is a measure of degree of stability or instability of a ship since the magnitudes of moments produced in a roll are directly proportional to this distance.

$$y_m = \frac{I}{V} \pm y_s \qquad (7\text{-}10)$$

where I = moment of inertia of ship's cross section at waterline about longitudinal axis through O, ft^4

V = volume of displaced liquid, ft^3

y_s = distance, ft, between centers of buoyancy and gravity when ship is in equilibrium

The negative sign should be used when the center of gravity is above the center of buoyancy.

7-5. Manometers

A manometer is a device for measuring pressure. It consists of a tube containing a column of one or two liquids that balances the unknown pressure. The basis for the calculation of this unknown pressure is provided by the height of the liquid column. All manometer problems may be solved with Eq. (7-4), $p = wh$. Manometers indicate h, the *pressure head*, or the difference in head.

The primary application of manometers is measurement of relatively low pressures, for which aneroid and Bourdon gages are not sufficiently accurate because of their inherent mechanical limitations. However, manometers may also be used in precise measurement of high pressures by arranging several U-tube manometers in series (Fig. 7-12c). Manometers are used for both static and flow applications, although the latter is most common.

Three basic types are used (shown in Fig. 7-11): piezometer, U-tube manometer, and differential manometer. Following is a brief discussion of the basic types.

The **piezometer** (Fig. 7-11a) consists of a tube with one end tapped flush with the wall of the container in which the pressure is to be measured and the other end open to the atmosphere. The only liquid it contains is the one whose pressure is being measured (the **metered liquid**). The piezometer is a sensitive gage, but it is limited to the measurement of relatively small pressures, usually heads of 5 ft of water or less. Larger pressures would create an impractically high column of liquid.

Example 7-3 ▪ The gage pressure p_c in the pipe of Fig. 7-11a is 2.17 psi. The liquid is water with $w = 62.4$ lb/ft^3. What is h_m?

$$h_m = \frac{p_c}{w} = \frac{2.17 \times 144}{62.4} = 5.0 \text{ ft}$$

For pressures greater than 5 ft of water, the **U-tube manometer** (Fig. 7-11b) is used. It is similar to the piezometer except that it contains an **indicating liquid** with a specific gravity usually much larger than that of the metered liquid. The only other criteria are that the indicating liquid should have a good meniscus and be immiscible with the metered liquid.

The U-tube manometer is used when pressures are either too high or too low for the piezometer. High pressures can be measured by arranging U-tube manometers in series (Fig. 7-12c). Very low pressures, including negative gage pressures, can be measured if the bottom of the U tube extends below the center line of the container of the metered liquid. The most common use of the U-tube manometer is measurement of the pressures of flowing water. In this application, the usual indicating liquid is mercury.

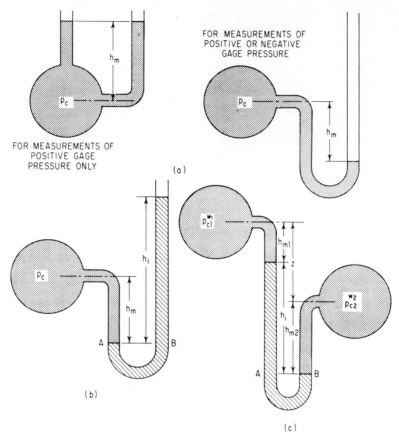

Fig. 7-11.   Basic types of manometers. (*a*) Piezometers; (*b*) U-tube manometer; (*c*) differential manometer.

A movable scale, as opposed to a fixed scale, facilitates reading the U-tube manometer. The scale is positioned between the two vertical legs and moved to adjust for the variation in distance h_m from the center line of the pressure vessel to the indicating liquid. This zero adjustment enables a direct reading of the heights h_i and h_m of the liquid columns. The scale may be calibrated in any convenient units, such as ft of water or psi.

The **differential manometer** (Fig. 7-11*c*) is identical to the U-tube manometer but measures the difference in pressure between two points. (It does not indicate the pressure at either point.) The differential manometer may have either the standard U-tube configuration or an inverted U-tube configuration, depending on the comparative specific gravities of the indicating and metered liquids. The inverted U-tube configuration (Fig. 7-12*b*) is used when the indicating liquid has a lower specific gravity than the metered liquid.

Example 7-4 ▪ A differential manometer (Fig. 7-11*c*) is measuring the difference in pressure between two water pipes. The indicating liquid is mercury (specific gravity = 13.6), h_i is 2.25 ft, h_{m1} is 9 in, and z is 1.0 ft. What is the pressure differential between the two pipes?

$$h_{m2} = h_i + h_{m1} - z = 2.25 + 0.75 - 1.0 = 2.0 \text{ ft}$$

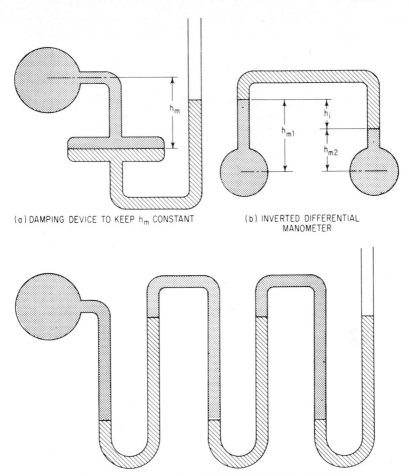

(a) DAMPING DEVICE TO KEEP h_m CONSTANT

(b) INVERTED DIFFERENTIAL MANOMETER

(c) U-TUBE MANOMETERS IN SERIES TO MEASURE HIGH PRESSURE

Fig. 7-12.   Manometers may contain a sump, as in (a) to damp flow disturbances; they may be shaped like an inverted U (b) for measuring pressures on liquids with low specific gravity; and they may be connected in series (c) for measuring high pressures.

The pressure at B, lb/ft^2, is

$$P_B = p_{c2} + w_2 h_{m2} = p_{c2} + 62.4 \times 2.0 = p_{c2} + 125$$

The pressure at A, lb/ft^2, is

$$P_A = p_{c1} + w_1 h_{m1} + w_i h_i$$

$$= p_{c1} + 62.4 \times 0.75 + 13.6 \times 62.4 \times 2.25 = p_{c1} + 1957$$

Since the pressure at A must equal that at B,

$$p_{c2} + 125 = p_{c1} + 1,957$$

$$p_{c2} - p_{c1} = 1832 \text{ lb/ft}^2 = 12.7 \text{ psi}$$

When small pressure differences in water are measured, if the specific gravity of the indicating liquid is between 1.0 and 2.0 and the points at which the pressure is being measured are at the same level, the actual pressure difference, when expressed in feet of water, is magnified by the differential manometer. For example, if the actual difference is 0.50 ft of water and the indicating liquid has a specific gravity of 1.40, the magnification will be 2.5; that is, the height of the liquid column h_i will be 1.25 ft of water. The closer the specific gravities of the metered and indicating liquids, the greater the magnification and sensitivity. This is true only up to a magnification of about 5. Above 5, the increased sensitivity may be deceptive because the meniscus between the two liquids becomes poorly defined and sluggish in movement.

Many factors affect the accuracy of manometers. Most of them, however, may be neglected in the majority of hydraulics applications since they are significant only in precise reading of manometers, such as might be required in laboratories. One factor, however, is significant: the existence of surges in the manometer caused by the pulsations and disturbances in the flow of water resulting from turbulence. These surges make reading of the manometer difficult. They may be reduced or eliminated by installing a large-diameter section, or sump, in the manometer, as shown in Fig. 7-12a. This sump will damp the pulsations and keep the distance from the center line of the conduit to the indicating liquid essentially at a constant value.

7-6. Fundamentals of Fluid Flow

For fluid energy, the law of conservation of energy is represented by the **Bernoulli equation:**

$$Z_1 + \frac{p_1}{w} + \frac{V_1^2}{2g} = Z_2 + \frac{p_2}{w} + \frac{V_2^2}{2g} \tag{7-11}$$

where Z_1 = elevation, ft, at any point 1 of flowing fluid above an arbitrary datum
Z_2 = elevation, ft, at downstream point in fluid above same datum
p_1 = pressure at 1, lb/ft^2
p_2 = pressure at 2, lb/ft^2
w = specific weight of fluid, lb/ft^3
V_1 = velocity of fluid at 1, ft/s
V_2 = velocity of fluid at 2, ft/s
g = acceleration due to gravity, 32.2 ft/s^2

The left side of the equation sums the total energy per unit weight of fluid at 1, and the right side, the total energy per unit weight at 2. Equation (7-11) applies only to an ideal fluid. Its practical use requires a term to account for the decrease in total head, ft, through friction. This term h_f, when added to the downstream side of Eq. (7-11), yields the form of the Bernoulli equation most frequently used:

$$Z_1 + \frac{p_1}{w} + \frac{V_1^2}{2g} = Z_2 + \frac{p_2}{w} + \frac{V_2^2}{2g} + h_f \tag{7-12}$$

The energy contained in an elemental volume of fluid thus is a function of its elevation, velocity, and pressure (Fig. 7-13). The energy due to elevation is the potential energy and equals WZ_a, where W is the weight, lb, of the fluid in the elemental volume and Z_a is its elevation, ft, above some arbitrary datum. The energy due to velocity is the kinetic energy. It equals $WV_a^2/2g$, where V_a is the velocity, ft/s. The pressure energy equals Wp_a/w, where p_a is the pressure lb/ft^2, and w is the specific weight of the fluid, lb/ft^3. The total energy, in the elemental volume of fluid, is

$$E = WZ_a + \frac{Wp_a}{w} + \frac{WV_a^2}{2g} \tag{7-13}$$

Dividing both sides of the equation by W yields the energy per unit weight of flowing fluid, or the *total head,* ft:

$$H = Z_a + \frac{p_a}{w} + \frac{V_a^2}{2g}$$ (7-14)

p_a/w is called **pressure head;** $V_a^2/2g$, **velocity head.**

As indicated in Fig. 7-13, $Z + p/w$ is constant for any point in a cross section and normal to the flow through a pipe or channel. Kinetic energy at the section, however, varies with velocity. Usually, $Z + p/w$ at the midpoint and the average velocity at a section are assumed when the Bernoulli equation is applied to flow across the section or when total head is to be determined. Average velocity, ft/s $= Q/A$, where Q is the quantity of flow, ft³/s, across the area of the section A, ft².

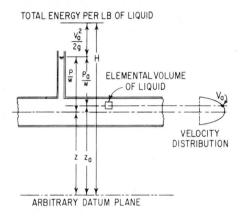

Example 7-5 ▪ Determine the energy loss between points 1 and 2 in the 24-in-diameter pipe in Fig. 7-14. The pipe carries water flowing at 31.4 ft³/s.

Average velocity in the pipe $= Q/A = 31.4/3.14 = 10$ ft/s. Select point 1 far enough from the reservoir outlet that V_1 can be assumed to be 0. Since the datum plane passes through point 2, $Z_2 = 0$. Also, since the pipe has free discharge, $p_2 = 0$. Thus substitution in Eq. (7-12) yields

$$30 + 20 + 0 = 0 + 0 + \frac{10^2}{64.4} + h_f$$

Fig. 7-13.   Energy in a liquid depends on elevation, velocity, and pressure.

where h_f is the friction loss, ft. Hence, $h_f = 50 - 1.55 = 48.45$ ft.

Note that in this example h_f includes minor losses due to the pipe entrance, gate valve, and any bends.

The Bernoulli equation and the variation of pressure may be represented graphically, respec-

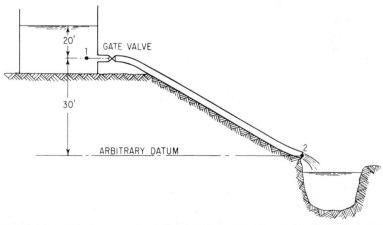

Fig. 7-14.   Flow from an elevated reservoir—application of the Bernoulli equation. (See Example 7-5.)

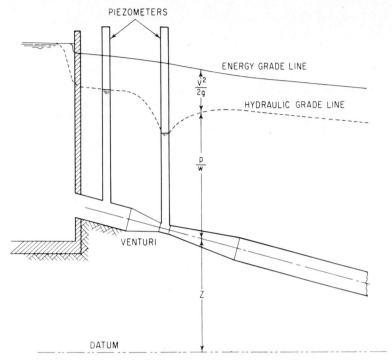

Fig. 7-15. Energy grade line and hydraulic grade line indicate variations in energy and pressure head, respectively, in a liquid as it flows along a pipe or channel.

tively, by energy and hydraulic grade lines (Fig. 7-15). The *energy grade line,* sometimes called the total head line, shows the decrease in total energy per unit weight H in the direction of flow. The slope of the energy grade line is called the **energy gradient** or friction slope. The **hydraulic grade line** lies a distance $V^2/2g$ below the energy grade line and shows the variation of velocity or pressure in the direction of flow. The slope of the hydraulic grade line is termed the **hydraulic gradient.** In open-channel flow, the hydraulic grade line coincides with the water surface, while in pressure flow, it represents the height to which water would rise in a piezometer (see also Example 7-7, Art. 7-9).

Momentum is a fundamental concept that must be considered in the design of essentially all waterworks facilities involving flow. A change in momentum, which may result from a change in either velocity, direction, or magnitude of flow, is equal to the impulse, the force F acting on the fluid times the period of time dt over which it acts. Dividing the total change in momentum by the time interval over which the change occurs gives the momentum equation, or impulse-momentum equation:

$$F_x = \rho Q \, \Delta V_x \tag{7-15}$$

where F_x = summation of all forces in X direction per unit time causing change in momentum in X direction, lb
   ρ = density of flowing fluid, lb $\cdot$ s^2/ft^4 (specific weight divided by g)
   Q = flow rate, ft^3/s
   ΔV_x = change in velocity in X direction, ft/s

Similar equations may be written for the Y and Z directions. The impulse-momentum equation often is used in conjunction with the Bernoulli equation [Eq. (7-11) or (7-12)] but may be used separately.

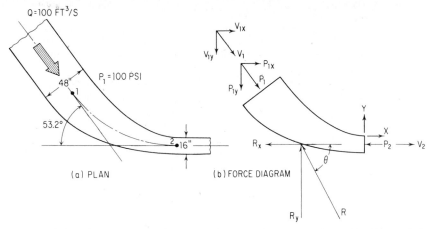

Fig. 7-16. Flow induces forces in a pipe at bends and at changes in size of section—application of momentum equation. (See Example 7-6.)

Example 7-6 ▪ Calculate the resultant force on the reducer elbow in Fig. 7-16. The pipe center line lies in a horizontal plane. The pipe reduces from 48 in in diameter to 16 in. The pressure at the upstream side of the reducer bend (point 1) is 100 psi, and the water flow is 100 ft³/s. (Neglect friction loss at the bend.)

Velocity at points 1 and 2 is found by dividing $Q = 100$ ft³/s by the respective areas: $V_1 = 100 \times 4/4^2\pi = 7.96$ ft/s and $V_2 = 100 \times 4/1.33^2\pi = 71.5$ ft/s.

With p_1 known, the Bernoulli equation yields the pressure p_2 at 2:

$$0 + 100 \times \frac{144}{62.4} + \frac{7.96^2}{2 \times 32.2} = 0 + \frac{p_2}{62.4} + \frac{71.5^2}{2 \times 32.2}$$

$$p_2 = 9500 \text{ lb/ft}^2$$

The total pressure force at 1 is $P_1 = p_1 A_1 = 181,000$ lb, and at 2, $P_2 = p_2 A_2 = 13,200$ lb.

Let R be the force, lb, exerted by the pipe on the fluid (equal and opposite in direction to the force against the pipe, which is to be determined). Then, the force F changing the momentum of the fluid equals the vector sum $P_1 - P_2 + R$. To find F, apply Eq. (7-15) first in the X direction, then in the Y direction, and determine the resultant of the forces:

In the X direction, since $\Delta V_x = -(7.96 \sin 53.2° - 71.5) = 65.1$ and the density $\rho = 62.4/32.2 = 1.94$,

$$F_x = 181,000 \cos 53.2° - 13,200 + R_x = 1.94 \times 100 \times 65.1$$

$$R_x = -82,600 \text{ lb}$$

In the Y direction, since $\Delta V_y = -(-7.96 \cos 53.2° - 0) = 4.78$,

$$F_y = -181,000 \sin 53.2° + R_y = 1.94 \times 100 \times 4.78$$

$$R_y = 145,700 \text{ lb}$$

The resultant $R = \sqrt{R_x^2 + R_y^2} = 167,500$ lb. It acts at an angle θ with the horizontal such that $\tan \theta = 145,700/82,600$, so $\theta = 60.5°$. The force against the pipe acts in the opposite direction.

7-7. Hydraulic Models

A model is a system whose operation can be used to predict the characteristics of a similar system, or prototype, usually more complex or built to a much larger scale. A knowledge of the laws governing the phenomena under investigation is necessary if the model study is to yield accurate quantitative results.

Forces acting on the model should be proportional to forces on the prototype. The four forces usually considered in hydraulic models are inertia, gravity, viscosity, and surface tension. Because of the laws governing these forces and because the model and prototype are normally not the same size, it is usually not possible to have all four forces in the model in the same proportions as they are in the prototype. It is, however, a simple procedure to have two predominant forces in the same proportion. In most models, the fact that two of the four forces are not in the same proportion as they are in the prototype does not introduce serious error. The inertial force, which is always a predominant force, and one other force are made proportional.

Ratios of the forces of gravity, viscosity, and surface tension to the force of inertia are designated, respectively, Froude number, Reynolds number, and Weber number. Equating the Froude number of the model and the Froude number of the prototype insures that the gravitational and inertial forces are in the same proportion. Similarly, equating the Reynolds numbers of the model and prototype insures that the viscous and inertial forces will be in the same proportion. And equating the Weber numbers insures proportionality of surface tension and inertial forces.

The **Froude number** is

$$\mathbf{F} = \frac{V}{\sqrt{Lg}} \tag{7-16}$$

where $\mathbf{F}$ = Froude number (dimensionless)
V = velocity of fluid, ft/s
L = linear dimension (characteristic, such as depth or diameter), ft
g = acceleration due to gravity, 32.2 ft/s^2

For hydraulic structures, such as spillways and weirs, where there is a rapidly changing water-surface profile, the two predominant forces are inertia and gravity. Therefore, the Froude numbers of the model and prototype are equated:

$$\mathbf{F_m} = \mathbf{F_p} \qquad \frac{V_m}{\sqrt{L_m g}} = \frac{V_p}{\sqrt{L_p g}} \tag{7-17a}$$

where subscript m applies to the model and p to the prototype. Squaring both sides of Eq. (7-17a) and grouping like terms yields

$$\frac{V_m^2}{V_p^2} = \frac{L_m}{L_p} \tag{7-17b}$$

Let $V_r = V_m/V_p$ and $L_r = L_m/L_p$. Then

$$V_r^2 = L_r \qquad V_r = L_r^{1/2} \tag{7-18}$$

The subscript r indicates ratio of quantity in model to that in prototype.

If the ratios of all the physical dimensions of a model to all the corresponding physical dimensions of the prototype are equal to the length ratio, the model is termed a **true model**. In a true model where the Froude number is the governing design criterion, the length ratio is the only variable. Once the length ratio has been set, all the physical dimensions of the model are fixed. The discharge ratio is determined as follows:

$$Q_r = V_r A_r \tag{7-19a}$$

Since $V_r = L_r^{1/2}$ and A_r = area ratio = L_r^2,

$$Q_r = V_r A_r = L_r^{5/2} \tag{7-19 b}$$

By this method all the necessary characteristics of a spillway or weir model can be determined.
The **Reynolds number** is

$$\mathbf{R} = \frac{VL}{\nu} \tag{7-20}$$

$\mathbf{R}$ is dimensionless, and ν is the kinematic viscosity of fluid, ft^2/s. The Reynolds numbers of model and prototype are equated when the viscous and inertial forces are predominant. Viscous forces are usually predominant when flow occurs in a closed system, such as pipe flow where there is no free surface. The following relations are obtained by equating Reynolds numbers of the model and prototype:

$$\frac{V_m L_m}{\nu_m} = \frac{V_p L_p}{\nu_p} \tag{7-21 a}$$

$$V_r = \frac{\nu_r}{L_r} \tag{7-21 b}$$

The variable factors that fix the design of a true model when the Reynolds number governs are the length ratio and the viscosity ratio.
The **Weber number** is

$$\mathbf{W} = \frac{V^2 L \rho}{\sigma} \tag{7-22}$$

where ρ = density of fluid, $\text{lb} \cdot \text{s}^2/\text{ft}^4$ (specific weight divided by g)
    σ = surface tension of fluid, lb/ft^2
The Weber numbers of model and prototype are equated in certain types of wave studies, the formation of drops and air bubbles, entrainment of air in flowing water, and other phenomena where surface tension and inertial forces are predominant. The velocity ratio is determined as follows:

$$\frac{V_m^2 L_m \rho_m}{\sigma_m} = \frac{V_p^2 L_p \rho_p}{\sigma_p} \tag{7-23 a}$$

$$V_r^2 = \frac{\sigma_r}{\rho_r L_r} \tag{7-23 b}$$

The fluid properties and the length ratio fix the design of a model governed by the Weber number.
In some cases, such as a morning-glory spillway, inertial, viscous, and gravity forces all have an important effect on the flow. In these cases it is usually not possible to have both the Reynolds and Froude numbers of the model and prototype equal. The solution to this type of problem is mostly empirical and may consist of an attempt to evaluate the effects of viscosity and gravity separately.
For the flow of water in open channels and rivers where the friction slope is relatively flat, model designs are often based on the Manning equation. The relations between the model and prototype are determined as follows:

$$\frac{V_m}{V_p} = \frac{(1.486/n_m) R_m^{2/3} S_m^{1/2}}{(1.486/n_p) R_p^{2/3} S_p^{1/2}} \tag{7-24}$$

where n = Manning roughness coefficient ($T/L^{1/3}$, T representing time)
   R = hydraulic radius (L)
   S = loss of head due to friction per unit length of conduit (dimensionless)
   = slope of energy gradient
For true models, $S_r = 1$, $R_r = L_r$. Hence,

$$V_r = \frac{L_r^{2/3}}{n_r} \qquad (7\text{-}25)$$

In models of rivers and channels, it is necessary for the flow to be turbulent. The U.S. Waterways Experiment Station has determined that flow will be turbulent if

$$\frac{VR}{\nu} \geqq 4000 \qquad (7\text{-}26)$$

where V = mean velocity, ft/s
   R = hydraulic radius, ft
   ν = kinematic viscosity, ft^2/s
If the model is to be a true model, it may have to be uneconomically large for the flow to be turbulent. Another problem also encountered in true models is surface tension. In a true model of a wide river where the depth may be only a fraction of an inch, the surface tension will distort the flow to such an extent that the model may be useless. To overcome the effect of surface tension and to get turbulent flow, the depth scale is often made much larger than the length scale. This type of model is called a **distorted model.**

The relations between a distorted model of a channel and a prototype are determined in the same manner as was Eq. (7-24). The only difference is that the slope ratio S_r equals the depth ratio d_r and the hydraulic-radius ratio is a function of the width ratio and depth ratio.

One type of model, called a movable-bed model, is used to study erosion and transportation of silt in riverbeds. Because the laws governing the transportation of material are not fully understood, movable-bed models are built largely on the basis of experience and give only qualitative results.

(J. E. A. John and W. L. Haberman, "Introduction to Fluid Mechanics," Prentice-Hall, Inc., Englewood Cliffs, N.J.)

PIPE FLOW

The term pipe flow as used in this section refers to flow in a circular closed conduit entirely filled with fluid. For closed conduits other than circular, reasonably good results are obtained in the turbulent range with standard pipe-flow formulas if the diameter is replaced by four times the hydraulic radius. But when there is severe deviation from a circular cross section, as in annular passages, this method gives values that are much too low. (J. F. Walker, G. A. Whan, and R. R. Rothfus, "Fluid Friction in Noncircular Ducts," *Journal of the American Institute of Chemical Engineers,* vol. 3, 1957.)

7-8. Laminar Flow

In laminar flow, fluid particles move in parallel layers in one direction. The parabolic velocity distribution in laminar flow, shown in Fig. 7-17, creates a shearing stress $\tau = \mu\, dV/dy$, where dV/dy is the rate of change of velocity with depth and μ is the coefficient of viscosity (see Viscosity, Art. 7-2). As this shearing stress increases, the viscous forces become unable to damp out disturbances, and turbulent flow results. The region of change is dependent on the fluid's velocity, density, and viscosity and the size of the conduit.

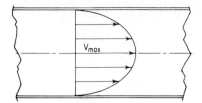

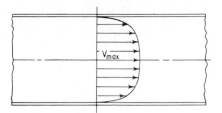

Fig. 7-17. Velocity distribution for laminar flow in a circular pipe is parabolic. Maximum velocity is twice the average velocity.

Fig. 7-18. Velocity distribution for turbulent flow in a circular pipe is more nearly uniform than that for laminar flow.

A dimensionless parameter called the Reynolds number has been found to be a reliable criterion for the determination of laminar or turbulent flow. It is the ratio of inertial forces to viscous forces, and is given by

$$\mathbf{R} = \frac{VD\rho}{\mu} = \frac{VD}{\nu} \tag{7-27}$$

where V = fluid velocity, ft/s
  D = pipe diameter, ft
  ρ = density of fluid, lb $\cdot$ s^2/ft^4 (specific weight divided by g, 32.2 ft/s^2)
  μ = viscosity of fluid lb $\cdot$ s/ft^2
  ν = μ/ρ = kinematic viscosity, ft^2/s
For a Reynolds number less than 2000, flow is laminar in circular pipes. When the Reynolds number is greater than 2000, laminar flow is unstable; a disturbance will probably be magnified, causing the flow to become turbulent.

In laminar flow, the following equation for head loss due to friction can be developed by considering the forces acting on a cylinder of fluid in a pipe:

$$h_f = \frac{32\mu LV}{D^2\rho g} = \frac{32\mu LV}{D^2 w} \tag{7-28}$$

where h_f = head loss due to friction, ft
  L = length of pipe section considered, ft
  g = acceleration due to gravity, 32.2 ft/s^2
  w = specific weight of fluid, lb/ft^3
Substitution of the Reynolds number yields

$$h_f = \frac{64}{\mathbf{R}} \frac{L}{D} \frac{V^2}{2g} \tag{7-29}$$

For laminar flow, Eq. (7-29) is identical to the Darcy-Weisbach formula Eq. (7-30) since in laminar flow the friction $f = 64/\mathbf{R}$.

(H. W. King and E. F. Brater, "Handbook of Hydraulics," McGraw-Hill Book Company, New York.)

7-9. Turbulent Flow

In turbulent flow, the inertial forces are so great that viscous forces cannot dampen out disturbances caused primarily by the surface roughness. These disturbances create eddies, which have both a rotational and translational velocity. The translation of these eddies is a mixing action that affects an interchange of momentum across the cross section of the conduit. As a result, the velocity distribution is more uniform, as shown in Fig. 7-18, than for laminar flow (Fig. 7-17).

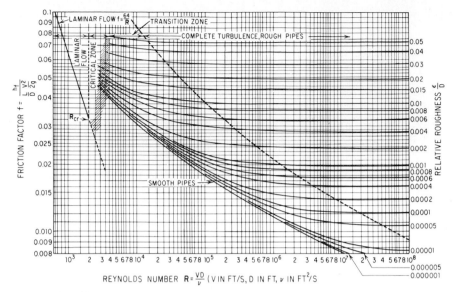

Fig. 7-19. Chart gives friction forces for flow in pipes.

For a Reynolds number greater than 2000 but to the left of the dashed line in Fig. 7-19, there is a transition from laminar to turbulent flow. In this region, there is a laminar film at the boundaries that covers some of the smaller roughness projections. This explains why the friction loss in this region has both laminar and turbulent characteristics. As the Reynolds number increases, this laminar boundary layer decreases in thickness until, at completely turbulent flow, it no longer covers any of the roughness projections. To the right of the dashed line in Fig. 7-19, the flow is completely turbulent, and viscous forces do not affect the friction loss.

Because of the random nature of turbulent flow, it is not practical to treat it analytically. Therefore, formulas for head loss and flow in the turbulent regions have been developed through experimental and statistical means. Experimentation in turbulent flow has shown that:

The head loss varies directly as the length of the pipe.

The head loss varies almost as the square of the velocity.

The head loss varies almost inversely as the diameter.

The head loss depends on the surface roughness of the pipe wall.

The head loss depends on the fluid's density and viscosity.

The head loss is independent of the pressure.

The Darcy-Weisbach formula, one of the most widely used equations for pipe flow, satisfies the above condition and is valid for laminar or turbulent flow in all fluids.

$$h_f = f \frac{L}{D} \frac{V^2}{2g} \tag{7-30}$$

where h_f = head loss due to friction, ft
  f = friction factor (see Fig. 7-19)
  L = length of pipe, ft
  D = diameter of pipe, ft

TABLE 7-3 Typical Values of Roughness for Use in the Moody Diagram (Fig. 7-19) to Determine f

	ϵ, ft
Steel pipe:	
Severe tuberculation and incrustation	0.03 –0.008
General tuberculation	0.008 –0.003
Heavy brush-coat asphalts, enamels, and tars	0.003 –0.001
Light rust	0.001 –0.0005
New smooth pipe, centrifugally applied enamels	0.0002–0.00003
Hot-dipped asphalt; centrifugally applied concrete linings	0.0005–0.0002
Steel-formed concrete pipe, good workmanship	0.0005–0.0002
New cast-iron pipe	0.00085

V = velocity of fluid, ft/s

g = acceleration due to gravity, 32.2 ft/s^2

It employs the Moody diagram (Fig. 7-19) for evaluating the friction factor f. (L. F. Moody, "Friction Factors for Pipe Flow," *Transactions of the American Society of Mechanical Engineers,* November 1944.)

Because Eq. (7-30) is dimensionally homogeneous, it can be used with any consistent set of units without changing the value of the friction factor.

Roughness values ϵ (ft) for use with the Moody diagram to determine the Darcy-Weisbach friction factor f are in Table 7-3.

The following formulas were derived for head loss in waterworks design and give good results for water-transmission and -distribution calculations. They contain a factor that depends on the surface roughness of the pipe material. The accuracy of these formulas is greatly affected by the selection of the roughness factor, which requires experience in its choice.

The Chezy formula holds for head loss in conduits and gives reasonably good results for high Reynolds numbers:

$$V = C\sqrt{RS} \tag{7-31}$$

where V = velocity, ft/s

C = coefficient, dependent on surface roughness of conduit

S = slope of energy grade line or head loss due to friction, ft/ft of conduit

R = hydraulic radius, ft

Hydraulic radius of a conduit is the cross-sectional area of the fluid in it divided by the perimeter of the wetted section.

Manning's formula: Through experimentation, Manning concluded that the C in the Chezy equation [Eq. (7-31)] should vary as $R^{1/6}$

$$C = \frac{1.486R^{1/6}}{n} \tag{7-32}$$

where n = coefficient, dependent on surface roughness. Substitution into Eq. (7-31) gives

$$V = \frac{1.486}{n}R^{2/3}S^{1/2} \tag{7-33a}$$

Upon substitution of $D/4$, where D is the pipe diameter, for the hydraulic radius of the pipe, the following equations are obtained for pipes flowing full:

$$V = \frac{0.590}{n}D^{2/3}S^{1/2} \tag{7-33b}$$

TABLE 7-4   Values of n for Pipes, to Be Used with the Manning Formula

Material of pipe	Variation		Use in designing	
	From	To	From	To
Clean cast iron	0.011	0.015	0.013	0.015
Dirty or tuberculated cast iron	0.015	0.035		
Riveted steel	0.013	0.017	0.015	0.017
Welded steel	0.010	0.013	0.012	0.013
Galvanized iron	0.012	0.017	0.015	0.017
Wood stave	0.010	0.014	0.012	0.013
Concrete	0.010	0.017		
Good workmanship			0.012	0.014
Poor workmanship			0.016	0.017

$$Q = \frac{0.463}{n} D^{8/3} S^{1/2} \qquad (7\text{-}33\,c)$$

$$h_f = 4.66 n^2 \frac{LQ^2}{D^{16/3}} \qquad (7\text{-}33\,d)$$

$$D = \left(\frac{2.159 Q n}{S^{1/2}} \right)^{3/8} \qquad (7\text{-}33\,e)$$

where Q = flow, ft^3/s.

Tables 7-4 and 7-11 (p. 7-52) give values of n for the foot-pound-second system. See also tables in Art. 8-6 for velocity and flow at various slopes.

The Hazen-Williams formula is one of the most widely used formulas for pipe-flow computations of water utilities, although it was developed for both open channels and pipe flow:

$$V = 1.318 C_1 R^{0.63} S^{0.54} \qquad (7\text{-}34\,a)$$

For pipes flowing full:

$$V = 0.55 C_1 D^{0.63} S^{0.54} \qquad (7\text{-}34\,b)$$

$$Q = 0.432 C_1 D^{2.63} S^{0.54} \qquad (7\text{-}34\,c)$$

$$h_f = \frac{4.727}{D^{4.87}} L \left(\frac{Q}{C_1} \right)^{1.85} \qquad (7\text{-}34\,d)$$

$$D = \frac{1.376}{S^{0.205}} \left(\frac{Q}{C_1} \right)^{0.38} \qquad (7\text{-}34\,e)$$

where V = velocity, ft/s
C_1 = coefficient, dependent on surface roughness
R = hydraulic radius, ft
S = head loss due to friction, ft/ft of pipe
D = diameter of pipe, ft
L = length of pipe, ft
Q = discharge, ft^3/s
h_f = friction loss, ft

The C_1 terms in Table 7-5 are in the foot-pound-second system.

The problem of flow in branching pipes illustrates the use of friction-loss equations and the hydraulic-grade-line concept.

TABLE 7-5   Values of C_1 in Hazen and Williams Formula

Type of pipe	C_1
Cast iron:	
New	All sizes, 130
5 years old	All sizes up to 24 in , 120
	24 in and over, 115
10 years old	12 in , 110
	4 in , 105
	30 in and over, 85
40 years old	16 in , 80
	4 in , 65
Welded steel	Values the same as for cast-iron pipe, 5 years older
Riveted steel	Values the same as for cast-iron pipe, 10 years older
Wood stave	Average value, regardless of age, 120
Concrete or concrete-lined	Large sizes, good workmanship, steel forms, 140
	Large sizes, good workmanship, wood forms, 120
	Centrifugally spun, 135
Vitrified	In good condition, 110

Example 7-7 ▪ Figure 7-20 shows a typical three-reservoir problem. The elevations of the hydraulic grade lines for the three pipes are equal at point D. The Hazen-Williams equation for friction loss [Eq. (7-34d)] can be written for each pipe meeting at D. With the continuity equation for quantity of flow, there are as many equations as there are unknowns:

$$Z_a = Z_d + \frac{p_D}{w} + \frac{4.727 L_A}{D_A^{4.87}} \left(\frac{Q_A}{C_A} \right)^{1.85} \tag{7-35 a}$$

$$Z_b = Z_d + \frac{p_D}{w} + \frac{4.727 L_B}{D_B^{4.87}} \left(\frac{Q_B}{C_B} \right)^{1.85} \tag{7-35 b}$$

$$Z_c = Z_d + \frac{p_D}{w} - \frac{4.727 L_C}{D_C^{4.87}} \left(\frac{Q_C}{C_C} \right)^{1.85} \tag{7-35 c}$$

$$Q_A + Q_B = Q_C \tag{7-36}$$

where p_D = pressure at D
w = unit weight of liquid

With the elevations Z of the three reservoirs and the pipe intersection known, the easiest way to solve these equations is by trying different values of p_D/w in Eqs. (7-35) and substituting the values obtained for Q into Eq. (7-36) for a check. If the value of $Z_d + p_D/w$ becomes greater than Z_b, the sign of the friction-loss term is negative instead of positive. This would indicate water is flowing from reservoir A into reservoirs B and C.

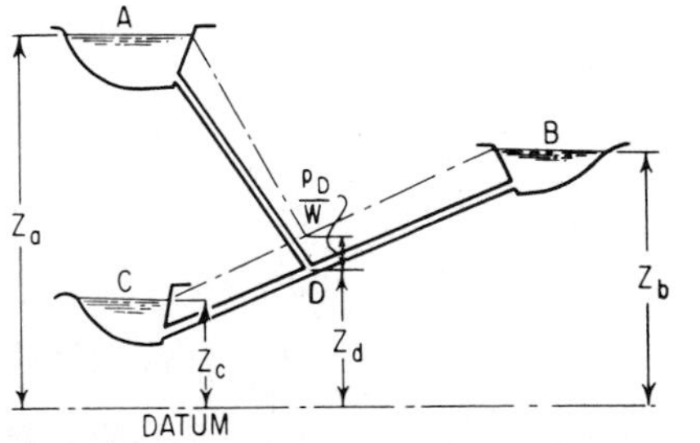

7-10 Minor Losses in Pipes

These are losses occurring in contractions, bends, enlargements, and valves and other pipe fittings. These losses can usually be neglected if the length of the pipeline is greater than 1500 times the pipe's diameter. However, in short pipelines, because these losses may exceed the friction losses, minor losses must be considered.

Fig. 7-20. Flow between three reservoirs. (See Example 7-7.)

Sudden Enlargement. ▪ The following equation for the head loss, ft, across a sudden enlargement of pipe diameter has been determined analytically and agrees well with experimental results:

$$h_L = \frac{(V_1 - V_2)^2}{2g} \tag{7-37}$$

where V_1 = velocity before enlargement, ft/s
V_2 = velocity after enlargement, ft/s
g = 32.2 ft/s^2

It was derived by applying the Bernoulli equation and the momentum equation across an enlargement.

Another equation for the head loss caused by sudden enlargements was determined experimentally by Archer. This equation gives slightly better agreement with experimental results than Eq. (7-37):

$$h_L = \frac{1.1(V_1 - V_2)^{1.92}}{2g} \tag{7-38}$$

A special application of Eq. (7-37) or (7-38) is the discharge from a pipe into a reservoir. The water in the reservoir has no velocity, so a full velocity head is lost.

Gradual Enlargement ▪ The equation for the head loss due to a gradual conical enlargement of a pipe takes the following form:

$$h_L = \frac{K(V_1 - V_2)^2}{2g} \tag{7-39}$$

where K = loss coefficient (see Fig. 7-21).

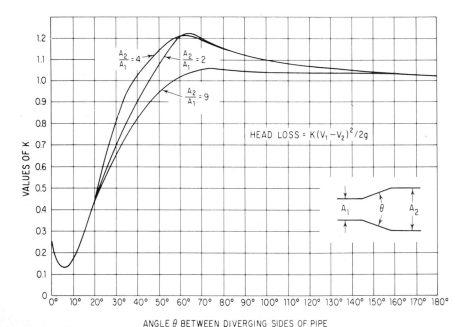

Fig. 7-21. Head-loss coefficients for a pipe with diverging sides depend on the angle of divergence.

TABLE 7-6 C_c for Contractions in Pipe Area from A_1 to A_2

A_2/A_1	0.1	0.2	0.3	0.4	0.5	0.6	0.7	0.8	0.9	1.0
C_c	0.62	0.63	0.64	0.66	0.68	0.71	0.76	0.81	0.89	1.0

Since the experimental data available on gradual enlargements are limited and inconclusive, the values of K in Fig. 7-21 are approximate. (A. H. Gibson, "Hydraulics and Its Applications," Constable & Co., Ltd., London.)

Sudden Contraction ▪ The following equation for the head loss across a sudden contraction of a pipe was determined by the same type of analytical studies as Eq. (7-37):

$$h_L = \left(\frac{1}{C_c} - 1 \right)^2 \frac{V^2}{2g} \qquad (7\text{-}40)$$

where C_c = coefficient of contraction (see Table 7-6)
V = velocity in smaller-diameter pipe, ft/s
This equation gives best results when the head loss is greater than 1 ft. Table 7-6 gives C_c values for sudden contractions, determined by Julius Weisbach ("Die Experiments-Hydraulik").

Another formula for determining the loss of head caused by a sudden contraction, determined experimentally by Brightmore, is

$$h_L = \frac{0.7(V_1 - V_2)^2}{2g} \qquad (7\text{-}41)$$

This equation gives best results if the head loss is less than 1 ft.

A special case of sudden contraction is the entrance loss for pipes. Some typical values of the loss coefficient K in $h_L = KV^2/2g$, where V is the velocity in the pipe, are presented in Table 7-7.

Bends and Standard Fitting Losses ▪ The head loss that occurs in pipe fittings, such as valves and elbows, and at bends is given by

$$h_L = \frac{KV^2}{2g} \qquad (7\text{-}42)$$

Table 7-8 gives some typical K values for these losses.

The values in Table 7-8 are only approximate. K values vary not only for different sizes of fitting but with different manufacturers. For these reasons, manufacturers' data are the best source for loss coefficients.

Experimental data available on bend losses cover a rather narrow range of laboratory experiments utilizing small-diameter pipes and do not give conclusive results. The data indicate the losses vary with surface roughness, Reynolds number, ratio of radius of bend r to pipe diameter D, and angle of bend. The data are in agreement that the head loss, not including friction loss, decreases sharply as the r/D ratio increases from zero to around 4 or 5. When r/D increases

TABLE 7-7 Coefficients for Entrance Losses

Pipe projecting into reservoir	$K = 0.80$
Sharp-cornered entrance	$K = 0.50$
Bellmouth entrance	$K = 0.05$
Slightly rounded entrance	$K = 0.25$

TABLE 7-8 Coefficients for Fitting
Losses and Losses at Bends

Fitting	K
Globe valve, fully open	10.0
Angle valve, fully open	5.0
Swing check valve, fully open	2.5
Gate valve, fully open	0.2
Closed-return bend	2.2
Short-radius elbow ($r/D \approx 1.0$)*	0.9
Long-radius elbow ($r/D \approx 1.5$)	0.6
45° elbow	0.4

*r = radius of bend; D = pipe diameter.

above 4 or 5, there is disagreement. Some experiments indicate that the head loss, not including friction loss in the bend, increases significantly with an increasing r/D. Some recent work by Ito, on smooth pipes, indicates that this increase is very slight and that above an r/D of 4, the bend loss essentially remains constant. (H. Ito, "Pressure Losses in Smooth Pipe Bends," *Transactions of the American Society of Civil Engineers,* series D, vol. 82, no. 1, 1960.)

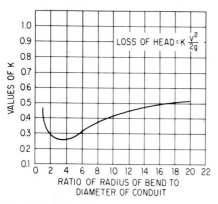

Fig. 7-22. Recommended values of head-loss coefficients K for 90° bends in closed conduits.

Because experiments have produced such widely varying data, bend-loss coefficients give only an approximation of losses to be expected. Figure 7-22 gives values of K for 90° bends for use with Eq. (7-42). (K. H. Beij, "Pressure Losses for Fluid Flow in 90° Pipe Bends," *Journal of Research, National Bureau of Standards,* vol. 21, July 1938.)

To obtain losses in bends other than 90°, Hinds suggested the following formula to adjust the K values given in Fig. 7-22:

$$K' = K \sqrt{\frac{\Delta}{90}} \qquad (7\text{-}43)$$

where Δ = the deflection angle, °.

The K' value may be used in place of K in Eq. (7-42).

7-11. Orifices

An orifice is an opening with a closed perimeter through which water flows. Orifices may have any shape, although they are usually round, square, or rectangular.

Discharge through an orifice may be calculated from

$$Q = Ca \sqrt{2gh} \qquad (7\text{-}44)$$

where Q = discharge, ft³/s
C = coefficient of discharge
a = area of orifice, ft²
g = acceleration due to gravity, ft/s²
h = head on horizontal center line of orifice, ft

Coefficients of discharge C are given in Table 7-9 for low velocity of approach. If this velocity is significant, its effect should be taken into account. Equation (7-44) is applicable for any head for which the coefficient of discharge is known. For low heads, measuring the head

TABLE 7-9 Smith's Cofficients of Discharge for Circular and Square Orifices with Full Contraction

Dia of circular orifices, ft				Head, ft	Side of square orifices, ft			
0.02	0.04	0.01	1.0		0.02	0.04	0.1	1.0
	0.637	0.618		0.4		0.643	0.621	
0.655	0.630	0.613		0.6	0.660	0.636	0.617	
0.648	0.626	0.610	0.590	0.8	0.652	0.631	0.615	0.597
0.644	0.623	0.608	0.591	1	0.648	0.628	0.613	0.599
0.637	0.618	0.605	0.593	1.5	0.641	0.622	0.610	0.601
0.632	0.614	0.604	0.595	2	0.637	0.619	0.608	0.602
0.629	0.612	0.603	0.596	2.5	0.634	0.617	0.607	0.602
0.627	0.611	0.603	0.597	3	0.632	0.616	0.607	0.603
0.623	0.609	0.602	0.596	4	0.628	0.614	0.606	0.602
0.618	0.607	0.600	0.596	6	0.623	0.612	0.605	0.602
0.614	0.605	0.600	0.596	8	0.619	0.610	0.605	0.602
0.611	0.603	0.598	0.595	10	0.616	0.608	0.604	0.601
0.601	0.599	0.596	0.594	20	0.606	0.604	0.602	0.600
0.596	0.595	0.594	0.593	50	0.602	0.601	0.600	0.599
0.593	0.592	0.592	0.592	100	0.599	0.598	0.598	0.598

*Hamilton Smith, Jr., "Hydraulics," 1886.

from the center line of the orifice is not theoretically correct; however, this error is corrected by the C values.

The coefficient of discharge C is the product of the coefficient of velocity C_v and the coefficient of contraction C_c. The coefficient of velocity is the ratio obtained by dividing the actual velocity at the *vena contracta* (contraction of the jet discharged) by the theoretical velocity. The theoretical velocity may be calculated by writing Bernoulli's equation for points 1 and 2 in Fig. 7-23.

$$\frac{V_1^2}{2g} + \frac{p_1}{w} + Z_1 = \frac{V_2^2}{2g} + \frac{p_2}{w} + Z_2 \qquad (7\text{-}45)$$

With the reference plane through point 2, $Z_1 = h$, $V_1 = 0$, $p_1/w = p_2/w = 0$, and $Z_2 = 0$, and Eq. (7-45) becomes

$$V_2 = \sqrt{2gh} \qquad (7\text{-}46)$$

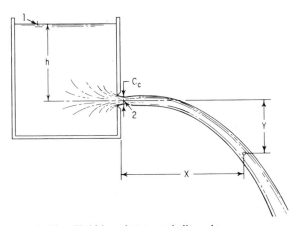

Fig. 7-23. Fluid jet takes a parabolic path.

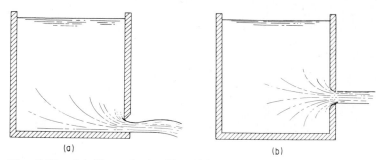

Fig. 7-24. (*a*) Sharp-edged orifice with partly suppressed contraction. (*b*) Round-edged orifice has no contraction.

The actual velocity, determined experimentally, is less than the theoretical velocity because of the energy loss from point 1 to point 2. Typical values of C_v range from 0.94 to 0.99.

The coefficient of contraction C_c is the ratio of the smallest area of the jet, the vena contracta, to the area of the orifice. Contraction of a fluid jet will occur if the orifice is square-edged and so located that some of the fluid approaches the orifice at an angle to the direction of flow through the orifice. This fluid has a momentum component perpendicular to the axis of the jet which causes the jet to contract. Typical values of the coefficient of contraction range from 0.61 to 0.67.

If the water entering the orifice does not have this momentum, the contraction is completely suppressed. Figure 7-24*a* is an example of a partly suppressed contraction; no contraction occurs at the bottom of the jet. In Fig. 7-24*b*, the edges of the orifice have been rounded to reduce or eliminate the contraction. With a partly suppressed orifice, the increased area of jet caused by suppressing the contraction on one side is partly offset because more water at a higher velocity enters on the other sides. The result is a slightly greater coefficient of contraction.

Submerged Orifices ▪ Flow through a submerged orifice may be computed by applying Bernoulli's equation to points 1 and 2 in Fig. 7-25.

$$V_2 = \sqrt{2g\left(h_1 - h_2 + \frac{V_1^2}{2g} - h_L\right)} \qquad (7\text{-}47)$$

where h_L = losses in head, ft, between 1 and 2.

Assuming $V_1 \approx 0$, setting $h_1 - h_2 = \Delta h$, and using a coefficient of discharge C to account for losses, Eq. (7-48) is obtained.

$$Q = Ca\sqrt{2g\,\Delta h} \qquad (7\text{-}48)$$

Values of C for submerged orifices do not differ greatly from those for nonsubmerged orifices. (For table of values of coefficients of discharge for submerged orifices, see H. W. King and E. F. Brater, "Handbook of Hydraulics," McGraw-Hill Book Company, New York.)

Discharge under Falling Head ▪ The flow from a reservoir or vessel when the inflow is less than the outflow represents a condition of falling head. The time required for a certain quantity of water to flow from a reservoir can be calculated by equating the volume of water that flows through the orifice or pipe in time dt to the volume decrease in the reservoir (Fig. 7-26):

$$Ca\sqrt{2gy}\,dt = A\,dy \qquad (7\text{-}49)$$

Solving for dt yields

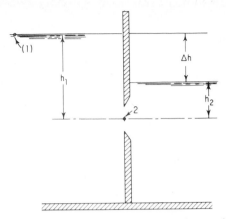

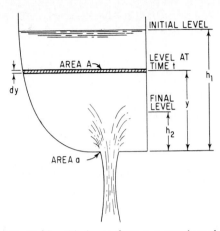

Fig. 7-25. Discharge through a submerged orifice.

Fig. 7-26. Discharge from a reservoir under falling head.

$$dt = \frac{A\ dy}{Ca\ \sqrt{2gy}} \tag{7-50}$$

where a = area of orifice, ft^2
A = area of reservoir, ft^2
y = head on orifice at time t, ft
C = coefficient of discharge
g = acceleration due to gravity, 32.2 ft/s^2

Expressing the area as a function of y $[A = F(y)]$ and summing from time zero, when $y = h_1$, to time t, when $y = h_2$, Eq. (7-50) becomes

$$t = \int_{h_2}^{h_1} \frac{F(y)\ dy}{Ca\ \sqrt{2gy}} \tag{7-51}$$

If the area of the reservoir is constant as y varies, Eq. (7-51) upon integration becomes

$$t = \frac{2A}{Ca\ \sqrt{2g}}\ (\sqrt{h_1} - \sqrt{h_2}) \tag{7-52}$$

where h_1 = head at the start, ft
h_2 = head at the end, ft
t = time interval for head to fall from h_1 to h_2, s

Fluid Jets ▪ Where the effect of air resistance is small, a fluid discharged through an orifice into the air will follow the path of a projectile. The initial velocity of the jet is

$$V_0 = C_v\ \sqrt{2gh} \tag{7-53}$$

where h = head on center line of orifice, ft
C_v = coefficient of velocity

The direction of the initial velocity depends on the orientation of the surface in which the orifice is located. For simplicity, the following equations were determined assuming the orifice is located in a vertical surface (Fig. 7-23). The velocity of the jet in the X direction (horizontal) remains constant.

$$V_x = V_0 = C_v\sqrt{2gh} \tag{7-54}$$

The velocity in the Y direction is initially zero and thereafter a function of time and the acceleration of gravity:

$$V_y = gt \qquad (7\text{-}55)$$

The X coordinate at time t is

$$X = V_x t = t C_v \sqrt{2gh} \qquad (7\text{-}56)$$

The Y coordinate is

$$Y = V_{\text{avg}} t = \frac{gt^2}{2} \qquad (7\text{-}57)$$

where V_{avg} = average velocity over period of time t. The equation for the path of the jet [Eq. (7-58)], obtained by solving Eq. (7-57) for t and substituting in Eq. (7-56), is that for a parabola:

$$X^2 = C_v^2 4hY \qquad (7\text{-}58)$$

Equation (7-58) can be used to determine C_v experimentally. Rearranging Eq. (7-58) gives

$$C_v = \sqrt{\frac{X^2}{4hY}} \qquad (7\text{-}59)$$

The X and Y coordinates can be measured in a laboratory and C_v calculated from Eq. (7-59).

Short Tubes ▪ When water flows from a reservoir into a pipe or tube with a sharp leading edge, the same type of contraction occurs as for a sharp-edged orifice. In the tube or pipe, however, the water contracts and then expands to fill the tube. If the tube is discharging at atmospheric pressure, a partial vacuum is created at the contraction, as can be seen by applying the Bernoulli equation across points 1 and 2 in Fig. 7-27. This reduced pressure causes the flow through a short tube to be greater than that through a sharp-edged orifice of the same dimensions. If the head on the tube is greater than 50 ft and the tube is short, the water will shoot through the tube without filling it. When this happens, the tube acts as a sharp-edged orifice.

For a short tube flowing full, the coefficient of contraction $C_c = 1.00$ and the coefficient of velocity $C_v = 0.82$. Therefore, the coefficient of discharge $C = 0.82$. Solving for head loss as a proportion of final velocity head, a K value for Eq. (7-42) of 0.5 is obtained as follows: The theoretical velocity head with no loss is $V_T^2/2g$. Actual velocity head is $V_a^2/2g = (0.82 V_T)^2/2g$

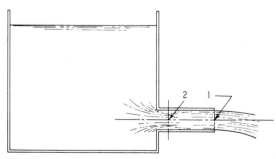

Fig. 7-27. Flow through a tube with a sharp-edged inlet.

$= 0.67 V_T^2/2g$. Head loss $h_L = 1.00 V_T^2/2g - 0.67 V_T^2/2g = 0.33 V_T^2/2g$. From $h_L = KV_a^2/2g$, where $V_a^2/2g$ is the actual velocity head,

$$K = 2gh_L/V_a^2 = (0.33 V_T^2 \times 2g)/(2g \times 0.67 V_T^2) = 0.5$$

For a reentrant tube projecting into a reservoir (Fig. 7-28), the coefficients of velocity and discharge equal 0.75, and the loss coefficient K equals 0.80.

Diverging Conical Tubes ▪ This type of tube can greatly increase the flow through an orifice by reducing the pressure at the orifice below atmospheric. Equation (7-60) for the pressure at the entrance to the tube is obtained by writing the Bernoulli equation for points 1 and 3 and points 1 and 2 in Fig. 7-29,

$$p_2 = wh \left[1 - \left(\frac{a_3}{a_2} \right)^2 \right] \tag{7-60}$$

where p_2 = gage pressure at tube entrance, lb/ft²
    w = unit weight of water, lb/ft³
    h = head on center line of orifice, ft
    a_2 = area of smallest part of jet (vena contracta, if one exists), ft²
    a_3 = area of discharge end of tube, ft²

Discharge is also calculated by writing the Bernoulli equation for points 1 and 3 in Fig. 7-29.

For this analysis to be valid, the tube must flow full, and the pressure in the throat of the tube must not fall to the vapor pressure of water. Experiments by Venturi show the most efficient angle θ to be around 5°.

7-12. Siphons

A siphon is a closed conduit that rises above the hydraulic grade line and in which the pressure at some point is below atmospheric (Fig. 7-30). The most common use of a siphon is the siphon spillway.

Flow through a siphon can be calculated by writing the Bernoulli equation for the entrance and exit. But the pressure in the siphon must be checked to be sure it does not fall to the vapor pressure of water. This is accomplished by writing the Bernoulli equation across a point of

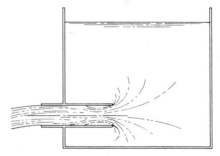

Fig. 7-28. Flow through a reentrant tube resembles that through a flush tube (Fig. 7-27), but head loss is greater.

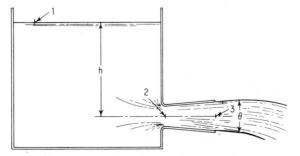

Fig. 7-29. Diverging conical tube increases flow through an orifice.

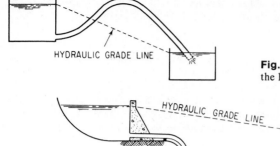

Fig. 7-30.   Siphon is a pipe that rises above the hydraulic grade line.

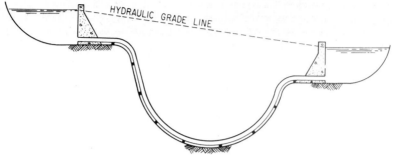

Fig. 7-31.   Sag pipe connects two reservoirs.

known pressure and a point where the elevation head or the velocity head is a maximum in the conduit. If the pressure were to fall to the vapor pressure, vaporization would decrease or totally stop the flow.

The pipe shown in Fig. 7-31 is also commonly called a siphon or inverted siphon. This is a misnomer since the pressure at all points in the pipe is above atmospheric. The American Society of Civil Engineers recommends that the inverted siphon be called a **sag pipe** to avoid the false impression that it acts as a siphon.

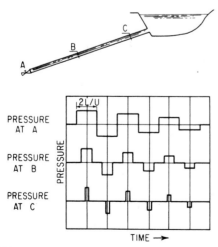

Fig. 7-32.   Variation of pressure with time along a penstock, for water hammer caused by instantaneous closure of a valve.

7-13. Water Hammer

Water hammer is a change in pressure, either above or below the normal pressure, caused by a variation of the flow rate in a pipe. Every time the flow rate is changed, either increased or decreased, it causes water hammer. However, the stresses are not critical in small-diameter pipes with flows at low velocities.

The water flowing in a pipe has momentum equal to the mass of the water times its velocity. When a valve is closed, this momentum drops to zero. The change causes a pressure rise, which begins at the valve and is transmitted up the pipe. The pressure at the valve will rise until it is high enough to overcome the momentum of the water and bring the water to a stop. This pressure buildup travels the full length of the pipe to the reservoir (Fig. 7-32).

At the instant the pressure wave reaches the reservoir, the water in the pipe is motionless, but at a pressure much higher than normal. The differential pressure between the pipe and the reservoir then causes the water in the pipe to rush back into the reservoir. As the water flows into the reservoir, the pressure in the pipe falls.

At the instant the pressure at the valve reaches normal, the water has attained considerable momentum up the pipe. As the water flows away from the closed valve, the pressure at the valve drops until differential pressure again brings the water to a stop. This pressure drop begins at the valve and continues up the pipe until it reaches the reservoir.

The pressure in the pipe is now below normal, so water from the reservoir rushes into the pipe. This cycle repeats over and over until friction damps these oscillations. Because of the high velocity of the pressure waves, each cycle may take only a fraction of a second.

The equation for the velocity of a wave in a pipe is

$$U = \sqrt{\frac{E}{\rho}} \sqrt{\frac{1}{1 + ED/E_p t}} \qquad (7\text{-}61)$$

where U = velocity of pressure wave along pipe, ft/s
E = modulus of elasticity of water, 43.2×10^6 lb/ft^2
ρ = density of water, 1.94 lb · s/ft^4 (specific weight divided by acceleration due to gravity)
D = diameter of pipe, ft
E_p = modulus of elasticity of pipe material, lb/ft^2
t = thickness of pipe wall, ft

Instantaneous Closure ▪ The magnitude of the pressure change that results when flow is varied depends on the rate of change of flow and the length of the pipeline. Any gradual movement of a valve that is made in less time than it takes for a pressure wave to travel from the valve to the reservoir and be reflected back to the valve produces the same pressure change as an instantaneous movement. Instantaneous closure:

$$T < \frac{2L}{U} \qquad (7\text{-}62)$$

where L = length of pipe from reservoir to valve, ft
T = time required to change setting of valve, s

A plot of pressure vs. time for various points along a pipe is shown in Fig. 7-32 for the instantaneous closure of a valve. Equation (7-63 a) for the pressure rise or fall caused by adjusting a valve was derived by equating the momentum of the water in the pipe to the force impulse required to bring the water to a stop.

$$\Delta p = \rho U \Delta V \qquad (7\text{-}63\,a)$$

In terms of pressure head, Eq. (7-63 a) becomes

$$\Delta h = \frac{U \Delta V}{g} \qquad (7\text{-}63\,b)$$

where Δp = pressure change from normal due to instantaneous change of valve setting, lb/ft^2
Δh = head change from normal due to instantaneous change of valve setting, ft
ΔV = change in the velocity of water caused by adjusting valve, ft/s

If the closing or opening of a valve is instantaneous, the pressure change can be calculated in one step from Eq. (7-63).

Gradual Closure ▪ The following method of determining the pressure change due to gradual closure of a valve gives a quick, approximate solution. The pressure rise or head change is assumed to be in direct proportion to the closure time:

$$\Delta h_g = \frac{t_i \, \Delta h}{T} = \frac{2L \, \Delta V}{Tg} \tag{7-64}$$

where Δh_g = head change due to gradual closure, ft
  t_i = time for wave to travel from the valve to the reservoir and be reflected back to valve, s
  T = actual closure time of valve, s
  Δh = head rise due to instantaneous closure, ft
  L = length of pipeline, ft
  ΔV = change in velocity of water due to instantaneous closure, ft/s
  g = acceleration due to gravity, 32.2 ft/s^2

Arithmetic integration is a more exact method for finding the pressure change due to gradual movement of a valve, and it can be readily programmed for a computer. Integration is a direct means of studying every physical element of the process of water hammer. The valve is assumed to close in a series of small movements, each causing an individual pressure wave. The magnitude of these pressure waves is given by Eq. (7-63). The individual pressure waves are totaled to give the pressure at any desired point for a certain time.

The first step in this method is to choose the time interval for each incremental movement of the valve. (It is convenient to make the time interval some submultiple of L/U, such as L/aU, where a equals any integer, so that the pressure waves reflected at the reservoir will be superimposed upon the new waves being formed at the valve. The wave formed at the valve will be opposite in sign to the water reflected from the reservoir, so there will be a tendency for the waves to cancel out.) Assuming a valve is fully open and requires T seconds for closing, the number of incremental closing movements required is $T/\Delta t$, where Δt, the increment of time, equals L/aU.

Once the time interval has been determined, an estimate of the velocity change ΔV during each time interval must be made, to apply Eq. (7-63). A rough estimate for the velocity following the incremental change is $V_n = V_o(A_n/A_o)$, where V_n is the velocity following a certain incremental movement, V_o the original velocity, A_n the area of the valve opening after the corresponding incremental movement, and A_o the original area of the valve opening.

The change in head can now be calculated with Eq. (7-63). With the head known, the estimated velocity V_n can be checked by the following equation:

$$V_n = \frac{V_o A_n}{A_o} \sqrt{\frac{H_o + \Sigma \, \Delta h}{H_o}} \tag{7-65}$$

where H_o = head at valve before any movement of valve, ft
$H_o + \Sigma \, \Delta h$ = total pressure at valve after particular movement; this includes pressure change caused by valve movement plus effect of waves reflected from reservoir, ft
  A_n = area of valve opening after n incremental closings; this area can be determined from closure characteristics of valve or by assuming its characteristics, ft^2

If the velocity obtained from Eq. (7-65) differs greatly from the estimated velocity, then that obtained from Eq. (7-65) should be used to recalculate Δh.

Two other widely used methods for solution of water-hammer problems are the Angus graphical method and the Allievi chart method. The Angus graphical method is simple yet can handle most complex problems that are encountered. The Allievi chart method is widely used; however, it has a relatively limited range of application. (See C. V. Davis and K. E. Sorensen, "Handbook of Applied Hydraulics," McGraw-Hill Book Company, New York.)

TABLE 7-10 Calculation of Pressure Due to Gradual Closing of Valve

Time, sec	Interval Δt_i, sec	ΔV, ft/s	V, ft/s	$Q = 78.5V$, ft³/s	Δh at valve, ft	$2\Delta h$ reflected from reservoir, ft	Total head $H + h_o$, ft	Valve area, A_n/A_o	Change in head from static, Col. (8) minus 1,000 ft
(1)	(2)	(3)	(4)	(5)	(6)	(7)	(8)	(9)	(10)
0			10.00	785			1,000	1.00	0
0.95	0.95	1.46	8.54	670	144		1,144	0.80	144
1.90	0.95	1.67	6.87	539	165		1,309	0.60	309
2.85	0.95	2.37	4.50	353	234	−288	1,255	0.40	255
3.80	0.95	2.36	2.14	168	233	−330	1,158	0.20	158
4.75	0.95	2.14	0	0	211	−180	1,189	0.00	189
5.70	0.95	Valve closed at time $t = 4.75$				−136	1,053		53
6.65	0.95					−242	811		−189
7.60						+136	947		−53
8.55						+242	1,189		189
9.50						−136	1,053		53

Example 7-8 ▪ The following problem illustrates the use of the preceding methods and compares the results: Steel penstock, length = 3000 ft, diameter = 10 ft, area = 78.5 ft², initial velocity = 10 ft/s, penstock thickness = 1 in, head at turbine with valve open = 1000 ft, and modulus of elasticity of steel = 43.2×10^8 lb/ft².

(For penstocks as shown in Fig. 7-32, thickness and diameter normally vary with head. Thus, the velocity of the pressure waves is different in each section of the penstock. Separate calculations for the velocity of the pressure wave should be made for each thickness and diameter of penstock to obtain the time required for a wave to travel to the reservoir and back to the valve.)

Velocity of pressure wave, from Eq. (7-61), is

$$U = \sqrt{\frac{E}{\rho}} \sqrt{\frac{1}{1 + ED/E_p t}}$$

$$= \sqrt{\frac{43.2 \times 10^6}{1.94}} \sqrt{\frac{1}{1 + 43.2 \times 10^6 \times 10 \times 12/(1 \times 43.2 \times 10^8)}}$$

$$= 3180 \text{ ft/s}$$

The time required for the wave to travel to the reservoir and be reflected back to the valve = $2L/U$ = 6000/3,180 = 1.90 s.

If closure time T of the valve is less than 1.90 s, the closure is instantaneous, and the pressure rise, from Eq. (7-63), is

$$\Delta h = \frac{U \Delta V}{g} = \frac{3180 \times 10}{32.2} = 990 \text{ ft}$$

Assuming T = 4.75 s, approximate equation (7-64) gives the following result:

$$\Delta h_g = \frac{t_i \Delta h}{T} = \frac{1.90 \times 990}{4.75} = 396 \text{ ft}$$

By arithmetic integration, the solution requires that the variation of the area of the valve with time be determined. For this example, the assumption will be made that the area varies linearly with time.

The next step is to choose a time interval: $t = 2L/2U = 0.95$ s. (This time interval is too large for exact calculations, but it illustrates the procedure.) Hence, valve closure comprises $4.75/0.95 = 5$ increments. A table can now be set up and computations made (Table 7-10). Assuming ΔV for the first incremental closing is 1.5 ft/s, Δh is calculated as follows:

$$\Delta h = \frac{U \, \Delta V}{g} = \frac{3180 \times 1.5}{32.2} = 148 \text{ ft}$$

Now check to see if the assumed ΔV was correct.

$$V_1 = \frac{V_o \times A_n}{A_o} \sqrt{\frac{H_o + \Sigma \, \Delta h}{H_o}}$$

$$= 10 \times 0.8 \sqrt{\frac{1000 + 148}{1000}} = 8.54 \text{ ft/s}$$

$$\Delta V = V_o - V_1 = 10 - 8.54 = 1.46 \text{ ft/s}$$

Recalculate Δh:

$$\Delta h = \frac{3180 \times 1.46}{32.2} = 144 \text{ ft}$$

This procedure for determining Δh is repeated for each incremental movement of the valve. Results are posted in Table 7-10. Column 7 is a total of all the wave pressures at the valve due to reflection. The waves for which heads are given in column 6 are reflected back from the reservoir and reach the valve $2L/U$ seconds after they were formed. The wave pressures were Δh when formed but are $-\Delta h$ when reflected back, the difference being $-2\Delta h$. The waves for which values are given in column 7 also are reflected back to the valve every $2L/U$ seconds and must be included. For example, for the fifth increment ($t = T = 4.25$), the reflected pressure is -2×234 from the third increment plus $+288$, a total of -180 ft. Note that the waves continue after the valve is closed.

For further information see C. V. Davis and K. E. Sorensen, "Handbook for Applied Hydraulics," McGraw-Hill Book Company, New York.

Surge Tanks ▪ It is uneconomical to design long pipelines for pressures created by water hammer or to operate a valve slowly enough to reduce these pressures. Usually, to prevent water hammer, a surge tank is installed close to valves at the end of long conduits. A surge tank is a tank containing water and connected to the conduit; the water column, in effect, floats on the line.

When a valve is suddenly closed, the water in the line rushes into the surge tank. The water level in the tank rises until the increased pressure in the surge tank overcomes the momentum of the water. When a valve is suddenly opened, the surge tank supplies water to the line when the pressure drops. The section of pipe (Fig. 7-33) between the surge tank and the valve must still be designed for water hammer; but the closure time to reduce the pressures for this section will be only a fraction of the time required without the surge tank.

Fig. 7-33.   Surge tank is placed near a valve on a penstock to prevent water hammer.

Although a surge tank is one of the most commonly used devices to prevent water hammer, it is by no means the only one. Various types of relief valves and air chambers are widely used on small-diameter lines, where the pressure of water hammer may be relieved by the release of a relatively small quantity of water.

PIPE STRESSES

7-14. Stresses Perpendicular to the Longitudinal Axis

The stresses acting perpendicular to the longitudinal axis of a pipe are caused by either internal or external pressures on the pipe walls.

Internal pressure creates a stress commonly called hoop tension. It may be calculated by taking a free-body diagram of a 1-in-long strip of pipe cut by a vertical plane through the longitudinal axis (Fig. 7-34). The forces in the vertical direction cancel out. The sum of the forces in the horizontal direction is

$$pD = 2F \tag{7-66}$$

where p = internal pressure, psi
D = outside diameter of pipe, in
F = force acting on each cut of edge of pipe, lb
Hence, the stress, psi, on the pipe material is

$$f = \frac{F}{A} = \frac{pD}{2t} \tag{7-67}$$

where A = area of cut edge of pipe, ft^2
t = thickness of pipe wall, in
From the derivation of Eq. (7-67), it would appear that the diameter used for calculations should be the inside diameter. However, Eq. (7-67) is not theoretically exact and gives stresses slightly lower than those actually developed. For this reason the outside diameter often is used (see also Art. 1-10).

Equation (7-67) is exact for all practical purposes when D/t is equal to or greater than 50. If D/t is less than 10, this equation will usually be quite conservative and therefore will yield an uneconomical design. For steel pipes, Eq. (7-67) gives directly the thickness required to resist internal pressure.

For concrete pipes, this analysis is approximate, however, since concrete cannot resist large tensile stresses. The force F must be carried by steel reinforcing. The internal diameter is used in Eq. (7-67) for concrete pipe.

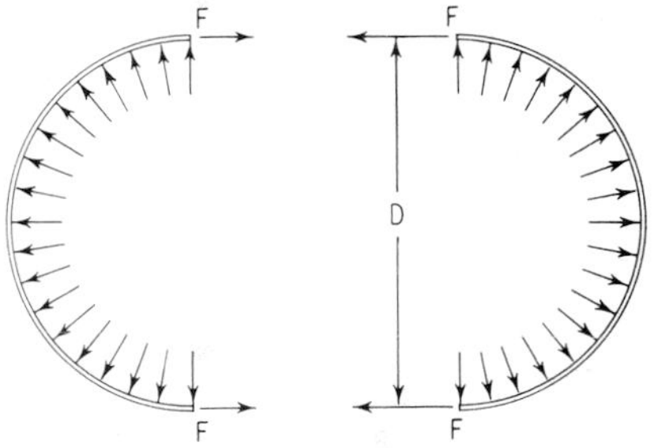

Fig. 7-34.   Internal pressure in a pipe produces hoop tension.

When a pipe has external pressure acting on it, the analysis is much more complex because the pipe material no longer acts in direct tension. The external pressure creates bending and compressive stresses that cause buckling problems. Equation (7-68) gives the thickness required for an empty steel pipe to resist buckling under uniform external pressure. (S. Timoshenko, "Strength of Materials," Van Nostrand Reinhold Company, New York. For calculation of pressure due to soil loading, see Art. 6-28.)

$$t = D \sqrt[3]{\frac{6p}{E}} \tag{7-68}$$

where t = shell thickness, in
D = diameter of pipe, ft
p = uniform external pressure, lb/ft^2
E = modulus of elasticity, psi

7-15. Stresses Parallel to the Longitudinal Axis

If a pipe is supported on piers, it acts like a beam. The stresses created can be calculated from the bending moment and shear equations for a continuous circular hollow beam. This stress is usually not critical in high-head pipes. However, thin-walled pipes usually require stiffening to prevent buckling and excessive deflection from the concentrated loads.

7-16. Temperature Expansion

If a pipe is subject to a wide range of temperatures, the stress due to temperature variation must be designed for or expansion joints provided. The stress, psi, due to a temperature change is

$$f = cE \, \Delta T \tag{7-69}$$

where E = modulus of elasticity of pipe material, psi
ΔT = temperature change from installation temperature
c = coefficient of thermal expansion of pipe material
The movement that should be allowed for, if expansion joints are to be used, is

$$\Delta L = Lc \, \Delta T \tag{7-70}$$

where ΔL = movement in length L of pipe
L = length between expansion joints

7-17. Forces Due to Pipe Bends

It is common practice to use thrust blocks in pipe bends to take the forces on the pipe caused by the momentum change and the unbalanced internal pressure of the water.

In all bends, there will be a slight loss of head due to turbulence and friction. This loss will cause a pressure change across the bend, but it is usually small enough to be neglected. When there is a change in the cross-sectional area of the pipe, there will be an additional pressure change that can be calculated with the Bernoulli equation (see Example 6, Art. 7-6). In this case, the pressure differential may be large and must be considered.

The force diagram in Fig. 7-35 is a convenient method for finding the resultant force on a

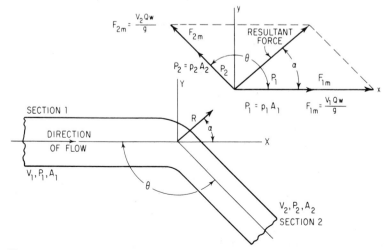

Fig. 7-35. Forces produced by flow at a pipe bend and change in diameter.

bend. The forces can be resolved into X and Y components to find the magnitude and direction of the resultant force on the pipe.

In Fig. 7-35:

V_1 = velocity before change in size of pipe, ft/s
V_2 = velocity after change in size of pipe, ft/s
p_1 = pressure before bend or size change in pipe, lb/ft^2
p_2 = pressure after bend or size change in pipe, lb/ft^2
A_1 = area before size change in pipe, ft^2
A_2 = area after size change in pipe, ft^2
F_{2m} = momentum of water in section 2 = $V_2 Q w / g$
F_{1m} = momentum of water in section 1 = $V_1 Q w / g$
P_2 = pressure of water in section 2 times area of section 2 = $p_2 A_2$
P_1 = pressure of water in section 1 times area of section 1 = $p_1 A_1$
w = unit weight of liquid, lb/ft^3
Q = discharge, ft^3/s

If the pressure loss in the bend is neglected and there is no change in magnitude of velocity around the bend, Eqs. (7-71) and (7-72) give a quick solution.

$$R = 2A \left(w \frac{V^2}{g} + p \right) \cos \frac{\theta}{2} \tag{7-71}$$

$$\alpha = \frac{\theta}{2} \tag{7-72}$$

where R = resultant force on bend, lb
α = angle R makes with F_{1m}
p = pressure, lb/ft^2
w = unit weight of water, 62.4 lb/ft^3
V = velocity of flow, ft/s
g = acceleration due to gravity, 32.2 ft/s^2
A = area of pipe, ft^2
θ = angle between pipes ($0° \leq \theta \leq 180°$)

Although thrust blocks are normally used to take the force on bends, in many cases the pipe material takes this force. The stress caused by this force is directly additive to other stresses along the longitudinal axis of the pipe. In small pipes, the force caused by bends can easily be carried by the pipe material; however, the joints must also be able to take these forces.

CULVERTS

A culvert is a closed conduit for the passage of surface drainage under a highway, a railroad, canal, or other embankment. The slope of a culvert and its inlet and outlet conditions are usually determined by the topography of the site. Because of the many combinations obtained by varying the entrance conditions, exit conditions, and slope, no single formula can be given that will apply to all culvert problems.

The basic method for determining discharge through a culvert is applying the Bernoulli equation between a point just outside the entrance and a point somewhere downstream. An understanding of uniform and nonuniform flow is necessary to understand culvert flow fully. However, an exact theoretical analysis, involving detailed calculation of drawdown and backwater curves, is usually unwarranted because of the relatively low accuracy attainable in determining runoff. Neglecting drawdown and backwater curves does not seriously affect the accuracy but greatly simplifies the calculations.

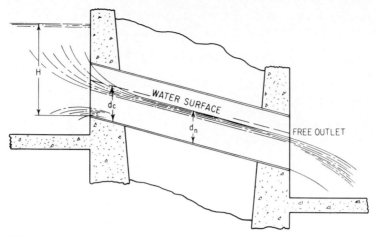

Fig. 7-36.   Flow through a culvert with free discharge. Normal depth d_n is less than critical depth d_c; slope is greater than critical. Discharge depends on type of inlet and head H.

7-18. Culverts on Critical Slope or Steeper

Entrance Submerged or Unsubmerged but Free Exit ▪ If a culvert is on critical slope or steeper, that is, the normal depth is equal to or less than the critical depth (Art. 7-23), the discharge will be entirely dependent on the entrance conditions (Fig. 7-36). Increasing the slope of the culvert past critical slope (the slope just sufficient to maintain flow at critical depth) will decrease the depth of flow downstream from the entrance. But the increased slope will not increase the amount of water entering the culvert because the entrance depth will remain at critical.

The discharge is given by the equation for flow through an orifice if the entrance is submerged, or by the equation for flow over a weir if the entrance is not submerged. Coefficients of discharge for weirs and orifices give good results, but they do not cover the entire range of entry conditions encountered in culvert problems. For this reason, charts and nomographs have been developed and are used almost exclusively in design. ("Handbook of Concrete Culvert Pipe Hydraulics," EB058W, Portland Cement Association.)

Entrance Unsubmerged but Exit Submerged ▪ In this case, the submergence of the exit will cause a hydraulic jump to occur in the culvert (Fig. 7-37). The jump will not affect the culvert discharge, and the control will still be at the inlet.

Entrance and Exit Submerged ▪ When both the exit and entrance are submerged (Fig. 7-38), the culvert flows full, and the discharge is independent of the slope. This is normal pipe flow and is easily solved by using the Manning or Darcy-Weisbach formula for friction loss [Eq. (7-33d) or (7-30)]. From the Bernoulli equation for the entrance and exit, and the Manning equation for friction loss, the following equation is obtained:

$$H = (1 + K_e)\frac{V^2}{2g} + \frac{V^2 n^2 L}{2.21 R^{4/3}} \tag{7-73a}$$

$$V = \sqrt{\frac{H}{(1 + K_e/2g) + (n^2 L/2.21 R^{4/3})}} \tag{7-73b}$$

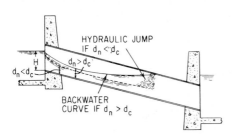

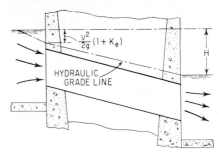

Fig. 7-37. Flow through a culvert with entrance unsubmerged but exit submerged. When slope is less than critical, open-channel flow takes place, and $d_n > d_c$. When slope exceeds critical, flow depends on inlet conditions, and $d_n < d_c$.

Fig. 7-38. With entrance and exit submerged, normal pipe flow occurs in a culvert, and discharge is independent of slope. The fluid flows under pressure. Discharge may be determined from Bernoulli and Manning equations.

where H = elevation difference between headwater and tailwater, ft
   V = velocity in culvert, ft/s
   g = acceleration due to gravity, 32.2 ft/s^2
   K_e = entrance-loss coefficient (Art. 7-20)
   n = Manning's roughness coefficient
   L = length of culvert, ft
   R = hydraulic radius of culvert, ft
This equation can be solved directly since the velocity is the only unknown.

7-19. Culverts with Slope Less than Critical

Critical slope is the slope just sufficient to maintain flow at critical depth (Art. 7-23).

Entrance Submerged or Unsubmerged but Free Exit ▪ For these conditions, depending on the head, the flow can be either pressure or open-channel.

The discharge, for the open-channel condition (Fig. 7-39), is obtained by writing the Bernoulli equation for a point just outside the entrance and a point a short distance downstream from the entrance. Thus,

$$H = K_e \frac{V^2}{2g} + \frac{V^2}{2g} + d_n \tag{7-74}$$

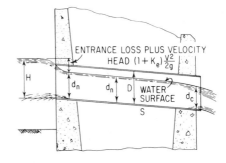

Fig. 7-39. Flow through culvert with free discharge and normal depth d_n greater than the critical depth d_c when the entrance is unsubmerged or slightly submerged. Open-channel flow occurs and discharge depends on head H, loss at entrance, and slope of culvert.

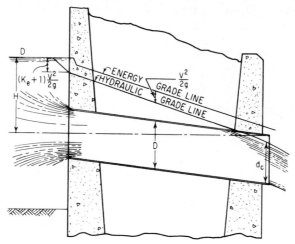

Fig. 7-40.   Flow through culvert with free discharge and normal depth d_n greater than critical depth d_c when entrance is deeply submerged. The culvert flows full. Discharge is given by equations for pipe flow.

The velocity can be determined from the Manning equation:

$$V^2 = \frac{2.2SR^{4/3}}{n^2} \tag{7-75}$$

Substituting this into Eq. (7-74) yields

$$H = (1 + K_e)\frac{2.2}{2gn^2}SR^{4/3} + d_n \tag{7-76}$$

where H = head on entrance measured from bottom of culvert, ft
  K_e = entrance-loss coefficient (Art. 7-20)
  S = slope of energy grade line, which for culverts is assumed to equal slope of bottom of culvert
  R = hydraulic radius of culvert, ft
  d_n = normal depth of flow, ft

To solve Eq. (7-76), it is necessary to try different values of d_n and corresponding values of R until a value is found that satisfies the equation. If the head on a culvert is high, a value of d_n less than the culvert diameter will not satisfy Eq. (7-76). This means the flow is under pressure (Fig. 7-40), and discharge is given by Eq. (7-73).

When the depth of the water is slightly below the top of the culvert, there is a range of unstable flow fluctuating between pressure and open channel. If this condition exists, it is good practice to check the discharge for both pressure flow and open-channel flow. The condition that gives the lesser discharge should be assumed to exist.

Short Culvert with Free Exit ▪ When a culvert on a slope less than critical has a free exit, there will be a drawdown of the water surface at the exit and for some distance upstream. The magnitude of the drawdown depends on the friction slope of the culvert and the difference between the critical and normal depths. If the friction slope approaches critical, the difference between normal depth and critical depth is small (Fig. 7-39), and the drawdown will not extend for any significant distance upstream. When the friction slope is flat, there will be a large difference between normal and critical depth. The effect of the drawdown will extend a greater distance upstream and may reach the entrance of a short culvert (Fig. 7-41). This drawdown of the water level in the entrance of the culvert will increase the discharge, causing it to be about the same as for a culvert on a slope steeper than critical (Art. 7-18). Most culverts, however, are on too steep a slope for the backwater to have any effect for an appreciable distance upstream.

Entrance Unsubmerged but Exit Submerged ▪ If the level of submergence of the exit is well below the bottom of the entrance (Fig. 7-37), the backwater from the submergence will not extend to the entrance. The discharge for this case will be given by Eq. (7-76).

If the level of submergence of the exit is close to the level of the entrance, it may be assumed that the backwater will cause the culvert to flow full and a pipe flow condition will result. The discharge for this case is given by Eq. (7-73).

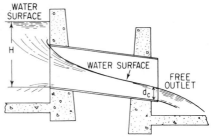

Fig. 7-41. Drawdown of water surface at free exit of a short culvert with slope less than critical affects depth at entrance and controls discharge.

When the level of submergence falls between these two cases and the project does not warrant a trial approach with backwater curves, it is good practice to assume the condition that gives the lesser discharge. ▪

7-20. Entrance Losses for Culverts

Flow in a culvert may be significantly affected by loss in head because of conditions at the entrance (Arts. 7-18 and 7-19). Following are coefficients of entrance loss K_e for some typical entrance conditions:

Sharp-edged projecting inlet	0.9
Flush inlet, square edge	0.5
Concrete pipe, groove or bell, projecting	0.15
Concrete pipe, groove or bell, flush	0.10
Well-rounded entrance	0.08

These values are for culverts flowing full. When the entrance is not submerged, the coefficients are usually somewhat lower. But because of the many unknowns entering into culvert problems, the values tabulated can be used for submerged or unsubmerged cases without sacrificing accuracy.

Example 7-9 ▪ *Given:* Maximum head above the top of the culvert = 5 ft, slope = 0.01, length = 300 ft, discharge Q = 40 ft³/s, n = 0.013, and free exit. *Find:* size of culvert.

Procedure: First assume a trial culvert; then investigate the assumed section to find its discharge. Assume a 2 × 2 ft concrete box section. Calculate Q assuming entrance control, with Eq. (7-44) for discharge through an orifice. The coefficient of discharge C for a 2-ft-square orifice is about 0.6. Head h on center line of entrance = 5 + ½ × 2 = 6 ft. Entrance area a = 2 × 2 = 4 ft².

$$Q = Ca\sqrt{2gh} = 0.6 \times 4\sqrt{64.4 \times 6} = 47.2 \text{ ft}^3/\text{s}$$

For entrance control, the flow must be supercritical and d_n must be less than 2 ft. First find d_n.

To calculate the hydraulic radius, assume the depth is slightly less than 2 ft, since this will give the maximum possible value of the hydraulic radius for this culvert.

$$R = \frac{\text{area of flow}}{\text{wetted perimeter}} = \frac{2 \times 2}{6} = 0.67 \text{ ft}$$

Application of Eq. (7-33a) gives

$$V = \frac{1.486}{n} R^{2/3} S^{1/2} = \frac{1.486}{0.013} \times 0.67^{2/3} \times 0.01^{1/2} = 8.76 \text{ ft/s}$$

$$d_n = \frac{Q}{V \times \text{width}} = \frac{47.2}{8.76 \times 2} = 2.69 \text{ ft}$$

Since d_n is greater than the culvert depth, the flow is under pressure, and the entrance will not control.

Since the culvert is under pressure, Eq. (7-73a) applies. But

$$H = 5 + 0.01 \times 300 = 8 \text{ ft}$$

(see Fig. 7-40). The hydraulic radius for pipe flow is $R = 2^2/8 = \frac{1}{2}$. Substitution in Eq. (7-73a) yields

$$8 = \frac{1.5 V^2}{2g} + 0.0575 V^2 = 0.0808 V^2$$

$$V = \sqrt{8/0.0808} = 9.95 \text{ ft/s}$$

$$Q = Va = 9.95 \times 4 = 39.8 \text{ ft}^3/\text{s}$$

Since the discharge of the assumed culvert section under the allowable head equals the maximum expected runoff, the assumed culvert would be satisfactory.

OPEN-CHANNEL FLOW

7-21. Basic Elements of Open Channels

Free surface flow, or open-channel flow, includes all cases of flow in which the liquid surface is open to the atmosphere. Thus, flow in a pipe is open-channel flow if the pipe is only partly full.

A **uniform channel** is one of constant cross section. It has **uniform flow** if the grade, or slope, of the water surface is the same as that of the channel. Hence, depth of flow is constant throughout. **Steady flow** in a channel occurs if the depth at any location remains constant with time.

The **discharge** Q at any section is defined as the volume of water passing that section per unit of time. It is expressed in cubic feet per second, ft^3/s, and is given by

$$Q = VA \tag{7-77}$$

where V = *average velocity,* ft/s
   A = cross-sectional *area* of flow, ft^2

When the discharge is constant, the flow is said to be **continuous** and therefore

$$Q = V_1 A_1 = V_2 A_2 = \cdots \tag{7-78}$$

where the subscripts designate different channel sections. Equation (7-78) is known as the continuity equation for continuous steady flow.

In a uniform channel, **varied flow** occurs if the longitudinal water-surface profile is not parallel with the channel bottom. Varied flow exists within the limits of backwater curves, within a hydraulic jump, and within a channel of changing slope or discharge.

Depth of flow d is taken as the vertical distance, ft, from the bottom of a channel to the water surface. The **wetted perimeter** is the length, ft, of a line bounding the cross-sectional area of flow, minus the free surface width. The **hydraulic radius** R equals the area of flow divided by its wetted perimeter. The **average velocity** of flow V is defined as the discharge divided by the area of flow,

$$V = \frac{Q}{A} \tag{7-79}$$

The velocity head H_V, ft, is generally given by

$$H_V = \frac{V^2}{2g} \tag{7-80}$$

where V = average velocity from Eq. (7-79), ft/s
      g = acceleration due to gravity, 32.2 ft/s^2

Velocity heads of individual filaments of flow vary considerably above and below the velocity head based on the average velocity. Since these velocities are squared in head and energy computations, the average of the velocity heads will be greater than the average-velocity head. The **true velocity head** may be expressed as

$$H_{Va} = \alpha \frac{V^2}{2g} \tag{7-81}$$

where α is an empirical coefficient that represents the degree of turbulence. Experimental data indicate that α may vary from about 1.03 to 1.36 for prismatic channels. It is, however, normally taken as 1.00 for practical hydraulic work and is evaluated only for precise investigations of energy loss.

The total energy per pound of water relative to the bottom of the channel at a vertical section is called the **specific energy head** H_e. It is composed of the depth of flow at any point, plus the velocity head at the point. It is expressed in feet as

$$H_e = d + \frac{V^2}{2g} \tag{7-82}$$

A longitudinal profile of the elevation of the specific energy head is called the **energy grade line,** or the **total-head line.** A longitudinal profile of the water surface is called the **hydraulic grade line.** The vertical distance between these profiles at any point equals the velocity head at that point.

Figure 7-42 shows a section of uniform open channel for which the slopes of the water surface S_W and the energy grade line S equal the slope of the channel bottom S_o.

Loss of head due to friction h_f in channel length L equals the drop in elevation of the channel ΔZ in the same distance.

7-22. Normal Depth of Flow

The depth of equilibrium flow that exists in the channel of Fig. 7-42 is called the normal depth d_n. This depth is unique for specific discharge and channel conditions. It may be computed by a trial-and-error process when the channel shape, slope, roughness, and discharge are known. A form of the Manning equation has been suggested for this calculation. (V. T. Chow, "Open-Channel Hydraulics," McGraw-Hill Book Company, New York.)

$$AR^{2/3} = \frac{Qn}{1.486S^{1/2}} \tag{7-83}$$

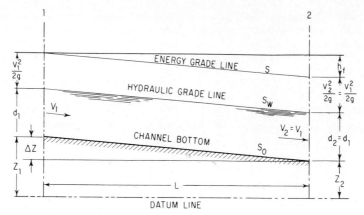

Fig. 7-42. Flow characteristics for uniform open-channel flow.

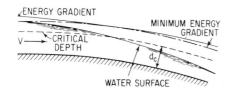

Fig. 7-43. Prismatic channel with gradually increasing bottom slope. Normal depth decreases downstream as slope increases.

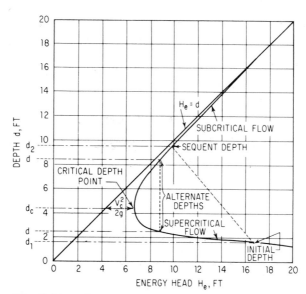

Fig. 7-44. Specific energy head H_e changes with depth for constant discharge in a rectangular channel of changing slope and is a minimum at critical depth.

where A = area of flow, ft^2
$\quad R$ = hydraulic radius, ft
$\quad Q$ = amount of flow or discharge, ft^3/s
$\quad n$ = Manning's roughness coefficient
$\quad S$ = slope of energy grade line or loss of head, ft, due to friction per lin ft of channel
$AR^{2/3}$ is referred to as a section factor. Determination of d_n for uniform channels is simplified by use of tables that relate d_n to the bottom width of a rectangular or trapezoidal channel, or to the diameter of a circular channel. (See, for example, H. W. King and E. F. Brater, "Handbook of Hydraulics," McGraw-Hill Book Company, New York.)

In a prismatic channel of gradually increasing slope, normal depth decreases downstream, as shown in Fig. 7-43, and specific energy first decreases and then increases as shown in Fig. 7-44.

The specific energy is high initially where the channel is relatively flat because of the large normal depth (Fig. 7-43). As the depth decreases downstream, the specific energy also decreases. It reaches a minimum at the point where the flow satisfies the equation

$$\frac{A^3}{T} = \frac{Q^2}{g} \tag{7-84}$$

in which T is the top width of the channel, ft. For a rectangular channel, Eq. (7-84) reduces to

$$\frac{d}{2} = \frac{V^2}{2g} \tag{7-85}$$

where $V = Q/A$ = mean velocity of flow, ft^3/s
$\quad d$ = depth of flow, ft
This indicates that the specific energy is a minimum where the normal depth equals twice the velocity head. As the depth continues to decrease in the downstream direction, the specific energy increases again because of the higher velocity head (Fig. 7-44).

7-23. Critical Depth

The depth of flow that satisfies Eq. (7-84) is called the *critical depth d_c*. For a given value of specific energy, the critical depth gives the greatest discharge, or conversely, for a given discharge, the specific energy is a minimum for the critical depth (Fig. 7-44).

In the section of mild slope upstream from the critical-depth point in Fig. 7-43, the depth is greater than critical. The flow there is called *subcritical flow,* indicating that the velocity is less than that at critical depth. In the section of steeper slope below the critical-depth point, the depth is below critical. The velocity there exceeds that at critical depth, and flow is *supercritical.*

Critical depth may be computed for a uniform channel once the discharge is known. Determination of this depth is independent of the channel slope and roughness since critical depth simply represents a depth for which the specific energy head is a minimum. Critical depth may be calculated by trial and error with Eq. (7-84), or it may be found directly from tables (H. W. King and E. F. Brater, "Handbook of Hydraulics," McGraw-Hill Book Company, New York). For rectangular channels, Eq. (7-84) may be reduced to

$$d_c = \sqrt[3]{\frac{Q^2}{b^2 g}} \tag{7-86}$$

where d_c = critical depth, ft
$\quad Q$ = quantity of flow or discharge, ft^3/s
$\quad b$ = width of channel, ft

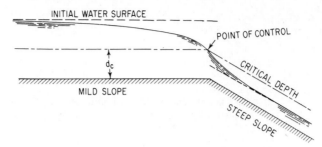

Fig. 7-45.   Change in flow stage from subcritical to supercritical occurs gradually.

Critical slope is the slope of the channel bed that will maintain flow at critical depth. Such slopes should be avoided in channel design because flow near critical depth tends to be unstable and exhibits turbulence and water-surface undulations.

Critical depth, once calculated, should be plotted for the full length of a uniform channel, regardless of slope, to determine whether the normal depth at any section is subcritical or super-critical. [As indicated by Eq. (7-85), if the velocity head is less than half the depth in a rectangular channel, flow is subcritical, but if velocity head exceeds half the depth, flow is super-critical.] If channel configuration is such that the normal depth must go from below to above critical, a hydraulic jump will occur, along with a high loss of energy. Critical depth will change if the channel cross section changes, so the possibility of a hydraulic jump in the vicinity of a transition should be investigated.

For every depth greater than critical depth, there is a corresponding depth less than critical that has an identical value of specific energy (Fig. 7-44). These depths of equal energy are called *alternate depths*. The fact that the energy is the same for alternate depths does not mean that the flow may switch from one alternate depth to the other and back again; flow will always seek to attain the normal depth in a uniform channel and will maintain that depth unless an obstruction is met.

It can be seen from Fig. 7-44 that any obstruction to flow that causes a reduction in total head causes subcritical flow to experience a drop in depth and supercritical flow to undergo an increase in depth.

If supercritical flow exists momentarily on a flat slope because of a sudden grade change in the channel (Fig. 7-53b), the depth will increase suddenly from the depth below critical to a depth above critical in a hydraulic jump. The depth following the jump will not be the alternate depth, however. There has been a loss of energy in making the jump. The new depth is said to be sequent to the initial depth, indicating an irreversible occurrence. There is no similar phenomenon that allows a sudden change in depth from subcritical flow to supercritical flow with a corresponding gain in energy. Such a change occurs gradually, without turbulence, as indicated in Fig. 7-45.

7-24. Manning's Equation for Open Channels

One of the more popular of the numerous equations developed for determination of discharge in an open channel is Manning's variation of the Chezy formula,

$$V = C\sqrt{RS}\qquad(7\text{-}87)$$

where R = hydraulic radius, ft
$\quad V$ = mean velocity of flow ft/s

S = slope of energy grade line or loss of head due to friction, ft/lin ft of channel

C = Chezy roughness coefficient

Manning proposed

$$C = \frac{1.486^{1/6}}{n} \tag{7-88}$$

where n is the coefficient of roughness in the earlier Ganguillet-Kutter formula (see also Art. 7-25). When Manning's C is used in the Chezy formula, the familiar Manning equation results:

$$V = \frac{1.486}{n} R^{2/3}S^{1/2} \tag{7-89}$$

Since the discharge $Q = VA$, Eq. (7-89) may be written

$$Q = \frac{1.486}{n} AR^{2/3}S^{1/2} \tag{7-90}$$

where A = area of flow, ft^2

Q = quantity of flow, ft^3/s

7-25. Roughness Coefficient for Open Channels

Values of the roughness coefficient n for Manning's equation (Art. 7-24) have been determined for a wide range of natural and artificial channel construction materials. Excerpts from a table of these coefficients taken from V. T. Chow, "Open-Channel Hydraulics," McGraw-Hill Book Company, New York, are in Table 7-11. Dr. Chow compiled data for his table from work by R. E. Horton and from technical bulletins published by the U.S. Department of Agriculture.

Channel roughness does not remain constant with time or even depth of flow. An unlined channel excavated in earth may have one n value when first put in service and another when overgrown with weeds and brush. If an unlined channel is to have a reasonably constant n value over its useful lifetime, there must be a continuing maintenance program.

Shallow flow in an unlined channel will result in an increase in the effective n value if the channel bottom is covered with large boulders or ridges of silt since these projections would then have a larger influence on the flow than for deep flow. A deeper-than-normal flow will also result in an increase in the effective n value if there is a dense growth of brush along the banks within the path of flow. When channel banks are overtopped during a flood, the effective n value increases as the flow spills into heavy growth bordering the channel. The roughness of a lined channel experiences change with age because of both deterioration of the surface and accumulation of foreign matter; therefore, the average n values given in Table 7-11 are recommended only for well-maintained channels. (See also Art. 7-9 and Table 7-4.)

7-26. Water-Surface Profiles for Gradually Varied Flow

Examples of various surface curves possible with gradually varied flow are shown in Fig. 7-46. These surface profiles represent backwater curves that form under the conditions illustrated in examples (a) through (r).

These curves are divided into five groups, according to the slope of the channel in which they appear (Art. 7-23). Each group is labeled with a letter descriptive of the slope: M for mild (subcritical), S for steep (supercritical), C for critical, H for horizontal, and A for adverse. The two dashed lines in the left-hand figure for each class are the *normal-depth line* N.D.L. and the

TABLE 7-11   Values of the Roughness Coefficient n for Use in the Manning Equation

	Min	Avg	Max
A. Open-channel flow in closed conduits			
1. Corrugated-metal storm drain	0.021	0.024	0.030
2. Cement-mortar surface	0.011	0.013	0.015
3. Concrete (unfinished)			
a. Steel form	0.012	0.013	0.014
b. Smooth wood form	0.012	0.014	0.016
c. Rough wood form	0.015	0.017	0.020
B. Lined channels			
1. Metal			
a. Smooth steel (unpainted)	0.011	0.012	0.014
b. Corrugated	0.021	0.025	0.030
2. Wood			
a. Planed, untreated	0.010	0.012	0.014
3. Concrete			
a. Float finish	0.013	0.015	0.016
b. Gunite, good section	0.016	0.019	0.023
c. Gunite, wavy section	0.018	0.022	0.025
4. Masonry			
a. Cemented rubble	0.017	0.025	0.030
b. Dry rubble	0.023	0.032	0.035
5. Asphalt			
a. Smooth	0.013	0.013	
b. Rough	0.016	0.016	
C. Unlined channels			
1. Excavated earth, straight and uniform			
a. Clean, after weathering	0.018	0.022	0.025
b. With short grass, few weeds	0.022	0.027	0.033
c. Dense weeds, high as flow depth	0.050	0.080	0.120
d. Dense brush, high stage	0.080	0.100	0.140
2. Dredged earth			
a. No vegetation	0.025	0.028	0.033
b. Light brush on banks	0.035	0.050	0.060
3. Rock cuts			
a. Smooth and uniform	0.025	0.035	0.040
b. Jagged and irregular	0.035	0.040	0.050

critical-depth line C.D.L. The N.D.L. and C.D.L. are identical for a channel of critical slope, and the N.D.L. is replaced by a horizontal line, at an arbitrary elevation, for the channels of horizontal or adverse slope.

There are three types of surface-profile curves possible in channels of mild or steep slope, and two types for channels of critical, horizontal, and adverse slope.

The M1 curve is the familiar surface profile from which all backwater curves derive their name and is the most important from a practical point of view. It forms above the normal-depth line and occurs when water is backed up a stream by high water in the downstream channel, as shown in Fig. 7-46a and b.

The M2 curve forms between the normal- and critical-depth lines. It occurs under conditions shown in Fig. 7-46c and d, corresponding to an increase in channel width or slope.

The M3 curve forms between the channel bottom and critical-depth line. It terminates in a hydraulic jump, except where a drop-off in the channel occurs before a jump can form. Examples of the M3 curve are in Fig. 7-46e and f (a partly opened sluice gate and a decrease in channel slope, respectively).

The S1 curve begins at a hydraulic jump and extends downstream, becoming tangent to a horizontal line (Fig. 7-46g and h) under channel conditions corresponding to those for Fig. 7-46a and b.

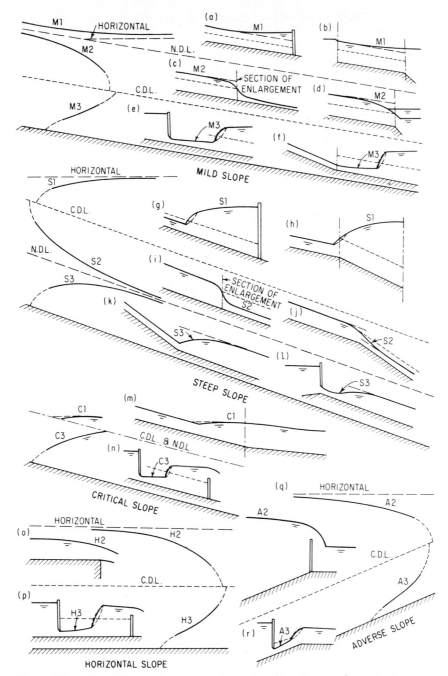

Fig. 7-46. Typical flow profiles for channels with various slopes. N.D.L. indicates normal-depth line; C.D.L., critical-depth line.

The S2 curve, commonly called a drawdown curve, extends downstream from the critical depth and becomes tangent to the normal-depth line under conditions corresponding to those for Fig. 7-46*i* and *j*.

The S3 curve is of the transitional type. It forms between two normal depths of less than critical depth under conditions corresponding to those for Fig. 7-46*k* and *l*.

Examples in Fig. 7-46*m* through *r* show conditions for the formation of C, H, and A profiles.

The curves in Fig. 7-46 approach the normal-depth line asymptotically and terminate abruptly in a vertical line as they approach the critical depth. The curves that approach the bottom intersect it at a definite angle but are imaginary near the bottom since velocity would have to be infinite to satisfy Eq. (7-77) if the depth were zero. The curves are shown dotted near the critical-depth line as a reminder that this portion of the curve does not possess the same degree of accuracy as the rest of the curve because of neglect of vertical components of velocity in the calculations. These curves either start or end at what is called a point of control.

A **point of control** is a physical location in a prismatic channel at which the depth of steady flow may readily be determined. This depth is usually different from the normal depth for the channel because of a grade change, gate, weir, dam, free overfall, or other feature at that location that causes a backwater curve to form. Calculations for the length and shape of the surface profile of a backwater curve start at this known depth and location and proceed either up or downstream, depending on the type of flow. For subcritical flow conditions, the curve proceeds upstream from the point of control in a true backwater curve. The surface curve that occurs under supercritical flow conditions proceeds downstream from the point of control and might better be called a downwater curve.

The point of control is always at the downstream end of a backwater curve in subcritical flow and at the upstream end for supercritical flow. This is explained as follows: A backwater curve may be thought of as being the result of some disruption of uniform flow that causes a wave of disturbance in the channel. The wave travels at a speed, known as its **celerity,** which always equals the critical velocity for the channel. If a disturbance wave attempts to move upstream against supercritical flow (flow moving at a speed greater than critical), it will be swept downstream by the flow and have no effect on conditions upstream. A disturbance wave is held steady by critical flow and moves upstream in subcritical flow.

When a hydraulic jump occurs on a mild slope and is followed by a free overfall (Fig. 7-52), backwater curves form both before and after the jump. The point of control for the curve in the supercritical region above the jump will be located at the vena contracta that forms just below the sluice gate. The point of control for the backwater curve in the subcritical region below the jump is at the free overfall where critical depth occurs. Computations for these backwater curves are carried toward the jump from their respective points of control and are extended across the jump to help determine its exact location. But a backwater curve cannot be calculated through a hydraulic jump from either direction. The surface profiles involved terminate abruptly in a vertical line as they approach the critical depth, and a hydraulic jump always occurs across critical depth. See Art. 7-32.

7-27. Backwater-Curve Computations

The solution of a backwater curve involves computation of a gradually varied flow profile. Solutions available include the graphical-integration, direction-integration, and step methods. Explanations of both the graphical- and direct-integration methods are in V. T. Chow, "Open-Channel Hydraulics," McGraw-Hill Book Company, New York.

Two variations of the step method include the direct or uniform method and the standard method. They are simple and widely used.

For step-method computations, the channel is divided into short lengths, or reaches, with

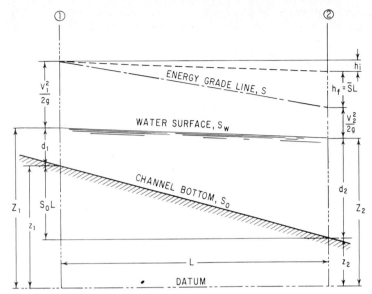

Fig. 7-47.   Channel with constant discharge and gradually varying in area.

relatively small variation. In a series of steps starting from a point of control, each reach is solved in succession. Step methods have been developed for channels with uniform or varying cross sections.

Direct step method of backwater computation involves solving for an unknown length of channel between two known depths. The procedure is applicable only to uniform prismatic channels with gradually varying area of flow.

For the section of channel in Fig. 7-47, Bernoulli's equation for the reach between sections 1 and 2 is

$$S_o L + d_1 + \frac{V_1^2}{2g} = d_2 + \frac{V_2^2}{2g} + \overline{S}L \tag{7-91}$$

where V_1 and V_2 = mean velocities of flow at sections 1 and 2, ft/s
  d_1 and d_2 = depths of flow at sections 1 and 2, ft
  g = acceleration due to gravity, 32.2 ft/s^2
  $\overline{S}$ = average head loss due to friction, ft/ft of channel
  S_o = slope of channel bottom
  L = length of channel between sections 1 and 2, ft

Note that $S_o L = \Delta z$, the change in elevation, ft, of the channel bottom between sections 1 and 2, and $\overline{S}L = h_f$, the head loss, ft, due to friction in the same reach. (For uniform, prismatic channels, h_i, the eddy loss, is negligible and can be ignored.) $\overline{S}$ equals the slope calculated for the average depth in the reach but may be approximated by the average of the values of friction slope S for the depths at sections 1 and 2.

Solving Eq. (7-91) for L gives

$$L = \frac{\left(d_2 + \dfrac{V_2^2}{2g} \right) - \left(d_1 + \dfrac{V_1^2}{2g} \right)}{S_o - \overline{S}} = \frac{H_{e2} - H_{e1}}{S_o - \overline{S}} \tag{7-92}$$

where H_{e1} and H_{e2} are the specific energy heads for sections 1 and 2, respectively, as given by Eq. (7-82). The friction slope S at any point may be computed by the Manning equation, rearranged as follows:

$$S = \frac{n^2 V^2}{2.21 R^{4/3}} \qquad (7\text{-}93)$$

where R = hydraulic radius, ft
$\quad n$ = roughness coefficient (Art. 7-25)

Note that the slope S used in the Manning equation is the slope of the energy grade line, not the channel bottom. Note also that the roughness coefficient n is squared in Eq. (7-93), and its value must therefore be chosen with special care to avoid an exaggerated error in the computed friction slope. The smaller the value of n, the longer the backwater curve profile, and vice versa. Therefore, the smallest n possible for the prevailing conditions should be selected for computation of a backwater curve if knowledge of the longest possible flow profile is required.

The first step in the direct step method involves choosing a series of depths for the end points of each reach. These depths will range from the depth at the point of control to the ending depth for the backwater curve. This ending depth is often the normal depth for the channel (Art. 7-22) but may be some intermediate depth, such as for a curve preceding a hydraulic jump. Depths should be chosen so that the velocity change across a reach does not exceed 20% of the velocity at the beginning of the reach. Also the change in depth between sections should never exceed 1 ft.

The specific energy head H_e should be computed for the chosen depth at each of the various sections and the change in specific energy between sections determined. Next, the friction slope S should be computed at each section from Eq. (7-93). The average of two sections gives the friction slope $\overline{S}$ between sections. Finally, the difference between $\overline{S}$ and slope of channel bottom S_o should be computed and the length of reach determined from Eq. (7-92).

These computations can be handled most conveniently in a table. The table should be arranged with separate columns for results of calculations and separate rows for each chosen depth (Table 7-12a).

TABLE 7-12a Direct Step Backwater-Curve Calculations for Example 7-10

Section	d, ft (1)	A, ft² (2)	W.P., ft (3)	R, ft (4)	$R^{4/3}$ (5)	V, ft/s (6)	$V^2/2g$, ft (7)
1	3.86	38.6	17.72	2.18	2.83	11.14	1.93
2	4.30	43.0	18.60	2.31	3.05	10.00	1.55
3	4.75	47.5	19.50	2.44	3.28	9.05	1.27
4	4.94	49.4	19.88	2.48	3.36	8.70	1.18

Section	H_e, ft (8)	ΔH_e, ft (9)	S (10)	$\overline{S}$ (11)	$S_o - \overline{S}$ (12)	L, ft (13)	ΣL, ft (14)
1	5.79		0.00389				
		−0.06		0.00340	−0.00140	43	43
2	5.85		0.00291				
		−0.17		0.00256	−0.00056	304	347
3	6.02		0.00222				
		−0.10		0.00211	−0.00011	909	1,256
4	6.12		0.00200				

Column 1 in Table 7-12a lists d, depth of flow, ft, arbitrarily assigned. Column 2 lists areas of flow, ft², corresponding to depth in column 1. Column 3 gives wetted perimeter, ft, corresponding to depth in column 1. For column 4, hydraulic radius R, ft, equals column 2 divided by column 3. In column 6, V is the mean velocity, ft/s, or discharge Q divided by column 2. The values in column 6 are used to compute the velocity head, ft, in column 7, which may be adjusted for turbulence if α is known [see Eq. (7-81)].

Column 8 gives specific energy H_e, ft, the sum of column 7 and column 1 [Eq. (7-82)]. Column 9 is the value in column 8 for a section minus that in the previous section ($H_{e2} - H_{e1}$). Column 10 lists friction slope S computed from Eq. (7-93), with known value of n, V as given in column 6, and $R^{4/3}$ from column 5. Column 11 is $\overline{S}$, the average friction between steps, which equals the arithmetic mean of the slope for a section in column 10 and the one computed for the previous section. Column 12 is obtained by subtracting column 11 from the known bottom slope. Column 13 gives length of reach, ft, between the consecutive sections, column 9 divided by column 12 [Eq. (7-92)]. Finally, column 14 lists distance from the section under consideration to the point of control, equal to the cumulative sum of the values in column 13.

Example 7-10. Direct Step Method for Computing Backwater Curves ▪ A 10-ft-wide rectangular channel with a slope S_o = 0.0020 and roughness factor n = 0.014 carries a discharge Q = 430 ft³/s. The channel ends in a free overfall as shown in Fig. 7-48. Determine the water-surface profile and the distance from the free overfall to the location where the backwater curve joins the normal-depth line.

Solution.   Critical depth d_c and normal depth d_n are found from Eqs. (7-84) and (7-83) to be 3.86 and 4.94 ft, respectively. Since the normal depth is greater than the critical depth, the slope is mild and flow is subcritical. Calculations for the backwater curve progress upstream from the point of control, which in this case is the critical depth (see Table 7-12a). Critical depth may be assumed to occur from $3d_c$ to $4d_c$, about 15 ft, upstream of the overfall (Art. 7-32). The water depth at the overfall is about $0.7d_c$ = 0.7 × 3.86 = 2.70 ft. The water-surface profile is of the M2 type, as shown in Fig. 7-48. The total distance from the overfall to the location where the curve joins the normal-depth line equals 1256 ft plus 15 ft, or 1271 ft.

Standard step method allows computation of backwater curves in both nonprismatic natural channels and nonuniform artificial channels as well as in uniform channels. This method involves solving for the depth of flow at various locations along a channel with Bernoulli's energy equation and a known length of reach.

A surface profile is determined in the following manner: The channel is examined for changes in cross section, grade, or roughness, and the locations of these changes are given station numbers. Stations are also established between these locations such that the velocity change between any two consecutive stations is not greater than 20% of the velocity at the former station. Data concerning the hydraulic elements of the channel are collected at each station. Computation of the surface curve is then made in steps, starting from the point of control and

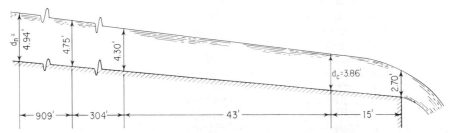

Fig. 7-48. Backwater curve for rectangular channel with free overfall (right) in Example 7-10.

progressing from station to station—in an upstream direction for subcritical flow and downstream for supercritical flow. The length of reach in each step is given by the stationing, and the depth of flow is determined by trial and error.

Nonprismatic channels do not have well-defined points of control to aid in determining the starting depth for a backwater curve. Therefore, the water-surface elevation at the beginning must be determined as follows: The step computations are started at a point in the channel some distance upstream or downstream from the desired starting point, depending on whether flow is supercritical or subcritical, respectively. Then, computations progress toward the initial section. Since this step method is a converging process, this procedure produces the true depth for the initial section within a relatively few steps.

The energy balance used in the standard step method is shown graphically in Fig. 7-47, in which the position of the water surface at section 1 is Z_1 and at section 2, Z_2, referred to a horizontal datum. Writing Bernoulli's equation [Eq. (7-11)] for sections 1 and 2 yields

$$Z_1 + \frac{V_1^2}{2g} = Z_2 + \frac{V_2^2}{2g} + h_f + h_i \tag{7-94}$$

where V_1 and V_2 are the mean velocities, ft/s, at sections 1 and 2; the friction loss ft, in the reach $(\overline{S}L)$ is denoted by h_f; and the term h_i is added to account for eddy loss, ft.

Eddy loss, sometimes called **impact loss,** is a head loss caused by flow running contrary to the main current because of irregularities in the channel. No rational method is available for determination of eddy loss, and it is therefore often accounted for, in natural channels, by a slight increase in Manning's n. Eddy loss depends mainly on a change in velocity head. For lined channels, it has been expressed as a coefficient k to be applied as follows:

$$h_i = k \left(\frac{V_1^2}{2g} - \frac{V_2^2}{2g} \right) = k \left(\Delta \frac{V^2}{2g} \right) \tag{7-95}$$

The coefficient k is 0.2 for diverging reaches, from 0 to 0.1 for converging reaches, and about 0.5 for abrupt expansions and contractions.

The total head at any section of the channel is given by

$$H = Z + \frac{V^2}{2g} \tag{7-96}$$

where Z equals the elevation of the channel bottom above the given datum plus the depth of flow d at that section.

The standard step method is most easily used if the computations are arranged in tabular form, similar to that used for the uniform step method (Table 7-12a). One form is shown in Table 7-12b.

In column 1, each section is identified by a station number, such as Station $2 + 80$ (280 ft from initial station). Column 2 gives water-surface elevation, ft, at the station. The first entry in this column is the known elevation of the water surface at the initial section. Subsequent entries are trial values, to be verified or rejected by the computations made in the remaining columns of the table.

Column 3 lists depth of flow, ft. This depth corresponds to the elevation in column 2, adjusted for S_oL (see Fig. 7-47). Column 4 shows area of flow, ft², corresponding to the depth in column

TABLE 7-12b Computation of a Flow Profile by the Standard Step Method

Station (1)	Z (2)	d (3)	A (4)	V (5)	$V^2/2g$ (6)	H (7)	R (8)	$R^{4/3}$ (9)	S (10)	$\overline{S}$ (11)	L (12)	h_f (13)	h_i (14)	H (15)

3. In column 5 is mean velocity, ft/s, equal to the given discharge Q divided by column 4. It yields velocity head, ft, for column 6. Addition of columns 6 and 2 then produces total head, ft, for column 7 [Eq. (7-96)].

Column 8 contains hydraulic radius R, ft, corresponding to the depth in column 3. Column 10 is friction slope S computed from Eq. (7-93), with V from column 5, $R^{4/3}$ from column 9, and n from Table 7-11. Column 11 lists $\overline{S}$, average friction slope for the reach; it equals the mean of the friction slope for a section in column 10 and that for the previous section. Column 12 gives length of reach, ft, between sections. These values are the differences in station numbers for the reach (column 1). Column 13 contains friction loss h_f in the reach, column 11 times column 12. For column 14, which lists eddy loss h_i in the reach, ft, a coefficient k is multiplied by the result obtained by subtracting the value for a section in column 6 from that for the previous section.

Finally, column 15 gives total head H, ft. This is obtained from Eq. (7-94), which after substitution of H from Eq. (7-96) becomes

$$H_1 = H_2 + h_f + h_i \tag{7-97}$$

where H_1 and H_2 equal the total head of sections 1 and 2, respectively. The value of total head computed from Eq. (7-97) must agree with the value of total head given in column 7 or the assumed water-surface elevation in column 2 is incorrect. Agreement is assumed if the two values of total head are within 0.1 ft in elevation. If the two values of total head do not agree (column 15 $\neq$ column 7), a new water-surface elevation must be assumed for column 2 and the computations repeated until agreement is obtained. The value that finally leads to agreement gives the correct water-surface elevation. This value should be underlined to indicate its acceptance, and the computations may then proceed for the next step.

Additional columns may be added to the table to give such incidental information as the invert elevation, bottom width of the channel, wetted perimeter, and change in velocity head between sections.

Backwater curves for natural river or stream channels (irregularly shaped channels) are calculated in a manner similar to that described for regularly shaped channels. However, some account must be taken of the varying channel roughness and the differences in velocity and capacity in the main channel and the overbank or flood plain portions of the stream channel. The most expeditious way of determining the backwater curves is to plot the channel cross section to a scale convenient for measurement of lengths and areas; subdivide the cross section into main channels and floodplain areas; and determine the discharge, velocity, and friction slope for each subarea at selected water-surface elevations. Utilizing the above data, determine the total discharge (the sum of the subarea discharges), the mean velocity (the total discharge divided by the total area), and α (the energy coefficient or coriolis coefficient to be applied to the velocity head). (See V. T. Chow, "Open-Channel Hydraulics," McGraw-Hill Book Company, New York.)

The backwater curve is usually started by assuming normal depth at a point some distance downstream from the start of the reach under analysis. Several intermediate cross sections should be taken between the point where normal depth is assumed and the start of the reach for which a detailed water-surface profile is required. This allows the intermediate sections to "dampen out" any minor errors in the assumed starting water-surface elevation.

The accuracy or validity of the water-surface profile is contingent on an accurate evaluation of the channel roughness and judicious selection of cross-section location. A greater number of cross sections generally enhances the validity of the water-surface profile; however, because of the extensive calculations involved with each cross section, their number should be limited to as few as accuracy permits.

The effect of bridges, approach roadways, bridge piers, and culverts can be determined using procedures outlined in V. T. Chow "Open-Channel Hydraulics," McGraw-Hill Book Company,

New York, and J. N. Bradley, "Hydraulics of Bridge Waterways," Hydraulics Design Series no. 1, 2d ed., U.S. Department of Transportation, Federal Highway Administration, Bureau of Public Roads, 1970.

7-28. Hydraulic Jump

This is an abrupt increase in depth of rapidly flowing water (Fig. 7-49). Flow at the jump changes from a supercritical to a subcritical stage with an accompanying loss of kinetic energy (Art. 7-23).

A hydraulic jump is the only means by which the depth of flow can change from less than critical to greater than critical in a uniform channel. A jump will occur either where supercritical flow exists in a channel of subcritical slope, as shown in Figs. 7-52 and 7-53*b*, or where a steep channel enters a reservoir. The first condition is met in a mild channel downstream from a sluice gate or ogee overflow spillway, or at an abrupt change in channel slope from steep to mild. The second condition occurs where flow in a steep channel is blocked by an overflow weir, a gate, or other obstruction.

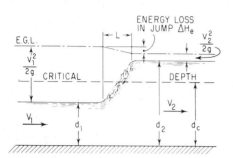

Fig. 7-49.   Hydraulic jump.

A hydraulic jump can be either stationary or moving, depending on whether the flow is steady or unsteady, respectively.

Depth at the jump is not discontinuous. The change in depth occurs over a finite distance, known as the length of jump. The upstream surface of the jump, known as the roller, is a turbulent mass of water, which is continually tumbling erratically against the rapidly flowing sheet below.

The depth before a jump is the initial depth, and the depth after a jump is the sequent depth. The specific energy for the sequent depth is less than that for the initial depth because of the energy dissipation within the jump. (Initial and sequent depths should not be confused with the depths of equal energy, or alternate depths.)

According to Newton's second law of motion, the rate of loss of momentum at the jump must equal the unbalanced pressure force acting on the moving water and tending to retard its motion. This unbalanced force equals the difference between the hydrostatic forces corresponding to the depths before and after the jump. For rectangular channels, this resultant pressure force is

$$F = \frac{d_2^2 w}{2} - \frac{d_1^2 w}{2} \tag{7-98}$$

where d_1 = depth before jump, ft
d_2 = depth after jump, ft
w = unit weight of water, lb/ft^3

The rate of change of momentum at the jump per foot width of channel equals

$$F = \frac{MV_1 - MV_2}{t} = \frac{qw}{g} (V_1 - V_2) \tag{7-99}$$

where M = mass of water, lb · s^2/ft
V_1 = velocity at depth d_1, ft/s
V_2 = velocity at depth d_2, ft/s
q = discharge per foot width of rectangular channel, ft^3/s

t = unit of time, s

g = acceleration due to gravity, 32.2 ft/s^2

Equating the values of F in Eqs. (7-98) and (7-99), and substituting $V_1 d_1$ for q and $V_1 d_1/d_2$ for V_2, the reduced equation for rectangular channels becomes

$$V_1^2 = \frac{gd_2}{2d_1}(d_2 + d_1) \tag{7-100}$$

Equation (7-100) may then be solved for the sequent depth:

$$d_2 = \frac{-d_1}{2} + \sqrt{\frac{2V_1^2 d_1}{g} + \frac{d_1^2}{4}} \tag{7-101}$$

If $V_2 d_2/d_1$ is substituted for V_1 in Eq. (7-100),

$$d_1 = \frac{-d_2}{2} + \sqrt{\frac{2V_2^2 d_2}{g} + \frac{d_2^2}{4}} \tag{7-102}$$

Equation (7-102) may be used in determining the position of the jump where V_2 and d_2 are known. Relationships may be derived similarly for channels of any cross section.

The head loss in a jump equals the difference in specific-energy head before and after the jump. This difference (Fig. 7-49) is given by

$$\Delta H_e = H_{e1} - H_{e2} = \frac{(d_2 - d_1)^3}{4d_1 d_2} \tag{7-103}$$

where H_{e1} = specific-energy head of stream before jump, ft

H_{e2} = specific-energy head of stream after jump, ft

The specific energy for free-surface flow is given by Eq. (7-82).

The depths before and after a hydraulic jump may be related to the critical depth by the equation

$$d_1 d_2 \frac{d_1 + d_2}{2} = \frac{q^2}{g} = d_c^3 \tag{7-104}$$

where q = discharge, ft^3/s per ft of channel width

d_c = critical depth for the channel, ft

It may be seen from this equation that if $d_1 = d_c$, d_2 must also equal d_c.

7-29. Jump in Horizontal Rectangular Channels

The form of a hydraulic jump in a horizontal rectangular channel may be of several distinct types, depending on the Froude number of the incoming flow $\mathbf{F} = V/(gL)^{1/2}$ [Eq. (7-16)], where L is a characteristic length, ft; V is the mean velocity, ft/s; and g = acceleration due to gravity, ft/s^2. For open-channel flow, the characteristic length for the Froude number is made equal to the **hydraulic depth** d_h.

Hydraulic depth is defined as

$$d_h = \frac{A}{T} \tag{7-105}$$

where A = area of flow, ft^2

T = width of free surface, ft

For rectangular channels, hydraulic depth equals depth of flow.

Various forms of hydraulic jump, and their relation to the Froude number of the approaching flow F_1, were classified by the U.S. Bureau of Reclamation and are presented in Fig. 7-50, (V. T. Chow, "Open-Channel Hydraulics," McGraw-Hill Book Company, New York.)

For $F_1 = 1$, the flow is critical and there is no jump.

For $F_1 = 1$ to 1.7, there are undulations on the surface. The jump is called an undular jump.

For $F_1 = 1.7$ to 2.5, a series of small rollers develop on the surface of the jump, but the downstream water surface remains smooth. The velocity throughout is fairly uniform and the energy loss is low. This jump may be called a weak jump.

For $F_1 = 2.5$ to 4.5, an oscillating jet is entering the jump. The jet moves from the channel bottom to the surface and back again with no set period. Each oscillation produces a large wave of irregular period, which, very commonly in canals, can travel for miles, doing extensive damage to earth banks and riprap surfaces. This jump may be called an oscillating jump.

For $F_1 = 4.5$ to 9.0, the downstream extremity of the surface roller and the point at which the high-velocity jet tends to leave the flow occur at practically the same vertical section. The action and position of this jump are least sensitive to variation in tailwater depth. The jump is well-balanced, and the performance is at its best. The energy dissipation ranges from 45 to 70%. This jump may be called a steady jump.

For $F_1 = 9.0$ and larger, the high-velocity jet grabs intermittent slugs of water rolling down the front face of the jump, generating waves downstream and causing a rough surface. The jump action is rough but effective, and energy dissipation may reach 85%. This jump may be called a strong jump.

Note that the ranges of the Froude number given for the various types of jump are not clear-cut but overlap to a certain extent, depending on local conditions.

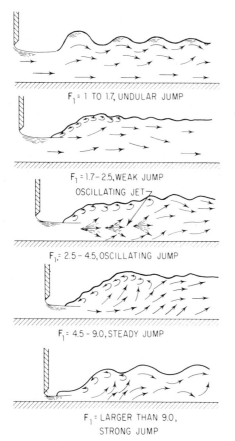

Fig. 7-50. Type of hydraulic jump depends on Froude number. *(After V. T. Chow, "Open-Channel Hydraulics," McGraw-Hill Book Company, New York.)*

7-30.   Hydraulic Jump as an Energy Dissipator

A hydraulic jump is a useful means for dissipating excess energy in supercritical flow (Art. 7-23). A jump may be used to prevent erosion below an overflow spillway, chute, or sluice gate by quickly reducing the velocity of the flow over a paved apron. A special section of channel built to contain a hydraulic jump is known as a **stilling basin.**

If a hydraulic jump is to function ideally as an energy dissipator, below a spillway, for example, the elevation of the water surface after the jump must coincide with the normal tailwater elevation for every discharge. If the tailwater is too low, the high-velocity flow will continue downstream for some distance before the jump can occur. If the tailwater is too high, the jump will be drowned out, and there will be a much smaller dissipation of total head. In either case, dangerous erosion is likely to occur for a considerable distance downstream.

The ideal condition is to have the *sequent-depth curve,* which gives discharge vs. depth after the jump, coincide exactly with the *tailwater-rating curve.* The tailwater-rating curve gives normal depths in the discharge channel for the range of flows to be expected. Changes in the spillway design that can be made to alter the tailwater-rating curve involve changing the crest length, changing the apron elevation, and sloping the apron.

Accessories, such as chute blocks and baffle blocks, are usually installed in a stilling basin to control the jump. The main purpose of these accessories is to shorten the range within which the jump will take place, not only to force the jump to occur within the basin but to reduce the size and therefore the cost of the basin. Controls within a stilling basin have additional advantages in that they improve the dissipation function of the basin and stabilize the jump action.

7-31.   Length of Hydraulic Jump

The length of a hydraulic jump L may be defined as the horizontal distance from the upstream edge of the roller to a point on the raised surface immediately downstream from cessation of the violent turbulence. This length (Fig. 7-49) defies accurate mathematical expression, partly because of the nonuniform velocity distribution within the jump. But it has been determined experimentally. The experimental results may be summarized conveniently by plotting the Froude number of the upstream flow F_1 against a dimensionless ratio of jump length to downstream depth L/d_2. The resulting curve (Fig. 7-51) has a flat portion in the range of steady jumps. The curve thus minimizes the effect of any errors made in calculation of the Froude number in the range where this information is most frequently needed. The curve, prepared by V. T. Chow from data gathered by the U.S. Bureau of Reclamation, was developed for jumps in rectangular channels, but it will give approximate results for jumps formed in trapezoidal channels.

For other than rectangular channels, the depth d_1 used in the equation for Froude number is the hydraulic depth given by Eq. (7-105).

7-32.   Location of a Hydraulic Jump

It is important to know where a hydraulic jump will form since the turbulent energy released in a jump can extensively scour an unlined channel or destroy paving in a thinly lined channel. Special reinforced sections of channel must be built to withstand the pounding and vibration of

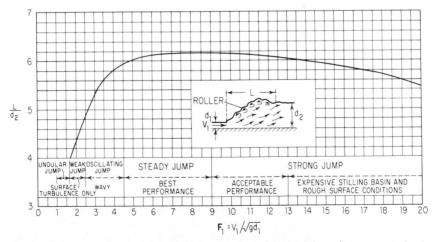

Fig. 7-51.   Length of hydraulic jump in a horizontal channel depends on sequent depth d_2 and Froude number of approaching flow. *(From V. T. Chow, "Open-Channel Hydraulics," McGraw-Hill Book Company, New York.)*

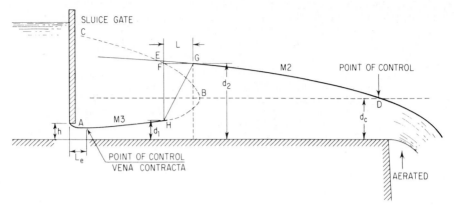

Fig. 7-52.   Graphical method for locating hydraulic jump before a sluice gate.

a jump and to provide extra freeboard for the added depth at the jump. These features are expensive to build; therefore, a great savings can be realized if their use is restricted to a limited area through a knowledge of the jump location.

The precision with which the location is predicted depends on the accuracy with which the friction losses and length of jump are estimated and on whether the discharge is as assumed. The method of prediction used for rectangular channels is illustrated in Fig. 7-52.

The water-surface profiles of the flow approaching and leaving the jump, curves AB and ED in Fig. 7-52, are type M3 and M2 backwater curves, respectively (Fig. 7-46e and c).

Backwater curve ED has as its point of control the critical depth d_c, which occurs near the channel drop-off. Critical depth does not exist exactly at the edge, as theory would indicate, but instead occurs a short distance upstream. The distance is small (from three to four times d_c) and can be ignored for most problems. The actual depth at the brink is 71.5% of critical depth, but it is normally assumed to be $0.7d_c$ for simplicity.

The point of control for backwater curve AB is taken as the depth at the vena contracta, which forms just downstream from the sluice gate. The distance from the gate to the vena contracta L_e is nearly equal to the size of gate opening h. The amount of contraction varies with both the head on the gate and the gate opening. Depth at the contraction ranges from 50 to over 90% of h. The depth of flow at the vena contracta may be taken as $0.75h$ in the absence of better information.

Jump location is determined as follows: The backwater curves AB and ED are computed in their respective directions until they overlap, using the step methods of Art. 7-27. With values of d_2 obtained from Eq. (7-101), CB, the curve of depths sequent to curve AB, is plotted through the area where it crosses curve ED. A horizontal intercept FG, equal in length to L, the computed length of jump, is then fitted between the curves CB and ED. The jump may be expected to form between the points H and G since all requirements for the formation of a jump are satisfied at this location.

If the downstream depth is increased because of an obstruction, the jump moves upstream and may eventually be drowned out in front of the sluice gate. Conversely, if the downstream depth is lowered, the jump moves to a new location downstream.

When the slope of a channel has an abrupt change from steeper than critical (Art. 7-23) to mild, a jump forms that may be located either above or below the grade change. The position of the jump depends on whether the downstream depth d_2 is greater than, less than, or equal to the depth d_1' sequent to the upstream depth d_1. Two possible positions are shown in Fig. 7-53.

It is assumed, for simplicity, that flow is uniform, except in the reach between the jump and the grade break. If the downstream depth d_2 is greater than the upstream sequent depth d_1', computed from Eq. (7-101) with d_1 given, the jump occurs in the steep region, as shown in Fig. 7-53a. The surface curve EO is of the S1 type and is asymptotic to a horizontal line at O. Line CB' is a plot of the depth d_1' sequent to the depth of approach line AB. The jump location is found by producing a horizontal intercept FG, equal to the computed length of the jump, between lines CB' and EO. A jump will form between H and G since all requirements are satisfied for this location. As depth d_2 is lowered, the jump moves downstream to a new position, as shown in Fig. 7-53b. If d_2 is less than d_1', computed from Eq. (7-102), the jump will form in the mild channel and can be located as described for Fig. 7-52.

7-33. Flow at Entrance to a Steep Channel

The discharge Q, ft^3/s, in a channel leaving a reservoir is a function of the total head H, ft, on the channel entrance, the entrance loss, ft, and the slope of the channel. If the channel has a slope steeper than the critical slope (Art. 7-23), the flow passes through critical depth at the entrance, and discharge is at a maximum. If the channel entrance is rectangular in cross section, the critical depth $d_c = \frac{2}{3}H_e$ [according to Eqs. (7-82) and (7-85)], where H_e is the specific energy head, ft, in the reservoir and datum is the elevation of the lip of the channel (Fig. 7-54a).

From $Q = AV$, with the area of flow $A = bd_c = \frac{2}{3}bH_e$ and the velocity

$$V = \sqrt{2g(H_e - d_c)} = \sqrt{\frac{64.4H_e}{3}}$$

the discharge for rectangular channels, ignoring entrance loss, is

$$Q = 3.087bH_e^{3/2} \qquad (7\text{-}106)$$

where b is the channel width, ft.

If the entrance loss must be considered, or if the channel entrance is other than rectangular, the inlet depth must be solved for by trial and error since the discharge is unknown. The procedure for finding the correct discharge is as follows:

A trial discharge is chosen. Then, the critical depth for the given shape of channel entrance is determined from appropriate tables, such as those in H. W. King and E. F. Brater, "Handbook of Hydraulics," McGraw-Hill Book Company, New York. Adding d_c to its associated velocity head gives the specific energy in the channel entrance, to which the resulting entrance loss is added. This sum then is com-

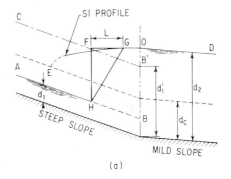

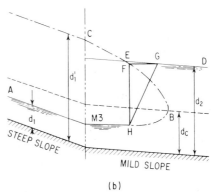

Fig. 7-53. Hydraulic jump may occur above (a), or below (b), or at a change of bottom slope of a channel.

pared with the specific energy of the reservoir water, which equals the depth of water above datum plus the velocity head of flow toward the channel. (This velocity head is normally so small that it may be taken as zero in most calculations.) If the specific energy computed for the depth of water in the reservoir equals the sum of specific energy and entrance loss determined for the channel entrance, then the assumed discharge is correct; if not, a new discharge is assumed, and the computations are continued until a balance is reached.

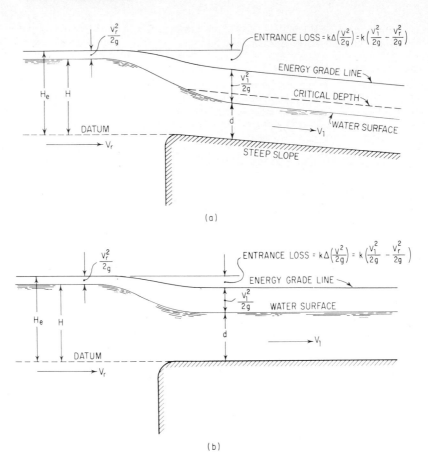

Fig. 7-54.   Entrance to (*a*) steep channel, (*b*) mild-slope channel from a reservoir.

A first trial discharge may be found from $Q = A \sqrt{2g(H_e - d)}$, where $(H_e - d)$ gives the actual head producing flow (Fig. 7-54). A reasonable value for the depth d would be $\frac{2}{3}H_e$ for steep channels and an even greater percentage of H_e for mild channels.

The entrance loss equals the product of an empirical constant k and the change in velocity head ΔH_v at the entrance. If the velocity in the reservoir is assumed to be zero, then the entrance loss is $k(V_1^2/2g)$, where V_1 is the velocity computed for the channel entrance. Safe design values for the coefficient vary from about 0.1 for a well-rounded entrance to slightly over 0.3 for one with squared ends.

7-34. Flow at Entrance to a Channel of Mild Slope

When water flows from a reservoir into a channel with slope less than the critical slope (Art. 7-23), the depth of flow at the channel entrance equals the normal depth for the channel (Art. 7-22). The entrance depth and discharge are dependent on each other. The discharge that results from a given head is that for which flow enters the channel without forming either a backwater or drawdown curve within the entrance. This requirement necessitates the formation of normal depth d since only at this equilibrium depth is there no tendency to change the discharge or to form backwater curves. (In Fig. 7-54*b*, *d* is normal depth.)

A solution for discharge at entrance to a <u>channel of</u> mild slope is found as follows: A trial discharge, ft³/s, is estimated from $Q = A \sqrt{2g(H_e - d)}$, where $H_e - d$ is the actual head, ft, producing flow. H_e is the specific energy head, ft, of the reservoir water relative to datum at lip of channel; A is the cross-sectional area of flow, ft²; and g is acceleration due to gravity, 32.2 ft/s². The normal depth of the channel is determined for this discharge from Eq. (7-83). (Tables in H. W. King and E. F. Brater, "Handbook of Hydraulics," McGraw-Hill Book Company, New York, may be used to advantage for this calculation.) The velocity head is computed for this depth-discharge combination, and an entrance-loss calculation is made (see Art. 7-33). The sum of the specific energy of flow in the channel entrance and the entrance loss must equal the specific energy of the water in the reservoir for an energy balance to exist between those points (Fig. 7-54b). If the trial discharge gives this balance of energy, then the discharge is correct; if not, a new discharge is chosen, and the calculations continued until a satisfactory balance is obtained.

7-35. Channel Section of Greatest Efficiency

If a channel of any shape is to reach its greatest hydraulic efficiency, it must have the shortest possible wetted perimeter for a given cross-sectional area. The resulting shape gives the greatest hydraulic radius and therefore the greatest capacity for that area. This can be seen from the Manning equation for discharge [Eq. (7-83)], in which Q is a direct function of hydraulic radius to the two-thirds power.

The most efficient of all possible open-channel cross sections is the semicircle. There are practical objections to the use of this shape because of the difficulty of construction, but it finds some use in metal flumes where sections can be preformed. The most efficient of all trapezoidal sections is the half hexagon, which is used extensively for large water-supply channels. The rectangular section with the greatest efficiency has a depth of flow equal to one-half the width. This shape is often used for box culverts and small drainage ditches.

7-36. Subcritical Flow Around Bends in Channels

Because of the inability of liquids to resist shearing stress, the free surface of steady uniform flow is always normal to the resultant of the forces acting on the water. Water in a reservoir has a horizontal surface since the only force acting on it is the force of gravity.

Water reacts in accordance with Newton's first law of motion: It flows in a straight line unless deflected from its path by an outside force. When water is forced to flow in a curved path, its surface assumes a position normal to the resultant of the forces of gravity and radial acceleration. The force due to radial acceleration equals the force required to turn the water from a straight-line path, or mV^2/r_c for m, a unit mass of water, where V is its average velocity, ft/s, and r_c the radius of curvature, ft, of the center line of the channel.

The water surface makes an angle ϕ with the horizontal such that

$$\tan \phi = \frac{V^2}{r_c g} \qquad (7\text{-}107)$$

The theoretical difference y, ft, in water-surface level between the inside and outside banks of a curve (Fig. 7-55) is found by multiplying $\tan \phi$ by the top width of the channel T, ft. Thus,

$$y = \frac{V^2 T}{r_c g} \qquad (7\text{-}108)$$

where the radius of curvature r_c of the center of the channel is assumed to represent the average curvature of flow. This equation gives values of y smaller than those actually encountered because of the use of average values of velocity and radius, rather than empirically derived values more representative of actual conditions. The error will not be great, however, if the depth of flow is well above critical (Art. 7-23). In this range, the true value of y would be only a few inches.

The difference in surface elevation found from Eq. (7-108), although it involves some drop in surface elevation on the inside of the curve, does not allow a savings of freeboard height on

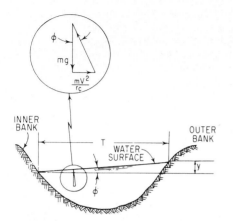

the inside bank. The water surface there is wavy and thus needs a freeboard height at least equal to that of a straight channel.

The top layer of flow in a channel has a higher velocity than flow near the bottom because of the retarding effect of friction along the floor of the channel. A greater force is required to deflect the high-velocity flow. Therefore, when a stream enters a curve, the higher-velocity flow moves to the outside of the bend. If the bend continues long enough, all the high-velocity water will move against the outer bank and may cause extensive scour unless special bank protection is provided.

Since the higher-velocity flow is pressed directly against the bank, an increase in friction

Fig. 7-55.   Water-surface profile at a bend in a channel with subcritical flow.

loss results. This increased loss may be accounted for in calculations by assuming an increased value of the roughness coefficient n

within the curve. Scobey suggests that the value of n be increased by 0.001 for each 20° of curvature in 100 ft of flume. His values have not been evaluated completely, however, and should be used with discretion. (F. C. Scobey, "The Flow of Water in Flumes," *U.S. Department of Argiculture, Technical Bulletin* 393.)

7-37. Supercritical Flow Around Bends

When water, traveling at a velocity greater than critical (Art. 7-23), flows around a bend in a channel, a series of standing waves are produced. Two waves form at the start of the curve. One is a positive wave, of greater-than-average surface elevation, which starts at the outside wall and extends across the channel on the line *AME* (Fig. 7-56). The second is a negative wave, with a surface elevation of less-than-average height, which starts at the inside wall and extends

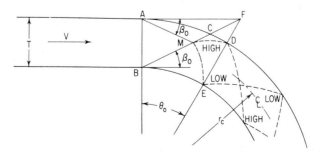

Fig. 7-56.   Plan view of supercritical flow around a bend in a channel.

across the channel on the line *BMD*. These waves cross at *M*, are reflected from opposite channel walls at *D* and *E*, recross as shown, and continue crossing and recrossing.

The two waves at the entrance form at an angle with the approach channel known as the wave angle β_o. This angle may be determined from the equation

$$\sin \beta_o = \frac{1}{\mathbf{F}_1} \tag{7-109}$$

where $\mathbf{F}_1$ represents the Froude number of flow in the approach channel [Eq. (7-16)] .

The distance from the beginning of the curve to the first wave peak on the outside bank is determined by the central angle θ_o. This angle may be found from

$$\tan \theta_o = \frac{T}{(r_c + T/2) \tan \beta_o} \tag{7-110}$$

where *T* is the normal top width of channel and r_c is the radius of curvature of the center of channel. The depths along the banks at an angle $\theta < \theta_o$ are given by

$$d = \frac{V^2}{g} \sin^2 \left(\beta_o \pm \frac{\theta}{2} \right) \tag{7-111}$$

where the positive sign gives depths along the outside wall and the negative sign, depths along the inside wall. The depth of maximum height for the first positive wave is obtained by substituting the value of θ_o found from Eq. (7-110) for θ in Eq. (7-111).

Standing waves in existing rectangular channels may be prevented by installing diagonal sills at the beginning and end of the curve. The sills introduce a counterdisturbance of the right magnitude, phase, and shape to neutralize the undesirable oscillations that normally form at the change of curvature. The details of sill design have been determined experimentally.

Good flow conditions may be insured in new projects with supercritical flow in rectangular channels by providing transition curves or by banking the channel bottom. Circular transition curves aid in wave control by setting up counterdisturbances in the flow similar to those provided by diagonal sills. A transition curve should have a radius of curvature twice the radius of the central curve. It should curve in the same direction and have a central angle given, with sufficient accuracy, by

$$\tan \theta_t = \frac{T}{2r_c \tan \beta_o} \tag{7-112}$$

Transition curves should be used at both the beginning and end of a curve to prevent disturbances downstream.

Banking the channel bottom is the most effective method of wave control. It permits equilibrium conditions to be set up without introduction of a counterdisturbance. The cross slope required for equilibrium is the same as the surface slope found for subcritical flow around a bend (Fig. 7-55). The angle ϕ the bottom makes with the horizontal is found from the equation

$$\tan \phi = \frac{V^2}{r_c g} \tag{7-113}$$

7-38. Transitions in Open Channels

A transition is a structure placed between two open channels of different shape or cross-sectional area to produce a smooth, low-head-loss transfer of flow. The major problems associated with design of a transition lie in locating the invert and determining the various cross-sectional areas so that the flow is in accord with the assumptions made in locating the invert. Many variables, such as flow-rate changes, wall roughness, and channel shape and slope, must be taken into account in design of a smooth-flow transition.

When proceeding downstream through a transition, the flow may remain subcritical or supercritical (Art. 7-23), change from subcritical to supercritical, or change from supercritical to subcritical. The latter flow possibility may produce a hydraulic jump.

Special care must be exercised in the design if the depth in either of the two channels connected is near the critical depth. In this range, a small change in energy head within the transition may cause the depth of flow to change to its alternate depth. A flow that switches to its subcritical alternate depth may overflow the channel. A flow that changes to its supercritical alternate depth may cause excessive channel scour. The relationship of flow depth to energy head can be shown on a plot such as Fig. 7-44.

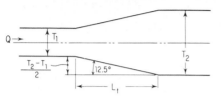

Fig. 7-57. Plan view of transition between two open channels with different widths.

To place a transition properly between two open channels, it is necessary to determine the design flow and calculate normal and critical depths for each channel section. Maximum flow is usually selected as the design flow. Normal depth for each section is used for the design depth. After the design has been completed for maximum flow, hydraulic calculations should be made to check the suitability of the structure for lower flows.

The transition length that produces a smooth-flowing, low-head-loss structure is obtained for an angle of about 12.5° between the channel axis and the lines of intersection of the water surface with the channel sides, as shown in Fig. 7-57. The length of the transition L_t is then given by

$$L_t = \frac{\frac{1}{2}(T_2 - T_1)}{\tan 12.5°} \qquad (7\text{-}114)$$

where T_2 and T_1 are the top widths of sections 2 and 1, respectively.

In design of an inlet-type transition structure, the water-surface level of the downstream channel must be set below the water-surface level of the upstream channel by at least the sum of the increase in velocity head, plus any transition and friction losses. The transition loss, ft, is given by $K(\Delta V^2/2g)$, where K, the loss factor, equals about 0.1 for an inlet-type structure; ΔV is the velocity change, ft/s; and $g = 32.2$ ft/s². The total drop in water surface y_d across the inlet-type transition is then $1.1[\Delta(V^2/2g)]$, if friction is ignored.

For outlet-type structures, the average velocity decreases, and part of the loss in velocity head is recovered as added depth. The rise of the water surface for an outlet structure equals the decrease in velocity head minus the outlet and friction losses. The outlet loss factor is normally 0.2 for well-designed transitions. If friction is ignored, the total rise in water surface y_r across the outlet structure is $0.8[\Delta(V^2/2g)]$.

Many well-designed transitions have a reverse parabolic water-surface curve tangent to the water surfaces in each channel (Fig. 7-58). After such a water-surface profile is chosen, depth and cross-sectional areas are selected at points along the transition to produce this smooth curve. Straight, angular walls usually will not produce a smooth parabolic water surface; therefore, a transition with a curved bottom or sides has to be designed.

The total transition length L_t is split into an even number of sections of equal length x. For Fig. 7-58, six equal lengths of 10 ft each are used, for an assumed drop in water surface y_d of 1 ft. It is assumed that the water surface will follow parabola AC for the length $L_t/2$ to produce a water-surface drop of $y_d/2$ and that the other half of the surface drop takes place along the

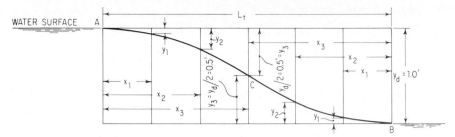

Fig. 7-58.   Profile of reverse parabolic water-surface curve for well-designed transitions.

parabola *CB*. The water-surface profile can be determined from the general equation for a parabola, $y = ax^2$, where y is the vertical drop in the distance x, measured from A or B.

The surface drops at sections 1 and 2 are found as follows: At the midpoint of the transition, $y_3 = ax^2 = y_d/2 = 0.5 = a(30)^2$, from which $a = 0.000556$. Then $y_1 = ax_1^2 = 0.000556(10)^2 = 0.056$ ft and $y_2 = ax_2^2 = 0.000556(20)^2 = 0.222$ ft.

7-39. Types of Weirs

A weir is a barrier in an open channel over which water flows. The edge or surface over which the water flows is called the *crest*. The overflowing sheet of water is the *nappe*.

If the nappe discharges into the air, the weir has *free discharge*. If the discharge is partly under water, the weir is *submerged* or drowned.

A weir with a sharp upstream corner or edge such that the water springs clear of the crest is a *sharp-crested weir* (Fig. 7-59). All other weirs are classed as *weirs not sharp-crested*. Sharp-crested weirs are classified according to the shape of the weir opening, such as rectangular weirs, triangular or V-notch weirs, trapezoidal weirs, and parabolic weirs. Weirs not sharp-crested are classified according to the shape of their cross section, such as broad-crested weirs, triangular weirs, and, as shown in Fig. 7-60, trapezoidal weirs.

The channel leading up to a weir is the *channel of approach*. The mean velocity in this channel is the *velocity of approach*. The depth of water producing the discharge is the *head*.

Sharp-crested weirs are useful only as a means of measuring flowing water. In contrast, weirs not sharp-crested are commonly incorporated into hydraulic structures as control or regulation devices, with measurement of flow as their secondary function.

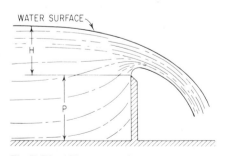

Fig 7-59.   Sharp-crested weir.

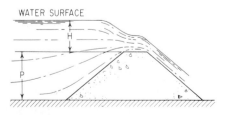

Fig. 7-60.   Weir not sharp-crested.

7-40. Sharp-Crested Weirs

Article 7-39 classifies weirs as sharp-crested or not sharp-crested. Discharge over a rectangular sharp-crested weir is given by

$$Q = CLH^{3/2} \qquad (7\text{-}115)$$

where Q = discharge, ft³/s
C = discharge coefficient
L = effective length of crest, ft
H = measured head = depth of flow above elevation of crest, ft
The head should be measured at least $2.5H$ upstream from the weir, to be beyond the drop in the water surface (surface contraction) near the weir.

Numerous equations have been developed for finding the discharge coefficient C. One such equation, which applies only when the nappe is fully ventilated, was developed by Rehbock and simplified by Chow:

$$C = 3.27 + 0.40\frac{H}{P} \qquad (7\text{-}116)$$

where P is the height of the weir above the channel bottom (Fig. 7-59) (V. T. Chow, "Open-Channel Hydraulics," McGraw-Hill Book Company, New York).

The height of weir P must be at least $2.5H$ for a complete crest contraction to form. If P is less than $2.5H$, the crest contraction is reduced and said to be partly suppressed. Equation (7-116) corrects for the effects of friction, contraction of the nappe, unequal velocities in the channel of approach, and partial suppression of the crest contraction and includes a correction for the velocity of approach and the associated velocity head.

To be fully ventilated, a nappe must have its lower surface subjected to full atmospheric pressure. A partial vacuum below the nappe can result through removal of air by the overflowing jet if there is restricted ventilation at the sides of the weir. This lack of ventilation causes increased discharge and a fluctuation and shape change of the nappe. The resulting unsteady condition is very objectionable when the weir is used as a measuring device. At very low heads, the nappe has a tendency to adhere to the downstream face of a rectangular weir even when means for ventilation are provided. A weir operating under such conditions could not be expected to have the same relationship between head and discharge as would a fully ventilated nappe. A V-notch weir should be used for measurement of flow at very low heads if accuracy of measurement is required.

End contractions occur when the weir opening does not extend the full width of the approach channel. Water flowing near the walls must move toward the center of the channel to pass over

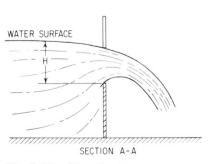
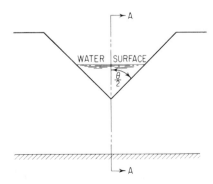

Fig. 7-61. V-notch weir.

the weir, thus causing a contraction of the flow. The nappe continues to contract as it passes over the crest, so below the crest, the nappe has a minimum width less than the crest length.

The effective length L, ft, of a contracted-width weir is given by

$$L = L' - 0.1NH \qquad (7\text{-}117)$$

where L' = measured length of crest, ft
N = number of end contractions
H = measured head, ft

If flow contraction occurs at both ends of a weir, there are two end contractions and $N = 2$. If the weir crest extends to one channel wall but not the other, there is one end contraction and $N = 1$. The effective crest length of a full-width weir is taken as its measured length. Such a weir is said to have its contractions suppressed.

7-41. Triangular or V-Notch Sharp-Crested Weirs

The triangular or V-notch weir (Fig. 7-61) has a distinct advantage over a rectangular sharp-crested weir (Art. 7-40) when low discharges are to be measured. Flow over a V-notch weir starts at a point, and both discharge and width of flow increase as a function of depth. This has the effect of spreading out the low-discharge end of the depth-discharge curve and therefore allows more accurate determination of discharge in this region.

Discharge is given by

$$Q = C_1 H^{5/2} \tan\frac{\theta}{2} \qquad (7\text{-}118)$$

where θ = notch angle
H = measured head, ft
C_1 = discharge coefficient

The head H is measured from the notch elevation to the water-surface elevation at a distance $2.5H$ upstream from the weir. Values of the discharge coefficient were derived experimentally by Lenz, who developed a procedure for including the effect of viscosity and surface tension as

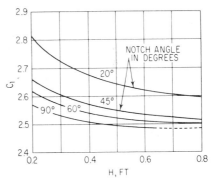

Fig. 7-62. Chart gives discharge coefficients for sharp-crested V-notch weirs. The coefficients depend on head and notch angle.

well as the effect of contraction and velocity of approach (A. T. Lenz, "Viscosity and Surface Tension Effects on V-Notch Weir Coefficients," *Transactions of the American Society of Civil Engineers,* vol. 69, 1943). His values were summarized by Brater, who presented the data in the form of curves (Fig. 7-62) (E. F. Brater and H. W. King, "Handbook of Hydraulics," McGraw-Hill Book Company, New York).

A V-notch weir tends to concentrate or focus the overflowing nappe, causing it to spring clear of the downstream face for even the smallest flows. This characteristic prevents a change in the head-discharge relationship at low flows and adds materially to the reliability of the weir.

7-42. Trapezoidal Sharp-Crested Weirs

The discharge from a trapezoidal weir (Fig. 7-63) is assumed the same as that from a rectangular weir and a triangular weir in combination.

$$Q = C_2 L H^{3/2} + C_3 Z H^{5/2} \qquad (7\text{-}119)$$

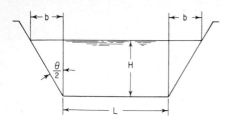

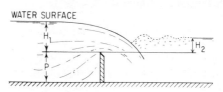

Fig. 7-63.   Trapezoidal sharp-crested weir.   **Fig. 7-64.**   Submerged sharp-crested weir.

where Q = discharge, ft^3/s
   L = length of notch at bottom, ft
   H = head, measured from notch bottom, ft
   Z = b/H [substituted for tan $(\theta/2)$ in Eq. (7-118)]
   b = half the difference between lengths of notch at top and bottom, ft
No data are available for determination of coefficients C_2 and C_3. They must be determined experimentally for each installation.

7-43. Submerged Sharp-Crested Weirs

The discharge over a submerged sharp-crested weir (Fig. 7-64) is affected not only by the head on the upstream side H_1 but by the head downstream H_2. Discharge also is influenced to some extent by the height P of the weir crest above the floor of the channel.

The discharge Q_s, ft^3/s, for a submerged weir is related to the free or unsubmerged discharge Q, ft^3/s, for that weir by a function of H_2/H_1. Villemonte expressed this relationship by the equation

$$\frac{Q_s}{Q} = \left[1 - \left(\frac{H_2}{H_1} \right)^n \right]^{0.385}$$ (7-120)

where n is the exponent of H in the equation for free discharge for the shape of weir used. (The value of n is $\frac{3}{2}$ for a rectangular sharp-crested weir and $\frac{5}{2}$ for a triangular weir.) To use the Villemonte equation, first compute the rate of flow Q for the weir when not submerged, and then, using this rate and the required depths, solve for the submerged discharge Q_s. (J. R. Villemonte, "Submerged-Weir Discharge Studies," *Engineering News-Record*, Dec. 25, 1947, p. 866.)

Equation (7-120) may be used to compute the discharge for a submerged sharp-crested weir of any shape simply by changing the value of n. The maximum deviation from the Villemonte equation for all test results was found to be 5%. Where great accuracy is essential, it is recommended that the weir be tested in a laboratory under conditions comparable with those at its point of intended use.

7-44. Weirs Not Sharp-Crested

These are sturdy, heavily constructed devices, normally an integral part of hydraulic projects (Fig. 7-60). Typically, a weir not sharp-crested appears as the crest section for an overflow dam or the entrance section for a spillway or channel. Such a weir can be used for discharge measurement, but its purpose is normally one of control and regulation.

The discharge over a weir not sharp-crested is given by

$$Q = CLH_t^{3/2}$$ (7-121)

where Q = discharge, ft^3/s
    C = coefficient of discharge
    L = effective length of crest, ft
    H_t = total head on crest including velocity head of approach, ft

The head of water producing discharge over a weir is the total of measured head H and velocity head of approach H_v. The velocity head of approach is accounted for by the discharge coefficient for sharp-crested weirs but must be considered separately for weirs not sharp-crested. Thus, for such weirs, Eq. (7-115) is rewritten in the form

$$Q = CL\left(H + \frac{V^2}{2g} \right)^{3/2}$$
(7-122)

where H = measured head, ft
    V = velocity of approach, ft/s
    $V^2/2g$ = H_v, velocity head of approach, ft, neglecting degree of turbulence given by Eq. (7-81)
    g = acceleration due to gravity, 32.2 ft/s^2

Since velocity and discharge are dependent on each other in this equation and both are unknown, discharge must be found by a series of approximations, which may be done as follows: First, compute a trial discharge from the measured head, neglecting the velocity head. Then, using this discharge, compute the velocity of approach, velocity head, and finally total head. From this total head, compute the first corrected discharge. This corrected discharge will be sufficiently accurate if the velocity of approach is small. But the process should be repeated, starting with the corrected discharge, where approach velocities are high.

The discharge coefficient C must be determined empirically for weirs not sharp-crested. If a weir of untested shape is to be constructed, it must be calibrated in place or a model study made to determine its head-discharge relationship. The problem of establishing a fixed relation between head and discharge is complicated by the fact that the nappe may assume a variety of shapes in passing over the weir. For each change of nappe shape, there is a corresponding change in the relation between head and discharge. The effect is most critical for low heads. A nappe undergoes several changes in succession as the head varies, and the successive shapes that appear with an increasing stage may differ from those pertaining to similar stages with decreasing head. Therefore care must be exercised when using these weirs for flow measurement to ensure that the conditions are similar to those at the time of calibration. (E. F. Brater and H. W. King, "Handbook of Hydraulics," McGraw-Hill Book Company, New York.)

Large weirs not sharp-crested often have piers on their crest to support control gates or a roadway. These piers reduce the effective length of crest by more than the sum of their individual widths because of the formation of flow contractions at each pier. The effective crest length for a weir not sharp-crested is given by

$$L = L' - 2(NK_p + K_a)H_t$$
(7-123)

where L = effective crest length, ft
    L' = net crest length, ft = measured length minus width of all piers
    N = number of piers
    K_p = pier-contraction coefficient
    K_a = abutment-contraction coefficient
    H_t = total head on crest including velocity head of approach, ft

(U.S. Department of the Interior, "Design of Small Dams," Government Printing Office, Washington, D.C. 20402.)

The pier-contraction coefficient K_p is affected by the shape and location of the pier nose,

TABLE 7-13 Pier-Contraction Coefficients

Condition	K_p
Square-nosed piers with corners rounded on a radius equal to about 0.1 of the pier thickness	0.02
Round-nosed piers	0.01
Pointed-nosed piers	0

TABLE 7-14 Abutment-Contraction Coefficients

Condition	K_a
Square abutment with headwall at 90° to direction of flow	0.20
Rounded abutments with headwall at 90° to direction of flow when $0.5H_d \geq r* \geq 0.15H_d$	0.10
Rounded abutments where $r* > 0.5H_d$ and headwall is placed not more than 45° to direction of flow	0

$*r$ = radius of abutment rounding.

thickness of pier, head in relation to design head, and approach velocity. For conditions of design head H, the average pier-contraction coefficients are as shown in Table 7-13.

The abutment-contraction coefficient K_a is affected by the shape of the abutment, the angle between the upstream approach wall and the axis of flow, the head in relation to the design head, and the approach velocity. For conditions of design head H_d, average coefficients may be assumed as shown in Table 7-14.

7-45. Submergence of Weirs Not Sharp-Crested

Spillways and other weirs not sharp-crested are submerged when their tailwater level is high enough to affect their discharge. Because of the surface disturbance produced in the vicinity of the crest, such a spillway or weir is unsatisfactory for accurate flow measurement.

Approximate values of discharge may be found by applying the following rules proposed by E. F. Brater: "(1) If the depth of submergence is not greater than 0.2 of the head, ignore the submergence and treat the weir as though it had free discharge. (2) For narrow weirs having a sharp upstream leading edge, use a submerged-weir formula for sharp-crested weirs. (3) Broad-crested weirs are not affected by submergence up to approximately 0.66 of the head. (4) For weirs with narrow rounded crests, increase discharge obtained by a formula for submerged sharp-crested weirs by 10% or more. Of the above rules, 1, 2, and 3 probably apply quite accurately, while 4 is simply a rough approximation."

7-46. The Ogee-Crested Weir

The ogee-crested weir was developed in an attempt to produce a weir that would not have the undesirable nappe variation normally associated with weirs not sharp-crested. A shape was needed that would force the nappe to assume a single path for any discharge, thus making the weir consistent for flow measurement. The ogee-crested weir (Fig. 7-65) has such a shape. Its crest profile conforms closely to the profile of the lower surface of a ventilated nappe flowing over a rectangular sharp-crested weir.

The shape of this nappe, and therefore of an ogee crest, depends on the head producing the discharge. Consequently, an ogee crest is designed for a single total head, called the design head

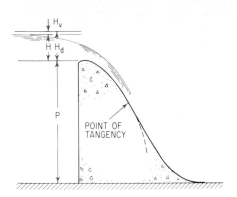

Fig. 7-65. Ogee-crested weir with vertical upstream face.

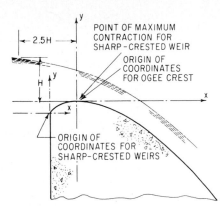

Fig. 7-66. Location of origin of coordinates for sharp-crested and ogee-crested weirs.

H_d. When an ogee weir is discharging at the design head, the flow glides over the crest with no interference from the boundary surface and attains near-maximum discharge efficiency.

For flow at heads lower than the design head, the nappe is supported by the crest and pressure develops on the crest that is above atmospheric but less than hydrostatic. This crest pressure reduces the discharge below that for ideal flow. (Ideal flow is flow over a fully ventilated sharp-crested weir under the same head H.)

When the weir is discharging at heads greater than the design head, the pressure on the crest is less than atmospheric, and the discharge increases over that for ideal flow. The pressure may become so low that separation in flow will occur; however, according to Chow, the design head may be safely exceeded by at least 50% before harmful cavitation develops (V. T. Chow, "Open-Channel Hydraulics," McGraw-Hill Book Company, New York).

The measured head H on an ogee-crested weir is taken as the distance from the highest point of the crest to the level of the water surface at a distance $2.5H$ upstream. This depth coincides with the depth measured between the upstream water level and the bottom of the nappe, at the point of maximum contraction, for a sharp-crested weir. This relationship is shown in Fig. 7-66.

Discharge coefficients for ogee-crested weirs are therefore determined from sharp-crested-weir coefficients after an adjustment for this difference in head. These coefficients are a function of the approach velocity, which varies with the ratio of height of weir P to actual total head H_t, where discharge is given by Eq. (7-122). Figure 7-67 for an ogee weir with a vertical upstream face gives coefficient C_d for discharge at design head H_d. (U.S. Department of the Interior, "Design of Small Dams," Government Printing Office, Washington, D.C. 20402. This manual and V. T. Chow, "Open-Channel Hydraulics," McGraw-Hill Book Company, New York, present methods for determining the shape of an ogee crest profile.) When the weir is discharging at other than the design head, the flow differs from ideal, and the discharge coefficient changes from that given in Fig. 7-67.

Figure 7-68 gives values of the discharge coefficient C as a function of the ratio H_t/H_d, where H_t is the actual head being considered and H_d is the design head.

If an ogee weir has a sloping upstream face, there is a tendency for an increase in discharge over that for a weir with a vertical face. Figure 7-69 shows the ratio of the coefficient for an ogee weir with a sloping face to the coefficient for a weir with a vertical upstream face. The coefficient of discharge for an ogee weir with a sloping upstream face, if flow is at other than the design head, is determined from Fig. 7-67 and is then corrected for head and slope with Figs. 7-68 and 7-69.

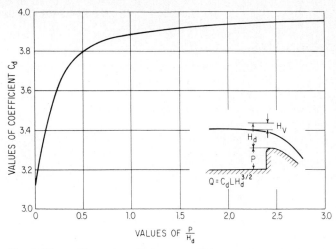

Fig. 7-67. Chart gives discharge coefficients at design head H_d for vertical-faced ogee-crested weirs. *(From "Design of Small Dams," U.S. Department of the Interior.)*

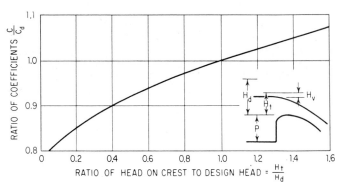

Fig. 7-68. Chart gives discharge coefficients for vertical-faced ogee-crested weirs at head H_t other than design head. *(From "Design of Small Dams," U.S. Department of the Interior.)*

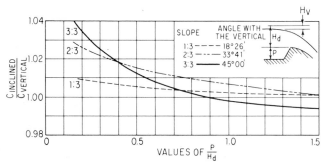

Fig. 7-69. Chart gives design coefficients at design head H_d for ogee-crested weirs with sloping upstream face. *(From "Design of Small Dams," U.S. Department of the Interior.)*

7-47. Broad-Crested Weir

This is a weir with a horizontal or nearly horizontal crest. The crest must be sufficiently long in the direction of flow that the nappe is supported and hydrostatic pressure developed on the crest for at least a short distance. A broad-crested weir is nearly rectangular in cross section. Unless otherwise noted, it will be assumed to have vertical faces, a plane horizontal crest, and sharp right-angled edges.

Figure 7-70 shows a broad-crested weir that, because of its sharp upstream edge, has contraction of the nappe. This causes a zone of reduced pressure at the leading edge. When the head H on a broad-crested weir reaches one to two times its breadth b, the nappe springs free, and the weir acts as a sharp-crested weir.

Discharge over a broad-crested weir is given by Eq. (7-115) since the velocity of approach was ignored in experiments performed to determine the coefficient of discharge. These coefficients probably apply more accurately, therefore, where the velocity of approach is not high. Values of the discharge coefficient, compiled by King, appear in Table 7-15 (H. W. King and E. F. Brater, "Handbook of Hydraulics," McGraw-Hill Book Company, New York).

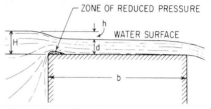

Fig. 7-70.   Broad-crested weir.

7-48. Weirs of Irregular Section

This group includes those weirs whose cross section deviates from typical broad-crested or ogee-crested weirs. Weirs of irregular section, fairly common in waterworks projects, are used as spillways and control structures. Experimental data are available on the more common shapes. (See, for example, H. W. King and E. F. Brater, "Handbook of Hydraulics," McGraw-Hill Book Company, New York.)

TABLE 7-15   Values of C in $Q = CLH^{3/2}$ for Broad-Crested Weirs

Measured head H, ft	Breadth of crest of weir, ft										
	0.50	0.75	1.00	1.50	2.00	2.50	3.00	4.00	5.00	10.00	15.00
0.2	2.80	2.75	2.69	2.62	2.54	2.48	2.44	2.38	2.34	2.49	2.68
0.4	2.92	2.80	2.72	2.64	2.61	2.60	2.58	2.54	2.50	2.56	2.70
0.6	3.08	2.89	2.75	2.64	2.61	2.60	2.68	2.69	2.70	2.70	2.70
0.8	3.30	3.04	2.85	2.68	2.60	2.60	2.67	2.68	2.68	2.69	2.64
1.0	3.32	3.14	2.98	2.75	2.66	2.64	2.65	2.67	2.68	2.68	2.63
1.2	3.32	3.20	3.08	2.86	2.70	2.65	2.64	2.67	2.66	2.69	2.64
1.4	3.32	3.26	3.20	2.92	2.77	2.68	2.64	2.65	2.65	2.67	2.64
1.6	3.32	3.29	3.28	3.07	2.89	2.75	2.68	2.66	2.65	2.64	2.63
1.8	3.32	3.32	3.31	3.07	2.88	2.74	2.68	2.66	2.65	2.64	2.63
2.0	3.32	3.31	3.30	3.03	2.85	2.76	2.72	2.68	2.65	2.64	2.63
2.5	3.32	3.32	3.31	3.28	3.07	2.89	2.81	2.72	2.67	2.64	2 63
3.0	3.32	3.32	3.32	3.32	3.20	3.05	2.92	2.73	2.66	2.64	2.63
3.5	3.32	3.32	3.32	3.32	3.32	3.19	2.97	2.76	2.68	2.64	2.63
4.0	3.32	3.32	3.32	3.32	3.32	3.32	3.07	2.79	2.70	2.64	2.63
4.5	3.32	3.32	3.32	3.32	3.32	3.32	3.32	1.88	2.74	2.64	2.63
5.0	3.32	3.32	3.32	3.32	3.32	3.32	3.32	3.07	2.79	2.64	2.63
5.5	3.32	3.32	3.32	3.32	3.32	3.32	3.32	3.32	2.88	2.64	2.63

Environmental engineers are concerned with works developed to protect and promote public health and to improve the environment. Their practice includes surveys, reports, designs, reviews, management, operation, and investigations of such works. They also engage in research in engineering sciences and such related sciences as chemistry, physics, and microbiology to advance the objectives of protecting public health and controlling environment.

Environmental engineering deals with treatment and distribution of water supply; collection, treatment, and disposal of wastewater; control of pollution in surface and underground waters; collection, treatment, and disposal of solid wastes; sanitary handling of milk and food; housing and institutional sanitation; rodent and insect control; control of atmospheric pollution; limitations on exposure to radiation; and other environmental factors affecting the health, comfort, safety, and well-being of people.

This section, which deals primarily with the handling of liquid wastes, covers such topics as sources of water pollution, sewer design and construction, and various methods of treating wastewater.

Section 8

Environmental Engineering

William T. Ingram
Consulting Engineer
Whitestone, NY

8-1. Prevention of Environmental Pollution

Because of public concern over continuous deterioration of the natural environment, Congress established the Environmental Protection Agency (EPA) and passed legislation to control disposal of solid wastes and pollution of water and air. The following legislation is of particular significance to environmental engineers.

National Environmental Policy Act ▪ All agencies of the Federal government and state and municipal agencies executing programs supported by Federal funds are required to carefully consider the environmental consequences of major actions, including proposed construction projects, and proposed legislation. The objectives are:

1. Fulfill the responsibilities of each generation as trustee of the environment for the succeeding generation.

2. Assure for all Americans safe, healthful, productive, and esthetically and culturally pleasing surroundings.

3. Attain the widest range of beneficial uses of the environment without degradation, risk to health or safety, or other undesirable and unintended consequences.

4. Preserve important historic, cultural, and natural aspects of our national heritage, and maintain, wherever possible, an environment that supports diversity and variety of individual choice.

5. Achieve a balance between population and resource use that will permit high standards of living and a wide sharing of life's amenities.

6. Enhance the quality of renewable resources and approach the maximum attainable recycling of depletable resources.

Clean Water Act (Federal Water Pollution Control Act) ▪ The objective is to restore and maintain the chemical, physical, and biological integrity of the nation's waters. Discharge of pollutants into navigable waters is to be eliminated as soon as possible. The objective will be met through a combination of Federal and state regulatory actions and Federal financial and advisory support of state and local activities. The act directs EPA to establish technology-based limitations and standards for industrial discharges. The states set water-quality standards for their waters. Control is achieved principally by issuance of permits by EPA or delegated states under the National Pollutant Discharge Elimination System. EPA also provides financial support for construction or modification of publicly owned sewage-treatment works and awards grants to states for preparation of water-quality management plans and for executing requirements of the act.

Safe Drinking Water Act ▪ EPA is required to establish regulations for public drinking water supplies. Primary regulations set maximum allowable levels for contaminants in drinking water and establish criteria for water treatment. Secondary regulations deal with taste, odor, and appearance of drinking water. Other regulations protect groundwater through controls over injection wells under the Underground Injection Control Program. EPA delegates primary responsibility for enforcement to the states and supports state programs with grants.

Resource Conservation and Recovery Act ▪ The objectives are to improve management of solid wastes, protect the environment and human health, and conserve valuable material and energy resources. The act also provides for state programs regulating hazardous wastes from generation to disposal, including disposal of industrial sludges containing toxic materials. The states regulate disposal of solid wastes on land in accordance with Federal criteria.

Marine Protection, Research and Sanctuaries Act ▪ EPA is required to protect the oceans from indiscriminate dumping of wastes and to designate safe sites for dumping. An objective is an ultimate halt in ocean dumping of wastes. The Corps of Engineers issues, subject to EPA approval, permits for dredging, filling of wetlands, or dumping of dredged material.

Superfund (Comprehensive Environmental Response, Compensation and Liability Act) ▪ The Federal government is authorized to remove and safely dispose of pollutants in hazardous waste sites and other facilities. The act establishes a Hazardous Waste Response Fund to pay for cleanup and damage claims. EPA designates substances that may present substantial hazards to public health or welfare or to the environment. The National Response Center should be notified of releases of hazardous substances.

Clean Air Act ▪ The objective is to protect public health and welfare from the harmful effects of air pollution. EPA promulgates National Ambient Air Quality Standards. To meet these standards, the states prepare State Implementation Plans and plans for enhancement of visibility and prevention of significant deterioration of air quality in areas where the standards have been attained. EPA also develops New Source Performance Standards, to reduce pollutant emissions, and National Emission Standards for Hazardous Air Pollutants, applicable to pollutants that will cause an increase in mortality or incapacitating illness. In addition, EPA sets limits on emissions from moving sources of air pollution.

8-2. Point Sources of Water Pollution

There are two major sources of water pollution: point sources and nonpoint sources. The former consists of sources that discharge pollutants from a well-defined place, such as outfall pipes of sewage-treatment plants and factories. Nonpoint sources, in contrast, cannot be located with such precision. They include runoff from city streets, construction sites, farms, or mines. Therefore, prevention of water pollution requires a mixture of controls on discharges from both point and nonpoint sources. Under the Clean Water Act (Art. 8-1), the EPA provides funding for planning control strategies under its Water Quality Management Program.

Domestic wastewater and industrial discharges are major point sources. The Clean Water Act and the Marine Protection, Research and Sanctuaries Act (Art 8-1) aim at elimination of discharge of pollutants in navigable waters and the ocean.

Wastewater is the liquid effluent of a community. This spent water is a combination of the liquid and water-carried wastes from residences, commercial buildings, industrial plants, and institutions, plus groundwater, surface water, or storm water.

Wastewater may be grouped into four classes:

Class 1 ▪ Effluents that are nontoxic and not directly polluting but liable to disturb the physical nature of the receiving water; they can be improved by physical means. They include such effluents as cooling water from power plants.

Class 2 ▪ Effluents that are nontoxic but polluting because they have an organic content with high oxygen demand. They can be treated for removal of objectionable characteristics by biological methods. The main constituent of this class of effluent usually is domestic sewage. But the class also includes storm water and wastes from dairy product plants and other food factories.

Class 3 ▪ Effluents that contain poisonous materials and therefore are often toxic. They can be treated by chemical methods. When they occur, such effluents generally are included in industrial wastes, for example, those from metal finishing.

Class 4 ▪ Effluents that are polluting because of organic content with high oxygen demand and, in addition, are toxic. Their treatment requires a combination of chemical, physical, and biological processes. When such effluents occur, they generally are included in industrial wastes, for example, those from tanning.

Domestic wastewater is collected from dwelling units, commercial buildings, and institutions of the community. It may include process wastes of industry as well as groundwater infiltration and miscellaneous waste liquids. It is primarily spent water from building water supply, to which have been added the waste materials of bathroom, kitchen, and laundry. (See Art. 8-15.)

Storm water is precipitation collected from property and streets and carrying with it the washings from surfaces.

Industrial wastes are primarily the specific liquid waste products collected from industrial processing but may contain small quantities of domestic sewage. Such wastes vary with the process and contain some quantity of the material being processed or chemicals used for processing purposes. Industrial cooling water when mixed with process water is also called industrial waste.

Industrial wastes, as distinguished from domestic wastes, are related directly to processing operations and usually are the liquid fraction of processing that has no further use in recovery of a product. These wastes may contain substances that, when discharged, cause some biological, chemical, or physical change in the receiving body of water.

Organic substances exert a biochemical oxygen demand (BOD) of relatively high proportion compared with domestic waste. It is not unusual in food processing to have wastes with a BOD of 1000 to 5000 mg/L or in the processing of edible oils to have 10,000 to 25,000 mg/L BOD.

The wastes may cause discoloration of a receiving stream, as in the release of dyes, or increase the temperature of the water, as in the case of a cooling tower or process-cooling water discharges.

Chemicals in the waste may be toxic to aquatic life, animals, or human populations using the water, or may in some way affect water quality by imparting taste or odor. Phenols introduced into water in the parts-per-billion range can produce such marked taste that the water becomes unusable for many purposes. Some chemicals may stimulate aquatic growth, and algae populations in the receiving stream may be increased. Some algae are detrimental to water quality since they too produce taste, odor, color, and turbidity in the water.

Industrial wastes that contain large quantities of solids may produce objectionable and dangerous levels of sludge on the bottom of a stream or along the banks. These add to the chemical, biological, and physical degradation of the stream. Discharges containing oil may render bathing beaches useless, interfere with nesting water fowl, and present extra problems of removal in water-treatment processes.

Wastes containing acids or alkalies may attack pier structures and water craft and produce serious toxic effects on fish life.

Some wastes interfere with the normal processes of wastewater treatment and may, if mixed with municipal waste, render the whole treatment process inoperative. Pretreatment of industrial wastes is often required to protect the sewers and treatment plant maintained by a municipal agency. Toxic pollutants are controlled by EPA General Pretreatment Regulations, which contain limits on specific substances discharged by various industries. Treatment of industrial wastes to the degree required to protect a receiving body of water is a requirement in all states; it may range from neutralization and other simple primary treatments to complete treatment or, in some instances, even an advanced stage of treatment to remove such pollutants as trace chemicals.

Combined wastes are the mixed discharge of domestic waste and storm water in a single pipeline. Industrial waste may or may not be found in a combined waste and can be carried apart from either in an industrial sewer.

(E. B. Besselievre and M. Schwartz, "Treatment of Industrial Wastes," and H. Azad, "Industrial Wastewater Management Handbook," McGraw-Hill Book Company, New York; N. L. Nemerow, "Liquid Wastes of Industry: Theories, Practices and Treatment," and R. L. Culp, G. M. Wesner, and G. L. Culp, "Advanced Wastewater Treatment," Van Nostrand Reinhold Company, New York.)

8-3. Types of Sewers

A sewer is a conduit through which wastewater, storm water, or other wastes flow. Sewerage is a system of sewers. Usually, it includes all the sewers between the ends of building-drainage systems and sewage-treatment plants or other points of disposal.

Sanitary sewers carry mostly domestic sewage. They may also receive some industrial wastes. But they are not designed for storm water or groundwater.

Storm sewers are designed specifically to convey storm water, street wash, and other surface water to disposal points.

Combined sewers are designed for both sewage and storm water. They cost less than separate sanitary and storm sewers, but disposal of the flow may create objectionable or hazardous conditions or involve costly treatment. A large flow of water from a storm may make adequate wastewater treatment impossible or increase its cost considerably.

Building sewers, or **house connections,** are pipes carrying sewage from the plumbing systems of buildings to a sewer or disposal plant. In urban areas, the flow goes to a **common sewer,** which serves abutting property. This conduit may be a **lateral,** one that receives sewage only from house sewers. A **submain,** or **branch,** sewer takes the flow from two or more laterals. A **main,** or **trunk,** sewer handles the flow from two or more submains or a submain plus laterals. An **outfall sewer** extends from the end of a collection system or to a treatment plant disposal point.

An **intercepting sewer** receives dry-weather flow and specific, limited quantities of storm water from several combined sewers. A **storm-overflow sewer** carries storm-flow excess from a main or intercepting sewer to an independent outlet.

A **relief sewer** is one built to relieve an existing sewer with inadequate capacity.

Usually, sewage or storm-water flow does not completely fill the conduit. But all sewers may be filled at some time and must be capable of withstanding some hydraulic pressure. Some types are always under pressure. **Force mains** flow full under pressure from a pump. **Inverted siphons,** conduits that dip below the hydraulic grade line, also flow full and under pressure.

8-4. Estimating Quantity of Sewage

Before a sewer is designed, the community or area to be served should be studied for the purpose of estimating the type and quantity of flow to be handled. Design should be based on the flow estimated at some future time, 25 to 50 years ahead, or at completion of the development. Also, the engineer must have, in advance, policy decisions on whether separate or combined sewers will be built.

The quantity and flow patterns of domestic sewage are affected principally by population and population increase; population density and density change; water use, water demand, and water consumption; industrial requirements; commercial requirements; expansion of service geographically; groundwater geology of the area; and topography of the area.

Since sewage consists primarily of wastewater, population and water consumption per capita are the more important factors. The quantity of sewage, however, generally is less than water consumption since some portion of water used for firefighting, lawn irrigation, street washing, industrial processing, and leakage does not reach the sewer. Some of these losses, however, may be offset by addition of water from private wells, groundwater infiltration, and illegal connections from roof drains. If the community to be served by the sewerage system already exists, the sewage flow may be estimated from the gallons per capita per day (gcd) of water being consumed. For a planned community, the estimate may be based on the gcd of water being consumed by an existing similar community. Table 8-1 lists reported flows for several large United States cities. Although sewage flow may range from 70 to 130% of water consumption, designers often assume the average sewage flow equal to the average water consumption or, for estimating purposes, 100 to 110 gcd.

(H. S. Azad, "Industrial Wastewater Management Handbook," and Metcalf & Eddy, Inc., "Wastewater Engineering," McGraw-Hill Book Company, New York; "Wastewater Treatment Design," MOP8, Water Pollution Control Federation, Washington, D.C.)

8-5. Rate of Wastewater Flow

Sewage flow varies with water use. But fluctuations tend to dampen out since there is a time lag from the time of water use to the time the flow reaches the sewer mains. Hourly, daily, and seasonal fluctuations affect design of sewers, pumping stations, and treatment plants.

TABLE 8-1 Municipal Discharges*

City	State	Population served	Design flow, mgd	Design, gcd
Bismarck	ND	37,000	4.95	133
Boise	ID	75,000	10	133
Bozeman	MT	21,000	5.2	247
Chicago	IL			
West S.W.		2,900,000	1200	413
Calumet		604,000	310	513
North Side		1,243,330	410	330
Cleveland	OH			
East		819,101	123	150
South		635,000	96	151
Denver	CO	1,023,200	98	96
Des Moines	IA	201,200	35	174
Detroit	MI	2,400,000	1290	538
Houston	TX			
North Side and 69th		460,000	55	120
Sims Bayou		359,463	48	134
Southwest		173,433	30	173
Indianapolis	IN			
S. Belmont Rd		539,108	120	223
Southport Rd		205,516	57	277
Jacksonville	FL	164,000	17.5	107
Kansas City	MO	418,000	85	203
Los Angeles	CA			
Hyperion		3,000,000	420	140
Terminal Island		115,000	14	122
Minneapolis	MN	434,000	218	502
New Orleans	LA	594,000	23	39
New York	NY			
Wards Island		1,270,000	180	142
Hunts Point		770,000	150	195
Bowery Bay		725,000	120	166
Tallman's Island		460,000	60	130
Twenty-sixth Ward		425,000	60	141
Owls Head		815,000	160	196
Newtown Creek		2,100,000	310	148
Oakwood Beach		105,000	15	143
Rockaway		389,000	30	77
Oklahoma City (South)	OK	218,900	30	137
Omaha	NB	280,000	18	64
Philadelphia	PA			
Northeast		1,240,000	175	141
Southwest		925,000	136	147
Portland	OR	377,800	100	265
Reno, Sparks	NV	110,000	20	182
Salt Lake City	UT	181,650	45	248
San Francisco	CA			
North Point		353,840	150	424
Richmond, Sunset		220,030	30	136
South East		177,450	37	208
Schenectady	NY	77,985	15.1	194
Seattle (West Point)	WA	494,000	125	253
St. Louis	MO			

TABLE 8-1 (*Continued*)*

City	State	Population served	Design flow, mgd	Design, gcd
Le May		849,783	240	282
Bistle Point		988,357	251	254
Washington	DC	1,780,000	240	135
Witchita	KS	275,000	45	164

*From Computer Run 1974: National Water Quality Inventory, app. C, vol. II, Office of Water Planning and Standards, EPA 440/9-74-001.

Daily and seasonal variations depend largely on community characteristics. In a residential district, greatest use of water is in the early morning; a pronounced peak usually occurs about 9 A.M., in the laterals. In commercial and industrial districts, where water is used all day, a peak may occur during the day, but it is less pronounced. At the outfall, the peak flow probably will occur about noon. Wherever possible, measure flow in existing sewers and at treatment plants to determine actual variations in flow.

Weekend flows may be lower than weekday. Also, industrial operations of a seasonal nature influence the seasonal average. The seasonal and annual averages often are about equal in May and June. The seasonal average may rise to about 125% of the annual average in late summer and drop to about 90% at winter's end.

Peak flows may exceed 300% of average in laterals and 200% of average at the treatment plant. Several state health departments require that laterals and submains be designed for a minimum of 400 gcd, including normal infiltration (see below), and main, trunk, and outfall sewers, for a minimum of 250 gcd, including normal infiltration, and any known substantial amounts of industrial waste.

Infiltration into Sewers ▪ Water may infiltrate sewers through poor joints, cracked pipes, walls of manholes, perforated manhole covers, and drains from flooded cellars. Sewers in wet ground with a high water table or close to streambeds will have more infiltration than sewers in other locations. Since infiltration increases the sewage load, it is undesirable. The sewer design should specify joints that will allow little or no infiltration, and the joints should be carefully made in the field.

Some specifications limit infiltration to 500 gal per day per inch diameter per mile. Often, enforcement agency specifications and requirements call for leakage tests. Some states limit the net leakage to 500 gal/day per inch diameter per mile for any section of the system.

Estimating Storm-Water Flow ▪ An estimate of the quantity of storm water flowing into sewers during or following a period of rain is necessary for their design. Preparation of the estimate requires knowledge of intensity and duration of storms, distances the water will travel before reaching the sewers, permeability and slope of the drainage area, and shape and size of the drainage area. Estimating by the rational method (empirical) incorporates these general considerations into one equation:

$$Q = CIA \tag{8-1}$$

where Q = peak runoff, ft^3/s
   A = drainage area, acres
   C = coefficient of runoff of area
   I = average rainfall rate, in/h of rain producing runoff
A common value for C used for residential areas with considerable land in lawn, garden, and shrubbery is 0.30 to 0.40. In built-up areas, C may be taken as 0.70 to 0.90.

The **time of concentration** is the time required for the maximum runoff rate to develop at a point in a sewer. At an inlet to a sewer, time of concentration equals inlet time, the theoretical time required for a drop of water to flow to the inlet from the most distant point of the area served by the inlet. The time of concentration for a point in the first sewer entered equals the inlet time plus the time of flow in the sewer to that point. Where branches connect to a sewer, the longest time of concentration for all the branches is used in design.

Inlet time may range from 5 min for a steep slope on an impervious area to 30 min for a slightly sloped city street. Time of flow in a sewer (assumed to be flowing full) may be taken as the length of the sewer to the point of concentration divided by the velocity of flow. Flood crest and storage time while the sewer is filling usually are neglected. The effect of this approximation is calculation of a larger rate of flow, which provides a safety factor in design.

Critical duration of rainfall on a watershed is the time required to develop maximum runoff and therefore equals time of concentration. Observations indicate that rainfall rate I is a function of storm duration t, minutes. Therefore, I for design of storm sewers may be estimated from rate-duration curves or formulas by substituting time of concentration for t.

Rainfall-intensity values are selected on the basis of frequency as well as duration of storms that have occurred in the vicinity. Rainfalls that are exceeded only once in 10 years are called 10-year storms, once in 20 years, 20-year storms, and so on. The designer has to decide for which frequency storm to design, and this involves a calculated risk combined with engineering judgment. For relatively inexpensive structures in residential areas, a 5-year storm may be used for design of storm sewers with reasonable safety. Where failure would endanger property, a 10-, 25-, or 50-year storm would be more conservative. A 50-year storm may be chosen if flooding would cause costly damage and disrupt essential activities. In such cases, cost-benefit studies may be made to guide selection of a suitable storm frequency.

Since storm-sewage flow is very large compared with dry-weather flow, combined sewers may be designed on the same basis as storm sewers. But cross sections and appurtenances of the sewers must be designed to handle the dry-weather flow efficiently.

(Metcalf & Eddy, Inc., "Wastewater Engineering," and V. T. Chow, "Handbook of Applied Hydrology," McGraw-Hill Book Company, New York; "Design and Construction of Sanitary and Storm Sewers," Manual 37, American Society of Civil Engineers.)

8-6. Sewer Design

Before a sewer system can be designed, the quantities of sewage to be handled and the rates of flow must be estimated. This requires a comprehensive study of the community or area to be served (Arts. 8-4 and 8-5). Then, a preliminary layout of the sewerage can be made. Also, pipe sizes, slopes, and depths below grade can be tentatively selected. Preliminary drawings should include a plan of the proposed system and show, in elevation and plan, location of roads, streets, water courses, buildings, basements, underground utilities, and geology. In addition, construction costs should be estimated.

After the preliminary design has been accepted, a survey should locate, in plan and elevation, all existing structures that may affect the design. Preferably, borings should be taken to determine soil characteristics along sewer routes and at sites for structures in the system. Physical characteristics of the area, including contours, should be shown on a topographic map. Scale may be 1 in to 200 ft, unless the number of details requires a larger scale. Contours at 5- or 10-ft intervals usually are satisfactory. Elevations of streets should be noted at intersections and abrupt changes in grade.

Sufficient cover should be placed over the tops of sewers to prevent damage from traffic loads. Also, the sewers should be below the frost level. Municipal and state regulations on cover should always be reviewed before a design for a specific location is undertaken.

The location of the sewers should be shown in elevation on profiles. Horizontal scale may be

1 in in 40 ft or 1 in in 100 ft, depending on the amount of detail. Vertical scale generally is 10 times the horizontal.

The final design should include a general map of the whole area showing location of all sewers and structures and the drainage areas; detailed plans and profiles of sewers showing ground levels, sizes of pipe and slopes, and location of appurtenances; detailed plans of all appurtenances and structures; a complete report with necessary charts and tables to make clear the exact nature of the project; complete specifications; and a confidential estimate of costs for the owner or agency responsible for the project.

Extensive plans require tabulation of data beginning at the upper end of the system and proceeding downstream from manhole to manhole. The addition to flow from connecting sewers should be included.

For combined sewers, provision also must be made for handling dry-weather or sanitary flow at proper velocities in sewers that may carry large quantities of water after a storm. Design is complicated by the need for diversion of waters not flowing to a sewage-treatment plant. Diversion structures should be located at or near water courses into which storm water may be discharged. The effects of discharging polluted water, a combination of sanitary sewage and storm water, should be fully investigated.

Approval of a supervising government agency, such as the state health department or Federal and state environmental protection agencies, usually must be obtained for the plans. Sewer designers should be familiar with requirements for sewers in the state in which work is to be done.

Design Flows ▪ Unless force mains are required because sewage must be pumped, or inverted siphons are necessary because of a drop in terrain, sewers usually are sized for open-channel flow. Maximum flow occurs when a conduit is not completely full. For example, for a circular pipe, maximum discharge takes place at about 0.9 of the total depth of the section. Sewers, however, should be designed to withstand some hydraulic pressure.

Sanitary sewers should be designed to carry peak design flow with a depth from one-half full for the smallest sewers to full. For storm stewers, common practice is to permit pipe to carry design flow at full depth.

Laterals may be designed for ultimate flow of the area to be served. Some designers size them to carry the expected flow while half full, to provide a safety factor. Submains may be designed for 10 to 40 years ahead. Trunk sewers may be planned for long periods, with provision made in design for parallel or separate routings of trunks of smaller size to be constructed as the need arises. The large sewers may be conservatively sized to carry the ultimate design flow while full. Appurtenances may have a different life since replacement of mechanical equipment will be necessary. Usually, they are designed for 20 to 25 years ahead, and a timetable of additions during that period is then scheduled in an overall improvement plan.

In general, flow may be assumed uniform in straight sewers. Velocity changes, however, will occur at obstacles and changes in sewer cross section and should be considered in making hydraulic computations.

Velocity Formulas ▪ Velocity of flow, ft/s, in straight sewers without obstructions may be estimated with satisfactory accuracy from the Manning formula

$$V = \frac{1.5}{n} R^{2/3} S^{1/2} \tag{8-2}$$

where n = coefficient dependent on roughness of conduit surface
R = hydraulic radius, ft = area, ft^2, of fluid divided by wetted perimeter, ft
S = energy loss, ft/ft of conduit length; approximately the slope of the conduit invert for uniform flow

TABLE 8-2 Slopes of Sewers, Ft/1000 Ft*

Velocity, ft/s	Hydraulic radius, ft							
	0.15	0.25	0.50	1.0	2.0	3.0	4.0	5.0
0.5	0.25	0.12	0.048					
1.0	0.96	0.48	0.19	0.077	0.03	0.02	0.01	
2.0	3.8	1.9	0.77	0.30	0.12	0.07	0.05	0.04
3.0	8.6	4.4	1.7	0.68	0.27	0.16	0.11	0.08
4.0	15.4	7.8	3.1	1.2	0.48	0.28	0.20	0.14
5.0	24.0	12.1	4.8	1.9	0.76	0.44	0.30	0.22
6.0	35.6	17.5	7.0	2.7	1.1	0.64	0.44	0.32
7.0	47.1	23.8	9.5	3.7	1.5	0.87	0.60	0.44
8.0	61.5	31.0	12.4	4.9	1.9	1.1	0.78	0.58

*From Manning formula [Eq. (8-3)] for $n = 0.013$. For other values of n, multiply slope given in table by $n/0.013$.

(See also Art. 7-9.) A common value for n is 0.013, suitable for well-laid brickwork, good concrete pipe, riveted steel pipe, and well-laid vitrified tile. Smaller values may be used for new, smooth pipe; but the roughness, and value of n, is likely to increase with age. For $n = 0.013$, the Manning formula becomes

$$V = 114R^{2/3}S^{1/2} \tag{8-3}$$

Table 8-2 lists slopes given by this formula for various velocities and hydraulic radii. The quantity of flow, ft^3/s, is given by

$$Q = AV \tag{8-4}$$

where A = cross-sectional area of flow, ft^2.

Minimum Velocity ▪ Velocity should be at least 2 ft/s in sanitary sewers to prevent settlement of solids. Slopes and cross sections of sewers should be chosen to given this velocity, or greater, for design flows. Greater velocities are desirable for storm and combined sewers because the flow may carry heavy sand and grit; 3 ft/s is desirable. Where sewers are sized for lower velocities than recommended minimums, provision for flushing and removal of obstructions should be made in the design.

Slopes ▪ Pipe slopes generally should exceed the minimum desirable for maintaining minimum velocity for design flow since actual flows, especially before a development reaches its ultimate size, may be much smaller than design flow. Actual velocity then may be less than the cleaning velocity. For example, suppose a circular pipe is sized and sloped to handle design flow when flowing full at 3 ft/s. This velocity will also be maintained when the pipe is flowing half full to full. But if the depth of flow drops to one-third the diameter, the velocity will decrease to about 2.4 ft/s; and at a depth 0.2 of the diameter, velocity declines to about 1.8 ft/s.

Table 8-3 gives the hydraulic characteristics of circular pipe. It enables the quantity and velocity of flow to be computed for a circular pipe flowing partly full, when the respective values for the pipe flowing full are known. The quantity, ft^3/s, for flow full may be estimated from

$$Q = \frac{0.463}{n} d^{8/3}S^{1/2} \tag{8-5}$$

and the velocity for flow full from

$$V = \frac{0.59}{n} d^{2/3}S^{1/2} \tag{8-6}$$

where d = inside diameter of pipe, ft.

Table 8-4 lists the quantities and slopes given by these formulas for various velocities and diameters. Information such as that in Tables 8-1 to 8-4 may be stored in computer memories for design use. Programs for application of the data are available commercially.

Minimum Pipe Size ▪ In many cities, 8 in is the minimum diameter of sewer permitted, and in large cities and metropolitan areas 10 in may be the minimum. In any case, pipe smaller than 6 in in diameter should not be used because of the possibility of stoppages.

Maximum Velocities ▪ High velocities in sewers also should be avoided because the solids carried in the flow may erode the conduit. A usual upper limit for sanitary sewers is 10 ft/s. For velocities in that range, though, lining at least the lower portion of the sewers with abrasion-resistant material, such as vitrified-tile blocks, is advisable. Maximum design velocities for storm sewers, however, may be much greater when such flows are likely to occur infrequently. Concrete channels have carried 40 ft/s without damage.

Energy Losses ▪ The assumption of uniform, open-channel flow in sewer design implies that the hydraulic grade line, or water surface, will parallel the sewer invert. This may quite often be true. But where conditions exist that change the slope of the water surface, the carrying capacity of the sewer will change, regardless of the constancy of the invert slope. This should be taken into account in hydraulic computations for flow near intersections of large sewers, any structure combining the flow from two or more sources, interchange of velocity and pressure head, and submerged outlets at outfalls.

In curved sewer lines, allowance must be made for larger energy losses than in straight sewers. One way of doing this is to increase the value of n by, say, 0.003 to 0.005.

TABLE 8-3 Hydraulic Characteristics of a Circular Pipe

Depth of flow	Partial area	Quantity, ft³/s, partly full	Velocity partly full
Inside diameter	Total area	Quantity, ft³/s, flowing full	Velocity flowing full
0	0	0	0
0.05	0.019	0.005	0.25
0.10	0.052	0.021	0.40
0.15	0.094	0.049	0.52
0.20	0.143	0.088	0.62
0.25	0.196	0.137	0.70
0.30	0.252	0.195	0.77
0.35	0.312	0.262	0.84
0.40	0.374	0.336	0.92
0.45	0.437	0.416	0.95
0.50	0.500	0.500	1.00
0.60	0.627	0.671	1.07
0.70	0.748	0.837	1.12
0.80	0.858	0.977	1.14
0.90	0.950	1.062	1.12
0.95	0.982	1.073	1.09
1.00	1.000	1.000	1.00

TABLE 8-4 Quantities, Velocities, and Slopes for Circular Sewers, Flowing Full*

Dia, in		Velocity, ft/s								
		0.5	1.0	2.0	3.0	4.0	5.0	6.0	7.0	8.0
8	Q†	0.17	0.35	0.70	1.1	1.4	1.8	2.1	2.4	2.8
	S‡	0.21	0.83	3.3	7.5	13.3	20.8	30.0	40.7	53.2
10	Q	0.27	0.55	1.1	1.6	2.2	2.7	3.3	3.8	4.4
	S	0.15	0.62	2.5	5.6	9.9	15.5	22.3	30.3	39.6
12	Q	0.39	0.79	1.6	2.4	3.1	3.9	4.7	5.5	6.3
	S	0.12	0.48	1.9	4.4	7.8	12.1	17.5	23.8	31.0
15	Q	0.61	1.2	2.5	3.7	4.9	6.1	7.4	8.6	9.8
	S	0.09	0.36	1.4	3.2	5.8	9.0	13.0	17.8	23.0
18	Q	0.88	1.8	3.5	5.3	7.1	8.8	10.6	12.4	14.2
	S	0.07	0.28	1.1	2.5	4.5	7.1	10.1	13.8	18.1
21	Q	1.2	2.4	4.8	7.2	9.6	12.0	14.4	17.8	19.2
	S	0.06	0.23	0.92	2.1	3.7	5.8	8.3	11.3	14.7
24	Q	1.6	3.1	6.3	9.4	12.6	15.7	18.8	22.0	25.2
	S	0.05	0.19	0.77	1.7	3.1	4.8	7.0	9.5	12.4
27	Q	2.0	4.0	8.0	11.9	15.9	19.9	23.9	27.9	31.9
	S	0.04	0.16	0.66	1.5	2.6	4.1	5.9	8.1	10.5
30	Q	2.5	4.9	9.8	14.7	19.6	24.5	29.4	34.4	39.3
	S	0.04	0.14	0.57	1.3	2.3	3.6	5.2	7.0	9.2
33	Q	3.0	5.9	11.9	17.8	23.8	29.7	35.7	41.7	47.6
	S	0.03	0.13	0.50	1.1	2.0	3.1	4.5	6.2	8.1
36	Q	3.5	7.1	14.1	21.2	28.3	35.4	32.4	49.5	56.6
	S	0.03	0.11	0.45	1.1	1.8	2.8	4.0	5.5	7.2
42	Q	4.8	9.6	19.2	28.9	38.4	48.1	57.7	67.3	76.9
	S		0.09	0.36	0.82	1.5	2.3	3.3	4.5	5.8
48	Q	6.3	12.6	25.2	37.7	50.3	62.8	75.4	88.0	101
	S		0.08	0.30	0.68	1.2	1.9	2.7	3.7	4.9
54	Q	8.0	15.9	31.8	47.7	63.6	79.5	95.4	111	127
	S		0.07	0.26	0.59	1.0	1.6	2.4	3.2	4.2
60	Q	9.8	19.6	39.2	58.8	78.5	98.1	118	137	157
	S		0.06	0.23	0.51	0.90	1.4	2.0	2.8	3.6
66	Q	11.9	23.8	47.6	71.3	95.1	119	143	166	190
	S		0.05	0.20	0.45	0.80	1.2	1.8	2.4	3.2
72	Q	14.1	28.3	56.5	84.7	113	141	170	198	226
	S		0.04	0.17	0.40	0.71	1.1	1.6	2.2	2.8
78	Q	16.6	33.2	66.4	99.5	133	166	199	232	266
	S		0.04	0.16	0.36	0.64	0.99	1.4	2.0	2.5
84	Q	19.2	38.5	77.0	115	154	192	231	270	308
	S		0.04	0.14	0.33	0.58	0.91	1.3	1.8	2.3
90	Q	22.1	44.2	88.4	133	177	221	265	309	353
	S		0.03	0.13	0.30	0.53	0.83	1.2	1.6	2.1
96	Q	25.1	50.3	101	151	201	252	302	352	402
	S		0.03	0.12	0.27	0.48	0.76	1.1	1.5	1.9
108	Q	31.8	63.6	127	191	254	318	381	444	508
	S		0.03	0.10	0.23	0.41	0.64	0.93	1.3	1.7
120	Q	39.2	78.5	157	236	314	392	471	549	628
	S		0.02	0.09	0.20	0.36	0.56	0.81	1.1	1.5

*From Manning formula [Eqs. (8-5) and (8-6)] for $n = 0.013$. For other values of n, multiply slopes given in the table by $n/0.013$; multiply quantities and velocities by $0.013/n$.

†Q = quantity of flow ft³/s.

‡S = slope, ft/1000 ft.

To account for the energy loss due to change in direction of sewers at manholes, the invert in the manhole may be dropped about 0.04 ft. If the sewer increases in size at the manhole, the crowns of the pipes may be kept aligned and the invert dropped accordingly, or the 0.8 depth points of the pipes may be set at the same elevation. The invert drop also may offset head losses due to size changes. Thus, it reduces the danger of the flow backing up and building up pressure.

Sewer Shapes ▪ In selection of a sewer shape, designers sometimes favor one that permits higher velocities at both small and large flows. For example, an egg shape, with the small end down, offers a rapidly decreasing cross-sectional area for decreasing flows. Since for a given quantity velocity is inversely proportional to area, velocity in an egg shape does not fall off so rapidly with decreasing flow as in other shapes. But cost of constructing such curved sections may be higher than that for simpler shapes. Often, a compromise shape is chosen, one that has favorable hydraulic characteristics and relatively low cost.

For this reason, circular sewers generally are used, especially for prefabricated conduit. This shape provides the maximum cross-sectional area for the volume of material in the wall and has fair hydraulic properties (Table 8-3). But because of the roundness, there is added cost in bedding circular pipe compared with shapes with a flat bottom.

Figure 8-1 shows some typical shapes that have been used for large reinforced concrete

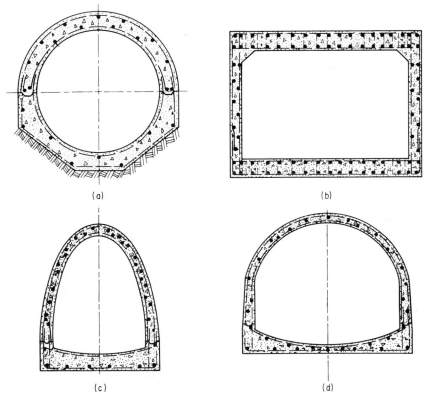

(a)   (b)

(c)   (d)

Fig. 8-1.   Some shapes used for large reinforced concrete sewers: (*a*) Circular; (*b*) rectangular; (*c*) semielliptical; (*d*) horseshoe.

sewers. The inverts usually are curved or incorporate a *cunette,* or small channel, to concentrate small flows to obtain desirable velocities.

Sewer Materials ▪ Sewers should be made of materials resistant to corrosion and abrasion and with sufficient strength to resist hydraulic pressure, handling, and earth and traffic loads with economy. Materials meeting these requirements include salt-glazed vitrified clay, reinforced concrete, cast iron, galvanized iron, brick, asbestos-cement, coated steel, bituminized fiber, and plastics formulated for the purpose. Sewer pipe is covered by Federal standards and specifications of the American Society for Testing and Materials and American Public Works Association.

(G. M. Fair, J. C. Geyer, and D. A. Okun, "Elements of Water Supply and Wastewater Disposal," John Wiley & Sons, Inc., New York; Metcalf & Eddy, Inc., "Wastewater Engineering," and H. W. King and E. F. Brater, "Handbook of Hydraulics," McGraw-Hill Book Company, New York; "Design and Construction of Sanitary and Storm Sewers," Manual 37, American Society of Civil Engineers.)

Loads on Sewers ▪ Sewers must be designed with adequate strength to withstand superimposed loads without crushing, collapsing, or through cracks. Usually, the loads are produced by earth pressure or loads transmitted through earth and may be assumed to be uniformly distributed.

Vertical earth loads on sewers may be estimated as indicated in Art. 6-28. Stresses in large sewers may be computed by elastic theory, and the sewers can be sized to resist these stresses. Standard culverts and sewer pipe generally may be selected with the aid of allowable-load tables prepared by the manufacturers.

8-7. Storm-Water Inlets

An inlet is an opening in a gutter or curb for passing storm-water runoff to a drain or sewer. In urban areas, inlets usually are positioned at street intersections to remove storm water before it reaches pedestrian crossings and so that water is never required to cross over the street crown to reach an inlet. If a block is more than 500 ft long, an inlet may be placed near the midpoint. Along rural highways, inlets generally are installed at low points. Spacings generally range from 300 ft for flat terrain and expressways to 600 ft. Often, however, the capacity of an inlet is increased by permitting some of the water to flow past to an inlet at a lower level. A common practice is to provide three inlets in each sag vertical curve, one at the low point and one at each side of it where the gutter elevation is about 0.2 ft higher. Several inlets also are necessary to reduce pondage where the drainage area would be too large for a single inlet in a valley.

Flow through an inlet is directed by a concrete or masonry enclosure to a pipe at the bottom (Fig. 8-2). The size of the enclosure generally is determined by the inlet length, which in turn is determined by the quantity of runoff to be drained, depth of water in the gutter at the inlet, and slope of the gutter. Runoff quantity can be estimated by use of the rational formula [Eq. (8-1)].

An inlet may be a curb opening, a gutter grating, or a combination of the two. Capacity of the curb-opening type when diverting 100% of gutter flow may be computed from

$$Q = 0.7L(a + y)^{3/2} \tag{8-7}$$

where Q = quantity of runoff, ft^3/s
L = length of opening, ft
a = depression in curb inlet, ft
y = depth of flow at inlet, ft

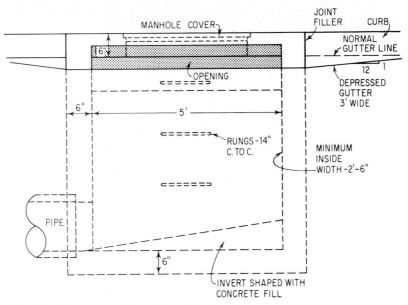

Fig. 8-2. Storm-water inlet with opening in curb.

In practice, the gutter may be depressed up to 5 in below the normal gutter line along the length of the inlet. Slope of the gutter commonly is 1 in in 12 in. The depression may extend up to 3 ft from the curb. Depth of flow in the gutter may be estimated from the Manning Formula.

Grate inlets should be placed with bars parallel to the flow. Length of opening should be at least 18 in to allow the flow to fall clear of the downstream end of the slot. For depths of flow up to 0.4 ft, capacity of inlet may be calculated from the weir formula

$$Q = 3Py^{3/2} \tag{8-8}$$

where P = perimeter, ft, of grate opening over which water may flow, ignoring the bars. For depths of flow greater than 1.4 ft, capacity may be computed from the orifice formula

$$Q = 0.6A \sqrt{2gy} \tag{8-9}$$

where A = total area of clear opening, ft^2
     g = acceleration due to gravity, 32 ft/s^2

At depths between 0.4 and 1.4 ft, neither formula may be applicable because of turbulence. A rough estimate may be made by using the smaller of the values of Q obtained from Eqs. (8-8) and (8-9).

Combination inlets are desirable, especially at low points, because the curb opening provides relief from flooding if the gate becomes clogged. If the gutter grate is efficient, the combination inlet will have a capacity only slightly greater than a similar inlet with grate alone. Hence, only the grate capacity should be depended on in designing a combination inlet.

Catch basins (Fig. 8-3) are inlets with enclosures that permit debris to settle out before the water enters the sewer. With good sewer grades and careful construction, however, catch basins are unnecessary because flow will be adequate to prevent debris from clogging the sewer. Also, since water trapped in catch basins may permit mosquitoes to hatch and may be a source of bad odors, simple inlets are preferable. Furthermore, catch basins are more expensive to maintain because they must be cleaned frequently.

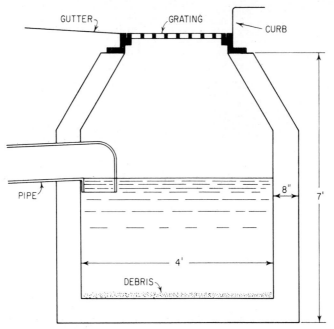

Fig. 8-3.   Catch basin with grating inlet in gutter.

("Design and Construction of Sanitary and Storm Sewers," Manual 37, American Society of Civil Engineers; Metcalf & Eddy, Inc., "Wastewater Engineering," McGraw-Hill Book Company, New York; "The Design of Storm-Water Inlets," Storm Drainage Research Committee, Johns Hopkins University, Baltimore, Md., 1956.)

8-8. Manholes

A manhole is a concrete or masonry enclosure for providing access to a sewer. The lower portion usually is cylindrical, with an inside diameter of at least 4 ft to allow adequate space for workers. The upper portion generally tapers to the opening to the street. About 2 ft in diameter, the opening is capped with a heavy cast-iron cover seated on a cast-iron frame. Figure 8-4a shows a typical manhole for sewers up to about 60 in in diameter, and Fig. 8-4b shows one type used for larger sewers.

Sewers are interrupted at manholes to permit inspection and cleaning. The flow passes through the manholes in channels at the bottom. Cast-iron rungs on the manhole walls enable workers to climb down to the sewers.

For sewers up to about 60 in in diameter, manholes are spaced 300 to 400 ft apart. They also are placed where sewers intersect or where there is a significant change in direction, grade, or pipe size. Since workers can walk through larger sewers, manholes for these may be spaced farther apart.

Drop manholes are used where one sewer joins another several feet below. The lower sewer enters the manhole at the bottom in the usual manner. The upper sewer, however, turns down sharply just outside the manhole and enters it at the bottom, where a channel feeds the flow to the main channel. To permit cleaning of the upper sewer from the manhole, the upper sewer also extends to the manhole at constant slope past the sharp drop through which the sewage

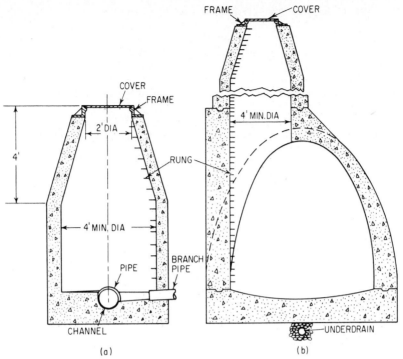

Fig. 8-4.   Concrete manholes: (*a*) For sewer under 60 in in diameter; (*b*) for larger sewer.

flows. Although the upper sewer could be brought down to the lower one more gradually, use of the drop manhole permits a more reasonable slope and thus saves considerable excavation. If the drop is less than 2 ft, however, a steeper slope for the upper sewer is usually more economical.

Where a large quantity of sewage must be dropped a long distance, a wellhole may be used. The fall may be broken by staggered horizontal plates in the shaft or by a well or sump at the bottom from which the sewage overflows to a lower level sewer. In a flight sewer, concrete steps break the fall.

Most street and highway departments and departments of public works have standard plans for manholes.

("Design and Construction of Sanitary and Storm Sewers," Manual 37, American Society of Civil Engineers.)

8-9. Sewer Outfalls

Type of outfall depends on quantity of sewage to be discharged, degree of treatment of the sewage, and characteristics of the disposal source. The outlet should be located to avoid pollution of water supplies and creation of a nuisance. Submerged outlets away from shore are preferable to discharge along a shore or bank, which may create an unsightly appearance and odors. Currents should be strong enough to prevent buildup of sludge near the outlet. It should be protected against scour by its location or suitable construction. A flap valve or automatically closing gate is desirable at the outlet to prevent entrance of water into the sewer during high-water stages.

Outfalls in tidal waters require special investigations to insure suitable dispersion of the sewage and to avoid floating wastes at the water surface. These outfalls often are constructed with a multiple discharge at the end, thus spreading the effluent over a large area and through a large volume of water. Depth of water over the outfall must be sufficient to accomplish dispersion before currents can transport the concentrated effluent streams shoreward, over shellfish beds, or into shallow water.

The outfall may be laid on the bottom. For protection against waves and scour, the pipe may be set in a trench or between two rows of piles and securely anchored.

An outlet discharging treated sewage into a small stream should be protected, by a concrete head wall and a concrete apron on the bank, against undercutting by the flow of the stream or sewage. A similarly protected outlet may be used at a river bank for storm-water discharge from a combined sewer. The dry weather flow may be carried farther out into the river through a small pipe along the bottom.

8-10. Inverted Siphons (Sag Pipes)

These are sewers that dip below the hydraulic grade line. They are used to avoid such obstructions as waterways, open-cut railways, subways, and extensive utility piping and structures. After passing under an obstruction, the pipe is brought to grade to permit open-channel flow in the continuation, to keep down the amount of cut and thus the cost of installing the sewer. The portion of the sewer below the hydraulic grade line flows full under pressure. Hence, it must

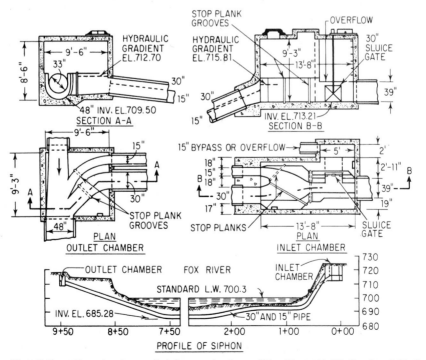

Fig. 8-5. Two-pipe inverted siphon at Appleton, Wis. *(From E. W. Steel and T. J. McGhee, "Water Supply and Sewerage," McGraw-Hill Book Company, New York.)*

have tight joints and be made of a material suitable for this job, and it must be designed for the maximum expected pressure.

To prevent solids from being deposited and obstructing an inverted siphon, it should be sized and sloped to keep flow velocities as much above 3 ft/s as feasible. Although experience has been good with a single pipe, 12 to 24 in in diameter, carrying flows with such velocities, a pipe big enough to handle the maximum flow at an adequate velocity may carry small flows at undesirably low speeds. In that case, two or more parallel pipes may be used instead of a single pipe.

An inlet chamber is constructed at the upstream end of the inverted siphon and an outlet chamber at the downstream end. These chambers may be concrete enclosures, which may be entered through manholes extending to grade. The inlet chamber for a multiple-pipe inverted siphon usually incorporates flow-regulating devices to control the flow to each pipe. As a safety measure, the inlet chamber may also incorporate a bypass, or overflow, pipe to relieve the inlet should the inverted siphon be overloaded or obstructed. In the outlet chamber, the inverts of the pipes merge into a single channel, which becomes the invert of the continuing sewer. Provision should be made in the chambers for cleaning and repairing for pipes and for draining them for these purposes. The designer should always investigate the hydraulic heads required in the inlet chamber to avoid surcharge on the upstream pipes.

Figure 8-5 shows a two-pipe inverted siphon, with stop planks for regulating flow in the inlet chamber. When the smaller pipe is full, sewage overflows a stop plank, to enter the larger pipe. Similarly, a three-pipe system may be used for a large combined sewer. The smallest pipe may be assigned the minimum dry-weather flow; a larger pipe, the excess up to a specified percentage of the maximum flow; and the largest pipe, the remainder of the flow. Built-in weirs may be used to regulate the flow to each pipe.

(E. W. Steel and T. J. McGhee, "Water Supply and Sewerage," McGraw-Hill Book Company, New York.)

8-11. Flow-Regulating Devices in Sewers

Sewerage systems often require some means for controlling flow, such as weirs, spillway siphons, and gates and valves. The devices may be used to divert flow from one conduit to another or to distribute flow among several pipes.

A common application is control of flow in a combined sewer when the discharge goes through a treatment plant. Treatment of maximum flow may not be economic, even if feasible, so flow to the plant is limited, usually to twice the dry-weather flow. For this purpose, a regulating device is installed in the sewer to permit the desired quantity to pass to the treatment plant. The excess flow is diverted into other conduits or discharged untreated into a waterway.

Side Weirs ▪ A simple device for such an application is a side weir, an overflow weir along the side wall of the combined sewer (Fig. 8-6a). Diversion, ft^3/s, may be estimated from the Engels formula:

$$Q = 3.32l^{0.83}h^{1.67} \tag{8-10}$$

where l = length of weir, ft
 h = depth of flow over weir at downstream end, ft

Siphon Spillways ▪ Although simple to construct, side weirs may not control flow as closely as desired. Siphon spillways (Fig. 8-6b) are more effective, especially for large flows. The outlet may be placed considerably below the inlet (differences in elevation up to 33.9 ft at sea level under standard atmospheric conditions may be used). Siphons operate under higher heads than weirs and permit much larger flows. Control is better because siphons can be constructed to start or stop discharge at any desired depth of flow in the combined sewer.

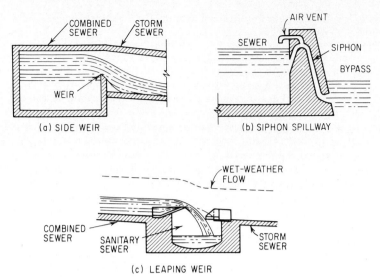

Fig. 8-6.   Flow-regulating devices for sewers.

Area, ft^2, of the siphon throat can be determined from

$$A = \frac{Q}{c\sqrt{2gh}}$$

(8-11)

where Q = discharge, ft^3/s
c = coefficient of discharge, which varies from 0.6 to 0.8
g = acceleration due to gravity = 32.2 ft/s^2
h = head, ft

For proper operation, the air vent should have an area of about $A/24$. The siphon inlet should be shaped to minimize entrance losses. The outlet should be completely submerged or sealed by the discharge.

Leaping Weirs ▪ A leaping weir, set in an invert of a combined sewer, permits low flows to drop through an opening into a sanitary sewer (Fig. 8-6c). Higher flows, having higher velocities, jump the opening and are discharged through a storm sewer. The opening may be made adjustable to correct for inaccuracies in computations based on theory.

Gates and Valves ▪ Diversion of flow also may be accomplished with float-actuated gates and valves. For example, low flows may be permitted to reach an outfall through an opening controlled by a gate. When water reaches a predetermined level in a float chamber, a float valve closes the gate to divert the water to a bypass. The designer must provide access to diversion chambers for cleaning since debris carried into the combined sewer will fill in the channel, clog the openings, and otherwise defeat the purpose of the flow-regulating device.

8-12.   Sewer-Construction Methods

Sewers usually are placed in trenches, but occasionally sewers may be constructed or installed in tunnels or laid at grade and covered with embankment.

In trench construction, the sewer line is located with respect to an offset line, laid out with transit and tape sufficiently far away to avoid disturbance. The trench then is marked or staked out on the ground and excavated. For small sewers, both the vertical and the horizontal positions of the conduit in the trench may be determined with the aid of a string set at a convenient elevation across batter boards straddling the trench at 25- or 50-ft intervals. For large sewers, key points should be located with transit and tape.

Trench excavation may be done by hand or with powered equipment. In rock, explosives should be avoided or used with great caution to avoid collapsing the trench or damaging nearby structures or utilities.

Experience generally will indicate whether the depth and type of soil require that the sides of the trench be sheeted and braced. If there is any doubt, sheeting should be used so as not to endanger workers. Sheeting methods (such as soldier beams and wood lagging) are applicable to trench construction. Unless this is forbidden by specifications to prevent possible failures, the sheeting may be salvaged as backfilling proceeds.

Water may be drained, except in quicksand, by leading it to sumps and pumping it out. Wellpoints may be necessary to prevent quicksand from forming in a sandy trench bottom or to dry out the bottom.

The support for the sewer should be shaped to the conduit bottom, whether the support be the subgrade, a granular fill, or a concrete cradle. In rock, excavation should be carried to a depth of one-fourth the conduit diameter below the bottom of the conduit, but not less than 4 in below. The space between the trench bottom and the conduit should be refilled with ¾-in gravel or lean concrete (1:4½ :9 mix), so that at least 120° of the pipe will be supported on it.

Pipelaying usually proceeds upgrade. Pipe is laid with bell ends upstream, to receive the spigots of subsequent sections. Grade usually is required to be within ½ in of that specified.

Joints between lengths of pipe usually are calked with a gasket of plastic, twisted hemp, or oakum and a filling of plastic, bitumen, or portland cement mortar (1:1 mix). Resilient joints are preferable to rigid types, which differential settlement may crack.

Feeder sewers come with Y or T stub branches for house sewer connections. If these connections are not made when the feeder is installed, a disk stopper is mortared into the bell of the stubs. Field notes should record the location of each branch so that it can be found when a connection has to be made in the future.

Backfilling should start as soon as possible. Earth should be placed and tamped evenly around the pipe to avoid disturbance of newly made joints and creation of high or unbalanced side pressures on the pipe. Material should be placed in layers not exceeding 6 in in thickness and tamped lightly until about 2 ft of fill covers the top of the pipe. Walking on the fill above the pipe should be prohibited until at least 1 ft of fill has been placed over the pipe.

The upper portion of the backfill should be heavily tamped, to reduce future settlement, if the surface over the trench is to be paved. Backfill must be carefully placed throughout, and materials that may permit excessive settlement should not be used.

For trenches in fields, the backfill need not be tamped. After all the material previously excavated from the trench has been replaced, the resulting mound may be left to settle naturally.

Large sewers in trenches generally are constructed of reinforced concrete, cast in long, reusable forms. Often, the invert is concreted first. Then, the forms for the upper portion are supported on the hardened invert concrete.

("Design and Construction of Sanitary and Storm Sewers," Manual 37, American Society of Civil Engineers.)

8-13. Pumping Stations for Sewage

Lift stations are used where it is necessary to pump sewage to a higher level. The installation may be underground or above grade, housed in a building. (For a discussion of sewage pumps, see Art. 8-14.)

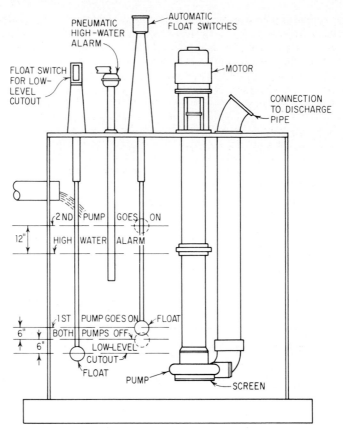

Fig. 8-7.   Small automatic sewage pumping station.

Most installations have at least two pumps. One is available as a standby, ready to take over if the first should fail. Main pumping stations should have at least three pumps; with the largest pump out of service, the other two should be able to handle the design flow. Several pumps with different capacities permit flexibility of operation. The smallest pump should be able to handle minimum flow. The others can be brought on-stream in succession as flow increases.

At a small pumping station, sewage may flow into a manhole or a tank. A horizontal pump may be installed in a "dry" compartment alongside the manhole; or a vertical pump, on the roof of the tank (Fig. 8-7). At a large pumping station, sewage flows into a wet well. The pumps may be installed above or in an adjacent dry well.

Often, the pumps operate automatically when the liquid in the wet well reaches a selected level. (See, for example, Fig. 8-7.) The motors may be started and stopped by switches operated by a float rod, which rises and falls with the liquid level. Usually, two sources of electric power are provided to insure continuity of operation. If no attendants are present in an automatic station, provision should be made for an alarm to be sounded and recorded at a remote station when a pump fails or the liquid level rises above a selected elevation.

Seepage in a dry well should be directed to a sump. It can be drained by one of the sewage pumps or by a special pump. This pump may also have a suction line to the wet well to drain it for cleaning and repair.

The wet well usually is small, to preclude septic action in the sewage; however, the well must be designed to handle maximum flow without flooding. The well should be vented to the outside, to prevent accumulation of odors. It may be divided into two interconnected compartments, which may be isolated for cleaning and repair by closure of a gate.

Pumps, though nonclogging, should be protected against debris in the sewage by a screen. For that purpose, a basket screen may be placed at the entering sewer or a bar screen ahead of the wet well.

("Design and Construction of Sanitary and Storm Sewers," Manual 37, American Society of Civil Engineers; "Design of Wastewater and Stormwater Pumping Stations," MOP No.FD-4, Water Pollution Control Federation, Washington, D.C.)

8-14. Sewage Pumps

Although sewage generally flows by gravity through conduit and treatment plants, pumping sometimes is required. Pumping may be the most economical means of conveying sewage past a hill, or the only way to get sewage from a cellar to a sewer at a higher level. Where desirable invert slopes would place a sewer far underground, making construction costs high, a more economical method is to raise the sewage in a pumping plant and then let it flow by gravity. Similarly, pumping may be necessary to give sufficient head for sewage to flow by gravity through a treatment plant.

"Nonclogging" centrifugal pumps are generally used. They are capable of passing solids with a maximum size of about 80% of the inside diameter of the pump suction and discharge pipes. These single-suction volute pumps may be bladeless or have two vanes. In some cases, grit chambers may be desirable ahead of pumps, to prevent accelerated wear in the pumps, and bar screens, perhaps mechanically cleaned, may be justified.

The pumps generally are driven by electric motors. Types preferred have a high efficiency over a wide range of operating conditions, but dependability is the most important characteristic. Also, slow-speed pumps are desirable for long life and less noise.

The shaft of the pump may be horizontal or vertical. Vertical pumps permit installation of motors above the pump pit, where they are less likely to be damaged by floods.

Sewage ejectors operated by compressed air are an alternative to nonclogging centrifugal pumps. In buildings where compressed air is available, such ejectors may be used as sump pumps.

In a commonly used type of sewage ejector, the sewage flows into a storage chamber until it is full. During this stage, air is exhausted from the chamber as the liquid level rises. A float rod closes the air exhaust and opens a compressed-air inlet. The compressed air forces the sewage up the discharge pipe. When the storage chamber is emptied, the float valve shuts the compressed-air valve and opens the air exhaust. Check valves in inlet and discharge pipes prevent back flow.

(T. G. Hicks, "Pump Application Engineering," and I. Karassik, "Pump Handbook," McGraw-Hill Book Company, New York.)

8-15. Characteristics of Domestic Wastewater

Usually, wastewater contains less than 0.1% of solid matter. Much of the flow looks like bath or laundry effluent, with garbage, paper, matches, rags, pieces of wood, and feces floating on top. Within a few hours at temperatures above 40°F, wastewater becomes stale. Later, it may become septic, often with the odors of hydrogen sulfide, mercaptans, and other sulfur compounds predominating. The more putrescible compounds there are in wastewater, the greater is its concentration or strength. In general, strength will vary with the amount of organic matter, water consumption per capita, and amount of industrial wastes.

Solids ▪ Total solids in wastewater comprise suspended and dissolved solids. About one-third of the total solids usually are in suspension. Suspended solids are those that can be filtered out on an asbestos mat. Usually more than half of these solids are organic material.

Suspended solids include settleable solids and colloids. Settleable solids precipitate out in sedimentation tanks in the usual detention periods. Colloids, mostly organic material, are smaller than 0.0001 mm in diameter and can remain in suspension indefinitely. They can pass through filter paper but are retained on a filtering membrane. Elimination of suspended solids from wastewater is desirable because they contain insoluble organic and inorganic pollutants and harbor bacteria and viruses.

Dissolved solids are the residue from evaporation after removal of suspended solids. Excessive dissolved solids can have adverse effects on living things, taste, irrigation, and water softness.

Solids also may be classified as volatile or fixed. The loss of weight when dried solids are burned is attributed to the volatile solids, which are considered to be organic material. The residue comprises fixed solids, which are assumed to be inorganic.

Organic Content ▪ The organic content of wastewater may be classified as nitrogenous and nonnitrogenous. Principal nitrogenous compounds include proteins, urea, amines, and amino acids. Principal nonnitrogenous compounds comprise soaps, fats, and carbohydrates.

Analyses of Sewage ▪ Tests are made on sewage to determine its strength, potential harmful effects in its disposal, and progress made in treating it. The most commonly made tests measure:

Suspended solids

Biochemical oxygen demand (BOD)

Amounts of ammonia, which decrease as sewage is treated

Nitrites and nitrates, which increase as sewage is treated

Dissolved oxygen

Ether-soluble matter, or fats and greases, which can form a heavy scum

pH value, which decreases, indicating greater acidity, as sewage becomes stale

Chemical oxygen demand, which approximates the total oxidizable carbonaceous content

Phosphorus, which can stimulate undesirable algae growths in lakes and streams

Heavy metals, such as mercury, silver, and lead, which are toxic

Total organic carbon (TOC), which may be determined in large laboratories or in industrial plants; small laboratories are not equipped with suitable apparatus to run the required test

Chlorine demand, the amount of chlorine added to sewage to produce a residual after a certain time, usually 15 min

Bacteria and other microorganisms

Coliform tests are usually required. Fecal coliform tests may be required when effluent is discharged into bathing or drinking waters or into tidal waters where shellfish are harvested.

Bacteria ▪ These may be aerobic, requiring air for survival; anaerobic, thriving without air; or facultative, carrying on with or without air. (Some may be pathogenic, causers of intestinal diseases. If they are present, the effluent may have to be chlorinated or otherwise treated to eliminate such bacteria, depending on the method of disposal.) Bacteria are useful in stabilizing wastewater, breaking it down into substances that do not decompose further.

Anaerobic bacteria are used in sludge digestion, the stabilization of organic material removed from sewage by sedimentation. Anaerobic stabilization takes longer than aerobic, is more sensitive to environmental conditions, and produces more disagreeable odors. Because the process is lengthy, it usually is not carried to complete stability but to a stage where further decomposition proceeds slowly. Stabilization is part of a cycle in which the products of decomposition become food for plants, then in turn food for people and animals, and finally are reconverted into wastes (Fig. 8-8a).

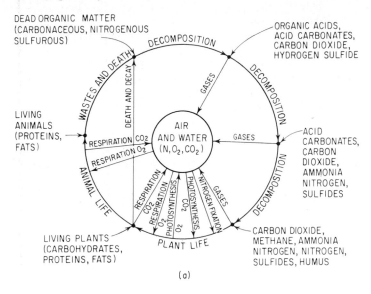

(a)

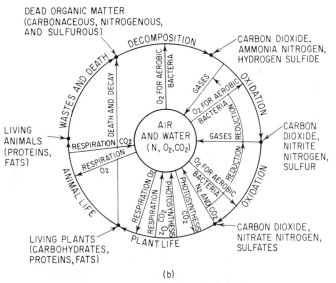

(b)

Fig. 8-8. Carbon, nitrogen, and sulfur cycles in (a) anaerobic decomposition; (b) aerobic decomposition. (*From E. W. Steel, "Water Supply and Sewerage," McGraw-Hill Book Company, New York.*)

Aerobic bacteria serve in self-purification of streams, trickling filters, and the activated-sludge method of treatment. In streams, oxygen may become available from several sources: absorption of air at the water surface; release by algae, which absorb carbon dioxide and release oxygen; and production by decomposition of such compounds as nitrates. In trickling filters, oxygen is supplied by allowing sewage to pass over filtering media while air circulates through the voids. In the activated-sludge process, oxygen is furnished by passing air through a mixture of sewage and previously activated sludge and by strongly agitating the mixture to dissolve air into the liquid. In aerobic stabilization also, decomposition occurs in steps and is part of a cycle (Fig. 8-8b). If the supply of oxygen is inadequate, however, anaerobic action will occur and disagreeable odors may be produced.

BOD and COD ▪ The amount of oxygen used during decomposition of organic material is the **biochemical oxygen demand (BOD)**. It is a measure of the amount of biodegradable organic material present. If the BOD of wastewater discharged into a stream or lake exceeds the oxygen content of that water, the oxygen will be used up and the stream or lake will become septic at the discharge area. Fish and aquatic plants cannot survive in such conditions.

BOD is determined by diluting a wastewater sample with water with known dissolved-oxygen content and storing the mixture for 5 days at 20°C. The oxygen content at the end of the period is measured, and the difference is reported as the BOD.

At the end of a period of t days at 20°C

$$BOD = O(1 - 10^{-K_1 t}) \tag{8-12}$$

where O = oxygen demand when $t = 0$ or at start of any oxidation period

K_1 = deoxygenation coefficient, usually about 0.1 for sewage, but may range from less than 0.05 to more than 0.2. For temperatures other than 20°C, multiply K_1 for 20°C by 1.047^{T-20}

T = temperature, °C

To obtain the initial oxygen demand at temperatures other than 20°C, multiply O for 20°C by $0.02T + 0.6$.

The load on a receiving body of water may be estimated from the size of contributing population. For example, the 5-day BOD, lb/day per capita, may be assumed as 0.2 for domestic wastewater, 0.3 for combined sewage, and 0.5 for combined sewage with large amounts of industrial wastewater.

Sometimes, the wastewater concentration is expressed as **population equivalent**, the number of persons required to create the total oxygen demand of the wastewater per day. For example, suppose domestic wastewater has a BOD of 5000 lb/day. The population equivalent then may be taken as $5000/0.2 = 25,000$ persons.

As an example of the use of BOD, consider a residential community of 100,000 persons producing a wastewater flow of 25 mgd to be disposed of in a river with no BOD and a dissolved-oxygen content of 10 ppm. Permissible oxygen content downstream is 6.5 ppm. What should the flow in the river be?

The total oxygen demand may be assumed to be $100,000 \times 0.2 = 20,000$ lb/day. Since a gallon of water weighs 8.33 lb, this for 25 mgd of wastewater is equivalent to

$$\frac{20,000}{25 \times 8.33} = 96 \text{ ppm or mg/L}$$

The required river flow Q, mgd, must supply this oxygen. Hence,

$$8.33(10 - 6.5)Q = 20,000 \quad \text{and} \quad Q = 686 \text{ mgd}$$

Some of the organic material, such as pesticides, in wastewater may not be biologically degradable. They are not measured by BOD. Some of these materials may have adverse long-term effects on living things and can create undesirable taste, odors, and colors in a receiving body of water. Chemical oxygen demand (COD) is a measure of the quantities of such materials present in the water. COD, however, as measured in a COD test, also includes the demand of biologically degradable materials because more compounds can be oxidized chemically than biologically. Hence, the COD is larger than the BOD. Treatments are available for removing COD and BOD from wastewater.

Relative stability is a measure of the amount of oxygen needed for stabilization of a sewage-treatment-plant effluent. Table 8-5 shows how relative stability varies with storage time at 20°C. The table indicates that the aerobic process is nearly completed after 20 days. If the time required to exhaust oxygen from an effluent is known, the relative stability given by Table 8-5 also is taken as the percent of the initial oxygen demand O that has been satisfied.

TABLE 8-5 Relative Stability of Treatment-Plant Effluent

Time at 20°C, or time required for decolorization of methylene blue, days	Proportion oxidized or relative stability, %	Time at 20°C, or time required for decolorization of methylene blue, days	Proportion oxidized or relative stability, %
0.5	11	8.0	84
1.0	21	9.0	87
1.5	30	10.0	90
2.0	37	11.0	92
2.5	44	12.0	94
3.0	50	13.0	95
4.0	60	14.0	96
5.0	68	16.0	97
6.0	75	18.0	98
7.0	80	20.0	99

The time can be determined by adding to a sample of an effluent a small amount of methylene blue, an aniline dye. On exhaustion of the oxygen in the sample, anaerobic bacteria become active. They release enzymes that remove the color from the dye. The time required at 20°C for this to take place may be used with Table 8-5 to determine the percent of organic material stabilized. For example, a sample that decolorizes in 5 days has a relative stability of 68%. Only 32% of the initial oxygen demand remains. Such an effluent may be stable enough to be discharged into a stream.

Since concentration and composition of wastewater vary considerably throughout a day, care must be taken to obtain a representative sample for each type of test. Sampling and analyses should be made as directed in Standard Methods for the Examination of Water and Sewage, American Public Health Association, 1015 18th St., N.W., Washington, D.C. 20036; American Water Works Association, 6666 Quincy Ave., W., Denver, Col. 80235; Water Pollution Control Federation, 3900 Wisconsin Ave., Washington, D.C. 20016.

(H. S. Azad, "Industrial Wastewater Management Handbook," Metcalf & Eddy, Inc., "Wastewater Engineering," E. W. Steel and T. J. McGhee, "Water Supply and Sewerage," and R. E. McKinney, "Microbiology for Sanitary Engineers," McGraw-Hill Book Company, New York; G. M. Fair, J. C. Geyer, and D. A. Okun, "Elements of Water Supply and Wastewater Disposal," John Wiley & Sons, Inc., New York; L. D. Benefield and C. W. Randall, "Biological Process Design for Wastewater Treatment," Prentice-Hall, Inc., Englewood Cliffs, N.J.)

8-16. Wastewater Treatment and Disposal

Because of the objectionable characteristics of raw sewage (Art. 8-15), disposal requires consideration of many factors, especially health hazards; odors, appearance, and other nuisance conditions; and economics. Rarely do conditions exist that permit low-cost disposal of raw sewage. Usually, some degree of treatment is necessary. Selection of the type and degree of treatment depends on the nature of the raw wastewater, effluent quality after treatment, initial cost of the treatment plant, costs of operation and maintenance, process reliability, capability for disposal of sludge produced, potential for air pollution from pollutants removed, treatment chemicals required, energy consumed in the process, space requirements for the treatment plant, and potential hazards within the plant and in the surrounding area if the plant should malfunction or during transport of materials to and from the plant.

Several methods are used for disposal of wastewater on land. Among them are oxidation ponds, or lagoons; irrigation; incineration; burial; composting; and dewatering and conversion into fertilizer.

Irrigation is of importance because it permits reclamation of the water content, to replenish the groundwater. Surface, flood, or subsurface irrigation may be used: *Surface irrigation* discharges wastewater on the ground. Part evaporates and part percolates into the ground, but a sizable amount remains on the surface and must be collected in surface drainage channels. For domestic sewage, the method is not efficient. A modification, *spray irrigation*, however, has been used successfully for some industrial wastes. *Flood irrigation* also discharges the sewage on the ground, but the sewage seeps down and is usually collected in underdrains. The soil acts as a filter and partly purifies the waste. But unless the sewage is treated before irrigation, odors and insects may be produced, the soil may become clogged by grease or soap, and surface and groundwater may become contaminated. Surface irrigation sometimes is used for watering and fertilizing crops. This application, however, may create potential health hazards unless treatment has stabilized and disinfected the effluent. Another form of irrigation, *subsurface irrigation*, often is used with cesspools (Art. 8-25) and septic tanks (Art. 8-24).

Self-Purification ▪ Wastewater, with or without treatment, has been disposed of by dilution in a natural body of water. Partial or complete treatment then takes place in the water. Sometimes self-purification occurs; more often, if the wastewater has not had adequate treatment, the body of water becomes polluted. It may be unsafe for water supply and swimming, may contaminate or kill fish and shellfish, and may produce odors and have an unpleasant appearance. Therefore, treatment consistent with the self-purification characteristics of the body of water is desirable and is usually required by law. Secondary treatment now is required in most states. Requirements for tertiary treatment may be imposed to protect stream-water quality.

In polluted water, decomposition of organic matter utilizes oxygen from the water. If there is an adequate supply of oxygen, the BOD may be satisfied while enough dissolved oxygen remains to support fish life. If not, anaerobic decomposition will occur (Art. 8-15); the water becomes septic and malodorous and unable to support fish life.

Unpolluted water usually is saturated with oxygen. Table 8-6 shows the amount of oxygen that fresh water can hold in solution at various temperatures. The saturation quantity also depends on the concentration of dissolved substances. Salt water, for example, holds about 80% as much oxygen as fresh water.

Oxygen deficit D is the difference between saturation content and actual content, ppm or mg/L. As oxygen is removed from the water, the loss is offset by absorption of atmospheric oxygen at the surface. The rate at which this reaeration occurs depends on deficit D, the amount of turbulence, and the ratio of volume of water to the surface area. At any time t, days,

$$D = \frac{K_1 O}{K_2 - K_1} (10^{-K_1 t} - 10^{-K_2 t}) + 10^{-K_2 t} D_o \qquad (8\text{-}13)$$

where K_1 = coefficient of deoxygenation [see Eq. (8-12)]

K_2 = reaeration coefficient, which ranges from 0.05 to 0.5 at 20°C, depending on depth,

TABLE 8-6   Solubility of Oxygen in Fresh Water at Sea Level

Temperature		Dissolved oxygen, ppm or mg per liter	Temperature		Dissolved oxygen, ppm or mg per liter
°C	°F		°C	°F	
1	33.8	14.23	16	60.8	9.95
2	35.6	13.84	17	62.6	9.74
3	37.4	13.48	18	64.4	9.54
4	39.2	13.13	19	66.2	9.35
5	41.0	12.80	20	68.0	9.17
6	42.8	12.48	21	69.8	8.99
7	44.6	12.17	22	71.6	8.83
8	46.4	11.87	23	73.4	8.68
9	48.2	11.59	24	75.2	8.53
10	50.0	11.33	25	77.0	8.38
11	51.8	11.08	26	78.8	8.22
12	53.6	10.83	27	80.6	8.07
13	55.4	10.60	28	82.4	7.92
14	57.2	10.37	29	84.2	7.77
15	59.0	10.15	30	86.0	7.63

velocity, and turbulence of the water. For temperatures other than 20°C, multiply K_2 by 1.047^{T-20}

T = temperature, °C

O = oxygen demand at $t = 0$, ppm or mg/L

D_o = oxygen deficit at point of pollution, or $t = 0$, ppm or mg/L

(H. W. Streeter, "The Role of Atmospheric Reaeration of Sewage-Polluted Streams," *Transactions, American Society of Civil Engineers,* vol. 89, p. 1355, 1926.)

In Fig. 8-9, the deoxygenation curve indicates the amount of dissolved oxygen remaining at any time as sewage with initial demand O stabilizes, if the supply of oxygen is not replenished. The reaeration curve shows the amount of new oxygen dissolved during the same period. The oxygen sag curve represents at any given time the dissolved oxygen present, the sum of the remaining oxygen after deoxygenation and the oxygen from reaeration. The oxygen deficit D, as given by Eq. (8-13), is the ordinate of the oxygen sag curve measured from the horizontal line representing oxygen content at saturation.

The lowest or critical point of the sag curve indicates the occurrence of minimum dissolved oxygen, or maximum deficit. The time at which this occurs may be calculated from

$$t_c = \frac{1}{K_1(f-1)} \log f \left[1 - \frac{D_o}{O}(f-1) \right] \tag{8-14a}$$

where $f = K_2/K_1$ = self-purification coefficient.

When $f = 1$,

$$t_c = \frac{0.434}{K_1} \left(1 - \frac{D_o}{O} \right) \tag{8-14b}$$

The critical deficit is given by

$$D_c = \frac{O}{f} 10^{-K_1 t_c} \tag{8-15}$$

The pollution load O that a stream may absorb depends on the value of D_c, coefficients f and K_1, and the initial deficit D_o. The allowable value of D_c usually is established by law. The initial deficit is determined by existing pollution. The coefficients may be estimated from tests on the wastewater and the receiving body of water, or values may be assigned based on experience. Seasonal variations in temperature and water level or stream flow affect the amount of oxygen the water can hold and the amount of water available for dilution. Hence, the most critical conditions usually occur during summer, when rainfall is low and temperatures are high.

Self-purification is slower in lakes than in streams because of the low rate of dispersion of sewage. With turbulence usually not present, mixing of water and sewage in lakes depends mostly on currents and wind. Outfalls should be designed to take advantage of conditions encouraging dispersion, to prevent sludge buildup at the discharge.

In estuaries, tides complicate dispersion. They carry pollutants back and forth many times. Salinity, density, and currents may change with time. These factors may also affect dispersion in ocean waters. Special care is necessary in outfall design to promote mixing and to take advantage of currents.

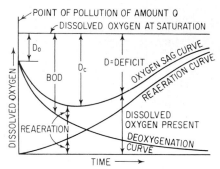

Fig. 8-9.   Curves show the variation in oxygen content of a stream below a point of pollution.

Quality Standards ▪ Present legal standards for water quality for recreation and water supply are not uniform. A typical standard may limit coliforms to an average of 10/mL; 5-day BOD to an average of 3 and a maximum of 6.5 mg/L; and phenols to a maximum of 0.001 mg/L. Dissolved oxygen may be required to be at least 5 and average 6.5 mg/L, while pH must be between 6.5 and 8.5.

Stream Capacity ▪ A rough approximation of the capacity of a stream to absorb a pollutional load may be based on the **dilution factor**, the ratio of amounts of diluting water to sewage. The significance of this factor is questionable. A few early studies indicated that nuisances would result with dilution factors for untreated sewage of less than 20 and might occur under unfavorable conditions with factors up to 40. These findings are no longer generally accepted. Use of Eq. (8-15) is preferred.

Types of Treatment ▪ The disposal problem generally makes some treatment of wastewater necessary.

Wastewater treatment is any process to which wastewater is subjected to remove or alter its objectionable constituents and thus render it less offensive or dangerous. Treatment may be classified as preliminary, primary, secondary, or tertiary or advanced, depending on the degree of processing.

Preliminary treatment or pretreatment may be the conditioning of industrial waste before discharge to remove or neutralize substances injurious to sewers and treatment processes, or it may be unit operations that prepare the wastewater for major treatment.

Primary treatment is the first and sometimes the only treatment of wastewater. This process removes floating solids and suspended solids, both fine and coarse. If a plant provides only primary treatment, the effluent is considered only partly treated.

Secondary treatment applies biological methods to the effluent from primary treatment. Organic matter still present is stabilized by aerobic processes.

Tertiary or complete treatment removes a high percentage of suspended, colloidal, and organic matter. The wastewater also may be disinfected.

Advanced waste treatment is any physical, chemical, or biological process that accomplishes a degree of treatment higher than secondary.

TABLE 8-7 Efficiencies of Wastewater Treatment Methods*

Type of treatment	% reduction		
	Suspended matter	BOD	Bacteria
Fine screens	5–20		10–20
Plain sedimentation	35–65	25–40	50–60
Chemical precipitation	75–90	60–85	70–90
Low-rate trickling filter, including presedimentation and final sedimentation	70–90+	75–90	90+
High-rate trickling filter, including presedimentation and final sedimentation	70–90	65–95	70–95
Conventional activated sludge, including presedimentation and final sedimentation	80–95	80–95	90–95+
High-rate activated sludge, including presedimentation and final sedimentation	70–90	70–95	80–95
Contact aeration, including presedimentation and final sedimentation	80–95	80–95	90–95+
Intermittent sand filtration, including presedimentation	90–95	85–95	95+
Chlorination:			
Settled sewage		†	90–95
Biologically treated sewage		†	98–99

*From E. W. Steel, "Water Supply and Sewerage," McGraw-Hill Book Company, New York.
†Reduction is dependent on dosage.

Efficiency of treatment depends on quality of plant design and operation and on type and strength of sewage. Table 8-7 lists efficiencies for commonly used treatment methods in terms of percent reduction of suspended solids, bacteria, and BOD.

Details of treatment methods are given in the following articles.

(H. S. Azad, "Industrial Wastewater Management Handbook," Metcalf & Eddy, Inc., "Wastewater Engineering," and E. W. Steel and T. J. McGhee, "Water Supply and Sewerage," McGraw-Hill Book Company, New York; R. L. Culp, G. M. Wesner, and G. L. Culp, "Advanced Wastewater Treatment," Van Nostrand Reinhold Company, New York; H. W. Parker, "Wastewater Systems Engineering," Prentice-Hall, Inc., Englewood Cliffs, N.J.; "Sewage Treatment Plant Design," Manual 36, "Sanitary Landfill," Manual 39, and "Glossary—Water and Wastewater Control Engineering," American Society of Civil Engineers.)

8-17. Wastewater Pretreatment

The purpose of pretreatment is to remove from wastewater coarse materials that may interfere with treatment, or do not respond to treatment, or may damage or clog pumps, pipes, valves, and nozzles. Various types of screening devices are used for this purpose. Generally, they are the first units in a treatment plant.

Racks are fixed screens composed of parallel bars, set vertically or sloped in the direction of flow, to catch debris. Coarse racks have spaces between the bars of 2 in or more. They usually are used for large plants to protect sewage pumps. Medium racks, used more frequently, have bar spacings of ½ to 1½ in. They may be fixed or movable. Movable racks are three-sided cages. Sewage enters through the open side and leaves through the bars. One cage is periodically hoisted to the surface for manual cleaning, while sewage passes through a second cage. Fixed-bar racks may be manually or mechanically cleaned. The bars may be curved to the horizontal at the top to facilitate cleaning.

While a minimum velocity of about 2 ft/s is desirable in the approach channel to prevent sediment from clogging it, velocity through a rack should be lower, perhaps 0.5 to 1 ft/s, so that objects should not be forced through. This requires enlargement of the conduit in the vicinity of the rack. To allow for head loss through a rack, the conduit bottom may be lowered below the rack 3 to 6 in.

Fine screens, with uniform-size openings or slots ⅛ in wide or less, have low efficiency for sewage treatment but are useful for removal of bulky and fibrous materials from industrial wastes. Generally, fine screens are movable and mechanically cleaned. Various types are used: rotating disk or drum, band, plate, or vibratory screens.

Screenings may be disposed of by burial, incineration, or digestion. Digestion of sludge proceeds normally when fine screenings are added in sludge-digestion tanks. In some treatment plants, screenings are passed through a grinder and returned to the flow, to settle out subsequently in a sedimentation tank. Screening and cutting are combined in such devices as comminutors, barminutors, and griductors. Their high-speed rotating edges cut through the sewage flow and chop and shred the solids, which then pass on to a sedimentation tank. Shearing-type units should be located after a grit chamber to prevent excessive wear of cutting edges.

Skimming tanks also may be placed ahead of sedimentation tanks. Skimmers remove oil and grease, which tend to form scum, clog fine screens, obstruct filters, and reduce the efficiency of activated sludge. Compressed air, applied through porous plates in the bottom of the tank, coagulates the grease and oil and causes them to rise to the surface. About 0.1 ft^3 of air is required per gallon. Detention period ranges from 5 to 15 min. About 2 mg/L of chlorine increases the efficiency of grease removal. After the effluent reaches the sedimentation tank, the coagulated material is removed with the scum or settled solids.

(E. Besselievre and M. Schwartz, "The Treatment of Industrial Wastes," Metcalf & Eddy, Inc., "Wastewater Engineering," L. Rich, "Wastewater Treatment Systems," and E. Schroeder, "Water and Wastewater Engineering Treatment," McGraw-Hill Book Company, New York.)

8-18. Sedimentation

At most wastewater-treatment plants, sedimentation is a primary treatment. In activated-sludge plants, sedimentation is required after oxidation. It also is used after oxidation of wastewater on trickling filters.

The major objective of sedimentation is removal of settleable solids. But often, some floating materials also are removed by clarifier, skimming devices built into sedimentation tanks. These processes occur while wastewater moves slowly through a settling basin.

Efficiency of a sedimentation tank depends on particle size, specific gravity, and settling velocity and on several other factors: concentration of suspended matter, temperature, surface area of the liquid, retention period, depth and shape of basin, baffling, total length of flow, wind, and biological effects. Density currents and short circuiting may negate theoretical detention computations. Improper baffling may reduce the effective surface area of the liquid and create dead or nonflow areas within the tank. In general, a settling tank of good design should have an efficiency in the upper range of that given in Table 8-7.

Settling velocity of a particle is a function of the specific gravity and diameter of the particle, and specific gravity and viscosity of fluid. Settling rates of particles larger than 200 μm are determined empirically. Sizes less than 200 μm settle in accordance with Stokes' law for drag of small settling spheres in a viscous fluid.

Theoretically, if the forward motion of the water is less than the vertical settling rate of all the particles, they will settle some distance below the surface in a given time interval while in the tank. After that period, if the surface layer of water were removed, it would contain no solids.

Surface settling rate, or overflow rate, gal/ft^2 of surface area per day, is a measure of the rate of flow through the basin when the rate of flow, ft^3/s, equals the surface area, ft^2, times the settling velocity ft/s, of the smallest particle to be removed. Hence, selection of an overflow or surface settling rate establishes a relationship between flow and area.

Detention period is the theoretical time water is detained in a basin. The average detention period is V/Q, where Q is the flow, mgd or ft^3/s, and V, the basin volume. Since most of the settleable solids will settle out in 1 to 2 h, long detention periods are not advantageous. In fact, they are undesirable because the wastewater may become septic.

The flowing-through period is the time required for wastewater to pass through the basin. This time may be estimated by adding sodium chloride (or dye) to the influent and testing the effluent for increase of chloride or checking for dye. The flowing-through period should be at least 30% of the theoretical detention period. Dye may be used to follow the flow pattern.

Grit chambers (Fig. 8-10) are settling basins used to remove coarse inorganic solids. They may also trap heavier organic material, such as seeds. Grit chambers are necessary with com-

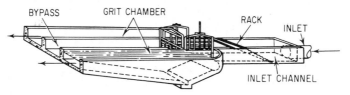

Fig. 8-10.   Grit chamber.

bined sewers if the flow is to be treated. The wet-weather flow usually contains sand and grit, which must be removed to prevent damage to pumps and interference with wastewater treatment.

Design of a grit chamber should insure settlement of all particles over 0.2 mm in size but should not remove organic solids. Flow should be fast enough to secure this result but without scouring solids already deposited. Scour will occur if the horizontal velocity, ft/s, of the sewage exceeds

$$v = 2.2 \sqrt{\frac{gd}{f}(s-1)} \qquad (8\text{-}16)$$

where f = roughness coefficient (Darcy formula for flow in pipes) for chamber (see Fig. 7-19)

   g = acceleration due to gravity, 32.2 ft/s^2

   d = particle diameter, ft

   s = specific gravity of particle

Usually, grit chambers are designed for a flow of about 1 ft/s. Flow may be controlled by specially shaped gates or weirs to keep velocity constant. The material settling out may be removed manually or mechanically. Also, devices may be added to mechanically cleaned units to wash most of the organic material out of the grit.

A plain sedimentation tank is a settling basin where sedimentation is not aided by coagulants and the settled solids, or sludge, are not retained for digestion. Generally, sludge and scum are removed mechanically. Any of several methods may be used to remove light, suspended material.

Flocculent suspensions have little or no settling velocity. Although they may occur in raw sewage, more frequently they are encountered when effluents from activated-sludge units undergo secondary settling. The suspensions may be removed by passing inflowing sewage upward through a blanket of the flocculent material (vertical-flow sedimentation tank). The objective is to produce a mechanical sweeping action in which small particles attach to larger particles, which then have sufficient weight to settle. Another removal method employs an inner chamber equipped with baffles that rotate and stir the liquid, to aid formation of larger, heavier floc. The same results also may be achieved by agitation with air. Some of the settled sludge is raised by air lift and mixed with the floc, to form a conglomerate with better settling characteristics. Variations utilizing the preceding principles have been introduced by several manufacturers, for example, the up-flow tube clarifier.

Design of a sedimentation tank should be based on the settling velocity of the smallest particle to be removed. Depth should be no larger than necessary for preventing scour and to accommodate cleaning mechanisms. Surface area of the liquid is more important than depth, so depth usually is held to 10 ft or less (at sidewalls). The surface-settling-rate requirement generally is 600 gal/(ft^2 · day) for primary treatment alone and 800 to 1000 for all other tanks. The detention period normally is 2 h. These three design parameters must be adjusted since each is dependent on the other for a given design flow (average daily flow for a plant).

Rectangular tanks are built in units with common walls. Width per unit ranges up to 25 ft. Minimum length should be at least 10 ft. The length-width ratio should not exceed 5:1. Final sizes may be determined by dimensions of available sludge-removal equipment.

Provision should be made for sludge removal on a regular schedule. If sludge is not removed, gasification occurs, and large blocks of sludge appear on the surface. These must then be broken up so that they will settle, or they must be removed by the scum-removal mechanism. In circular tanks (Fig. 8-11) radial blades scrape the bottom to move the sludge to a central sludge hopper. In rectangular tanks, the hopper is located near the inlet end since the heaviest sludge accumulation occurs in that region. Blades moving along the bottom against the flow of sewage push the sludge to the hopper. In some tanks, the same blades may be lifted to the surface and, traveling with the sewage flow, move scum to the outlet end. There, the scum may be trapped by a baffle until taken out by a scum-removal device.

Many mechanical aids for use with sedimentation tanks are available commercially. Manufacturers' literature should be carefully studied and specifications written to insure procurement of equipment exactly meeting design requirements.

Actual flowing-through time is influenced by inlet and outlet construction. For circular tanks, inlets are submerged, at the center (Fig. 8-11). Sewage rises inside a baffle extending downward, to still the currents. The outlet device nearly always is a circumferential weir adjusted to level after installation. The weir may be sharp-edged and level or provided with V notches about 1 ft or less apart. The notches permit more constant flow since they are less affected by local differences in weir elevation and surface tension. For rectangular tanks, inlets

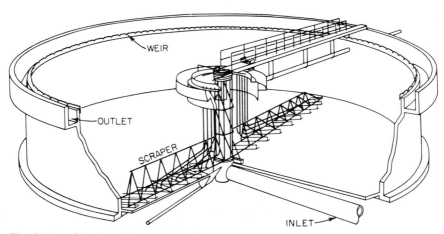

Fig. 8-11. Circular sedimentation tank.

also may be submerged, but at one end. More often, the sewage is brought to a trough that has a weir extending the width of the tank. The flow then moves forward with less short circuiting. At the outlet, to provide enough weir length, a launder is used. This consists of a series of fingerlike shallow conduits set at water level and receiving flow from both sides. Each finger is connected to a common discharge trough. Normal weir loading should not exceed 10,000 gal/ft of weir per day in small plants, or 15,000 in units handling more than 1 mgd.

Chemical precipitation sometimes is used to improve the effluent from sedimentation. The process is similar to that for water clarification. The high cost of chemicals and the intermediate grade of treatment obtained with chemicals have kept the process from general use. Chemical precipitation has, however, been found useful in specialized treatment. Phosphorus removal, preparation of sludges for filtration or dewatering, and removal of trace metals are examples of such treatment.

Alum, ferric chloride, ferric sulfate, lime, sodium aluminate, ferrous chloride, ferrous sulfate, and polyelectrolytes are chemicals used to expedite precipitation. The coagulation resulting is, actually, the result of a complex group of reactions involving the hydrolysis products of the added chemicals. Effectiveness of the various chemicals depends on the conditions under which they are used and the types of wastes.

There has to be an optimum pH and an optimum dosage for efficient wastewater coagulation. Consequently, dosages are often determined by trial (jar tests). Measurement of zeta potential (an electrical potential related to particle stability and hence useful in controlling coagulation) and of phosphate content is also desirable.

Design requirements include rapid mixing, mixer-blade peripheral speeds less than 5 ft/s, control of slurry concentration, minimum sludge blanket levels, and controlled horizontal movement of clearer water by launder or weir spacing and by weir overflow rate control.

(Metcalf & Eddy, Inc., "Wastewater Engineering," McGraw-Hill Book Company, New York; G. M. Fair, J. C. Geyer, and D. A. Okun, "Elements of Water Supply and Wastewater Disposal," John Wiley & Sons, Inc., New York; "Wastewater Treatment Plant Design," MOP 8, Water Pollution Control Federation, Washington, D.C.; R. L. Culp, G. M. Wesner, and G. L. Culp, "Advanced Wastewater Treatment," Van Nostrand Reinhold Company, New York; H. W. Parker, "Wastewater Systems Engineering," Prentice-Hall, Inc., Englewood Cliffs, N.J.; "Water Treatment Plant Design," American Water Works Association, Denver, Colo.)

8-19. Wastewater Filtration

Secondary treatments frequently employ oxidation to decompose and stabilize the putrescible matter remaining after primary treatments. Filtration is one of these secondary treatments. Others include the activated-sludge process, oxidation ponds, and irrigation. These oxidation methods bring organic matter in wastewater into immediate contact with microorganisms under aerobic conditions. In filtration, the microorganisms coat the filtering media. As the wastewater flows through, adsorption occurs, and most of the organic materials are removed by contact with the coating. The organisms decompose organic nitrogen compounds and destroy carbohydrates. Efficiency of the method, as measured by reduction of BOD, is high (Table 8-7).

Intermittent sand filters are sand beds, usually 2½ to 3 ft deep, with underdrains for collecting and carrying off the effluent. Settled wastewater, the effluent from a sedimentation tank, is applied to the sand surface in intermittent doses. A rest period between doses allows time for air to assist in oxidation of the organic matter. Application rates generally range from 20,000 gallons per acre per day (gad) to 125,000 gad when the filters serve as a secondary treatment. Rates may go as high as 0.5 million gallons per acre per day (mgad) for tertiary treatments.

Sand for an intermittent filter should have a uniformity coefficient of 5 or less; 3.5 is preferred. (**Uniformity coefficient** is the ratio of the sieve size that will pass 60% of the material to the effective size of the sand. **Effective size** is the size, mm, of the sieve that passes 10%, by

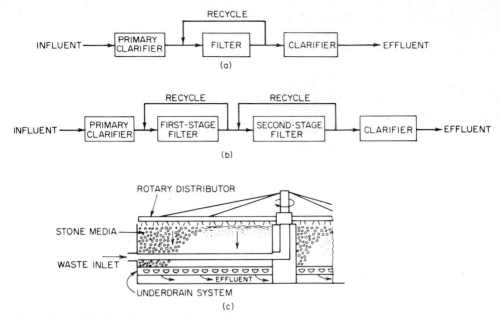

Fig. 8-12. Trickling filters supply bacteria for consumption of organic matter in wastewater. (*a*) Schematic of process with single-stage filtration. (*b*) Schematic of process with two-stage filtration. (*c*) Cross section of trickling filter with a rotary distributor of wastewater.

weight, of the sand.) The effective size of the sand should be between 0.2 and 0.5 mm. A bed of gravel 6 to 12 in thick usually underlies the sand.

A mat of solids forms on the filter surface and must be removed periodically. Generally, the mat can be scraped off when dry, but occasionally the top 6 in or so of the filter material must be replaced.

In winter, there is danger that the sand surface will freeze. To keep the filter in operation, the bed may be ridged on 3-ft centers to support the ice while the wastewater flows underneath it.

Granular filters may be adapted from potable backwash sand filter types to the treatment of secondary effluents from wastewater-treatment plants. The design should provide 6- to 8-h runs with terminal head loss less than 10 ft. Filtration rates above 3 gal/(min · ft^2) lead to high terminal head loss and short runs. Experience gained in filtration of potable water should be used to support designs of this type. ("Water Treatment Plant Design," American Water Works Association, Denver, Colo.)

Trickling filters are beds of coarse aggregate over which settled wastewater is dropped or sprayed and through which the wastewater trickles to under drains (Fig. 8-12). Filter media include gravel, crushed rock, ceramic chapes, slag, redwood slats, or plastics. Stone and crushed rock that do not fragment, flour or soften on exposure to sewage are widely used. Generally, rock sizes are kept between 2- and 4-in nominal diameter. Underdrains collect and carry off the effluent. Filters may be ventilated through the underdrain system or by other means, to supply air to the aerobic organisms that grow on the media surfaces.

Since suspended solids can clog filters, sedimentation of the wastewater is desirable before it is fed to the filters. When, however, a waste, such as milk waste, contains a concentration of dissolved solids, it may be applied directly to a filter. In that case, preaeration is desirable, so that the waste contains some dissolved oxygen.

In time, oxidized matter breaks away from the filter media and is flushed from the filter with the effluent. Hence, the effluent is passed through a secondary settling basin, or **clarifier**. Design of these basins is similar to that of primary sedimentation tanks. Efficiency, or percent reduction of BOD, of a trickling filter generally is measured for both the filter and final sedimentation.

Trickling filters are classified as standard or low-rate, high-rate, and controlled.

Standard filters were introduced in the United States early in the twentieth century. They consisted of an underdrained bed of stones, 6 to 8 ft deep. Settled wastewater was distributed over the surface through fixed nozzles. Later, the fixed nozzles were superseded by a rotary distributor. This type of distributor has two or four radial arms supported on a center pedestal (Fig. 8-12c). Jets of sewage from nozzles on the arms cause rotation. Thus, the filter surface is sprayed as the arms revolve. Dosing, as a result, is intermittent, though the interval between doses is short, often not more than 15 s. A distributor may be kept rotating continuously by feeding the nozzles from a weir box or a dosing tank, with siphons or pumps. To accommodate rotary distribution, standard filters are built round in plan.

These low-rate filters are dosed at a rate of 1 to 4 mgad, substantially lower than that for high-rate filters. Loading also may be expressed in terms of 5-day BOD, lb/acre-ft · day. Some state health departments limit the load on a standard filter to 400 to 600 lb/acre-ft · day. The approximate load w to be applied to a filter, lb/day · acre-ft of filter volume, when the BOD of the sewage is known and a limit is specified for the BOD of the effluent, may be computed from

$$ w = 13,840 \left(\frac{B}{A - B} \right)^2 \tag{8-17} $$

where A = 5-day BOD of the influent, mg/L
$\quad\ B$ = specified maximum BOD of effluent, mg/L

High-rate filters receive a load three or more times greater than that usually applied to standard filters. Usual rate is about 20 mgad, but rates from 9 to 44 mgad have been used. Some state health departments limit the load to 2000 to 5000 lb of BOD per acre-ft per day.

Such high rates are feasible because the effluent is recirculated through the filter (Fig. 8-12a). Recirculation reduces the load on the filter, seeds the media continuously with organisms, allows continuous dosage, offsets fluctuations in sewage flow, and reduces odors by freshening the influent. Several recirculation alternatives may be used. For example, part of the filter effluent may be returned directly to the filter. (Proponents of this method of recirculation claim direct return intensifies biological oxidation.) Or part of the effluent of the filter or the final clarifier may be combined with the influent to the primary sedimentation tank. Sometimes, dual recirculation is used: The filter effluent is returned to the primary sedimentation tank, while part of the final clarifier effluent is sent back through the filter. In some cases, the sludge from the final clarifier is recirculated through the primary clarifier.

Two-stage filtration (Fig. 8-12b) may be used when a better effluent is desired than can be obtained from a single filter. For this purpose, two filters are connected in series. Various recirculation methods may be used in this case also.

The recirculation ratio, or ratio of returned effluent to sewage influent, ranges from 1:1 to about 5:1. At each passage, the amount of BOD removed decreases because response to treatment decreases. If the ratio of the decrease per passage to the BOD is given by k, then the number of effective passages of sewage through a filter may be computed from

$$ F = \frac{1 + R}{(1 + kR)^2} \tag{8-18} $$

where R = recirculation ratio. Under normal conditions, k may have a value of about 0.1.

The approximate load, lb of BOD per day per acre-ft of filter volume, to be applied to a single-stage high-rate filter or the first filter of a two-stage system, when the BOD of the sewage is known and a limit is specified for the BOD of the effluent, may be computed from

$$w = 13,840F \left(\frac{B}{A - B} \right)^2 \tag{8-19}$$

where A = 5-day BOD of influent , mg/L
  B = specified maximum BOD of effluent, mg/L
The approximate load for a second-stage filter may be estimated from

$$w = 13,840 \left(\frac{B_1}{A_1} \right)^2 \left(\frac{B_2}{A_2 - B_2} \right)^2 F \tag{8-20}$$

where A_1 = 5-day BOD of influent of first-stage filter, mg/L
  B_1 = specified maximum BOD of effluent of first-stage filter, mg/L
  A_2 = 5-day BOD of influent of second-stage filter, mg/L
  B_2 = specified maximum BOD of effluent of final clarifier, mg/L
  F = number of effective passages through second-stage filter
Equations (8-17) to (8-20) are based on formulas recommended by a committee of the National Research Council ("Sewage Treatment at Military Institutions," *Sewage Works Journal,* vol. 18, no. 5, p. 794. September 1946).

Sewage is sprayed over high-rate filters by rotary distributors or by a motor-driven disk that rains sewage continuously and uniformly over the surface. Hence, the filters are built circular.

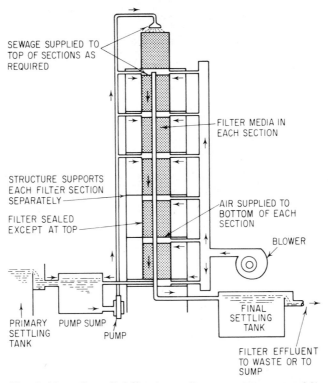

Fig. 8-13. Controlled filtration applies wastewater to tops of filter sections installed in sequence vertically. Each filter is sealed, except at the top, and has liquid inlets and outlets and an air inlet.

Controlled filters consist of sectionalized units combined into a deep filter. The loading rate with no recirculation is 10 to 12 times that for low-rate filters.

Essentials of this type of filter include sectional design, means for introduction and distribution of controlled quantities of wastewater to top or upper sections of the filter, means for introduction of controlled quantities of air under each section of filter, temperature control between 15 and 30°C, and nonabsorbing filter media of sufficient uniformity to provide both media surface and void space (Fig. 8-13).

For domestic wastes having BOD values that do not limit the rate of absorption of oxygen, hydraulic loadings may be used as a primary design parameter according to the equation

$$ n = C \left[\frac{V'(1 + R)}{Q} \right]^k \tag{8-21} $$

where n = fraction of BOD remaining
  C = constant
  V' = total filter volume, thousands of ft^3
  Q = daily flow, mgd
  R = recirculation ratio
  k = constant

Figure 8-14 can be used to select the constants C and k when $R = 0$. The V'/Q value may be read directly as the reciprocal of the filter hydraulic application rate L_H, million gal per 1000 ft^3/day (mgtcfd) since

$$ \frac{V'}{Q} = \frac{1}{L_H} \tag{8-22} $$

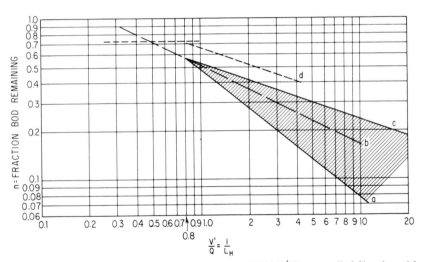

Fig. 8-14.   Curves represent the equation $n = C(V'/Q)^k$ for controlled filtration with no recirculation, where n is the fraction of BOD remaining; V' is the total filter volume, 10^3ft^3; Q is the daily flow; and C and k are constants.

Curve	C	k
a	0.48	−0.795
b	0.51	−0.482
c	0.52	−0.343
d	0.65	−0.343

When Eq. (8-22) is used in industrial-waste treatment, allowance must be made for organic loading and treatability of individual process wastes. Hence, it is advisable to develop pilot-plant information on filter application before final design.

Hydraulic surface loadings should always be greater than 70 mgad, to provide continuous washing or scouring of the filter. Unlike high-rate and low-rate filters, application of wastewater must be continuous.

Equations have been developed by engineers concerned with design and performance of trickling filters. These equations include the Velz formula (1948), Schulz formula (1960), Eckenfelder formula (1963), and Galler and Gotaas formula (1965). Each formula incorporates the influences that the investigators believed to be of primary importance.

(Metcalf & Eddy, Inc., "Wastewater Engineering," McGraw-Hill Book Company, New York; "Filtering Materials for Sewage Treatment Plants," Manual 13, and "Sewage Treatment Plant Design," Manual 36, American Society of Civil Engineers; G. M. Fair, J. C. Geyer, and D. A. Okun, "Water and Wastewater Engineering," John Wiley & Sons, Inc., New York; "Wastewater Treatment Plant Design," MOP8, Water Pollution Control Federation, Washington, D.C.)

8-20. Activated-Sludge Processes

An activated-sludge process is a biological treatment in which a mixture of wastewater and a sludge of microorganisms is agitated and aerated and from which the solids are subsequently removed and returned to the aeration process as required.

Passing air bubbles through wastewater coagulates colloids and grease, satisfies some of the BOD, and reduces ammonia nitrogen a little. Aeration also may prevent wastewater from becoming septic in a following sedimentation tank. But if wastewater is mixed with previously aerated sludge and then aerated, as is done in activated-sludge methods, the effectiveness of aeration is considerably improved. Reduction of BOD and suspended solids in the conventional activated-sludge process, including presettling and final sedimentation, may range from 80 to 95% and of coliforms, from 90 to 95% (Table 8-7). Furthermore, cost of constructing an activated-sludge plant may be competitive with other types of treatment plants producing comparable results. Unit operating costs, however, are relatively high.

The activated-sludge method is a secondary biological treatment employing oxidation to decompose and stabilize the putrescible matter remaining after primary treatments. Other oxidation methods include filtration, oxidation ponds, and irrigation. These oxidation methods bring organic matter in wastewater into immediate contact with microorganisms under aerobic conditions.

In a conventional activated-sludge plant (Fig. 8-15a), incoming wastewater first passes through a primary sedimentation tank. Activated sludge is added to the effluent from the tank, usually in the ratio of 1 part of sludge to 3 or 4 parts of settled sewage, by volume, and the mixture goes through an aeration tank. In that tank, atmospheric air is mixed with the liquid by mechanical agitation, or compressed air is diffused in the fluid by various devices: filter plates, filter tubes, ejectors, and jets. In either method, the sewage thus is brought into intimate contact with microorganisms contained in the sludge. In the first 15 to 45 min, the sludge adsorbs suspended and colloidal solids. As the organic matter is adsorbed, biological oxidation occurs. The organisms in the sludge decompose organic nitrogen compounds and destroy carbohydrates. The process proceeds rapidly at first, then falls off gradually for 2 to 5 h. After that, it continues at a nearly uniform rate for several hours. Generally, the aeration period ranges from 6 to 8 or more hours.

The aeration-tank effluent goes to a secondary sedimentation tank, where the fluid is detained, usually from 1½ to 2 h, to settle out the sludge. The effluent from this tank is completely treated and, after chlorination, may be safely discharged.

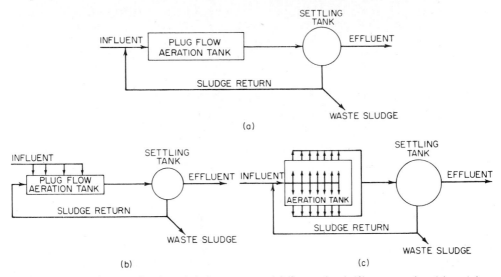

Fig. 8-15.   Schematics of activated-sludge processes: (*a*) Conventional; (*b*) step aeration; (*c*) complete mix. *(From "Environmental Pollution Control Alternatives: Municipal Wastewater," Environmental Protection Agency, Cincinnati, Ohio.)*

About 25 to 35% of the sludge from the final sedimentation tank is returned for recirculation with incoming sewage. Sludge should not be detained in the tank. Frequent removal (at intervals of less than 1 h) or continuous removal is necessary to avoid deaeration.

Overflow rates for final sedimentation normally range from about 800 gal/(ft^2 · day) for small plants to 1000 for plants of over 2-mgd capacity. Weir loadings preferably should not exceed 10,000 gal/(lin ft · day). When tank volume required exceeds 2500 ft^3, multiple sedimentation tanks are desirable.

Multiple aeration tanks are required when total tank volume exceeds 5000 ft^3. Aeration tanks in which compressed air is used generally are long and narrow. To conserve space, the channel may be turned 180° several times, with a common wall between the flow in opposing directions. An air main is generally run along the top of the wall to feed diffusers (Fig. 8-16*a* and *b*) or porous plates (Fig. 8-16*c*) along its length. The air sets up a spiral motion in the liquid as it flows through the tanks. This agitation reduces air requirements.

Width of channel ranges from 15 to 30 ft. Depth is about 15 ft.

Dissolved oxygen should be maintained at 2 ppm (mg/L) or more. Air requirements normally range from 0.2 to 1.5 ft^3/gal of wastewater treated. Most state authorities require a minimum of 1000 ft^3 of air per lb of applied BOD per day.

Mechanical aeration may be done in square, rectangular, or circular tanks, depending on the mechanism employed for agitation. In some plants, the fluid may be drawn up vertical tubes and discharged in thin sheets at the top, or the liquid may pass down draft tubes while air is bubbled through it. In both methods, agitation at the surface produced by the movement of the liquid increases aeration. Detention periods generally are longer, 8 h or more, than for tanks with diffused air.

Several modifications of the activated-sludge method, seeking to improve performance or cut costs, are in use. These include modified, activated, tapered, step, and complete-mix aeration, and the Kraus, biosorption, and bioactivation processes.

Modified aeration decreases the aeration period to 3 h or less and holds return sludge to a low proportion. Results are intermediate between primary sedimentation and full secondary treatment.

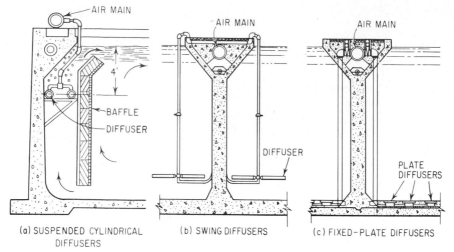

(a) SUSPENDED CYLINDRICAL DIFFUSERS

(b) SWING DIFFUSERS

(c) FIXED–PLATE DIFFUSERS

Fig. 8-16.   Air main atop aeration-tank walls supplies air to diffusers in adjoining channels in which the mixture of activated sludge and sedimentation-tank effluent flows.

Activated aeration places aeration tanks in parallel. The activated sludge from one final sedimentation tank or group of such tanks is added to the influent of the aeration tanks. Other sludge is concentrated and removed. With much less air, results are better than with modified aeration.

Tapered aeration differs from conventional in that air diffusers are not uniformly spaced. Instead, more diffusers are placed near the inlet end of the aeration tanks than near the outlet. The theory is that oxygen demand is greater near the inlet, and so the efficiency of the treatment should be improved if more air is supplied there. However, results depend on degree of longitudinal mixing, rate of sludge return, and characteristics of recirculated matter, for example, air content of sludge or mixed liquor.

Step aeration adds sewage at four or more points in an aeration tank (Fig. 8-15b). Each increment reacts with sludge already present in the tank. Thus, air requirements are nearly uniform throughout the tank. **Complete-mix aeration** (Fig. 8-15c) obtains better results by dispersing the influent wastewater as uniformly as possible along the entire length of the aeration basin, to produce a uniform oxygen demand throughout. **Extended aeration** is similar, but the wastewater is aerated for 24 h instead of the conventional 6 to 8 h.

The Kraus process adds to the sewage an aerated mixture of activated sludge and material from sludge digester tanks. The **biosorption** process mixes sewage with sludge preaerated in a separate tank. The **bioactivation process** uses primary sedimentation, a trickling filter, and short secondary sedimentation, then adds activated sludge and passes the mixture through aeration and final sedimentation tanks.

Excellent results have been obtained by substituting oxygen for air in the activated-sludge process. For efficient use of the oxygen, the aeration tanks may be covered. The oxygen may then be recirculated through several stages, entering the first stage of the process and flowing through the oxygenation basin with the wastewater being treated. Pressure under the tank cover is close to atmospheric and enough to maintain control and prevent backmixing of successive stages. Within each stage, mixing may be achieved with surface aerators or a submerged rotating sprayer. Pure oxygen permits use of smaller tanks, and oxygenation time may be 1½ to 2 h instead of the conventional 6 to 8 h. The activated sludge produced settles more easily and is easier to dewater than that from conventional processes.

Activated-sludge plants should be closely controlled for optimum performance. This requires frequent checking of the sludge content of the mixed liquor. Solids usually are limited to 1500 to 2500 ppm (mg/L) in diffused-air plants and about 1000 ppm when mechanical agitation is used. Settling characteristics of the sludge are indicated by the Mohlman index:

$$\text{Mohlman index} = \frac{\text{volume of sludge settled in 30 min, \%}}{\text{volume of suspended solids, \%}} \tag{8-23}$$

A good settling sludge has an index below 100. An alternative measure is the sludge density index, 100 divided by the Mohlman index. Operating control may be maintained by holding a constant mixed-liquor suspended-solids (MLSS) or volatile-suspended-solids (MLVSS) concentration, by holding a constant ratio of food to microorganisms (F : M), or by holding a constant mean cell residence time (MCRT). The latter may be the simplest because only suspended solids concentration in the aeration basin and in the waste activated sludge need be measured.

Sludge age is another important factor. It is the average time that a particle of suspended solids remains under aeration. Sludge age is measured by the ratio of dry weight of sludge in the aeration tank, lb, to the suspended-solids load, lb/day, of the incoming wastewater. In a well-operated activated-sludge plant, sludge age is 3 to 5 days. But it may be only 0.3 days for a modified process that is well operated.

(Metcalf & Eddy, Inc., "Wastewater Engineering," and L. Rich, "Wastewater Treatment Systems," McGraw-Hill Book Company, New York; "Sewage Treatment Plant Design," Manual 36, American Society of Civil Engineers; G. M. Fair, J. C. Geyer, and D. A. Okun, "Water and Wastewater Engineering," John Wiley & Sons, Inc., New York.)

8-21. Contact Stabilization

This is a secondary treatment similar to the activated-sludge method. Contact stabilization (Fig. 8-17) also uses air diffusion to supply oxygen and keep a suspension containing microorganisms thoroughly mixed with incoming sewage. In addition, active growths of microorganisms are maintained on plates of impervious material, such as asbestos cement, suspended in the mixing liquor of the aeration tank. Slime growth forms on the plates, and liquid passing by furnishes the organisms on the plates with nutrients. The organisms decompose organic nitrogen compounds and destroy carbohydrates.

Plates may be fixed or rotate about a horizontal axis. As they rotate, biological growth adheres to them and is alternately immersed in waste liquid and exposed to the air. This alternation insures an aerobic condition for growth.

The aeration period in contact aerators may be 5 h or more. Aeration usually is preceded by 1 h of preaeration of the raw sewage and return sludge before primary settling. The load on the contact aerator is based on two factors: pounds per day of BOD per 1000 ft² of contact surfaces

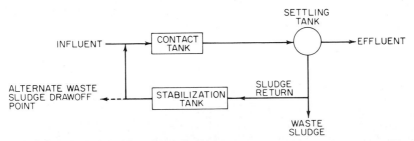

Fig. 8-17. Schematic of the contact stabilization process. *(From "Environmental Pollution Control Alternatives: Municipal Wastewater," Environmental Protection Agency, Cincinnati, Ohio.)*

(6.0 or less) and pounds per day of BOD per 1000 ft^2/h of aeration (1.2 or less). About 1.5 ft^3 of air per gal of flow is required. Overall plant efficiency may be about 90% BOD removal, with a higher percentage removal of suspended solids.

(H. W. Parker, "Wastewater Systems Engineering," Prentice-Hall, Inc., Englewood Cliffs, N.J.)

8-22. Sludge Treatment and Disposal

Sludge comprises the solids and accompanying liquids removed from wastewater in screening and treating it. Solids are removed as screenings, grit, primary sludge, secondary sludge, and scum. Often sludge treatment is necessary to make possible safe, economical disposal of these wastes. The treatment to be selected depends on quantity and characteristics of the sludge, nature and cost of disposal, and cost of treatment.

Screenings are putrescible and offensive. They may be disposed of by burning, burial, grinding and return to sewage, or grinding and transfer to sludge digester. The quantity of screenings is variable and dependent on sewage characteristics. Coarse screenings may range from 0.3 to 5 ft^3/million gal. Fine screenings may range from 5 to 35 ft^3/million gal.

Sand and other gritty materials also may be present in widely varying amounts. Normally, the volume will be between 1 and 10 ft^3/million gal.

Sludge varies in quantity and characteristics with the characteristics of the sewage and plant operations. Usually, more than 90% is water containing suspended solids with a specific gravity of about 1.2. Roughly, there may be about 0.20 lb of these solids per capita daily in sanitary sewage; 0.22 lb if a moderate amount of industrial wastes is present; 0.25 lb in effluents of combined sewers if considerable industrial wastes are present; and 0.32 to 0.36 lb if the sewage contains ground garbage also.

Primary sludge, derived from sedimentation tanks or the influent of digestion chambers of Imhoff tanks, is putrescible and odorous. It is composed of gray, viscous identifiable solids and has a moisture content of 95% or more. Primary treatment of 1 million gal of sewage may produce about 2500 gal of this sludge.

Trickling filter sludge is black or dark brown, granular or flocculent, and partly decomposed. It is not highly odorous when fresh. Moisture content may be about 93%. Passage of 1 million gal of sewage through a trickling filter may produce about 500 gal of this sludge.

Activated sludge is dark to golden brown, granular or flocculent, and partly decomposed. It has an earthy odor when fresh. Moisture content may be about 98%. Influent to an activated-sludge plant may yield about 13,500 gal of waste sludge per million gal.

Chemical-precipitation sludge may have a solids content more than double that of sludge from primary sedimentation. Normally, chemical precipitation from 1 million gal of sewage will yield about 5000 gal of sludge with moisture content of 95%.

Digested sludge, from septic, Imhoff, or separate digestion tanks, is very dark in color and has a homogeneous texture. When wet, it has a tarry odor. Roughly, treatment of 1 million gal of sewage will produce 800 gal of digested sludge with a moisture content of about 90%.

The sludges removed in wastewater treatment may contain as much as 97% water. The objective of sludge treatment is to separate the solids from the water and return that water to a wastewater-treatment plant for processing. Sludge treatment may require:

1. **Conditioning.** Sludge is treated with chemicals or heat so that the water may be readily separated.

2. **Thickening.** Removal of as much water as possible by gravity or flotation.

3. **Stabilization.** Processes known as sludge digestion are employed to stabilize (make less odorless and less putrescible) the organic solids in the sludge so that they can be

handled or used as soil conditioners without creating a nuisance or health hazard.

4. Dewatering. Further removal of water by drying the sludge with heat or suction.

5. Reduction. The solids are converted into a stable form by incineration or wet oxidation processes.

Sludge conditioning may employ any of several available methods to facilitate separation of the water from the solids in sludge. One method is to add a coagulant, such as ferric chloride, lime, or organic polymers, which cause the solids to clump together. Another method is to first grind the sludge and then heat it to between 350 and 450°F under pressures of 150 and 300 psi in a reactor. Under these conditions, the water contained in the solids is released. The sludge is fed from the reactor to a settling tank, where the solids are concentrated before the dewatering step. Still another conditioning method is to apply heavy doses of chlorine under pressures of 30 to 40 psi.

Sludge thickening usually is accomplished in one of two ways: settlement, or gravity thickening, or flotation thickening. Simple and inexpensive, gravity thickening is essentially a sedimentation process, employing a tank similar in appearance and action to a circular clarifier used in primary and secondary sedimentation (Fig. 8-18a). Best results are obtained with sludges from primary wastewater treatment. In flotation thickening (Fig. 8-18b), air is injected into the sludge under pressures of 40 to 80 psi. Containing large amounts of dissolved air, the sludge flows into an open tank. There, under atmospheric pressure, the dissolved air

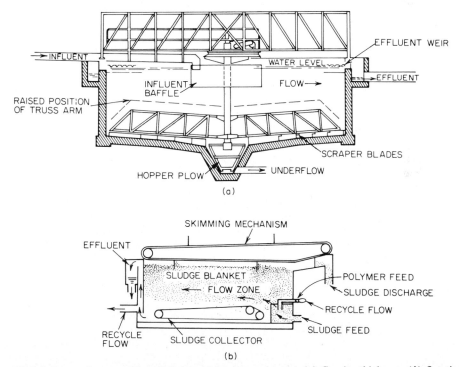

Fig. 8-18.   Cross sections of sludge-thickening equipment: (*a*) Gravity thickener; (*b*) flotation thickener.

comes out of solution as minute air bubbles. These attach themselves to solids in the sludge and float them to the surface, where a skimming mechanism removes them. This method is effective on activated sludge, which is difficult to thicken by gravity.

Sludge digestion is the biological decomposition of organic matter, which makes up about 70% of total solids, by weight, in sludge. The process results in partial gasification, liquefaction, and mineralization of the solids. It can be applied to treatment-process sludges other than chemical sludges and those containing substances toxic to sludge organisms, such as cyanides and chromium. Advantages of sludge digestion include production of a stable, inoffensive sludge (if the process is continued long enough); 35 to 45% reduction of suspended solids; 55 to 75% reduction in dry weight of volatile matter; reduction in moisture content; and production of a sludge from which water may be more easily removed. The digested sludge may be used as a soil conditioner and weak fertilizer under certain conditions. Furthermore, gases produced during digestion may be used as fuel. (If the sludge is to be dewatered and incinerated, digestion is not usually employed.) Digestion may be anaerobic, performed in closed tanks devoid of oxygen, or aerobic, with air injected into the sludge.

For anaerobic sludge digestion, sludges are transferred to separate digestion tanks, unless Imhoff-type tanks or septic tanks are used. While sludge decomposes in a digester, fresh sludge is added periodically. Anaerobic bacteria attack the carbohydrates first, forming organic acids. After this initial acid fermentation, acid digestion occurs. Organisms living in the acid environment attack the organic acids and nitrogenous matter. Then, a period of digestion, stabilization, and gasification takes place, in which the anaerobic bacteria feed on proteins and amino acids. Volatile acids are reduced, and the pH rises. In the final stage, methane fermentation occurs, with methane as the principal gaseous product. Speed of digestion is indicated by the rate of gas formation. Periodic removal of liquefied matter, excess liquor (or supernatant liquor), and digested solids makes room for fresh sludge.

Supernatant liquor, the liquid fraction in a digester, is high in solids and biochemical oxygen demand. It has an offensive odor. Withdrawn from a digester in small quantities at a level where the liquor contains relatively few solids, it is disposed of by insertion in the influent to a primary sedimentation tank.

Sludge-gas production under good operating conditions is about 12 ft³/lb of volatiles destroyed. The gas is 60 to 70% methane, 20 to 30% carbon dioxide, plus minor amounts of other gases, including hydrogen sulfide. Fuel value of sludge gas usually ranges from 600 to 700 Btu/ft³. The gas may be used at the treatment plant to operate auxiliary engines and provide heat for sludge-heating systems. Excess gas is burned.

All stages of the anaerobic process proceed simultaneously in the tank. Mingling of well-digested sludge with fresh sludge provides balance. If the pH holds between 7.2 and 7.4, conditions for digestion will be most favorable. Once achieved, balance may usually be maintained if addition of fresh solids is held to less than 4%, by weight, of the solids in the tank.

Speed of anaerobic digestion depends on temperature (Fig. 8-19). In conventional sludge digestion, as illustrated in Fig. 8-20a (mesophilic range), 100°F is the optimum temperature. Between 110 and 140°F (thermophilic range), thermophilic, or heat-loving, bacteria become active and speed digestion even more, with an optimum temperature of 130°F. Tanks usually are heated to hasten digestion (Fig. 8-20b).

Most states have established schedules of capacity requirements for digestion tanks, depending on type of sludge and whether or not the tanks are heated. Typical requirements set a capacity, ft³ per capita, for heated tanks of 2 to 3 for primary sludges, 3 to 4 for mixtures of primary and standard-filter sludges, and 4 to 6 for activated sludge or mixtures of primary and high-rate filter sludges. Capacities of unheated tanks should be twice as great for each type of sludge.

Sludge-digestion tanks may be circular or rectangular in plan. They generally provide a means of manipulating the sludge. The system also may include preheater and heater equipment, recirculation pumps with sludge suction at several levels, supernatant-liquor drawoff at

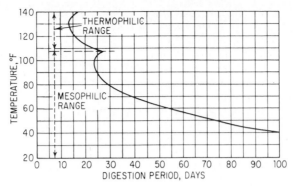

Fig. 8-19. Digestion period decreases with increasing temperature, reaching a minimum in the mesophilic range at about 100°F and in the thermophilic range at about 130°F.

several levels, gas dome or collector, stirring mechanism, sludge rakes, and drawoff. The tank cover may be floating or fixed. With a fixed cover, when fresh sludge is added to a tank kept full, an equal volume of supernatant liquor must be removed. Addition of sludge creates currents, as a result of which the liquor being removed may carry off some of the sludge. A floating cover allows the liquor to be withdrawn before or after the fresh sludge enters the tank.

In multistage anaerobic digestion, two or more digesters are placed in series. The sludge drawoff of each is fed to a subsequent one, and digested sludge is removed from the last (Fig. 8-21). The system provides flexibility in manipulating and mixing sludges and in controlling supernatant liquor. Also, it may be possible to use a smaller tank than required for single-stage operation or, for a given-size tank, to retain solids longer. In two-stage digestion, good results may be obtained if less than 20%, by volume, of material transferred from the first to the second tank is the best-digested sludge and more than 80% is supernatant liquor with the lowest solid content.

In aerobic digestion, organic sludges are aerated in an open tank similar to an activated-sludge aeration tank. The process provides about the same reduction in solids as the anaerobic process but is more stable in operation and recycles fewer pollutants to the wastewater-treatment plant. Aerobic digestion, however, has higher power costs and does not produce fuel gases.

Stabilization of primary and secondary wastewater sludges for reuse as sludge conditioners

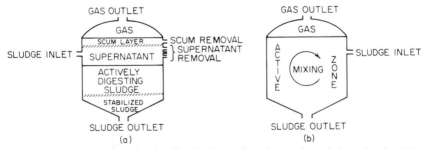

Fig. 8-20. Sludge digestion. (*a*) Standard-rate digestion—unheated, detention time 30 to 60 days, loading 0.03 to 0.10 lb of volatile suspended solids per ft³ · day, intermittent feeding and withdrawal, and stratification. (*b*) High-rate digestion—heated to between 85 and 95°F, detention 15 days or less, loading 0.10 to 0.50 lb of volatile suspended solids per ft³ · day, continuous or intermittent feeding or withdrawal, and homogeneous.

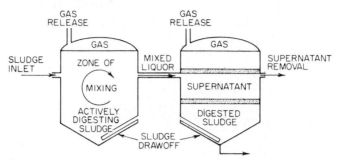

Fig. 8-21.   Two-stage anaerobic digestion.

may be accomplished by composting. In this process, the sludge is mixed with a bulking material, such as wood chips or refuse. Placed in piles or windrows about 7 ft high, the mixture undergoes biological action that stabilizes the sludge and heats it sufficiently to kill most disease-causing organisms in it. The composting takes about 3 weeks, after which the mixture usually is cured for about another month before its reuse.

Sludge Dewatering ▪ Before disposal, digested sludge from relatively small treatment plants may be concentrated in drying beds. Area needed for this purpose is about 2 to 3 ft² per capita (about three-fourths as much if the beds are covered). Beds consist of up to 12 in of coarse sand over 12 to 18 in of gravel. The natural earth bottom is sloped to underdrains, usually spaced about 30 ft apart. A bed may be from 20 to 30 ft wide and up to 125 ft long. It may be bounded or separated from an adjacent bed by a concrete wall extending about 15 in above the sand surface.

The bed is dosed with sludge to a depth of 9 to 12 in and allowed to drain and dry. A well-digested, granular sludge drains easily and reduces to a depth of 3 to 4 in when dry (60 to 70% moisture content). Sludge removed from the bed has little or no odor. It may be used as a weak fertilizer or landfill.

Sludge processing may be required if the sludge is to be disposed of by other methods. One sludge-processing method is elutriation, or washing of sludge with plant effluent. This removes undesirable amino-ammonia nitrogen and reduces or eliminates the need for conditioning chemicals. After settlement, the washed sludge is drawn off for conditioning and filtration.

As an alternative, lime or ferric chloride may be used to prepare sludge for vacuum filtration. For relatively large treatment plants, mechanical dewatering systems are advantageous because they are more compact and more controllable. Such systems include vacuum filtration, centrifuging, and pressure filtration.

Vacuum filtration reduces moisture content to about 80%. Filter cakes are easier to handle than the digested sludge from digesters. (In some plants, raw sludge is conditioned and processed on various filters without digestion. Such sludge is offensive and is handled in the same manner as screenings.)

Filter rates range from 2.5 lb/(ft² · h) of dry solids for fresh or digested activated sludge to 8 for primary digested sludge. Usually, a vacuum filter is a hollow drum that rotates slowly about a horizontal axis in a basin of sludge (Fig. 8-22). The filter is covered with wire, plastic, or cotton cloth or with flexible, metal, springlike coils. A vacuum in the compartmented interior of the drum holds sludge against the cover and separates water from the solids. As the drum rotates, a blade scrapes the cake into a conveyor or the cake is dislodged by release of the vacuum when the filter fabric passes off the drum over small rollers. The filtrate is returned to wastewater influent or to elutriators.

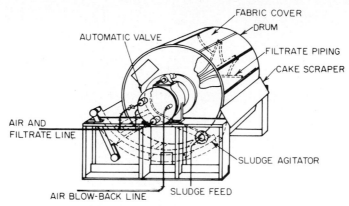

Fig. 8-22.   Rotary drum vacuum filter.

Centrifuge dewatering of sludge is accomplished in a horizontal drum rotated at 1600 to 2000 rpm. Sludge is pumped into the centrifuge and injected with polymers for sludge conditioning. As the drum turns, the solids are spun to the outside of the drum and removed by a conveyer (Fig. 8-23). Costs and results are similar to those obtained with vacuum filtration.

Pressure filtration is accomplished by pumping sludge at pressures up to 225 psi through filters attached to a series of plates. The plates are supported in a frame between a fixed and a moving end. When sludge is forced into the chambers between plates, the liquid passes through the filters while the solids are retained. When the filter chambers fill up with solids, the sludge feed is stopped. The filter cake is dislodged by shifting the moving end so that the plates can be moved. Pressure filtration provides the driest cake obtained by mechanical dewatering methods, produces a clear filtrate, and often reduces chemical conditioning costs.

Sludge Soil Conditioning ▪ Because sludge from municipal wastewater treatment contains some essential plant nutrients, it can be used as a fertilizer or soil conditioner. For that purpose, however, it is desirable that the sludge be first stabilized. It is often also dewatered.

Some cities apply liquid sludge to croplands. This eliminates dewatering costs but requires transporting of large amounts of sludge, and for health reasons, the sludge cannot be used for root crops or crops eaten raw. In the Chicago area, crops fertilized with liquid sludge include corn, soybeans, and winter wheat.

In some cases, sludge is dried in high-heat flash driers to reduce the volume substantially. Flash driers operate by mixing a portion of dried sludge with incoming wet sludge cake and introducing a high-velocity high-temperature gas stream. The dried material is separated from the gas in a cyclone separator and moved to storage. If a refuse incinerator is located at the wastewater-treatment site, it can provide heat for sludge drying.

Sludge Reduction ▪ If sludge is not to be used as a soil conditioner and if a landfill disposal site is not available, the sludge may be reduced to a more innocuous and more easily handled form by incineration, chemical oxidation, or wet oxidation.

During incineration, the moisture in the sludge is completely evaporated and the organic solids are burned to a sterile ash. Digested as well as undigested sludge, however, may be disposed of by incineration. Auxiliary heat is needed because the moisture content of the filter cake is high. Gas, including digester gas, oil, or coal, may be used as fuel.

In the past, incinerators used to burn sludge were multiple-hearth. Fed initially to the top

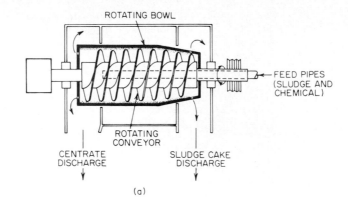

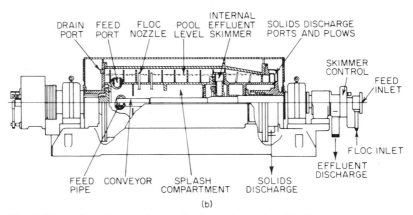

Fig. 8-23.   Centrifuge equipment for dewatering sludge: (*a*) Continuous countercurrent, solid-bowl, screw-conveyor-discharge centrifuge. (*b*) Concurrent-flow, solid-bowl, conveyor-discharge centrifuge.

hearth, sludge is pushed down to the next hearth by agitator arms as it dries. The heat drives off water and volatile gases, which are ignited by the high temperature. To avoid excessive odors, the temperature should be maintained at 1500°F or more. Ash residue, if it meets state standards, may be used for fill or cover on sanitary landfill. Flue gases are passed through a scrubber to limit air pollution.

If digester gas is not available for fuel, cost of sludge incineration may be high. As an alternative, filter cake may be mixed with solid wastes and burned in a municipal incinerator, if it adjoins the treatment plant.

A fluidized-bed incinerator is an alternative.

Chemical oxidation of sludge is suitable for use at small- to medium-size treatment plants. The process provides batch treatment of sludge with chlorine, which results in formation of biologically inert compounds. Most of the free available chlorine reacts to form hydrochloric (HCl) or hypochlorous (HOCl) acids. The HOCl subsequently breaks down into HCl with release of nascent oxygen O. Both HOCl and O are strong oxidants.

The raw sludge is pumped through a macerator to reduce particle size. The sludge then is mixed with conditioned sludge in the ratio of 3.8 gal of recirculated sludge per gal of raw sludge

(Fig. 8-24). Chlorine is added to the mixture in dosages ranging from 600 to 4800 mg/L, depending on the type of sludge. For domestic primary sludge, dosage usually ranges between 1500 and 2000 mg/L of sludge added. A recirculation pump feeds the mixture to a reaction tank, where the reaction with chlorine takes place almost instantaneously. A portion of the flow is fed to a second reaction tank, and the remainder is recirculated. The recirculation aids in mixing and provides efficient utilization of the chlorine. A pressure control pump at the discharge end of the second reaction tank maintains from 30- to 40-psi pressure on the system.

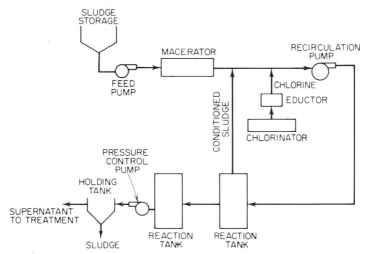

Fig. 8-24.   Schematic of process for chemical oxidation of sludge.

The sludge dewaters easily in conventional drying beds. Bed filtrate is returned to the primary settling tank. The sludge is sterile and may be disposed of by incineration or landfill, except for limitations on heavy metals.

Wet oxidation utilizes the principle that a combustible organic material can be oxidized in the presence of water in liquid form at temperatures between 250 and 700°F. Thus the process is suitable for reduction of sludges difficult to dewater. In the wet oxidation process, the sludge is ground and then fed into a reactor where temperatures about 500°F and pressures of 1000 to 1700 psi are maintained. The high pressure prevents conversion of the water to steam. Air is injected into the sludge to speed oxidation. The liquid and oxidized solids are separated by settling, vacuum filtration, or centrifuging. In addition to eliminating the dewatering step, the process has the advantage over incineration of a lower potential for air pollution because oxidation takes place in water without producing exhaust gases and ash; but the process creates a liquid very high in BOD, phosphorus, and nitrogen, and this liquid must be returned to the wastewater-treatment plant.

When wastewater treatment involves the coagulation-sedimentation process, large volumes of chemical sludges are produced. Generally, the sludges may be dewatered and disposed of in the same manner as the organic sludges from secondary treatment. If lime is the coagulant used, however, the lime may be recovered for reuse. For that purpose, the sludge is dewatered by one of the processes described previously and then subjected to recalcining in a multiple-hearth or fluidized-bed incinerator. Recalcining drives off water and carbon dioxide and leaves a residue of lime, which can be collected for reuse.

(Metcalf & Eddy, Inc., "Wastewater Engineering," McGraw-Hill Book Company, New

York; G. M. Fair, J. C. Geyer, and D. A. Okun, "Water and Wastewater Engineering," John Wiley & Sons, New York; "Sewage Treatment Plant Design," Manual 36, American Society of Civil Engineers; W. F. Ettlich et al., "Operations Manual—Sludge Handling and Conditioning," Environmental Protection Agency, Cincinnati, Ohio; M. J. Satriana, "Large-Scale Composting," Noyes Data Corp., Park Ridge, N.J.; "Sludge Thickening," MOP 15 no. FD-1, Water Pollution Control Federation, Washington, D.C.)

8-23. Imhoff Tanks

Developed by Karl Imhoff in Germany for the Emscher sewage district, this type of tank has been widely used in the United States since 1907 for primary treatment of sewage. The tank permits both sedimentation and sludge digestion to take place. Sludge comprises the settled solids in wastewater, and sludge digestion is the anaerobic decomposition of organic matter in sludge (Art. 8-22).

Efficiency of Imhoff tanks is about the same as for plain sedimentation tanks. Imhoff effluents are suitable for treatment in trickling filters. Sludge digestion, however, may proceed much more slowly in an Imhoff tank than in a separate digester. In an Imhoff tank, sludge digestion takes place without heat. Since rate of digestion decreases with drop in temperature (Fig. 8-19), lack of temperature control is a disadvantage, especially in regions where winters are cold.

Imhoff sludge has a tarlike odor and a black, granular appearance. It is dense. When withdrawn from a tank, it may have a moisture content of 90 to 95%. It dries easily, and when dry, it is comparatively odorless. It is an excellent humus but not a fertilizer.

Imhoff tanks are compartmented (Fig. 8-25). Sedimentation occurs in an upper, or flowing-through, chamber. Sludge settles into a lower chamber for digestion. To facilitate transfer of the settling solids, the flowing-through chamber has a smooth, sloping bottom (about 60° with the horizontal) with a slot at the lowest level. After particles pass through the slot, they are trapped in the lower chamber. Their path is obstructed either by overlapping walls at the slot, as shown in the cross section in Fig. 8-25, or by a triangular beam with an apex just below the slot.

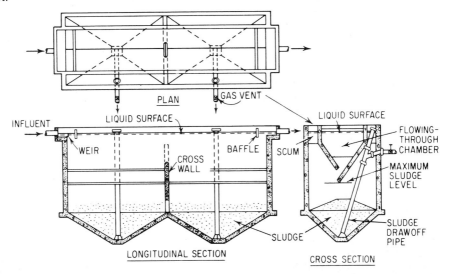

Fig. 8-25.   Imhoff tank permits sedimentation of wastewater in upper compartments, sludge digestion in lower.

As digestion proceeds in the lower chamber, scum is formed by rising sludge in which gas is trapped. The scum is directed to a scum chamber and gas vent alongside the upper chamber. As gases escape, sludge sinks back from the scum chamber to the lower chamber. (The gas vents occasionally may give off offensive odors.) The scum chamber should have a surface area 25 to 30% of the horizontal surface of the digestion chamber. Vents should be at least 24 in wide. And top freeboard should be at least 2 ft to contain the scum. If foaming occurs at a gas vent, it can be knocked down with a water jet from a hose.

In the digestion chamber, sludge settles to the sloped bottom. After sufficient time has elapsed for anaerobic decomposition, the sludge is removed through drawoff pipes. Since the height of a tank usually is 30 to 40 ft, the sludge can be expelled under the hydraulic pressure of the liquid in the tank. Ordinarily, sludge withdrawals are made twice a year. With such a schedule, the digestion chamber may be designed for a capacity of 3 to 5 ft^3 per capita of connected wastewater load. If, however, sludge removal is less frequent, or if industrial wastes with large quantities of solids are present in the wastewater, the capacity should be greater. Some chambers have been constructed with capacities up to 6.5 ft^3 per capita.

Large tanks are provided with means for reversing flow in the upper chamber. Since sedimentation generally is largest near an inlet, flow reversal permits a more even distribution of settled solids over the digestion chamber.

Detention period in the upper chamber usually is about 2½ h. Surface settling rate generally is 600 gal/(ft^2 · day). The weir overflow rate normally does not exceed 10,000 gal/lin ft of weir per day. Velocity of flow is held below 1 ft/s.

Length-width ratios of Imhoff tanks range from 3:1 to 5:1. Depth to slot is about equal to the width.

Multiple units are preferable to a single large tank. Sometimes it also is expedient to set two flowing-through chambers above one digestion chamber.

8-24. Septic Tanks

Like Imhoff tanks (Art. 8-23), septic tanks permit both sedimentation and sludge digestion. But unlike Imhoff tanks, septic tanks do not provide separate compartments for these processes. While undergoing anaerobic decompositions, the settled sludge is in immediate contact with wastewater flowing through the tank.

Septic tanks have limited use in municipal treatment. Their effluents are odorous, high in biochemical oxygen demand, and dangerous because of possible content of pathogenic organisms. Septic tanks, however, are widely used for treatment of wastewater from individual residences. Such tanks also are used by isolated schools and institutions and for treatment of sanitary wastewater at small industrial plants.

The tanks have a capacity of about 1 day's flow, plus storage capacity for sludge. Design of residential tanks generally is based on 75 gal of sewage per person per day or 150 gal per bedroom per day. If bedrooms are used as a criterion, allowance should be made for future conversion of some rooms into bedrooms. If garbage grinders may be used, tank capacity should be increased (Table 8-8). Most states set a minimum capacity of 500 gal for a single tank. Some states require a second compartment of 300-gal capacity, separated from the first compartment by a vertical partition. The partition has a horizontal slot, about 6 in high, to permit passage of effluent from the first compartment.

A septic tank may be constructed of coated metal or reinforced concrete and should be watertight. It should have a minimum liquid depth of 4 ft. Length of a rectangular tank may be about twice the width. Cast-in-place concrete tanks should be at least 6 in thick, unless completely reinforced. The top slab, at least, should be reinforced to support 150 psf. The tank top should be between 12 and 24 in below finished grade. An opening at least 16 in in diameter should be provided for a manhole. The underside of the tank top should be at least 1 in above

TABLE 8-8 Minimum Capacities of Septic Tanks

Bedrooms	Persons	Liquid capacity, gal	
		Without garbage grinders*	With garbage grinders†
2 or fewer	4	500	750
3	6	600	900
4	8	750	1,000

*Add 150 gal for each bedroom over 4.
†Add 250 gal for each bedroom over 4.

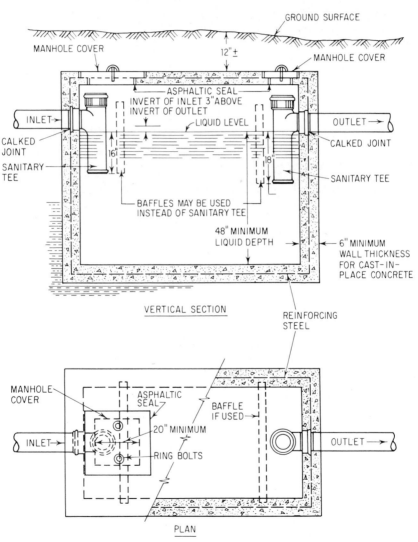

Fig. 8-26. Septic tank permits sedimentation and sludge digestion in the same compartment.

the tops of partitions and baffles. The invert of the inlet pipe should be at least 1 in, preferably 3 in, above the invert of the outlet. When the length of a tank exceeds 9 ft, two compartments should be used. Figure 8-26 shows a typical tank.

Residential septic tanks usually are buried in the ground and forgotten until the system gives trouble because of clogging or overflow. Actually, sludge should not be permitted to accumulate to a depth greater than that indicated in Table 8-9.

TABLE 8-9   Allowable Sludge Accumulation in Septic Tanks

Tank capacity, gal	Distance from bottom of outlet device to top of sludge, in , for liquid depth of	
	4 ft	5 ft
500	16	21
600	13	18
750	10	13
900	7	10
1,000	6	8

Commercial scavenger companies are available for sludge removal in most areas. Using a tank truck equipped with pumps, they remove the contents of a septic tank and cart them to a sewer manhole or a treatment plant for disposal. In rural areas, the sludge may be buried in an isolated site.

Municipal and institutional septic tanks are designed to hold 12- to 24-h flow, plus stored sludge. For camps for 40 or more persons, septic tanks should have a liquid capacity of at least 25 gal per person served. For day schools, the capacity may be two-thirds as large.

For residential units, the main vent for the house plumbing normally provides adequate ventilation. For large septic tanks, however, separate vents for the tanks are desirable.

Septic-tank effluent may be disposed of in a leaching cesspool (Art. 8-25) or a tile field. The latter consists of lines of open-jointed tile or perforated pipe laid in trenches 18 to 30 in deep. The lines receive the effluent from a distribution box, which distributes the liquid equally. From the box, the lines spread out, so that they are at least 6 ft apart. Lines should be of equal length, but none should be over 60 ft long.

Laid on a slight slope, not more than 1/16 in/ft, the tile or pipe is firmly set in a bed of crushed stone or washed gravel. The aggregate should extend 12 in below and 2 in above the conduit. The effluent, discharging from the openings, disperses over the entire trench bottom and seeps into the ground. The size of the tile field should be determined from the results of soil-percolation tests (Table 8-10).

At least two soil-percolation tests should be made in the area of the tile field. To perform a test, dig a hole 8 in in diameter or 12 in square. It should extend 6 in below the trench bottom or about 30 in below the final ground surface. Place 2 in of coarse sand or fine gravel in the bottom of the hole. Presoak the hole by filling it with water several hours before the test and again at the time of test and allowing the water to seep away. Remove any soil that falls into the hole. Pour clean water to a depth of 6 in in the hole. Record the time, minutes, required for the water to drop 1 in. Repeat until the time for the water to drop from the 6- to 5-in levels is about the same for two successive tests. Use the results of the last test as the stabilized rate. Alternative percolation-test methods have been developed for use where peculiar soil conditions exist.

Lots with less than 10 ft of soil above a rock formation usually are not suitable for construction of both wastewater systems and well-water supplies because of contamination hazards. Tile

TABLE 8-10 Suggested Sizes of Tile Fields for Septic-Tank Effluent

Soil-percolation rate, min*	Sewage application, gal per ft² per day	Trench width, in	Trench length, lin ft, for sewage loads, gal/day, of			
			300	450	600	1,000
0–5	2.4	24	63	94	125	209
6–7	2.0	24	73	110	146	244
8–10	1.7	36	59	88	118	196
11–15	1.3	36	77	116	154	256
16–20	1.0	36	95	143	191	317
21–30	0.8	36	125	188	250	417
31–45†	0.6	36	167	250	334	555
46–60†	0.4	36	250	375	500	834

*Time for 1-in drop in water level in soaked hole.
†If the percolation rate exceeds 60 min, the system is not suitable for a tile field. A rate over 30 min indicates borderline suitability for soil absorption; special care should be used in design and construction.

fields should not be constructed under driveways. The fields should be more than 100 ft away from any source of water supply, 20 ft from house foundation walls, and 10 ft from property lines. Trench bottoms should be at least 2 ft above groundwater, 5 ft above rock. Roof, footing, and basement drains should not be connected to septic tanks or they will be overloaded with water not requiring treatment. Water from roof gutters and other storm water should be routed away from the tile field. This water would saturate the soil and interfere with proper operation of the field.

Where soil is impervious or nearly so, an underdrained tile field may be used. This, in reality, is a buried sand filter placed below the tile drainage system. The drainage tile is laid in trenches filled with gravel or other porous media. Underdrains at the bottom collect and convey the effluent to a central collection point. There, the waste may be either drained out by gravity, chlorinated and discharged to a body of water, or pumped to a discharge point.

("Sewage Disposal Systems for the Home," Part III, *Bulletin* 1, Department of Health, State of New York, Albany, N.Y.; "Studies on Household Sewage Disposal System," Parts 1 to 3, Robert A. Taft Sanitary Engineering Center, U.S. Public Health Service; F. Wright, "Rural Water Supply and Sanitation," McGraw-Hill Book Company, New York; W. J. Jewell and R. Swan, "Water Pollution Control in Low-Density Areas," University of Vermont.)

8-25. Cesspools and Seepage Pits

A cesspool is a lined and covered hole in the ground into which sewage is discharged. It is used only when a sewerage system is not available. It may be watertight or leaching. A watertight cesspool retains wastewater until it is removed, by pumps or buckets. This type of cesspool is used only where no drainage into surrounding soil or rock is permitted. A leaching cesspool allows wastewater to seep into the surrounding ground.

Seepage pits of similar construction may be used to supplement tile fields (Art. 8-24) or instead of such fields where conditions are favorable. The pits also may be used in series with cesspools or septic tanks, to drain overflow liquid into the surrounding soil. Results are similar to those obtained with septic tanks (Art. 8-24).

Use of a leaching cesspool for direct disposal should be restricted to a small family in a remote location where there is absorptive soil and no danger of groundwater pollution. Leaching cesspools and seepage pits should never be attempted in clay soils.

The bottom of a seepage pit should be at least 2 ft above groundwater and 5 ft above rock. Lots with less than 10 ft of soil above a rock formation generally are not suitable for construction of both seepage pits and well-water supplies because of contamination hazards. Pits should be located more than 100 ft from a source of water supply, 20 ft from buildings, and 10 ft from property lines. Clear distance between two pits should be at least two times the diameter of the larger pit.

Size of seepage pit should be determined on the basis of 75 gal per person per day or 150 gal per bedroom per day. When bedrooms are used as a criterion, allowance should be made for future conversion of some rooms into bedrooms. The pit lining should be open-jointed or perforated to permit liquid to leak out. Wall area should be large enough to allow the soil to absorb the liquid without the pit overflowing. The required wall area, or effective absorption area, should be determined from soil-percolation tests (Table 8-11).

TABLE 8-11   Suggested Absorption Areas for Seepage Pits

Soil-percolation rate, min*	Sewage application, gal per ft² per day	Required absorption area, ft², for sewage loads, gal/day, of			
		300	450	600	1,000
0–5	3.2	94	141	188	313
6–10	2.3	130	196	261	435
11–15	1.8	167	250	334	555
16–20	1.5	200	300	400	666
21–30	1.1	273	409	545	911
31–45†	0.8	375	562	750	1,250
46–60†	0.5	600	900	1,200	2,000

*Time for 1-in drop in water level in soaked hole.
†If the percolation rate exceeds 60 min, the system is not suitable for a seepage pit. A rate over 30 min indicates borderline suitability for soil absorption; special care should be used in design and construction.

Percolation tests for seepage pits are the same as for tile fields (Art 8-24). The tests should be made, however, at half the depth and at the full estimated depth of the seepage pit. A larger excavation may be made for the upper portion of the hole, to facilitate execution of the test.

When the required absorption area has been obtained from Table 8-11, the outside diameter and effective depth of pit may be obtained from Table 8-12. The lining generally is made of concrete block or precast-concrete sections. Thickness should be at least 8 in. With rectangular block, the bottom should not be more than 10 ft below grade; with interlocking block, not

TABLE 8-12   Seepage-Pit Dimensions* for Required Absorption Area, Ft²

Depth, ft	Outside diameter, ft							
	5	6	7	8	9	10	11	12
3	47	57	66	75	85	94	104	113
4	63	75	88	101	113	126	138	151
5	79	94	110	126	141	157	173	188
6	94	113	132	151	169	189	207	226
7	110	132	154	176	197	220	242	263
8	126	151	176	201	225	252	276	302
9	141	170	198	226	254	283	310	339
10	157	189	220	251	282	314	346	377
11	173	207	242	276	310	346	380	415
12	188	226	263	302	339	377	415	453

*Outside diameter and effective depth. Bottom area excluded from computations.

more than 15 ft. For deeper pits, the lining should be structurally designed to resist saturated-earth pressures. The top should have a watertight manhole and concrete cover.

Coarse gravel should be placed in the bottom of the pit to a depth of 6 in. Backfill around the lining in the absorption area should be clean crushed stone or gravel, 1½ to 2 in in diameter, to a thickness of at least 6 in. A 2 in-thick layer of straw should be placed on top of the gravel before soil is backfilled.

When a seepage pit is used at the end of a tile field, the pit wall should be at least 6 ft from the end of the trench. The pipe connecting the end of the line with the pit should have tight joints.

(Sewage Disposal for the Home," Part III, *Bulletin* 1, Department of Health, State of New York, Albany, N.Y.)

9-1. Types of Surveys

Surveying is the science and art of making the measurements necessary to determine the relative positions of points above, on, or beneath the surface of the earth or to establish such points. Surveying continues to undergo important changes.

Plane surveying neglects curvature of the earth and is suitable for small areas.

Geodetic surveying takes into account curvature of the earth. It is applicable for large areas, long lines, and precisely locating basic points suitable for controlling other surveys.

Land, boundary, and cadastral surveys usually are closed surveys that establish property lines and corners. The term *cadastral* is now generally reserved for surveys of the public lands. There are two major categories: *retracement surveys* and *subdivision surveys*.

Topographic surveys provide the location of natural and artificial features and elevations used in map making.

Route surveys normally start at a control point and progress to another control point in the most direct manner permitted by field conditions. They are used for surveys for railroads, highways, pipelines, and so on.

Construction surveys are made while construction is in progress to control elevations, horizontal positions and dimensions, and configuration. Such surveys also are made to obtain essential data for computing construction pay quantities.

As-built surveys are postconstruction surveys that show the exact final location and layout of civil engineering works, to provide positional verification and records that include design changes.

Hydrographic surveys determine the shoreline and depths of lakes, streams, oceans, reservoirs, and other bodies of water. **Sea surveying** covers surveys for port and off-shore industries and the marine environment, including measurement and marine investigations by ship-borne personnel.

Solar surveying includes surveying and mapping of property boundaries, solar access easements, positions of obstructions and collectors, determination of minimum vertical sun angles, and other requirements of zoning boards and title insurance companies.

Section 9

Surveying

Russell C. Brinker
Visiting Professor of Civil Engineering
New Mexico State University

Satellite surveying provides positioning data and imagery, which is received by equipment, stored, and automatically verified on tape in selected data coordinates with each satellite pass. Dopler and global positioning are used as standard practice in remote regions and on subdivided lands.

Global Positioning System (GPS) utilizes a constellation of 24 high-altitude navigational satellites positioned in three orbital planes and spaced so that an operator of specialized equipment can receive signals from at least six satellites at all times.

Inertial surveying systems acquire coordinate data obtained by use of a helicopter or ground vehicle. Inertial equipment now coming into use has a dramatic impact on the installation of geodetic and cadastral control.

Photogrammetric surveys utilize terrestrial and aerial photographs or other sensors that provide data and can be a part of all the types of surveys listed in the preceding.

9-2. Surveying Organizations

The National Ocean Survey (NOS), through a component, the National Geodetic Survey (NGS), Department of Commerce, is responsible for extending leveling begun by the U.S. Coast Survey in 1856 and the vertical control system. National networks now include 235,000 horizontal and 500,000 vertical control points. Other past and present vertical datums that may have to be considered by surveyors include the International Great Lakes Datum (IGLD) of 1955 (agreed upon by the United States and Canada for that area), the Mississippi River datum, and numerous others formerly used in local regions.

The present vertical reference system, based on a 1929 adjustment extending through Alaska and Central America, is being readjusted with the aid of huge computers. Work on the new National Geodetic Vertical Datum (NGVD) was begun in 1978 and will be completed in 1987, to provide new descriptions and improved (changed) elevations of 480,000+ bench marks. Tidal data and bench-mark sheets issued prior to Nov. 28, 1980, will be corrected.

Other activities of the NGS include triangulation, preparation of nautical and aeronautical charts, photogrammetric surveys, tide and current studies, and collection of magnetic data.

The United States Geological Survey, Department of the Interior, is engaged in mapping and remapping the country, among other assignments. It prepares the commonly used 7½ × 15 min quadrangles.

The Bureau of Land Management, Department of the Interior, replaced the General Land Office established in 1812. It has jurisdiction over the surveys and sales of the public lands and prepares and publishes the "Manual of Surveying Instructions" for the survey of the public lands of the United States.

9-3. Units of Measurement

Units of measurement used in past and present surveys are:

For construction work: feet, inches, fractions of inches

For most surveys: feet, tenths, hundredths, thousandths

For National Geodetic Survey control surveys: meters, 0.1, 0.01, 0.001 m

The most-used equivalents are:

1 meter = 39.37 in (exactly) = 3.2808 ft

1 rod = 1 pole = 1 perch = 16½ ft

1 engineer's chain = 100 ft = 100 links

1 Gunter's chain = 66 ft = 100 Gunter's links (lk) = 4 rods = ⅟₈₀ mi

1 acre = 100,000 sq (Gunter's) links = 43,560 ft^2 = 160 rods^2 = 10 sq (Gunter's) chains
        = 4046.87 m^2 = 0.4047 hectares

1 rood = ¼ acre = 40 rods^2 (also local unit = 5½ to 8 yd)

1 hectare = 10,000 m^2 = 107,639.10 ft^2 =   acres

1 arpent = about 0.85 acre, or length of side of 1 square arpent (varies)

1 statute mile = 5280 ft = 1609.35 m

1 mi^2 = 640 acres

1 nautical mile (U.S.) = 6080.27 ft = 1853.248 m

1 fathom = 6 ft

1 cubit = 18 in

1 vara = 33 in (Calif.), 33⅓ in (Texas), varies

1 degree = ⅟₃₆₀ circle = 60 min = 3600 s = 0.01745 rad

sin 1° = 0.01745241

1 rad = 57°17′44.8″ or about 57.30°

1 grad (grade) = 1/400 circle = 1/100 quadrant = 100 centesimal min
             = 10^4 centesimals (French)

1 mil = 1/6,400 circle = 0.05625°

1 military pace (milpace) = 2½ ft

9-4. Significant Figures

These are the digits read directly from a measuring device plus one digit that must be estimated and therefore is questionable. For example, a reading of 654.32 ft from a steel tape graduated in tenths of a foot has five significant figures. In multiplying 798.16 by 37.1, the answer cannot have more significant figures than either number used; i.e., three in this case. The same rule applies in division. In addition or subtraction, for example, 73.148 + 6.93 + 482, the answer will have three significant figures, all on the left side of the decimal point.

Small, hand-held and large computers now available provide 10 or more digits, but carrying computation results beyond justifiable significant figures leads to false impressions of precision.

9-5. Measurement of Distance

Reasonable precisions for different methods of measuring distances are

Pacing (ordinary terrain): ⅟₅₀ to ⅟₁₀₀.

Taping (ordinary steel tape): ⅟₁₀₀₀ to ⅟₁₀,₀₀₀. (Results can be improved by use of tension apparatus, transit alignment, leveling.)

Base line (invar tape): ⅟₅₀,₀₀₀ to ⅟₁,₀₀₀,₀₀₀.

Stadia: ⅟₃₀₀ to ⅟₅₀₀ (with special procedures).

Subtense bar: ⅟₁₀₀₀ to ⅟₇₀₀ (for short distances, with a 1-s theodolite, averaging angles taken at both ends).

Electronic Distance Measurement (EDM) devices have been in use since the middle of the twentieth century—the geodimeter since 1948 (using the speed of light); the tellurometer since 1957 (employing microwaves); and the electrotape (with radio-frequency signals). These devices now have many competitors. (See Table 9-12, Art. 9-14.)

Slope Corrections ▪ In slope measurements, the horizontal distance $H = L \cos x$, where L = slope distance and x = vertical angle, measured from the horizontal—a simple hand calculator operation. Table 9-1 (conveniently carried in a field book to give a "feel" for the corrections) supplies an alternate method for determining reductions from slope to horizontal distances, and shortened lengths where one end of a tape is off line.

In measuring horizontal distances, the following corrections must be added:

For nonstandard temperature,

$$C_T = 0^{r} \qquad {}^{-}(T - 68)L \tag{9-1}$$

TABLE 9-1 Correction for Slope Distance or Tape Off Line

Elevation difference of ends or off-line distance, ft	Correction, ft		Elevation difference of ends or off-line distance, ft	Correction, ft	
	Tape length, ft			Tape length, ft	
	100	300		100	300
0.0	0.000	0.000	1.8	0.016	0.005
0.5	0.001	0.000	2.0	0.020	0.007
0.6	0.002	0.001	3.0	0.045	0.015
0.8	0.003	0.001	4.0	0.080	0.027
0.9	0.004	0.001	5.0	0.125	0.042
1.0	0.005	0.002	6.0	0.180	0.060
1.1	0.006	0.002	7.0	0.245	0.082
1.2	0.007	0.002	8.0	0.320	0.107
1.3	0.008	0.003	9.0	0.405	0.135
1.4	0.010	0.003	10.0	0.500	0.167
1.5	0.011	0.004	15.0	1.131	0.375
1.6	0.013	0.004	20.0	2.020	0.667

Note: Values by formula $C = d^2/2L$; for slopes greater than 10%, $C = d^2/2L + d^4/8L^3$, where d is the elevation difference or off-line distance.

where L = measured distance, ft

T = temperature at time of measurement, °F

For example, if the measured distance is 8785.32 ft when the temperature is 80°F, $C_T = (0.0000065)(80 - 68)(8785.32) = 0.69$ ft, and the true distance is 8786.01 ft. If the measured distance is 4721.30 ft when the temperature is 35°F, $C_T = -1.01$ ft, and the true distance is 4720.29 ft. In general, if $L = 1000$ ft and $T - 68 = 1°$, $C_T = 0.0065$ ft.

For incorrect tape length,

$$C_t = \frac{(\text{actual tape length} - \text{nominal tape length})L}{\text{nominal tape length}} \tag{9-2}$$

For nonstandard tension,

$$C_p = \frac{(\text{applied pull} - \text{standard tension})L}{AE} \tag{9-3}$$

where A = cross-sectional area of tape, in²

E = modulus of elasticity = 29,000,000 psi for steel

For sag correction between points of support, ft,

$$C = -\frac{w^2 L_s^3}{24 P^2} \tag{9-4}$$

where w = weight of tape per foot, lb

L_s = unsupported length of tape, ft

P = pull on tape, lb

Sources and Types of Error ▪ There are three sources of error in taping—instrumental, natural, and personal—and nine general types of errors. Table 9-2 lists the types of errors and their sources and classifies them as systematic or accidental.

TABLE 9-2 Types, Sources, and Classification of Taping Errors

Type of error	Source*	Classification†	Departure from standard to produce 0.01-ft error for a 100-ft tape
Tape length	I	S	0.01 ft
Temperature	N	S or A	15°F
Tension	P	S or A	15 lb
Sag	N, P	S	7⅜ in at center as compared with support throughout
Alignment	P	S	1.4 ft at one end, or 8½ in at midpoint
Tape not level	P	S	1.4 ft
Interpolation	P	A	0.01 ft
Marking	P	A	0.01 ft
Plumbing	P	A	0.01 ft

*I = instrumental, N = natural, P = personal.
†S = systematic, A = accidental

All errors in Table 9-2 produce, in effect, an incorrect tape length. Therefore, only four basic tape problems exist: *measuring* a line between fixed points with a tape too long or too short, and *laying out* a line from one fixed point with a tape too long or too short. A simple one-line sketch (Fig. 9-1) with tick marks for nominal and actual tape lengths is a foolproof method for deciding whether to add or subtract the correction in any case.

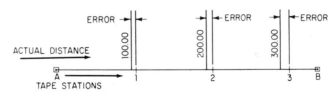

Fig. 9-1.   Cumulative error from measuring with a tape that is too long.

In base-line measurements with steel or invar tapes (three or more tapes should be used on different sections of the line), corrections are applied for inclination; temperature; nonstandard length of tape, for both full and partial tape lengths; and reduction to sea level.

9-6. Leveling

A few definitions introduce the subject:

Vertical Line ▪ A line to the center of the earth from any point. Commonly considered to coincide with a plumb line.

Level Surface ▪ A curved surface that, at every point, is perpendicular to a plumb line through the point.

Level Line ▪ A line in a level surface, therefore a curved line.

Horizontal Plane ▪ A plane perpendicular to the plumb line.

Horizontal Line ▪ A straight line perpendicular to the vertical.

Datum ▪ Any level surface to which elevations are referred, such as mean sea level, which is most commonly used; also called datum plane although not actually a plane.

Mean Sea Level (MSL) ▪ The average height of the surface of the sea. MSL was established originally over a 19-year period, for all tidal stages, at 21 United States coastal stations and 5 in Canada. The basic National Geodetic Vertical Datum net is being connected to all accessible primary tide and water-level stations.

Orthometric Correction ▪ This is a correction applied to preliminary elevations due to flattening of the earth in the polar direction. Its value is a function of the latitude and elevation of the level circuit.

Curvature of the earth causes a horizontal line to depart from a level surface. The departure, C_f, ft; or C_m, m, may be computed from

$$C_f = 0.667M^2 = 0.0239F^2 \qquad (9\text{-}5\,a)$$

$$C_m = 0.0785K^2 \qquad (9\text{-}5\,b)$$

where M, F, and K are distances in miles, thousands of feet, and kilometers, respectively, from the point of tangency to the earth.

Refraction causes light rays that pass through the earth's atmosphere to bend toward the earth's surface. For horizontal sights, the average angular displacement (like the sun's diameter) is about 32 min. The displacement, R_f, ft, or R_m, m, is given approximately by

$$R_f = 0.093M^2 = 0.0033F^2 \qquad (9\text{-}6\,a)$$

$$R_m = 0.011K^2 \qquad (9\text{-}6\,b)$$

To obtain the combined effect of refraction and curvature of the earth, subtract R_f from C_f or R_m from C_m.

Differential leveling is the process of determining the difference in elevation of two points. The procedure involves sighting with a level on a ruled rod set on a point of known elevation (backsight or plus sight), then on the rod set on points (or intermediate points) whose elevations are to be determined (foresights). These elevations equal the height of instrument minus the foresight reading. The height of instrument equals the known elevation plus the backsight reading. For accuracy, the sum of backsight and foresight distances should be kept nearly equal.

Elevations commonly are taken to 0.01 ft in engineering surveys and to 0.001 m in precise National Geodetic Survey work.

Table 9-3 shows a typical left-hand page of open-style notes. In closed-style (condensed) notes, B.S., H.I., F.S., and elevation values are placed on the same line, thereby saving space (which is cheap in a field book) but reducing the clarity of steps for beginners. The right-hand page contains bench-mark descriptions, sketches, date of survey, names of survey-party members, and information on the weather, equipment used, and other necessary remarks.

As noted in Brinker, Austin, and Minnick, "Note forms for Surveying Measurements," Landmark Enterprises, Rancho Cordova, Cal.: The critical importance of field notes is sometimes neglected. If any of the five main features used in evaluating notes—accuracy, integrity, legibility, arrangement and clarity—is absent, delays, mistakes and increased costs in completing field work, computations, and mapping result.

Pushbutton (mechanical) notes for field measurements of angles and distances, as well as reduction of slope distances and computation of coordinates, are now being recorded in various types of Data Collectors. They are displayed and the data automatically recorded by pushing buttons. Reading and transcribing errors are thus eliminated, both in the field and office where the data collector automatically transfers the field notes to a surveying calculator for processing. The results then go to a printer, which makes working plots and convenient page-width printouts.

TABLE 9-3   Differental Leveling Notes

	Differential Leveling—BM Civil to BM Dorm				
Station	B.S.	H.I.*	F.S.	Elev. †	Dist.
BM Civil				100.00	
	4.08	104.08			175
TP 1			0.20	103.88	180
	6.09	109.97			160
BM Dorm			4.32	105.65	155
	10.17		4.52		670
	4.52				
	5.65				
BM Dorm				105.65	
	4.37	110.02			165
TP 2			6.14	103.88	165
	0.93	104.81			170
BM Civil			4.80	100.01	175
	5.30		10.94		675
			5.30		
			5.64		
	Elev. Diff. = 5.64 ft				
	Loop Closure = 0.01 ft				

*Height of instrument (H.I.) = elevation + backsight (B.S.).
†Elevation = H.I. − foresight (F.S.).

Data collectors do not completely replace the field book, which still is used to record backup information, including sketches and notes to show station identification for the permanent project. Actually, since only a small part of total field time is occupied in recording measurements in a field book, the important time-saving advantage of a data collector is gained in the office and drafting room.

A usable tool for notekeepers is photography. With a reasonably priced, reliable and lightweight camera a photographic record of monuments set or found, and other field evidence to the survey can be prepared.

Profile leveling determines the elevations of points at known distances along a line. When these points are plotted, a vertical section through the earth's surface is shown. Elevations are taken at full stations (100 ft) or closer in irregular terrain, at breaks in the ground surface, and at critical points such as bridge abutments and road crossings. Profiles are generally plotted on special paper with an exaggeration of from 5:1 up to 20:1, or even more, so that elevation differentials will show up better. Profiles are needed for route surveys, to select grades and find earthwork quantities. Elevations are usually taken to 0.01 ft on bench marks and 0.1 ft on the ground.

Reciprocal leveling is used to cross streams, lakes, canyons, and other topographic barriers that prevent keeping backsights and foresights balanced. On each side of the obstruction to be crossed, a plus sight is taken on the near rod and several minus sights on the far rod. The resulting differences in elevation are averaged to eliminate the effects of curvature and refraction, and inadjustment of the instrument. Even though a number of minus sights are taken for averaging, their length may reduce the accuracy of results.

Borrow-pit or cross-section leveling produces elevations at the corners of squares or rectangles whose sides are dependent on the area to be covered, type of terrain, and accuracy desired. For example, sides may be 10, 20, 40, 50, or 100 ft. Layout can be made by tape alone, or transit and tape. Contours can be located readily, topographic features not so well. Quantities of material to be excavated or filled are computed, in cubic yards, by selecting a grade elevation, or final ground elevation, computing elevation differences for the corners, and substituting in

$$Q = \frac{nxA}{108} \qquad (9\text{-}7)$$

where n = number of times a particular corner enters as part of a division block
$\quad x$ = difference in ground and grade elevation for each corner, ft
$\quad A$ = area of each block, ft^2

Cross-section leveling also is the term applied to the procedure for locating contours or taking elevations on lines at right angles to the center line in a route survey.

Three-wire leveling is a type of differential leveling with three horizontal sighting wires in the level. Upper, middle, and lower wires are read to obtain an average value for the sight, check the precision of reading the individual wires, and secure stadia distances for checking lengths of backsights and foresights. The height of instrument is not needed or computed. The National Geodetic Survey has long used three-wire leveling for its control work, but more general use is now being made of the method.

Grade designates the elevation of the finished surface of an engineering project and also the rise or fall in 100 ft of horizontal distance; for example, a 4% grade (also called gradient). Note that since the common stadia interval factor is 100, the difference in readings between the middle and upper (or lower) wire represents ½ ft in 100 ft, or a ½% grade.

Types of levels in general use are listed in Table 9-4.

TABLE 9-4 Types of Levels

Type	Use
Hand level	Rough work. Sights on ordinary level rod limited to about 50 ft because of zero- to 2-power magnification
Engineer's level, Wye or Dumpy	Suitable for ordinary work (third- or fourth-order). Elevations to 0.01 ft without target
Tilting level	Faster, more accurate sighting. Good for third-, second-, or first-order work depending upon refinement
Self-leveling, automatic levels	Fast, suitable for second-order and third-order work
Precise level	Very sensitive level vials, high magnification, tilting, other features

Note: Instruments are arranged in ascending order of cost.

Special construction levels include the Blout & George Laser Tracking Level (which can search a 360° horizontal plane and lock on a pocket-size target), the Dietzgen Laser Swinger, Spectra-Physics Rotolite Building Laser, and AGL Construction Laser. Some laser instruments are available for shaft plumbing and setups inside large pipe lines.

9-7. Vertical Control

The National Geodetic Survey has added over 650,000 km of first- and second-order level lines since 1929 to provide vertical control for all types of surveys. NGS furnishes descriptions and elevations of bench marks free upon request. The relative accuracy C, mm, required between directly connected bench marks for the three orders of leveling are:

First-order: $C = 0.5\sqrt{K}$ for Class I and $0.7\sqrt{K}$ for Class II

Second-order: $C = 1.0\sqrt{K}$ for Class I and $1.3\sqrt{K}$ for Class II

Third-order: $C = 2.0\sqrt{K}$

where K is the distance between bench marks, km.

9-8. Magnetic Compass

A magnetic compass consists of a magnetized needle mounted on a pivot at the center of a graduated circle. The compass is now used primarily for retracement purposes and checking, although some surveys not requiring precision are made with a compass; for example, in forestry and geology. American transits have traditionally come with a long compass needle, whereas on optical instruments the compass is merely an accessory, and therefore the instruments can be smaller and lighter.

A small weight is placed on the south end of the needle in the northern hemisphere to counteract the dip caused by magnetic lines of force. Since the magnetic poles are not located at the geographic poles, a horizontal angle (declination) results between the axis of the needle and a true meridian. East declination occurs if the needle points east of true (due) north, west declination if the needle points west of true north.

The National Geodetic Survey publishes a world chart every fifth year showing the positions of the agonic line, isogonic lines for each degree, and values for annual variation of the needle. The **agonic line** is a line of zero declination; i.e., a magnetic compass set up on points along this line would point to true north as well as magnetic north. For points along an isogonic line, declination should be constant, barring local attraction.

Table 9-5 lists the periodic variations in the declination of the needle that make it unreliable. In addition, local attraction resulting from power sources, metal objects, and so on may

TABLE 9-5 Periodic Variations in Declination of Magnetic Needle

Variation	Remarks
Secular	Largest and most important. Produces wide unpredictable swings over a period of years, but records permit comparison of past and present declinations
Daily (diurnal)	Swings about 8 min per day in the U.S. Relatively unimportant
Annual	Periodic swing amounting to less than 1 min of arc; it is unimportant
Irregular	From magnetic storms and other sources. Can pull needle off more than a degree

produce considerable error in bearings taken with a compass. If the source of local attraction is fixed and constant, however, angles between bearings are correct, even though the bearings are uniformly distorted.

The Brunton compass or pocket transit has some of the features of a sighting compass, a prismatic compass, a hand level, and a clinometer. It is suitable for some forest, geological, topographical, and preliminary surveys of various kinds.

A common problem today is the conversion of past magnetic bearings based upon the declination of a given date to present bearings with today's declination, or to true bearings. A sketch, such as Fig. 9-2, showing all data with pencils of different colors, will make the answer evident.

9-9. Bearings and Azimuths

The direction of a line is the angle measured from any reference line, such as a magnetic or true meridian. Bearings are angles measured from the north and the south, toward the east or the west. They can never be greater than 90° (Fig. 9-3).

Bearings read in the advancing direction are forward bearings; those in the opposite direction are back bearings. Computed bearings are obtained by using a bearing and applying a direct, deflection, or other angle. Bearings, either magnetic or true, are used in rerunning old surveys, in computations, on maps, and in deed descriptions.

An azimuth is a clockwise angle measured from some reference line, usually a meridian. Government surveys use geodetic south as the base of azimuths. Other surveys in the northern hemisphere may employ north. Azimuths are advantageous in topographic surveys, plotting, direction problems, and other work where omission of the quadrant letters and a range of angular values from 0 to 360° simplify the work.

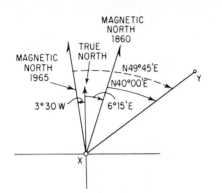

Fig. 9-2.   Magnetic bearing of a line *XY* in a past year is found by plotting magnetic north for that year with respect to true north.

9-10. Horizontal Control

All surveys require some kind of control, be it a base line or bench mark, or both. Horizontal control consists of points whose positions are established by traverse, triangulation, or trilateration. The National Geodetic Survey has established control monuments throughout the country and tabulated aziumths, latitude and longitude, statewide coordinates, and other data for them. Surveys on the statewide coordinate system have increased the number of control points available to all surveyors.

Traverses ▪ For a traverse, the survey follows a line from point to point in succession. The lengths of lines between points and their directions are measured. If the traverse returns to the point of origin, it is called a closed traverse. The United States–Canada boundary, for example,

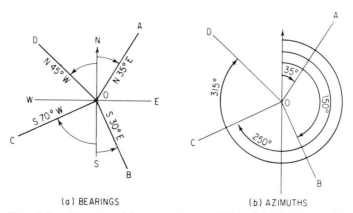

(a) BEARINGS   (b) AZIMUTHS

Fig. 9-3.   Direction of lines may be specified by (*a*) bearings or (*b*) azimuth.

was run by traverse. In contrast, the boundary of a construction site would be surveyed by a closed traverse. Permissible closures for traverses that make a closed loop or connect adjusted positions of equal-order or higher-order control surveys are given in Table 9-6.

Transit-tape traverses provide control for areas of limited size as well as for the final results on property surveys, route surveys, and other work. Stadia traverses are good enough for small-area topographic surveys when tied to higher control. Faster and more accurate traversing may be accomplished with electronic distance-measuring devices and with theodolites with direct readings to seconds and much lighter than the older-type, bulky transits.

Reasonable ratios of error for different types of property surveys, depending on land values, are shown in Table 9-7.

TABLE 9-6   Permissible Traverse Closures*

Traverse order	Max permissible closure after azimuth adjustment	Max azimuth closure at azimuth checkpoint	
		Sec per station	Sec†
First order	1:100,000	1.0	$2\sqrt{N}$
Second order			
Class I	1:50,000	1.5	$3\sqrt{N}$
Class II	1:20,000	2.0	$6\sqrt{N}$
Third order			
Class I	1:10,000	3.0	$10\sqrt{N}$
Class II	1:5,000	8.0	$30\sqrt{N}$

*National Geodetic Survey.
†N = number of stations

TABLE 9-7   Acceptable Ratio of Error for Various Surveys

Wasteland	1/500
Ordinary farmland	1/1,000
Small community	1/2,000–1/5,000
Small city	1/5,000–1/10,000
Metropolitan area	1/10,000–1/20,000

Triangulation ▪ In triangulation, points are located at the apexes of triangles, and all angles and one base line are measured. Additional base lines are used when a chain of triangles, quadrilaterals, or central-point figures is required (Fig. 9-4). All other sides are computed and adjustments carried from the fixed base lines forward and backward to minimize the corrections. Angles used in computation should exceed 15°, and preferably 30°, to avoid the rapid change in sines for small angles.

Chains of triangles are unsuitable for high-precision work since they do not permit the rigid adjustments available in quadrilaterals and more complicated figures. Quadrilaterals are advantageous for long, relatively narrow chains; polygons and central-point figures for wide systems and perhaps for large cities, where stations can be set on tops of buildings.

Strength of figure in triangulation is an expression of relative precision possible in the system based on the route of computation of a triangle side. It is independent of the accuracy of observations and utilizes the number of directions observed, conditions to be satisfied, and rates of

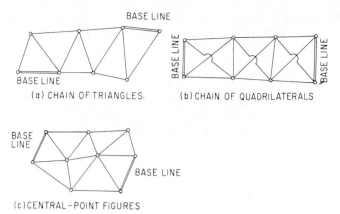

Fig. 9-4.   Triangulation chains.

changes for the sines of distance angles. Triangulation stations that cannot be occupied require additional computation for reduction to center in obtaining coordinates and other data.

Permissible triangulation closures for the three orders of triangulation specified by the National Geodetic Survey are given in Table 9-8 and specifications for base-line measurements in Table 9-9.

TABLE 9-8   Triangulation Closures

Specification item	First order	Second order		Third order	
		Class I	Class II	Class I	Class II
Avg triangle closure, s	1.0	1.2	2.0	3.0	5.0
Max triangle closure, s	3.0	3.0	5.0	5.0	10.0

TABLE 9-9   Specifications for Base-Line Measurements

Order	Max standard error of base
First	1/1,000,000
Second	
Class I	1/900,000
Class II	1/800,000
Third	
Class I	1/500,000
Class II	1/250,000

Trilateration has replaced triangulation for establishment of control in many cases, such as photogrammetry, since the development of electronic measuring devices. All distances are measured and the angles computed as needed.

9-11. Stadia

Stadia is a method of measuring distances by noting the length of a stadia or level rod intercepted between the upper and lower sighting wires of a transit, theodolite, or level. Most transits and levels have an interval between stadia wires that gives a vertical intercept of 1 ft on a rod 100 ft away. A stadia constant varying from about ¾ to 1¼ ft (usually assumed to be 1 ft) must be added for older-type external-focusing telescopes. The internal-focusing short-length telescopes common today have a stadia constant of only a few tenths of a foot, and so it can be neglected for normal readings taken to the nearest foot.

Figure 9-5 clearly shows stadia relationships for a horizontal sight with the older-type external-focusing telescope. Relationships are comparable for the internal-focusing type.

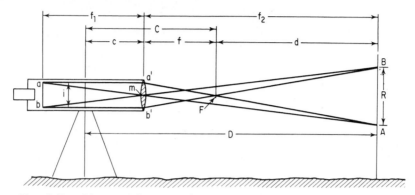

Fig. 9-5.   Distance D is measured with an external-focusing telescope by determining interval R intercepted on a rod AB by two horizontal sighting wires a and b.

For horizontal sights, the stadia distance, ft (from instrument spindle to rod), is

$$D = R \frac{f}{i} + C \qquad (9\text{-}8)$$

where R = intercept on rod between two sighting wires, ft
$\quad f$ = focal length of telescope, ft (constant for specific instrument)
$\quad i$ = distance between stadia wires, ft
$\quad C = f + c$
$\quad c$ = distance from center of spindle to center of objective lens, ft
C is called the stadia constant, although c and C vary slightly.

The value of f/i, the stadia factor, is set by the manufacturer to be about 100, but it is not necessarily 100.00. The value should be checked before use on important work, or when the wires or reticle are damaged and replaced.

For inclined sights (Fig. 9-6) the rod is held vertical, as indicated by a rod level or other means because it is difficult to assure perpendicularity to the sight line on sloping shots. Reduction to horizontal and vertical distances is made according to formulas, such as

$$H = 100R - 100R \sin^2 \alpha + C \qquad (9\text{-}9)$$
$$V = 100R(\tfrac{1}{2} \sin 2\alpha) \qquad (9\text{-}10)$$

where H = horizontal distance from instrument to rod, ft
$\quad V$ = vertical distance from instrument to rod, ft
$\quad \alpha$ = vertical angle above or below level sight .

A Beaman arc on transits and alidades simplifies reduction of slope sights. It consists of an H scale and a V scale, both graduated in percent, with spacing based on the stadia formulas. The H scale gives the correction per 100 ft of slope distance, which is subtracted from $100R + C$ to get the horizontal distance. A V-scale index of 50 for level sights eliminates minus values in determining vertical distance. Readings above 50 are angles of elevation; below 50, angles of

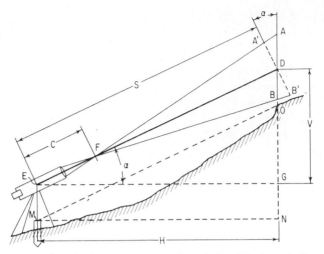

Fig. 9-6. Stadia measurement of vertical and horizontal distances *V* and *H* is done by reading with a telescope rod intercept *AB* and vertical angle α.

depression. Each unit above or below 50 represents 1-ft difference in elevation per 100 ft of sight. By setting the *V* scale to a whole number, even though the middle wire does not fall on the height of the instrument, you need only mental arithmetic to compute vertical distance. The *H* scale is read by interpolation since the value generally is small and falls in the area of wide spaces.

As an illustration, to determine the elevation of a point *X* from a setup at point *Y*, compute elevation *X* = elevation *Y* + height of instrument + (arc reading − 50)(rod intercept) − reading of middle wire.

Some self-reducing tachymeters have curved stadia lines engraved on a glass plate, which turns and appears to make the lines move closer or farther apart. A fixed stadia factor of 100 is used for horizontal reduction, but several factors are required for elevation differences, depending on the slope.

Stadia traverses can be run with direct or azimuth angles. Distances and elevation differences should be averaged for the foresights and backsights. Elevation checks on bench marks are necessary at frequent intervals to maintain reasonable precision.

Poor closures in stadia work usually result from incorrect rod readings rather than errors in angles. A difference of 1 min in vertical angle has little effect on horizontal distances and produces a correction smaller than 0.1 ft for sights up to 300 ft.

Stadia distances, normally read to the nearest foot, are assumed to be valid within about ½ ft. For the same line and lateral error on a 300-ft shot, sin α = ½/300 = 0.00167 and α = 5.7 min. Thus for stadia sights up to 300 ft, comparable distance-angle precision is obtained by reading horizontal angles to the nearest 5 or 6 min. This can be done by estimation on the scale without using the vernier graduations. (See Fig. 9-7.)

Closely approximate answers to many problems in surveying, engineering, mechanics, and other fields can be computed mentally by memorizing the sin of 1′ = 0.00029 (or roughly 0.0003), and sin 1° = 0.01745 (about 0.01¾). For sines of angles from 0° to 10°, the numerical values increase almost linearly. The divergence from true value at 10° is only ½%; at 30° just 4½%—high for surveying but within, say, some design load estimates. Values of tangents found by multiplying the tangent of 1° by other angular sizes diverge more rapidly but still are off only 1% at 10°.

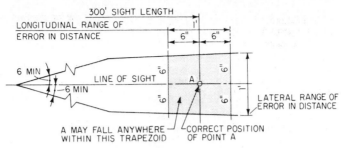

Fig. 9-7.   Comparable precision of angles and stadia distances.

9-12. Planetable Surveying

This time-tested method was the primary one used by the U.S. Geological Survey and others to prepare topographic maps before introduction of photogrammetry. Planetable surveying still finds many applications in civil and mining engineering, geology, agriculture, forestry, archeology, and military mapping as well as in field checking photogrammetric maps *(ground-truth surveys)*. It generally is the preferred method for surveying small tracts at large scales.

Planetable surveying has many advantages and disadvantages, as shown below.

Advantages of the planetable:

Features are mapped while in full view.

Contours and irregular lines are accurately sketched, with half the number of sights needed for transit stadia.

No notes need be taken.

Less total time (fieldwork plus office) is required.

Setups can be made at desirable points without advance determination by using two- or three-point location.

Self-reducing alidades and parallel rulers can be used.

Disadvantages of the planetable:

Plotted control is needed in advance of fieldwork.

Good weather is needed.

More field time is required.

Brush or wooded areas hamper sketching.

Elbow-height table setups give low sight clearance.

Computations are needed immediately for plotting.

Many awkward items must be carried.

Lengths and angles must be scaled if required later.

Difficulty in keeping the table level and oriented results from leverage caused by a light pressure on the outer area of the board.

Considerable experience is essential.

Planetable surveying enables maps to be drawn partly or completely in the field as measurements are made. The method is especially suitable for mapping topography. For a planetable survey, a hard, flat surface that can be adjusted to be level is set up within the area to be mapped. Map paper is fastened to the surface for recording measurements diagrammatically. A surveying instrument, called an alidade, is placed on the plane table and used to sight on a stadia rod and for drawing map lines.

Two basic kinds of table are used: a small traverse table with peep-sight alidade and fixed-leg tripod without leveling head, obviously appropriate only for rough work; and the standard planetable board, usually 24 by 31 in, set on a tripod having either the National Geodetic Survey four-screw leveling head or the Johnson ball-and-socket head.

Planetables are oriented by a declinator, by backsighting as with a transit, or by resection. Permanent backsights (towers, lone trees, fixed signals) enable the person working with the instruments to check orientation frequently without interrupting the rodperson's movements.

A traverse is run by orienting the table, sighting the next point, drawing a line along the alidade blade and plotting the stadia length, then moving to the plotted point and repeating the process. An average distance and elevation are obtained from the foresight and backsight. Adjusting the vertical-arc index to read zero when the bubble is centered is an ever-present problem because the table goes off level.

Topographic details are located by radiation or intersection. Short distances can be measured with a cloth tape for large-scale maps. The intersection method (graphical triangulation) is appropriate for long sights taken from two planetable stations—or three for checking—to inaccessible points. Elevations of inaccessible points may be determined from vertical angles and scaled map distances.

The stepping method (Fig. 9-8) also is used in rough terrain to save difficult travel for the rodperson. To find the elevation of the top of a steep cliff, rod intercept *ab* is taken for a level reading at the base of the cliff. Next, the lower hair is set on the ground and an identifiable spot noted where the upper wire hits the bluff. The lower hair then is moved up to the same spot and

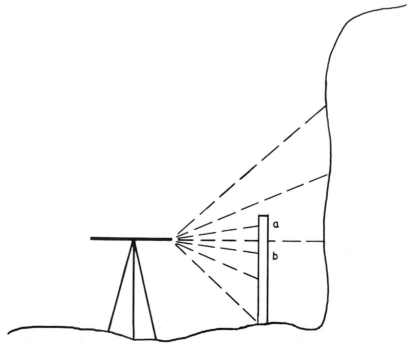

Fig. 9-8.   Stepping method determines elevations on a steep slope.

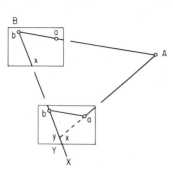

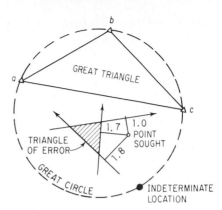

Fig. 9-9.   Two-point resection orients plane-table at new station Y when two stations A and B already plotted are visible.

Fig. 9-10.   Lehmann method orients plane-table at a new station when three stations already plotted are visible.

the process repeated until the upper hair is near the summit. A final partial "step" is estimated. The number of steps times the rod intercept, plus the final estimated number of feet, gives an *approximate* height of the cliff. The method can also be used with a transit or theodolite.

In **resection,** orientation of the table at positions not yet identified on the map is done by either the two- or three-point method. In two-point location, a direction toward the unselected next point X is plotted as in Fig. 9-9. After setting up at any selected spot on the projected line *bx,* the table is oriented using that line. By sighting to the known plotted point A, which preferably is at an angle of 60 to 90° with *bx,* the setup location is fixed at point *y,* the inter-section of *bx* and *aA* extended.

The three-point location method determines the planetable position after the table is set up at a place from which three or more prominent, plotted control signals can be seen. (In the past, navigation and planetable procedures used directions, but electronic distance-measuring devices solve the problem with lengths.) Arcs drawn with radii equal to the measured distances and using the plotted signal points as centers give the desired point. A check is obtained if the three arcs intersect at a unique point.

Various solutions are available on the planetable, such as the tracing-paper method, three-arm protractor location, and Lehmann's method. All give a strong solution of the position (point sought) if it is inside the great triangle (Fig. 9-10) or away from the great circle through the three control points. An indeterminate solution results if the point sought is on or very close to a great circle through the control points. The size of the triangle of error in Fig. 9-10 depends on how well the table was oriented by estimation to start the process, and the mapping scale.

When lines drawn at the new station to the three control points do not intersect at a point, three simple rules are used to find the point. (A second, or possibly a third, application of the trial method may be necessary.)

1. The point sought is inside the triangle of error if the station occupied is within the great triangle.

2. The point sought is either to the right or the left (when facing the control points) of all three resection lines drawn from the signals.

3. The point sought is always distant from each of the three resection lines in proportion to the distances from the respective signals to the planetable station. In Fig. 9-10, using propor-tional estimated distances to the three signals, perpendiculars are drawn by trial from the resection lines until they intersect at a unique point, the point sought.

9-13. Topographic Surveys

Topographic surveys are made to locate natural and constructed features for mapping purposes. By means of conventional symbols, culture (bridges, buildings, boundary lines, and so on), relief, hydrography, vegetation, soil types, and other topographic details are shown for a portion of the earth's surface.

Planimetric (line) maps define natural and cultural features in plane only. **Hypsometric maps** give elevations by contours, or less definitely by means of hachures, shading, and tinting.

Horizontal and vertical control of a high order is necessary for accurate topographic work. Triangulation, trilateration, traversing, and photogrammetry furnish the skeleton on which the topographic details are hung. A level net must provide elevations with closures smaller than expected of the topographic traverse and side shots. For surveys near lake shores or slow-moving streams, the water surface on calm days is a continuous bench mark.

Seven methods are used to locate points in the field, as listed in Table 9-10. The first four require a *base line* of known length. An experienced instrument person selects the simplest method considering both fieldwork and office work involved.

TABLE 9-10   Methods for Locating Points in the Field

Method	Principal Use
1. Two distances	Short taping, details close together, trilateration
2. Two angles	Graphical triangulation, plane table
3. One angle, adjacent distance	Transit and stadia
4. One angle, opposite distance	Special cases
5. One distance, right-angle offset	Route surveys, curved shorelines or boundaries
6. String lines from straddle hubs	Referencing hubs for relocation
7. Two angles at point to be located	Three-point location for plane table, navigation

A **contour** is a line connecting points of equal elevation. The shoreline of a lake not disturbed by wind, inlet, or outlet water forms a contour. The vertical distance (elevation) between successive contours is the contour interval. Intervals commonly used are 1, 2, 5, 10, 20, 25, 40, 50, 80, and 100 ft, depending on the map scale, type of terrain, purpose of the map, and other factors.

Methods of taking topography and pertinent points on the suitability of each for given conditions are given in Table 9-11 opposite.

9-14. Electronic Surveying Devices and Other Equipment

Since introduction of the first electronic distance measuring instruments (EDMI), numerous improved EDMI with increased efficiency, ease of operation, greater range, and higher precision have been developed. Correction factors can be dialed in to eliminate calculations required previously. Various models are suitable for staking and layout on construction and route surveys. Displayed distances to reflectors change in 2 s as a *rod* is moved on line.

Theodolites have also entered the electronic age. Horizontal and zenith angles are read electronically, displayed, and, along with distances, transmitted to a hand-held field computer without manual input. Then, by proper programming, the instrument computes horizontal distance, difference in elevation, average angle, latitude and departure, deviation, and so forth.

TABLE 9-11   Methods of Obtaining Topography

Method	Suitability
Transit and tape	Accurate, but slow and costly. Used where accuracy beyond plotting precision is desired
Transit and stadia	Fast, reasonably accurate for plotting purposes. Contours by direct (trace-contour) method in gently rolling country, or by indirect (controlling-point) system where high, low, and break points are found in rugged terrain, or on uniform slopes, and contours interpolated
Planetable	Plotting and checking in field. Good in cluttered areas of many details. Contours by direct or indirect method. Now generally replaced by photogrammetry for large areas. Used to check photogrammetric maps
Coordinate squares	Better for contours than culture. Elevations at corners and slope changes interpolated for contours. Size of squares dependent on area covered, accuracy desired, and terrain. Best in level to gently rolling country
Offsets from center line or cross sectioning	On route surveys, right-angle offsets taken by eye or prism at full stations and critical points, along with elevations, to get a cross profile and topographic details. Contours by direct or indirect method. Elevations or contours recorded as numerator, and distance out as denominator
Photogrammetry	Fast, cheap, and now commonly used for large areas covering any terrain, where ground can be seen. Basic control by ground methods, some additional control from photographs

For surveying computations, fast and accurate programmable computers are within the budget of even small surveying offices, or a terminal can be shared. Or services of computer service companies may be utilized. Desk-size computers for surveying, tied in to hand-held field recorders, may be used to calculate data electronically for plotting from field notes to final printout.

Table 9-12 lists most EDMI presently on the market, with their manufacturer, name, range, accuracy, weight, and year of introduction. Several *space* surveying instruments are in a special class. The portable Magnavox MX 1502 Georeceiver Satellite Surveyor acquires, evaluates, displays, and records data without operator attention or a land survey. Three-dimensional computations are made on the spot with 1-m accuracy at distances up to 1000 km between two or more instruments. Use of a 10-site network adjustment provides 300-mm accuracy for site separations up to 1000 km.

The Spanmark Inertial Survey System transported in a helicopter or terrain vehicle produces computer printouts of final calculated coordinate data with identification number, latitude, longitude, northing, easting, and elevation. The Motorola Mini-Ranger Satellite Surveying System establishes geodetic horizontal and vertical positions for precise local survey projects using the U.S. Navy's Transit Satellite system.

TABLE 9-12 Some Electronic Distance-Measuring Instruments*

Manufacturer and instrument	Range, m			Accuracy	Weight, kg, inst. + yoke	Date introduced, month/year	Remarks
	Number of prisms						
	1	3	More than 3				
AGA Geodimeter							
Geodimeter 116	600	1,600	2,600	$\pm$(5 mm + 5 ppm)	4.1	3/80	
110	1,000	2,000	5,000	$\pm$(5 mm + 5 ppm)	4.1	1/80	
120	1,300	2,300	5,000	$\pm$(5 mm + 5 ppm)	4.1	9/79	
112	1,600	2,900	6,000	$\pm$(5 mm + 3 ppm)	4.1	1/80	
14A	6,000	8,000	14,000	$\pm$(5 mm + 5 ppm)	4.0	3/80	
Benchmark, Inc. Surveyor I	1,600	2,000		$\pm$(5 mm + 2 ppm)	2.3‡	11/79	
Cubic Western Data							
Minitape HDM-70	900	1,600		$\pm$(5 mm +10 ppm)	5.0	2/78	
Hewlett-Packard							
3805A	600	1,600		$\pm$(7 mm + 7 ppm)	7.7	1/74	
3808A†	3,000	6,000	10,000	$\pm$(5 mm + 1 ppm)	9.0	6/78	
3810A		1,600		$\pm$(5 mm + 5 ppm)	12.1		Total station
3810B	2,000	5,000	8,000	$\pm$(5 mm + 1 ppm)	12.0		Total station
3820A	1,000	3,000	5,000	$\pm$(5 mm + 5 ppm)	9.6		Total station
Kern Instruments DM502	1,200	2,000		$\pm$(5 mm + 5 ppm)	1.6‡	5/80	Total station

Keuffel & Esser AutoRanger A	250	500		± 2 mm	3.0	9/79	
AutoRanger	1,000	2,000		±(5 mm + 6 ppm)	3.0	3/76	
AutoRanger II	1,600	3,600		±(5 mm + 6 ppm)	3.0	1/79	
AutoRanger S	1,600	3,600		±(5 mm + 6 ppm)	3.0	6/78	
UniRanger	3,200	4,800	10,000	±(10 mm + 2 ppm)	9.1	10/78	
Ranger V†	8,000	16,000	25,000	±(5 mm + 2 ppm)	18.1	10/76	
Rangemaster III†	16,000	24,000	64,000	±(5 mm + 1 ppm)	25	5/79	
Lietz Company Red 1A	1,600	2,500	3,200	±(5 mm + 5 ppm)	3.5	3/80	
Precision International							
Super Beetle	1,400	2,000	3,000	±(5 mm + 5 ppm)	2.5		
Citation CI-450	1,600	2,300	4,000	±(5 mm + 5 ppm)	3.1	3/80	
Topcon							
Guppy CTS-100	1,000				5.0		Total station
DM-S1	1,200	1,600	2,000	±(5 mm + 5 ppm)	2.8	1/81	
DM-C3	1,800	2,500	3,600	±(5 mm + 5 ppm)	2.8	10/80	
Wild Heerbrug							
DI-4 Distomat	1,000	1,600	2,000	±(5 mm + 5 ppm)	1.9‡	3/80	
DI-10		2,000		±(10 mm)	3.1‡		
Tachymat TC1							Total station
Zeiss							
Elta 2	1,600	2,000	4,000	±(10 mm + 2 ppm)	13.5		Total station
Elta 4	1,000	1,500	3,000	±(5 mm + 2 ppm)	6.5		Total station

*Under normal conditions. All infrared except those marked by †, which are laser type.

‡No yoke.

Index